Reliability of Semiconductor Lasers and Optoelectronic Devices

Woodhead Publishing Series in Electronic and Optical Materials

Reliability of Semiconductor Lasers and Optoelectronic Devices

Edited by

Robert W. Herrick

Osamu Ueda

ELSEVIER

WP
WOODHEAD PUBLISHING
An imprint of Elsevier

Woodhead Publishing is an imprint of Elsevier
The Officers' Mess Business Centre, Royston Road, Duxford, CB22 4QH, United Kingdom
50 Hampshire Street, 5th Floor, Cambridge, MA 02139, United States
The Boulevard, Langford Lane, Kidlington, OX5 1GB, United Kingdom

British Library Cataloguing-in-Publication Data
A catalogue record for this book is available from the British Library

Library of Congress Cataloging-in-Publication Data
A catalog record for this book is available from the Library of Congress

ISBN: 978-0-12-819254-2 (print)

ISBN: 978-0-12-819255-9 (online)

For information on all Woodhead Publishing publications
visit our website at https://www.elsevier.com/books-and-journals

Publisher: Matthew Deans
Acquisitions Editor: Kayla Dos Santos
Editorial Project Manager: Isabella C. Silva
Production Project Manager: Surya Narayanan Jayachandrans
Cover Designer: Miles Hitchen

Typeset by MPS Limited, Chennai, India

Dedication

This book is dedicated to the late Dr. Robert G. Waters, my first and most important reliability mentor. Without him, I never would have gone down this very rewarding career path.

Robert W. Herrick

Contents

List of contributors **xi**
Preface **xiii**
Acknowledgments **xv**

1 Introduction to optoelectronic devices **1**
Robert W. Herrick and Qiang Guo
1.1 Introduction 1
1.2 Optoelectronic applications 2
1.3 Principles of operation for optoelectronic components 11
1.4 Method of fabrication 25
1.5 Critical metrics 31
1.6 Laser and light-emitting diode reliability 34
1.7 New technology developments 36
1.8 Summary 41
References 41

2 Reliability engineering in optoelectronic devices and fiber optic transceivers **47**
Robert W. Herrick
2.1 Reliability engineering organizations and management 47
2.2 Developing a product: design for reliability 49
2.3 Developing a test plan: standards-based testing versus customized testing 53
2.4 Test engineering and data collection 67
2.5 Data collection and analysis 69
2.6 Failure analysis 80
2.7 Physics of failure and common failure mechanisms 80
2.8 Product release, ongoing monitoring, and postrelease support 82
2.9 Conclusion 83
References 84

3 Case studies in fiber optic reliability **89**
Robert W. Herrick
3.1 Introduction 89
3.2 Group 1: issues that were caught before product release 90
3.3 Group 2: cases that had reliability issues from the start, but were not detected in the reliability qualification 98

3.4 Group 3: parts that were reliable after release, but later developed reliability issues during high-volume manufacturing 104
3.5 Conclusion 111
References 111

4 Materials science of defects in GaAs-based semiconductor lasers 113
Kunal Mukherjee
4.1 Introduction 113
4.2 Characteristics of simple point defects in GaAs 114
4.3 Dislocations in GaAs 122
4.4 Epitaxial integration of GaAs-based materials on silicon 143
4.5 Recombination-enhanced processes 152
4.6 Outlook 166
References 167

5 Grown-in defects and thermal instability affecting the reliability of lasers: III–Vs versus III-nitrides 177
Osamu Ueda and Shigetaka Tomiya
5.1 Introduction 177
5.2 Grown-in defects in III–Vs semiconductors for optoelectronics 177
5.3 Influence of grown-in defects on device reliability and degradation in III-V based optoelectronics 193
5.4 Grown-in defects in III-nitrides 202
5.5 Composition-modulated structures and ordered structures in III-V based optoelectronics 214
5.6 Thermal instability in III-nitrides 226
5.7 Conclusion 232
References 233

6 Reliability of lasers on silicon substrates for silicon photonics 239
Justin C. Norman, Daehwan Jung, Alan Y. Liu, Jennifer Selvidge, Kunal Mukherjee, John E. Bowers and Robert W. Herrick
6.1 Introduction 239
6.2 Early-stage prospective technologies 240
6.3 Heterogeneous lasers on silicon 242
6.4 Heteroepitaxial lasers grown on silicon substrates 246
6.5 Reliability results for quantum dot lasers grown on silicon substrates 250
6.6 Degradation mechanisms in quantum dot lasers on silicon 259
6.7 The path forward and future improvements for quantum dot lasers grown on silicon substrates 265
6.8 Conclusion 267
Acknowledgments 268
References 268

7 Degradation mechanisms of InGaN visible LEDs and AlGaN UV LEDs **273**
C. De Santi, A. Caria, F. Piva, G. Meneghesso, E. Zanoni and M. Meneghini
7.1 Introduction 273
7.2 Degradation of InGaN visible LEDs 273
7.3 Degradation of AlGaN UV LEDs 292
7.4 Conclusions 303
References 304

Index **313**

List of contributors

John E. Bowers Quintessent, Inc., Santa Barbara, CA, United States; Institute for Energy Efficiency, University of California, Santa Barbara, CA, United States

A. Caria Department of Information Engineering, University of Padova, Padova, Italy

C. De Santi Department of Information Engineering, University of Padova, Padova, Italy

Qiang Guo Intel Corporation, Rio Rancho, NM, United States

Robert W. Herrick Intel Corporation, Santa Clara, CA, United States

Daehwan Jung Korea Institute of Science and Technology, Seoul, South Korea

Alan Y. Liu Quintessent, Inc., Santa Barbara, CA, United States

G. Meneghesso Department of Information Engineering, University of Padova, Padova, Italy

M. Meneghini Department of Information Engineering, University of Padova, Padova, Italy

Kunal Mukherjee Department of Materials Science and Engineering, Stanford University, Stanford, CA, United States

Justin C. Norman Quintessent, Inc., Santa Barbara, CA, United States

F. Piva Department of Information Engineering, University of Padova, Padova, Italy

Jennifer Selvidge Institute for Energy Efficiency, University of California, Santa Barbara, CA, United States

Shigetaka Tomiya Sony Corporation, Atsugi, Japan; Tokyo Institute of Technology, Yokohama, Japan

Osamu Ueda Meiji University, Tama-ku, Kawasaki, Japan

E. Zanoni Department of Information Engineering, University of Padova, Padova, Italy

Preface

Semiconductor laser reliability is a subject with great depth and complexity, but one where most practitioners must become comfortable making decisions based on incomplete knowledge. Unlike many elements of optoelectronic design where modeling can predict performance with high accuracy, and measurements can be made with many significant digits of precision, semiconductor laser reliability is governed by flaws that are ill-understood, very hard to see, and often hard to measure. What is known about the subject is usually shrouded in secrecy, under the cloak of "company proprietary" information, so little of what is known gets published or shared with the larger technical community. Furthermore, there are only a few "reliability physics" graduate programs, and a few other "reliability statistics" programs worldwide, so there are almost no "reliability engineering graduates" to be hired. Over 99% of the practitioners of the field worldwide are taught through an apprentice system, often making expensive mistakes in the process, and failing to use industry best practices even after they have been on the job for several years. We hope this book can help to articulate many of the best practices and "demystify" the discipline of reliability engineering for those entering the profession. Hopefully it will also explain many of the reasons why the reliability standards and reliability qualifications are structured the way they are today, so that practitioners are not just executing the qualifications by rote, but understand the purpose of each test.

In this book, we have brought together several of the best-known teams in the world to cover many key areas of the subject. The book is organized into two main parts. The first two chapters are intended for new entrants into the field. While there are a small number of books addressing the subject of semiconductor laser reliability, all of them are advanced in their treatment, and require many years of practice in the field before they can be understood. The large majority of those who approach me for advice on the subject are not capable of reading and understanding the existing volumes. In this book, we have addressed that gap in the market with two entry-level chapters. The first is a simple introduction to optoelectronics with a focus on the major applications that are likely to dominate market growth in the next decade. This is written in a way that does not require understanding of solid-state physics, or "energy band theory," and is intended to be accessible to anyone with a good high school science background. The second chapter is an introduction to the principles of reliability engineering. Other specialized books cover selected elements of reliability engineering, such as reliability calculations, burn-in optimization, or accelerated aging. However, this chapter is the first of its kind to cover *all* elements of the practice, from hiring and organization, to building a reliability test capability, to supporting released products. It is extensively cross-referenced, to where the reader is directed to the best of the existing other volumes for assistance on advanced topics. The third chapter is also unique: a set of 10 case studies in reliability that make it far easier to understand the principles of reliability engineering, and how they apply in practice. By looking at cases from companies (primarily those that are no longer in the market sector), it gives readers a "peek behind the curtain" to see how others have tried to solve hard reliability problems, and in some cases, where they could have done better. This third chapter should be both accessible and of interest to all readers of the book, both beginners and experts.

The second half of the book is intended for advanced readers. Chapter 4, Materials science of defects in GaAs-based semiconductor lasers, covers the material science of dislocations in optoelectronics, which is a very difficult subject to understand, and is not well-covered in material science references. Chapter 5, Grown-in defects and thermal instability affecting the reliability of lasers: III–Vs vs. III-nitrides, is a comprehensive chapter on defects in III-V's and III-N's by two of the most famous authors in the field, and provides experimental detail on the most important aging mechanisms from a material science characterization viewpoint. Chapter 6, Reliability of lasers on silicon substrates for silicon photonics, deals with the very cutting-edge topic of reliability of lasers on silicon, including the issue of heteroepitaxy (where III-V compounds such as GaAs or InP are grown on silicon substrates). And finally, Chapter 7, Degradation mechanisms of InGaN visible LEDs and AlGaN UV LEDs, deals with the physics of degradation of visible and UV LEDs, by a group with unique access to cutting-edge methods of studying the physics of degradation, and decades of experience doing so.

We believe this new volume will be highly instructive to readers at a wide range of expertise levels. But hopefully this book will be just a starting point, and readers will dive into the references, and learn more about the subjects that interest them most. Feel free to reach out to Professor Ueda or myself if you have questions or feedback.

Robert W. Herrick

Acknowledgments

I often reflect on how grateful I am for the many mentors who helped me in my professional development, and want to take a few minutes to thank several of them for what they taught me, before discussing acknowledgments that are specific to this book. My first reliability mentor is the most important on this list: the late Dr. Robert G. Waters. Bob was a co-worker of mine at McDonnell Douglas from 1987 to 1992, working at the Opto-Electronic Center in Elmsford, New York, and doing ground-breaking work in the physics of failure of semiconductor lasers for space-based laser communication systems. He never hesitated to share what he knew, or to take a call. Even with the limited tools available back in the late 1980s, he was able to develop ingenious experiments to understand the root causes of sudden and gradual failure in laser diodes, culminating in elimination of sudden failures by the development of compressively strained quantum well lasers. Discovering the secrets of semiconductor laser failure is a remarkably difficult challenge, given the atomic scale of the defects, the paucity of published literature, and the patience required to wait for experimental results, but he was able to open the "black box" and tease out many of the secrets using a multidisciplinary approach, and shared his excitement for the breakthroughs with me. He not only published some of the best papers ever written on the subject, but chaired many of the conference sessions where the breakthroughs were being first announced. I feel fortunate to have such an excellent role model to have followed, and for the very satisfying career that it led me to. I would never have followed a career path in reliability without the inspiration he provided, which is why I have dedicated this book to his memory.

To mention the most important other mentors (in chronological order), Joe Levy was my manager at my first job in McDonnell Douglas, and taught me much of what I know about semiconductor lasers. I owe him a debt of gratitude for encouraging me to take risks, to go beyond my comfort zone, and get into device simulation and design, wafer processing, characterization, mask design, and other fields that were outside both my job responsibilities and my knowledge base. When I decided to go back to school to get a PhD, Professor Larry Coldren brought me in to UCSB, funded my PhD work, and made me part of his VCSEL group for the first couple of years of graduate studies, and I will always be grateful to him for those opportunities. Professor Pierre Petroff taught me most of what I know about Materials Science, and was an excellent research advisor for the second half of my PhD. Brent Wahl hired me in to Hewlett Packard, and taught me much of what I know about transceiver reliability (including much of what is in Chapter 2, Reliability engineering in optoelectronic devices and fiber optic transceivers). Kevin Weitsman was our Department Manager, and one of the most capable and amazing people I have ever worked with. I also owe him a debt of gratitude for what he taught me about the business side of managing a quality department. Saurin Shah has been a manager or co-worker of mine for most of the past 15 years, and he deserves the biggest share of the credit for building the quality department for Intel SPPD out of almost nothing in his first year on the job. Saurin's manager, the late Pinder Mathru, had the vision for what our group could become, and helped put me in front of many important connections throughout Intel, and played an important part in our current success—I just wish he were around to see how his vision has been fulfilled. Finally, I would like to thank my many excellent co-workers and managers along the way.

I benefitted from many others I enlisted for help in providing feedback and editing assistance on the chapters I wrote. The most helpful of these was from my co-worker Felipe Vallini; his honest feedback was very beneficial, and helped me to re-organize two of the chapters. I also want to thank my co-author Qiang Guo for reading and commenting on all of the chapters in this book, as well as his help with the first three chapters. He also managed the figure permissions for several of the chapters, and found and shared many great background articles. Andra Hioki also provided a very careful reading of Chapter 1, Introduction to optoelectronic devices, and I appreciate her copy editing help. I also enlisted family and friends without engineering degrees to read the first chapter to assess its suitability for newcomers to the field—thank you to Andy Herrick, John Herrick, and John South for your careful reading and suggestions.

At Elsevier, I would like to thank Kayla Dos Santos, who originally approached Osamu and I in 2018 and invited us to put together a book proposal on the subject of laser reliability, and helped shepherd the proposal through Elsevier management. I also would like to thank Isabella Silva for program managing the book and its production. And I would also like to thank my co-editor Osamu Ueda for also providing inspiration early in my career with his excellent published work, and for his collaboration in the past decade where I have had the chance to work together with him on many conferences and publications. Finally, I would like to thank all of the authors of this book for the work they have put into their chapters, and sharing what they know with the readers of this book. I hope this work can help to make the process of reliability engineering a less mysterious and more approachable discipline.

Robert W. Herrick

Chapter 1

Introduction to optoelectronic devices

Robert W. Herrick[1] *and Qiang Guo*[2]
[1]Intel Corporation, Santa Clara, CA, United States, [2]Intel Corporation, Rio Rancho, NM, United States

1.1 Introduction

The laser was invented in 1960, and in the years since then it has played a rapidly growing part of our lives, carrying data across the Internet, powering laser printers, barcode scanners in supermarkets, and driving imaging and proximity sensors such as computer mice. The light-emitting diode (LED) has gone past being a simple indicator lamp in its early life, to powering most colored lighting and traffic signals, and is gradually taking over the majority of the lighting market, as well as most backlighting for TVs and computer displays. Optoelectronics play an integral role in many of our daily activities, in more ways than most people appreciate. The purpose of this chapter is to give an introduction to how lasers and LEDs work, how they are fabricated, and a few of their key applications.

In the past decade (the 2010s), three large markets developed in optoelectronics, each of which resulted in billions of new semiconductor optoelectronic devices. The first was the large-scale replacement of conventional incandescent and fluorescent lighting with solid-state lighting, based on LEDs. The second was the development of the mega-scale cloud data centers, which are powered by fiber optic transceivers, where transceivers and other fiber optics now account for more than a third of the total investment. The third, and most recent, is the explosive growth of sensors for facial or gesture recognition, such as the “FaceID” sensor first used in the Apple iPhone X. Looking forward to the 2020s, laser-based range mapping (LIDAR, short for "light detection and ranging") for autonomous driving and other advanced driver-assistance systems will add a fourth market that is expected to also ramp up quickly into the billions of devices created and sold.

All of these will create hundreds of thousands of new jobs for those designing, fabricating, testing, and assembling those systems [1–5]. While we expect that those designing the devices will continue to be trained through the same time-consuming and thorough process in engineering schools around the world, many more in the fabrication, test, assembly, market development, and management roles will transfer in from successful careers in other fields, without the benefit of a graduate-school education in optoelectronics. It is that need for retraining and continuous education that this chapter seeks to address, with an introduction that has a lower barrier to entry than previous

Reliability of Semiconductor Lasers and Optoelectronic Devices. DOI: https://doi.org/10.1016/B978-0-12-819254-2.00010-2

tutorials. We will attempt to emphasize ease-of-understanding over rigorous or in-depth treatments, and introduce a minimum of theory and no equations. For many, we hope this will be enough to better understand decisions being made in engineering meetings. For others, it can help provide a larger context before reading other chapters in this book, or in other books or papers on optoelectronic design and fabrication. Some of the most popular treatments elsewhere include Coldren [6], Verdeyen [7], Kasap [8], and Culshaw [9], and these are recommended for those looking for greater depth, with the time and inclination to dive into the equations and physics.

We assume that the reader has a background in college-level physics, and basic engineering concepts. Unlike other treatments of the subject, we do not assume previous training in semiconductor devices or solid-state physics. We cover a few basics of laser design and operation, with a particular emphasis on elements important to the manufacturing and test process and (in keeping with this book) with device reliability. This chapter does not require sequential reading, so we encourage readers to skim the chapter, and pick and choose just the sections that look relevant to their industry, and to their job.

1.2 Optoelectronic applications

In the following section, we introduce three optoelectronic applications that have achieved over $1B per year in revenue: InGaN LEDs for high-efficiency lighting, lasers for sensing arrays, and lasers for data communication. On top of the fiber-optic network backbone that was built up in previous decades, which will be discussed in Section 1.3, these represent the three most important new optoelectronic markets of the past decade.

1.2.1 InGaN-based light-emitting diodes for high-efficiency lighting

The most efficient type of general lighting now available is based on blue InGaN LEDs with phosphors to down-convert most of the photons to other lower-energy wavelengths to provide broad-spectrum white light. In many jurisdictions, traditional incandescent lighting has now been banned as being too inefficient and wasteful of energy, and for contributing to climate change. Fluorescent lighting, while far more efficient than incandescent, is less efficient than LED lighting, and also has a shorter lifetime, and environmental concerns related to the disposal of the mercury used in fluorescent bulbs. In Fig. 1.1, we show images of LED-powered bulbs [10,11].

LED lighting really took off starting in the late 1990s, with the invention of efficient InGaN LEDs. Professor Shuji Nakamura received the 2014 Nobel Prize in Physics for the development of the method of growth used to produce these devices. It took another couple of decades for it to ramp up to ever more powerful LED chips, and to scale up production volumes and cut costs. At the time of this writing,

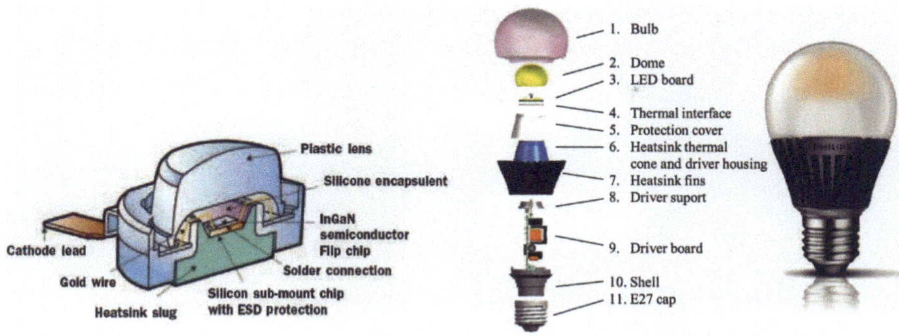

Figure 1.1 (Left) An image of the construction of a high-brightness white LED package; (right) a common low-cost LED lamp.

Source: (left) Copyright 2005 Society of Photo-Optical Instrumentation Engineers (SPIE) & Lumileds LLC from M.G. Craford, LEDs for solid state lighting and other emerging applications: status, trends, and challenges, in: Proc. of the SPIE, vol. 5941, Sep. 2005, p. 594101, https://doi.org/10.1117/12.625918; (right) [2018] IEEE. Reprinted, with permission, from X. Perpiñà et al., Thermal management strategies for low- and high-voltage retrofit LED lamp drivers, IEEE Trans. Power Electron. 34 (4) (2019) 3677–3688, https://doi.org/10.1109/TPEL.2018.2853119.

an 800-lumen bulb (the most popular type for indoor lighting) is expected to last 20 years, with an annual energy cost only $1.20, leading to $6.08 annual energy savings when compared to the incandescent bulbs it replaced [12]. Based on a US Department of Energy forecast, LED lighting is projected to make up over 99% of the total residential lighting market by 2030 in the United States [13].

1.2.2 Lasers for sensing arrays

In 2017, Apple introduced "Face ID" to its phones, which was powered by a vertical-cavity surface-emitting laser (VCSEL) array with 219 VCSEL emitters [14]. The system is capable of creating a 3-D map of the face of each user, which is used as a security biometric system for sign-on to the phone. Other gesture recognition systems use a similar engine. Production of these sensors reached 197 million units in 2018 [15]. With the addition of these smartphone applications to the previous base, the total VCSEL market reached US$0.738B in 2018 [4,16]. As with many consumer products, it is not uncommon for each dollar of spending on lasers in the supply chain to enable 100 dollars of value/consumer spending in the final product (e.g., on DVD or CD lasers in disc players). Images of the dot projector that powers the FaceID are shown in Fig. 1.2 [17].

1.2.3 Lasers for data- and telecommunications

Nearly everyone is aware that low-cost telecommunication is powered by fiber optic communications. However, fewer people are aware of the fact that around 45% of the cost of modern cloud data centers is the cost of "optics," including both the fiber, and

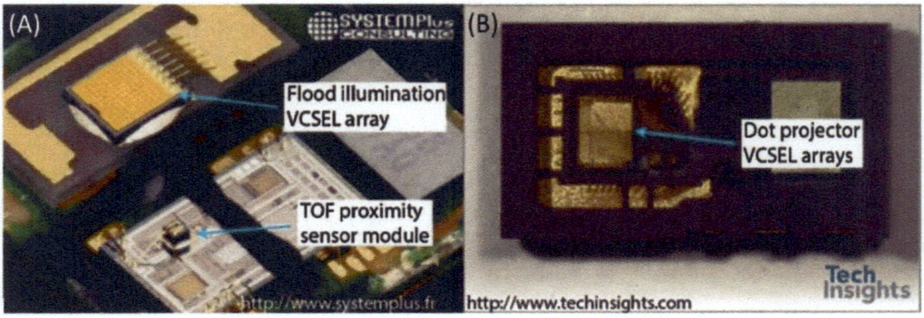

Figure 1.2 Image of (A) face recognition module and (B) vertical-cavity surface-emitting laser-based "dot projector arrays."

Source: Reprinted from A. Liu, P. Wolf, J.A. Lott, and D. Bimberg, Vertical-cavity surface-emitting lasers for data communication and sensing, Photonics Res. 7 (2) (2019) 121–136, https://doi.org/10.1364/PRJ.7.000121, copyright 2019. With permission from Chinese Laser Press (CLP).

the components we will discuss in this section. It is useful to understand when communications are made by electronics over high-speed cables, and when it is more advantageous to make them over optical fiber. A simple summary would be to say that where it is possible to do communications through electrical signals alone, that is the normal industry practice. However, there are real limitations, in both speed and link length, to high-speed electronic communications, as shown in Fig. 1.3. Above the combination of speeds and distances shown (where the product of distance in meters and speed exceeds approximately 100 m-Gbps) [18], the signal degradation is severe, the power required is prohibitive, or both. In those cases, it is worth the additional expense of adding an "electrical/optical and optical/electrical" converter to the link. Photons do not suffer from the problems high-speed electrical links do with capacitance or electromagnetic radiation—they are able to pass through fiber optic cables with much lower degradation per meter traveled, and require less energy for longer links.

To take advantage of fiber optic communication, a pair of core components are needed: one to convert high-speed electronic data into high-speed optical data (E-O), as well as a companion converter of optical signals back to electrical signals (O-E). The former E-O converter is known as a "transmitter" or Tx, while the latter, O-E, is known as a "receiver" or Rx. These components were first deployed in the Bell Telephone network back in 1977. In modern application, the Tx and Rx are combined into a single unit called a "transceiver" with a tightly defined set of mechanical, electrical, optical, and software specifications that are part of a "multisource agreement" or "MSA" document that assures interchangeability. The most popular types of transceivers are made by dozens of companies, and have seen the same sort of rigorous price competition seen in other electronics commodity industries, such as hard drives or RAM memory modules. The net outcome has been international phone calls that used to cost dollars per minute are now too cheap to meter, and free streaming video is available to homes and cell phones.

The transceiver, which is approximately the size of a finger, normally contains a few optical components, as well as a number of integrated circuits (ICs). A typical

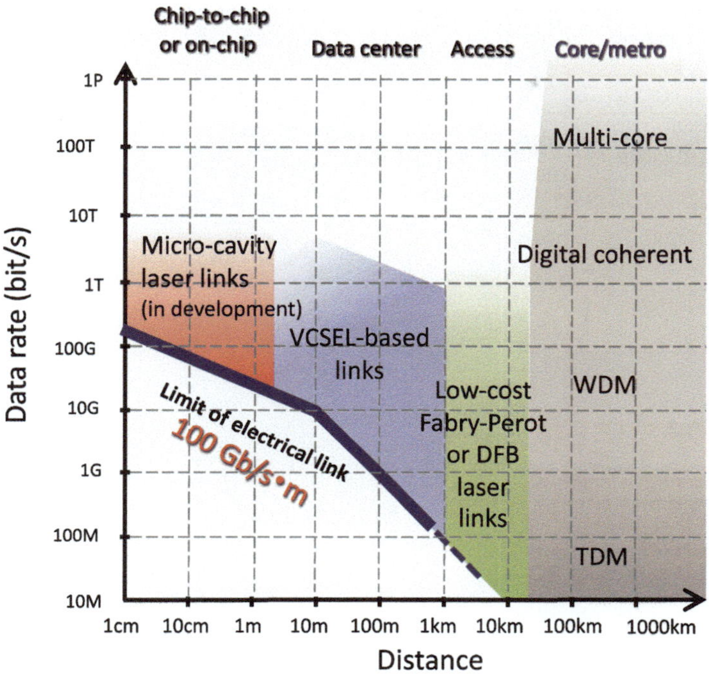

Figure 1.3 Graph of data rate vs. link distance. Copper links are the most cost effective for low speeds and short links (lower left), but fiber optics are used for longer or higher speed links. After Reference [19].

layout is shown in Fig. 1.4 [20]. Of special interest is the optical subassembly [or transmitter optical sub assembly (TOSA) on the Tx side]. The receiver optical components tend to be less challenging, although the ICs that are used in the receiver are at least as challenging as those on the Tx side. The TOSA in turn usually provides three functions. First, it must provide a source to generate photons. Second, it monitors the average light out, to allow adjustment to a predetermined set point in response to degradation or changes in the thermal environment (although some very simple designs may dispense with this and operate an "open loop"). Third, it needs a way to turn the stream of light on and off rapidly—either using a "modulator," or directly by powering the laser up and down.

Infrared wavelengths (especially 1310 and 1550 nm) are typically used in optical communication due to the available laser sources and low transmission loss of optical fiber. Fig. 1.5 shows that this region can be further subdivided into multiple wavelength bands [21]. The "O-band," at 1310 nm, gained popularity because it is at the dispersion minimum, and thus works best for many multimode fiber applications. The O-band is also widely used within data centers and for shorter distance communication. The C-band at 1550 nm is widely used for long-haul optical communication because of a mature erbium-doped fiber amplifier (EDFA) technology and the lowest transmission loss in this band. Near the C-band, the L-band and the

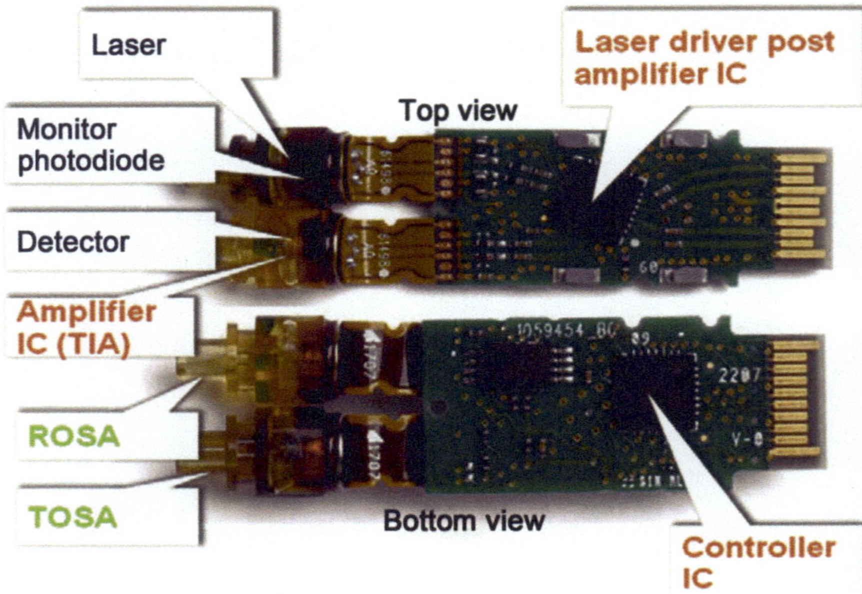

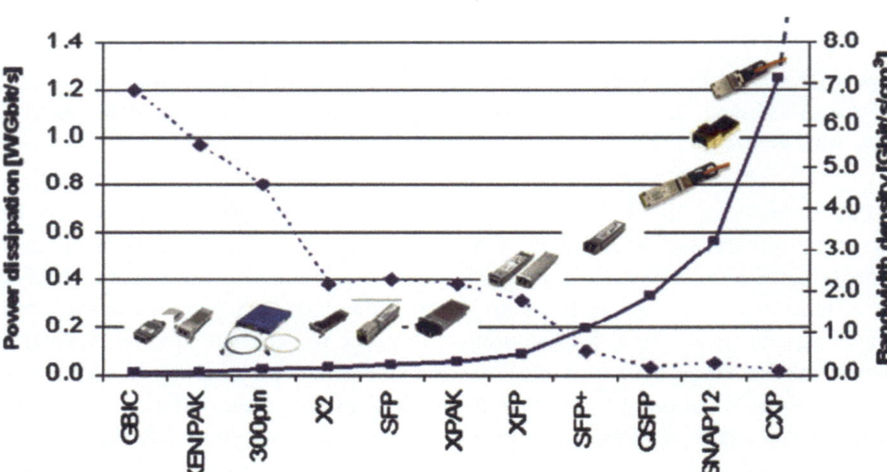

Figure 1.4 Photo of a fiber optic transceiver, including key internal components on both the top and bottom of the PC board (upper part of the figure). The laser is embedded in the subassembly labeled "TOSA." In the graph at the bottom, the transceivers are sorted by bandwidth density, with the older types on the left, and newer ones on the right. The dotted line shows the substantial reductions that have been obtained in terms of reduced power dissipation per bit, and the solid line shows how bandwidth density has increased.

Source: Reprinted from J.S. Eng and C. Kocot, Optical Fiber Telecommunications VIA: Chapter 14. VCSEL-Based Data Links, Elsevier Inc., 2013, copyright (2013), with permission from Elsevier.

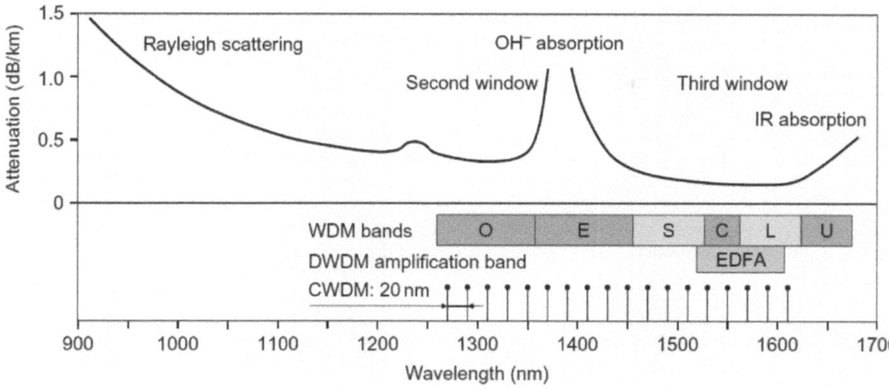

Figure 1.5 Communication bands used for medium- and long-distance optical communication.

Source: Reprinted from K. Grobe, Chapter 5—Optical wavelength-division multiplexing for data communication networks, in: C. DeCusatis (Ed.), Handbook of Fiber Optic Data Communication (4th Ed.), Academic Press: Oxford, 2013, pp. 85–122, copyright (2013), with permission from Elsevier.

S-band (so-named because they are slightly longer or shorter in wavelength than the original "center" band) also show sufficiently low absorption loss to support an increased number of channels. Thus, wavelength-division multiplexing (WDM) can be applied across these multiple bands to boost transmission capacity. WDM is a technology which can fully utilize the bandwidth of optical fiber by transmitting optical signals using different wavelengths at the same time. It can be divided into coarse wavelength division multiplexing (CWDM) and dense wavelength division multiplexing (DWDM). The main difference between CWDM and DWDM is channel spacing. CWDM has wider spacing between channels (e.g., 20 nm), whereas DWDM has very close channel spacing below 1 nm. Although DWDM allows more data per fiber, it has higher cost and power consumption.

1.2.4 The history of the laser

The idea of the laser dates back to the 1930s, and it is widely believed that all the technology elements were available at that time. Many of the most important inventions and technologies are shown in the timeline in Fig. 1.6. However, the theory was misunderstood at the time, and it was thought that the necessary "population inversion" (explained later) required a "negative temperature," and was thus impossible to obtain. It was not until successful demonstration of microwave amplification of stimulated emission of radiation in the late 1950s that it was thought that optical stimulated emission might be possible, including by Townes and Schalow [22,23]. Theodore Maiman at Hughes Research Labs demonstrated the first laser in 1960. The first semiconductor lasers came a couple of years later in 1962, and the first visible LED was also demonstrated by Nick Holonyak at General Electric in 1962, as shown in the timeline in Fig. 1.6. The first large-volume laser applications

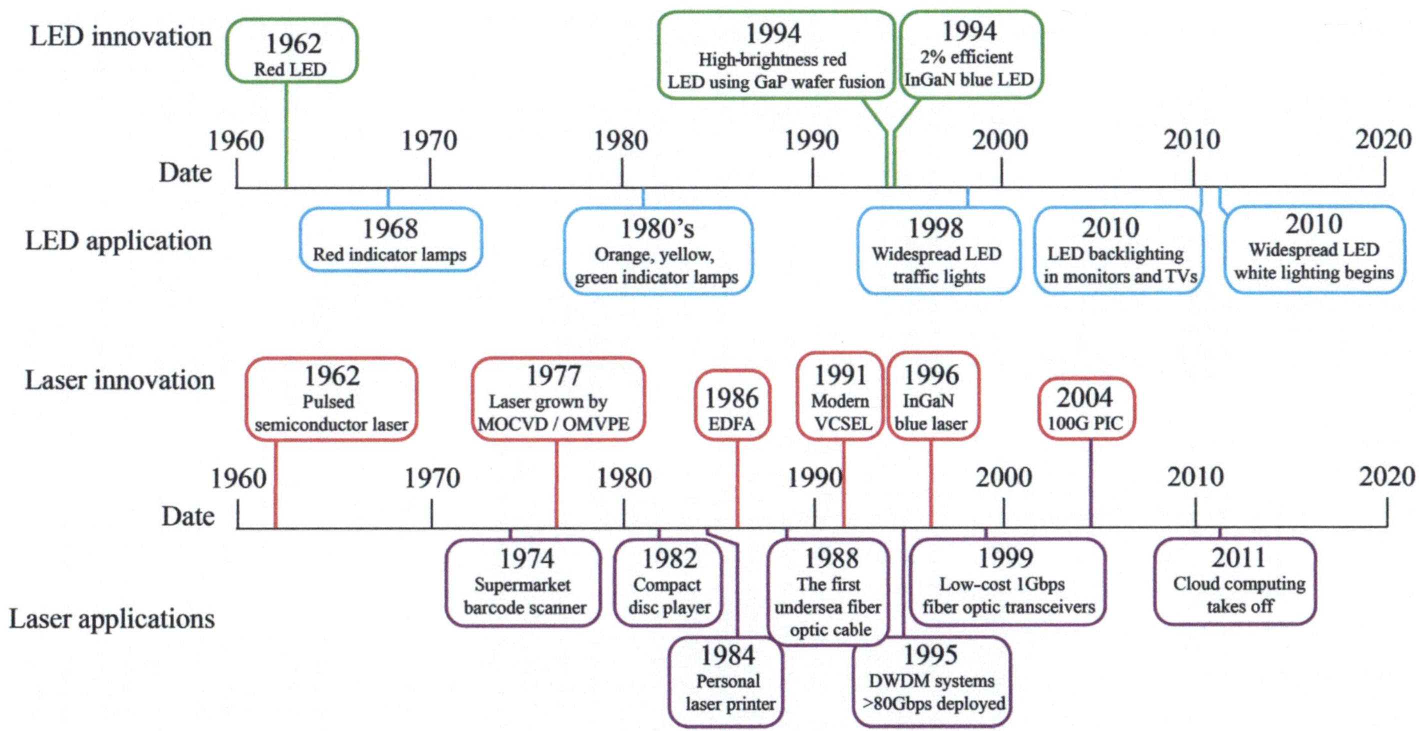

Figure 1.6 Timeline showing many of the most important technologies and applications for the LED (top) and for the laser (bottom). The details of many of the innovations are discussed later in this chapter. *DWDM*, dense wavelength division multiplexing; *EDFA*, erbium-doped fiber amplifier; *LED*, light-emitting diode; *MOCVD*, metal organic chemical vapor deposition; *OMVPE*, organo-metallic vapor phase epitaxy; *VCSEL*, vertical-cavity surface-emitting laser.

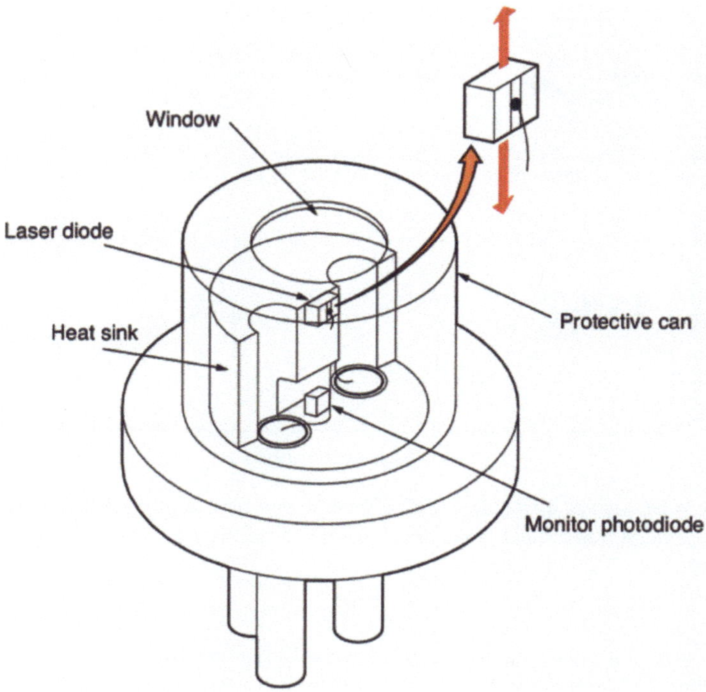

Figure 1.7 CD laser assembly in a "transistor outline can" (TO can) package.

Source: After Newport Corporation, Laser Diode Technology. https://www.newport.com/t/laser-diode-technology, 2019 (accessed 26.10.19), permission to use granted by Newport Corporation.

were material processing using carbon dioxide (CO_2) lasers beginning in 1967, and helium-neon (HeNe) lasers for barcode scanners in retail scanning systems in the 1980s. Telecommunication lasers for fiber optics started being produced in small volumes for ultra-long-haul submarine communication in 1988 [24,25], but they cost a few thousand dollars each at the time.

The first low-cost laser was the gallium arsenide (GaAs)-based 780 nm CD laser, which exceeded 10 million units produced in 1987 [26], and is pictured in Fig. 1.7 [27]. The cost per unit fell to just a few dollars each by 1990. Of course, CD lasers and the DVD lasers that superseded them have been largely replaced by cloud delivery of video content (also made possible by optoelectronics), as the storage and data communication to stream the files on demand have become so cheap on a cost-per-bit basis that it has become less expensive than the previous procedure of printing and shipping discs to transfer the information.

The next major chapter in the development of the semiconductor laser was the development of the vertical cavity laser, which enabled even lower cost, lower power consumption through a very small pumped area, and improved beam quality. VCSELs were first suggested by Professor Kenichi Iga, and low-temperature pulsed

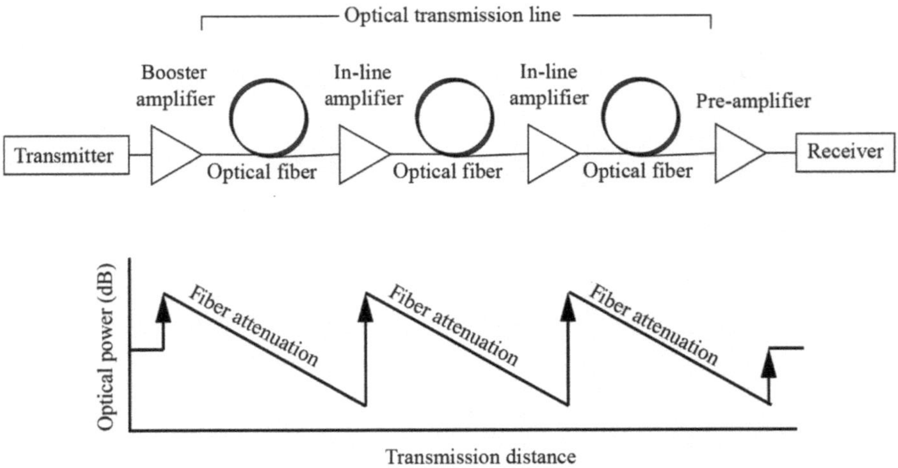

Figure 1.8 Schematic of a fiber link with erbium-doped fiber amplifiers to restore the signal periodically, which was critical in the development of "too cheap to meter" phone calls and data transmission for Internet service.

operation was demonstrated in the laboratory [28]. The first practical VCSEL was demonstrated by Jack Jewell of Bell Labs in 1991, followed shortly by Professor Larry Coldren's group at UCSB the same year. By the late 1990s, companies such as Hewlett Packard had introduced the first "low-cost" VCSEL-based gigabit fiber optic transceivers (for $300 each, while previously gigabit per second transceivers typically cost $5000 each), which enabled data center and storage server communications, and were instrumental in the dot-com explosion of the late 1990s.

Another important development was the high-power 980 nm laser for pumping of EDFAs, pictured in Fig. 1.8, which were the main focus of the telecom bubble of 1998–2001 [25,29]. At the time, telecommunication was moving to "DWDM," where 40–80 different wavelength "channels" could be sent down the same fiber, with different information encoded on each one. However, the information attenuated sufficiently within 80–100 km which previously had needed to go through splitting, detection, and retransmission, which required large capital investment of very expensive components. However, EDFA-based systems which simply restored the signal amplitude periodically through optical amplification were found to be capable of transmitting the information once, and crossing the entire distance (across the Atlantic Ocean, for example) with no repeaters.This created huge reductions in the cost per bit that were essential for the development of the Internet and unmetered phone communication [24,25,30]. Finally, in the past decade, we have seen widespread replacement of incandescent and fluorescent lighting with solid-state LED lighting, and also the rise of large-scale cloud computing powering social media and remote meetings. We will have more to say about these technologies in applications in the sections to come. For additional laser history information, refer to Hecht's excellent books and papers [22,31–33].

1.3 Principles of operation for optoelectronic components

In this section, we discuss first LEDs, then lasers, and finally photodiodes which are used in receivers. Lasers and LEDs have many common design features, although obtaining lasing is more demanding, and additional features are included in the laser.

1.3.1 Light emission

While a full discussion of the principles of emission from semiconductor lasers is beyond the scope of this chapter, we wish to briefly discuss the phenomena of "spontaneous emission" and "stimulated emission," which are the bases of LED and laser emission, respectively. While neon signs and gas lasers have mostly been replaced by LED signs and diode lasers, we still introduce emission from gas plasmas because we hope the emission from isolated atoms will be more familiar to readers, and also since it forms the basis for the more complex "band theory" that is necessary for understanding semiconductor optoelectronic devices.

1.3.1.1 Light emission in gas plasmas: a review

In middle school science, or basic chemistry, the concept of the Bohr atom is introduced, with atomic orbitals. The neon atom, which we will be focusing on in this section and the next, is a noble gas, which has both the first and second "shells" fully populated, with a total of 10 electrons. If neon gas is "excited," for example by a plasma, or by a strong light, the electrons can be "lifted" to orbitals farther away from the nucleus, which are called "excited states," and have additional potential energy. After residing there for up to a few billionths of a second (i.e., a few nanoseconds), the electron is likely to decay back to the ground state, and in the process of relaxing back, it will give off a photon in stimulated emission. While there are a broad range of transitions possible, we are most familiar with the red emission seen in neon signs, as seen in Fig. 1.9 [34].

1.3.1.2 Stimulated emission in gas lasers

"Laser" is an acronym which is short for "light amplification by stimulated emission of radiation," so we wish to spend a moment explaining stimulated emission here. As shown in Fig. 1.10, if a passing photon comes close enough to the excited atom, it will encourage (stimulate) the excited atom to drop the electron from the excited state to a lower state, emitting an accompanying second photon in the process, and amplifying the optical electromagnetic emission in the process. In doing so, the second photon will be: (1) at the same wavelength as the stimulating wave; (2) in the same polarization; (3) in the same direction of travel; and (4) in the same phase as the stimulating wave. These attributes are important in providing the characteristics that we value in lasers, of being monochromatic, directional, and coherent. A photo of a helium-neon laser (He-Ne), emitting at 632.8 nm is shown on the right-hand side of Fig. 1.10 [35]. Millions of this type of laser were sold for point-of-sale barcode scanning in supermarkets in the 1980s and 1990s, before being phased out for other types of low-cost imaging systems, or diode-laser-based barcode

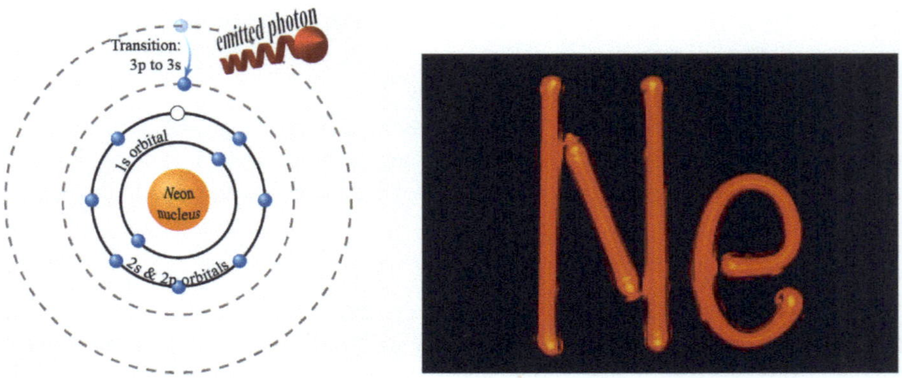

Figure 1.9 Spontaneous emission (left) results when an excited electron gives up potential energy to return back to a lower state, and emits a photon in the process. A neon sign (right) emits across many red transitions from the 3p orbitals to the 3s orbitals.

Source: Neon sign image from Pslawinski, Neon. https://commons.wikimedia.org/wiki/File:NeTube.jpg, 2005.

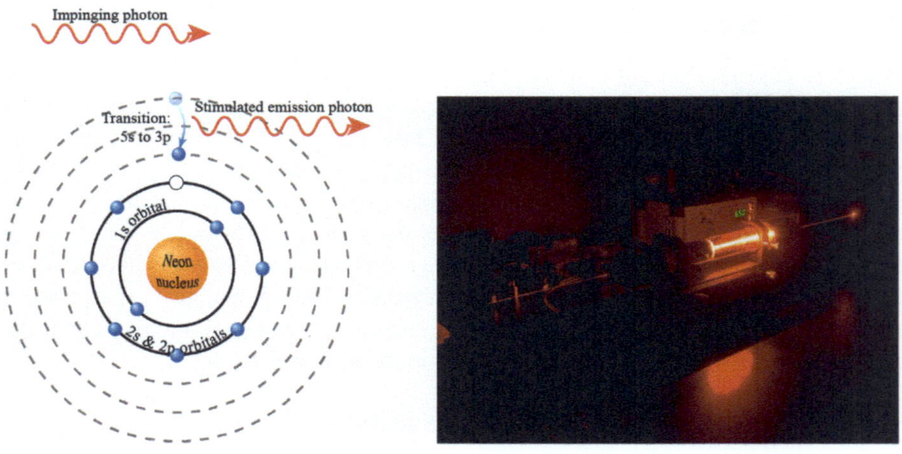

Figure 1.10 Stimulated emission prompts the atom to release a second photon as it passes close enough to prompt the electron to drop down from the excited state (left); a helium-neon laser tube emitting at the 632.8 nm wavelength (right).

Source: HeNe laser image from T. Markstein, Deutsch: Helium-Laser in Betrieb (from https://commons.wikimedia.org/wiki/File:Henelaser.jpg), 2009.

scanners. The laser properties, such as beam quality and divergence for He-Ne laser are superior, but cost, ruggedness, and lifetime are inferior to the modern alternatives.

1.3.1.3 Spontaneous and stimulated emission from semiconductors

In semiconductor crystals, the close proximity of the atoms means that instead of having sharp, well-defined spectral lines like a gas lamp would, we have smearing of the occupancy levels into "bands," with the filled lower states we saw above being similar to the normally full "valance band" in an unexcited semiconductor, and the upper states in the atom mapping to the "conduction band" that excited electrons can be lifted into. By forming a LED or a semiconductor laser diode, we can create structures that are sandwiches of different materials where recombination takes place in very thin layers at the diode's electrical junction, allowing electrical excitation by simply forward biasing the diodes. For lasers, these thin layers enable us to populate most of the excited states in the conduction band, and empty most of the states in the valance band, with currents as low as a few milliamperes, which in turn enables devices like a VCSEL-based computer mouse to run off of a single set of batteries for over a year. We will discuss device design further in the next section.

1.3.2 Light-emitting diodes

LEDs started out as rugged "indicator lamps." The first visible (red) LED was invented by Nick Holonyak in 1962 when he was at General Electric [36,37]. The first LEDs were constructed from a single alloy from the group III and group V columns of the periodic table (GaAsP, in the case of this first LED). However, it was quickly realized that more efficient emitters required a "sandwich" of a low-energy semiconductor, surrounded by higher energy barriers around it to contain the carriers, as we explain in the next section.

1.3.2.1 Carrier confinement

One of the characteristics of semiconductor materials is that they possess a "band gap" or "energy gap." In the case of some wafers, such as those made from gallium phosphide (GaP), you can actually see it, as the wafers appear to be tinted orange. For photons with energy below the band gap (e.g., infrared, red, and orange photons in the case of GaP substrates), the material appears transparent. For light above the band gaps (e.g., violet, blue, green, or yellow photonics for GaP), the material is highly absorptive. This gap can be seen qualitatively in Fig. 1.11, and quantitatively

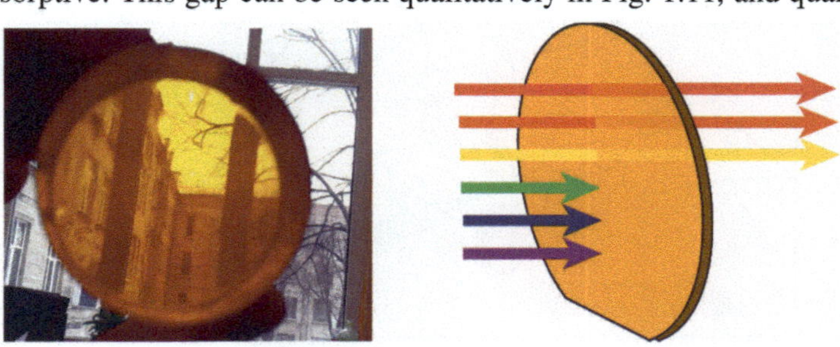

Figure 1.11 Photo of a zinc selenide wafer (left); note that you can see through it, but that it provides an orange tint. Higher energy photons are absorbed in zinc selenide substrates (right), while lower energy photons pass through it with low absorption.

Source: Reprinted from I. Dafinei, et al., Production of 82Se enriched Zinc Selenide (ZnSe) crystals for the study of neutrinoless double beta decay, J. Cryst. Growth 475 (2017) 158–170, copyright (2018), with permission from Elsevier.

in Fig. 1.12 [38,39]. With sufficient electrical excitation in a LED or diode laser, the semiconductor material will emit light near the edge of this cutoff. Note that energy and wavelength are inversely proportional, so materials with shorter wavelength emission require wider band gap materials.

Much of the design and engineering of LEDs and laser diodes comes from choosing from the limited palette of materials that both (1) can be economically grown on readily available, low-cost substrates, and (2) have acceptable power

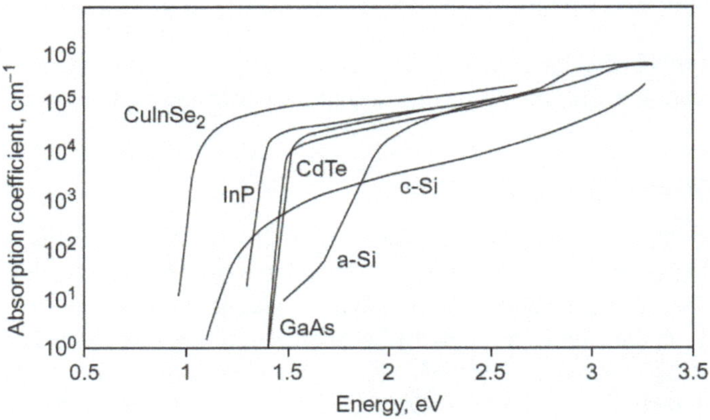

Figure 1.12 Absorption coefficient, α, as a function of photon energy for several semiconductors shows very low absorption for photons with energy below the band gap, with absorption that rises by several orders of magnitude for photons with energy above the band gap.

Source: Reprinted from T. Markvart and L. Castañer, Chapter IA-2—Semiconductor materials and modelling, in: A. McEvoy, T. Markvart, and L. Castañer (Eds.), Practical Handbook of Photovoltaics (second ed.), Academic Press: Boston, 2012, pp. 33–62, copyright (2007), with permission from Elsevier.

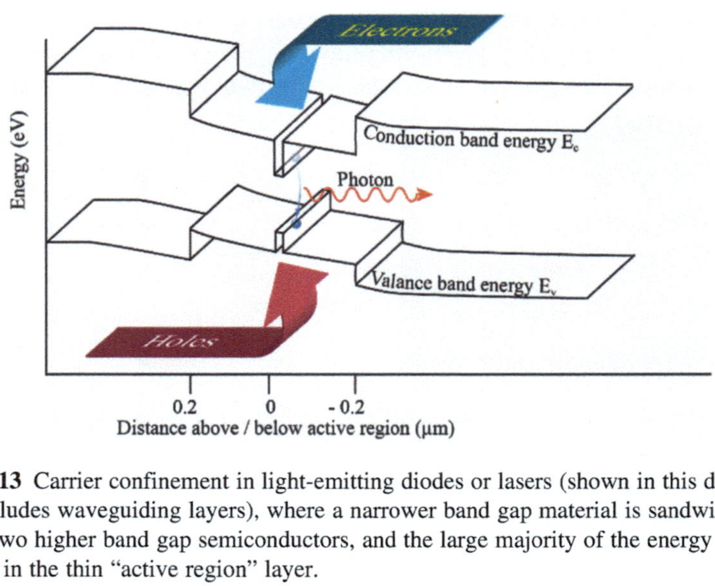

Figure 1.13 Carrier confinement in light-emitting diodes or lasers (shown in this diagram, which includes waveguiding layers), where a narrower band gap material is sandwiched between two higher band gap semiconductors, and the large majority of the energy is deposited in the thin “active region” layer.

conversion efficiency (PCE) at the desired wavelength of emission. LEDs typically sandwich a lower band gap energy material between higher energy barriers on the side, trapping the carriers in a sheet, as shown in Fig. 1.13.

1.3.2.2 Light extraction

One major challenge for LEDs has been that the semiconductor materials have a very high index of refraction, where most photons generated would normally undergo total internal reflection, making it difficult to escape the device, as shown in Fig. 1.14 [40]. Many strategies have been employed to come up with novel geometries to improve the ratio of photons that escape on the first or second pass, rather than being lost to

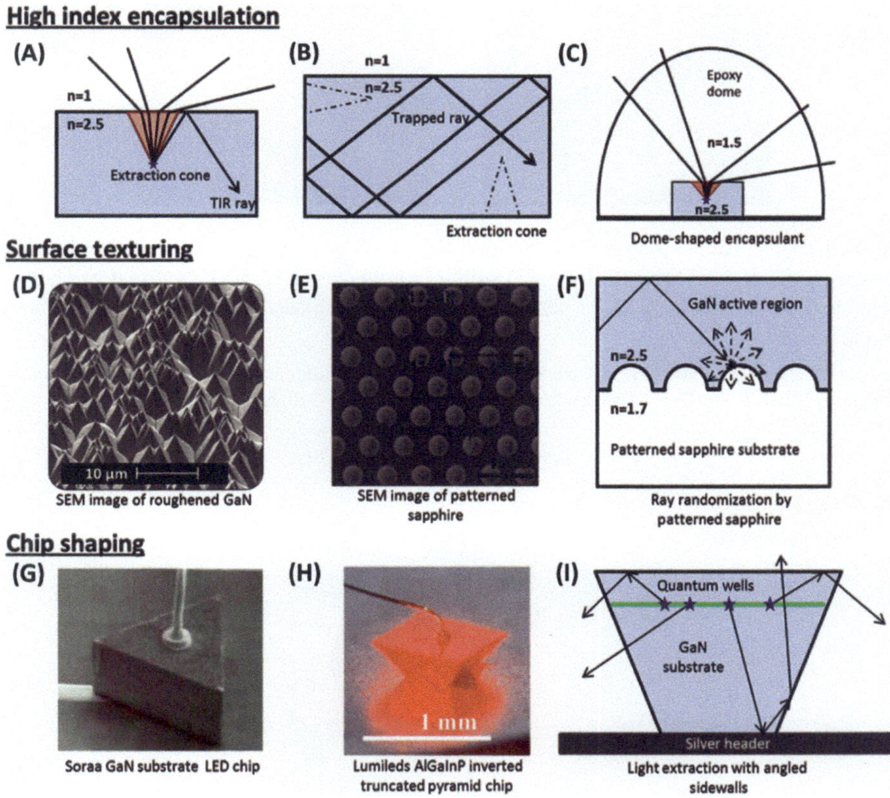

Figure 1.14 Summary of techniques used to enhance light extraction in LED chips. These techniques include (A)–(C) high index encapsulation to increase the critical angle for extraction, (D)–(F) surface texturing such as roughening or patterning to randomize light, and (G)–(I) chip shaping to redirect rays at each bounce and increase the probability of extraction on the next attempt.

Source: Reprinted by permission from Springer from T.-Y. Seong, J. Han, H. Amano, and H. Morkoç (Eds.), III-Nitride Based Light Emitting Diodes and Applications. Springer, Netherlands, 2013 (copyright 2017), figure 9.4.

absorption from repeated internal reflection. Another problem previously was that red and orange AlGaAs LEDs were grown on a highly absorbing GaAs substrate. This problem was solved by replacing the absorbing substrate with a transparent GaP substrate, as shown in Fig. 1.15 [41,42].

1.3.2.3 Heat extraction

While indicator lamps may have only limited optical power needs, other applications such as general lighting want to get a large amount of light (hundreds of

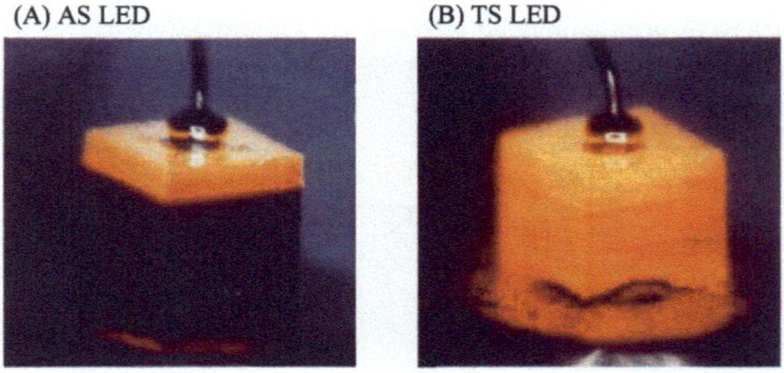

Figure 1.15 (A) Amber AlGaInP LED with a GaP window layer and absorbing GaAs substrate (AS). (B) Amber AlGaInP LED with a GaP window layer and a transparent GaP substrate (TS) fabricated by wafer bonding. Conductive Ag-loaded die-attach epoxy can be seen at bottom.

Source: Reprinted, with permission, from D.A. Steigerwald et al., Illumination with solid state lighting technology, IEEE J. Sel. Top. Quantum Electron. 8 (2) (2002) 310–320, https://doi.org/10.1109/2944.999186, IEEE (2002).

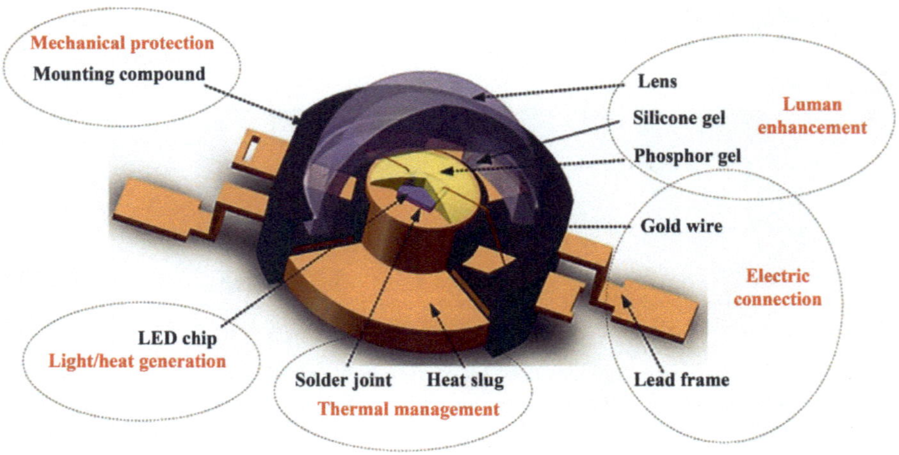

Figure 1.16 Construction of a high-power light-emitting diode.

Source: Reprinted from X. Luo, R. Hu, S. Liu, and K. Wang, Heat and fluid flow in high-power LED packaging and applications, Prog. Energy Combust. Sci. 56 (2016) 1–32, https://doi.org/10.1016/j.pecs.2016.05.003, copyright (2016), with permission from Elsevier.

lumens, requiring many watts of power) out of the smallest possible diode, in order to maximize the number of die per wafer, and minimize the cost per die. Electrical power of several watts per device in volumes of less than 1 mm^3 cause junction temperature increases of over 10°C, which then results in substantial drops in PCE. Overheating is managed with good heat sinks that spread the heat out to keep the device cool, as shown in Fig. 1.16 [43].

1.3.2.4 "Rollover" or "droop"

To minimize the cost of each lamp, manufacturers wish to be able to deliver the light from only a single LED to minimize cost, and ideally of a small size to allow a large number of chips per wafer to be fabricated. However, when putting several watts of power per chip into something which could be no larger than a grain of sand, self-heating lowers the PCE substantially. If you push enough current through the chip, efficiency drops of 30% or larger will be observed, as shown in Fig. 1.17. The debate about the root cause has raged for more than 10 years, with competing claims that the limiting factor is competing nonradiative ("Auger") recombination, electron overflow, or nonuniformities in the active layer thickness and composition [44–47]. Researchers are continuing to work on improving the maximum power per chip, but in the meantime, lamps are made with a larger number of inexpensive chips for

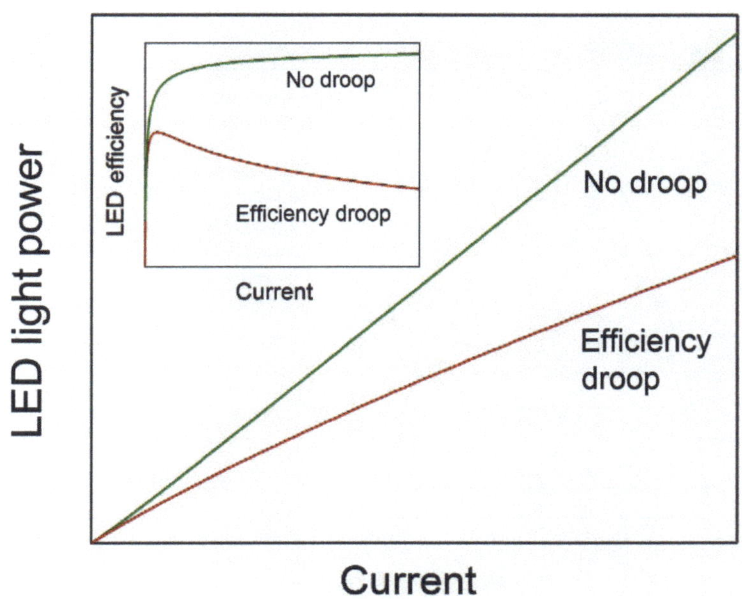

Figure 1.17 Illustration of light-emitting diode efficiency droop, where the output power deviates further from the ideal the more the bias current is increased.

Source: Reprinted from J. Piprek, Efficiency droop in nitride-based light-emitting diodes, Phys. Status Solidi A 207 (2010) 2217–2225, https://doi.org/10.1002/pssa.201026149, copyright 2010. With permission from John Wiley and Sons.

applications where the light does not need to be focused from a single source, or with more elaborate heat sinking and chip designs where it does matter.

1.3.2.5 "The green gap"

More than most people realize, we have a limited number of materials that are appropriate for making LEDs or semiconductor lasers. For example, there is no such thing as a yellow or orange semiconductor laser at the time of this writing—the best method available for now is to make an infrared laser with twice the wavelength, and use a nonlinear frequency doubler to get a yellow output beam—but this is only available for approximately one hundred times the cost of comparable red laser sources at this writing, and limited to specialty markets.

LEDs are less demanding than semiconductor lasers, and available in a wider range of wavelengths, but still have limits on power and conversion efficiency in many different wavelengths. For longer wavelengths, yellow, orange, and red are grown using AlGaInP on GaAs substrates, with electron overflow limiting the power available for the yellow and orange wavelengths. For the short wavelengths, violet, blue, and green are grown using GaN/InGaN, with green being less than half the efficiency of the blue LEDs at this writing. This phenomenon results in what is termed the "green gap," as shown in Fig. 1.18 [48]. The reasons for this are beyond

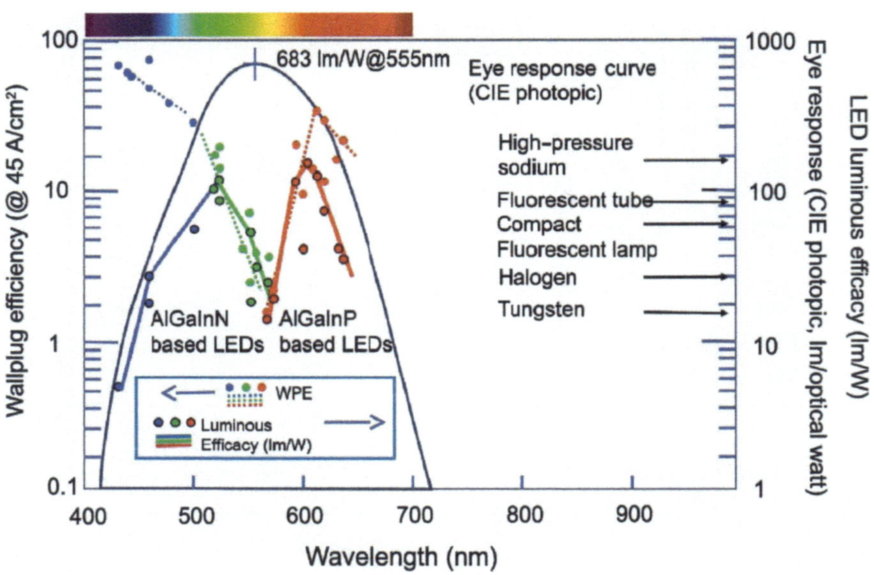

Figure 1.18 Plot of power conversion efficiency vs wavelength, showing reduced efficiency for green, yellow, and orange wavelengths—this is referred to as "the green gap."

Source: Reprinted from C. Weisbuch, Historical perspective on the physics of artificial lighting, Comptes Rendus Phys. 19 (3) (2018) 89–112, https://doi.org/10.1016/j.crhy.2018.03.001, copyright (2018), with permission from Elsevier.

the scope of this chapter, but for those interested in learning, see the Reference [49], with an understanding that the root causes require much more technical background to understand.

1.3.3 Lasers

1.3.3.1 What additional design features exist with lasers that are not present with an light-emitting diode?

While many of the design elements used to make the LED are also used in the laser, we also need to add a number of additional features to get laser emission from semiconductor lasers. We briefly discuss three of them in the section below.

Adding a "population inversion"

For lasing to occur, the excited (upper) state needs to be heavily populated, and the ground (lower) state needs to be depopulated. A certain minimum current density will typically need to be obtained before this criterion can be met, which creates a typical "threshold current." In addition, the gain must overcome optical losses and emitted power—another reason why below the threshold current there is no lasing output. We can reach the lasing threshold more easily by using very thin gain layers, which minimize the volume of semiconductor material we need to pump.

Adding waveguiding

In the LED, light radiates out in all directions equally. For lasers, however, we want to direct the light in one direction. To do so, we also need optical confinement with waveguiding. For those familiar with fiber optic cables [50], light is channeled down a particular direction, using total internal reflection to keep the light in a high index of the refraction core, that is bounded by a low-index-of-refraction cladding layer. Most lasers available today are "in-plane lasers" (IPLs) or "edge emitters" (as opposed to VCSELs, which we address later in the section). For edge emitters, the layer structure confines the light in the vertical ("transverse") direction, while ridge etching is usually the method of confining it from side to side (the "lateral" direction) as shown in Fig. 1.19. Since the vertical confinement is usually only on the order of

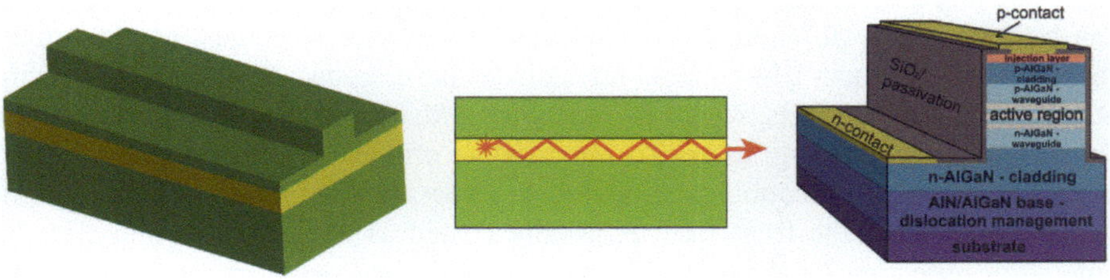

Figure 1.19 Ridge waveguide structure confines light vertically with a high index waveguiding layer shown in yellow, sandwiched between lower index layers above and below shown in green (left and center). AlGaN ridge waveguide laser with different epitaxial layers labeled (right).

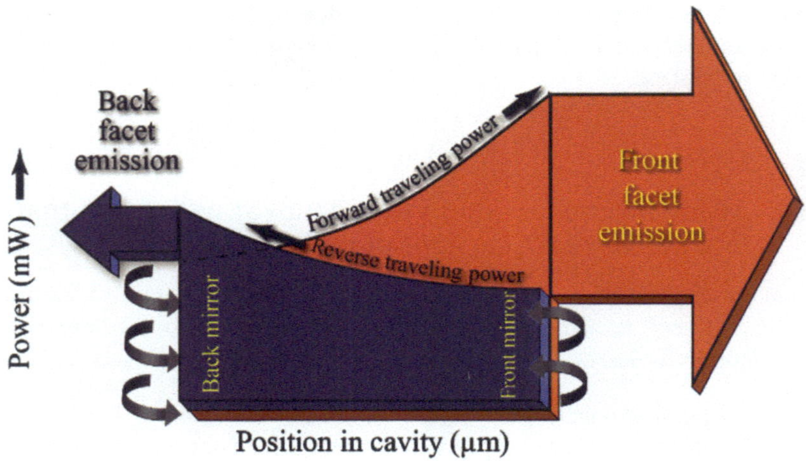

Figure 1.20 Drawing of how laser power varies along the cavity in a Fabry-Perot semiconductor laser, with simple reflecting mirrors at both ends of the cavity. Power rises as it passes through the gain medium, and then drops as a portion of the power is extracted at the emitting facets.

half a micron, and the lateral mode width is usually a few microns, we typically get an elliptical (nonround) and astigmatic beam, which requires additional optics to circularize later. This is one disadvantage that is not an issue for the VCSEL.

Adding mirrors

If we only added waveguiding and some stimulated emission, we would create a super-luminescent LED (SLED), which is a variant of an LED with greater directionality, and reduced spectral width. However, it would not be capable of lasing. The next element we must add is a resonant cavity, which provides positive feedback for lasers. This makes sure a sizeable portion of the power seeds continuing stimulated emission—in fact, no more power should be removed per round-trip than the gain of the laser can replace, or the power will drop. The resonant cavity can be realized by using reflective facets or gratings in semiconductor lasers. This can be done a few different ways. The easiest is to take advantage of the natural facet reflectivity: most III–V-based laser diodes have around 30% facet reflectivity at uncoated cleaved facets, which is usually more than sufficient for most materials for making edge-emitting lasers. To improve the efficiency, cleaved facets can be coated with a high-reflection back mirror, to direct much or nearly all of the light to the front facet. The front facet is usually coated as well, both to optimize the amount of reflection for the best efficiency at the desired output point, as well as to protect the facet from corrosion and to stabilize it during aging. The circulating power is shown in Fig. 1.20. One consequence of the simple cleaved-facet approach is that multiple different wavelengths will be supported in lasing, which can be disadvantageous for some applications.

More complex lasers use "gratings" which reflect a small amount of power at intervals of half a wavelength—these can force the laser to operate with only a single wavelength, and can also allow better control of the wavelength and power as a function of temperature.

1.3.3.2 Vertical-cavity surface-emitting lasers

Most semiconductor lasers operate with light moving along a micron or two below the top surface of the wafer—these lasers are referred to as "in-plane lasers" (IPLs), "edge-emitting lasers," or "edge-emitters" (as shown earlier in Fig. 1.19 or later in Fig. 1.31). A very popular alternative is the "VCSEL," where the light is emitted perpendicular to the surface of the wafer. VCSELs are widely used for powering the Internet backbone in short-reach fiber optic transceivers for links of just a few hundred meters, operating at 850 nm. Another major application is as a highly efficient directed light source for battery-powered computer mice. More recently, volumes have ramped up for use as illuminators for gesture recognition and facial recognition, such as in Apple's "Face ID." In each case, the advantage of the VCSEL is much lower power consumption due to a much smaller pumped area compared to edge-emitting lasers. Typical VCSELs might have an 8 μm diameter pumped area, for a total area of 50 μm^2, which is about an order of magnitude smaller than most edge-emitters.

VCSELs work by controlling the reflection through variations in growth layer composition. Most VCSELs at the time of this writing have 850 nm emission, and are grown with the GaAs/AlGaAs system, so variations in the aluminum mole fraction allow the designer to get a small reflection at each layer transition [51]. Back mirror reflectivities of 99.9% + and front reflectivities of 99.5% are typical—due to the short cavity, many more round trips are typical before the average photon would be expected to escape.

Most VCSELs are "oxide VCSELs" that have a thin "oxide aperture layer" of AlGaAs that is nearly pure AlAs, and is oxidized to convert it to Al_2O_3 in a furnace with a high-temperature flow of water vapor. This results in a controllable lateral oxidation of 15–20 μm around all exposed sides of the post. The oxide aperture that this creates performs two functions: both confining the current laterally and waveguiding light to control the lateral mode of the laser. The construction and modal flow of a typical VCSEL are shown in Fig. 1.21.

While a detailed comparison of the advantages of VCSELs versus edge emitters is outside the scope of this section, the main advantages of VCSELs are low power consumption for applications that only require a few milliwatts of power, easy fabrication and testing, and a circular emission pattern that can easily be focused. The disadvantages are higher thermal resistance, which limits output power, and limited wavelengths where the materials can support lasing. Materials must have not only excellent power conversion properties that would also be required for an edge emitter, but have very low optical loss when doped, good thermal conductivity, and high index contrast between alternating DBR materials.

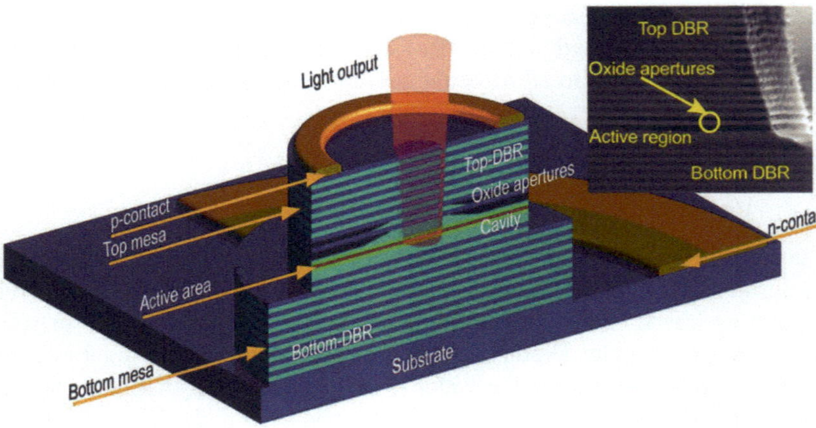

Figure 1.21 A typical gallium arsenide-based vertical-cavity surface-emitting laser (VCSEL) structure.

Source: Reprinted from A. Liu, P. Wolf, J.A. Lott, and D. Bimberg, Vertical-cavity surface-emitting lasers for data communication and sensing, Photonics Res. 7 (2) (2019) 121–136, https://doi.org/10.1364/PRJ.7.000121, copyright 2019. With permission from Chinese Laser Press (CLP).

1.3.3.3 Direct modulation

In data communication, the most popular modulation format is simple amplitude modulation, where a 0 is encoded as a low power, and a 1 is encoded as high power. To ensure a balanced average power, 64b/66b or 128b/132b encoding is used to allow multiple different versions of each 64 or 128 bit data string with two to four additional "coding bits" added to the 64 or 128 "payload bits.". The different versions of the encoded words can be chosen to keep the average number of 1's and 0's equal—this encoding prevents long unbroken strings of 1's or 0's, by allowing four versions of each word, including some with more 0's than 1's even for a data word of all 1's, for example.

The modulation speed has continued to increase over time, providing huge reductions in cost-per-bit, as the 10 Gbps transceivers eventually were available at the same price as the 1 Gbps transceivers had been less than 10 years earlier. This trend means that every few years, either the modulation rates must increase by another 30%–100%, or a different encoding format with more bits per period must be adopted.

For much of the history of fiber optic communication, simple "direct modulation" has been used, where the laser is ramped between low and high power, as shown in Fig. 1.22. However, the very fast ramping up and down has become more and more challenging, as it exceeds the natural limits of the device, as is discussed in Section 1.5.5.

1.3.4 Modulators

An increasingly popular alternative to powering the laser light source on and off directly is to leave the laser at a fixed power, and to then send the light through a

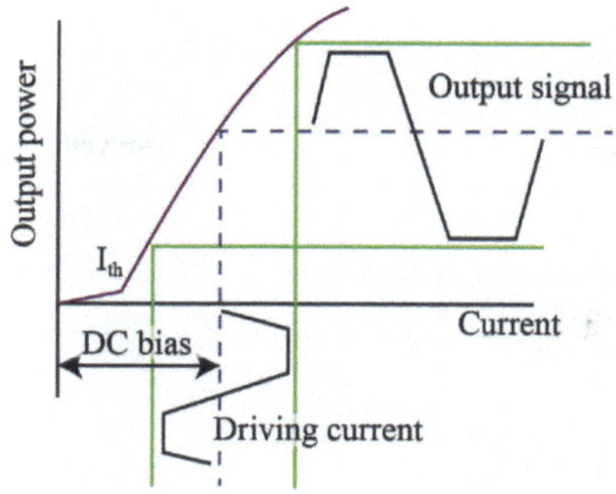

Figure 1.22 Direct modulation of the laser diode uses a high-speed AC swing superimposed on a DC bias to get a high-speed optical output.

"modulator" that is capable of attenuating the light to encode the data optically [52]. These modulators work through one of two main principles. The first is to absorb the light from the laser to encode a "zero" bit, for example in an electro-absorption modulator. The second is to destructively interfere the light using either a dual-path modulator, called a Mach-Zehnder interferometer, or through a ring modulator where the light is circulating around multiple times.

1.3.5 Photodetectors

The transceiver is a device that converts high-speed electrical signals to optical signals on one side of the transceiver, and back from optical to electrical signals on the other side of the device. The optoelectronic device used to convert photons back to an electrical signal is the photodiode.

1.3.5.1 Photodiodes

While the photoelectric effect was first discovered more than 100 years ago, modern photodiodes use the same principles that are used in solar panels to convert sunlight into electrical power. The current versus voltage curves are shown in Fig. 1.23. In solar cells, a combination of voltage and current is chosen by the controlling device (the "inverter") to maximize the power yield, and it operates in the lower right quadrant of Fig. 1.23. For optical communication devices, speed rather than power harvesting is the goal, so it operates in the lower left quadrant of Fig. 1.23, with a "reverse bias" of the diode. As can be seen from the curves, for these devices, little enhancement in signal is obtained by increasing the reverse electrical bias, but it does affect capacitance and rise time.

For short-wavelength 850 nm VCSEL-based links, early photodetector systems used both silicon and GaAs-based photodiodes. Modern 25 and 40 Gpbs 850 nm photodetectors use GaAs-based photodiodes. For longer wavelength 1310 and 1550 nm photodiodes, they generally use InGaAs-based photodiodes grown on InP substrates.

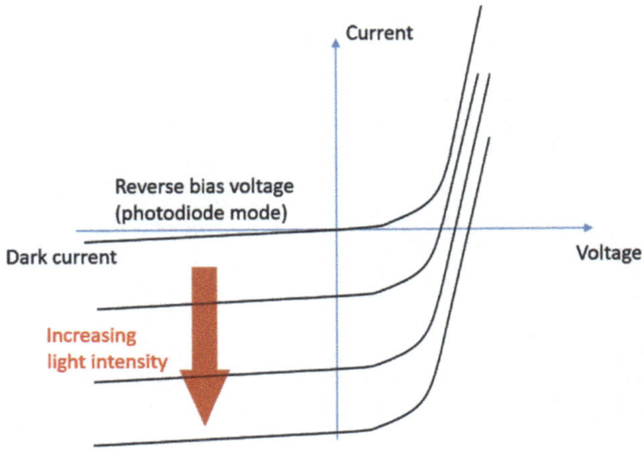

Figure 1.23 "Current–voltage" (I–V) curves show how a photocurrent is generated in a solar cell or photodetector when hit by photons. Solar cells operate in the lower-right quadrant, with positive voltage, and photodetectors operate in the lower left quadrant, with a modest reverse bias.

Photodiodes can be characterized by their responsivity, which is a number citing how many milliamps output a photodiode gives per milliwatt input. This ratio in turn is proportional to the quantum efficiency (how many electrons are collected vs. recombined before collecting). Commonly available designs provide 80%–90% of the maximum theoretical photocurrent that an ideal photodetector would provide, which is to say, 80%–90% of the photons can be converted into carriers and collected.

1.3.5.2 Avalanche photodiodes

In fiber optic links where there is high optical loss (e.g., long-haul systems), the optical power is quite weak at the receiver, and avalanche photodiodes (APDs) are able to provide cleaner signal amplification than electronic amplifiers can. While APDs are more difficult to design and fabricate, and more temperamental to bias and control, in a side-by-side comparison, systems that use them require one-fifth to one-tenth the optical power to complete a low-error-rate optical link [53]. This can be a substantial advantage in terms of increasing allowable link length for systems between repeaters. An APD takes the same absorption we saw for the photodiode, and adds a "multiplication" or "gain" stage, where a cascade of 5–20 electron–hole pairs can be generated by each electron the photodiode section creates, with lower numbers being typical for high-speed communications [54]. Higher multiplication factors (or "gains") of 100–1000 electrons per photon are possible for sensitive instrumentation that needs less of a high-speed response, but where used, these high gains allow single photons to be detected. The design and operation of a low-gain APD are shown in Fig. 1.24.

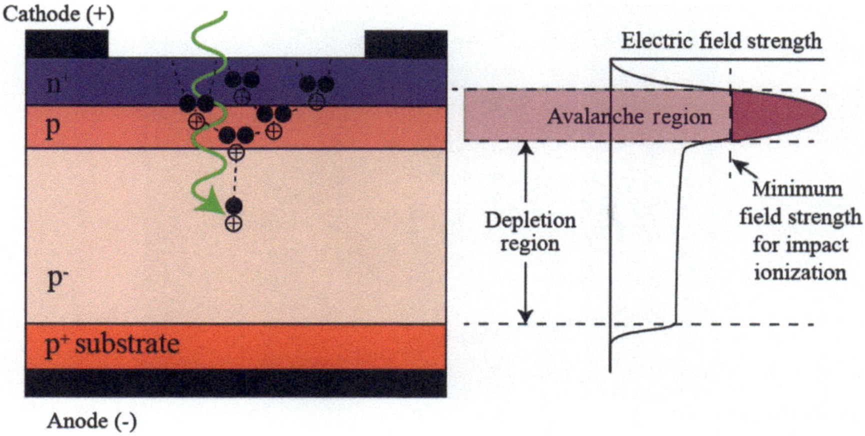

Figure 1.24 Operation of an avalanche photodiode, showing a photon (in green) being absorbed near the middle, and creating an electron–hole pair. The electron is attracted to the positive cathode at the top, and creates (in this example) a cascade of six electrons through impact in the avalanche region where the electric field exceeds the impact ionization voltage (after References [55,56]).

1.4 Method of fabrication

In this section, we give a brief overview of how lasers and LEDs are fabricated. The fabrication of simple devices can be cleanly divided into two sections: the growth of the epitaxial wafers, and wafer fabrication in the cleanroom. More complex devices may require selective regrowths where additional epitaxial growth is done after cleanroom patterning. After fabrication is complete, devices are tested, broken or sawn into individual units, and mounted in a package. Each device also goes through some form of reliability screening to eliminate weak devices and stabilize the device before it gets certified to be incorporated into its final system (e.g., transceiver or LED luminaire).

1.4.1 Epitaxial growth

Epitaxial growth is the manufacturing process where the starting material for lasers, LEDs, or photodiodes is created. It uses a large multi-million-dollar piece of equipment called an "epitaxial reactor," as shown in Fig. 1.25, to perform "epitaxial growth" of the desired semiconductor layers onto a "substrate" wafer that is used as a starting growth template. The substrate, at this writing, is usually a few hundred dollars each, and the resulting "epitaxial wafer" is usually sold for a thousand to a few thousand dollars for semiconductor lasers. For LEDs, the much larger volume, and lower costs of the inputs result in both substrate and finished wafer costs being more than an order of magnitude lower. Most devices are created using a process called metal organic chemical vapor deposition (MOCVD) or organo-metallic vapor

Figure 1.25 High-volume MOCVD reactor, with graphite platen that holds the substrate wafers at the right inside the glove box that is used for substrate loading under controlled atmosphere, and the gas handling equipment on the left. The footprint of this system is 1.2×6 m.

Source: Courtesy AIXTRON.

phase epitaxy (OMVPE) [57]—note these are both just the two most popular names for exactly the same growth method, and there is no difference between the two. In those growth processes, metal-organic compounds such as group III precursors trimethyl-aluminum (TMA) and trimethyl-gallium (TMG) are combined with group V gases such as arsine (AsH_3) over a very hot rotating substrate. The heat is sufficient to break the molecular bonds, and encourage the deposition of III–V compounds or alloys on the original starting substrate that was loaded into the reactor. By varying the ratio of the different gases, one can form the desired combination of layers to meet the desired vertical design of the device structure. For example, by varying the ratios of TMA to TMG, one can form AlAs, GaAs, or $Al_xGa_{1-x}As$ alloys anywhere between those two endpoints. By doing this, one can control items like emission wavelength, where carriers will be captured and held, and waveguiding properties for edge emitters or reflection for VCSELs. This concept is shown in Fig. 1.26 for a simplified AlGaInP red laser structure. Modern growth methods are capable of making rapid transitions between compositions within a few atomic layer thicknesses (i.e., within a transition thickness less than 1 nm) between one material and the next. MOCVD enabled the growth of efficient "quantum well" lasers and LEDs, with active layer thicknesses of less than 15 nm (which show quantum confinement effects). Forcing all the carriers into very thin layers reduces the pumped volume to the absolute minimum, thereby lowering the threshold current. (Common lasers used in data communication have an energy density of around 10 GW/cm^3 [58–60]).

Most engineers working in the field regard epitaxial growth as the most challenging part of device manufacturing, and the step most critical to making a reliable device. III–V compounds are still very sensitive to even part-per-billion level

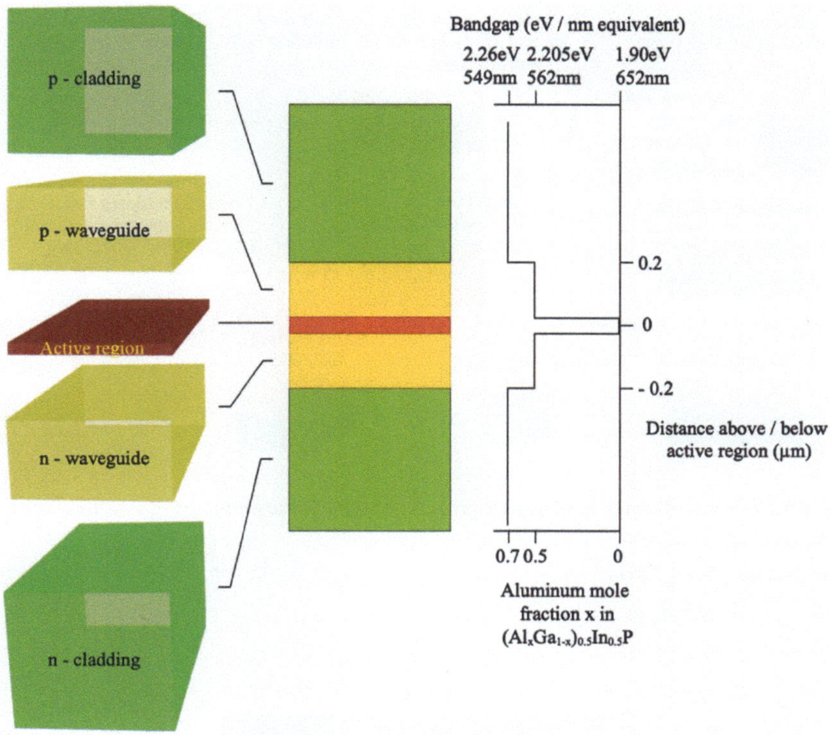

Figure 1.26 Sample red laser growth combines a few different alloy compositions to control material band gap and waveguiding in the structure.

impurities. Even with very pure gases, the design space is complex enough that the point defect concentration (of undesired interstitials and vacancies) or items like oxygen incorporation can be modified by subtleties such as growth temperature, the ratio of the group V to group III precursors, growth rate, and other factors. Because of the complexity, it is considered something of a "black art," where the outcome is difficult to predict based on the inputs, and is usually dialed in empirically based on past successful designs, or through iterative trials that require device fabrication and testing, followed by additional rounds of refinement, to reach a successful growth recipe.

1.4.2 Wafer fabrication

Once a wafer has been grown in an epitaxial growth reactor, and passes the initial characterization tests, it moves into wafer fabrication. While the epitaxial growth determines the device's vertical (transverse) structure, the wafer fabrication process determines the device's horizontal (lateral and longitudinal) structure. This fabrication is done through a series of photomasks. At a minimum, the wafer processing

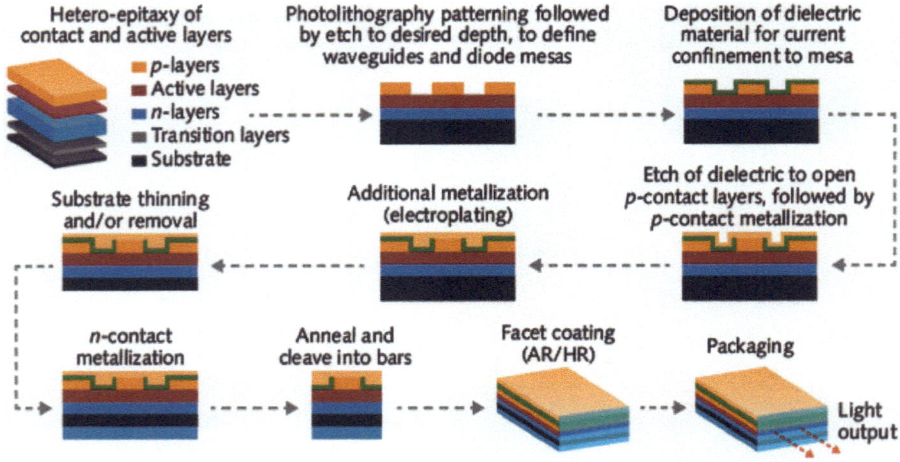

Figure 1.27 Typical wafer frabrication processing flow for a simple edge-emitting laser diode.

Source: Reprinted from the January 2018 edition of Laser Focus World, copyright 2019 by Endeavor Business Media, LLC.

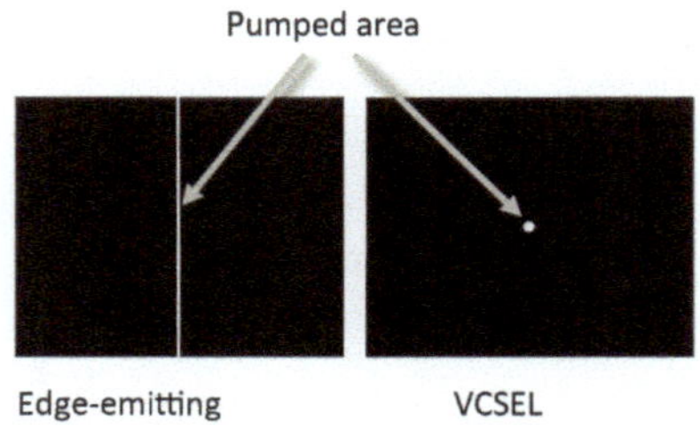

Figure 1.28 Typical die sizes for low-cost lasers are 200 × 250 μm, which is the smallest size that is practical to handle. However, even at that size, only a percent or less of the area is electrically pumped for most semiconductor lasers.

will define where the current will flow on the chip, and put down metal to make electrical contact with the devices. In most cases, the metal is patterned to make discrete contact pads. For most lasers, transverse waveguiding will also be added through patterning to form a ridge waveguide. For some lasers, the longitudinal mode control may be introduced through the addition of a "feedback grating." A process flow for a simple wafer-fabrication process is shown in Fig. 1.27.

While it is desired to maximize the die count per wafer to minimize device cost, practical limits to being able to handle devices have set a minimum device dimension of around 200 μm on a side. As a result, only a few percent of the area of the die is electrically active, as can be seen in Fig. 1.28. The remaining area is only present for mechanical purposes (supporting bond pads) and thermal purposes (heat spreading).

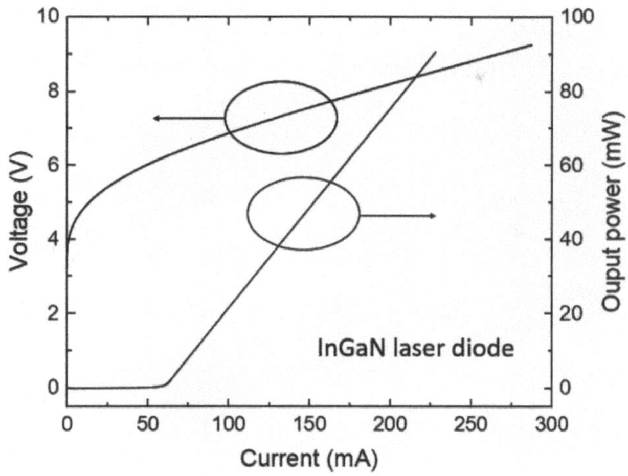

Figure 1.29 L–I–V curves of an InGaN laser diode.

1.4.3 Wafer test

After wafer fabrication is completed, devices go for a wafer test, where 100% of the devices are tested to see which devices meet specifications. The most universal test for lasers or LEDs is the "L–I" curve shown in Fig. 1.29, which looks at light out vs. current in, as well as "V–I" (bias vs. current). Certain features of these curves are extracted using software, and compared against upper and/or lower pass–fail limits. If any of the parameters for any of the limits fail, the device gets "inked out"—either physically or in a software log that provides a list of good and bad devices.

For some types of devices, additional testing may be needed, either on a 100% basis, or a sampling basis. Common examples include spectral testing to verify that the device is single longitudinal mode (i.e., it has only a single lasing wavelength), or high-speed testing to verify that a sample meets the rise and fall time requirements.

Most edge-emitters may only be testable after singulation into bars (1-cm width is common) and facet coating. After the bad die are inked out, they are then cleaved in the other direction, usually into 40 individual 250 μm-wide die, and moved into packaging.

1.4.4 Singulation and packaging

It is common to get up to 50,000 devices from a 3 inch diameter wafer, if the device size is small enough (e.g., 250×200 μm, resulting in 20 die/mm^2). The devices can be broken up into individual die using either "cleaving," where they break along clean lines similar to the way glass sheets are scribed and broken, or sawing. They are usually held on semi-sticky films during this singulation process, and the film is slightly stretched to a larger dimension to provide some space

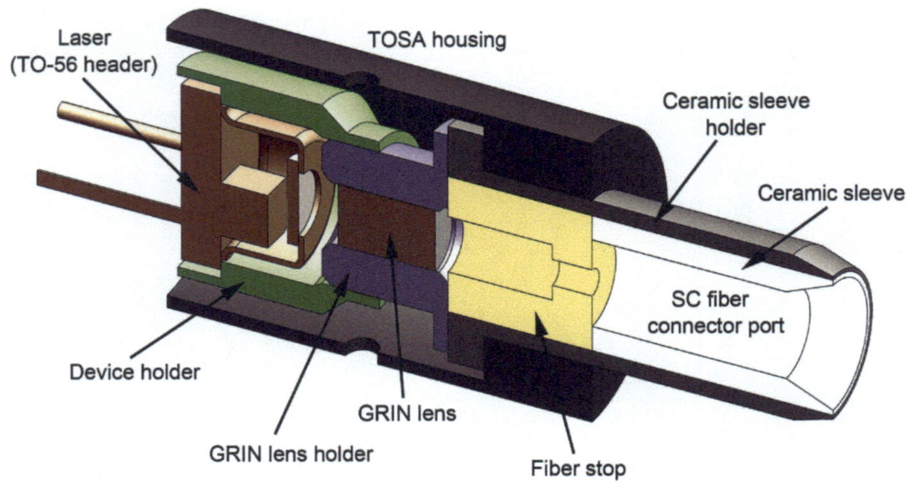

Figure 1.30 Cutaway of view of the components used in a typical data communications transmitter optical sub-assembly (TOSA).

Source: Reprinted with permission from K.P. Jackson and C.L. Schow, VCSEL-based transceivers for data communications, in: R. Michalzik (Ed.), VCSELs: Fundamentals, Technology and Applications of Vertical-Cavity Surface-Emitting Lasers, Springer: Berlin, Heidelberg, 2013, pp. 431–448, copyright 2013.

between die after the singulation process is completed. The good die are usually picked directly from this film, using a vacuum pickup needle on the top, and a "poking" needle on the bottom to help push the die off the film. They can then either be mounted in the package directly, or mounted on an intermediate sticky pack called a "gel pak" that has enough adhesion that it can go through shipping without worrying about the die coming off the pack.

After that, the die are mounted in a package. The package should, at a minimum, provide a secure platform for the die to be mounted, and electrical feedthroughs. The optoelectronic die can be attached to the package with adhesive (often a silver-filled epoxy) or alternately a solder. The die is electrically connected with the feedthroughs using wirebonding, in most cases, although some future die may use bump bonding to make electrical connections with the package. The package should also allow the die to be held securely relative to the optical fiber connections to avoid drift or misalignment over time. An example of a common package is shown in Fig. 1.30 [61]. The traditional construction of fiber optic transceivers involved the integration of transistor outline (TO) cans with top windows to allow the light through. The lids were made with glass, metal, and ceramic, in a totally hermetic package that prevented any risk of corrosion over time. The transmitter cans are referred to as a "TOSA," and the receivers with the packaged photodiodes are referred to as "ROSA" cans. Newer, lower cost transceiver assemblies use plastic only, which does not protect against atmospheric humidity, and thus requires die that are more rugged and can resist corrosion through additional design features such as dielectric oxide or nitride coatings applied during wafer fabrication.

Alternately, silicon photonics allow the integration of a large number of components on a single assembly that might otherwise require dozens of individual optical components.

1.4.5 Burn-in

Nearly all lasers and LEDs go through an accelerated aging process to stabilize the devices, called "burn-in." Burn-in seeks to force devices with infant mortality issues to quickly fail within a 1–2-day period, while not removing useful life from good devices. The burn-in, plus other quality control measures, can add substantially to the cost of the product, but is usually necessary to reach acceptable field failure rates, and is standard for all data communication and telecommunication applications. The burn-in aging would usually be done from 85°C to 150°C, and 2–4 × normal use current, depending on how rapidly infant mortality issues are accelerated by light, current, or thermal acceleration. (More light is present at the lower temperatures, and more thermal acceleration at the higher temperatures.)

Burn-in is usually done on packaged assemblies such as TOSAs, which in turn could be mounted into burn-in boards. However, more recently, lower cost methods, such as submount-level burn-in or wafer-level burn-in, are becoming more popular. Wafer-level burn-in involves testing all devices before aging, aging the entire wafer at once (or possibly in a few passes), and then testing again. Note that most burn-ins seek to distinguish between good devices that are stable (less than 2% change), and bad devices that are degrading (over 5% change during burn-in is a common upper cutoff). Because there is far more than this level of natural variation in incoming parameters, burn-in requires pre-and-post testing, and the pass or fail determination is off of the *change* in parameters, which is also termed a "delta specification."

Photodiodes generally do not need burn-in for most applications, due to much lower current densities. The large majority of types of photodiodes have very low or zero field failure rates, with the primary cause of failure being packaging-related issues such as broken wirebonds.

1.5 Critical metrics

A laser is separated from incoherent light sources by three main differences: a tightly confined beam with low divergence, narrow spectral width (monochromatic), and coherence. We discuss each of these elements first, and then other key metrics for optoelectronic devices.

1.5.1 Beam divergence

An incoherent emitter, in its purest form such as that from an incandescent bulb, emits uniformly in all directions. This property may be useful for lighting up a

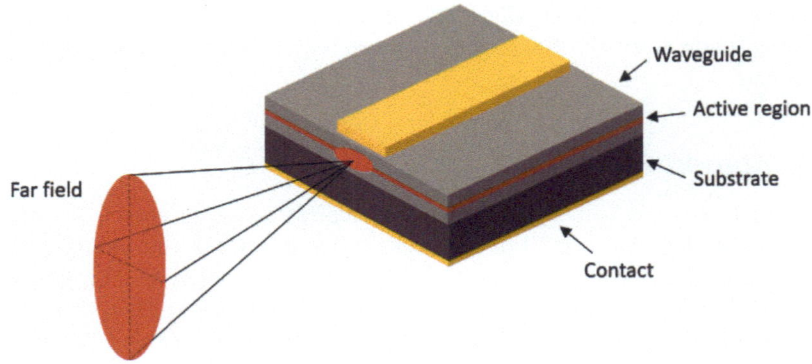

Figure 1.31 Drawing of a conventional edge emitter, showing the oval far field pattern, with fast divergence in the transverse (vertical) axis, and slower divergence in the lateral (horizontal) axis.

room, but is detrimental for display devices such as a projector, designed to direct all the light in just a few percent of that solid angle, or other applications that want to direct the energy into a tight beam or focus onto a small spot. Popular semiconductor laser designs naturally provide energy conversion at over 10% PCE with beam output characteristics that allow focusing of the beam to a near-diffraction-limited spot size. As we saw earlier, edge-emitting semiconductor lasers are inferior in natural beam characteristics due to their asymmetric shape. The beam divergence can reach 25 degrees along its vertical direction (fast axis), but only around 8 degrees along its horizontal axis (slow axis), as we can see in Fig. 1.31. This elliptical beam shape is not desirable for focusing. Typically, focusing the beam would require optics with crossed cylinder lenses, or complex molded optics required to ensure a high beam quality, and to shape the beam into something we should all be familiar with in the red laser pointers that can be purchased for <$10 today. By contrast, the VCSEL has a low beam divergence angle of less than 12 degrees due to its symmetric optical output, which allows a single low-cost spherical lens to focus the beam. Among other types of lasers, solid-state crystal lasers (such as the popular Nd:YAG laser) have a very low beam divergence angle, which is only 0.019 degrees. However, solid-state lasers and gas lasers with lower beam divergence have other disadvantages, such as high cost, bulky size, ruggedness, and high power consumption.

1.5.2 Single mode versus multimode

The second key indicator of a laser vs. other light emitters is that a laser is monochromatic, while other lamps are "polychromatic," meaning that they emit within a range of wavelengths that spread over dozens or even hundreds of nanometers. While lasers can emit multiple wavelengths simultaneously ("multilongitudinal mode"), the peaks usually fall within a few nm of each other for all of the included lasing peaks, and for the purposes of many applications, that is still

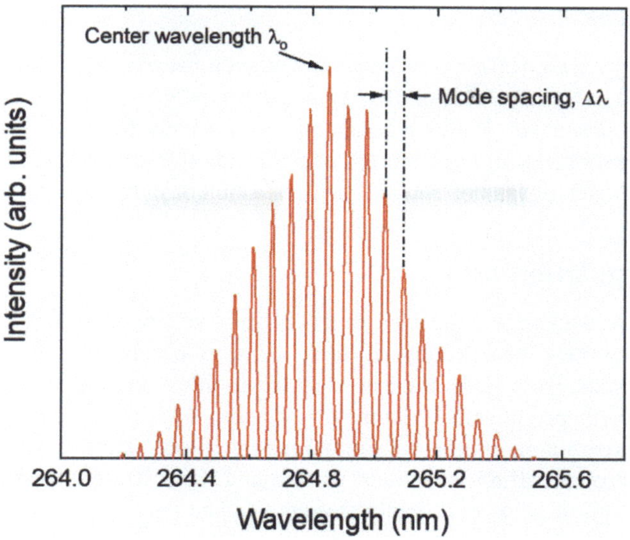

Figure 1.32 Spectra of a multimode AlGaN MQW laser, showing over a dozen modes lasing simultaneously.

"monochromatic." See Fig. 1.32 for an example of a spectrum from a multilongitudinal laser spectrum.

However, for fiber optic communications, these lines can travel at slightly different speeds through fiber optic cables and cause dispersion, thus degrading the signal the further it travels, as some of the light from later bits "catches up" with the light from earlier ones. As mentioned earlier, this problem can be avoided by building in wavelength features such as a wavelength grating that provides high optical loss for all wavelengths other than the desired primary wavelength, to make a single wavelength "distributed feedback" laser.

1.5.3 Coherence length

The third and final main indicator of laser emission is coherence, with the photons all being in phase with one another, allowing applications like holography, or coherent communication. In these applications, constructive and destructive interference is used. A key indicator is "coherence length," which is actually a more refined indicator of how monochromatic and phase-stable the light is. This indicates how far you could send the light out and back, and still be able to usefully perform coherent mixing. For example, a typical target for coherent fiber optic communication systems made today is 35–300 kHz linewidth, which corresponds to a 0.5–4 km coherence length in free space [62].

1.5.4 Power

Optical power and conversion efficiency are common metrics today. Most semiconductor lasers used today have at least 10% conversion efficiency of the electrical power in to the optical power out. Power requirements vary widely from a few milliwatts (for communications or sensing) to hundreds of milliwatts (for applications like optical disc writing or LiDAR) to watts of power (for illumination or industrial applications).

1.5.5 Modulation rate

The fiber optic data communications industry follows a path similar to Moore's law of introducing new generations of fiber optic transceivers with a 20% speed increase per year. Each generation is expected to allow not only a density increase (more bandwidth per physical volume), but also a 50% reduction in price per bit, and a 30% reduction in energy consumption per bit [63]. These improvements have taken international calling and video on demand from being expensive luxuries a few decades ago to being practically free today. The way that increase in transmission rate is being done, however, has started to change, from simply increasing the modulation rate to using multilevel or coherent modulation. Some additional terminology is useful here. Data communication rate, or the number of bits per second, is just the product of the baud rate (how many signal periods per second), times the number of bits per baud period.

While the number of bits per baud was traditionally only 1 (thus making bit rate and baud rate interchangeable in discussion), schemes to increase it to 2 or more bits per baud are becoming more popular. As mentioned in Section 1.3.3.3, increasing the modulation rate has become progressively more difficult over the past decade. Therefore, as an alternate, in the short term, multilevel analog modulation is being incorporated for an increasing number of next-generation interconnects, such as "PAM4," where 2 bits can be sent in each time slot, with four levels being allowed, as seen in Fig. 1.33 [64]. Long term, more bits can be sent in each slot using coherent modulation, which we will discuss further in the final "upcoming technologies" in Section 1.7.1.

1.6 Laser and light-emitting diode reliability

While this chapter is in a book focused on the reliability of lasers and LEDs, we offer a brief introduction for completeness, to make this a standalone chapter. Obviously, more detail can be found in this book and the other books listed in the references.

1.6.1 Reliability qualification

During the product design and qualification phase, proposed new products go through reliability testing, to detect any potential design weaknesses. As detailed in Chapter 2, Reliability engineering in optoelectronic devices and fiber optic transceivers, testing takes place on the subcomponent level (e.g., lasers and photodiodes),

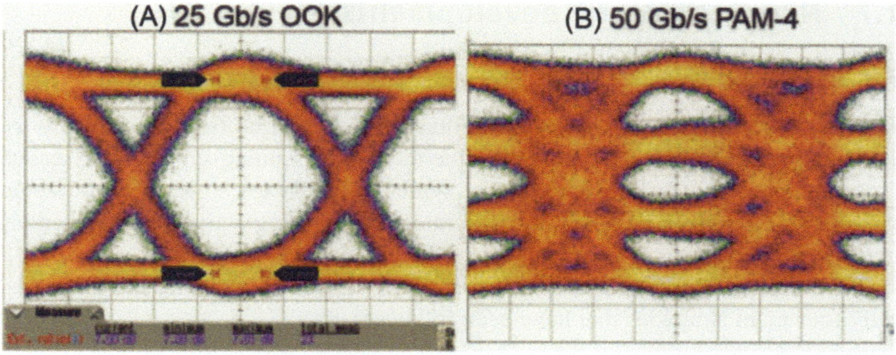

Figure 1.33 Typical on–off keying (OOK) (left) has only a high and low level and only 1 bit/baud period, while PAM4 coding (right) has intermediate levels, and encodes 2 bits/baud. In this case, both have the same baud rate of 25 Gb/s, and the same duration of 40 ps/bit period or signal unit.

Source: Reprinted, with permission, from C. Xiong, D. Gill, J. Proesel, J. Orcutt, W. Haensch, and W.M.J. Green, A monolithic 56 Gb/s CMOS integrated nanophotonic PAM-4 transmitter, in: 2015 IEEE Optical Interconnects Conference (OI), Apr. 2015, pp. 16–17, https://doi.org/10.1109/OIC.2015.7115665 (2015) IEEE.

the component level (transmitter or receiver assemblies), and the system level (e.g., transceiver, lamp, or system). The lower level tests allow higher acceleration levels to be reached, and larger quantities to be tested, with higher measurement accuracy, while the system-level tests allow full high-speed functionality and packaging issues to be included in the testing. Particularly for system-level testing, standards-based testing is commonplace, for example, the use of the Telcordia GR-468 standard for qualifying fiber optic transceivers.

1.6.2 Quality control

For products that have passed their reliability qualification, a number of precautions are used to control quality and reliability. In addition to the burn-in on 100% of devices, a 100% visual inspection is performed to allow the removal of any devices with cracks, scratches, or other handling or fabrication damage. While the burn-in is only intended to detect "infant mortality" (early-life failure) issues, and be the equivalent of 6–24 months of normal use, many manufacturers also carry out a harsher destructive sample test on 20–100 devices per wafer (out of the 50,000 die per wafer). This destructive test is intended to ensure that no reduction in wearout lifetime has occurred; it is the equivalent of 10–40 years of normal operation. If any failures occur in this test, the wafer will fail the wafer qualification test, and not move on to manufacturing. Finally, a sample of transceivers or finished assemblies will periodically (every 1–3 months) be reliability tested to ensure that the original build standard present during the reliability qualification has been maintained, and that no degradation has occurred due to unintentional changes in incoming part quality or build methods. Strict change control and careful documentation are also required to prevent intentional changes without proper qualification testing.

1.7 New technology developments

1.7.1 Fiber optics

As mentioned in Section 1.5.5, the information transfer rate (measured in bits per second) is the product of the baud rate (the number of sending symbols per second) times the number of bits transmitted per symbol. We discuss some future strategies to increase the overall data transmission rate that are being actively researched and implemented today.

To increase the amount of data sent through a fiber optic cable, there are basically four main strategies that have been followed:

1. Increase the baud rate by speeding up modulation, and thus reducing the bit period to increase the number of bits sent per second. As discussed earlier, it is difficult with today's technology to exceed 50 Gbaud, due to electrical limits, so the strategies below are preferred by most researchers as alternates. (However, because some 100 and even 200 Gbaud results have been demonstrated in labs and published [65,66], we can't rule out this option in the longer term, even if it is not currently favored in roadmaps.)
2. Send multiple wavelengths of light down the same fiber, which is known as WDM, as discussed in Section 1.2.3. This strategy has been widely used since the late 1990s, and for the purpose of this discussion, we will assume that it has been fully exploited in long-distance communication, but is becoming more popular in shorter reach communication.
3. Use more channels side-by-side per package. One option is to use a fiber ribbon rather than a single fiber. This option drives up fiber cost, but because the transceiver is simpler and cheaper than other options, it is popular as an alternative for short links. Another similar option is to have multiple cores in a single fiber, which is referred to as spatial division multiplexing.
4. Increase the number of bits sent per symbol, using either multiple bits per symbol, as mentioned for encoding schemes like PAM4, or better yet, phase encoding coherent communication.

Coherent communication encodes information with so-called "quadrature-amplitude modulation" (QAM). Fig. 1.34 shows a common form of 16-QAM used in optical systems. Two PAM-4 signals in quadrature phase ($\pi/2$ degrees apart) are

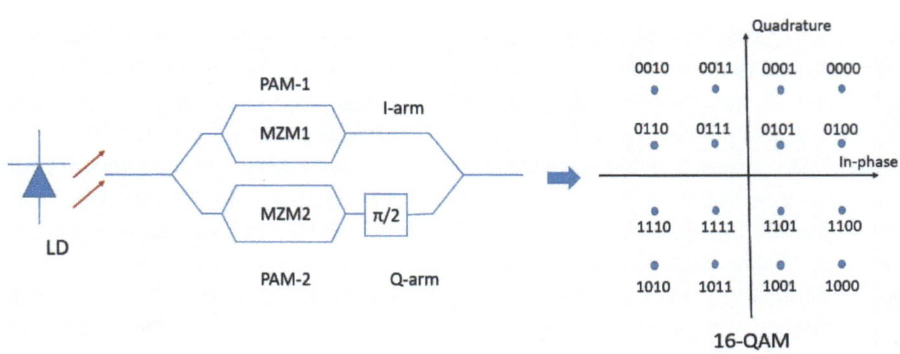

Figure 1.34 Diagram of how we can encode coherent signals with two interferometers (left), and the constellation of 16 points, each carrying 4 bits of data on it (right).

generated by using two separate Mach-Zehnder modulators. Thus, this modulation method has both phase and amplitude freedoms. 16-QAM allows twice the amount of information to be embedded within each wavelength channel compared with PAM-4 (4 vs 2 bits/baud). Coherent modulation is used today for long-distance communication (> 1000 km), and is starting to be phased in for Metro communication (10–100 km) over the next 5 years, and possibly in the data center and shorter-reach applications after that. Therefore many in the industry expect that this technology will be a major part of the future. One problem with the current implementation is power consumption: for longer links, the signal is substantially degraded through fiber dispersion, and requires power-hungry digital signal processing chips to recover the original phase information. However, it is thought that analog coherent communications links can avoid the power consumption of DSPs for shorter link lengths [67], and that coherent communication may well be the most cost-effective format 5–10 years from now.

A final technology that is being introduced is called co-packaging, as shown in Fig. 1.35 [68]. Fiber optic communication boxes today usually have the switch electronics in the center of the router, and the fiber optics at the edge of the box. When the data rates are low, this distance between the transceiver and the switching chip is not an issue. However, as the baud rate increases, power consumption from capacitive and inductive loading becomes an issue, along with signal degradation as the signals travel 10–30 cm from the transceiver to switch ASICs. It is thought that by co-packaging the optical communications chip right inside the switch ASIC package, power consumption savings of 30% or more can be obtained, along with a more compact design. Two challenges with co-packaging are:

1. Maintainability—Current boxes allow easy swapping of failed fiber optic transceivers, which are hot pluggable, right out of the front panel, while co-packaged optics will require disassembling the box to replace transceivers, and some configurations use soldered rather than socketed transceivers. This issue might be overcome through redundant channels in the fiber optics on co-packaged transceivers.
2. Heat removal—Because current designs have the cool air coming over the transceivers at the edge of the box, case temperatures are typically 35°C–55°C; by contrast, temperatures of 80°C–110°C are often discussed for the co-packaged transceivers that are mounted on the switch ASIC.

A compromise that is sometimes discussed is having the laser mounted at the edge or externally, with all other fiber optic components being co-packaged, since the components other than the laser do not tend to have challenges meeting reliability targets at high temperatures.

1.7.2 The future of optoelectronic devices

1.7.2.1 Photonic integrated circuits

While photonic integrated circuits (PICs) were discussed and even built in the 1980s, complex PICs with large numbers of different types of components only really started being made in the last two decades. Two platforms are widely

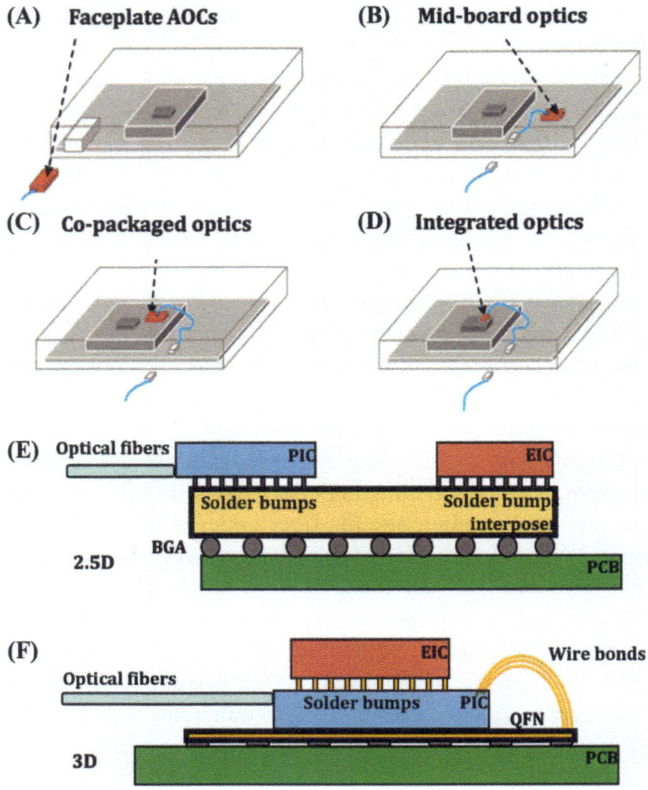

Figure 1.35 Integrated optics (D) minimizes signal degradation relative to conventional faceplate optics (A). Electronics and optics can be integrated as shown in the lower cross-sections (E) and (F). Options (B) and (C) show some "stepping stone" options that reduce power consumption and signal degradation without requiring the packaging technology development that full integrated optics solution require.

Source: Reprinted with permission from Q. Cheng, M. Bahadori, M. Glick, S. Rumley, and K. Bergman, Recent advances in optical technologies for data centers: a review, Optica 5 (11) (2018) 1354, https://doi.org/10.1364/OPTICA.5.001354, © 2018. The Optical Society of America.

available: InP-based PICs and silicon photonic circuits. The former has the advantage of having a range of lasers that can easily be grown on the InP substrate. However, silicon photonics seem to be increasing in relative market share, and have excellent components for everything except the laser, with good ways to integrate lasers through heterogeneous integration with InP, or couple external lasers in using fiber optic cables or direct mounting of lasers onto the silicon photonic wafers.

1.7.2.2 LiDAR

Self-driving cars are one of the biggest upcoming opportunities in the market [69–73]. These autonomous vehicles are typically outfitted with three types of

Figure 1.36 LiDAR devices such as those shown (left) enable a 3-D map (right) to be formed by computing the range to each reflection.

Source: Courtesy Velodyne.

sensors: multiple cameras, radar, and LiDAR, or "light detection and ranging." LiDAR-enabled detection systems enable the car to quickly establish distance and velocity to every object larger than 3 cm, at a range of up to 200 m. While current LiDAR systems for vehicles cost over US \$10,000, for the systems to become mainstream, LiDAR units will have to drop to a price of under US\$1000. Volumes are expected to exceed 42 million units per year by 2035, for a total market of over US\$28 billion. Examples of commercially available LiDAR systems, and the brightness and distance data cloud, are shown in Fig. 1.36.

1.7.2.3 Quantum dot lasers

While quantum dot lasers have been discussed for many years, and were the focus of many research groups in the 1990s, they finally appear to be reaching such a state of maturity and performance advantages over quantum well lasers that they may fill more than a small commercial niche. For native substrate quantum dot lasers, the two greatest advantages appear to be improved ability to operate at high temperatures, and resistance to back-reflection-induced instability. In addition to quantum dots on native III–V substrates such as GaAs or InP, quantum dot lasers grown on silicon substrates have many advantages for both integration with silicon photonics and cost reduction. This topic is the subject of Chapter 6, Reliability of lasers on silicon substrates for silicon photonics, and the papers in the references, to which the reader is referred for more detail. Many quantum dot laser designs look similar to quantum well lasers, as shown in Fig. 1.37 [74].

1.7.2.4 Photonic band gap/plasmonics

To meet future green data center standards, energy consumption for each bit should be reduced to less than 1 pJ/bit level. Photonic crystal lasers and plasmonic lasers may offer a long-term technical solution. These nanoscale lasers can confine photons into a small volume and achieve an ultra-low threshold lasing. Thus, they have a high direct modulation rate of over 100 GHz, low power consumption, and a

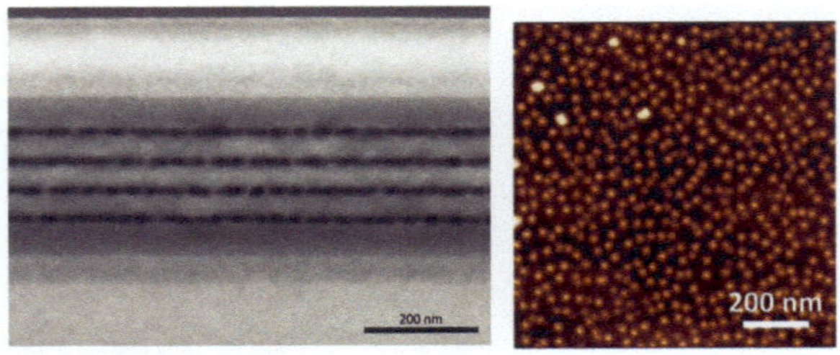

Figure 1.37 The TEM image of a four-layer quantum dot laser is shown on the left; on the right, an AFM scan of one layer of dots shows the density of the dots, with ~ 600 dots/μm^2.

Source: Reprinted from J.C. Norman, D. Jung, Y. Wan, and J.E. Bowers, Perspective: the future of quantum dot photonic integrated circuits, APL Photonics 3 (3) (2018) 030901, https://doi.org/10.1063/1.5021345, with the permission of AIP Publishing.

Figure 1.38 Photo of a blue laser in a TO can (left). Lasers like this can be used for display, printing, or chemical processing applications, with lenses to collimate and focus the light. Phosphor plates after it can convert the blue light to white light for illumination. Headlights based on blue lasers are now starting to be used in premium vehicles (right), and are expected to become more widespread over the next 5 years. *TO*, transistor outline.

Source: Left image adapted from a photo courtesy Osram; right image courtesy BMW.

small footprint, making them a good candidate for on-chip lasers used for interconnects in future data centers [75].

1.7.2.5 High-efficiency lighting and display markets

For high-efficiency lighting, the market in the developed world will eventually saturate, as legacy incandescent and fluorescent lighting is replaced with high-efficiency LED-based lighting. After that, increasing penetration into developing countries will continue for a decade or more. Improved color quality from LED lighting will be an increasingly valued metric. The blue light from the Sun has a strong effect on our body clocks, and lighting that helps to mimic that daily cycle (more blue light during the day, and less in the evening or at night) will become more widely available, along with better quality lights with improved color-rendering index values. High color gamut displays powered by three-color LEDs, or three-color lasers, will allow more accurate color rendering, and are now starting to be used in digital cinemas, but will become more affordable for home theaters and the consumer display and TV market. Micro-LED displays (with millions of

LEDs directly emitting) are another new technology with a lot of promise [76], although many production cost challenges need to be overcome before it can be available outside of premium price tiers. Another novel laser application will be blue lasers for high-efficiency lighting, as seen in Fig. 1.38. Powerful blue lasers can be used to pump white phosphors in a much tighter area than can be obtained with LEDs. These laser-driven light sources produce focused light with high directionality and contrast, allowing for applications such as projectors or car headlights. Further, the blue lasers themselves can be far smaller than LED chips of a comparable power, allowing for lower chip costs.

1.8 Summary

In summary, photonic devices have provided many important benefits over the past half-century, and are certain to become far more important over the next half-century. They are growing much faster than the technology market as a whole, and offer answers to many other problems that we have not had a chance to touch on here, such as biosensing, Internet of Things, optical computing, imaging, and display applications. In addition, many novel structures, such as organic photonics, graphene, or photonic band gap structures have promise for reduced cost or improved performance over what is available today. Higher levels of integration in silicon photonics or increasingly complex PICs are expected to continue to lower the cost-per-bit for optical communications for at least the next few decades, and will support more data-intensive applications like directing autonomous vehicles or artificial intelligence-based smart computing. We hope this overview of how the various technologies work, and are built, has been useful, and refer the reader to the references if more detail in any of the specific technologies is desired.

References

[1] E. Mills, Job creation and energy savings through a transition to modern off-grid lighting, Energy Sustain. Dev. 33 (2016) 155–166. Available from: https://doi.org/10.1016/j.esd.2016.06.001.

[2] PR_OPTICAL_TRANSCEIVERS_INDUSTRY_STATUS_YOLEGROUP_June2020.pdf. [Online]. Available: http://www.yole.fr/iso_upload/News/2020/PR_OPTICAL_TRANSCEIVERS_INDUSTRY_STATUS_YOLEGROUP_June2020.pdf (accessed 15.06.20).

[3] PR_LIDAR_IndustryUpdate_YOLE_April2019.pdf. [Online]. Available: http://www.yole.fr/iso_upload/News/2019/PR_LIDAR_IndustryUpdate_YOLE_April2019.pdf (accessed 15.06.20).

[4] PR_BIOMETRICS_CONSUMER_IndustryOverview_YOLE_Dec2018.pdf. [Online]. Available: http://www.yole.fr/iso_upload/News/2018/PR_BIOMETRICS_CONSUMER_IndustryOverview_YOLE_Dec2018.pdf (accessed 15.06.20).

[5] Mounier, E. and J.-L. Malinge, Yole Market report on Photonic Integrated Circuits and Silicon Photonics (PIC market to increase from $4B in 2018 to $19B in 2024). http://www.yole.fr/PhotonicIC_SiPhotonics_MarketUpdate_Intel.aspx#.XugTqPJlBBw (accessed 15.06.20).
[6] L.A. Coldren, S.W. Corzine, M.L. Mashanovitch, Diode Lasers and Photonic Integrated Circuits, 2nd ed., Wiley, Hoboken, N.J, 2012.
[7] J.T. Verdeyen, Laser Electronics, 3rd ed., Pearson, Englewood Cliffs, N.J., 1995.
[8] S.O. Kasap, Optoelectronics & Photonics: Principles & Practices, 2nd ed., Pearson, Boston, 2012.
[9] B. Culshaw, Introducing Photonics, Cambridge University Press, Cambridge; New York, NY, 2020.
[10] M.G. Craford, LEDs for solid state lighting and other emerging applications: status, trends, and challenges, in: Proc. of the SPIE, vol. 5941, Sep. 2005, p. 594101, https://doi.org/10.1117/12.625918.
[11] X. Perpiñà, et al., Thermal management strategies for low- and high-voltage retrofit LED lamp drivers, IEEE Trans. Power Electron. 34 (4) (2019) 3677–3688. Available from: https://doi.org/10.1109/TPEL.2018.2853119.
[12] U.S. Department of Energy, How Energy-Efficient Light Bulbs Compare with Traditional Incandescents, Energy.gov. https://www.energy.gov/energysaver/save-electricity-and-fuel/lighting-choices-save-you-money/how-energy-efficient-light (accessed 23.10.19).
[13] J. Penning, K. Stober, V. Taylor, and M. Yamada, Energy Savings Forecast of Solid-State Lighting in General Illumination Applications, Navigant Consulting Inc., Washington, DC (United States), DOE/EE-1467, Sep. 2016. https://doi.org/10.2172/1374119.
[14] Chin Han Lin, Weiping Li, and Xiaofeng Fan, "Creating arbitrary patterns on a 2-D uniform grid VCSEL array", U.S. Patent 10,153,614B1, issued Dec 11, 2018, and assigned to Apple, Inc.
[15] P. Huang, Adoption Rate of 3D Sensing in Smartphones is Estimated at 13.1% in 2018, while Apple Remains the Key Driver, Says TrendForce, TrendForce Corp., 2018. https://press.trendforce.com/press/20180307-3073.html (accessed 24.10.19).
[16] VCSELs - Market and Technology Trends 2019, i-Micronews. https://www.i-micronews.com/products/vcsels-market-and-technology-trends-2019/ (accessed 24.10.19).
[17] A. Liu, P. Wolf, J.A. Lott, D. Bimberg, Vertical-cavity surface-emitting lasers for data communication and sensing, Photonics Res. 7 (2) (2019) 121–136. Available from: https://doi.org/10.1364/PRJ.7.000121.
[18] A.V. Krishnamoorthy, et al., Progress in low-power switched optical interconnects, IEEE J. Sel. Top. Quantum Electron. 17 (2) (2011) 357–376. Available from: https://doi.org/10.1109/JSTQE.2010.2081350.
[19] T. Aihara et al., 56-Gbit/s Operations of Mach-Zehnder Modulators Using 300-μm-long Membrane InGaAsP Phase Shifters and SiN waveguides on Si, in: Optical Fiber Communication Conference (OFC) 2019 (2019), paper M4A.3, Mar. 2019, p. M4A.3, https://doi.org/10.1364/OFC.2019.M4A.3.
[20] J.S. Eng, C. Kocot, Optical Fiber Telecommunications VIA: Chapter 14. VCSEL-Based Data Links, Elsevier Inc, 2013.
[21] K. Grobe, Chapter 5—Optical wavelength-division multiplexing for data communication networks, in: C. DeCusatis (Ed.), Handbook of Fiber Optic Data Communication, 4th Ed., Academic Press, Oxford, 2013, pp. 85–122.
[22] J. Hecht, Beam: The Race to Make the Laser, 1st ed., Oxford University Press, Oxford; New York, 2005.
[23] J.L. Bromberg, The Laser in America, 1950-1970, MIT Press, 1991.

[24] From the 50th anniversary keynote presented at OFC 2020 by David Welch March 10, 2020, and archived at: https://www.ofcconference.org/library/images/ofc/PDF/2020/50-Years-of-Light-Speed-Connections.pdf.
[25] G.P. Agrawal, Optical communication: its history and recent progress, in: M.D. Al-Amri, M. El-Gomati, M.S. Zubairy (Eds.), Optics in Our Time, Springer International Publishing, Cham, 2016, pp. 177−199.
[26] O. Kumagai, M. Ikeda, M. Yamamoto, Application of laser diodes to optical disk systems: the history of laser diode development and mass production in three generations of optical disk systems, Proc. IEEE 101 (10) (2013) 2243−2254. Available from: https://doi.org/10.1109/JPROC.2013.2275017.
[27] Newport Corporation, Laser Diode Technology. https://www.newport.com/t/laser-diode-technology, 2019 (accessed 26.10.19).
[28] Y. Suematsu, K. Iga, Semiconductor lasers in photonics, J. Light. Technol. 26 (9) (2008) 1132−1144. Available from: https://doi.org/10.1109/JLT.2008.923615.
[29] P.J. Winzer, D.T. Neilson, A.R. Chraplyvy, Fiber-optic transmission and networking: the previous 20 and the next 20 years [Invited], Opt. Express 26 (18) (2018) 24190−24239. Available from: https://doi.org/10.1364/OE.26.024190.
[30] P. Winzer, Scaling optical fiber networks: challenges and solutions, Opt. Photonics News 26 (2015). Available from: https://doi.org/10.1364/OPN.26.3.000028.
[31] J. Hecht, Short history of laser development, Opt. Eng. 49 (9) (2010) 091002. Available from: https://doi.org/10.1117/1.3483597.
[32] J. Hecht, Laser Pioneers, 1st ed., Academic Press, Boston, 1991.
[33] J. Hecht, City of Light: The Story of Fiber Optics, Rev. ed., Oxford University Press, 1999.
[34] Pslawinski, Neon. Source: https://commons.wikimedia.org/wiki/File:NeTube.jpg, 2005.
[35] T. Markstein, Deutsch: Helium-Laser in Betrieb (from https://commons.wikimedia.org/wiki/File:Henelaser.jpg). 2009.
[36] N. Holonyak, S.F. Bevacqua, Coherent (visible) light emission from Ga(As1 − xPx) junctions, Appl. Phys. Lett. 1 (4) (1962) 3.
[37] M.G. Craford, From Holonyak to Today, Proc. IEEE 101 (10) (2013) 2170−2175. Available from: https://doi.org/10.1109/JPROC.2013.2274911.
[38] I. Dafinei, et al., Production of 82Se enriched Zinc Selenide (ZnSe) crystals for the study of neutrinoless double beta decay, J. Cryst. Growth 475 (2017) 158−170.
[39] T. Markvart, L. Castañer, Chapter IA-2—Semiconductor materials and modelling, in: A. McEvoy, T. Markvart, L. Castañer (Eds.), Practical Handbook of Photovoltaics, 2nd Ed., Academic Press, Boston, 2012, pp. 33−62.
[40] T.-Y. Seong, J. Han, H. Amano, H. Morkoç (Eds.), III-Nitride Based Light Emitting Diodes and Applications, Springer, Netherlands, 2013.
[41] D.A. Steigerwald, et al., Illumination with solid state lighting technology, IEEE J. Sel. Top. Quantum Electron. 8 (2) (2002) 310−320. Available from: https://doi.org/10.1109/2944.999186.
[42] F.A. Kish, R.M. Fletcher, Chapter 5 AlGalnP light-emitting diodes, in: G.B. Stringfellow, M.G. Craford (Eds.), Semiconductors and Semimetals, vol. 48, Elsevier, 1997, pp. 149−C3.
[43] X. Luo, R. Hu, S. Liu, K. Wang, Heat and fluid flow in high-power LED packaging and applications, Prog. Energy Combust. Sci. 56 (2016) 1−32. Available from: https://doi.org/10.1016/j.pecs.2016.05.003.
[44] J. Piprek, Efficiency droop in nitride-based light-emitting diodes, Phys. Status Solidi A 207 (2010) 2217−2225. Available from: https://doi.org/10.1002/pssa.201026149.

[45] J. Cho, E.F. Schubert, J.K. Kim, Efficiency droop in light-emitting diodes: challenges and countermeasures, Laser Photonics Rev. 7 (3) (2013) 408–421. Available from: https://doi.org/10.1002/lpor.201200025.
[46] Y.C. Shen, G.O. Mueller, S. Watanabe, N.F. Gardner, A. Munkholm, M.R. Krames, Auger recombination in InGaN measured by photoluminescence, Appl. Phys. Lett. 91 (14) (2007) 141101. Available from: https://doi.org/10.1063/1.2785135.
[47] J. Iveland, L. Martinelli, J. Peretti, J.S. Speck, C. Weisbuch, Direct measurement of Auger electrons emitted from a semiconductor light-emitting diode under electrical injection: identification of the dominant mechanism for efficiency droop, Phys. Rev. Lett. 110 (17) (2013) 177406. Available from: https://doi.org/10.1103/PhysRevLett.110.177406.
[48] C. Weisbuch, Historical perspective on the physics of artificial lighting, Comptes Rendus Phys. 19 (3) (2018) 89–112. Available from: https://doi.org/10.1016/j.crhy.2018.03.001.
[49] T.D. Moustakas, R. Paiella, Optoelectronic device physics and technology of nitride semiconductors from the UV to the terahertz, Rep. Prog. Phys. 80 (10) (2017) 106501. Available from: https://doi.org/10.1088/1361-6633/aa7bb2.
[50] I.S. Amiri, et al., Introduction to photonics: principles and the most recent applications of microstructures, Micromachines 9 (9) (2018). Available from: https://doi.org/10.3390/mi9090452.
[51] L.A. Coldren, R.S. Geels, S.W. Corzine, J.W. Scott, Efficient vertical-cavity lasers, Opt. Quantum Electron. 24 (2) (1992) S105–S119. Available from: https://doi.org/10.1007/BF00625818.
[52] B. Wang, et al., Modulation on silicon for datacom: past, present, and future, Prog. Electromagn. Res. 166 (2019) 27.
[53] J.C. Campbell, Optoelectronic Devices for Fiber Optics, vol. IEEE Online Course EDP013, 2009.
[54] J.C. Campbell, Recent advances in avalanche photodiodes, J. Light. Technol. 34 (2) (2016) 278–285. Available from: https://doi.org/10.1109/JLT.2015.2453092.
[55] M. Diepold, et al., Improved x-ray detection and particle identification with avalanche photodiodes, Rev. Sci. Instrum. 86 (5) (2015) 053102. Available from: https://doi.org/10.1063/1.4921195.
[56] Semiconductors_module_02.pdf. [Online]. Available: https://learnabout-electronics.org/Downloads/Semiconductors_module_02.pdf (accessed 08.08.20).
[57] R.M. Biefeld, D.D. Koleske, J.G. Cederberg, The science and practice of metal-organic vapor phase epitaxy (MOVPE), Handbook of Crystal Growth, Elsevier, 2015, pp. 95–160.
[58] A typical calculation might be 13mA peak in an 8um disc for a VCSEL, wtih 3x8nm-thick QWs, or 126mA peak in a 2 x 120um edge emitter, with 8 x 7nm-thick QWs. Both give about 10GW/cm^3, or 10mW/um, which is as high as anything in transistors (see transistor papers by Eric Pop), but includes much more minority carrier recombination.
[59] E. Pop, Energy dissipation and transport in nanoscale devices, Nano Res. 3 (3) (2010) 147–169. Available from: https://doi.org/10.1007/s12274-010-1019-z. Mar.
[60] E. Pop, S. Sinha, K.E. Goodson, Heat generation and transport in nanometer-scale transistors, Proc. IEEE 94 (8) (2006) 1587–1601. Available from: https://doi.org/10.1109/JPROC.2006.879794.
[61] K.P. Jackson, C.L. Schow, VCSEL-based transceivers for data communications, in: R. Michalzik (Ed.), VCSELs: Fundamentals, Technology and Applications of Vertical-Cavity Surface-Emitting Lasers, Springer, Berlin, Heidelberg, 2013, pp. 431–448.

[62] Coherence length, Wikipedia. Apr. 24, 2020 [Online]. Available: https://en.wikipedia.org/w/index.php?title = Coherence_length&oldid = 952867496 (accessed 22.06.20).

[63] C.D.A.C. Inc, Trends in Silicon Photonics for Fiber Optic Communication. http://www.photonicsspectra-digital.com/photonicsspectra/january_2019/MobilePagedArticle.action?articleId = 1456634 (accessed 16.06.20).

[64] C. Xiong, D. Gill, J. Proesel, J. Orcutt, W. Haensch, and W.M.J. Green, A monolithic 56 Gb/s CMOS integrated nanophotonic PAM-4 transmitter, in: 2015 IEEE Optical Interconnects Conference (OI), Apr. 2015, pp. 16–17, https://doi.org/10.1109/OIC.2015.7115665.

[65] M. He, et al., High-performance hybrid silicon and lithium niobate Mach–Zehnder modulators for 100 Gbit s −1 and beyond, Nat. Photonics 13 (2019) 1. Available from: https://doi.org/10.1038/s41566-019-0378-6.

[66] P.O. Weigel, et al., Bonded thin film lithium niobate modulator on a silicon photonics platform exceeding 100 GHz 3-dB electrical modulation bandwidth, Opt. Express 26 (18) (2018) 23728–23739. Available from: https://doi.org/10.1364/OE.26.023728.

[67] C.L. Schow and K. Schmidtke, INTREPID: Developing Power Efficient Analog Coherent Interconnects to Transform Data Center Networks, in: Optical Fiber Communication Conference (OFC) 2019, San Diego, California, 2019, p. M4D.9, https://doi.org/10.1364/OFC.2019.M4D.9.

[68] Q. Cheng, M. Bahadori, M. Glick, S. Rumley, K. Bergman, Recent advances in optical technologies for data centers: a review, Optica 5 (11) (2018) 1354. Available from: https://doi.org/10.1364/OPTICA.5.001354.

[69] SupplierInsight, EDITORIAL – Demand for Lidar systems to rise further with growing focus on self-driving vehicles – SupplierInsight. https://supplierinsight.ihsmarkit.com/news/5239955/editorial-demand-for-lidar-systems-to-rise-further-with-growing-focus-on-self-driving-vehicles, May 08, 2017 (accessed 26.10.19).

[70] How LiDAR Technology Enables Autonomous Cars to Operate Safely, Velodyne Lidar. https://velodynelidar.com/newsroom/how-lidar-technology-enables-autonomous-cars-to-operate-safely/, Sep. 20, 2018 (accessed 26.10.19).

[71] LiDAR drives forwards, Nat. Photonics, 12, 8, pp. 441–441, Aug. 2018, https://doi.org/10.1038/s41566-018-0235-z.

[72] J. Hecht, Lidar for self-driving cars, Opt. Photonics News (2018) 28–33.

[73] T. Simonite, Self-driving cars rely on sensor technology not ready for the mass market, MIT Technology Review. https://www.technologyreview.com/s/603885/autonomous-cars-lidar-sensors/ (accessed 26.10.19).

[74] J.C. Norman, D. Jung, Y. Wan, J.E. Bowers, Perspective: the future of quantum dot photonic integrated circuits, APL Photonics 3 (3) (2018) 030901. Available from: https://doi.org/10.1063/1.5021345.

[75] S. Matsuo, T. Kakitsuka, Low-operating-energy directly modulated lasers for short-distance optical interconnects, Adv. Opt. Photonics 10 (3) (2018) 567–643. Available from: https://doi.org/10.1364/AOP.10.000567.

[76] MicroLED-Info | The Micro-LED Experts. https://www.microled-info.com/ (accessed 15.11.20).

Chapter 2

Reliability engineering in optoelectronic devices and fiber optic transceivers

Robert W. Herrick
Intel Corporation, Santa Clara, CA, United States

2.1 Reliability engineering organizations and management

2.1.1 Why does it matter? Why do we need reliability testing?

Semiconductor lasers have a remarkably difficult job, since they require extremely high carrier recombination rates to be able to get past the lasing threshold, and these same conditions predispose them to rapid aging if they are not properly designed and manufactured. Furthermore, the materials that are most popular (and often the lowest cost) for making lasers are those that are not inherently reliable, but require special precautions and monitoring. Product expectations continue to increase as data centers use larger numbers of fiber optic links in each rack, but are not planning on allocating more maintenance personnel to repair failing parts. As we shall see in Chapter 3, Case studies in fiber optic reliability, significant downside risks are also present, where compensation payments can be orders of magnitude larger than the profits obtained for the components sold. A higher-than-expected reliability failure rate also leads to customer dissatisfaction, removal from the approved vendor list, and lost sales. The purpose of this chapter is to describe some of the basic steps for building a strong reliability program, and establishing quality and reliability that exceed customer expectations. A strong understanding of how to design for reliability, and how to validate the design and manufacturing process, is essential to produce parts that meet or exceed the reliability target.

2.1.2 Staffing a reliability department

Reliability engineering requires a remarkably broad range of specialties to support the different types of work listed in this chapter. One pitfall is to assume a single or a few reliability specialists can perform all of the tasks listed in Fig. 2.1. Reliability engineering requires multiple specialties that need different academic backgrounds and work experience, and it is a "team sport," not an activity where a few strong engineers can do the entire set of operations. Fig. 2.2 shows an organization chart for a typical quality department. The quality department is the one tasked with

Reliability of Semiconductor Lasers and Optoelectronic Devices. DOI: https://doi.org/10.1016/B978-0-12-819254-2.00003-5

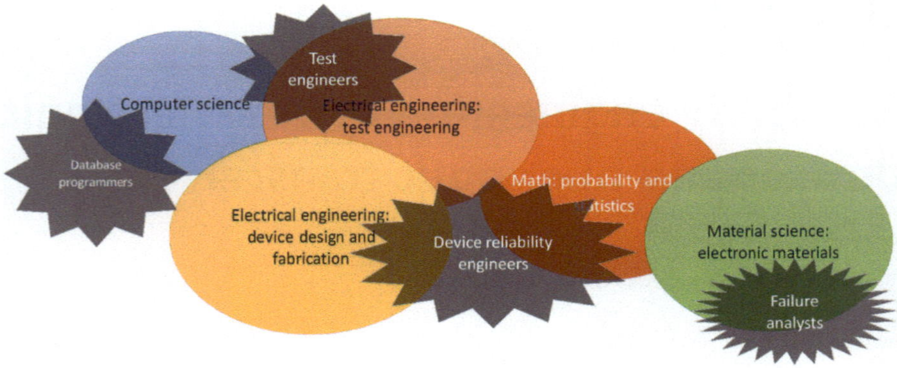

Figure 2.1 This diagram shows the academic backgrounds (ovals) and job descriptions (stars) of team members on the reliability team. One of the challenges is that the work is multidisciplinary, and most team members will be trained in only one or two of the academic disciplines, but will need to know many of them to operate at the highest levels.

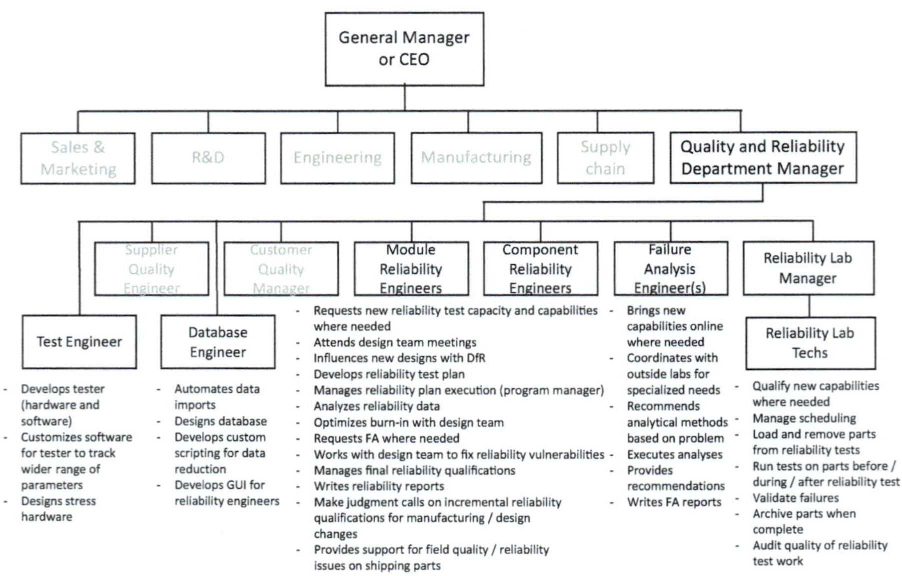

Figure 2.2 Organization chart of typical quality and reliability department roles and responsibilities.

maintaining product quality at or above customer and company expectations, and minimizing warranty and compensation costs. The quality department also acts as the internal voice of the customer, who is not directly present at the table when tradeoffs between cost, performance, and quality and reliability are being made. Fig. 2.3 shows a simplified flowchart for the work in the different functions in the reliability department. While management is very aware of the function and

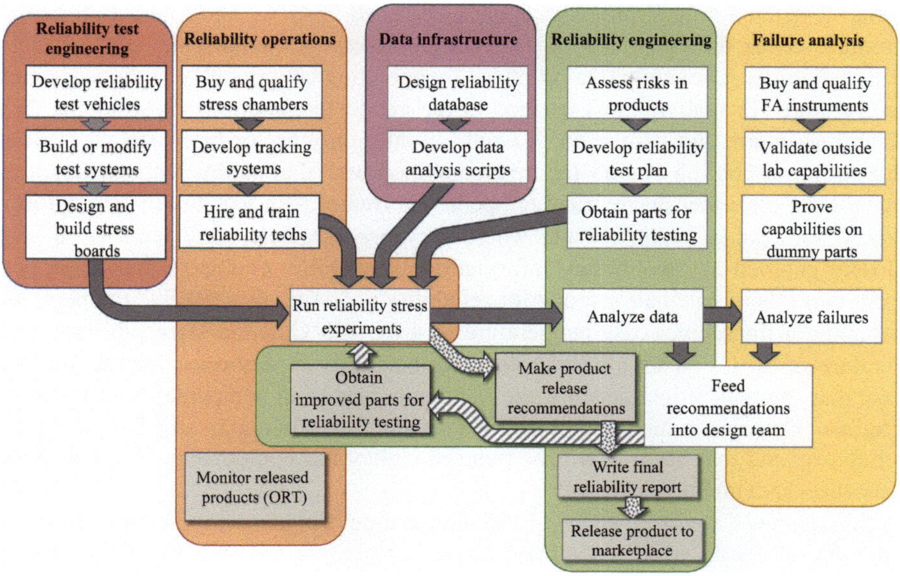

Figure 2.3 Simplified flow for work of the reliability team. Tasks are organized by different functional areas within the department.

importance of reliability operations and reliability engineering, in many companies, reliability test engineering support and data infrastructure support are areas of underinvestment. The return on investment for automating data rollup, or the potential for lower cost with test automation are less obvious, and test engineering is often understaffed, or combined with test engineering for product test, where the reliability test needs have trouble competing for resources.

Recruiting and training staff with the right background is usually challenging, since there are only a few programs in the world specializing in reliability engineering, and most work goes unpublished and is carefully guarded as company proprietary knowledge. Further, relatively little funded university research has been performed in understanding device aging, so much of what is known is understood only from an empirical basis, without the same level of underlying knowledge that is common in design-based disciplines.

2.2 Developing a product: design for reliability

2.2.1 Collecting information on customer expectations

Designing for reliability (DfR) is the process of "building in reliability" from the start. The first step is to understand customer needs, in terms of both "musts" and "wants." The three most critical customer requirements for the reliability engineer to gather are:

1. *Range of operating temperatures*. For most optoelectronic products, the standard data sheet "commercial-grade" temperature range is 0°C–70°C, although

some "extended-range" products cover −20°C to +85°C, and "industrial-grade" products cover the range −40°C to +85°C. However, for most modern products, laser reliability is severely compromised by operating continuously at the maximum temperature. In reality, most designs never get to temperatures close to the data sheet maximum, and if they do, it would usually only be for brief periods such as due to fan or cooling system failures, combined with the highest ambient temperatures. Data on actual deployment temperatures (rather than worst case) are unfortunately not readily available, and where they *are* available, they tend to be very closely guarded as "proprietary information." Reliability engineers thus often perform two sets of calculations: one for reliability operating at a "likely" temperature, assuming a transceiver case temperature of 40°C−50°C, and a laser substrate temperature of 50°C−60°C; a second set of calculations is performed for the unlikely case that someone might actually use it near the maximum temperature for its entire lifetime. Customers should be informed that active cooling is still beneficial for reliability, and that lifetime will be reduced if they run components at or near their data sheet maximum year-round.

2. *Expected "deployment time."* Another source of uncertainty is the question of how long products will be deployed in the field before they will be replaced with the next generation of parts, or what the expected lifetime is. At the low end, some consumer products might only be expected to be used for only a few years, or only turned on occasionally (think, for example, of an optical scanner hooked up to a home computer that only scans a few documents a month). In the middle of the range are fiber optic transceivers, where the deployment life can be as little as 3 years before they are replaced with the next generation, although 5−7 years is very common, and design for a 10-year lifetime is still a widely used target in the industry. At the high end, automotive, and some telecom, infrastructure calls for a 20-year lifetime. Here, depending on the design, there may be additional cost to designing for the longest life (especially at the highest temperature), so product marketing is expected to collect requirements from customers, and help in providing targets.

3. *Maximum allowed failure rate*. The acceptable failure rate, of course, always depends on the target application. At the least-demanding end might be flashing LED pendants that are given away as promotional prizes, where a few percent failure is fine, if that helps keep the cost low. In the middle, fiber optic transceivers usually have a failure rate target of <500 defects per million units shipped (dppm), or 0.05% failed, over the first year deployed. At the most demanding end, automotive, avionics, or military, are often looking for <50 dppm, and are willing to pay a high price for extensive screening to eliminate the last few failures, even if the screening costs more than the device itself. In most cases, customer feedback on existing products is the best indication. Where this does not exist, as in startups, employees with experience in similar market segments may be able to provide guidance. One of the best clarifying questions is to ask customers about where most of their failures come from. If the optoelectronics is one of the top three, chances are they will be interested in getting something better if it can be obtained without too much additional cost. However, if it is not in the top 5 or top 10, they might be

more interested in changes that drive cost down, rather than additional reliability improvements. For example, as shown in Fig. 2.4, a common data networking switch includes a large application-specific integrated circuit for switching with a 2%–4% failure rate over its 5-year deployment lifetime, a hard drive with a 1%–5% failure rate, and 32 transceivers. If the target for the 32 transceivers is to have a failure rate <1.5% over a 5-year lifetime (to avoid dominating the failure pareto), then the target failure rate would be 1.5%/32 = 0.047%, or 469 dppm per transceiver, which aligns well with the industry targets we mentioned earlier in this section, at least for early failures (during the first 1–6 months). An additional target would be for long-term reliability, where 300 "FITs" (short for "failures in time", or ~0.3%/year failure rate) is cited as a target, although it is common for transceivers to have failure rates 1–2 orders of magnitude lower than this.

2.2.2 Designing in margin and product tradeoffs

After customer requirements are collected, the next step in the design process is writing the "engineering requirements document," with all the specifications for the individual subcomponents. If the design allows the product to work even when the bias current doubles (100% increase criterion), gradual wearout will take far longer than a design where a 20% increase causes device failure, so building in the aging margin is another factor that should be considered during the design phase. One of the pitfalls is that most parameters in optoelectronic performance are improved by doing the opposite of what is good for reliability. The designer wants more optical output power (high bias), from the smallest laser possible (for both cost and power consumption reasons). By contrast, the reliability engineer wants low temperature, the lowest light output power requirement possible, and the current spread over a larger device for lower current density and lower thermal resistance. A larger laser will also in most cases have more margin for degradation, since it will be operating at a lower fraction of its peak power. A balance must be struck between the design and reliability, and the product designers need to be made explicitly responsible for

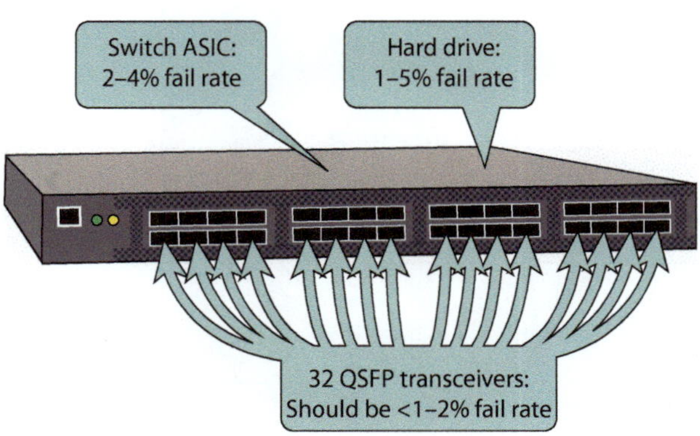

Figure 2.4 Drawing of a popular Ethernet switch design, with tree-key component reliability failure numbers listed for the duration of its deployment lifetime (usually 5–7 years).

meeting reliability targets, not just performance targets. In this way, we design-in reliability by increasing the strength of the product, and reducing the anticipated stresses on it.

2.2.3 Continuous improvement

Most of the optoelectronics industry has achieved very, very low failure rates through continuous improvement over decades of building ever more sophisticated products, and applying the learning from the previous generation to the next one. A strong failure analysis laboratory, and a "corrective action process" to apply anything learnt, are both key in getting this improvement. In Fig. 2.5, the iterative process of designing, testing, and improving the product is shown, with the target of <500 dppm (i.e., <1/20th of 1% failing) as the goal. Another important factor is a strong document control system, where the manufacturing process is well-documented, and changes to the process need to be qualified and approved through a sign-off list.

2.2.4 Ongoing reliability test

Another important part of any reliability program is an "ongoing reliability test" or ORT program. In this program, transceivers are randomly sampled every month or every quarter, and run through a few of the reliability tests from the GR-468 industry standard that is discussed in Section 2.3.4 [usually high temperature operating life (HTOL), 85/85, and temperature cycling], to verify that unexpected changes in the supplier part quality or build process have not taken place since the original GR-468 qualification. If failures are observed, failure analysis to determine the root cause and corrective action would be a first step, in addition to stepping up sampling. Sampling

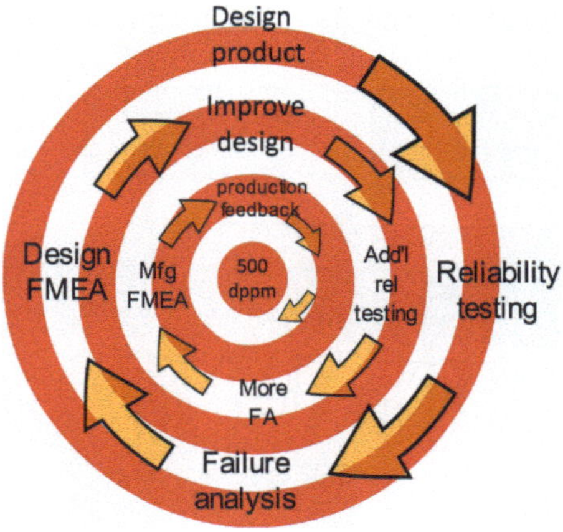

Figure 2.5 The iterative process of continuous improvement that is necessary to get to the very low failure rates obtained by the optoelectronic industry.

is more frequent for the first 6–12 months of production, and is only reduced if no failures are seen over a prolonged period. If failures are observed later, it may automatically trigger increased sampling for the next 6 months. In addition, module-level burn-in (a 24 or 48 hours HTOL on every module) is common for the first few thousand modules. Another industry-standard procedure is to temperature cycle all transceivers for 15–20 cycles before the final test, to weed out all parts with poor-quality solder joints or substandard discrete components.

2.2.5 Failure mode and effects analysis

In failure mode and effects analysis (FMEA), a large team representing all specialties meets for many hours, and creates tables of all opportunities for assembly or incoming quality defects that could lead to failure. Each opportunity is scored from 1–10 for three items: severity, detectability, and probability. It is common for thorough FMEA reports to have hundreds of items. After the items are scored, the three are multiplied together to get a "risk priority number" (RPN) score from 1 to 1000, where the highest score would be a very severe defect that was impossible to detect with tests in the factory, and was very probable to happen. The spreadsheet is sorted by RPN, and the team then focuses on what can be done to reduce the items with the highest RPN scores. The team looks for opportunities to design out the potential failure modes with the highest rankings, or where that is not possible, to reduce the probability and improve detection with additional screening tests. The new RPN scores after preventive action are reassessed, and the process is repeated to continuously improve the product until the assessed risk meets the targets [1].

Some companies with excellent reliability have a weak or nonexistent FMEA process, and rely on incremental product development using platforms they understand well, and do not change much from one generation to the next. But for totally new products, where there is little baseline, this process is a useful tool. This is particularly important for low-volume items like space-based deployment where every satellite could end up being unique, and only a single opportunity exists to get it right.

2.3 Developing a test plan: standards-based testing versus customized testing

2.3.1 Accelerated aging: an overview

People are intuitively familiar with certain types of reliability failures and aging. Incandescent light bulbs fail, car tires wear out, and rubber left out in the sun ages and cracks, as shown in Fig. 2.6A. That aging can be accelerated using solar concentrators such as those shown in Fig. 2.6B. Such concentrators allow us to get many years' worth of outdoor aging in just a few months of accelerated testing, and allow us to assess how long such items can go before failing, and whether additives to resist weathering are effective [2,3].

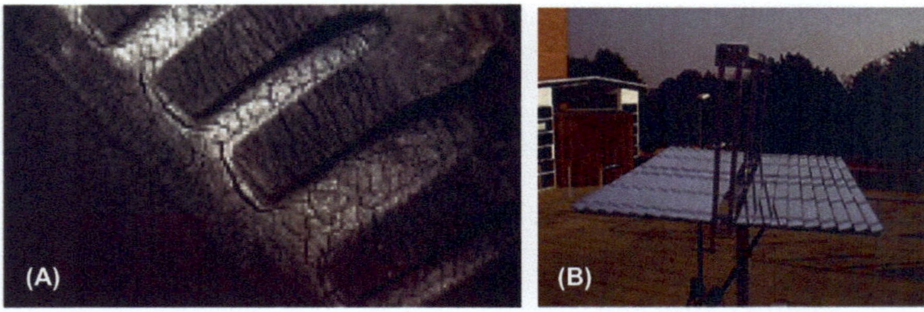

Figure 2.6 (A) Many types of rubber or polymers lose strength and crack and weaken in the presence of ultraviolet light and moisture. This type of aging can be accelerated with a solar concentrator mounted on a tracker (B), that focuses $>10\times$ the concentration of UV on the samples held at the focus of the parabola at the top of the tracker.

Source: (A) Copyright 2018, IEEE, Reprinted with permission from S. Nappi and G. Marrocco, Inkjet-printed RFID-skins for the detection of surface defects, in: 2018 2nd URSI Atlantic Radio Science Meeting (AT-RASC), May 2018, pp. 1–4, https://doi.org/10.23919/URSI-AT-RASC.2018.8471335; (B) reprinted from J.I. Rosell, X. Vallverdú, M.A. Lechón, and M. Ibáñez, Design and simulation of a low concentrating photovoltaic/thermal system, Energy Convers. Manage. 46 (18) (2005) 3034–3046, https://doi.org/10.1016/j.enconman.2005.01.012, copyright 2005, with permission from Elsevier.

Methods of accelerating aging, and a knowledge of the acceleration factors, are at the heart of reliability engineering [4,5]. The acceleration factor is how many equivalent hours of normal stress the part ages for every hour on accelerated stress. For qualification, we typically are looking to qualify a part as being suitable for 10 years of real-world use in about 6 weeks (1000 hours) of accelerated aging. For engineering experiments or production screens, we do not want to wait 6 weeks, so a higher acceleration factor is desired to get the answer in about 1 week. Table 2.1 summarizes a few common cases. "Burn-in" is a nondestructive screen used to degrade weak parts sufficiently to be able to detect and remove them.

2.3.2 Acceleration factors

While we will discuss acceleration models in greater detail below, the concept of the "acceleration factor" is introduced before moving on to the next section. The goal of reliability testing often requires us to subject components to the equivalent of 10 years or more of real-world use degradation, in 6 weeks or less of aging, by stressing the part in various ways as discussed in the next section (e.g., increased heat, mechanical stress, etc.). We attempt to build a quantitative model that tells us each hour of our accelerated stress is equal to some number of hours of normal-use stress [6,7]. This is referred to as an "acceleration factor," abbreviated as "AF". The equation expresses it as:

$$T_{eq\_use} = T_{stress} \times AF \tag{2.1}$$

Table 2.1 Typical test durations and acceleration factors, depending on the purpose of the reliability test.

Test purpose	Test duration		Acceleration factor		Equivalent use duration
Qualification	1000 h	×	100	=	100,000 h (11 years)
Production screen	96–250 h	×	1000	=	10–30 years
Burn-in	24–48 h	×	200–400	=	0.5–2 years

where the equivalent aging time under use conditions (T_{eq_use}) equals the accelerated stress time (T_{stress}) times the acceleration factor (AF). For most optoelectronic components, an acceleration factor of 100 × or more is desired, since we want to subject the components to >10 years (approximately 100,000 hours) in 6 weeks (1000 hours) or less. We come back to this topic, and cover it in greater detail in the following section.

2.3.3 Categories of reliability testing

Most reliability tests fit into one of four categories:

2.3.3.1 Accelerated aging with temperature and bias

This is the mainstay for device testing, such as for laser reliability. Operation at high temperatures effectively "softens" the lattice, allowing defects to move more easily through it. Increased current or bias also speeds up the degradation through a number of mechanisms (in fact, many types of aging will not take place at all without current or bias). A subset of this type of reliability testing is aging *without* bias, referred to as "high-temperature storage," or "HTS," as might be seen in shipment alone. The lowest biased stress that is often used is referred to as HTOL, where the maximum data sheet temperature, and typical use current or voltage are applied (e.g., 80°C, and normal operating current). This typically would be no acceleration over worst case stress, and 3–20 × acceleration over normal-use stress. One common gap with HTOL reliability testing is that even if it successfully passes with no failures, it only is the equivalent of 1 or 2 years of normal stress, so it provides no information about long-term wearout lifetime; in addition the number of equivalent device hours is limited unless a very large sample size is used. With higher temperatures (125°C–150°C), and higher currents (often up to 3 × normal use current), it is possible to get acceleration factors of 1000 × or more, and to build an acceleration model with a variety of different temperatures and currents, as discussed later.

2.3.3.2 Corrosion aging

Corrosion is a second reliability concern that is most often a problem for photodiodes. The way it is accelerated is with increased temperature and relative humidity. Acceleration factors of ~100 × over the worst case environments (outdoor in the tropics) can be obtained using an "85/85 chamber" (with parts at 85°C, and 85% relative humidity) [8,9]. Higher accelerations cannot be obtained at atmospheric pressure, due to being above the boiling point, but with a more expensive pressurized chamber, higher temperatures and accelerations can be obtained. Common stresses are shown in Table 2.2. One caution with pressurized testing conditions is that it has been seen to artificially embrittle many common plastics used in optoelectronics, and will result in failures that will not be seen even with extended corrosion testing at 85/85 conditions, so caution is warranted when such failures are observed not to over-reject them without first confirming failure in 85/85, which should always be run in parallel for qualifications.

Both biased and storage versions of corrosion testing are available, and ideally, both should be tried. The biased version can drive galvanic corrosion [10,11], if that is important for your structure. However, the bias can also cause self-heating, and drive off moisture, which lowers the relative humidity and thus the stress. Therefore low current is often specified in many testing requirements (e.g., only 1.2 × threshold current is called for in GR-468 when testing lasers [12], well below normal operation conditions which are usually at 3 × threshold current or higher).

Common acronyms for corrosion testing include temperature with humidity and bias (THB) and temperature with humidity unbiased (THuB) for 85/85 chambers. Damp heat with bias (DHB) and damp heat (without bias, "DH") are other popular equivalent abbreviations for the same atmospheric pressure testing condition. With pressure vessel chambers, pressure pot (PP), biased highly accelerated stress test (BHAST), and (unbiased) highly accelerated stress test (HAST) are common abbreviations.

Traditional optoelectronics were made with hermetically sealed enclosures that surround the components, using metal, glass, and ceramic to protect the components from moisture. However, a more recent trend is to reduce cost using assemblies that do not use hermetic cans: such reduced-cost parts are referred to as "nonhermetic" parts or designs. Another change over the past 20 years has been the replacement of traditional data center cooling with evaporative cooling in order to reduce the energy consumption used for climate control. This change exposes parts to both higher temperatures and higher humidity than were used in traditional "ASHRAE recommended temperature/humidity profiles" [13]. The shift to nonhermetic assemblies, and the shift to higher data center humidity and temperatures have both increased the risk of corrosion, and device designs have had to be improved to survive the increased corrosion stresses.

Acceleration models include the product of three factors: a thermal acceleration that follows a traditional Arrhenius equation; a humidity term that includes many possible forms to empirically fit the data; and a bias term. Refer to Osenbach's paper for more details [11], but note that while many of the humidity fitting methods fit his data, they diverge by orders of magnitude in their prediction for the predicted long-term corrosion rate, and would require longer-term acceleration model validation to confirm what the correct model is.

Table 2.2 Commonly used environmental conditions and acceleration factors for accelerated corrosion testing.

Goal	Type	Temp (°C)	Humidity (% R.H.)	Pressure (Atm abs)		Test duration		Acceleration factor		Equivalent use duration
Qualification	Ambient pressure	85°C	85%	1.0		1000 h	×	100	=	100,000 h (11 years)
Engineering/ production screen	High pressure	130°C	85%	2.25		96–250 h	×	1000	=	10–30 years

Another consideration is the use environment. We can break deployment environments into three broad categories: (1) controlled environment in temperate climates (most of the developed world, average 8 g/m^3 of H_2O vapor over the course of a year); (2) controlled environment in the tropics (e.g., Singapore, Miami, average 15 g/m^3); and (3) outdoor tropics (28 g/m^3). The acceleration factors in Table 2.2 and requirement for uncontrolled environments (UNC) in GR-468 are oriented toward category 3, but for category 2 deployments, we expect a three times lower corrosion rate than category 3, and category 1 was an additional two times lower corrosion rate than category 2. See figure 3 in the author's Electronic Component Technology Conference paper on the subject for more details [9]. For an additional discussion of device corrosion, refer to the references on electronics [8,14], or vertical-cavity surface-emitting lasers (VCSELs) [9,15,16]. Another type of corrosion testing that is not commonly used is "mixed flowing gas" testing which uses a mixture of corrosive gases to simulate what would happen in areas with high air pollution. Salt spray testing is useful for other applications that are deployed outdoors, but is not commonly used for optoelectronics corrosion testing.

2.3.3.3 Mechanical aging

Because many assemblies are built from dissimilar materials, cracking and delamination are major concerns, which we refer to as "mechanical failures" that often show up as open circuits where the device contacts break off from the printed circuit board. Less frequently, mechanical stress can induce defects within the semiconductors themselves. The primary workhorse for detecting such problems is "temperature cycle testing," abbreviated as "T/C." In such testing, the assembly would be cycled back and forth between a low temperature such as −40°C, and a high temperature limited by the data sheet maximum for storage or operation (usually +70°C to +100°C). The cycles typically have a 10 minutes minimum dwell time at the top and bottom of the cycle, and a 10°C/minute ramp rate, so each cycle might take 40–60 minutes. A total of 500 cycles are required by the GR-468 standard for fiber optic transceiver reliability testing [12], and validate the integrity of the construction and the stability of the fiber alignment.

Thermal shock testing can speed up testing considerably. Both liquid-to-liquid and air-to-air chambers exist. Liquid-to-liquid chambers are used for devices that can be submerged in inert fluids such as silicone oils, and dwell times might be as little as 1 minute for small packaged parts. Air-to-air units have an elevator that can move between a hot and cold chamber in just a few seconds, but typical dwell times would be 10 minutes. Either way, fatigue cycling drives failure. A good example paper is one published by Arnold et al. at the DfR Solutions website, and shown in Fig. 2.7 [17]. As might be expected, as the magnitude of the mechanical stress is increased, the number of cycles required to reach failure decreases. The measure in question is the change in

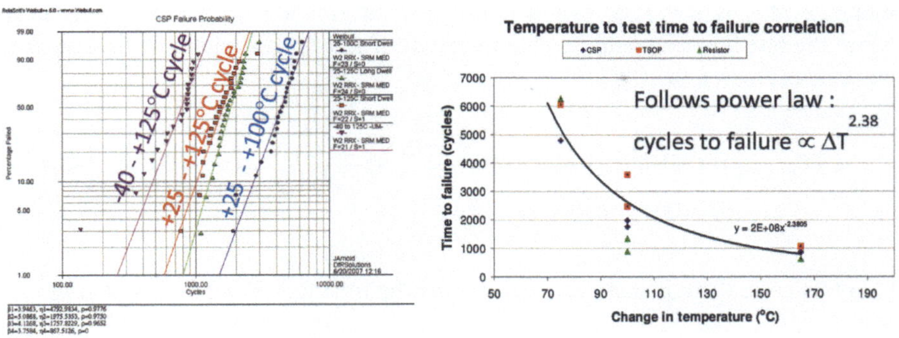

Figure 2.7 Weibull plots of the experiment (left), and plot of the median number of cycles to failure vs. temperature swing (right), showing that failures happen with fewer cycles when subjected to larger and more stressful temperature cycles. Courtesy of Dr. Craig Hillman at DfR Solutions.

Source: J. Arnold, N. Blattau, C. Hillman, C. Park, and K. Sweatman, Reliability testing of Ni-modified Sncu And Sac305 – accelerated thermal cycling, DfR soultions website. https://www.dfrsolutions.com/hubfs/DfR_Solutions_Website/Resources-Archived/Publications/2008-2010/2008_03_ni_modified_sncu_thermal_cycling.pdf.

temperature ΔT, which is the difference between the maximum and minimum temperatures used in the environmental chamber.

The Coffin Manson acceleration model for temperature cycling states that the acceleration factor between the test condition (such as those shown in Fig. 2.7), and the use condition follow the formula:

$$\mathrm{AF} = \left(\frac{\Delta T_{\mathrm{acc}}}{\Delta T_{\mathrm{use}}}\right)^n$$

where AF is the acceleration factor, ΔT_{acc} is the temperature swing from minimum to maximum for the reliability testing, ΔT_{use} is the temperature swing the device will experience in the field, and n is an empirically determined exponent that varies from 2 to 10, with the lowest values being observed on softer materials like solders, and the highest values being observed on brittle materials like ceramics [18]. For most fiber optic equations, where empirical values are not available, it is common to assume values as high as $n = 3$, or to use $n = 2$ for the most conservative assumptions.

Two examples serve to show both how the equation works, and also what elevates the risk. The first example is of a typical transceiver deployment in a high-volume cloud data center, using data published by the Open Compute Project [19]. These data show that the data center is ~5°C hotter during the day than at night. For this example, we combine it with data from the test boards shown in Fig. 2.5 [17], where the number of cycles to 1% failure ($\mathrm{CTF}_{01\_\mathrm{acc}}$) is approximately 1400 cycles for the mildest 25°C–100°C cycle, and the value of n for Figure 11 of reference [17] is $n = 2.38$. We will use this information to calculate how many cycles the test boards would be able to endure if left in that relatively benign cloud data center environment.

$$\begin{aligned} CTF_{01\_use} &= CTF_{01\_acc} \bullet AF = 1400\ \text{cy} \cdot \left(\frac{100^\circ C - 25^\circ C}{37^\circ C - 32^\circ C}\right)^{2.38} = 1400\ \text{cy} \cdot \left(\frac{75^\circ C}{5^\circ C}\right)^{2.38} \\ &= 1400\text{cy} \cdot 630 = 882\ \text{kcy} \end{aligned}$$

In this case, there is one cycle per day, so the time to 1% failure is given by:

$$T_{01} = \frac{CTF_{01\_use}}{\#\text{ of cycles per year}} = \frac{881,506\text{cy}}{365.25\frac{\text{cy}}{\text{year}}} = 2,413\ \text{years}$$

Our conclusion is that for the small excursion thermal cycles, there is very little fatigue, and we have a large margin of safety.

Our second example is of one where parts are power cycled twice per day, and the critical solder joints experience a 40°C swing every time the power is cycled, putting the solder joints under considerable stress. In this case, the equation is:

$$\begin{aligned} CTF_{01\_use} &= 1400\text{cy} \cdot \left(\frac{100^\circ C - 25^\circ C}{60^\circ C - 20^\circ C}\right)^{2.38} = 1400\text{cy} \cdot \left(\frac{75^\circ C}{40^\circ C}\right)^{2.38} \\ &= 1400\text{cy} \cdot 4.46 = 6250\text{cy} \end{aligned}$$

In this exercise, $T_{01} = \frac{CTF_{01\_use}}{\#\text{ of cycles per year}} = \frac{6,250\text{cy}}{2 \cdot 365.25\frac{\text{cy}}{\text{year}}} = 8.6\ \text{years}$

Since many deployments are in the 5–7-year range, this would normally be considered an insufficient margin of reliability, and we would either need to advise potential customers that the part should not be intentionally power cycled, or we would have to improve the part. Two ways of improving it would include more robust solder joints, that could increase the number of cycles to failure, or better thermal management that reduced the heating when the part is turned on.

Another type of mechanical stress testing that would not obviously be mechanical in nature is low-temperature operating life testing. In most electronic assemblies, the electronics or optoelectronics uses semiconductors with low coefficients of thermal expansion (CTE = 3–6 ppm/°C), surrounded by (or even embedded in) plastics with much higher CTE (20–30). They are cured at temperatures that are usually above the maximum use temperature. As temperature declines, the high-CTE materials apply greater compressive stress, and become harder and less yielding, which can result in failure as the encapsulant "squeezes" the component during normal use. For example, plastic LEDs embedded in clear epoxy are vulnerable to failure during aging at low temperatures [20], and so this is an important part of most LED reliability qualification tests. This type of testing is common for LEDs, but not widely used in transceivers, which usually would just go through "low-temperature storage" testing.

Another class of tests that some might put in the previous "corrosion" category are "powered temperature cycle", "cyclic moisture resistance" (e.g., MIL-STD-883, section 1004.7), or "humidity freeze". These are usually only relevant to parts that are going to see operation in nontemperature-controlled conditions, such as a telecommunications remote terminal or on a cellphone tower. They seek to saturate the interior of the part with moisture using a high-temperature, high-humidity exposure, and then cause it to condense as temperature is ramped down, sometimes even below freezing. Parts can fail when powered on if films of water are present internally, or due to cracking during the freeze cycle if water is present internally.

Another test, colloquially referred to as the "popcorn test," is done to determine the "moisture sensitivity level" (MSL) during assembly, which indicates how long parts can remain outside of packaging with moisture barriers and desiccant inside, before they go through the surface reflow soldering oven. Sensitive parts could absorb some moisture, only to have it expand to steam and explode like "popcorn" as the part goes through the surface reflow soldering operation. Test standards such as JEDEC Standard 22-A113D spell out the soak conditions and durations the parts must be able to tolerate prior to going through a reflow oven to obtain each higher (and more robust) MSL. These may not be relevant to pluggable optoelectronics like transceivers, but are important for on-board optoelectronics.

Two final types of testing that are frequently used are blowing dust tests and liquid immersion tests. Blowing dust tests check whether assemblies are adequately sealed to prevent dust buildup, which can obscure the light path. Immersion tests are used in consumer-grade optoelectronics, and many different classes are offered, from rain or sweat resistance at the low end, to submersion to a specific depth at the more demanding end.

2.3.3.4 Electrostatic discharge testing

While this is not strictly a type of reliability testing, it is normally a required part of new product qualification, to understand how robust the part is to different forms of electrostatic discharge (ESD). Most assembly houses are certified to a certain standard of ESD control (usually to 100 V human body model and charge device model). Parts that cannot pass at these levels of ESD test stress would need to either be re-engineered, or special modifications would need to be made on the assembly line to give tighter electrostatic controls. Additional requirements are that more exposed parts of the transceiver need to be more rugged, and air-to-air discharge at 15–25 kV is also often required where fibers would be plugged in and exposed to ESD during system configuration.

2.3.4 Reliability test standards

There are two fundamental approaches to reliability testing. The first is standards-based testing, which is the subject of this section, where little or no degradation is expected on the tests. The second is a more customized reliability testing at higher acceleration levels, where parts are aged to failure, and a reliability model is built up, which is discussed in Section 2.3.3.

For standards-based testing, relatively mild stresses are applied in many of the legs, and no parts are expected to fail. In most of the standards, a suite of several different types of tests are called for. If any failures are seen in any of the tests, the part fails the qualification, and failure analysis and corrective action are required, with the improved parts being required to repeat the test without failures to meet the goal. Standards-based testing informs designers about major design deficiencies, and infant failure modes, but generally says nothing about anticipated wearout lifetime, and only provides a ceiling on random failure rate. It is used for items like fiber optic transceivers that do not lend themselves to more accelerated testing, as is discussed later in this chapter.

The most-used reliability test standard for fiber optic transceivers is the Telcordia GR-468 test standard [12]. A typical table of stresses and sample sizes is shown in Table 2.3. Note that the plans will vary somewhat, depending on whether the devices are hermetically sealed or not, and also whether they are intended for "controlled office" deployment in environmentally controlled data centers, or in the more demanding uncontrolled (UNC) deployment environment, such as equipment boxes mounted outdoors with no climate control. A common GR-468 test would require ~88 parts, where most are deployed in parallel, but some shorter tests can be run sequentially. Such a qualification would take 3–5 people working for approximately 9 months to execute the qualification if no failures were encountered (although most of those people can also support another project in parallel during most phases). However, it is not common for a new transceiver design to pass all legs without failures on its first attempt, so it is important to start testing early, and to leave contingency time in the schedule and budget for repeats of several of the tests. Where failures are observed, failure analysis is critical to determine the cause of the failure, and allow the design team to determine how the design needs to be improved to pass on the next try. Most companies distinguish between early rounds of informal reliability testing, and the formal qualification on parts where the final design and manufacturing process is used. In the next section, we discuss subcomponents such as the laser or LED, where we want not just binary "pass/fail" standards data, but quantitative information on the random failure rate and wearout lifetime.

Solid-state LED lighting for a long time used company-specific test and analysis methods that could not be directly compared, but more recently, the IES-LM 80 standard for gradual degradation [21,22] and test method ANSI/IES TM-21 [23], as well as other standards for random failure such as IES-TM-26 [24], have provided a standardized baseline, and some are even required for certifications such as the US Energy Star label [25].

Automotive optoelectronics are a relatively small market at the time of this writing, but have significant potential for growth, both for items like LIDAR that are to be used in autonomous driving, as well as for in-vehicle data communication buses. Automotive parts need to be built in facilities with IATF-16949 certification, and the dominant test standard is AEC-Q102 [26].

Table 2.3 Sample GR-468 qualification plan for a pluggable fiber optic module.

Heading	Test	Reference	Condition	Test sample size	Failures allowed
Mechanical integrity	Mechanical shock	MIL-STD-883 Method 2002	500G, 1 ms, 5 times/axis	11	0
	Vibration	MIL-STD-883 Method 2007	Random, multiaxis, 20G rms, 20–2000 Hz, 2 h	11	0
	Thermal shock	MIL-STD-883 Method 1011	−40°C to +85°C, 15 cycles	11	0
Endurance	Accelerated aging (HTOL)	—	70°C case temperature, nominal voltage, 5000 h	22	0
	Temperature cycling	—	−40°C to +85°C, measured at 100 and 500 cycles	11	0
	Low-temperature storage	—	−20°C, 72 h, no bias	11	0
	Damp heat storage (humidity aging)	IEC Pub. 68-2-3 Test condition: Ca Non-Operating	85°C/85%RH, 2000 h	11	0
	Damp heat operating (humidity aging)	MIL-STD-202G Method 103B or IEC Pub 68-2-3	85°C/85%RH, 2000 h, 3.3 V	11	0
	Cyclic moisture resistance	MIL-STD-883E, Method 1004.7	7-step, 24 h cycle; +65°C to −10°C	11	0
Other	Electrical connector mating test		200 matings	11	0
	Optical connector mating test		200 matings	11	0
	Optical connector sideload test		2 kg sideload without damage	11	
	Blowing dust test	FOTP-35	6 modules with dust caps; 6 modules mated with fibers	6 + 6	0
	Mixed flowing gas corrosion test	EIA-364-65	6 modules with dust caps; 6 modules mated with fibers	6 + 6	

HTOL, high temperature operating life.

2.3.5 Developing a custom reliability test program for subcomponents

Standards-based testing can provide a diverse range of standardized stresses for finished assemblies such as fiber optic transceivers. But for the thermal + bias stress, the HTOL test has both low acceleration (AF = 1–10 ×), and relatively small sample sizes (usually 22 parts). By contrast, normal recommendations are to test subcomponents like lasers at acceleration factors of 100–1000 × , and use sample sizes of at least a few thousand devices during the qualification as a whole. In this way, we can collect measurable data on both random failure rate and wearout lifetime. A typical allocation of device reliability testing would put 70% of the parts on test at the maximum validated stress condition, to get quick answers to questions about whether design changes, epitaxial growth changes, or process changes helped or hurt reliability, and also to build up a trend chart of how reliability varies over time, even for a nominally fixed design. Another 20% might be allocated to alternate stresses to build an acceleration model, with the final 10% being allocated to "very long duration" tests at much lower stresses, to validate the aging rate closer to actual use condition. Note that while the HTOL module leg of the GR-468 test provides only the equivalent of a year or two of normal-use aging, the other legs of the GR-468 test provide better coverage with the corrosion and temperature cycling tests providing the equivalent of over 10 years of normal aging, and better detection of potential issues.

2.3.5.1 Device acceleration models

Two approaches can be to compute acceleration factors. For mature technologies, or early in the development process, the reliability engineer needs to rely heavily on previously published acceleration model values, and erring on the side of conservative values (e.g., using the lower published values of activation energy for calculating thermal acceleration). Over time, new experiments can be run to validate or refine the values for the specific design in development. For most semiconductor lasers or LEDs, the overall acceleration factor is modeled as the product of thermal acceleration and current acceleration, as shown in the following equation:

$$\mathrm{AF} = \left(\frac{I_{\mathrm{acc}}}{I_{\mathrm{use}}}\right)^{n}\left[\exp\left(\frac{E_a}{k_B}\left(1/T_{j,\mathrm{use}} - 1/T_{j,\mathrm{acc}}\right)\right)\right]$$

The right side of this equation covers thermal acceleration following the familiar Arrhenius law, and the sensitivity to temperature is characterized by an "activation energy," E_a. Common values of E_a are 0.6–0.8 eV for wearout aging, but lower values of 0.35 eV are normally assumed for random failures. The current acceleration is modeled on the left by a power law fit, with the accelerated current divided by the use current, taken to an empirically determined power n, where n usually varies from $n = 1$–3, and is usually taken to be $n = 2$ for VCSELs.

The acceleration matrix would normally be set up as an "inverse L," where a minimum of three temperatures are run, at the same maximum current, and a minimum of three currents would be run, at the same maximum temperature [27]. After the data are collected, the three different temperature cells at the same current can be analyzed first with an Arrhenius plot (of lifetime vs $1/T$, as shown below in Fig. 2.8), to determine the activation energy. Then the data versus current can be plotted to determine the power law applicable for the current acceleration. For improved accuracy, adjustments can be made to the current cell results to correct for junction temperature heating, either in advance by adjusting the stress temperatures to compensate, or mathematically by adjusting the lifetime data to compensate for the different junction temperatures seen in the high-current cells. For a more detailed discussion of semiconductor laser reliability testing, see the author's chapter on VCSEL reliability, where the concepts discussed are broadly applicable to all semiconductor lasers, not just to VCSELs [28].

2.3.5.2 Acceleration risk

One discussion that is always a subject of concern is the question of how valid accelerations are to use condition. This can work in either direction, for different reasons. The reliability engineers are concerned about the model being too optimistic. An example of this is where failure mechanisms have very high activation energy at highly accelerated conditions, but a lower activation energy mechanism dominates at use condition. This risk is also called out in GR-468, as shown on the left side of Fig. 2.8. The way to cover this risk is to reserve at least 10% of your capacity for longer-term tests, and prove that the same acceleration model follows down to at least 4 × extension in lifetime vs. the most accelerated cell, and preferably >10 × extension with multiple different cells.

On the other side, the design engineers or program managers are worried about overly pessimistic projections from a new failure mechanism being introduced at the high stresses used in reliability testing that are not present at use condition. An example of such a case is for plastic-encapsulated semiconductor devices: reliability testing at temperatures over 150°C causes wirebonding failures due to the black plastic encapsulant exceeding the glass transition temperature. It melts, and ends up pulling wirebonds off the die, as shown on the right side of Fig. 2.8. Where this

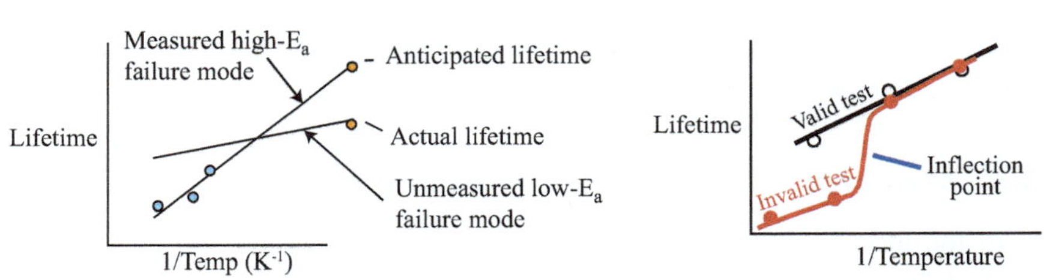

Figure 2.8 Acceleration can be either too optimistic (left), or too pessimistic (right)—see text for discussion.

concern is valid, a discontinuity is seen in the acceleration model, with a dramatically different model at high stresses from what is seen at low stresses. Failure analysis is another important indicator—if a different cause of failure or signature is seen at high temperatures than at low temperatures, that is another indicator that the acceleration model may not be valid across the range of conditions. However, for most types of devices, the questions below are good ones to ask to see if the acceleration in question is valid or not.

1. Is there a strong correlation between which parts fail first with high acceleration, and which fail first at more moderate conditions, when comparing above-average and below-average designs or production lots? That is to say, does the rank ordering from best batches of parts to worst batches hold to a similar order for the different acceleration conditions?
2. Is there a predictable acceleration model?
3. Is the failure mechanism the same in both cases?
4. Is there an abrupt "discontinuity" over the range of stresses (e.g., glass transition temperature)?

While it is common for design engineers to question the validity of accelerated testing (see case study 6 in Chapter 3: Case studies in fiber optic reliability), a change in failure modes across normally used stresses with AFs $<1000\times$ is less common than is widely believed. However, for AFs $>1000\times$, we have found limits to the correlation with the lasers we have tested, and junction temperatures above 200–220°C and current densities $>2-4\times$ normal-use current have generally pushed into the range where the correlation starts to fall apart.

2.3.5.3 Limitations in reliability testing

For every type of technology, there are always limits to acceleration. While it is great to get answers for reliability in the shortest time possible, the devices the author has worked with usually cannot be accelerated to greater than acceleration factors of a few thousand, or qualification lifetimes equivalent to 10–40 years of normal aging (to cover the full distribution of strength) in less than 100 hours or so. At currents and temperatures above this level, the correlation between predicted lifetime at ultra-high acceleration (say, a day or two to failure) and a high acceleration (1–2 weeks to failure) is not strong, and the failure signature is often different. If this were not the case, the practice of reliability testing would be very different, as it would allow quick testing on wafer probers in a serial manner. Any attempts to validate higher acceleration testing should be checked out across a range of failure modes, and pass all of the tests outlined in the previous section.

2.3.6 Maverick versus wearout reliability failure modes

One of the most important concepts in dealing with reliability issues is to understand that there are two ways in which your product might fail to meet its reliability target. This is shown in Fig. 2.9. The first way is that maverick failures could occur. In this case, most of the parts are highly reliable, but some small fraction of the

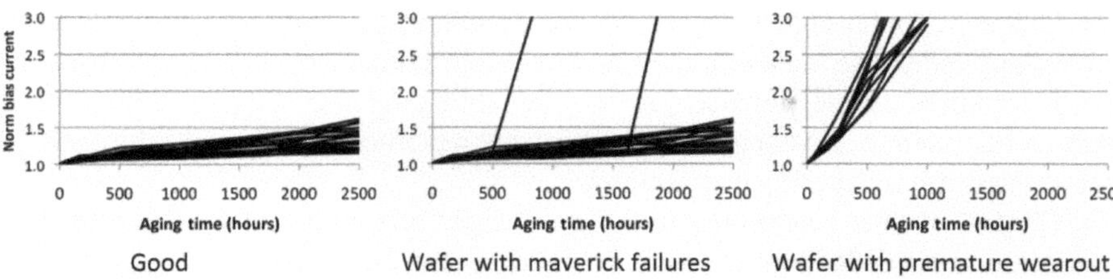

Figure 2.9 Results of reliability testing can show trends of increasing electrical power required to maintain the desired output light out. In the left case, a long lifetime for all parts is seen; in the center, a long lifetime for most parts, but with abrupt failures from a few, and at the right a short lifetime for all parts is seen.

parts fail (often very abruptly, with little warning). Two examples would be random contamination of wirebond pads that causes poor adhesion, or a cracked capacitor on a circuit board. In such cases, sample testing is not useful without prohibitively large sample sizes. For example, the target failure rate in the fiber optics industry might often be lower than 100 parts per million (ppm), or less than 0.01%, so 22,000 parts would need to be tested to verify that fewer than that number of failures were present with 90% confidence. To eliminate random failures, the industry carries out 100% visual inspection of parts to eliminate defective units, and 100% burn-in to identify and eliminate weak units.

The second way in which parts can fail is to have premature wearout lifetime. An example of this would be a batch of lasers grown in a reactor with an oxygen leak, that fail in one-tenth the normal expected time (a few years rather than >10 years). In this case, short nondestructive burn-in testing is inadequate to detect the reduced lifetime. The industry has responded by carrying out wafer qualification testing, and taking a sample from every wafer or every growth, and aging with very high acceleration factors to the equivalent of >10 years to verify that a wafer is good before allowing it to be built into devices.

2.4 Test engineering and data collection

2.4.1 Subcomponent-, component-, and system-level reliability testing

In performing reliability testing on a product like a fiber optic transceiver, coverage will be incomplete for a number of reasons, leaving potential flaws unexposed. Such flaws include:

1. Low-acceleration factors due to thermal limitations. Transceivers typically are built with glues that "melt" (i.e., have a glass transition temperature) when aged much above the data sheet maximum temperature, thus preventing the use of thermal acceleration.

2. Quantity limitations. Transceivers are expensive and complex to manufacture in the prototype phase, meaning that most reliability tests use 11, 22, or 44 transceivers. By contrast, good practice for laser reliability testing calls for thousands of lasers to be tested in the course of qualification, which can easily and affordably be done using normal experimental methods.

3. Control of bias conditions. In the transceiver, the laser and photodetector are often in an automatic control loop, so there is no direct control for aging—for example, you cannot age a laser at 3 × normal-use bias. However, if the components are aged individually, the aging racks can control the temperature and current for each device.

4. Detection of device parameters. At the component level, things like photodiode leakage current, photodiode breakdown voltage, and other parameters that might change well in advance of total failure can be detected. However, these would require special test stations, and such changes could never be detected when embedded in the transceiver itself.

5. Detection of small levels of drift. For transceivers, one of the limiting factors is fiber coupling repeatability when inserting fibers into the connector on the front of the transceiver. Usually, the variation is ± 1 dB ($\pm 26\%$). By contrast, repeatability of $< \pm 0.5\%$ is possible at the component level with properly designed test equipment, an improvement of almost two orders of magnitude. We can thus detect small, gradual drifts in performance that would be lost in the noise at the module level.

Due to these limitations, all fiber optic transceivers have their components (laser, photodetector, and integrated circuits) tested and qualified separately from the transceiver itself. Higher acceleration factors are available through aging infrastructure that can achieve high-acceleration factors, and subtle aging trends can be more easily measured and graphed during the applied stresses. Independent component qualification reports are written for each of the major subcomponents, particularly the laser and photodetector. However, transceiver testing is also still needed, both to validate the assembly techniques, as well as the way all the components function together to meet the optical communication standards.

2.4.2 Building and running a reliability test lab

While it is quite easy to outsource reliability testing for integrated circuits, optoelectronics are a smaller market that does not have the critical mass to have outsource suppliers at the time of this writing. The need to integrate photodetectors into the measurements of light source degradation, and the need to use constant current stress for light sources rather than constant voltage stress, are the main two challenges. The outsourced labs may be the best solution for early-stage startups, but are more expensive over time if a significant number of tests over a few years are going to be run, and therefore most companies usually purchase all the environmental chambers needed to run their own tests in-house. The level of effort required to establish reliability stress capabilities should not be underestimated, and normally should be a few man years for most labs, plus a few more for data handling in a reliability database. At least a few reliability technicians will also be needed to set up the reliability experiments, and run and monitor them.

When building a reliability test lab, there are two options: automated test systems with in situ monitoring versus simpler systems that only do stress but cannot do characterization. The automated option costs more in capital (usually \$300–500 per laser), but has very low data collection labor costs. The nonautomated option can be lower for capital (\$50–150 per laser), and allows for more complex testing, but requires more technician labor to run a series of stress tests (e.g., do read points at 0, 168, 500, 1000, and 2000 hours, and set up the system in the stress oven for each test leg). The reliability engineer and their management must make the decision based on the available capital budget, technician staffing levels, test engineering availability internally, and which parameters are being monitored. For example, if spectral measurements or high-speed measurements were desired, they are prohibitively expensive to integrate into in situ measurements in a rack.

2.5 Data collection and analysis

In this section we discuss how we collect and analyze reliability data. The challenge in this is not obvious in advance and is underappreciated. For this reason, plans to automate data collection and analysis only are usually put into place after the difficulty in handling, reducing, and manipulating the data to prepare reports becomes a significant engineering burden, and the return-on-investment for implementing a reliability database system becomes evident, usually after at least a few engineers are needed full time to prepare all the reports.

2.5.1 Data collection and storage

At its simplest, reliability data could involve running a part through a system test before putting it on accelerated aging, and then retesting at multiple test points, and simply looking for failures where it exceeded its pass/fail limits. However, most reliability engineers are looking for more than that, and want to plot quantitative performance trend data, and not just binary pass/fail data. A few decades ago, this might have been handled with spreadsheets and labor-intensive analysis, but today, extensive use of databases, scripting, and improved features in data analysis programs have streamlined the labor required to do data analysis, for those willing to make the investment in storing the data in a database system. The labor required to set up systems to handle anything more than small amounts of reliability data normally take at least a few man-years of effort—work that is rarely planned for or expected, but is a must for scaling the business. Surprisingly, there are no "out-of-the-box solutions" in this space, even though the requirements for all customers are very similar.

2.5.2 Data reduction

Reliability engineering requires the reduction of large amounts of data, in a number of steps to an ever-more-concise set of data that tells you whether the devices are likely to meet the lifetime and random failure rate targets that your customer expects them to meet. In Fig. 2.10, we see the "data pyramid," where each level up reduces the stored data volume by approximately 10 × from its source below. As detailed in Fig. 2.11, the lowest level data ("level 1 data" or "L1 data") are raw data from an individual device, tested at an individual time. At the next highest level, in L2 data we have extracted a single key parameter, such as required bias current to meet the target optical power out, or optical power out for a fixed current in, and then we plot it versus time for each device. For L3 data, we set a failure criterion, such as acceptable absolute power, or percentage degradation allowed, and look at how long each device takes to reach it, giving us a metric like "time to failure" for each device, or perhaps how much degradation each one shows at the end of a specified test. In L4 data, rather than make statements about the reliability of each individual device, the "lot" as a whole is analyzed, with metrics like "mean time to failure," "median time to failure," or "time to 10% failure" (T_{10}). The entire lot is tested at a single stress condition, and is of the same design, and the same source material, and we can then compare different designs or different acceleration conditions. The final level of the pyramid (L5) is the full model for the product, including distributions of strength, and response to temperature and bias. We can then make a prediction of what the distribution of failures will look like for the product under deployment, either at a likely temperature and bias, or a worst case temperature and bias.

Note that for data rollups, normal practice is to start by looking at the level 2 data for each lot under test first, which are the stability indicators for each individual device. If the data look as expected, little further scrutiny is usually needed, and the L2 data can be reported out to the team in a routine update. However, if the data do not follow expected trends, or if failures or rapid degradation are observed, then "reject validation" is needed. For transceivers, this would include inspecting the optical ports for dust contamination, retesting the transceivers to see if the results were repeatable, and also checking the control samples. For lasers, this

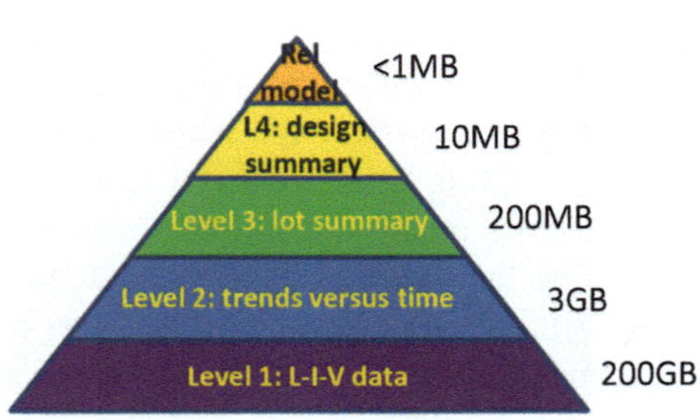

Figure 2.10 Pyramid of data, where each step pulls out a few key details from the more complete data in the level below, to allow us to summarize reliability trends.

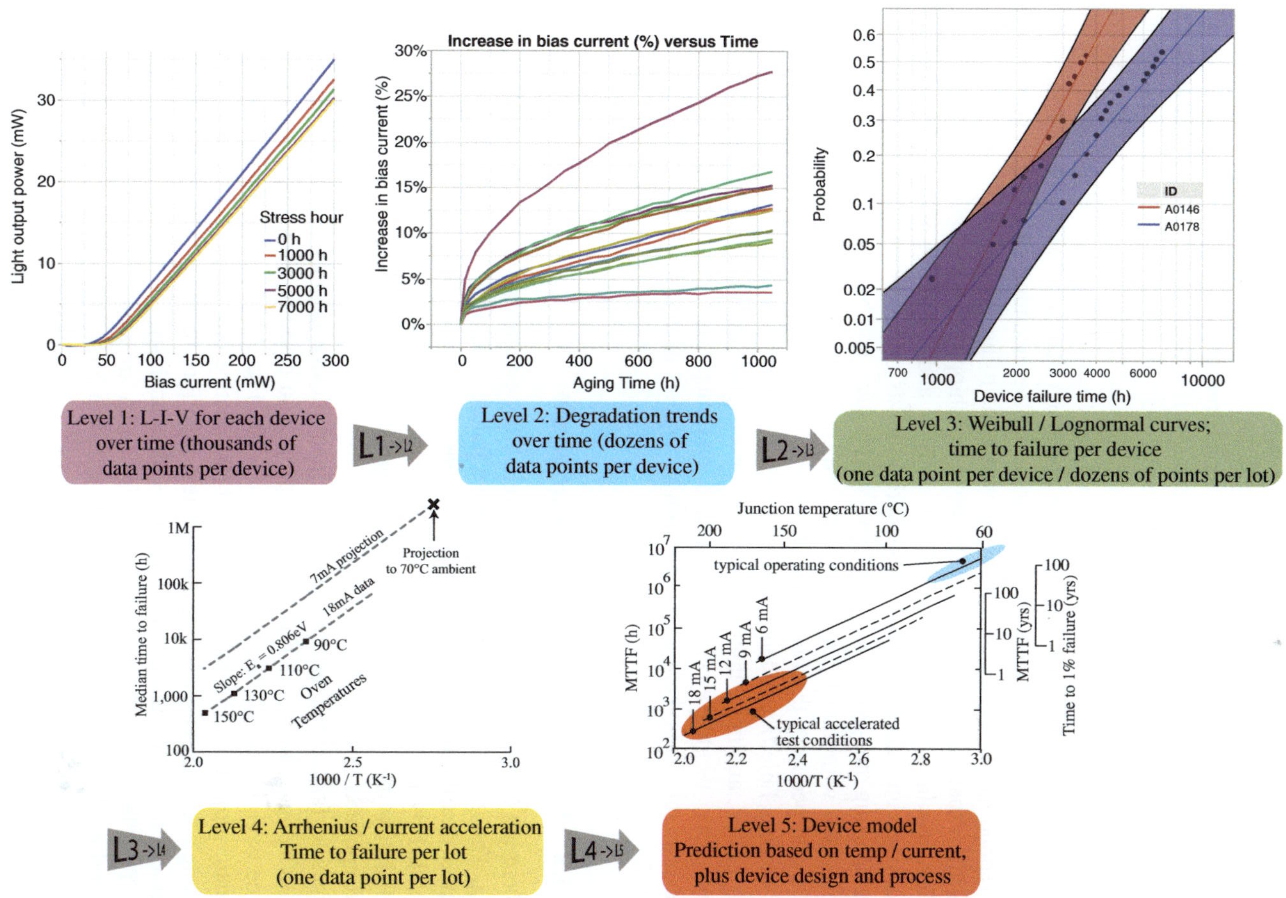

Figure 2.11 Raw data are collected in level 1, and data analysis is performed at each step to reduce it to more concise indicators.

would involve going back to level 1 data, and looking at how the L-I and I-V curves have changed over time, to see if the changes affect threshold current, device resistance, slope efficiency, etc., and whether they change rapidly or gradually. Another important step, especially if this failure could cause a device to fail its qualification, is to see if it could have been screened out based on burn-in pass/fail parameters, or tighter incoming specifications. If it is an outlier in terms of absolute parameters, or in the shifts observed during burn-in, this may help in allowing the devices to be eliminated by tightening the allowed specifications. Note that in reject validation, some types of failures may not be counted. Take the example of a laser under reliability test that fails due to a wirebond that falls off the chip. This could happen due to contamination on the surface of the incoming wafer, or nonoptimized wirebonding parameters. If the laser was built using the same process and same facilities that are intended for production, and shipped to customers, then significant effort would be expended on failure analysis and corrective action. However, in many circumstances, special "engineering builds" are used to build reliability test parts, using nonproduction facilities, which may not be optimized. In this case, unless the wirebond issues were seen frequently, or were due to a chip issue (such as metal peeling), the failure might be removed from both the numerator (number of devices failed) and denominator (number of devices tested), since it is a packaging failure, and not a laser chip failure. Other reliability lab-handling mistakes, like bent pins, would be another example that might not be counted as a failure. Records should be kept of these issues in support of customer audits, as some will ask for more details.

2.5.3 Key terms in reliability analysis

There are a number of key terms that reliability engineers use, that are worth defining. The most common is the cumulative failure rate, abbreviated as F(t), or cumulative distribution of failures. This is the fraction of devices that have failed, divided by the number of devices that started the test. The first step in analyzing reliability data is to convert device degradation trends into times to failure, or extrapolated times to failure if they have not failed yet. Note that using extrapolated times to failure can be misleading, as many device degradation trends accelerate with aging, as shown in Fig. 2.12A, so projection based on the aging rate at an early time (e.g., before 2000 h for the plot shown in Fig. 2.12A) would give an overoptimistic projection. For this reason, the best practice is to use a combination of high enough test stress, and long enough test duration, to drive at least a third of the devices to actual failure before doing the data analysis.

The "level 2" indicators shown in Fig. 2.12A are analyzed. In this case, a failure criterion of 2 dB increase in threshold current (a 58.5% allowable increase) was set, as shown by the dashed red line. When the threshold trends exceed that line, they are marked as failures, and the time to failure is noted in a table. This can be done automatically in some reliability software. The time to failure data is collected in a spreadsheet or database, and then the distribution is calculated by specialized reliability analysis software such as JMP or Weibull++ [29,30]. After calculating F(t)

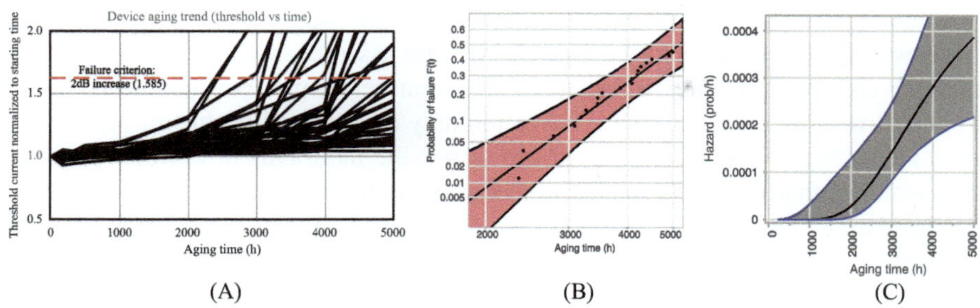

Figure 2.12 (A) Raw laser data are shown on the left, and when parts exceed the dashed red line, they are noted at that time as being failed. The data are compiled in the lognormal plot (B) with cumulative failure *F*(t) versus time. The hazard rate λ(t) is shown in (C), where for this condition, there is a nearly failure-free region for the first 1500 h, although the confidence limits are wide for this small sample size.

as shown in Fig. 2.12B, the next most important terms are the derivative terms that give the probability of failure per unit time, which are readily available from the same reliability software, or can be calculated based on the fitted parameters discussed in the next section. One of the calculated terms is called the "probability density function" (abbreviated as PDF), which is also referred to in equations as *f*(t). This is the failure rate per unit of time, from the original population. Another quantity that is probably the most familiar to everyone is the "hazard rate," abbreviated as λ(t). This is the failure rate per unit time divided by the surviving population (rather than the original population). The two quantities are very close where less than a few percent have failed, but when the majority of parts have failed, *f*(t) usually will drop over time, while the hazard rate λ(t) will continue to increase. Examples of typical laser *F*(t) and λ(t) graphs are shown in Fig. 2.12B and C, respectively. For a more detailed discussion, consult any reliability text such as Nash [31], or the NIST reliability handbook [32].

2.5.4 Reliability models

While several different fitting functions are offered in common reliability software, just a few are popular in optoelectronics. The two that are most commonly used are the lognormal distribution and the Weibull distribution. A third distribution is the exponential distribution, which is a special case of the Weibull distribution with the shape parameter $\beta = 1$, meaning that the failure rate is neither increasing nor decreasing. The exponential model (or "constant hazard rate model") for random failures is often assumed for electronic assemblies. It is best used when: (1) the wearout lifetime is long enough to be much longer than the normal-use lifetime, where we can ignore wearout, and (2) the failure rate is actually declining with time, as is the case with most infant mechanisms, to where assuming a constant hazard rate of $\beta = 1$ is actually more conservative than reality. In this case, the mean time to failure from the exponential distribution can give you information on how many parts are likely to fail over the first few years of deployment. While these estimates work well for complex electronic assemblies with extensive fielded data, the estimates that are published for new transceiver products tend to be overly conservative estimates that give information only on the level of due diligence from previous testing, since no one can afford to test the thousands or tens of thousands of assemblies that would be needed to gather likely numbers of failures.

The Weibull and lognormal models (in their most common forms) have two free parameters. One parameter adjusts the "slope" or "angle" of the fit line, and tells us how much variation in failure time there is. This derives from the spread in strength of the parts. One surprise for most device designers is how large this number is for typical semiconductor lasers, particularly when variation across multiple wafers is factored in. This parameter is called the "shape factor" β (beta) in the Weibull distribution, and the sigma μ (mu) in the lognormal distribution. Since the formulas work very differently, larger values of beta ($\beta > 3$) or smaller values or sigma ($\sigma < 0.6$) are generally considered to be fairly "tight" wearout distributions. Early failures, with declining failure rates, would correspond to $\beta < 0.8$ or $\sigma > 2$ for the Weibull or lognormal distributions, respectively.

The second adjustable parameter shifts the line right or leftward, and corresponds to the strength of the mean part. In the Weibull distribution this is referred to as the "characteristic lifetime," or the Greek letter "eta" (η) or "alpha" (α), depending on the source. It corresponds to when 63.2% of the devices have failed, usually in units of hours. In the lognormal distribution, the natural log of the median failure time is used, and referred to with the Greek letter "mu" (μ). In the example shown in Fig. 2.12, $\mu = 8.544$ (which is the natural log of the median time to failure of 5136 hours), and the spread or slope is given by $\sigma = 0.39$, which is a reasonably tight distribution for semiconductor lasers. These values, along with known acceleration factors, can be used in typical spreadsheet programs with functions such as "lognormdist" and "loginv" to make projections.

There are two approaches to deciding which distribution to use. If little is known about the physics of failure, the user can simply try both, preferably with multiple large data sets, and see which is the better fit, both by looking at residual error, and quality of fit to the eye, especially for early failures. The second approach is to use past knowledge, and consideration of physics of failure. The lognormal distribution typically fits better where cumulative damage governs device failure. The Weibull distribution typically is used for "weakest link" situations where any link of the chain failing could cause the entire device/system to fail. In general, the Weibull distribution will give a more pessimistic projection of time to early failures when fit to a given data set. For most semiconductor lasers, the lognormal function fits best. An example of that is shown in Fig. 2.13 for some early published laser data [33–35], where the data were fit to both distribution types, but the lognormal is obviously the better fit in this case.

2.5.5 Stress–strength model

While it is considered an advanced topic, and not often used in optoelectronics reliability engineering, it is worth considering what happens to our "good" devices as

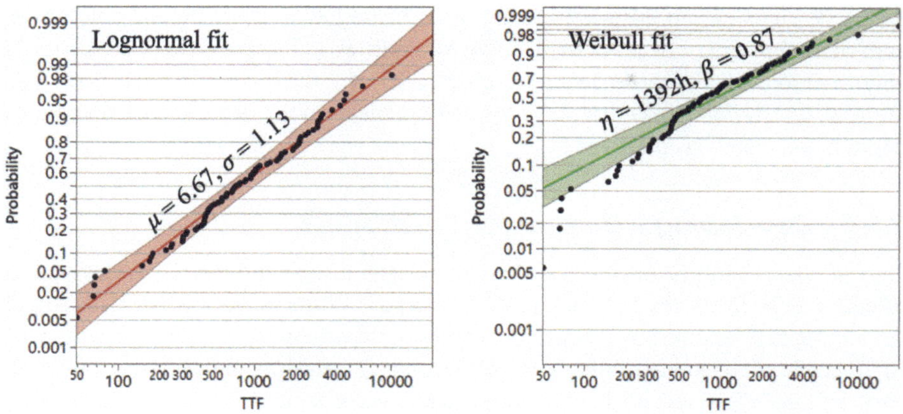

Figure 2.13 Laser reliability data fitted to lognormal (left) and Weibull (right) models.

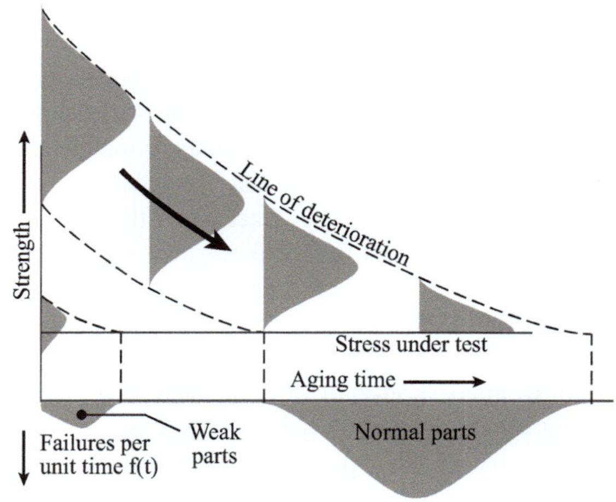

Figure 2.14 The strength of the lasers drops as they age as shown with the vertical plots; when the strength falls to the stress level they are tested or used at, failures are observed. The horizontal plot at the bottom shows how the failures map out over time, with a small number of weak parts that are removed by the burn-in, and a larger number of normal parts that start to fail after the wearout lifetime is exceeded.

they age, before failure is observed. Our devices start out with all devices hopefully having a significant margin of strength above the use stress they are being subjected to by operation. As they age, strength gradually declines, as shown in Fig. 2.14. When strength falls below the level of use stress, failure occurs.

If failure is gradual, we can easily track it, fit it, and make projections or extrapolations. But in some cases, abrupt or catastrophic failure is observed, and cannot normally be predicted or modeled by monitoring performance, as it may be totally stable before it fails. It is in this case that the stress–strength model may become useful. We can set up destructive (one-time) testing at a few different time points, and chart the dropping level of residual strength as aging progresses, to better be able to predict it. For more detail see chapter 13 of Prof. Meeker's book, *Statistical Methods for Reliability Data* [5], or the JMP manual on the use of their "destructive degradation" analysis platform.

2.5.6 Developing a burn-in for the product

One of the reliability engineer's most important responsibilities is to develop an effective burn-in for the new product. The purpose of the burn-in is twofold: first, it should identify and remove weak parts that would otherwise have failed within the first year in the customer's application. Second, it stabilizes the performance of the part, although this second item is only a benefit for some types of products, since others have little or no drift in burn-in.

The burn-in usually is the equivalent of 0.5–4 years' worth of normal use. The tradeoff is between having too short a burn-in, and letting bad devices escape because the burn-in was too brief to purge them, and having too long a burn-in, where the cost of burn-in was higher than needed without any real benefit. The ideal way of deciding what burn-in time to use would be to start with a proposed burn-in, and then monitor the failures at numerous intervals, such as a quarter, half, one, and two times the proposed duration, and see how many failures are observed at each of those points. By doing this, you can see what the shape is for the left side of the bathtub curve, and project what fraction of the failures can be removed for each proposed duration, and what the "leak through rate" would be if the burn-in were shortened. If the shorter rate only lets through failures in the 10–50 parts per million (ppm) rate, this might be acceptable in some applications. One challenge to collecting useful data is that many types of lasers may only have failure rates that are $<0.1\%$, so many thousands of parts need to be tested to collect statistically significant data on the time to failure for weak parts. In such cases, it may be necessary to do the experiment on production parts to get a large enough sample size with tens of thousands of parts. In such a case, the user could start with a "double burn-in," seeing how many fail in the first half vs. the second half. One other general piece of guidance is that there is a choice between high-temperature burn-ins, or ones at somewhat lower temperatures. Parts susceptible to dark-line defects (DLDs) (which include many of the most frequently used lasers) have DLDs driven by spontaneous emission that optically pumps those defects out at the edge that are otherwise too far away from the electrically pumped region [36]. Burn-ins at very high temperature (e.g., 125°C–150°C) reduce spontaneous emission, and thus do not drive DLD growth outside the pumped area as effectively during the burn-in, so many burn-ins operate at lower temperatures (e.g., 85°C–100°C) where at least weak lasing and stronger spontaneous emission are possible.

Where the burn-in falls short of its goals, three additional steps can be taken. The first is to use variable limits rather than fixed limits. For example, a burn-in with fixed limits might say all parts that change by more than $\pm 5\%$ should be screened out; variable limits might calculate the median and sigma, and remove everything outside 2 or 3 sigma of the median. This helps to remove the variability that is encountered on a lot-to-lot basis. The second is to use a multistep burn-in. In this case, you can separate the initial stabilization drift from the stabilization after burn-in, by analyzing the change in the second burn-in period, as shown in Fig. 2.15. In typical cases, after stabilization, lasers should change by less than $\pm 2\%$ (and possibly as little as $\pm 1\%$ depending on the test system repeatability). Parts that degrade by more than that amount can be singled out and removed. For some types of lasers, particularly AlGaInAs-based 1310 nm lasers or high power emitters, a low-temperature "optical burn-in" with high lasing power may be needed in addition to a thermal burn-in, since other types of defects may not be caught with the thermal burn-in [37].

Finally, the burn-in process, like other parts of the process, need to be tightly controlled, and the reliability engineer needs to be in the approval loop for any proposed changes to the burn-in, and analyze the impact of that change to the failure rate of shipped products. For a more detailed discussion of burn-in, Finn Jensen's treatment is my favorite on the subject [38].

2.5.7 Calculating random failure rates

One item customers will ask for is an estimate of random failure rate. This is often expressed in "failures in time" ("FITs"), as mentioned earlier, where 1 FIT is the fraction failing per hour times 10^9. (The units are more easily thought of as parts per million failing per thousand hours.) Three approaches are used to answer this question, as outlined in Telcordia spec SR-332 [39].

1. *Parts count method*: The authors of Telcordia SR-332 have collected large amounts of data on electronic assemblies to provide conservative baselines on how likely each component in an assembly is to fail. These are available both in paper

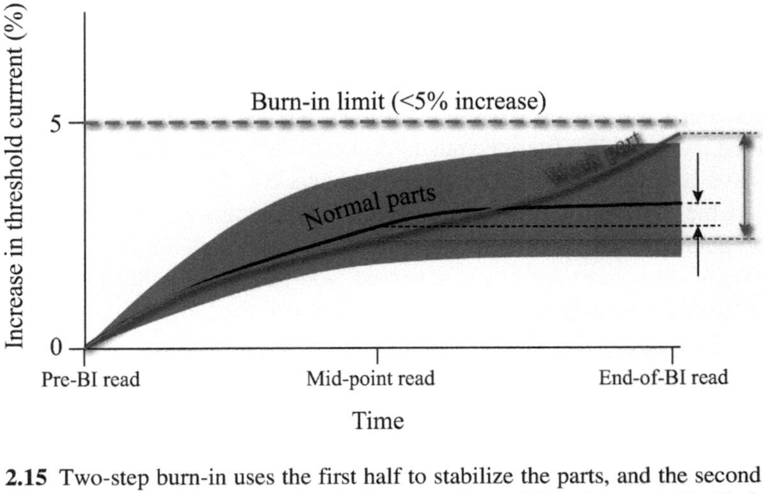

Figure 2.15 Two-step burn-in uses the first half to stabilize the parts, and the second half to verify stability. Typical parts (shown in black) change by $<1\%$ in the second half of the burn-in, while the weak part (shown in purple) changes by 3%.

tables, as well as in commercial software. The reliability engineer would load the bill of materials for all the components that go into the transceiver or other assembly (e.g., 13 resistors × 0.5 FITs per resistor + 4 capacitors × 1.2 FITs per capacitor, etc.). The random failure rates can then be added up to provide the estimated failure rate for the entire assembly. In most cases, while the electronics may be mature components that are listed in SR-332's tables, the optoelectronic components like the semiconductor lasers will be too novel to be covered, and the engineer will need to calculate that separately, and add it in manually. In general, the parts count method is extremely conservative, and provides an estimate of the failure rate that can be an order of magnitude or more above the actual field failure rate. But the numbers still can be compared, and give useful information on how complex the technologies are, which lets people know what to expect.

2. *Field failure method*: The most accurate method of determining the random failure rate is a backward-looking one, that can only usefully be applied to products deployed in the field for a year or more. To compute this rate, you would collect data on the number of the product that had been shipped each month, and use it to compute the number of device-hours that parts had been deployed. (It is normal to make an assumption about how long it takes between shipment and the start of customer usage, where 3 months is a typical value used.) From the "reverse logistics" employees who handle warranty returns, you can obtain data on the number of returns or field failures. The FIT rate is then just the number of failed units divided by the number of billion field device-hours accumulated.

There are two major problems with this approach. First, for new products, no reliability data have been accumulated, so this method is backward looking, and cannot be used for predictions except on mature products. Second, most customers will not bother to return every failed part, so it is common for this method to undercount the actual field failure rate by up to an order of magnitude. If part of the customer base is known to have a policy of returning every field failure, and requesting a failure analysis report on it, then the method can be computed using just their returns divided by just the deployed part-hours to that specific customer, to get a more accurate calculation.

3. *Laboratory test method*: The third method that can be used to calculate the random failure rate uses data from the reliability laboratory. Where numerous qualification rounds have been run, the number of device hours can once again be computed, by adding up the number of devices tested, times the duration of each individual test, times the assumed acceleration factor between the test condition and normal deployment. Even if no failures have been observed, you can use the chi-squared method to compute the upper bound on the likely failure rate with the formula:

$$\lambda = \frac{\chi^2/2}{\text{AF} \cdot \text{number of devices} \cdot \text{test duration}}$$

where λ is the failure rate, χ is a probability function dependent on confidence limits and number of observed failures that can be looked up in a table or in Excel, and "AF" is the effective acceleration factor of the reliability test. (With no failures, and 90% confidence limits, $\chi^2/2 = 2.305$ [40].)

Usually, the desired FIT rate by the customer is orders of magnitude lower than what can be proven by this method, since the quantities of parts tested rarely exceeds hundreds, and you would need to have tested 10,000–100,000 parts to collect an accurate failure rate. Thus this method only provides an upper bound to the possible undetected failure rate, and is best used to assess how far along or how thorough the reliability qualification has been.

2.5.8 Writing reliability reports

Reporting of reliability results takes two forms: internally, the reliability engineer will periodically present the status with a presentation of slides to the team, showing the results obtained. If the reliability tests passed without event, a simple status update to that effect may be all that is required. However, if failures were observed, then reject validation will be required, along with failure analysis, data mining on the affected parts and affected lot (including looking at the raw data), and more frequent meetings with the engineering staff to discuss the validity of the failures, and how to address the failure mode if applicable.

Once the part can consistently pass, and meets all reliability requirements for product release, the reliability engineer will start writing a reliability report. The report should include:

1. A description of the product under test;
2. The reliability plan that was set up to qualify the part;
3. The results of the reliability tests;
4. Analysis of the reliability data;
5. Estimates of the reliability under different conditions, and either a reliability model if acceleration factors have been empirically established, or assumed values if they have not;
6. A summary of the reliability qualification status. For example, that the part finished the qualification and is released to production, or that the part completed the preliminary qualification, but is still undergoing extended reliability testing, etc.;
7. Recommendations on limitations, if any, of the part, if they fall inside the established data sheet limits. For example, a part might allow −40°C to +85°C operation, but with the intention that operation over 60°C was only something intended during short periods during cooling system failures, and that the recommended long-term operating temperature was at or below 60°C.

When the reliability data sheet is completed, it is customary to have it reviewed by a change control board, including engineering, product managers, marketing, and other leadership, before it is issued as a formal released document that can be shared with customers, either on the company website, or by request with customers who have a nondisclosure agreement with the supplier.

2.6 Failure analysis

Failure analysis (or "FA") refers to the process of determining how and why a device failed. A strong failure analysis capability is critical to understanding why failures occurred, and how to eliminate them. Without it, there is no real underlying understanding of why the device fails in the way it does, and it is difficult to know what to change. FA also can often greatly reduce the learning time, since you can look for the defects immediately before and after a suspected step, and see if that step indeed caused the damage to the device. An important goal of failure analysis should be to look for telltale "fingerprints" or unique signatures seen in each failure mode that can be collected quickly in-house, but have been validated by deeper root cause techniques such as transmission electron microscopy. For a more in-depth discussion of failure analysis, refer to our earlier books and articles [41,42]. Samples of the techniques used can also be seen in Professor Mukherjee's chapter in this book (Chapter 4: Material science of defects in semiconductors).

Failure analysis is a critical part of the continuous improvement process discussed earlier, and the FA manager or the reliability department manager needs to be capable of judging whether the proposed corrective action is likely to actually prevent failure occurrences. They also need to be capable of persuading operations/manufacturing or design to make the changes necessary to improve product quality, even if (as is often the case), the proposed change may add to product cost, or hurt yield or performance. The tradeoff between the competing interests is a balancing act of getting optimal cost, performance, and reliability, and the quality department management plays a part in getting the balance of the parameters customers value most.

2.7 Physics of failure and common failure mechanisms

For semiconductor lasers, common failure mechanisms should be divided into gradual and abrupt failure mechanisms. Gradual aging can occur due to things like point defect migration increasing recombination centers in the active region [43]. Abrupt aging is often due to DLDs due to fast-growing climb dislocation networks that propagate from crystal defects or mechanical damage to the semiconductor, such as shown in Fig. 2.16.

One element that is underappreciated is the importance of materials selection for laser emitters. As can be seen in Table 2.4, different materials have differing levels of susceptibility to propagating DLDs, as are also briefly discussed in Chapter 4, Material science of defects in semiconductors, and Chapter 5, Grown-in defects and thermal instability affecting the reliability of lasers: III−Vs versus III-nitrides, of this book. Unfortunately, the large majority of laser products sold today are in the most vulnerable category, due to the lower cost and ease of manufacturing for that category. They are vulnerable to failure unless careful precautions are taken, such as 100% burn-in, visual inspection, and destructive sampling for lot qualification. Fortunately, with the right precautions, even lasers in category 3 have frequently

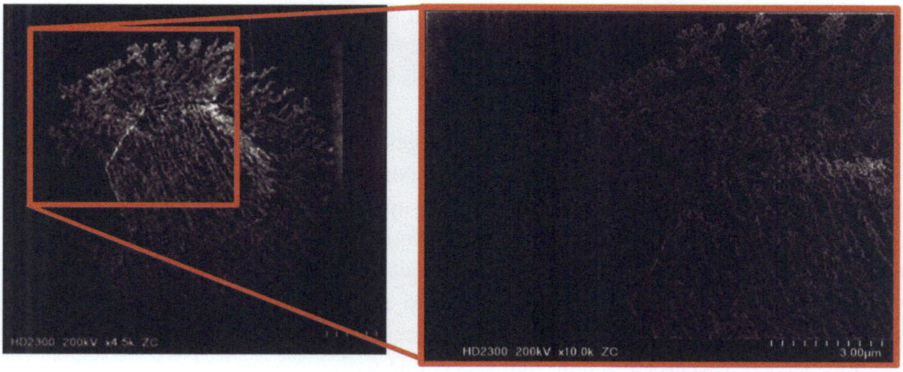

Figure 2.16 Transmission electron microscope (TEM) image of a failed vertical cavity laser (VCSEL), showing an extensive "dark-line-defect" that caused its failure.

Table 2.4 Potential laser emitter materials can be broken into three categories, with varying levels of vulnerability to dark-line defect growth.

#	Strength	Propagating dark-line defects?	Example materials
1	"Rugged"	Nonpropagating	InGaAsP on InP (1.3−1.55 μm); InGaN
2	"Semirugged"	Pinned by compressive strain	InGaAs on GaAs (> 920 nm)
3	"Normal"	Propagating	GaAs (850 nm); AlGaInAs on InP (1310 nm); InGaAsP on GaAs (0.8−0.9 μm); ZnSe

had failure rates that are very low, with <10 ppm after 5 years being obtained for most products.

The light source in many transceivers is a laser whose output is modulated on and off by direct modulation, from a low state not far above threshold, to a very high state with current densities >20 kA/cm^2 and energy densities >10 GW/cm^3 in the active region. At high temperatures, this can cause relatively rapid wearout rates. Cleaved facet lasers can experience issues such as corrosion behind the facet [44], premature failure, or damage from nonoptimized cleaving and coating, particularly on ridge lasers. These can be addressed in a few ways: one is to fabricate "window lasers," which are widely used for costly high-power lasers, but due to the additional processing required, this is something that is not widely used for data communications. Another way that will probably become more popular is to move from direct modulated lasers to externally modulated lasers (EMLs), with the laser kept at a constant power, and the power controlled with an external modulator that is in a separate section of the chip. EMLs can also help with the gradual aging rate, since a longer laser with lower current density can supply the power, because very high current densities to get fast rise and fall times are no longer needed, nor are small devices to get lower capacitance.

Another laser material with unique challenges is AlGaInAs lattice-matched to InP. This material provides better high-temperature performance than InGaAsP-based lasers, and has enabled uncooled transceiver operation up to (and over) 85°C ambient temperatures. It also generally has excellent wearout lifetime, but unfortunately is more vulnerable to sudden failure due to DLDs or intracavity melting. It includes an unusual sensitivity to failure due to optically activated aging modes in some cases. Special screening has helped [37], but additional lot acceptance and a highly optimized process are also needed to keep the failure rate acceptable with most cleaved ridge lasers.

For fiber optic transceivers, failure is often due to a large drop in laser power, but where that is not the cause, additional concerns are often due to mechanical issues such as drifting fiber alignment, or mechanisms that can be seen in electronic assemblies such as broken solder joints.

2.8 Product release, ongoing monitoring, and postrelease support

In this section, we cover the product life cycle, product release, and reliability support after release for fiber optic transceivers. For other products, the process is likely to be quite different, although those companies are likely to have adopted some of the same structure to oversee their product development process. Commercial product development is controlled by product release gates, which must be passed to gain continued development funding. To advance past a gate, the product must meet a number of criteria, including meeting performance, cost, and yield criteria. Certain levels of reliability test performance also must be met. The most important gate is the product release, or commercial availability, which is only granted after the product has completed its full GR-468 reliability qualification, and meets all product requirements. The suppliers start shipping "beta samples" (preproduction units) of transceivers approximately a year before the product release milestone is reached, to allow customers to qualify products under development, and see how they work in their equipment.

Before and shortly after product release, more intense testing and reliability screening at the module level will often take place. This might include module-level HTOL burn-in for 48 hours on every transceiver. This sort of enhanced screening is usually phased out as volume ramps up over 6 months, as long as the fallout from the screens being low. In addition, after the product passes its GR-468 reliability qualification, it goes into a periodic monitoring, sometimes referred to as "ORT/ORM" (on-going reliability test/ongoing reliability monitoring), as discussed in Section 2.2.4.

Occasionally, it will be necessary to change subcomponent suppliers, or make changes to the design to support cost reductions. When this needs to be done, the supplier will issue a process change notice, or "PCN," informing the customer of the change. Customers may spend $100,000 or more on qualifying a new transceiver design from a particular supplier the first time, and thus ask for both significant advance notice of such changes, and also that such changes are infrequent (preferably no more than once per year). Where many changes are needed, they ask that they be combined into a single PCN.

Late in the product life cycle, volumes drop for legacy transceivers, and supporting a large number of older transceivers may become costly, especially since such mature parts have low margins. Suppliers may decide to stop producing the parts, and offer the customers the opportunity to do a "final one-time buy," to purchase the inventory they need to maintain their systems for the anticipated remaining service life before upgrading to more modern technology.

The level of support required from reliability engineers for released products will vary widely depending on the reliability of the technology, the shipping volume, and how demanding the application is. In smaller organizations, the same reliability engineer who qualified the product can support it after release, while in larger organizations, a separate branch of the department may be responsible for monitoring the reliability of released products.

2.9 Conclusion

2.9.1 Future challenges in optoelectronic reliability engineering

There are many challenges that remain for reliability engineers. LED luminaires will continue to try to produce more power from smaller chips. Higher modulation speeds require higher operating power in data communication devices, which makes it more difficult to meet wearout lifetime requirements. Continued growth in cloud data centers and higher levels of integration will require even lower defect rates per laser to keep repair rates low. One item that might ease these requirements is systems that allow flexible or adaptive routing of data packets, where data can be shunted around downed links, allowing a "fail in place" strategy that does not require maintenance when link failure occurs. Data centers have gone from being aggressively cooled a few decades ago, to using more natural air cooling and evaporative cooling, that requires higher temperatures and humidity levels, which puts more stress on components [13]. Low-cost "nonhermetic" transceivers attempt to avoid failures due to corrosion without the benefit of hermetic enclosures using metal, ceramic, and glass. This usually requires additional encapsulation steps of vulnerable devices such as photodiodes. Coherent and tunable laser applications required better stability and lower noise over the life of the part. Photonic integrated circuits often have hundreds of thousands of channels in a single assembly, and thus require lower failure rates per channel than past applications. Finally, embedded copackaged optics involve mounting transceivers on top of chip packages, where temperatures are estimated to be as high as 110°C, creating challenges involved in higher operating temperatures.

2.9.2 Chapter summary and take-aways

In this chapter, we have covered many challenges involved in reliability engineering. One key is understanding whether the failure mode you are concerned about is a wearout failure mode, or due to maverick failure mechanisms that only apply to a small subpopulation of defective devices. The latter is usually more challenging to detect and measure, but often is what drives the product failure rate. We have covered how to assess customer requirements before starting, and how accelerated reliability testing allows us to determine expected product durability in a fraction of the time that the product is supposed to last for. Different types of tests, including temperature and bias testing, corrosion, and mechanical, were discussed, along with acceleration models for each. In Section 2.5, we showed how data for semiconductor lasers are collected, and reduced, to support reliability models, and make decisions about whether one product was better or worse than another, or how much margin it has over the product requirements.

2.9.3 Suggestions for resources for further information

While we have included many references in this chapter for details on specific points, it is worth highlighting a few more general references that are appropriate for the reader looking for more detail than a brief introductory chapter such as this can include. For reliability data analysis, Franklin Nash's first book *Estimating Device Reliability* [31]. The author worked at Bell Labs, and qualified some of the first subsea optical communication systems. Finn Jensen's book *Electronic Component Reliability*, though out of print, remains another good treatment that goes into much more detail on many of the topics touched on here [7]. McPherson also covers many of the same topics in his advanced treatment [6]. The book edited by Ueda and Pearton [45] is one of the newer books on the subject of optoelectronics reliability, and includes two tutorial chapters that I wrote on failure analysis and VCSEL reliability. Of course, other chapters in this book, especially the next chapter (with the case studies) are also good resources for those interested in reliability. In terms of websites, two useful sites are the NIST reliability handbook [46], and the Weibull.com website [47]. A few short courses are also available: Professor William Meeker of Iowa State University teaches the best reliability analysis workshop I have attended [48]; reliability short courses are also available at conferences such of the Optical Fiber Conference, for a more detailed discussion of GR-468 qualification of fiber optic transceivers [49].

References

[1] C. Carlson, Effective FMEAs: Achieving Safe, Reliable, and Economical Products and Processes using Failure Mode and Effects Analysis, 1 edition, Wiley, Hoboken, N.J, 2012.

[2] S. Nappi and G. Marrocco, Inkjet-printed RFID-skins for the detection of surface defects, in: 2018 2nd URSI Atlantic Radio Science Meeting (AT-RASC), May 2018, pp. 1–4. Available from: https://doi.org/10.23919/URSI-AT-RASC.2018.8471335.

[3] J.I. Rosell, X. Vallverdú, M.A. Lechón, M. Ibáñez, Design and simulation of a low concentrating photovoltaic/thermal system, Energy Convers. Manage. 46 (18) (2005) 3034–3046. Available from: https://doi.org/10.1016/j.enconman.2005.01.012.
[4] W.B. Nelson, Accelerated Testing: Statistical Models, Test Plans, and Data Analysis, 1 edition, Wiley-Interscience, Hoboken, N.J, 2004.
[5] W.Q. Meeker, L.A. Escobar, Statistical Methods for Reliability Data, 1 edition, Wiley-Interscience, New York, 1998.
[6] J.W. McPherson, Reliability Physics and Engineering: Time-To-Failure Modeling, 3rd ed., Springer International Publishing, 2019.
[7] F.B. Jensen, [(Electronic Component Reliability)], John Wiley and Sons Ltd, 1996.
[8] J.W. Osenbach, Corrosion-induced degradation of microelectronic devices, Semicond. Sci. Technol. 11 (2) (1996) 155–162. Available from: https://doi.org/10.1088/0268-1242/11/2/002.
[9] R.W. Herrick, Reliability of fiber optic datacom modules at Agilent Technologies, in: Electronic Components and Technology Conference, 2002. Proceedings 52nd, 2002, pp. 532–539.
[10] D.G. Ivey, J. Luo, S. Ingrey, R. Moore, I. Woods, Galvanic corrosion effects in InP-based laser ridge structures, J. Electron. Mater. 27 (2) (1998) 89–95. Available from: https://doi.org/10.1007/s11664-998-0195-z.
[11] J.W. Osenbach, T.L. Evanosky, Temperature-humidity-bias-behavior and acceleration model for InP planar PIN photodiodes, J. Light. Technol. 14 (8) (1996) 1865–1881. Available from: https://doi.org/10.1109/50.532025.
[12] GR-468 - Reliability Optoelectronic Devices Used I Telcordia. https://telecom-info.telcordia.com/site-cgi/ido/docs.cgi?ID = SEARCH&DOCUMENT = GR-468 (accessed 13.06.20).
[13] L. Li and J. Xue, Reliability of microelectronics packaging in the era of EnergyWise and Borderless networks, in: Reliability Physics Symposium (IRPS), 2010 IEEE International, 2010, pp. 440–445.
[14] S.K. Kurtz, D.S. Peck, (1989). *ISBN 089006-350-8* in: Emiliano Pollino (Ed.), Microelectronic Reliability, Vol. II: Integrity Assessment and Assurance, Pergamon, Artech House, Norwood, Massachusetts, 1989.
[15] A. Dafinca, et al., Reliability and degradation of oxide VCSELs due to reaction to atmospheric water vapor, Vertical-Cavity Surface-Emitting Lasers XVII, 8639, 2013, p. 86390L. Available from: http://doi.org/10.1117/12.2001195.
[16] R.W. Herrick, A. Dafinca, P. Farthouat, A.A. Grillo, S.J. McMahon, A.R. Weidberg, Corrosion-based failure of oxide-aperture VCSELs, IEEE J. Quantum Electron. 49 (12) (2013) 1045–1052. Available from: https://doi.org/10.1109/JQE.2013.2285572.
[17] J. Arnold, N. Blattau, C. Hillman, C. Park, and K. Sweatman, Reliability testing of Ni-modified Sncu And Sac305 – accelerated thermal cycling, DfR soultions website. https://www.dfrsolutions.com/hubfs/DfR_Solutions_Website/Resources-Archived/Publications/2008-2010/2008_03_ni_modified_sncu_thermal_cycling.pdf.
[18] H. Cui, Accelerated temperature cycle test and Coffin-Manson model for electronic packaging, in Annual Reliability and Maintainability Symposium, 2005. Proceedings, Jan. 2005, pp. 556–560. Available from: https://doi.org/10.1109/RAMS.2005.1408421.
[19] A. Chakravarty, S. Giridharan, M. Kelly, A. Poojary, and V. Zeng, Characterizing large-scale production reliability for 100G optical interconnect in Facebook data centers, in Frontiers in Optics 2017, Washington, D.C., 2017, p. JW4A.64. Available from: https://doi.org/10.1364/FIO.2017.JW4A.64.
[20] F.M. Steranka, High Brightness Light Emitting Diodes, Academic Press, 1998.

[21] ANSI/IES LM-80-20: Measuring LED luminous flux and color maintenance—ANSI Blog, The ANSI Blog, May 20, 2020. https://blog.ansi.org/2020/05/ansi-ies-lm-80-20-measure-led-luminous-color/ (accessed 13.06.20).
[22] Measuring luminous flux and color maintenance of LED packages, arrays and modules – Illuminating Engineering Society. https://www.ies.org/product/measuring-luminous-flux-and-color-maintenance-of-led-packages-arrays-and-modules/ (accessed 13.06.20).
[23] Technical memorandum: projecting long term lumen, photon, and radiant flux maintenance of LED light sources – Illuminating Engineering Society. https://www.ies.org/product/projecting-long-term-lumen-photon-and-radiant-flux-maintenance-of-led-light-sources/ (accessed 14.06.20).
[24] Methods for projecting catastrophic failure rate of LED packages – Illuminating Engineering Society. https://www.ies.org/product/methods-for-projecting-catastrophic-failure-rate-of-led-packages/ (accessed 14.06.20).
[25] ENERGY STAR Lamps V2_0 Program Requirements.pdf. [Online]. Available from: https://www.energystar.gov/sites/default/files/ENERGY%20STAR%20Lamps%20V2_0%20Program%20Requirements.pdf (accessed 14.06.20).
[26] AEC_Q102_Initial_Release_Final.pdf. [Online]. Available from: http://www.aecouncil.com/Documents/AEC_Q102_Initial_Release_Final.pdf (accessed 14.06.20).
[27] 8.3.1.4. Accelerated life tests. https://www.itl.nist.gov/div898/handbook/apr/section3/apr314.htm (accessed 13.06.20).
[28] R.W. Herrick, Reliability and degradation of vertical-cavity surface-emitting lasers, in: O. Ueda, S.J. Pearton (Eds.), Materials and Reliability Handbook for Semiconductor Optical and Electron Devices, Springer, New York, NY, 2013, pp. 147–205.
[29] Weibull++ 2020, 2020. Weibull++, a Reliasoft reliability analysis package. (Accessed 30 Nov 2020).
[30] 15.2, 2020. JMP, a software package by SAS software. (Accessed 30 Nov 2020).
[31] F.R. Nash, Estimating Device Reliability: Assessment of Credibility, Springer, 1993.
[32] 8.1.6. What are the basic lifetime distribution models used for non-repairable populations? https://www.itl.nist.gov/div898/handbook/apr/section1/apr16.htm (accessed 14.06.20).
[33] W.B. Joyce, K.-Y. Liou, F.R. Nash, P.R. Bossard, R.L. Hartman, Methodology of accelerated aging, AT&T Tech. J. 64 (3) (1985) 717–764. Available from: https://doi.org/10.1002/j.1538-7305.1985.tb00446.x.
[34] W.B. Joyce, R.W. Dixon, R.L. Hartman, Statistical characterization of the lifetimes of continuously operated (Al,Ga)As double-heterostructure lasers, Appl. Phys. Lett. 28 (11) (1976) 684–686. Available from: https://doi.org/10.1063/1.88622.
[35] F. Nash, Table 7.2, and Figs 7.16 & 7.21., in: Estimating Device Reliability: Assessment of Credibility, Kluwer Academic Publishers, Boston, 1993, pp. 145, 150, 155.
[36] R.G. Waters, Diode laser degradation mechanisms: a review, Prog. Quantum Electron. 15 (3) (1991) 153–174.
[37] R.W. Herrick, C.B. Roxlo, T.H.O. Sjolund, and T. Sudo, Methods for testing lasers using optical burn-in, US8067949B2, Nov. 29, 2011.
[38] F. Jensen and N.E. Petersen, Burn-in. Wiley New York, 1988.
[39] SR-332—Reliability Prediction Procedure | Telcordia. https://telecom-info.telcordia.com/site-cgi/ido/docs.cgi?ID = SEARCH&DOCUMENT = SR-332& (accessed 14.06.20).
[40] F.B. Jensen, Table 9.8, in: Electronic Component Reliability, Chichester, UK; New York, NY, 1995, p. 219.
[41] O. Ueda, R.W. Herrick, Failure analysis of semiconductor optical devices, in: O. Ueda, S.J. Pearton (Eds.), Materials and Reliability Handbook for Semiconductor Optical and Electron Devices, Springer, New York, NY, 2013, pp. 19–53.

[42] O. Ueda, Reliability and Degradation of III-V Optical Devices, Artech House Publishers, Boston, 1996.

[43] O. Ueda, Degradation of III–V opto-electronic devices, J. Electrochem. Soc. 135 (1) (1988) 11C–22C. Available from: 10.1149/1.2095535.

[44] W.J. Fritz, L.B. Bauer, and C.S. Miller, Analysis of aluminum gallium arsenide laser diodes failing due to nonradiative regions behind the facets, in Reliability Physics Symposium, 1989. 27th Annual Proceedings., International, 1989, pp. 59–64, doi: 10.1109/RELPHY.1989.36318.

[45] O. Ueda, S.J. Pearton (Eds.), Materials and Reliability Handbook for Semiconductor Optical and Electron Devices, Springer-Verlag, New York, 2013.

[46] 8. Assessing Product Reliability. https://www.itl.nist.gov/div898/handbook/apr/apr.htm (accessed 14.06.20).

[47] Life Data Analysis Reference Book - ReliaWiki. http://reliawiki.org/index.php/Life_Data_Analysis_Reference (accessed 14.06.20).

[48] William Q Meeker, *Department of Statistics*. https://www.stat.iastate.edu/people/william-meeker (accessed 14.06.20).

[49] SC347 | OFC, https://www.ofcconference.org/en-us/home/program-speakers/short-courses/reliability-and-qualification-of-fiber-optic-compo/ (accessed 14.06.20).

Chapter 3

Case studies in fiber optic reliability

Robert W. Herrick
Intel Corporation, Santa Clara, CA, United States

3.1 Introduction

Much of the rest of the material in this book has general recommendations ("do this," "don't do that"). I have tried to include the rationale for the recommendations where possible. However, the application of the information is often hard to see without specific examples. It also does not show what happens when you push the envelope—and it is often hard to follow all recommendations, all of the time.

I personally learn best through case studies or concrete examples, which show what happened where the recommendations were *not* followed, or short-cuts were taken, and negative consequences occurred as a result. While the case studies here are all of products that are no longer offered on the market, and from companies no longer in these business lines, the lessons contained are still relevant today, and I continue to see many similar mistakes from other companies in optoelectronics in the past decade as well.

Most of the companies in these stories were very successful companies with good overall track records on quality and reliability. Many got better after going through these cases. I hope the lessons learned can be taken to heart by those reading this chapter, even if they have not personally lived through them. I also hope it is clear from some of the examples that making the wrong decision in a quality and reliability setting can often have a very high cost associated with it. The median case study involved a financial loss of US $10–20 million.

I have organized this chapter in three major groups of case studies, by the "product life cycle" phase that the reliability problem occurred in. The first four cases had issues that were present from the start, and were detected before they ever were released to production. The next four cases also had issues that were present from the start, but were not detected until some time after product release. The final two cases deal with products that were reliable at the time of product release, and for some time after that, but had issues that came up later in the course of production. Within each group, the cases are ranked in order of ascending financial impact.

Reliability of Semiconductor Lasers and Optoelectronic Devices. DOI: https://doi.org/10.1016/B978-0-12-819254-2.00005-9

3.2 Group 1: issues that were caught before product release

In this group, issues were detected during the reliability qualification, and as such, never reached the field in large numbers. This avoided expensive recalls, which can often amount to many times the cost of the product, and over 100 times the cost of the defective subcomponent. But in spite of that, case study 4 in this group is probably still the highest financial loss episode in this chapter, so early detection alone is not sufficient to protect a company from the financial loss of an unsuccessful product development effort.

3.2.1 Case 1: "Hey, you cooked my VCSELs!"

Situation: A transceiver customer started seeing a number of fiber optic transceivers with early failures. They shipped them back to the transceiver manufacturer for failure analysis, which showed evidence of premature laser wearout, showing the same symptoms seen from long-term accelerated aging tests. The deliveries were from wafers shipped before a new wafer reliability monitoring had been put in place, so their reliability was suspect. A number of conference calls were arranged, with the vertical cavity surface emitting laser (VCSEL) supplier educating the customer on potential VCSEL reliability issues. Given the evidence of wearout failure after only a few weeks of deployed use, the supplier asked questions about the deployed environment. Getting the history of failed parts is always critical, and the customer should be asked to provide information on the deployed time and temperature, which ideally come from downloading the "digital diagnostics monitoring" or DDM data. The customer carried out an investigation, and found on further examination that there was a problem with the reliability screening setup these transceivers had gone through immediately before being rejected and returned. As was common practice, 100% of their 19″ rack-mount equipment was screened by running in a large "shake and bake" HASS chamber [1–4] by their subcontractor, where random vibration was applied while the parts were supposed to be temperature cycled multiple time between 10°C and 65°C for about 3 h. This type of screening has been very effective at detecting and eliminating poor solder joints in rack-mount equipment. Unfortunately, the duct work on the chambers in question was connected incorrectly (hopefully not in as obvious a way as shown in Fig. 3.1!), to where the exhaust ducting was connected to the input of the chambers, and temperatures were measured to exceed 100°C (with some estimates going as high as 140°C). This was well in excess of the maximum temperature the transceiver was designed to tolerate. The transceiver designers had not designed any reasonable upper bounds to the drive current—so 30 mA or more average current could be supplied by the current driver at high temperatures to a VCSEL normally designed to run at 10 mA or less. This was in a futile attempt to maintain a fixed optical power out—the optical power drops rapidly above 90°C from virtually all lasers. The combination of high temperatures and high current created an acceleration factor of over 1000 ×, and appeared to be responsible for the premature failures. Fortunately, a relatively small number of parts were impacted, and the failure rate was relatively low.

Figure 3.1 Cartoon of the "shake and bake chamber" from this case study, and misrouted piping that drove the temperature far above the original specifications.

In a related case, a different customer was intentionally "measuring margin" on all transceivers being considered by doing reliability testing on transceiver samples. They ran the transceivers at 150°C for a few days, and observed failures. Again, this was due to the very high maximum drive current the bias circuit provided. In this case, the customer took it as a source of concern, and inferred that this transceiver manufacturer had a "lack of reliability margin."

As a corrective action, circuit design and transceiver setup guidelines were changed to reduce the maximum current obtainable at high temperatures to a much lower value, that would not cause permanent damage with just a few hours of aging.

Business impact: Fortunately, the business impact was fairly minimal in the first case since it was caught quickly, and did not affect more than a few hundred transceivers, and also since the customer bore a significant share of the blame, the supplier did not have to pay for the replacement of the parts. It is not known if any business was lost in the case of the poor comparison at 150°C.

What they did right: The VCSEL supplier had a complete set of reliability test data, and sufficient failure analysis (FA) signature library to identify the returns as wearout failure, which would normally be very unusual to see from nothing more than customer burn-in or qualification. Together they were able to identify the problem as excessive temperature and current stress.

Opportunity for improvement: The laser driver circuit was set up to keep the link up at all costs (including permanent damage to the laser). It should have been set up instead to shut down the circuit if it got well beyond the data sheet limits, or certainly where it was damaging the parts with over-bias. The "shake and bake" highly accelerated stress screening (HASS) chamber should of course have had closed-loop monitoring to prevent overtemperature problems.

Summary of lessons learned: Think about what will happen to your part if it is used outside the data sheet parameters. Have failure analysis capabilities ready to go before the part is released to manufacturing. Collect FA data on reliability test parts to establish a "library of failure signatures."

3.2.2 Case 2: new integrated circuit packaging creates problems

Situation: A transceiver company was developing a next-generation transceiver with a much smaller form factor, and a narrower shell. In this case, smaller integrated circuit (IC) packages were needed to fit the smaller PC board layout. A search of packaging vendors revealed that although some research was under way to address these needs, nothing was likely to be commercially available for a few years—the company involved would have to develop their own smaller packaging or go to a much more expensive "double decker" 2-sided printed circuit (PC) board solution.

The transceiver manufacturer was able to work with suppliers to develop their own smaller package, and the schedule for the part progressed nicely. A large and thorough reliability test matrix was executed at the earliest opportunity. Customers were lining up for sample transceivers to test out their next-generation routers; a few thousand transceivers were delivered to them as "beta samples." As the reliability test results came in, a few failures were observed in the temperature cycle test. At first it was not clear whether these were just test artifacts, or something that people should worry about. They probably did not failure analyze the parts in question quickly enough (it took a few weeks), but they were no different than most businesses in this regard—analyzing customer returns come ahead of reliability test failures in priority. They did quickly repopulate the temp cycle cell, with an even larger number of parts, and rerun it to see if the results were reproducible—unfortunately, they were, and additional failures were observed when rerunning the test. Failure analysis showed that the failure mode was due to an open circuit in the small integrated circuit package that was the novel design, where the wirebond had broken off at the heel, as shown in Fig. 3.2. They started doing additional higher volume testing, at higher stresses (liquid-to-liquid thermal shock) on the ICs.

At this point, a few field failures came back from the beta customers, who had deployed a few thousand transceivers at this point. The field failures showed the same failure mechanism as had been seen from the temperature cycle tests. A few hundred parts were temperature cycled over a slightly smaller temperature range to see if the failures would still occur over a less severe regime—they did. Using the Coffin-Mason model discussed in Chapter 2, Reliability Engineering in Optoelectronic Devices and Fiber Optic Transceivers (see also refs [5,6]), projections were used to extrapolate how rapidly failures would be observed under normal use. The summary of these

Figure 3.2 Drawing of stress on wirebond from encapsulant (left); SEM of wirebond heel break from encapsulant pull force (center); projection of failures for different cycle sizes based on temperature cycle stress data. *SEM*, scanning electron microscope.

conclusions was that failures would be observed in significant numbers for parts that were shut down on a daily basis, but should be very rare for "always on" parts. Fortunately, parts in deployed use (including the "beta" parts) were rarely shut down more than a few times a year, and were deployed only in temperature-controlled data centers. It was decided to cut off beta part shipments until the problem was solved, but not to issue a recall. In hindsight, this proved to be the correct decision, as very few returns from fielded parts were subsequently observed through the rest of the normal deployed life of 3–5 years. However, the decision to cut off beta parts and delay release was painful to customers, since it also meant their next-generation routers were delayed by an equal amount, since no alternate suppliers of this smaller footprint transceiver were yet available.

A 4-month period followed of intensive product development to go to package materials with a lower coefficient of thermal expansion (CTE) encapsulant. This solved the problem, as was quickly verified using unmounted packages, and then in transceiver modules that were built up with the new package. The product was released, albeit behind the original schedule. It went on to be a very successful product, with a very low return rate.

Business impact: This was mostly borne by the end customers who had to push out the new product introduction by 1–2 quarters.

What they did right: Almost everything. Thorough reliability test matrix, decisions to cut off beta shipments, risk assessment for fielded modules, totally transparent with customers about risk, accurate future risk assessment, long-term fix effective.

Opportunity for improvement: In hindsight, perhaps there was an opportunity to catch the packaging problem sooner by testing the packaged ICs themselves with more severe stress months earlier. FA on reliability test failures was delayed a few weeks, and identification of the problem took over a month.

Summary of lessons learned: Run a reliability test matrix as early as possible, and pay attention when you see failures. Cut off shipments when there is a problem. Do not release to production before it is ready. Always look for CTE-matched materials whenever possible, and be prepared for problems when materials are mismatched.

"Please break my parts, and please do it now!"—a note on the importance of prequalification reliability testing

One of the more memorable conversations I had was with the director of transceiver product development when I was still relatively new to the job. I was assigned as the transceiver reliability engineer on one of his projects, which was a new platform, with many novel design features. While we were not yet ready to start the formal product qualification, since the design had not yet been finalized, he asked me to find opportunities to do prequalification reliability testing. Such testing was often more severe in terms of stress than the qualification testing, but was intended to show product weaknesses. It also included tests on subassemblies used in the transceiver with unproven parts.

He asked me a question: "Of the dozens of transceiver products I've seen released to market over the past decade, do you know how many have been released on schedule?" After I hazarded a few incorrect guesses, he gave me the answer: "Not one." And the reason for schedule delays? Always the same: "failures on the reliability test matrix which require redesign and requalification before product release." I thought (and still think) this history is remarkable, especially since it came from a well-managed company with great products. It was not uncommon for a product to have to go through three or four "test–fail–redesign" cycles before passing cleanly and getting released to production. My take away: the reliability engineer (and the team leaders responsible for making sure things happen) really do want to figure out how to do reliability testing at the absolute earliest opportunity, even if it will not count toward final product qualification, even if the part is still undergoing design changes, and even if it means allocating the parts to internal testing rather than to marketing for customer sampling. Delays not only cost money in terms of having to rerun qualifications, but you lose the chance to get designed in early at key accounts if you fall behind the competition.

3.2.3 Case 3: great mean lifetime, but a wide distribution of lifetime

A company went to produce a semiconductor laser for use in fiber optic transceivers. Failures occurred during the reliability tests, including very early in the tests, but they were not large in number. Even after being on lifetest many thousands of hours at very accelerated conditions, most of the lasers were still running with no measurable degradation. Reliability analysis showed that it met the two metrics set out for the product: mean time-to-failure (MTTF) and time to 1% failure (also known as T_{01}) were both projected to exceed 10 years. Management moved to release the laser to production.

However, a more experienced reliability engineer was asked to review the work of the component reliability engineer that had originally been assigned to the product. It turned out that the distribution of failures implied a very wide distribution of strength, and without additional work, unacceptable field failures would be expected. Although the time to 1% failure was indeed okay (albeit marginal), the design team had not considered factors like the time to 0.1% failure, as shown in Fig. 3.3. And the instantaneous failure rate (IFR) $f(t)$ and hazard rate $\lambda(t)$ were over 10 times what one key customer required, as shown in Fig. 3.4. The management of the product development group was disappointed, and felt that new (and unnecessary) more demanding requirements were now being added just weeks before a planned product release milestone. The MTTF and 1% goals were made with an assumption that a certain distribution (sigma <1 for a lognormal distribution, or beta >2 for a Weibull distribution) would be met, as was the case for traditional VCSELs and other lasers, as indicated with the red lines in Figs. 3.3 and 3.4. These assumptions are nearly always met for most types of semiconductor lasers. In this

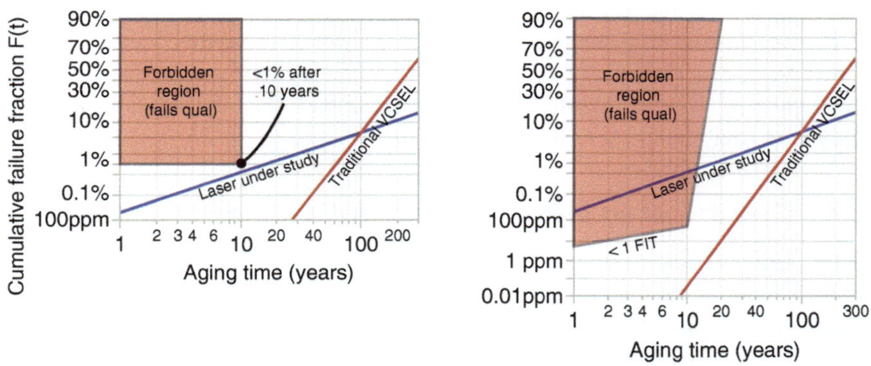

Figure 3.3 Cumulative failure requirements as originally given to the design team (left), and as actually required by key customers (right). The distribution of failures should not cross the red region for any of the projected aging life to be accepted by the customer.

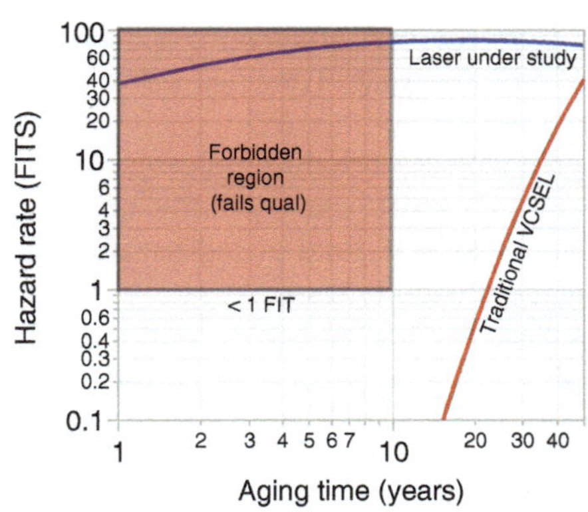

Figure 3.4 Hazard rate (per unit time) for the projected reliability at use condition. The larger number of early failures observed on the life test for the laser being qualified meant a projected failure rate of 40–80 FITs, well above the 1 FIT that is normally projected for early life.

case, the “hidden assumptions” were not met, and this discovery did not come out until the product design was ready to be frozen. The team needed to work another 6 months to improve the product fabrication process, wafer qualification screening, and laser burn-in, before it could meet the revised reliability specifications. If those specifications were known earlier, they possibly would have been able to make the improvements far earlier, and meet the original timeline.

Business impact: No negative impact to customers; the product release was delayed an additional 6 months to allow the team time to improve wafer screening and burn-in procedures.

What they did right: Good reliability test program which had sufficient data to make an high confidence assessment, at a wide range of test conditions.

Opportunity for improvement: Ignored early failures, because they were too focused on MTTF and T_{01}. Original reliability engineer did not have sufficient formal training in reliability engineering. Used the wrong reliability metrics in defining success.

Summary of lessons learned: Early failures, and IFR during the anticipated deployment life, matter more than MTTF.

3.2.4 Case 4: locked into a “no-burn-in” product definition

A large semiconductor company got into fiber optic transceivers because it fit well with their focus in communications electronics. They had a strong record as a provider of high-quality systems, and had long eliminated the need for any burn-in or wafer qualification on their semiconductor integrated circuits. With these ICs, the presence of “burn-in” or “wafer qualification” was seen as evidence of inadequate engineering. Clearly, it was preferable to have quality “designed in” rather than “screened in” through a burn-in. Doing otherwise (if designing out a failure mode is possible—a very, very big “if”) is usually a path to higher cost and lower product quality.

The executives at the semiconductor company hired many people with laser and transceiver experience from the best companies and research labs. They invested heavily; in the range of \$10 m a year. They had a world-class failure-analysis program, and extensive reliability data.

They pushed forward toward product release, and built a large lot for manufacturing verification. It failed the reliability qualification. An investigation showed that all of the failures came from a single wafer (out of more than a dozen wafers used to populate the reliability test).

At this point, it was clear that the executives who had set the ground rules in the first place overreached what was possible with the technology. Clearly, wafer screens and burn-in *were* needed—gallium arsenide (GaAs) is a very vulnerable material, after all, with vulnerability to dark-line defects (DLDs), as seen in Fig. 3.5. But now, 3 years into the program, it was a little late to be developing burn-ins and wafer screens. You need data on parts that have been put through the proposed screen, then aged a year or more under accelerated conditions to prove out the validity of the screen. Faced with years of additional work, the company withdrew, and invested in nonoptical products closer to its core competency.

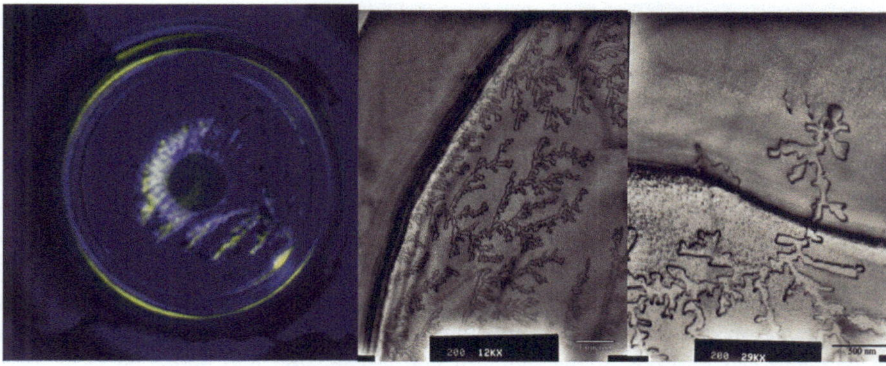

Figure 3.5 Laser scanning image of a similar 850 nm VCSEL, showing a dark-line defect (DLD) network originating at the edge of the mesa, and propagating in to the emitting area in the middle. TEM of edge (center) showing dendritic climb dislocations; TEM of point of origin (right), with cracks originating the climb dislocations. *VCSEL*, vertical cavity surface emitting laser; *TEM*, transmission electron microscope.

It is not to say it is not possible to make a product with no burn-in. I personally am optimistic that strained quantum-well lasers [7,8], which resist DLD arrays, might be capable of being used without burn-in for some less-demanding applications, particularly if other screens were first added. But I would never propose releasing a product without burn-in, or making that the baseline. The normal process would be to collect data on hundreds of thousands of units, document the failure rate through the burn-in process, and show that so few units fail burn-in that it is unneeded. This process would also include reject validation and failure analysis of the few units that did fail, to understand the failure mechanisms responsible. By following this process, (1) the customer is always protected; (2) the product gets released as early as possible, regardless of whether or not a burn-in is needed; and (3) the elimination of burn-in gets moved off the critical path to a "nice-to-have" cost-reduction goal. A wafer qualification screen (with a destructive sample being tested) would probably be required regardless, since semiconductor lasers are always vulnerable to epitaxial growth irregularities, and normal epitaxial wafer characterization is not capable of detecting most of the potential epitaxial reliability excursions.

Business impact: Hard to estimate, but probably at least \$50–100 m in direct investment lost, after the program had to be canceled.

What they did right: Excellent staffing, a high level of investment, many staff with laser and optoelectronics background. Good reliability test program that collected sufficient data at a range of different accelerations to allow high-quality decisions to be made, if given reasonable goals and objectives. An FA program that was best-in-class for its time. The company had a record of excellent quality systems. The reliability program uncovered unacceptable reliability, and parts never reached customers or failed in the field. Management made the correct decision to cancel the product rather than "start over" when it realized the very serious error that had been made early on.

Opportunity for improvement: Flaunted decades of conventional wisdom of the optoelectronics industry by claiming they would get rid of burn-in and wafer qualification in a vulnerable technology that *needs* both types of screens to have acceptable reliability.

Ignored numerous failures from reliability tests along the way that showed that the part *did* have random failures and that it was *not* inherently reliable. Looking back after the program was canceled, many wished they had not "swept the few failures under the carpet" as "oddballs" or "one-off-failures" or "odd wafers." This is a very common mistake to ignore reliability failures.

Summary of lessons learned: Put too many screens in, then reduce them slowly as confidence in the technology builds and data are collected. If you want to flaunt the conventional wisdom of the industry you are in, you had better be very, very certain that what people have been doing is unjustified or wrong—a more likely explanation is that you simply do not understand *why* they are doing what they are. While it is not uncommon for other areas of engineering to call for "scheduled breakthrough," and "meeting never-before-seen metrics," it is hard to schedule breakthroughs in reliability, and betting on it or requiring it frequently results in \$10 M + schedule delays

3.3 Group 2: cases that had reliability issues from the start, but were not detected in the reliability qualification

The next group of cases had issues that were not detected as part of the product qualification, varying from obvious to fairly challenging or obscure issues. The final case is a good reminder that the most important job of the quality department is to protect the company they work for against wearout failures that could go undetected for years before detection, only to result in massive recalls or lawsuits. Rapid infant failure problems or quality issues that show up at delivery or within a month will be reported to operations, which will be forced to react quickly. But if the reliability department is not taking the lead on assessing the risk of premature wearout failure, then it is likely no one else is either.

3.3.1 Case 5: reliability monitor fails to detect ongoing reliability shortcomings

Situation: A product with 850 nm GaAs-based fiber optic communication links had a relatively high field failure rate (0.1−1%/year). A few customers suffered highly visible field failures in their applications, and complained. Reliability engineers went back to the monthly reliability monitor data, and pointed out to customers that "no failures have *ever* been observed on our reliability monitor." In this case, 11 parts per month were sampled and then aged at the offshore manufacturing site for 1000 h at accelerated temperature and current. However, additional samples were pulled, and tested at the onshore engineering site, using different experimental equipment, and it showed a significant number of failures, with numbers consistent with the field failure rate. Now it was getting interesting—why were the results of the reliability monitor inconsistent with those of the "special experiments"—had the parts become less reliable over time? The offshore devices from the last several years, after passing the routine reliability monitor tests, had been carefully archived from many years back, and were pulled out of storage, shipped to the engineers responsible, and moved into additional reliability testing at the engineering site. Parts from years of production were put on a large reliability test, and showed failure rates similar to customer experience and to the recent engineering reliability test results. Now the reliability monitor test results offshore were starting to look like the problem. The reliability test lab at the manufacturing site was audited, and current meters were used to check the driver circuits. A problem was found. The instructions for setting up the reliability test said, "adjust the current regular 10-turn precision trimpot so that the voltage drop across the monitor resistor (which was in series with the devices) is 2.00 V"—this would correspond to the correct amount of current (i.e., the 100 mA specified in the instructions for this product). What the technicians were actually doing, however, was setting the current *across the entire chain* of eight devices + the resistor to 2.00 V, as shown in Fig. 3.6. Because of the nonlinear characteristic of diodes, this change meant that the current, instead of being 100 mA, was only a microamp—basically, the part was unbiased, and the results were no different than would be expected for unbiased high-temperature storage testing.

In response, a more thorough audit of procedures was undertaken, and retraining. However, that did not address the issue that had been uncovered: the underlying

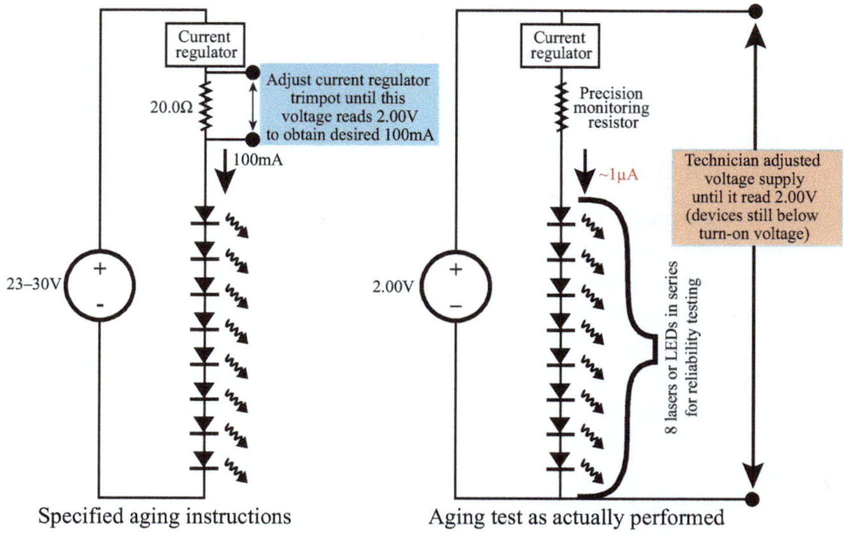

Figure 3.6 Schematic of the intended reliability test setup (left), and the test as it was actually performed (right), with devices biased below their turn-on bias.

reliability of the light-emitting diode (LED) part did not meet customer expectations, even for normal lots. A program to design a totally new device with much higher reliability at the same wavelength was started—and never succeeded. (Note, though, that different products such as 1310 nm links were highly reliable although slightly more expensive—they would have been better choices for the applications in question.) After this, marketing people were far more careful in the reliability claims they made for the low-cost product, and steered people wanting high-reliability parts away from this and similar low-cost parts.

Business impact: A few highly visible failures at key customers, damage to reputation, significant management and engineering time.

What they did right: Followed up on customer complaints, had reliability monitors, archived them afterwards, allowing continued reliability testing.

Opportunity for improvement: They were not aging the reliability monitor devices at the manufacturing site as intended; lack of training and supervision to assure that the test was being performed in accordance with the documented procedure. The documentation lacked photos or drawings of where to audit the voltage on the board, which would have made the mistake less likely. Oversold the parts as suitable for high-reliability applications.

Summary of lessons learned: Do not throw reliability specifications over the wall, and leave them for test engineers, operations engineers, and technicians to interpret and execute without feedback. The reliability engineer should spend enough time in the lab auditing to be familiar with all operations done in the lab, and be able to vouch for the details of their technicians' work, and of the equipment stress calibration, when meeting with internal and external customers. Do not use low-cost parts in high-reliability applications. GaAs-based parts are easy to make, but require a lot of vigilance to make with very low failure rates, and are vulnerable to DLDs.

3.3.2 Case 6: "But the part's never going to operate at 85°C, 85% relative humidity !!!"

Situation: A high-speed, high-power transistor was being qualified for use in wireless communication systems. The part repeatedly failed 85°C, 85% relative humidity (R.H.) reliability testing, sometimes in as little as a few hundred hours, which fell far short of the usual expected benchmark of surviving at least 1000 h of stress. The reliability engineers called on the part designers to improve the design; a few minor changes were tried, but unfortunately, all were unsuccessful. The product release date was slipping, and no progress was being made toward release. The design team objected that the stresses being used were unrealistic, and demanded a retest under "worst-case, real-life" stresses of 40°C, 90% R.H. The reliability engineer agreed, and a new reliability test was started at these lower stress conditions. The part passed 1000 h of testing, and the decision was made to release it to high-volume manufacturing. Within a year, a substantial number of failures were observed in ship-based maritime wireless communication systems that were deployed in tropical and subtropical locations in the Pacific Ocean, where the ambient temperatures would be 30°C–40°C, with high humidity, but for much longer than the 6-week reliability test had covered. The failure mechanism was corrosion based, with an identical FA signature to what had been seen in 85/85 testing. The customer demanded (and got) a full recall, accompanied by compensation payments for consequential damages to cover the equipment replacement costs. A redesign effort was begun, with the goal of passing at least 500 h of 85/85 with no failures from a 44-unit population. Future new products which failed 85/85 were delayed a year or more until the problem was fixed.

Business impact: At least a few million dollars in lost sales, damage control efforts, and compensation payments. Fortunately, the terrestrial customers did not see widespread failures, and no recall was needed for the majority of deployments.

What they did right: They did 85/85 testing in the first place. They tried (at least for a little while) to improve the parts before release. They did improve the parts after the field failure crisis.

Opportunity for improvement: I blame this primarily on the reliability engineer and his management. When challenged on the validity of accelerated testing, they were not able to defend it. With sufficient technical background, they could have justified *why* 85/85 testing was used, how it mapped to real-world stresses, past acceleration models in semiconductors, and a valid path toward a custom acceleration model for this product if the R&D team was proposing to disregard the early failures in 85/85 testing. In the absence of any of those things, escalating toward higher level reliability engineers who were capable of such support would have been appropriate. Abdicating the responsibility for reliability qualification plan to R&D engineers with no reliability engineering knowledge or experience proved to be an expensive mistake, and also harmed the brand's reputation with a few key customers.

Summary of lessons learned: Reliability engineers need to understand how and why industry standards were established, and the acceleration models that underlie them, especially if they are proposing to modify the pass/fail criteria for product release.

3.3.3 Case 7: broken design rules + no reliability testing = disaster

A new business group within a large conglomerate was created to pursue a promising new product, which involved custom-integrated circuits in a novel product. The product was developed quickly, and started gaining market share quickly. It had some custom ICs, but at the time, fabs did not have well-established design libraries, and design checking software had not been invented.

Unfortunately, the managers who started up the group did not have much experience with the quality and reliability function—they had always taken that part of their previous group for granted. The conglomerate they worked for was also highly decentralized, and had no real corporate quality or reliability standards, so they were not really given any central guidance or help either.

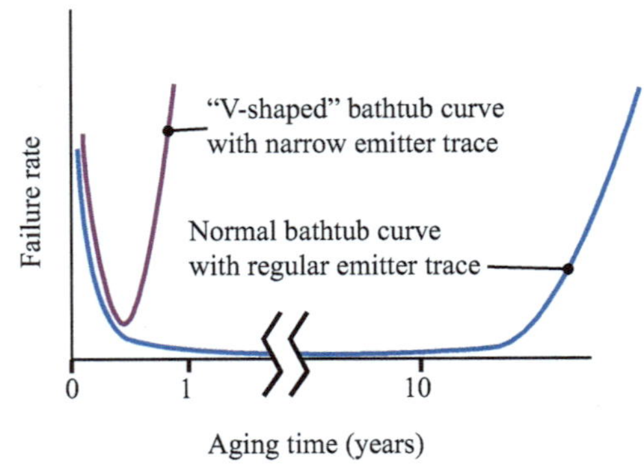

Figure 3.7 Bathtub curve (hazard rate vs. time) showed wearout occurring after only 6 months with this cases product, versus expected lifetimes of >10 years.

After their product had been on the market about 6 months, a few failures started coming back. After 8 months, lots of failure started coming back. They went to some of the quality gurus at the company. The gurus were shocked that no reliability testing had been done prior to product release, but figured "better late than never." So they put together highly accelerated reliability testing, which showed that the product had an extremely short wearout lifetime, with end-of-life failures *expected* to begin after about half a year. Further investigation revealed that the custom IC used the recommended trace width for the bipolar junction transistor (BJT) base on the emitter (which was supposed to be 10× wider, and carrier ~100× more current). The much higher than anticipated current density from this design error resulted in early failures due to electromigration (Fig. 3.7).

The conglomerate realized that virtually every system would fail while still under warranty, and the losses would be huge. The reputation of this new product line was now tarnished, probably unrecoverably. The management of this new group was clearly in over their heads. The conglomerate executives took quick action to shut down the new group, which was disbanded.

What they did right: Calling in experienced reliability engineers and failure analysts after problems were observed. Closing the group down.

Opportunity for improvement: No reliability testing, no design rule checking, no quality or reliability department present until after failures were observed.

Summary of lessons learned: Failure to do an adequate job on quality can result in financial losses orders of magnitude larger than the potential profits, and can destroy an otherwise successful business.

3.3.4 Case 8: compact disk lasers in gigabaud link modules

Situation: For many years, fiber optic transmitters with speeds of 10 Mb/s or 100 Mb/s were the fastest you could buy for prices under $1000. Then people wanted to go to 1 Gb/s, and LED sources were no longer sufficient—the decay times were just too slow to get past the 622 Mb/s node. Lasers fast enough for

1 Gb/s links were available for telecom sources, but they were over $1000 each, and transceiver prices were approximately $5000 each. Where would an affordable laser be found that could go into a low-cost datacom transceiver? The cheapest lasers you could buy, as of the early 1990s, were compact disk (CD) lasers (the lasers used in CD players). These were 780 nm aluminum gallium arsenide (AlGaAs) stripe lasers. They could be purchased for less than a dollar each, before reliability screening costs were included. Some reliability testing was done, and while significant degradation was observed, the reliability still looked good enough to meet customer needs. No real "reliability models" existed, or estimates of wear-out lifetime, since the CD player market had a much less demanding application than data communications. After all, the CD lasers were only intended for a few hundred hours of use—or perhaps a few thousand hours by a true audiophile, versus 50,000–100,000 h for a transceiver. The products were released to production, and tens of thousands of modules were sold (Fig. 3.8).

After about 2 years, field reliability problems started to become an issue. It looked like for many wafers, the lifetime was only about 3 years before significant numbers of failures would start occurring. CD lasers show much more rapid degradation than the VCSELs that were later used in this same application. The CD lasers, being based on AlGaAs, and being facet emitters, suffered from facet oxidation and microloop formation near the facet as they aged [9]. There *are* ways to prevent facet oxidation, and they were used on high-priced 980 nm pump lasers for erbium-doped fiber amplifiers, but they would be prohibitively expensive to apply to CD lasers, and are certainly not used in that application. But the modules were said to fail not because of the overall laser degradation, but rather due to the self-pulsation characteristics the CD lasers are known for [10], which changed as the CD laser aged, and moved from being well above the modulation frequency, to being close enough to interfere with the intended 1063 Mb/s modulation. For most laser reliability, only DC testing is required pre- and postaging, but this is a rare example where high-frequency testing after aging would have exposed this issue where DC characterization did not.

Figure 3.8 A photo of a 1063 Mbps GLM transceiver bottom (left) and top (right). *GLM*, gigabaud link modules.

After the gigabaud link modules (GLM) with CD lasers had been in the field a few years, epidemic failure from many of those modules occurred, with a few percent having failed, and a wearout failure rate that was increasing every month. Because the lasers were not suitable to this application, a recall of all transceivers per system had to be initiated. Because the $3 CD laser was not suitable for the application, a large percentage of the $300 transceivers needed to be replaced, with service visits that often cost thousands of dollars in travel costs and system down-time. As an aside, the ability to make good on compensation for defective products is one reason start ups or offshore companies have a very hard time getting adopted by large customers: the small companies do not have the financial standing to pay for a large recall campaign that will soak up hundreds or thousands of times the profits made from the original sales. The only saving grace was that the GLM system was a "pluggable" module, which made upgrading to more reliable VCSEL-based modules easier and cheaper than it would have been with soldered modules.

Business impact: Bankrupted at least one transceiver supplier, and nearly bankrupted others.

What they did right: Some reliability testing before deployment, did recall when it was clear the product would fail in large numbers.

Opportunity for improvement: Initial reliability testing inadequate to verify wearout reliability; laser was not suitable to application; no lot-to-lot wafer monitoring after product release to detect and screen out weak batches of lasers. No epidemic failure clause in contract; supplier too small to make good on costs incurred. No consideration of high-frequency modulation characteristics, and how those might change over time.

Summary of lessons learned: The component business is a very risky one financially, with recall costs of up to $100-1000\times$ above profits earned. Make sure the risks are low before you start shipping. Laser reliability is often the riskiest part, so make sure you have been especially diligent with laser reliability testing.

3.4 Group 3: parts that were reliable after release, but later developed reliability issues during high-volume manufacturing

The third and final group in this chapter is of parts that were appropriately qualified, and were reliable when released. The first case shows the importance of regular sampling and monitoring to detect excursions, since they show how the quality of inputs can vary, and should not be assumed to remain as reliable as they were at the time of product release. The final case study shows that a conservative approach is often needed in product changes—many of the most serious reliability qualification issues I have seen were ones where I was confident that the reliability risk was low, so "delta qualification" of product changes is important to do before moving forward with design or recipe changes.

3.4.1 Case 9: the dangers of copper contamination

Introduction: the risk from copper contamination

Copper is famous for its mobility and ease of diffusion in semiconductors, which is something that I find surprising, given that its atomic radius is relatively large. However, extensive precautions must be taken in silicon semiconductors to keep copper out of the transistor area, with a "damascene barrier" being used. No such barrier layers are used yet in semiconductor lasers, so other precautions must be taken to prevent copper ions from getting in the metallization, or copper ions from migrating into the device from the packaging. Copper contamination can speed up wearout degradation by acting as a recombination center. In large enough concentrations, it can form gettering centers that nucleate ⟨100⟩ climb dislocation networks, thereby creating maverick failures as well. Most of the three case studies here are for examples where copper was inadvertently embedded in the p-metal on top of the device, but other cases that we do not have room to cover include copper from nearby areas migrating laterally through the surface water layers that exist on normal nonhermetic products. The copper ions then migrate down with the assistance of the electric field, and pile up near the active region (usually quantum wells), as shown in Fig. 3.9. There are other contaminants that are also undesired recombination centers; some of them have even caused problems by being present in the metal-organic chemical vapor deposition (MOCVD) source gases and being grown into the crystal. But copper has a special role, both because it is capable of diffusing in after the fact even when it is not grown in, and because it is commonly used in heat sinks, copper pillars for packaging, and semiconductor fabrication equipment. This comes from the fact that copper has very high thermal conductivity (the highest of any metallic element other than silver). So let us go into a few of the many case studies I have seen where copper has caused problems—ask around, perhaps you could also find a few that have happened where you work!

3.4.1.1 Case 9a: The flaming power supply

GaAs-based LEDs were made for an opto-isolator. They were reliability tested and passed. The parts were shipped to customers, who used them in a large

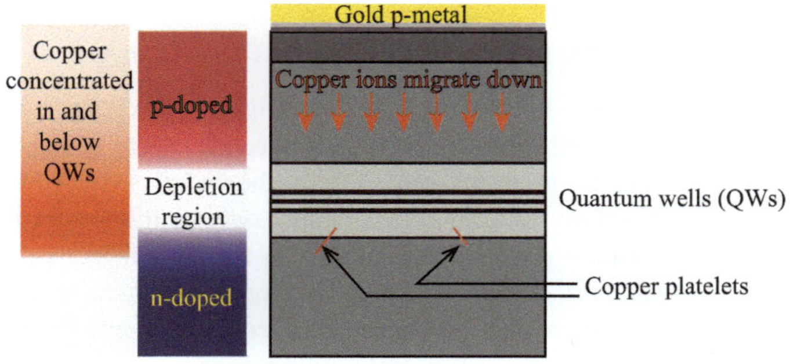

Figure 3.9 Cross-sectional drawing of copper migration from the p-metals on top to the bottom of the depletion region, where it forms small platelets.

equipment rack as a component in their power supply. Failure started occurring, and when they did, the failures were quite dramatic. The power supplies would catch on fire! The customers were angry, and demanded a recall. An investigation was launched. How could parts which had passed the reliability test fail in the field? Was this a maverick failure mode, on an undetected wearout failure mode? Failure analysis started to show symptoms of copper. Copper results in significant reverse bias leakage—but then, a lot of things can cause leakage. The parts were baked for a few hours at 150°C and the leakage "went away." This is consistent with the copper that is concentrated at the active layer diffusing away and being distributed over a wider area with far lower concentration in the active region. Then the parts were operated under normal conditions overnight, and efficiency started to drop again, and reverse leakage started to go back up. This was due to the applied forward bias field moving the copper ions back to the active region.

Then the parts were analyzed by energy dispersive X-ray spectrometry (EDX) (good for high levels of contamination), and a significant fraction of copper was observed in the p-metal contact on the top of the device. An investigation followed, and it was determined that the electron beam used to melt the gold in the evaporator had gone off of the gold target, and hit either the shutter or the copper on the edge of the pocket. Corrective action was taken to make sure this would not happen again, and compensation was paid to the customer for the field failures.

And what about the customer? They were partially at fault as well. Perhaps it is okay to have no redundancy in your power supply, but designing it so it catches on fire if the optocoupler LED fails? This sort of thing is easily avoided if you have any sort of failure mode and effects analysis.

Business impact: A few hundred thousand dollars in compensation.

What they did right: They had reliability monitors that they ran on batches routinely. Good failure analysis.

Opportunity for improvement: Needed more "fool proofing" in the evaporator; reliability monitor insufficient to catch this failure mode due to too few check points; customer's design increased impact of failures beyond just simple nonfunctionality.

3.4.1.2 Case study 9b: copper contamination in p-metal target

Another laser manufacturer with a strong wafer qualification procedure noticed that all recent wafers were failing reliability testing. Daily meetings were called, and correlations done between reliability performance and when the wafers were grown, and when they were processed. The firm was ISO certified, and used the framework to drive continuous improvement in their product. Weaknesses in the system that came up in the investigation were carefully documented, and many workers and managers filed corrective action reports for problems that had been observed, or preventative action reports for issues where key safeguards were lacking. Secondary ion mass spectrometry (SIMS) studies were done to look for contaminants in the epitaxial growth. This manufacturer had been very lucky in growth, and had few problems related to contaminated

MOCVD gas sources or reactor leaks. Therefore they had no SIMS baseline data to be able to compare to (most firms might send every 50th or 100th wafer out as a periodic monitoring check). They also did not even know which species to look for in requesting a SIMS scan. However, all of this quickly proved to be irrelevant, since they were able to narrow it down to a certain date in wafer fabrication, and proved it was uncorrelated to wafer growth date.

Simultaneously, failure analysis was done on the lasers. This firm did not have much experience with copper diffusion, but was able to find, using transmission electron microscope and EDX, that copper was present in high concentrations along the lower edge of the depletion region, where it would form thin copper platelets on the ⟨111⟩ plane. The identification of the source of the degradation as being copper was key. Now they just had to find out where the copper was coming from to eliminate the source. The pressure was intense, since they had little inventory, and shipments were on hold.

Many wet solutions were sent out for chemical analysis to see if copper was present. Samples of the p- and n-metals were going to be taken. The whole process was cut short when someone pulled the gold target out of the evaporator, and noticed the back of the "puck" was copper colored (see Fig. 3.10 for an example of what a similar thermal evaporator looks like). It appeared it had been grossly overheated near the edge of the copper pocket, and copper had alloyed into the gold. The puck was replaced (sold to a gold reclamation service), and the source was recharged with fresh gold, this time in a graphite boat/crucible so the previous alloying with the pocket could not be repeated. In a few weeks, new material was on reliability test, and it passed with very stable reliability characteristics. After about a month on shutdown, parts were delivered for an at-risk build, and the parts went on to pass the wafer qualification test.

Business impact: Perhaps a million dollars in lost sales for the customer, scrap and labor costs for the supplier.

What they did right: Excellent problem-solving approach gave a very quick problem resolution time of just a few weeks. Excellent data systems allowed

Figure 3.10 Electron beam evaporation is done in holders made of copper with multiple pockets (left). The photo on the right shows gold pellets in a tungsten crucible liner, and titanium pellets in a graphite crucible liner. Crucible liners are not universally used, but were implemented as a corrective action in this case study, after having the gold alloy directly with the copper from the evaporator pocket in the case study discussed above.
Photo on the right is reproduced from [11] A. Sarangan, Physical and chemical vapor deposition, in Nanofabrication, first ed., CRC Press, Boca Raton, 2016, pp. 53–98, copyright 2016, by permission of CRC Press.

problem to be quickly isolated. FA approach was effective and rapid. Communication with the customer was good. None of the bad parts ever reached the customer, thanks to an effective screening system. Good corrective action system for continuous improvement.

Opportunity for improvement: Inventory levels too low, causing supply disruption. (Semiconductor lasers are *not* a good product for "just in time" production!) Too few baselines were available for their epitaxy—seemed oblivious to the possibility of problems, even though epitaxial issues were not at fault in this instance.

3.4.1.3 Case study 9c: no supplier purity checks

A laser manufacturer had a problem with lasers failing the reliability test. They had suffered problems with copper contamination before, due to p-metal contamination, and immediately suspected the cause. The p-metal melt was analyzed, and 1% copper contamination was detected. The raw source pellets were examined, and were also found to have approximately 1% copper. The manufacturer was shocked—they were paying a premium price for "semiconductor-grade" or "research-grade" metals, and expected 99.99% or 99.999% (4N or 5N) pure materials, so they had not anticipated this very high level of impurity as being possible. They were perhaps justified in thinking no incoming inspection would be required. They had been using a special type of alloyed source pellets with p-dopant mixed into the gold, which was available from only a single supplier. While it was possible to separately buy the constituent elements and try to mix them into a homogeneous alloy, this was fraught with difficulty. The metal supplier claimed the 1% copper contamination was "within acceptable limits" according to their own specifications, and their documentation evidently made no real claims for the purity of their materials.

Business impact: $10 m in lost sales for the quarter due to having to scrap a couple of months of production.

What they did right: Screening test caught the problem—no bad parts were shipped to customers. Source of problem was found very quickly, good customer communication.

Opportunity for improvement: In hindsight, it appears that incoming purity checks from their supplier would have been warranted. Many other laser manufacturers require a "certificate of compliance" accompany every shipment before they accept incoming material, where a sample has been run through an independent certified analytical lab. The experts that I have talked to recommend requiring copper be ≤ 10 ppm in laser metalization. There are two sources of gold: "virgin gold," direct from mines, or "reclaimed gold," that comes from sources like jewelry, or chemically stripped off from plated items. Since 14-karat jewelry contains $>10\%$ copper, it is very difficult to reduce the concentration by more than $10{,}000\times$ to get to <10 ppm of reclaimed sources, and metal suppliers using "virgin gold" are more appropriate for most laser manufacturing. Using a nonstandard p-contact metal reduced their supplier options, lack of second source options for VCSELs or metal suppliers also hurt.

3.4.2 Case 10: high burn-in failure rate foretells early wearout failure

Situation: In the 1990s, there was a clear distinction between "telecom" and "datacom" parts. Telecom transceivers were ultra-high reliability parts, costing thousands of dollars each, built from expensive materials. Datacom transceivers were focused on being low-cost (typically <\$300 each) ways of establishing 100 Mb/s links over hundreds of meters, usually in applications where 1000 FITS ($\sim 1\%$ failure per year) was acceptable—the rack of parts serving data up to the transceiver was far more likely to fail than the transceiver itself. Over time, the electronics in the rack has become far more reliable. More importantly, the volume of data being transported had increased hugely, and applications with hundreds of transceivers in a box were starting to be seen for the first time. Over time, transceiver failure rates had dropped by $1-2$ orders of magnitude, and the prices have gone down to as low as \$30 a transceiver.

One of the changes that had to take place to get from 1000 FITS to 50 FITS was qualification of the transmitters going into the transceivers. In the 1990s, doing lot-based qualification was seen as cost-prohibitive—here is the story of how one manufacturer decided they *should* be doing 100% wafer testing.

Previously, a laser design was established, it was reliability tested for a few thousand hours at some mild acceleration factor, and if it passed, the design was frozen. No further lot-based qualification or reliability sampling was required. The rationale was "we are using the same recipe in the same reactor, so everything produced by this process should therefore be similarly reliable." Unfortunately, anybody familiar with epitaxial growth recognizes the weaknesses in this argument. Parts per million contamination in the precursor organometallic sources (even parts per billion of some species) can result in early laser failure, and these levels are undetectable to the gas suppliers providing the source gases used in the growth. Reactor leaks can occur, allowing oxygen contamination, which reduces part lifetime. The hydrogen purifiers can fail to function as intended. High epitaxial defect substrates can be shipped from the substrate supplier, or high particulates can fall off the reactor leading to epi defect counts thousands of times greater than those on the qualification wafers. Any experienced epitaxial growth engineer could probably add at least another half dozen possibilities.

However, these possibilities were not factored in for this case, or thought to be low enough that they did not justify the additional expense and complexity of adding routine wafer qualification monitors. Other safeguards were present, including a100% burn-in for all parts before shipment. In the burn-in, every laser was aged for a few days, at an acceleration condition that is the equivalent of $6-12$ months of worst-case normal use. If the burn-in yield was low ($>15\%$ yield loss) the wafer would normally be scrapped in favor of higher-yielding wafers. So parts with high amounts of degradation would be noticed and prevented from entering the supply chain. However, moderate levels of problems were not detectable. Parts with lifetimes of a few years could pass burn-in, and then go on to achieve "epidemic failure." Usually, large customers require suppliers to sign epidemic failure clauses on contracts. "Epidemic failure" (which might be defined as >2%–4% failing over a 5- or 10-year lifetime) triggers the contract clause, and requires the supplier to pay for a recall, including the cost of consequential damages. The cost of the recall could be 1–2 orders of magnitude above the revenue generated by the parts, and 2–3 orders of magnitude above the profit generated by the parts. It is this huge asymmetry between gains and potential losses that accounts for the great conservatism in reliability testing in fiber optic components.

On the laser in question, a minor design change was proposed to improve device yield. It was approved by the reliability engineer, who agreed that it was a minor change with low risk. No one thought much about it until high burn-in yield loss started showing up in production a few months later. The production engineers did the logical thing: they took the parts passing burn-in, and ran them through burn-in again, just to make sure that the ones that had passed burn-in were reliable, and no additional fallout would occur. They found that the failure was not just a parametric transient that drifted initially and then stabilized, as they had hoped. As shown in Fig. 3.11, a significant portion of the parts run through the second time would fail; and of the passing parts run through burn-in a third time, a significant portion of *those* would fail. Now it was time to switch from a one-time burn-in test being repeated multiple times, to a 500 h reliability test. It showed that *all* parts would fail early. There was no "good population" to be screened out. The executives in charge took the right action and stopped production for 4 months while the problem was identified. If this seems long, there were two complicating factors: first, this was prior to databases being available, and any attempt to run correlations between test data and wafer failure rate were slow and manual. The second problem causing the long shut-off was low inventory. It is standard practice in the industry to have at least 3 months of inventory, since the time to detect reliability problems is long (>1 month), and the time to solve them can be even longer (at least 3 weeks, and often up to 3 months). Epitaxial growth is a black art, with success being dependent on many factors that are below measurement thresholds, and thus success is not

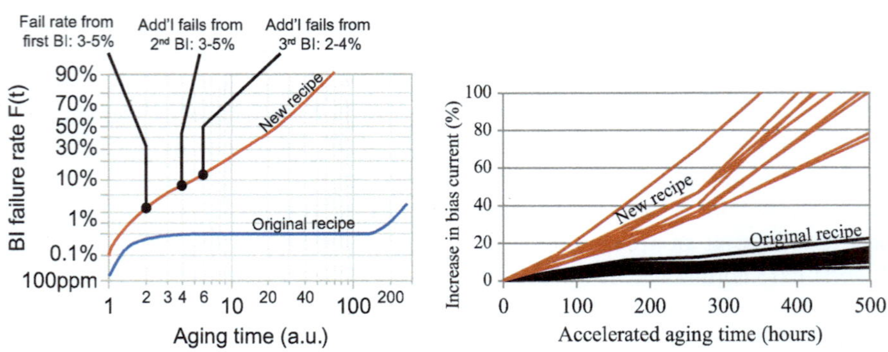

Figure 3.11 Burn-in failure rate trend versus burn-in time (left) show continuing fallout with repeated burn-in for the new recipe; parametric trends (right) show premature wearout for the new recipe.

readily predictable. Fixing problems involves "try–test–try something else" cycles, with an indeterminate number of cycles being required to solve a given problem.

They had to cut off all shipments, who qualified and switched to competitors on this industry-standard transceiver. When they fixed the problem, and resumed shipments, they had to drop the price by 40% to regain market share. In the wake of this, they chose to institute 100% wafer qualification, where every wafer was destructively tested for the equivalent of >10 years worst-case use life to verify that the wearout lifetime of the part was sufficient. Future parts were always required to be dual-source, so that if a similar shut-down of the primary source were encountered, transceivers could be shipped using an alternate laser supplier.

Business impact: Tens of millions in lost sales, plus a multiple of that once switching to competitors and reduced selling prices are accounted for.

What they did right: In spite of questionable reliability screening beforehand, very few bad parts ever reached customers. Corrective action after the fact would prevent recurrence of the problem in the future.

Opportunity for improvement: The lack of a wafer-qualification or indeed any systematic ongoing reliability test at the transmitter (LED or laser) level was a major gap in being able to go to today's optical communication markets that have more demanding reliability targets than those in the early 1990s did. (Note that the transceiver high-temperature operating life that the supplier did monthly is *not* capable of detecting this sort of laser issue.)

Key take-aways: Be realistic about whether something is "guaranteed by design." If it is not maybe there should be a wafer or lot qualification test in place to prevent it from reaching customers. Keep a decent buffer inventory. Have a secondary source of supply for key items.

3.5 Conclusion

If there is one single conclusion to this chapter, it should be just how expensive failure to do proper reliability engineering can be, and how even companies with fairly strong reliability teams can still make expensive mistakes if they overlook some of the key industry practices. Understanding what those practices are, and why they came into being, is key to avoiding these losses.

References

[1] G.K. Hobbs, Accelerated Reliability Engineering: HALT and HASS, Wiley, 2000.

[2] K.A. Gray, J.J. Paschkewitz, Next Generation HALT and HASS: Robust Design of Electronics and Systems, 1 edition, Wiley, Chichester, UK; Hoboken, NJ, 2016.

[3] H.W. McLean, HALT, HASS, and HASA Explained: Accelerated Reliability Techniques, Revised Edition, Revised edition, ASQ Quality Press, Milwaukee, Wis, 2009.

[4] "Highly accelerated life test," *Wikipedia*. April 02, 2020 [Online]. https://en.wikipedia.org/w/index.php?title = Highly_accelerated_life_test&oldid = 948724768. (Accessed June 13, 2020).
[5] J. Arnold, N. Blattau, C. Hillman, C. Park, K. Sweatman, "Reliability Testing of Ni-Modified SnCu and SAC305—Accelerated Thermal Cycling," *DfR soultions website*. https://www.dfrsolutions.com/hubfs/DfR_Solutions_Website/Resources-Archived/Publications/2008-2010/2008_03_ni_modified_sncu_thermal_cycling.pdf.
[6] H. Cui, "Accelerated temperature cycle test and Coffin-Manson model for electronic packaging," in *Annual Reliability and Maintainability Symposium, 2005. Proceedings*, Jan 2005, pp. 556−560, doi:10.1109/RAMS.2005.1408421.
[7] R.G. Waters, S.L. Yellen, N.F. Ruggieri, D.P. Bour, Inhibited dark-line defect formation in strained InGaAs/AlGaAs quantum well lasers, IEEE Photonics Technol. Lett. Inst. Electr. Electron. Eng. 2 (8) (1990).
[8] R.G. Waters, Diode laser degradation mechanisms: a review, Prog. Quantum Electron. 15 (3) (1991) 153−174.
[9] W.J. Fritz, L.B. Bauer, C.S. Miller, "Analysis of aluminum gallium arsenide laser diodes failing due to nonradiative regions behind the facets," in *Reliability Physics Symposium, 1989. 27th Annual Proceedings, International*, 1989, pp. 59−64, doi:10.1109/RELPHY.1989.36318.
[10] C. DeCusatis, Handbook of Fiber Optic Data Communication: A Practical Guide to Optical Networking, Academic Press, 2011.
[11] A. Sarangan, Physical and chemical vapor deposition, Nanofabrication, 1st edition, CRC Press, Boca Raton, 2016, pp. 53−98.

Chapter 4

Materials science of defects in GaAs-based semiconductor lasers

Kunal Mukherjee
Department of Materials Science and Engineering, Stanford University, Stanford, CA, United States

4.1 Introduction

The reliability of semiconductor lasers is an interdisciplinary subject. The confluence of high photon densities, carrier concentrations, electric fields, stresses, with dissimilar materials, yields a range of unusual defect phenomena that control the degradation of lasers. Thanks to decades of collaborative work between electrical engineers, solid-state physicists, chemists, and materials scientists, we now understand some of these processes, resulting in select laser designs rated to work for tens to hundreds of thousands of hours in extreme conditions. Extending this reliability to all material systems and upcoming integration schemes continues to be critical. In this chapter, we introduce the basic properties of defects in semiconductors and how they limit laser reliabilty.

We have taken the difficult decision to limit the scope of our discussion to the cubic III–V semiconductor GaAs family of materials; these have provided early evidence for degradation phenomena and remain technologically relevant. This chapter is organized into sections on (1) point defects, (2) dislocations, (3) heteroepitaxy of GaAs materials for integration with silicon, and (4) recombination-enhanced defect reactions (REDRs). These do not have to be read in order, but the ideas developed on point defects and dislocations are selected because they become relevant in GaAs on silicon integration and REDRs. Many concepts are borrowed from metals and silicon, and likewise can be extended to the InP family of lasers. The crystal structure and bonding in the nitride semiconductors make them quite different from cubic semiconductors, and interesting in their own right. We will not cover them here, although they are covered elsewhere in this book, in Chapter 5, Grown-in defects and thermal instability affecting the reliability of lasers: III–Vs versus III-nitrides, by Dr. Tomiya.

Reliability of Semiconductor Lasers and Optoelectronic Devices. DOI: https://doi.org/10.1016/B978-0-12-819254-2.00007-2

4.2 Characteristics of simple point defects in GaAs

Point defects in semiconductors are deviations in the arrangement of single or a few atoms in a crystal. Vacancies (a missing atom in the crystal) and interstitials (an atom in the gaps between the other atoms in the crystal) are two types of simple point defects. Point defects are an area of active research even in mature semiconductors and their electrical properties continue to be uncovered. As all semiconductors have point defects, it is important to know how many point defects and what types exist in any particular device. This is not a trivial task.

An important technique to measure the density of point defects, especially vacancies, is positron annihilation spectroscopy [1]. We will not go into details of this technique here, however there is good agreement between experiment and thermodynamic theory when crystals have been allowed to "equilibrate" with their surroundings for hours to days with well calibrated external vapor pressures of constituent species (that define the so-called chemical potential of the species). On the other hand, devices grown with techniques like molecular beam epitaxy (MBE) and metal-organic chemical vapor deposition (MOCVD) may be far from equilibrium, [2] and although device growers like to keep track of parameters such as group-V/group-III flux or flow ratios, these do not translate easily to the chemical potential on the crystal surface. Additionally, temperatures change rapidly during growth and cool down and the entire device stack is not in equilibrium with the vapor species in the growth chamber. This makes estimating the point defect concentration in working devices challenging. If the crystal is cooled down rapidly after growth at high temperature, point defects in the crystal are unable to equilibrate with the surface of the crystal and an excess (or in some rare cases, a deficiency) of point defects are locked-in; such a point defect population is termed metastable. This "out-of-equilibrium" concentration of point defects may provide the thermodynamic driving force for many of the point-defect-related failure modes in devices that are then kinetically enhanced by non-radiative carrier recombination (Section 4.5).

In this section, we ignore the difficulties of experimentally measuring the point defect concentration in devices. Rather, we aim to give growers and device designers a view into the thermodynamics of point defects gleaned from bulk crystals that are equilibrated over a long period of time, building intuition about what might be happening during thin film growth. The key take away is that the concentration of point defects can depend quite sensitively on the temperature, chemistry, as well as doping in the crystal. This dependence on doping is unique to semiconductors and can end up being the dominant effect, more important than V/III ratios, etc.

There are six simple native point defects in gallium arsenide. They are the gallium and arsenic vacancies (V_{Ga}, V_{As}), interstitials (I_{Ga}, I_{As}), and antisites (Ga_{As}, As_{Ga}). We introduce a few important concepts on the formation of these simple point defects in GaAs, avoiding discussion of point defect complexes involving more than one type of atom. Here, our coverage will go to the extent to which it becomes possible to appreciate point defect mediated dislocation climb (Section 4.3.5), an important failure mechanism in lasers. With the gallium vacancy as an example, we focus on two topics not commonly discussed in the context of laser reliability—the Fermi and quasi Fermi level effects. These may lead to large nonequilibrium populations of point defects either during crystal growth or device operation. We refer the reader to the book by McCluskey and Haller for a complete treatment of point defects in semiconductors [3]. Volume 38 in the Semiconductors and Semimetals series focuses on imperfections in GaAs-like materials, and remains a great resource despite being a few decades out of date.

4.2.1 Point defect thermodynamics and the Fermi level effect

At finite temperatures, there will always be point defects in semiconductors. The equilibrium concentration of point defects is also one where the Gibbs free energy of the crystal G_V is the lowest. By minimizing the Gibbs free energy of a crystal, one can show that the equilibrium concentration of a particular type of point defect, n, at a given temperature T and pressure P is given by:

$$n = N \exp\left(-\frac{G_V}{k_B T}\right) = N \exp\left(\frac{S_V}{k_B}\right) \exp\left(-\frac{H_V}{k_B T}\right). \tag{4.1}$$

Here, N is the total number of sites in the crystal, S_V is the change in the vibrational entropy of the crystal upon the introduction of a single point defect, H_V is the formation enthalpy of the point defect, and k_B is the Boltzmann constant. From this point onward, two approximations are common in the defect calculation literature. First, S_V is difficult to calculate but is estimated to have a value of about a few k_B and is ignored. Second, the formation enthalpy given by $H_V = E + PV$ is approximated as E. Here, E is the formation energy of the defect (includes chemical potential terms) while the pressure−volume term, PV, is ignored for ambient pressures. We will see later that the pressure can indeed be sizable near dislocations and will lead to some alterations in the formation enthalpy (Section 4.3.3). This equation is sufficient to estimate the concentration of point defects in simple metals. As an example, $E = 1\text{eV}$ for a simple vacancy in copper and one can accordingly estimate that 1 cm^3 of copper with about 10^{22} sites in the crystal has only 10^5 vacancies at room temperature, but that increases to about 10^{19} (one in 1000 atoms) near its melting point of 1085 °C. This simple estimate would suggest that semiconductors that are synthesized much below their melting temperature should have even fewer point defects. We now show that this is not the case.

The first factor controlling the formation energy of a point defect E in a compound semiconductor like GaAs is the chemical potential of relevant atomic and molecular species set by the environment (or growth conditions) of the crystal [4]. This is known to crystal growers who routinely vary parameters such as the group V/group III fluxes. However, we reiterate that it is not trivial to quantitatively link the reported group III and V fluxes in MBE or MOCVD growth to the chemical potential of the species in equilibrium with the crystal used for thermodynamic calculations and these calculations are best suited for bulk crystals. In thermal equilibrium, the vapor phase pressure above the crystal, the composition (nonstoichiometry or deviation from 1:1 Ga to As ratio) of the crystal, and the concentration of various point defects are all set by mutual dependencies [5]. In other words, adjusting any one of these parameters affects the others. For a two-component (Ga, As) and two-phase (solid crystal, vapor) situation, Gibbs' phase rule suggests there is only one free parameter that can fix the point defect concentration in the crystal for a given temperature. We set this to be the pressure of As_4 (gas) as experimentally it is most convenient to alter the vapor phase pressure above the crystal. One such defect equilibrium reaction may be considered:

$$\frac{1}{4}\mathrm{As}_4(g) \leftrightarrow V_{Ga}^0 + \mathrm{As}_{As}^0. \tag{4.2}$$

Here, an arsenic tetramer in vapor is in equilibrium with neutral gallium vacancies (V_{Ga}^0) and arsenic atoms $\left(\mathrm{As}_{As}^0\right)$ in the crystal. The law of mass action for chemical reactions tells us that the concentration of neutral gallium vacancies must therefore be proportional to the partial pressure of arsenic vapor as $\left(p_{As_4}\right)^{1/4}$. Likewise, the concentration of neutral arsenic vacancies is $\left(p_{As_4}\right)^{-1/4}$. Indeed, this is borne out in the experiment of Pearson and Chiang (Fig. 4.1A), where they use depth-dependent Hall effect

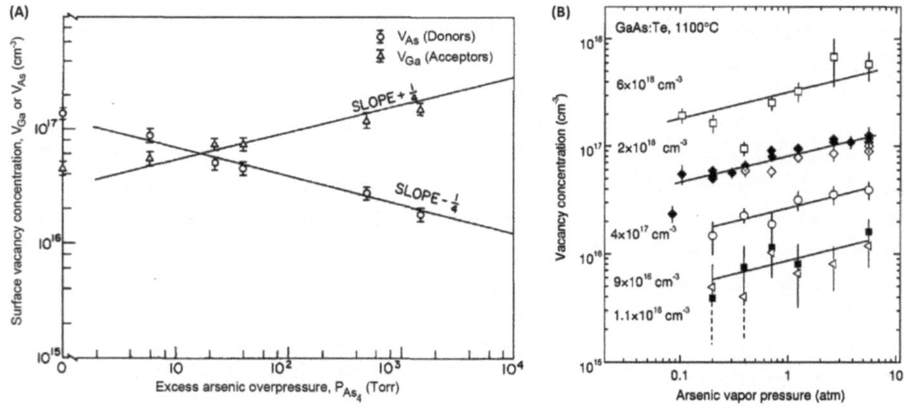

Figure 4.1 (A) Equilibrium concentration of surface vacancies of gallium and arsenic in a GaAs crystal as a function of arsenic (As_4) overpressure (800°C, 24 h), inferred from Hall effect measurements. The solid lines are power law fits with exponents of + 1/4 (gallium vacancy) and − 1/4 (arsenic vacancy), respectively. (B) Gallium vacancy concentration in Te-doped GaAs as a function of the As_4 vapor pressure (1100°C, 24 h) was inferred from positron annihilation spectroscopy. Lines are fits to a power law with $n = (0.25 \pm 0.02)$. The gallium vacancy concentration is strongly linked to the background n-type carrier concentration.

Source: (A) Reprinted from S.Y. Chiang and G.L. Pearson, Properties of vacancy defects in GaAs single crystals, J. Appl. Phys. 46 (7) (1975) 2986–2991, https://doi.org/10.1182/blood-2018-06-848705, with the permission of AIP Publishing; (B) reprinted with permission from J. Gebauer et al., Determination of the Gibbs free energy of formation of Ga vacancies in GaAs by positron annihilation, Phys. Rev. B 67 (23) (2003) 235207, https://doi.org/10.1103/PhysRevB.67.235207, copyright 2003, by the American Physical Society.

measurements find such as power-law dependence on the arsenic overpressure of the concentration of gallium and arsenic vacancies in a bulk crystal of GaAs (equilibrated at 800°C for 24 h) [6]. Likewise, Gebauer et al. envisage a $(p_{As_4})^{1/4}$ power law exponent for gallium vacancies in Te-doped (n-type) GaAs samples (Fig. 4.1B), this time measured using positron annihilation spectroscopy [7]. What is interesting is that Gebauer et al. do this for different Te-doping levels as well, and a dramatic effect of the gallium vacancy concentration on the n-type carrier concentration is seen. This happens because the formation energy E depends not on the type of point defect and the chemical potential of concerned species that are in equilibrium with the crystal (in this case As_4), but also on the Fermi level controlled charge state of the defect [8–10]. The takeaway here is that rather than E being a fixed number like in copper, it may drop or increase significantly in a semiconductor if the Fermi-level moves because defects in semiconductors can become charged. In fact, the doping may be one of the strongest levers to control the point defect concentration in a device; we show why this is the case using the gallium vacancy in GaAs as an example.

Device designers are very familiar with Fermi-Dirac statistics controlling dopant ionization (activation) in semiconductors. The ratio of the singly ionized to unionized concentration of acceptors $\frac{N_A^-}{N_A^0}$ is given by [11]

$$\frac{N_A^-}{N_A^0} = \frac{1}{g}\exp\left(-\frac{E_A - E_F}{kT}\right), \tag{4.3}$$

where g is the degeneracy factor (due to spin, valleys, etc.), E_F is the Fermi level, and E_A is the acceptor ionization level. The same analysis is applicable for point defects that may become charged. A neutral gallium vacancy, denoted as V_{Ga}^0, can act as an acceptor by transferring up to three holes to the valence band and becoming negatively charged V_{Ga}^{3-} in the process. Ignoring degeneracy factors for now, we have a situation analogous to Eq. (4.3):

$$\frac{c(V_{Ga}^{3-})}{c(V_{Ga}^0)} = exp\left(-\frac{E_{a,1} + E_{a,2} + E_{a,3} - 3E_F}{kT}\right). \tag{4.4}$$

Here $E_{a,i}$ is the i-th ionization level, also called the charge transition level. The meaning of $E_{a,i}$ may be understood more simply as the Fermi energy level at which there is a transition in the charge state of the defect. Both E_F and $E_{a,i}$ are commonly referenced to the valence band edge. For defects that can have multiple charge states such as the gallium vacancy, there are multiple values of ionization energies corresponding to a change in charge state of the vacancy from $0/-1$, $-1/-2$, and $-2/-3$, etc. This is why we write the formation energy of a charged point defect with charge z as:

$$E_V^z = E_V^0 - \left(zE_F - \sum E_{a,i}\right), \tag{4.5}$$

with E_V^0 being the formation energy of the neutral defect (previously denoted E) and the summation running from 0 to z. The values E_V^0 and $E_{a,i}$ can be determined by experiment or theory. Fig. 4.2A shows a recent calculation of the ionization levels of a number of simple point defects in GaAs using density functional theory [12]. The three charge transition levels of V_{Ga} can be seen near the valance band. E_F is then set by charge neutrality between charged point defects, charged extrinsic dopants, and free electrons and holes. One can see that if the Fermi level is close to the conduction band due to n-type doping (say at 1.2 eV), the gallium vacancy is predominantly present in the V_{Ga}^{3-} acceptor state and this would compensate for some of the n-type doping. The formation energy of V_{Ga}^{3-} is likely to be very low in this case as the system can lower the energy of its population of electrons by expending energy to create gallium vacancies and then recovering much of this defect formation energy by transferring three electrons to it. Likewise, Fig. 4.2A informs us that arsenic vacancies (if any) will exist in the V_{As}^{1-} state, arsenic antisites (if any) will be neutral, etc. for this value of the Fermi level.

Fig. 4.2B and C shows a calculation of the formation energy of simple point defects in GaAs as a function of the location of the Fermi level for two extreme conditions of arsenic-rich and gallium-rich equilibria, respectively. One sees that the two figures only differ in the vertical offset of the various lines. This is the role of the chemical potential. The magnitude and sign of the slopes of these lines correspond to the charge state of the defect (e.g., acceptor defects have a negative slope). One sees that n-type GaAs grown under arsenic-rich growth conditions most favor the arsenic antisite and gallium

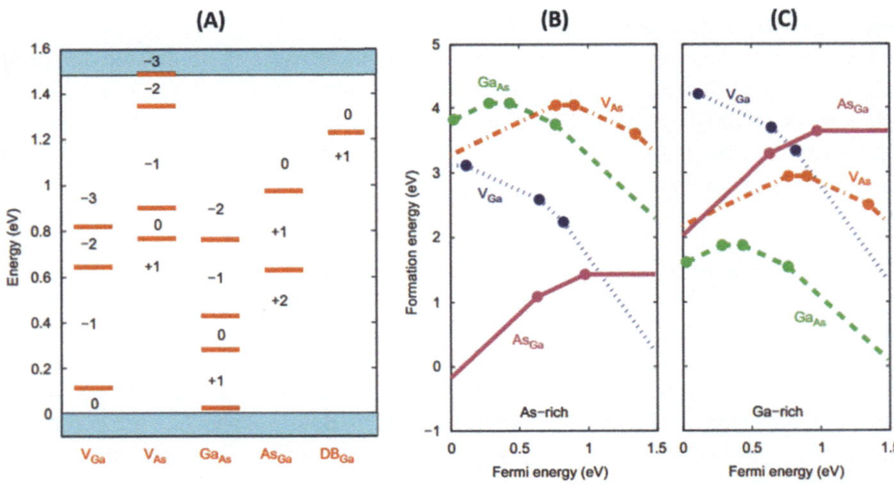

Figure 4.2 (A) Calculated charge transition levels of various intrinsic defects in GaAs, including dangling bonds (DB), formation energies vs. Fermi energy for the Ga vacancy (V_{Ga}, dotted, blue), the As vacancy (V_{As}, dash-dotted, red), the Ga antisite (Ga_{As}, dashed, green), and the As antisite (As_{Ga}, solid, magenta) in GaAs for (B) As-rich and (C) Ga-rich conditions. Discontinuities in the slopes occur at charge transition levels.

Source: Reprinted from H.-P. Komsa and A. Pasquarello, Comparison of vacancy and antisite defects in GaAs and InGaAs through hybrid functionals, J. Phys. Condens. Matter 24 (4) (2012), 045801, https://doi.org/10.1088/0953-8984/24/4/045801, with permission from Elsevier.

vacancies, the formation energy of the latter comes close to zero! Likewise, one can see that the formation energy of arsenic vacancies is very high when the crystal is grown under arsenic-rich conditions. Although this chart does not show interstitial defects, their formation energies are also very high.

4.2.2 Negative temperature coefficient of gallium vacancies in GaAs

Combining the effect of chemical potential and charge state of the defect, the concentration of gallium vacancies with charge state z in GaAs is given by [13]

$$c\left(V_{Ga}^{z-}\right) = N\left(\frac{p_{As_4}}{B_{As_4}}\right)^{1/4} \exp\left(-\frac{\left(E_V^z - TS_v\right)}{kT}\right). \tag{4.6}$$

B_{As_4} is the ideal gas pressure constant that links the partial pressure of a gas to a free energy. Evaluating the V_{Ga}^{3-} concentration in n-type GaAs as a function of temperature reveals an interesting phenomenon [13,14]. One notices that the concentration of V_{Ga}^{3-} decreases with increasing temperature for intermediate n-type doping. This result is counterintuitive at first because it is generally thought that there should be more point defects as the temperature is raised, however, such a situation has been confirmed by experiment in GaAs [7]. Fig. 4.3 shows that there is a notable increase in the experimentally measured gallium vacancy concentration

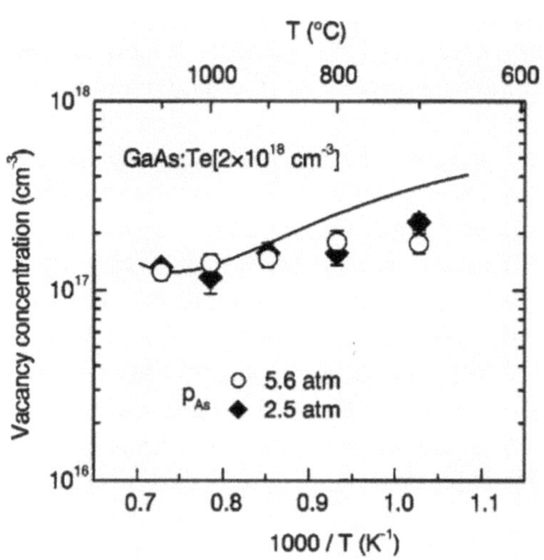

Figure 4.3 Negative temperature dependence of the gallium vacancy concentration in Te-doped GaAs ([Te] = 2×10^{18}/cm^3) for As_4 vapor pressures of 5.6 and 2.5 atm, inferred from position annihilation spectroscopy. The solid line is a calculation for a gallium vacancy charge of −3.

Source: Reprinted with permission from J. Gebauer et al., Determination of the Gibbs free energy of formation of Ga vacancies in GaAs by positron annihilation, Phys. Rev. B 67 (23) (2003) 235207, https://doi.org/10.1103/PhysRevB.67.235207, copyright 2003, by the American Physical Society.

upon cooling down the crystal from 1100°C to 700°C—the solid line is a fit to Eq. (4.6). This anomaly arises due to the Fermi-level moving closer to the conduction band as the crystal is cooled down and the sample becomes more n-type. Thus, the numerator in the exponential term in Eq. (4.6) changes to more than counter the change in the denominator. Unfortunately, it becomes difficult to continue measuring this effect down to room temperature as the crystal takes increasingly longer to equilibrate with the surrounding vapor due to slow point defect diffusion from the surface to the bulk. Nevertheless, it confirms the rather unusual behavior of the gallium vacancy in GaAs and illustrates the importance of both chemistry as well as doping in setting the point defect concentration. We also had an ulterior motive to discuss this anomalous effect over many others. We revisit the negative temperature dependence of the gallium vacancy in Section 4.5 in discussing a driving force for recombination-enhanced dislocation climb.

4.2.3 Relevance to MOCVD and MBE growth

MOCVD and MBE growth involve multiple steps of precursor dissociation, transport, surface reactions, adatom diffusion, and incorporation into the crystal. Usually, only one of these steps is rate limiting, and fortunately, all the others occur at near equilibrium. For example, in MBE growth, the arrival rate of group III adatoms is considered rate-limiting (and controls the growth rate), while the key process of a chemical reaction of vapor-phase precursors into the solid crystal occurs near equilibrium [15]. The same is true for MOCVD. At typical GaAs growth temperatures, the growth rate in MOCVD is controlled by diffusion of precursors through the boundary layer and not surface reactions. Thus, while MBE and MOCVD are nonequilibrium processes, thermodynamics has been applied quite successfully, for example, to estimate composition of ternary alloys from input V/III ratios and precursor flows [16,17]. This suggests that it should be possible to calculate the concentration of point defects using known values of the defect formation energies. Ichimura and Wada calculate the various point defects in GaAs as a function of input V/III ratios for both MBE and MOCVD [18,19]. In Fig. 4.4, we show their calculations for a growth temperature of 600°C. Overall, they find that MBE growth is slightly gallium-rich in comparison to MOCVD growth, with gallium antisites being the primary native impurity in the former as compared to arsenic antisites in the latter. It is understood that these estimates are highly sensitive to the assumed values of charge states of the defects, the background carrier concentration (intrinsic in this case), and the neutral defect formation energies.

4.2.4 Quasi Fermi level effect

We have seen that the position of the Fermi level controls the population of charged point defects in a crystal at equilibrium. However, during steady-state device operation, it is the quasi Fermi levels that yield the steady-state concentration of electrons and holes in the active region of a laser. It is interesting to think about how quasi Fermi levels might deviate the steady-state population of the point defects. Would

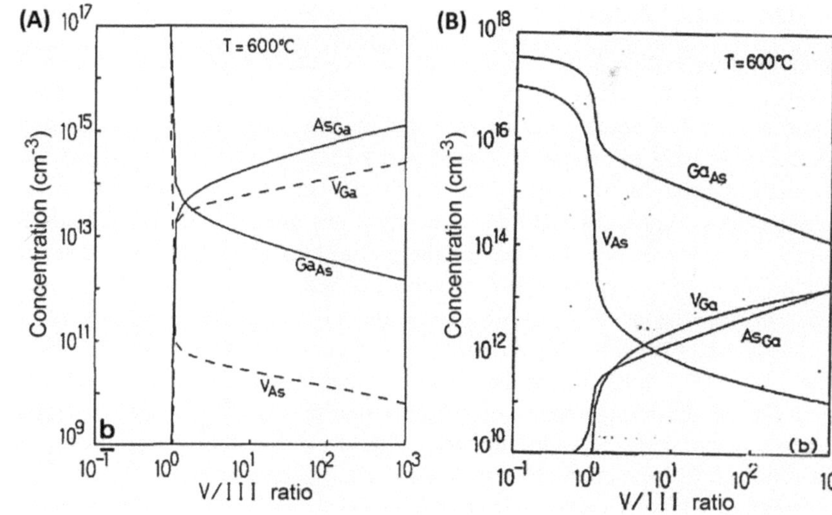

Figure 4.4 Calculated concentration of native point defects in undoped GaAs in (A) MOCVD and (B) MBE growths at 600°C. The V/III ratios are calculated for an input Ga pressure of 5×10^{-5} atm for (A) and incident Ga pressure of 2×10^{-6} Torr for (B).

Source: (A) Reprinted from M. Ichimura and T. Wada, Calculation of native defect concentrations in GaAs grown by organometallic vapor-phase epitaxy, J. Cryst. Growth 115 (1) (1991) 479–483, https://doi.org/10.1016/0022-0248(91)90790-C with permission from Elsevier; (B) reprinted from M. Ichimura and T. Wada, Calculation of point defect concentrations in GaAs grown by molecular beam epitaxy, J. Appl. Phys. 72 (3) (1992) 1200–1202, https://doi.org/10.1063/1.351806, with the permission of AIP Publishing.

it be the hole or electron quasi Fermi levels that now control the population of a particular point defect?

Inquiry into this idea has gained some experimental traction in wide band gap materials [20]. The quasi-Fermi level swing is so large in these materials that research groups are now attempting to illuminate the crystal with above band gap light *during growth* to alter the formation energy of point defects. Alberi and Scarpulla show that both the electron and hole quasi Fermi levels may contribute to the formation energy of a particular point defect, weighted by the details of the electron and hole emission and capture rates [21]. As an example, an acceptor-type state like the gallium vacancy in GaAs may become negatively charged by either capturing an electron from the conduction band or emitting a hole to the valence band. The position of the defect ionization level in the gap, electron and hole capture coefficients, free carrier concentration, etc., determines which of the two possibilities dominates. Quasi Fermi levels are not yet captured in laser degradation models but only growth models. However, the amount of excess carriers in a laser cavity far exceeds what is possible by illumination during growth and the electron quasi Fermi level in a laser reaches even higher than the conduction band minimum. We expect this topic to garner attention in the near future.

4.2.5 DX and EL2 defects

We conclude this section with a discussion of two other experimentally identified point defects in GaAs that go beyond the structurally 'simple' defects discussed so far. These are the DX center and the EL2 defect. The properties of the compensating DX center [22,23] in n-type AlGaAs and the EL2 defect [24,25] responsible for semi-insulating behavior in arsenic-rich GaAs are very well known, but these are both point defect complexes and their concentrations are difficult to calculate from the simple thermodynamics we used for the gallium vacancy. The DX center itself is a silicon atom that was previously a substitutional donor on a gallium site that becomes slightly displaced to form a silicon interstitial–gallium vacancy complex with acceptor-like behavior [26]. The role of DX centers in laser degradation is not resolved [27]. DX centers themselves may not cause nonradiative recombination, and there are contradicting reports on whether the density of DX centers decreases or increases in optically degraded material. The deep donor EL2 trap on the other hand is reported to be nonradiative in nature [28,29]. Interestingly, the EL2 defect is only seen in MOCVD, vapor phase epitaxy (VPE), and bulk crystals grown under arsenic-rich conditions. The likely origin is an arsenic antisite that displaces to form an arsenic interstitial with a gallium vacancy. The EL2 density is said to increase in optically degraded material [30]. Perhaps related to this, EL2 defects are also created after plastic deformation (dislocation motion) of GaAs, directly shown by site-specific electron beam-induced current and scanning deep-level transient spectroscopy on an indented site [31]. Farvacque et al. suggest that deformation processes lead to the formation of gallium and arsenic divacancies ($V_{Ga}-V_{As}$) [32], whereas Figielski et al. suggest that the dislocation itself getters and transports excess arsenic [31]. At the moment, we do not have sufficient evidence to directly implicate a particular point defect (either simple or complex) in GaAs-based laser degradation. We hope that this situation will change in the near future.

4.3 Dislocations in GaAs

Dislocations are one-dimensional (line) defects in crystals that play a key role in the plastic (irreversible) deformation of materials [33–35]. Plastic deformation in materials is accomplished by the motion of a large number of dislocations. This is energetically favored over a sudden concerted slipping of an entire plane of atoms. We liken this process to the difficulty of dragging a rug over a floor as opposed to pushing several small ripples out across the rug, both capable of moving the rug but the latter is easier. There are a number of references on the properties of dislocations in metals and semiconductors; we suggest the pioneering introductory book by Weertman and Weertman, the comprehensive reference by Hirth and Lothe, and for a focus on semiconductors, the recent book by Holt and Yacobi [36–38]. Dislocations in strongly bonded semiconductors like GaAs have key differences from their softer counterparts in metals. Unlike metals, semiconductors have bandgaps and the disruption in the crystal periodicity at the dislocation leads to electronic states in the gap. Further, the high penalty for dangling bonds at the core of the dislocation leads to an altered core structure comprised of reconstructed bonds (analogous to the tendency for semiconductor surfaces to reconstruct). This leads to dramatic changes in the anticipated electronic states introduced by the dislocation.

The purpose of this section is to place important characteristics of dislocations in the context of GaAs. We focus on the elementary steps involved in dislocation motion, glide and climb, in GaAs. These properties matter towards reducing the threading dislocation density (TDD) of GaAs/Si templates (Section 4.4) and are also those that contribute to laser degradation by becoming enhanced following nonradiative carrier recombination (Section 4.5).

4.3.1 Basic properties of dislocations

Dislocations are characterized by a Burgers vector and a line direction vector (which changes as a function of position if the dislocation is curved). Together, they set how planes of atoms around the dislocation are misaligned. Following one of many conventions, we may identify the Burgers vector $\vec{b}$ by choosing a sense for the line direction, $\vec{\xi}$, and then constructing a Burgers' circuit around the dislocation. Fig. 4.5A shows an example using the right-hand rule for determining the

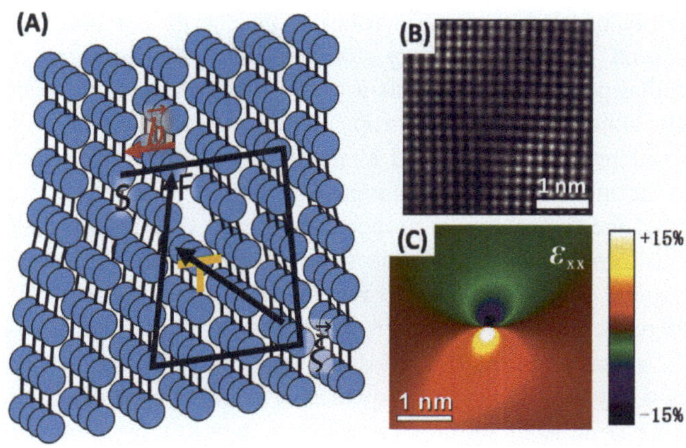

Figure 4.5 (A) A Burgers circuit showing circuit closure failure around an edge dislocation (such a circuit would close in a perfect crystal). The Burgers vector $\vec{b}$ can be defined as the vector joining the finish (F) to start (S) points of this circuit when drawn in the sense following a right-hand rule for the line direction $\vec{\xi}$ (clockwise here, since we have arbitrarily chosen the dislocation to be going into the page), and $\vec{b} \perp \vec{\xi}$ for an edge dislocation.
(B) High-resolution TEM image of an edge dislocation in a gold crystal. (C) Corresponding strain field (ε_{xx} component) experimentally determined from an analysis of the atomic displacements in the HRTEM image.

Source: (B) and (C) Reprinted from C.W. Zhao, Y.M. Xing, C.E. Zhou, and P.C. Bai, Experimental examination of displacement and strain fields in an edge dislocation core, Acta Mater. 56 (11) (2008) 2570–2575, https://doi.org/10.1016/j.actamat.2008.01.045, with permission from Elsevier.

sense of the closed Burgers circuit (thumb pointing along the line direction). The finish-to-start rule requires drawing around a dislocation what would be an otherwise closed circuit in a perfect crystal. The dislocation results in this circuit not closing, and the vector connecting the finish to the start points of this open circuit defines the Burgers vector. The angle between the Burgers vector and the line direction determines the character of the dislocation, taking values such as 90° (edge type), 0° (screw type), or other angles (mixed type). A single meandering or curved dislocation can have all of these characters at once even though each short segment of the dislocation has one character. Finally, the Burgers vector is constant along the line of the dislocation, and this remains the case even after the dislocation moves in the crystal—it is a fixed property of the dislocation.

Analytical expressions for the displacement, strain field and self-energy associated with the character of the dislocation and elastic moduli of the crystal have been derived [37], with special consideration for the properties of the core where linear elasticity breaks down. Fig. 4.5B shows a TEM image of an edge dislocation in gold clearly showing the extra half plane of atoms [39]. Fig. 4.5C shows one component of the corresponding strain field calculated from this image using a technique known as geometric phase analysis. Compressive strain is seen in the region above the half plane and tensile strain below the half plane. The strain fields are long ranged, meaning dislocations may interact attractively or repulsively with other dislocations, surfaces, point defects, etc. to minimize their energy. A dislocation in covalently bonded semiconductors has an energy per unit length of a few eV/plane of atoms, part of the energy stored in the strain field and the rest in the core itself. Unlike point defects, this is a high formation energy and creating a dislocation hardly contributes to the entropy of the crystal. This is why dislocations never exist spontaneously in a crystal as point defects do. Rather, they are created in response to thermomechanical or epitaxial stresses.

The energy of the dislocation is proportional to $\left|\vec{b}\right|^2$. Thus, two dislocations may fuse together to form a third dislocation with a Burgers' vector that is a sum of the individual Burgers' vectors (similar to Kirchhoff's laws for electrical current), but only if it also reduces the overall energy. This is referred to as Frank's rule after Charles Frank, a pioneer in the field of dislocations and also the first to suggest dislocations play a key role in crystal growth as a spiral source of atomic steps [40]. Finally, there are strict rules for how dislocations may or may not terminate at internal and external interfaces. These are often overlooked or misunderstood and have led to erroneous claims about the efficacy of dislocation filters among others; we highly recommend the review by McMahon et al. for a clear description of how dislocations may or may not terminate in thin film growth [41].

4.3.2 Structure of dislocations

The GaAs zincblende crystal structure has a face-centered cubic (FCC) lattice with a two-atom basis of As (0,0,0) and Ga (1/4,1/4,1/4). Therefore, we start with a basic description of features of dislocations in the simple FCC crystal and then introduce the impact of the two-atom basis. In an FCC crystal like copper with a lattice constant "a," the {111} planes have the highest atom packing density and the ⟨110⟩ direction on the {111} planes is the shortest distance between two atoms. Fig. 4.6A shows a top-down view of this close-packed (111) plane, and the shortest distance is highlighted by the red arrow and this is usually also the Burgers vector. This sets the lowest energy dislocations to lie on the {111} family of planes, oriented along one of the 12 ⟨110⟩ directions and the Burgers vectors are a/2 ⟨110⟩. This type of dislocation is a "perfect" dislocation as the Burgers vector translates lattice points. Common types of dislocations in FCC crystals are 60° dislocations (refers to the angle between $\vec{\xi}$ and $\vec{b}$) and screw dislocations $\left(\vec{\xi}\,||\,\vec{b}\right)$ [42].

A dislocation in an FCC crystal is able to spontaneously reduce its self-energy by dissociating into two partial dislocations that have to be connected by an intrinsic stacking fault. Fig. 4.6A shows the Burgers vector of the two partial dislocations that are a vector sum of the perfect dislocation according to Frank's rule. We obtain the following Burgers vectors for the two partial dislocations:

$$\frac{a}{2}[110] \rightarrow \frac{a}{6}[121] + \text{stackingfault} + \frac{a}{6}\left[21\overline{1}\right] \quad (4.7)$$

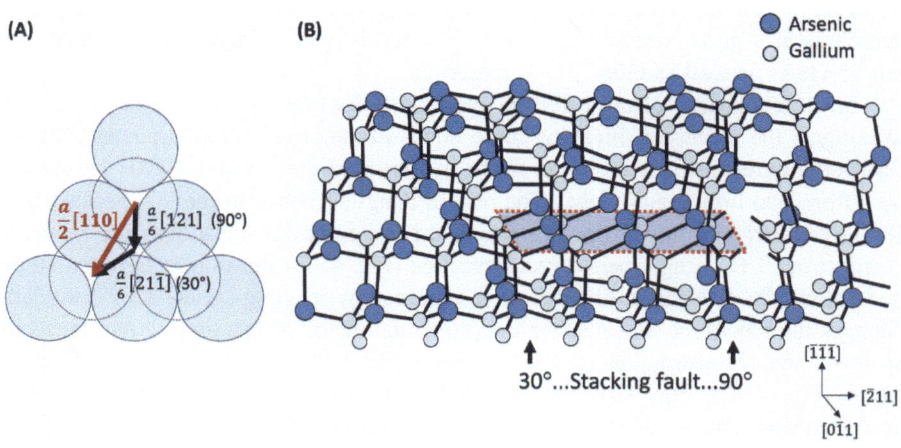

Figure 4.6 (A) Arrangement of two layers of atoms on the (111) plane of a FCC crystal. The shortest distance between any two atoms on this plane, corresponding to the Burgers vector of a dislocation, is a/2⟨110⟩. Frank's rule implies that such a dislocation dissociates into a 30° and a 90° partial dislocation. Here, the line direction is assume to be the horizontal axis. (B) Schematic of a 60° dislocation in the zincblende crystal structure of GaAs belonging to the glide set with a gallium core (β dislocation) dissociated into two partial dislocations. The shaded area indicates the stacking fault between partial dislocations and dangling bonds of the unreconstructed cores are also seen. The line direction of this dislocation is into the page.

Source: (B) Adapted from P.M. Petroff, Defects in III-V compound semiconductors, Semiconductors and Semimetals, vol. 22, part A, Academic Press, 1985, pp. 379–403.

The two partial dislocations now have 90° (edge) and 30° characters, respectively. High-resolution TEM images of dislocations in semiconductors have confirmed this dissociated nature along with a few nanometer wide stacking fault.

The two-atom basis of GaAs leads to a crystal comprised of alternating stacked {111} planes of gallium and arsenic. Therefore dislocations in GaAs with a line direction of ⟨110⟩ and a Burgers vector with an edge component (anything that is not a screw dislocation) have a core structure that is either all arsenic atoms or all gallium atoms. Note that in rare cases when the line direction is ⟨100⟩, the core has both arsenic and gallium atoms. Viewed along the [110] direction, Fig. 4.6B shows that the {111} planes alternate between closely spaced (glide-set) ones with three times as many bonds as the widely spaced (shuffle-set) ones. While intuition leads one to believe that forming a dislocation on a widely spaced plane costs less energy as the bonds have to distort less, the situation is complicated by the aforementioned tendency for dislocations in FCC systems to dissociate. Partial dislocations in GaAs connected by a stacking fault have been seen by TEM and this stacking fault must exist on the closely spaced planes [43]. This has led the scientific community to conclude that dislocations in GaAs form along the closely spaced (glide-set) {111} planes (at least at high temperatures). In this glide-set configuration, dislocations with an arsenic-terminated core are called α-dislocations and those with gallium-terminated cores are called β-dislocations. These two types of dislocations end up relieving strain in orthogonal directions, while also having very different glide velocities at a given stress/temperature. The result is asymmetries in strain relaxation in GaAs as well as other III–V materials.

The closely spaced planes offer the opportunity for bond reconstruction in the core to mitigate the penalty of breaking more bonds in the closely spaced planes. Fig. 4.7 shows the unreconstructed core structure of two partial dislocations [90° (edge) and 30°] from two orthogonal views. The row of dangling bonds in the 90° partial dislocation is clearly visible. There is a strong driving force for these dangling bonds to "reconstruct." The single-period reconstruction (Fig. 4.7B) shows how terminal atoms bond along the dislocation line direction, eliminating dangling bonds. This drastically changes the electronic and mechanical properties of the dislocation. Often, there are errors in the reconstruction (reconstruction defects) that result in isolated dangling bonds (Fig. 4.7C). Fig. 4.7E shows a 30° partial dislocation reconstruction with a double period, and a corresponding reconstruction defect is shown in Fig. 4.7F. As we see later, dislocations also possess short excursions, known as kinks and jogs, that interfere with reconstruction and yield additional dangling bonds.

4.3.3 *Thermal glide of dislocations*

Dislocations may extend or move in response to stress by a process known as glide. A majority of phenomena involving the motion of dislocations in semiconductors involve their glide, be it dislocation nucleation, multiplication of one dislocation into many, or annihilation of two or more dislocations. This motion does not involve addition or removal of atoms and is constrained to lie on a glide plane that contains both the Burgers vector and the line direction, shown in Fig. 4.8A (the glide plane is

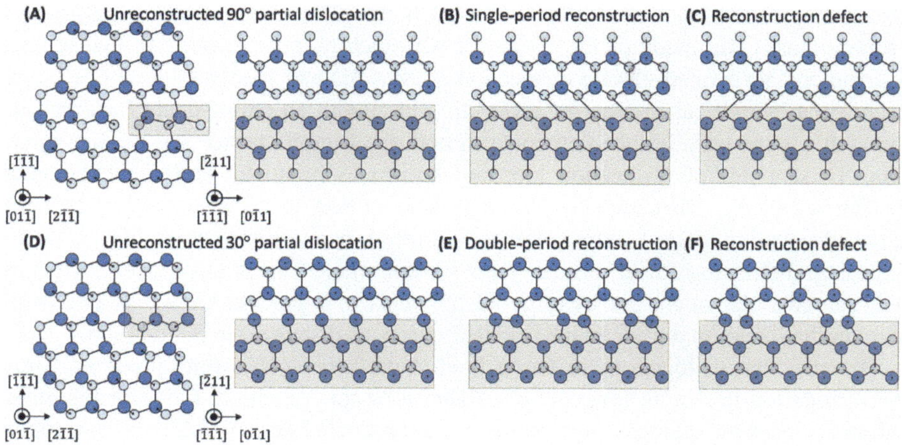

Figure 4.7 Orthogonal views of the (A) 90° and (D) 30° partial dislocations in the zincblende structure. The dangling bonds at the dislocation core undergo reconstruction; two different periods of reconstruction are shown. (B) Single-period reconstruction in 30° partial and (E) double-period reconstruction in 90° partial (bottom row). Finally, we also show reconstruction defects in these partial dislocations, also known as antiphase boundaries (or soliton defects), where dangling bonds remain (C) and (F). These are thought to play a role in the electronic properties of the dislocation.

Source: Adapted from Beckman et al., J. Phys. Condens. Matter 14 (2002) 12673–12680.

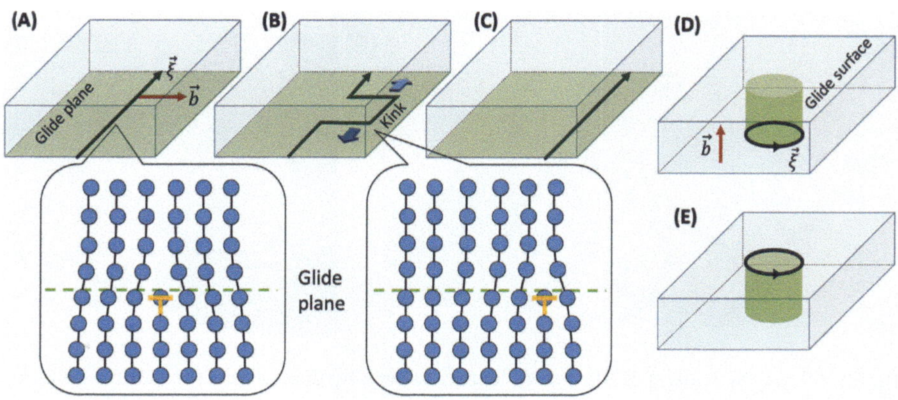

Figure 4.8 Sequence of elementary steps that result in dislocation glide. (A) The glide plane (green) is defined as the plane containing both the line direction $\vec{\xi}$ and the Burgers vector $\vec{b}$. A sketch of the atomic structure of such a dislocation is shown. (B) A short excursion of the dislocation known as a kink grows in the glide plane. This kink grows laterally over time and (C) eventually leads to the entire dislocation gliding across by distance of $|\vec{b}|$. (D) The cylindrical glide surface for a dislocation loop with an out-of-plane Burgers vector (prismatic loop) is shown. (E) This loop may only glide vertically along this surface.

shown in green). Dislocation glide is easiest if the glide plane also happens to be a primary (easy) slip plane of the crystal family, such as {111} family in GaAs; such dislocations are termed glissile dislocations. Fig. 4.8D and E shows how glide has to occur on a cylindrical surface for dislocation loop with an out-of-plane Burgers vector as this is the only surface that contains both the Burgers vector and the line direction. These types of dislocation loop are known as prismatic loops and the glide motion is called a prismatic slip. One may wonder how to define a glide plane for a screw dislocation since $\vec{\xi}$ is parallel to $\vec{b}$. Indeed, screw dislocations may glide on any plane belonging to the primary slip system where slip is easy, even switching between two equivalent planes to avoid obstacles in a process known as cross-slip. Motion that is not in a glide plane is called climb, as discussed in Section 4.3.5.

We saw that dislocations in GaAs have a preferred ⟨110⟩ line direction. Thus, the dislocation lies in an low energy valley along this direction. Glide of the dislocation requires the dislocation to move to a neighboring energy valley by surmounting an energy barrier known as the primary Peierls barrier. It is energetically unfeasible for the entire dislocation to make a concerted jump over this barrier all at once. Instead, dislocation glide is accomplished by a small atomic-scale unit of excursion of the dislocation line in the direction of the Burgers vector (staying in the glide plane) called a double kink or a kink-pair making this jump. Fig. 4.8A–C shows this process. Once the double kink is formed, it expands in size over time along the line direction and eventually the entire dislocation glides a distance equal to the magnitude of the Burgers vector. This expansion requires kinks moving over smaller secondary Peierls barriers as seen in Fig. 4.9A. Fig. 4.9B shows a TEM image of a 60° dislocation in silicon dissociated into 90° and 30° partial dislocations connected by a stacking fault [44]. Kinks in both partial dislocations are seen,

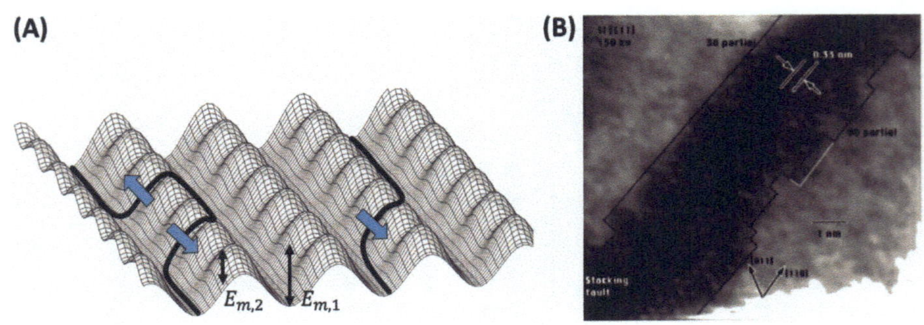

Figure 4.9 (A) Schematic of the energy landscape seen by dislocations (black lines) during glide. A double kink (left dislocation) and single kink (right) are shown as the dislocation, the formation of which required overcoming the primary Peierls barrier ($E_{m,1}$) in a small segment. Subsequent expansion or lateral migration of these kinks requires overcoming the smaller secondary Peierls barriers ($E_{m,2}$). (B) HR-TEM image of a dissociated 60° dislocation in silicon, showing kinks (outlined in black) in each of the partial segments.

Source: (A) Adapted from Cai et al. Dislocations in solids, Chapter 64; (B) Reprinted from H.R. Kolar, J.C.H. Spence, and H. Alexander, Observation of moving dislocation kinks and unpinning, Phys. Rev. Lett. 77 (19) (1996) 4031–4034, https://doi.org/10.1103/PhysRevLett.77.4031, copyright 1996 by the American Physical Society.

although the 30° partial dislocation appears to have far fewer, suggesting a higher kink formation energy. Overall, the motion of dislocations in semiconductors is strongly regulated by the breaking and rejoining of strong directional bonds at kinks, requiring the dislocation to thermally surmount energy barriers that are $1-3$ eV for each atomic step, this energy is split between the primary and secondary Peierls barriers. This high energy barrier for dislocation motion is why most semiconductors are brittle at room temperature.

4.3.3.1 Glide velocity

Dislocations glide in response to stress. The Peach–Koehler equation [45] succinctly captures the force per unit length acting on a dislocation due to mechanical stress tensor σ, depending on both the Burgers vector $\vec{b}$ and the line direction $\vec{\xi}$ as:

$$\frac{F}{L} = (\sigma\, b) \times \xi \tag{4.8}$$

This stress may be an external stress such as that from semiconductor lattice constant mismatch, device processing, or it may be an internal stress, for example, from other dislocations nearby. The component of the Peach–Koehler force in the direction of the Burger's vector gives the glide force. The glide velocity v of dislocations in semiconductors is represented phenomenologically as [46]

$$v = v_0 \left(\frac{\tau_{eff}}{\tau_0} \right)^m \exp\left(-\frac{E_t}{kT} \right), \tag{4.9}$$

where v_0 is a constant, and τ_0 is a normalizing constant for τ_{eff}, the effective stress on the dislocation that is in part determined using the Peach–Koehler formula. We clarify this term further for the specific case of heteroepitaxy in Section 4.4.1. The power law exponent m is experimentally seen to range between 1 and 2 and the term E_t is related to the energetics of kinks formation and migration. Dislocation glide in GaAs is a strongly thermally activated process, practically frozen at temperatures below 300°C. Specifically, the values m and E_t range between 1.6–1.8 and 1.3–1.4 eV, respectively, for α (As-core), β (Ga-core), and screw dislocations in undoped GaAs. However, v_0 can take drastically different values, leading to quite different velocities for the three types of dislocations [47]. For the same resolved shear stress, α-dislocations glide about two orders of magnitude faster than β- and screw-dislocations in undoped GaAs [48].

4.3.3.2 Diffusive glide

The model of Hirth and Lothe describes the glide of dislocations in response to stress in covalently bonded semiconductors [37]. Researchers have used this model to extract parameters such as the heights of the Peierls barriers [49]. The key idea is that the barriers to kink formation and migration are high in these materials.

Therefore, in the absence of stress, motion of kinks follows a purely diffusive process, akin to random walk, leading to no net motion. In the presence of stress, motion in one direction is only slightly biased; this largely resembles a diffusive process with a small element of drift. This situation is analogous to the motion of electrons in a crystal under a small electric field where the drift velocity of the electron is much smaller than the thermal velocity. At this point, we note that there are two regimes for this diffusive glide related to the way kinks move [50]. These subtleties play an important role in understanding recombination-enhanced dislocation glide in Section 4.5. For slow migration of kinks in long dislocations, the *kink-collision* regime predicts that two kinks of opposite signs will collide and annihilate before the entire dislocation has moved over to the next valley. In this case, the glide velocity does not depend on the length of the dislocation (a property that can be measured by TEM) but only on the mean free path of the kink (i.e., the average distance a kink travels before colliding with another kink). The second mode corresponds to the *kink-collisionless* case where kinks migrate along the dislocation length so rapidly that they do not encounter any other kink during this process. The glide velocity in this mode is proportional to the length of the dislocation. This is because kinks may conceivably form randomly at any point of the dislocation, and the longer the dislocation, the more sites there are for kinks to be formed and the dislocation glide would be faster. So far, the glide velocity of short dislocations in GaAs indeed appears to be proportional to the length of the dislocation according to Caillard et al., who measured the mean free path of kinks to be greater than 3 μm [49]. This suggests a kink-collisionless case. However, reports of a third mechanism of pinning by unknown localized obstacles has also been reported [50]. This has been inferred from jerky motion of dislocations and the dislocation appears to bow around what may be called pinning points. Clearly, more work is needed to elucidate microscopic aspects of dislocation glide in semiconductors.

4.3.4 Point defect and solute atmospheres around dislocations

Discussions on the mechanical and electronic properties of a dislocation as well as laser reliability frequently invoke the segregation of atomic impurities around dislocations. The key idea is that segregation of atoms of dissimilar sizes may reduce the overall energy of the system by interaction with the dislocation's strain field. Smaller substitutional impurities (such as nitrogen in GaAs or aluminum in InP) may segregate near the compressive side and larger substitutional atoms (indium in GaAs) or interstitial atoms segregate to the tensile regions around the dislocation. Fig. 4.10A and B shows a sketch of this process for an edge dislocation. It is well known that the segregation of impurities at dislocations can pin dislocations, hardening both steels and semiconductors (at high temperatures). It is therefore a compelling prospect that lasers may be made dislocation-tolerant by intentionally introducing impurities to pin dislocations from gliding (or climbing). We briefly discuss the energetics of such a process and estimate what concentration of point defects or impurities might exist in this way around a dislocation due to purely elastic interactions (without electrostatics) [51,52].

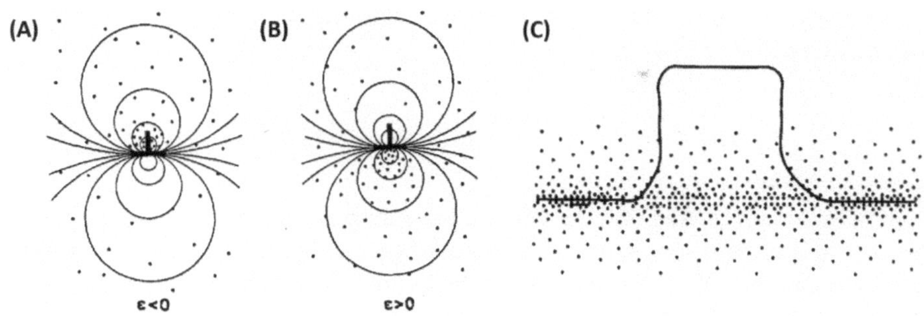

Figure 4.10 (A) and (B) Sketches of contours of constant hydrostatic pressure around an edge dislocation (half plane upwards) (compare with Fig. 4.5C). Regions of compression attract smaller impurity atoms ($\varepsilon < 0$) while regions of tension attract larger atoms ($\varepsilon > 0$). (C) A half loop breaking away from an atmosphere of impurities. This half loop can expand laterally in time and eventually free the entire dislocation.

Source: (A) Kirkby, Reprinted, with permission, from P.A. Kirkby, Dislocation pinning in GaAs by the deliberate introduction of impurities, IEEE J. Quantum Electron. 11 (7) (1975) 562–568, https://doi.org/10.1109/JQE.1975.1068634, copyright 1975 by IEEE. (B) Reprinted from B.W. Dodson, The effect of cottrell locking on structural stability in strained-layer semiconductor alloy structures, J. Cryst. Growth 92 (1) (1988) 222–232, https://doi.org/10.1016/0022-0248(88)90453-8, copyright 1988, with permission from Elsevier.

In considering the Gibbs free energy of point defects in Section 4.2, we had ignored the pressure–volume (PV) term corresponding to volume changes due to the introduction of the point defect. Solute atoms near dislocations may be treated similarly if we bring back the PV term. In the most general sense, interstitial defects are strong sources of hydrostatic compression, whereas vacancies are moderate sources of hydrostatic tension. These volume changes lead to elastic interactions between point defects and the stress fields of the dislocation. Following Lothe and Hirth [53], the external change in the volume ΔV of a solid upon introduction of a vacancy or solute atom is given by

$$\Delta V = (V_S - V_A)\left\{\frac{1 + (4\mu/3B)}{1 + (4\mu'/3B')}\right\}, \tag{4.10}$$

where V_S and V_A are the atomic volumes of the vacancy/solute atom and host/solvent. μ, μ' and B, B' are the shear and bulk moduli, respectively, of the host and impurity material. This solid may experience external pressures such as a misfit stress from epitaxy, thermal expansion mismatch-induced stress, or other processing-induced stresses. Additionally, there may be internal stresses originating from other dislocations, interfaces, etc. If these external and internal pressures with hydrostatic components P and p, respectively, work to create this vacancy, the external work done by this pressure modifies the Gibbs free energy of vacancy formation G_f to

$$\Delta G_0 = G_f - (P + p)\Delta V. \tag{4.11}$$

Analogus to Eq. (4.6), this leads to the equilibrium concentration of vacancies as:

$$c_0 = \left(\frac{1}{V_A}\right)\exp\left(-\frac{G_f - (P+p)\Delta V}{kT}\right) \quad (4.12)$$

It is important to note that the internal pressure p is spatially varying, for example due to a distribution of dislocations, and hence so will the vacancy concentration. We can see what this equilibrium looks like for an edge dislocation with a Burgers vector b_e. The hydrostatic component of p for an edge dislocation is: [37]

$$p = \frac{\mu b_e}{3\pi}\frac{1+\nu}{1-\nu}\frac{\sin\theta}{r}, \quad (4.13)$$

where ν is the Poisson's ratio. The key term in this equation is the sinusoidal function (ranging from -1 to $+1$) as this leads to a rise in concentration of vacancies in the compressive regions and fewer in tensile regions. This can be extended to understand if dissimilarly sized solute atoms (such as indium in GaAs) segregate near a dislocation. One may evaluate if solute atoms are a source of tension or compression in GaAs by comparing covalent tetrahedral radii of the atoms of the host (solvent) and the solute ($r_{Ga} = 124$ pm, $r_{In} = 144$ pm). The equilibrium concentration of solute atoms c_0 is found as:

$$c_0 = c_s\exp\left(-\frac{\beta\sin\theta}{rkT}\right), \quad (4.14)$$

where c_s is the mean concentration of the solute atom and $\beta = \frac{\mu b_e}{3\pi}\frac{1+\nu}{1-\nu}(V_s - V_A)$.

Dodson calculates this impurity dislocation interaction β for indium in InGaAs to around 0.4 eV-Å, which compared to $kT = 0.1$eV should lead to significant indium segregation. For a sufficiently strong interaction leading to solute concentrations in the percent level near the dislocation, Fermi-Dirac statistics replaces the Boltzmann statistics used here [54]. Using the ideas discussed here, Kirkby popularized the idea that certain impurties could be used to pin dislocations from moving [51].

4.3.4.1 Breakaway stress

Once impurties have segregated around the dislocation and lowered its overall energy, movement of the dislocation is constrained by the impurity cloud. At low temperatures, the impurity cloud is immobile and dislocation can move only if a double kink (or half loop) is able to break away from the impurity and continue to grow, in time freeing the dislocation of the impurity (Fig. 4.10C). Dodson estimates that while such half loops form easily in metals, they have a very high barrier of formation in semiconductors, and dislocations should be effectively pinned by impurities.

4.3.4.2 Experimental evidence

While the preceeding discussion gives the impression that solute segregegation near dislocations in semiconductors is understood, the reality is that there are significant discrepancies between experiment and theory. The pinning of dislocations by Kirkby's suggestion of intentional segregation of impurities to the dislocation in semiconductor lasers has not yet been verified. The most precise experimental determination of solute concentration around a dislocation comes to us from atom probe tomography (APT) in metals [55]. Fig. 4.11 shows an outstanding example of an APT reconstruction of a boron-containing iron–aluminum alloy with high lateral spatial resoltion where the dislocation can be seen directly in the lattice planes, a Burgers circuit reveals an edge type dislocation [56]. An atmosphere of boron atoms is segregated at the dislocation core. Unfortunately, even here there is poor agreement between the expected concentration and location of boron atoms with experiment, with the surprising finding that an interstitial boron atom segregates more than in just the tensile strained regions around the core [57]. Qualitatively similar discrepencies are seen in semiconductor samples where segregation appears to be radially symmetric, notably arsenic atoms decorating dislocations loops in silicon [58] and silver decorating dislocations in PbTe [59]. The difference between typical metallic samples studied by APT and a laser structure is that the former are bulk samples that have been allowed to equilibrate over long periods of time at high temperatures allowing for segregation to proceed to completion. Dodson estimates that bulk diffusion is too slow at typical growth temperatures in semiconductors. This, when taken together with the rapid growth rates of semiconductors, simply may not allow for the formation of an atmosphere of impurties around dislocations in lasers [52]. Recently, atmospheres of germanium and indium were seen around a dislocation in GaAs in a Ge/GaAs/InGaAs heterostructure by APT. Here,

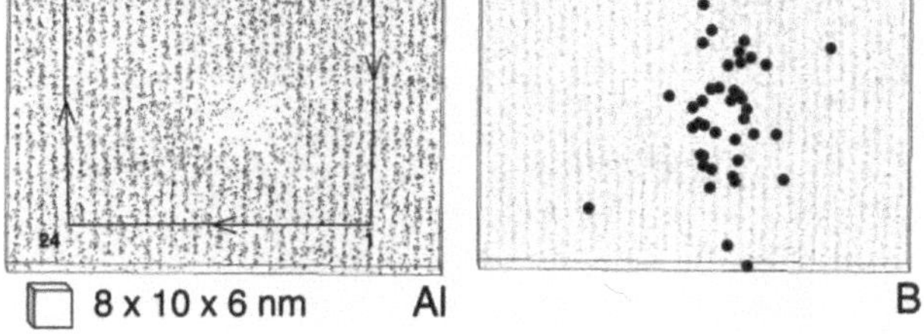

Figure 4.11 Maps of aluminum (small dots; A) and boron (big dots; B) measured by atom probe tomography showing the presence of a boron-enriched Cottrell atmosphere along an edge dislocation in a Fe-Al alloy along a small slice (thickness = 6 nm) cut perpendicularly to the dislocation. Imaged planes are almost perpendicular to the ⟨001⟩ depth axis. The Burgers circuit indicates the presence of an edge dislocation.

Source: Reprinted from E. Cadel, A. Fraczkiewicz, and D. Blavette, Atomic scale observation of Cottrell atmospheres in B-doped FeAl (B2) by 3D atom probe field ion microscopy, Mater. Sci. Eng. A 309–310 (2001) 32–37, https://doi.org/10.1016/S0921-5093(00)01688-9, with permission from Elsevier.

the atmosphere was formed via rapid short-circuit or pipe diffusion along the dislocation from between the layers rather than lateral segregation [60]. In other cases, atmospheres most likely have to form via surface diffusion and stay locked in after. Neither mechanism to circumvent slow bulk diffusion has been explored in detail. There is no doubt that impurities segregate to dislocations and influence their properties, but it remains difficult to quantify this and devise ways to intentionally modulate the environment around the dislocation in lasers.

4.3.5 Thermal climb of dislocations

4.3.5.1 Overview

The second mode of dislocation motion involves flux of atoms toward or away from the dislocation core leading to the motion of the dislocation out of the glide plane. This type of motion is called climb. Dislocation climb is an important degradation mode in laser reliability and continues to be an area of active research. Consider the edge dislocation in Fig. 4.12A acted upon by a tensile biaxial stress component σ_{xx}. One can see from Eq. (4.8) that the total force on the dislocation is purely in the y-direction with no force resolved along the glide plane (green). Hence, the edge dislocation cannot glide. On the other hand, this force could make

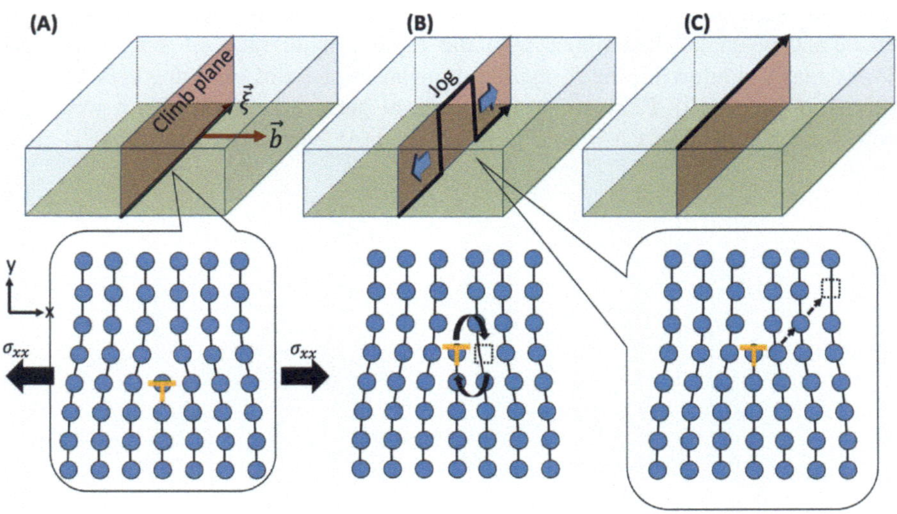

Figure 4.12 Schematic showing an edge dislocation emitting vacancies to undergo climb under the influence of tensile stress. (A) The climb plane (red) is defined as orthogonal to the Burgers vector $\vec{b}$. (B) The edge dislocation emits a vacancy, creating an excursion in the climb plane called a jog. This vacancy may diffuse away from the dislocation core if conditions permit. This jog may expand laterally by continuing to emit vacancies, and (C) eventually the entire dislocation has climbed up a distance equal to the jog height.

the dislocation climb upwards along a plane that is orthogonal to the Burgers vector (climb plane shown in red). Climb motion is more complicated than glide because the vertical motion of the dislocation will also create a vacancy (Fig. 4.12B) (or absorb an interstitial) in the crystal in order to keep the total number of atomic sites in the crystal constant. Likewise, downward motion of the dislocation due to compressive biaxial stress would lead to absorption of vacancies (or emission of interstitials). Climb proceeds via small excursions in the climb plane known as jogs (Fig. 4.12B). These jogs expand laterally by absorbing or emitting point defects, eventually leading to the entire dislocation to move up by a jog height (Fig. 4.12C).

Continued climb of such a dislocation leads to an imbalance in the number of point defects near the dislocation and builds up what is known as an osmotic pressure that can quite effectively counter the externally applied stress and halt climb. In Fig. 4.12B, we show that climb-generated vacancies may diffuse away from the dislocation core if there is a vacancy concentration gradient that permits this, thereby lowering the osmotic pressure and allowing for continued dislocation climb. This explains why dislocation climb is seen in metals at high temperatures as point defects need thermal activation to diffuse toward or away from the dislocation. At lower temperatures, dislocation climb may occur slowly in special cases, like our example in Fig. 4.12, where the stress is applied such that it cannot cause the dislocation to glide.

The diffusion of point defects in the bulk of III–V semiconductors is, however, sluggish even at growth temperatures. Hence, dislocation climb is secondary in importance to glide in semiconductors, and indeed, there are only a handful of studies on purely thermally activated dislocation climb in semiconductors [61–63]. Yet, it is useful to understand what dislocation climb is and the driving force behind it to appreciate the more complex recombination-enhanced dislocation climb (Section 4.5) where point defect diffusion is enhanced to the point of facilitating dislocation climb even at low temperatures.

In semiconductor lasers, it is not only an external stress that drives dislocation climb but also the osmotic pressure. In most cases, thin films of semiconductors grown at high-temperature have an out-of-equilibrium number of point defects when cooled down to room temperature. Perhaps point defects near the surfaces of such a crystal may be able to equilibrate, but those in the bulk of the semiconductor crystal may not be able to. Let us examine how a nonequilibrium concentration of point defects, a purely chemical driving force, leads dislocations to climb.

4.3.5.2 Osmotic force

If an edge dislocation has a Burgers vector, b_e, corresponding to an extra half plane, this half plane can extend or contract via jogs in the climb plane (contrasted with kinks that are in the glide plane). We had seen that elastic interactions between vacancies and dislocations lead to a spatially varying but equilibrium (steady state) concentration c_0. If the actual (out-of-equilibrium) vacancy concentration is c, we may, following Lothe and Hirth [53], define a chemical potential for vacancies using this spatially varying reference as:

$$\overline{G} = kTlog\left(\frac{c}{c_0}\right). \tag{4.15}$$

This value is positive for a supersaturation of the vacancy. Now consider an edge dislocation acted upon by a simple external biaxial tensile stress σ_{xx}. This stress may have origins in crystal lattice-mismatch, device packaging stress, etc. The component of the Peach–Koehler force perpendicular to the glide plane gives the elastic climb force per unit length from this stress; in this case, in the upward direction.

$$\frac{F_{y\_elastic}}{L} = \sigma_{xx}b_e \tag{4.16}$$

On the other hand, we can calculate the downward osmotic force per unit length on this dislocation by considering how much the free energy changes if the dislocation climbs down by absorbing vacancies with a volume V_A. Here, each vacancy absorbed leads to a change of the free energy by $\overline{G}$, and the number of such vacancies absorbed per unit length if the dislocation climbs a distance Δy is given by $\frac{b_e \Delta y}{V_A}$. This yields:

$$\left(\frac{F_{y\_osmotic}}{L}\right)\Delta y = -\left(\frac{b_e \Delta y}{V_A}\right)\overline{G}, \tag{4.17}$$

leading to,

$$\frac{F_{y\_osmotic}}{L} = -\frac{kTb_e}{V_A}log\left(\frac{c}{c_0}\right). \tag{4.18}$$

In the presence of both stress and a nonequilibrium vacancy concentration, the total force on the edge dislocation is:

$$\frac{F_y}{L} = \frac{F_{y\_elastic}}{L} + \frac{F_{y\_osmotic}}{L} = \sigma_{xx}b_e - \frac{kTb_e}{V_A}\log\left(\frac{c}{c_0}\right) \tag{4.19}$$

We can now see how external stress may counteract the osmotic force from a nonequilibrium point defect by setting the overall force on the dislocation to zero. Thus, a biaxial tensile stress opposes the osmotic force from a supersaturation of vacancies. Likewise, a compressive stress opposes that due to excess interstitials. Hopgood et al. propose this latter point as to why compressively strained InGaAs quantum-well lasers are reliable against climb mediated by absorption of interstitials (or emission of vacancies) [64]. Hirth and Lothe show how to generalize this discussion for mixed dislocations, where only the edge component of the Burgers vector of the dislocation contributes to climb [37].

4.3.5.3 Climb in the two-atom basis of GaAs

Unlike simple cubic metals with a monoatomic basis, the formation of a dislocation jog in GaAs requires interaction with two atoms belonging to the Ga- and As-sublattices. Moreover, these interactions must occur in sequence to maintain the polarity of this ordered crystal. For the most generic case involving only vacancies and interstitials, Kirkby works out the osmotic force as follows [51]. As before, if the dislocation moves down a distance Δy, the net number of point defects absorbed per unit length of the dislocation is given by:

$$\frac{b_e \Delta y L}{V_A} = n(\mathrm{V_{As}}) - n(\mathrm{I_{As}}) = n(\mathrm{V_{Ga}}) - n(\mathrm{I_{Ga}}) \tag{4.20}$$

With n(X) and $\overline{G}(X)$ referring to the number of the particular type of point defect X and the free energy associated with that defect, the change in free energy ΔG is given by:

$$\Delta G = \overline{G}(\mathrm{V_{As}})n(\mathrm{V_{As}}) + \overline{G}(\mathrm{V_{Ga}})n(\mathrm{V_{Ga}}) - \overline{G}(\mathrm{I_{As}})n(\mathrm{I_{As}}) - \overline{G}(\mathrm{I_{Ga}})n(\mathrm{I_{Ga}}) \tag{4.21}$$

To see this in practice, we follow Petroff and Kimerling who discuss the osmotic force for dislocation climb in GaAs grown under gallium-rich conditions of liquid phase epitaxy (LPE). In LPE growth, it may be feasible for the crystal to have an excess (a supersaturation) of gallium interstitials and arsenic vacancies. Fig. 4.13 shows a sketch of an arsenic-terminated 60° dislocation in GaAs, with a gallium interstitial atom nearby. Consider the following sequence of events. The gallium atom diffuses to the dislocation core and attaches momentarily (Fig. 4.13A). In this intermediate state, the polarity of the dislocation is imbalanced and this is energetically unfavorable. An arsenic atom from a lattice-site detaches and immediately joins the dislocation core to convert it back to an arsenic-terminated core. This completes the climb process by moving the dislocation core upwards (Fig. 4.13B and C). Importantly, note how this climb step also effectively absorbs a single gallium interstitial atom and expels a single arsenic vacancy. Thus we have the change in free energy of point defects as [65]:

$$\Delta G = \overline{G}(\mathrm{V_{As}}) - \overline{G}(\mathrm{I_{Ga}}), \tag{4.22}$$

which using Eq. (4.18) leads to:

$$\frac{F_{y\_osmotic}}{L} = -\frac{kTb_e}{V_A}\log\left(\frac{c_0(\mathrm{I_{Ga}})c(\mathrm{V_{As}})}{c(\mathrm{I_{Ga}})c_0(\mathrm{V_{As}})}\right). \tag{4.23}$$

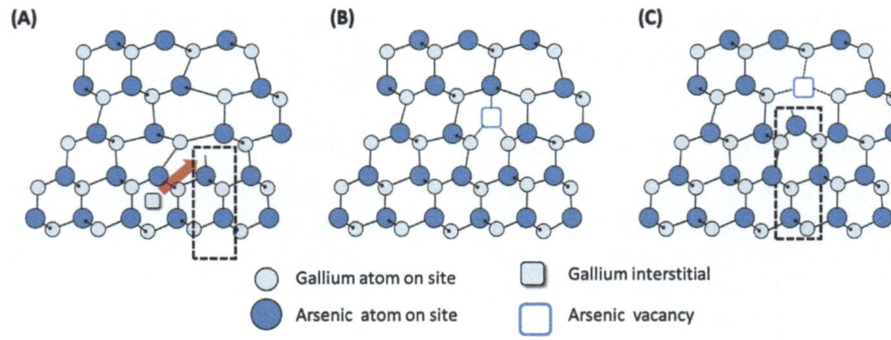

Figure 4.13 Sketch of the sequence of events in a climb process in GaAs. (A) A gallium interstitial diffuses toward and attaches to an arsenic terminated dislocation. (B) This creates a new arsenic vacancy site in the dislocation half plane that is (C) rapidly occupied by an arsenic atom, leaving behind an arsenic vacancy in the bulk. The net result is that the dislocation has climbed up one unit in the crystal, and also converted a gallium interstitial into an arsenic vacancy.

Source: Adapted from P.M. Petroff and L.C. Kimerling, Dislocation climb model in compound semiconductors with zinc blende structure, Appl. Phys. Lett. 29 (8) (1976) 461–463, https://doi.org/10.1063/1.89145.

If we know the undesaturation or oversaturation of point defects in a sample, this expression allows us to calculate what the osmotic force is on a dislocation segment that causes it to climb. More importantly, it can then inform on the type and magnitude of external stress (either tensile or compressive) that may be applied to counter dislocation climb.

4.3.5.4 Other considerations

Dislocation climb often leads to unique structures not possible via glide. For example, a dislocation loop is able to expand or contract due to climb by interacting with point defects (Fig. 4.14A and B). This is a key mechanism by which dissolved point defects coalesce to form dislocation loops that are often seen in semiconductor devices. It is possible to identify if the dislocation loop is formed out of vacancies or interstitials using sophisticated TEM analyses. Likewise, a screw dislocation can develop into a helix. This may be seen in the sketch in Fig. 4.14C and D where the climb plane (orthogonal to both $\vec{\xi}$ and $\vec{b}$) is shown. The dislocation must climb along this plane, although these climbed segments may subsequently move out of this plane via prismatic slip. The sense of winding of the helix provides clues about whether climb involves vacancy absorption or emission (or interstitial emission or absorption). Fig. 4.14E shows a TEM image of a helical dislocation from the emitter-base region of a silicon bipolar junction transistor [66]. Here, it is suggested that the diffusion of phosphorus from the emitter also brings with it a flux of silicon interstitials that leads to a supersaturation and hence dislocation climb. Such structures are not only visually stunning but also provide evidence in failure analysis of devices that a climb process

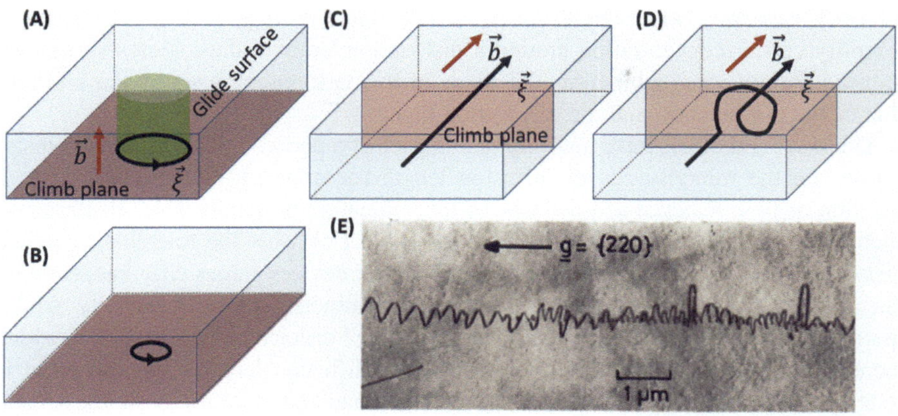

Figure 4.14 (A) The climb plane (red) of a prismatic dislocation loop with Burgers vector $\overline{b}$ out of plane of the dislocation loop. (B) Such a dislocation expands or contracts by climb following interactions (absorption/emission) with point defects. (C) The climb plane for a screw dislocation and (D) helical structures are generated following climb in a screw dislocation. The sense of winding of the helix can inform on the sense of climb (and can narrow down the type of point defect interaction). (E) TEM image of a screw dislocation in the emitter-base region of a bipolar junction transistor in silicon that has undergone climb resulting in a tightly wound helical structure. The individual windings can undergo glide rearrangement along the line direction resulting in an evenly spaced coil.

Source: (E) Reproduced is from H. Strunk, U. Gösele, and B.O. Kolbesen, Interstitial supersaturation and climb of misfit dislocations in phosphorus-diffused silicon, J. Microsc. 118 (1) (1980) 35–39, https://doi.org/10.1111/j.1365-2818.1980.tb00243.x, Copyright 1980, with permission from John Wiley and Sons.

has occurred and that one must look into point defects. We will revisit these topics in recombination-enhanced dislocation climb in Section 4.5.

4.3.6 Electrical and optical properties

4.3.6.1 Overview

Dislocations influence majority and minority charge carriers due to their line charge via electrostatic interactions, their strain fields via the deformation potential and piezoelectric coupling, and by trap states in the band gap [67,68]. This leads to reduced carrier mobility akin to impurity scattering as well as reduced minority carrier lifetimes. A dislocation line is often phenomenologically modeled as a cylindrical region with reduced minority carrier lifetime with respect to the bulk [69]. Another related method involves treating the dislocation as a line defect with a line recombination velocity (or a cylinder with surface recombination velocity) [70,71]. This reduced lifetime or the line recombination velocity is ascribed to nonradiative Shockley-Read-Hall recombination that describes recombination via single trap states. Uniform injection of minority carriers into the semiconductor leads to a lower minority carrier concentration around the dislocation and a net current of minority carriers toward the dislocation. For light-emitting devices, the reduced minority carrier concentration around a dislocation leads to dark spots or lines via reduced radiative recombination, the shape of the dark region is set by the intersection of the p-n junction and the dislocation line.

Donolato and others have analyzed the impact of a periodic array of straight dislocations on the minority carrier diffusion length (or minority carrier lifetime) as a function of bulk material and dislocation recombination properties [71,72]. Based on such analyses, there are reports tackling the subject of what the tolerable threading dislocation densities are for various minority carrier devices across different semiconductors. In general, semiconductors with a lower minority carrier diffusivity exhibit more tolerance to dislocations. There is a "tolerable" dislocation density below which there is minimal impact on the minority carrier diffusion length. This threshold is roughly $10^6/cm^2$ for GaAs solar cells (not lasers) and serves as the basis for several design goals for heteroepitaxial III–V/Si solar cells [70,73]. Interestingly, this threshold is around $10^8-10^9/cm^2$ for In GaN, which has a smaller bulk diffusivity as well as carrier localization [74]. This simple analysis is one reason why nitride LEDs have high radiative efficiencies despite large threading dislocation densities. We note that the dislocation density is assumed to be fixed here and hence this analysis does not apply to cases of recombination-enhanced dislocation motion that leads to an increasing dislocation density as a function of time. This brings us to the second reason for the success of nitride LEDs despite dislocations—their exceptional reliability to recombination-enhanced processes is not available to the softer GaAs-based materials. As discussed in Section 4.5.3, for semiconductor lasers, even a single dislocation can give rise to a "dark line defect" network that will kill a GaAs laser.

4.3.6.2 *Origin of traps*

Heteroepitaxy of III–Vs on mismatched substrates, such as GaAs-on-silicon growth for integrated lasers, has been a topic of active research, covered in detail in Section 4.4 and Chapter 6, Reliability of lasers on silicon substrates for silicon photonics. It has become increasingly difficult to reduce the threading dislocation density (TDD) in III-V/Si within the constraints of maintaining a scalable and cost-effective growth process. This motivates a study of what sets the recombination strength of a dislocation in the hope that it might provide a complementary pathway to improved optoelectronic devices. Initial models of dislocations often assumed that if they have line directions and Burgers vectors yielding edge components, they will have dangling bonds. In the earliest work on the electrical properties of dislocations in semiconductors, Shockley proposed that such a line of dangling bonds would have one electron each and correspond to a half-filled band in the gap leading to metallic behavior along the core. This was shown not to be true for covalently bonded semiconductors where the energetic cost of a single dangling bond is high. Therefore most dislocations in semiconductors have their cores reconstructed —forming bonds (such as dimers) along the line direction [75]. While these bonds are not ideal, especially for compound semiconductors, and deviate from the tetrahedral coordination (glide-set α-dislocations will form As–As bonds and β-dislocations form Ga–Ga bonds), this reconstruction should once again lead to the formation of bonding and antibonding orbitals, not necessarily with a gap as large as the host material, but forming shallow levels. Shallow electron or hole traps should not lead to the reported high dislocation recombination strengths. Sitch et al. find that both 90° α [As(glide-set)] and β [Ga(glide-set)] dislocations are strongly reconstructed, possessing no mid-gap states but only shallow levels (the latter, near the valence band acts as a hole trap for β dislocations) [76]. On the other hand, Beckman et al. calculated that both the reconstructed 30° and 90° α-dislocations lead to mid-gap states [77]. Fig. 4.15A–D show their results for 90° dislocations with both single- and double-period reconstruction (see Section 4.3.1). Along with shallow levels created near the conduction band, a trap state is seen close to mid-gap for α-dislocations. They find that the β-dislocation partials introduce states

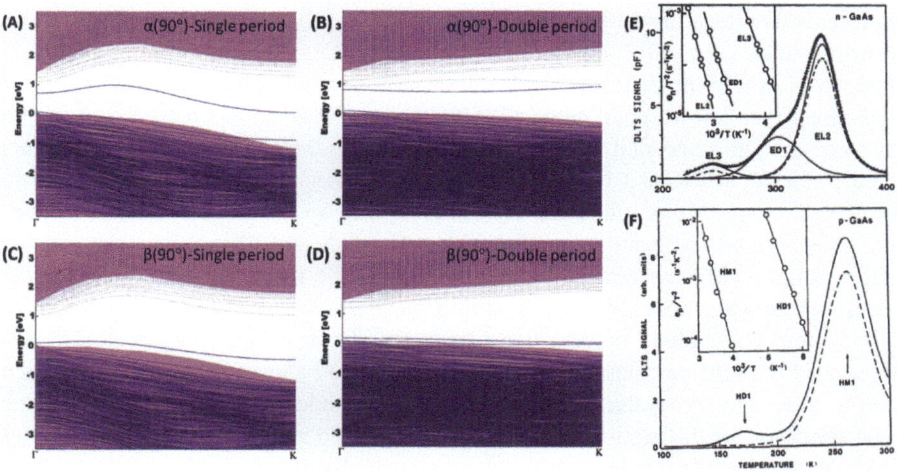

Figure 4.15 Electronic structure calculations for reconstructed (see Fig. 4.7) arsenic-core (α) and gallium-core (β) partial dislocations in GaAs in the glide-set within the local density approximation to density functional theory. α(90°) partial dislocation with a (A) single- and (B) double-period reconstructions, both showing shallow states near the conduction band along with a partially filled state near the mid-gap. β(90°) partial dislocation in the (C) single- and (D) double-period reconstructions, both introducing partially filled states just above the valence band. (E) DLTS spectra of an n-type GaAs sample with conventional bulk distributed electron traps (EL3 and EL2) along with a dislocation-related trap (ED1) with an activation energy of 0.68 eV. (F) DLTS spectra of p-type GaAs showing a conventional bulk distributed hole trap (HM1) and a dislocation-related hole trap (HD1) with an activation energy of 0.37 eV.

Source: (A)–(D) Reprinted from S.P. Beckman and D.C. Chrzan, Dislocation cores and their electronic states: partial dislocations in GaAs, Phys. B Condens. Matter, 340–342 (2003) 1001–1004, https://doi.org/10.1016/j.physb.2003.09.189, with permission from Elsevier; (E) and (F) reprinted from T. Wosiński and T. Figielski, Energy levels and electrical activity of dislocation electron states in GaAs, Acta Phys. Pol. A 83 (1) (1993) 51–58, https://doi.org/10.12693/APhysPolA.83.51, with permission from the Polish Physical Society.

above the valence band. Their structural models, which set the electronic properties, have impressive agreement with HRTEM images [78,79]. Recent calculations by Park et al. disagree with the results of Beckman et al., suggesting that only the 90° α single-pair reconstruction lead to deep gap states, not the double-pair reconstruction [80]. They go on to propose a new quadruple period reconstruction as the lowest energy state of the 90° β dislocation. Clearly, our understanding of the electronic structure of dislocations continues to evolve.

It has been challenging to experimentally quantify the electronic effects of dislocations in bulk materials as most processes that generate dislocations (such as mechanical bending) may also increase the number of point defects. Indeed, there is an unresolved controversy about whether the electrical effects of dislocations are due more to the dangling bonds of the dislocation core or to the cloud of impurities and defects that surround the dislocation line in a cylinder surrounding it—it could be both. Nevertheless, the linear arrangement of trap states along a dislocation (as opposed to randomly distributed point defects) leads to signatures in state-filling measurements like deep level transient spectroscopy (DLTS), making them ideally suited to probe dislocations [81]. Fig. 4.15E and F shows DLTS spectra for n-type and p-type GaAs, where the dislocation-related electron and hole traps (ED1 and HD1) are seen along with known point defects [78]. Such measurements on n-type GaAs reveal an electron trap associated with dislocations that lie 0.68 eV below the conduction band (0.58 eV in InGaAs [82]). Measurements varying the α/β dislocation density ascribe this trap to α-As(glide-set) dislocations. Likewise, a hole trap 0.37 eV above the valence band has been associated with β-Ga(glide-set) dislocations using p-type GaAs samples. This assignment agrees well with the reduction of glide activation energies following nonradiative recombination, discussed in detail in Section 4.5.2. The capture coefficients of both these traps are about $10^{14}/cm^2$. These results also agree well with the calculations by Beckman described above. As an aside, an extension to this logic shows that the more ionic a semiconductor becomes, the less likely it becomes for the dislocation to reconstruct and should naturally lead to deeper levels from dangling bonds. This is indeed the case for wider gap materials.

In addition to traps natively associated with straight/reconstructed dislocations, there is evidence for traps at kinks, jogs, or reconstruction defects. Other traps arise from the atmosphere of point defects and impurities that are bound to the core by elastic or electrostatic interactions; elastic interactions are discussed in Section 4.3.4. Notably in GaAs, the impact of high-temperature annealing and quenching vs. slow cooling has exposed variability in the recombination strength of the dislocation by cathodoluminescence mapping [83]. This was explained by suggesting that quenching the sample prevents the point defect atmosphere from forming to its equilibrium level and leads to a reduction in the dislocation recombination strength.

4.3.6.3 A connection between electronic states and glide velocity

Dislocations in compound semiconductors have further complexities that have an impact on devices. We mention, for example, asymmetry in the glide velocity of α

(As-core) and β (Ga-core) dislocations of the glide-set that leads to unidirectional cracking in GaAs/Si films. The origins of what sets differences in glide velocities are worth looking into. The more reconstructed the core of a dislocation is, the more it becomes difficult to nucleate a double kink that is key to glide. Sitch et al. note that β-dislocations are more strongly reconstructed (Ga−Ga bonds) than α-dislocations (As−As bonds) [76]. This might partially explain why α-dislocations have a higher glide velocity than β-dislocations. However, this is not the complete picture. The electronic states associated with the dislocation also manifest themselves via a doping dependence of the glide velocity. Specifically, α-dislocations slow down in GaAs upon both p-type and n-type doping, whereas β-dislocations slow down going from p-type to n-type. This is possible because kinks play an important role in both the creation of electronic states in the gap as well as the glide process. Hirsch explains this by suggesting that a double kink on dislocations may act as amphoteric defects, creating both localized acceptor and donor states in the gap, and thus acquiring a charge depending on the Fermi level [84]. Thus the concentration of kinks on a dislocation may be controlled by the Fermi level, analogous to the Femi level effect in determining the concentration of charged point defects based on their charge transition levels (Section 4.2). This presents an opportunity to tune the mechanical properties of a crystal by electronic means.

4.3.6.4 Optical properties of dislocations

Individual dislocations along with their strain fields might at first appear too small to appreciably interact with the large μm-sized mode of a photon and result in below-bandgap absorption. Sub-bandgap absorption resembling an Urbach tail (absorption coefficients of 10^2-10^3/cm at telecom wavelengths), however, has been reported in GaAs samples with high dislocation densities (10^7-10^8/cm^2) [85]. This is ascribed to band bending at the dislocation due to localized charge or a piezoelectric field arising from strain. As a closing note, we mention a curious property of dislocations in silicon. They emit sub-bandgap light from localized states (four prominent emission lines are identified: D1−D4) [86]. Indeed infrared LEDs with external efficiencies of 0.1% have been made from mechanically deformed silicon [87]. Nevertheless, dislocation recombination is not expected to be fast enough to compete with band-to-band recombination in direct gap semiconductors like GaAs.

4.4 Epitaxial integration of GaAs-based materials on silicon

We summarize certain key aspects of heteroepitaxial growth of III−V/Si in this section, relating it to the basic properties of dislocations that we have discussed previously. The ability to integrate dissimilar materials by epitaxy has tremendous utility in engineering new energetic landscapes for electrons and holes. These have enabled band gap tuning, enhanced carrier capture, population inversion, leakage blocking, etc. The role of heterostructures and strain in improving electronic devices can hardly be overstated. Yet, sometimes, the dissimilarity extends too far and the heterostructure becomes plagued with crystal defects, particularly dislocations. The focus of this section is the generation of dislocations due to a mismatch in lattice constant, although defects may also be introduced in otherwise closely lattice-matched systems due to unfavorable growth modes, interface reactions, or thermal expansion mismatch, to name a few.

4.4.1 Critical thickness for dislocation nucleation

The misfit strain between a film and a substrate is defined as:

$$\in = \frac{a_{film} - a_{substrate}}{a_{substrate}}. \tag{4.24}$$

This strain is initially accommodated entirely elastically by a change in the bond lengths and angles between the atoms in the film. The grown-in (preexisting) dislocations of the substrate participate in the initial mode of plastic accommodation of some of this strain. It is possible to define a "critical" thickness of such a strained film above which it becomes energetically favorable to extend a vertically threading dislocation across an interface to create a misfit dislocation via glide [88]. We emphasize that the descriptors "threading" and "misfit" only clarify the line directions; if one remembers the preferred Burgers vectors for dislocations in the GaAs crystal we can see that misfit dislocations are 60° dislocations while threading dislocations may range from screw to 60°. Also note that this glide process has to follow all the rules described in Section 4.3.3, namely there has to be a non-zero force on the dislocation due to lattice-mismatch strain along its glide plane that is set by the Burgers vector and line direction, and this glide plane should be a primarily slip plane for the crystal. Fig. 4.16A shows a plan-view TEM image of a threading dislocation extending to form a misfit segment in $In_{0.1}Ga_{0.9}As$/GaAs as the layer crosses the critical thickness [89,90]. At the critical thickness, the self-energy of this additional length of misfit dislocation (see Section 4.3.1) is compensated by a permanent reduction in the long-range strain in the film, and hence a reduction in the energy of the epilayer + dislocation system as a whole. Strained quantum well lasers, unless designed otherwise, stay well within this critical thickness limit to avoid creating misfit dislocations in the active region [91,92].

The treatment of Freund [93] and Nix [94] is well suited to the discussion of critical thickness and driving forces for glide of the threading dislocation. Consider the (001)-oriented growth of a misfitting film with thickness h, such as InGaAs on GaAs where 60° dislocations with Burgers vector $\vec{b}$ glide to relieve stress, as shown in Fig. 4.16B. We look at how a single threading dislocation glides to lay down a misfit segment at the interface. The primary (111) slip system is inclined to the growth direction, hence only some resolved component of the biaxial shear stress σ acts to make the dislocation glide. With the external

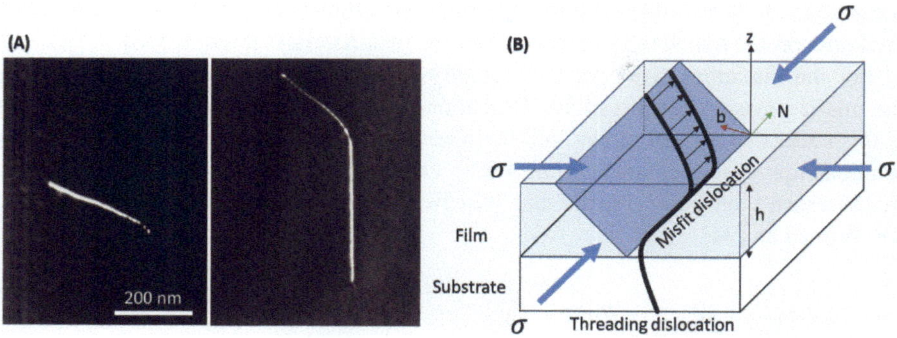

Figure 4.16 (A) TEM image of a threading dislocation at the onset of misfit dislocation generation via glide in GaAs/$In_{0.1}Ga_{0.9}As$. (B) Schematic of the glide of a threading dislocation upon application of biaxial stress σ, laying down a segment of misfit dislocation at the film/substrate interface. The critical thickness is achieved when the force on the threading dislocation exceeds the line tension of the misfit dislocation segment.

Source: (A) Reprinted from J. Zou, D.J.H. Cockayne, and B.F. Usher, Misfit dislocations and critical thickness in InGaAs/GaAs heterostructure systems, J. Appl. Phys. 73 (2) (1993) 619–626, https://doi.org/10.1063/1.353372, with the permission of AIP publishing. (B) Adapted from W.D. Nix, Mechanical properties of thin films, Metall. Trans. A, 20 (11) (1989) 2217, https://doi.org/10.1007/BF02666659.

stress acting upon the dislocation to make it glide by a unit length, the work done on the film W_{layer} is given by:

$$W_{layer} = \frac{\sigma b h}{2}. \tag{4.25}$$

The length of the threading dislocation is assumed to stay constant but there is now a unit length of misfit dislocation at the interface. The energy of this unit length of 60° dislocation, W_{60}, lying at the interface between the film and substrate is given by: [90]

$$W_{60} = 0.95\frac{b^2}{4\pi(1-\nu)}\frac{2\mu_f\mu_s}{\left(\mu_f + \mu_s\right)}\ln\left(\frac{\beta h}{b}\right). \tag{4.26}$$

Here ν is the Poisson's ratio and β is a numerical constant about unity (Nix et al. use $\beta = 0.755$), reflecting the spatial extent of the strain field. The expression for W_{60} looks more complicated than what was described in the discussion in Section 4.3.1 because the strain field of the dislocation is split between the substrate and film with different shear moduli μ_s and μ_f, respectively, and has mixed edge and screw character. Equating these two energy terms yields the well-known critical thickness criterion of Matthews [95]. From a force perspective, at this film thickness, the line tension of the misfit dislocation exactly balances the misfit stress on the threading dislocation, leading to no net or excess stress that would lead to dislocation motion. Often, researchers use simple design rules to stay within the critical thickness and avoid the formation of misfit dislocations, limiting the (dimensionless) lattice mismatch-thickness product $\varepsilon h < 0.2$ nm works well for InGaAs/GaAs systems [96].

For the threading dislocation to actually glide to lay down a misfit dislocation, the misfit stress on the threading dislocation needs to exceed the line tension, and it is this excess stress that enters into the phenomenological dislocation glide velocity relationships described in Section 4.3.2. This is an important difference between measurements on bulk crystals and thin films. This excess stress τ_{eff} [97] can be determined as [94]

$$\tau_{eff} bh\sqrt{\frac{3}{2}} = W_{layer} - W_{60}. \tag{4.27}$$

It is understood that epilayers can be grown above the critical thickness in systems where dislocations do not experience any resolved shear along a primary slip system or experience kinetic barriers to glide (Peierls barriers or other obstacles, sluggish nucleation). This is the case for c-plane oriented GaN. On the other hand, when a primary slip plane is active, the high growth temperatures of semiconductors imply that the formation of misfit dislocations from grown-in (preexisiting) dislocations in the substrate closely follows the critical thickness. In most cases, the low density of grown-in dislocations of most commercially available III–V substrates is insufficient to relieve even small amounts of epilayer strain. A commercial GaAs wafer has a dislocation density that is less than 500/cm^2. Thus, even if all these preexisting dislocations glide in response to epilayer strain above critical thickness, most of the epilayer does not experience strain relief. Thus, average relaxation measured by techniques such as X-ray diffraction may show that the films are completely strained even beyond the critical thickness, even though a low density of misfit dislocations may be seen after careful TEM, and may prove detrimental to devices. Continued growth above critical thickness eventually leads to the formation of threading dislocations that are orders of magnitude higher in number than that present in the starting substrate. Dislocation nucleation at surfaces and multiplication at sources (more likely) are key processes here, although it has been difficult to unequivocally identify the precise mechanism in many cases [98,99]. A design rule of $\varepsilon h < 0.8$ nm is specified in the InGaAs/GaAs system to keep existing threading dislocations gliding but minimize the copious nucleation of new dislocations [96].

4.4.2 Dislocations formed in the heteroepitaxy of GaAs on Si

The lattice constant of GaAs is 4% larger lattice constant than that of silicon. While the critical thickness of a planar layer GaAs on Si calculated from Eq. 4.27 less than 1 nm, the growth proceeds actually via 3D nucleation of tiny islands that eventually coalesce [100]. The complex nature of island coalescence leads to threading dislocations, sometimes because of the mismatch but often because of subunit cell misalignments in the position of atoms between neighboring islands [101]. Optimized growth (two-temperature step, [102] use of substrate miscut or annealing to obtain double-steps [103]) of 0.5–1 μm of GaAs on Si typically leads to the formation of 10^9 threading dislocations/cm^2 and the film is nearly relaxed at growth temperature. It is easy to see that this large number of threading dislocations is a geometric consequence of the short segments of misfit dislocations that arise due to island growth mode. An analysis of the character of misfit dislocations at the GaAs/Si reveals a fair number of misfit edge dislocations (90° dislocations) in thicker films [104]. As the glide plane for edge dislocations does not experience resolved shear for (001) biaxial strain (i.e., they are sessile), these edge dislocations are thought to arise by fusion of two 60° dislocations according to Frank's rule. We briefly review strategies to reduce this 10^9 threading dislocations/cm^2 along with their rationale from the perspective of dislocation dynamics.

4.4.2.1 Increase film thickness

Simply growing thicker layers of highly mismatched semiconductors Si leads to a decline in the threading dislocation density (TDD). Fig. 4.16A shows this process for four different combinations, InAs/GaAs, GaAs/Ge/Si, GaAs/InP, and InAs/InP [105]. A $1/h$ scaling law captures the experimentally observed trend in TDD as a function of film thickness [106]. The continued growth of GaAs does not lead to any glide force on threading dislocations. Yet, as these threading dislocations are constrained to $\{111\}$ glide planes and are inclined to the surface, there is a finite probability of two threading dislocations approaching each other. The reaction of two threading dislocations may lead to an energetically favorable (note that Frank's rule is not satisfied for all combinations of Burgers vectors) product of either a single dislocation (fusion) or no dislocation (annihilation). If the threading dislocations are strictly static, the likelihood of two threading dislocations exactly intersecting is vanishingly small. Thankfully, nearly intersecting threading dislocations that could fuse or annihilate impart an attractive force on each other that causes them to glide toward each other to facilitate the reaction. One may wonder about the equally vanishing likelihood that two threading dislocations find themselves on the same glide plane to react. In reality, threading dislocations are able to change their line directions to become screw-like, enabling them to cross-slip on to an orthogonal $\{111\}$ glide plane and intersect [107]. The distance over which the attractive force between two such threading dislocations overcomes the Peierls stress barrier σ_P sets the interaction radius and can be estimated in the low growth temperature limit from a force balance as $r_{int} = \mu b/2\pi\sigma_P$. Speck et al. calculated this so-called annihilation radius to be around 50–100 nm for GaAs and this agrees with experimental fits to the $1/h$ scaling law [106]. It should be possible for dislocations to attract or repel each other over longer distances at higher growth temperatures.

Interestingly, Frank's rule in an FCC lattice (GaAs, Si, etc.) dictates that the fusion reaction and product must involve two glissile and a sessile dislocation [108]. As an example, if two glissile (g) threading dislocations react, they must form a sessile (s) dislocation ($g + g \rightarrow s$). On the other hand, a glissile dislocation reacts with a sessile dislocation to form another glissile dislocation ($g + s \rightarrow g$). This implies that the sample will never run out of glissile dislocations during processes involving fusion reactions, the ratio of glissile to sessile dislocations tends to stay equal. The same is not true for the hexagonal closed packed lattice (GaN, ZnO, etc.), where the sample may run out of all glissile dislocations and prevent further defect reduction [108]. This point is less relevant to strain-relaxation in conventional in c-plane oriented InGaN/GaN quantum-wells where the threading dislocations are largely vertical (only slightly inclined, if at all) and there is no resolved shear due to lattice-mismatch or thermal stresses along the primary slip system of GaN, but it does become relevant for growth along other orientations.

The nature of the scaling law predicts that a reduction in TDD faces diminishing returns with increasing film thickness. Indeed, the dislocation density is reduced to only $10^8/cm^2$ after a 10 μm thick epilayer of GaAs (Fig. 4.17A). In GaAs films grown above ~10 μm, the TDD reduces even less sharply, deviating from the $1/h$ scaling law. In any case, it is impractical to rely on thickness alone as a means to reduce TDD beyond a certain value. Ward et al. remark that for an interaction radius in the range of tens of nm, simply growing thicker films is unlikely to lead to threading dislocation densities $< 10^5/cm^2$ [108]. A more successful approach to reducing the dislocation density lies in effectively increasing this interaction radius to beyond that determined by geometric parameters. We can achieve this by making

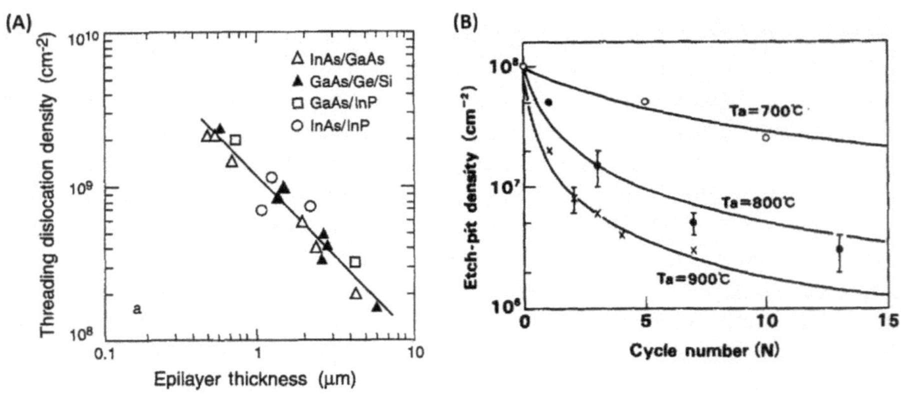

Figure 4.17 (A) Decrease in threading dislocation density seen as a function of increasing the epilayer thickness for different material systems. Threading dislocations interact with each other only within an interaction radius that is on the order of a few tens of nm. This method is most useful in the initial stages of growth, but practical limitations beyond 1−2 μm require additional filtering techniques. (B) Reduction in threading dislocation density in 2 μm GaAs films on Si as a function of thermal cyclic anneal parameters (cycled between high-temperature Ta and room temperature for N cycles), estimated from etch pit density measurements. The solid lines are a fit to a theory of fusion (2 dislocation → 1 dislocation) and annihilation (2 dislocations → 0 dislocations).

Source: (A) Reprinted from P. Sheldon, K.M. Jones, M.M. Al-Jassim, and B.G. Yacobi, Dislocation density reduction through annihilation in lattice-mismatched semiconductors grown by molecular-beam epitaxy, J. Appl. Phys. 63 (11) (1988) 5609−5611, https://doi.org/10.1063/1.340343, with the permission of AIP Publishing. (B) Reprinted from M. Yamaguchi, M. Tachikawa, Y. Itoh, M. Sugo, and S. Kondo, Thermal annealing effects of defect reduction in GaAs on Si substrates, J. Appl. Phys. 68 (9) (1990) 4518, https://doi.org/10.1063/1.346156, with the permission of AIP Publishing.

threading dislocations glide under stress to increase the probability that two dislocations come within the annihilation radius. This may be achieved either using thermal expansion or compositional stress. We discuss this aspect in detail in the following two subsections [109].

4.4.2.2 Thermal cyclic anneals

We saw that the annihilation radius for threading dislocations plays a key role in determining the rate of the $1/h$ scaling law. A mismatch in the coefficient of thermal expansion (CTE) between GaAs and Si (ΔCTE = 3 ppm/K) has been very effectively exploited for this [110]. Cycling the temperature between a low and high temperature limits is the most effective method for a significant reduction in TDD in GaAs/Si films. Using etch pit density measurements, Yamaguchi et al. showed an impressive reduction of nearly three orders of magnitude in TDD in 2 μm GaAs/Si films (Fig. 4.17B) [110]. We assume that sufficient misfit dislocations exist at the GaAs/Si interface to relieve lattice-mismatch strain at the growth temperature (say T_{growth} = 600°C). Increasing the temperature after growth leads to greater lattice-expansion in GaAs with respect to the silicon substrate. This imparts compressive stress in the GaAs film as the silicon substrate is much thicker than the film. The amount of arsenic overpressure available in the system to avoid the formation of Ga-droplets limits the high temperature setpoint T_{high}; 700°C–750°C is close to the limit for MBE and MOCVD films. This increase in strain is quite small (<0.1%) but causes μm-thick films to exceed the critical thickness after a certain ΔT ($T_{high} - T_{growth}$). The increase in annealing temperature leads to a lateral glide of threading dislocations in the film, relaxes the strain, and results in a modified (higher) density of misfit dislocations at the GaAs/Si interface. Researchers hope that during this lateral glide, more threading dislocations interact with each other and react. The interaction radius r_{int} itself does not change for low stresses, only the probability of encountering other threading dislocations increases. Following this step, the substrate is cooled down to temperatures in the range of T_{low} = 400°C – 500°C. Here, the GaAs epilayer contracts more than the Si substrate and leads to the build up of tensile strain in GaAs. The film once again crosses critical thickness, leading to the backward glide of threading dislocations. Cooling the films any further is not useful as dislocation glide is thermally activated and becomes sluggish below 300°C and could lead to cracking. Thermal cyclic anneal becomes impractical beyond 10^7/cm^2 for most *in situ* annealed structures.

4.4.2.3 Dislocation filter layers

In cases where it might be impractical to anneal the GaAs layers at high temperatures for extended periods, lattice-mismatch strain using III–V ternary alloys is used to initiate the lateral motion of threading dislocations in an effort to get them to fuse or annihilate. The misfitting strain obtained by compositional mismatch can be far greater than that from thermal mismatch and thus needs to be introduced carefully. The misfitting layer certainly needs to be above the critical thickness to induce threading dislocation glide, yet the strain should be low enough to not lead to the formation of new threading dislocations by multiplication. The strain-thickness product design rules mentioned preciously help here. Several researchers have shown up to an order of magnitude reduction in the TDD using misfitting layers. Popular designs for GaAs/Si films include single composition GaAsP or InGaAs layers on GaAs, and the so-called strained-layer superlattices (SLS) with each layer above the critical thickness (assuming the previous layer is completely relaxed) [111,112]. Ward et al. measured a dislocation interaction radius of 40 nm for the InAlAs/GaAs system [108]. A systematic study of defect reduction by thickness alone followed by SLS by George et al. reveals a weaker $1/h^{0.5}$ dependence on thickness (as opposed to $1/h$ mentioned by Speck et al.) and interaction radii that is TDD dependent. Fig. 4.18A shows their structure. Each set of 5-period SLS consisting of 10nm-$In_{0.18}Ga_{0.82}As$/10 nm-GaAs pairs yields defect reduction, but the top filters (where the TDD is lower) yield greater filtering ability (Fig. 4.18B). Currently, the glide of threading dislocations from strain induced by both thermal cyclic anneals and mismatched layers appear to be effective. Yet, neither has resulted in the unambiguous demonstration of GaAs/Si with TDD below $10^6/cm^2$. It is unlikely that practical (time, cost) dislocation filters will yield threading dislocation densities much below $10^6/cm^2$, thus a better strategy is to design resilient devices that can tolerate this level of threading dislocations [113].

4.4.2.4 Geometric filters

A distinct approach to blanket growth is aspect ratio trapping (ART) [114–116]. This method is performed via the selective growth of GaAs on Si in trenches through openings in a dielectric mask. The openings have a high aspect ratio and this leads to all tilted threading dislocations terminating at the dielectric boundary. While an aspect ratio of 1.43 is the geometric minimum required to trap threading dislocations on {111} planes, typical implementations use significantly greater aspect ratios. Results from IMEC have shown TDD in GaAs/Si in the mid-$10^6/cm^2$ for much thinner films compared to blanket growth approaches without the need for thermal or strained layer filters [115]. Despite the promise of ART, device designs have to adapt to this unique geometry of tall and slender fins of high-quality semiconductor material and are not as versatile as the blanket films discussed previously. Simple ART designs also do not prevent the propagation of planar defects such as stacking faults. A blanket film may be produced if the crystal is allowed to grow outside of the trenches and coalesce with material from neighboring trenches. This process is called epitaxial lateral overgrowth (ELOG), although studied initially for GaAs on Si [117,118], today it is commonly associated with GaN on sapphire [119]. More recently, ELOG has shown promise in cubic semiconductors with both demonstrations of buffers of GaAs on Si for lasers that have comparable threading dislocation densities as those with blanket filters discussed previously [120,121]. The mechanism of how ELOG filters threading dislocations needs to be considered carefully. McMahon et al. discuss the transition from ART to ELOG in Fig. 4.19 from the perspective of selective area growth, where oxide stripes define the trenches and growth

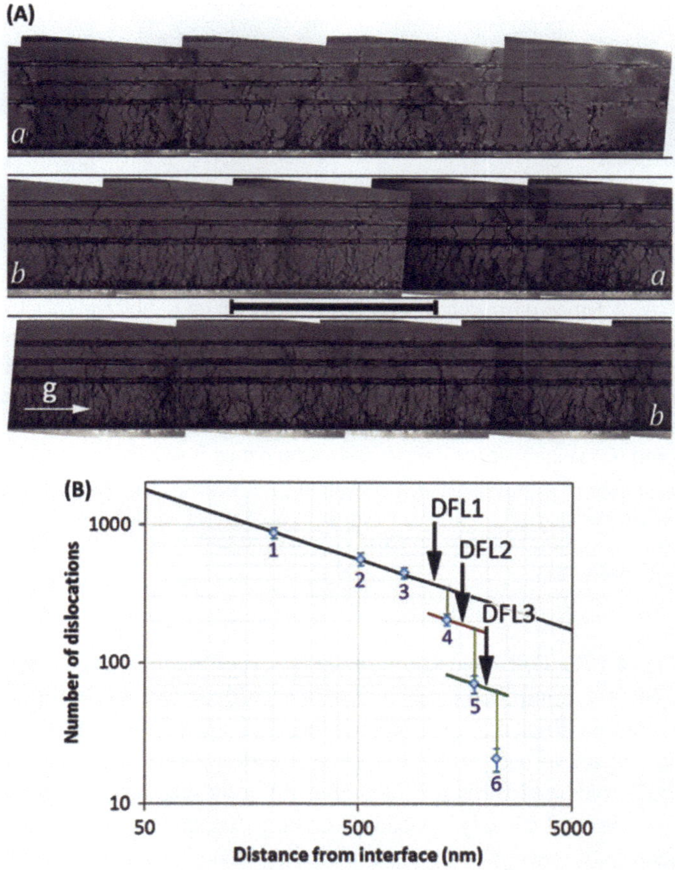

Figure 4.18 (A) Montage of 13 bright field $g = 220$ TEM cross-section images (total width 48 μm) of a GaAs/Si layer with three sets of [(5 ×)10 nm $In_{0.18}Ga_{0.82}As$ – (4 ×)10 nm GaAs] strained layer superlattice defect filter layers (DFLs) grown to reduce the threading dislocation density. The images join at edges a–a and b–b. The scale bar is 5 μm.
(B) Log–log plot showing threading dislocation counts at the six positions in the sample. A fit to the first three points is shown using a power law $\rho_{TD} = Ah^{-0.5}$.

Source: Reprinted from I. George, F. Becagli, H.Y. Liu, J. Wu, M. Tang, and R. Beanland, Dislocation filters in GaAs on Si, Semicond. Sci. Technol. 30 (2015) 114004, https://doi.org/10.1088/0268-1242/30/11/114004, copyright 2015, with permission from IOP Publishing, Ltd.

occurs exclusively in the trenches [41]. In Fig. 4.19A, we see what is akin to an ART process, with the dislocation (dashed red line) allowed to terminate at the oxide stripe. At this stage, the top of the "epipillar" can indeed be free of dislocations. However, if the epipillars are allowed to grow out of the trench and coalescence into a blanket, akin to ELOG, the blocked threading dislocations should, in principle, re-emerge on the other side of the oxide stripe (Fig. 4.19B). This is in contrast to the incorrect assumption (Fig. 4.19E) that dislocations can be terminated at interior

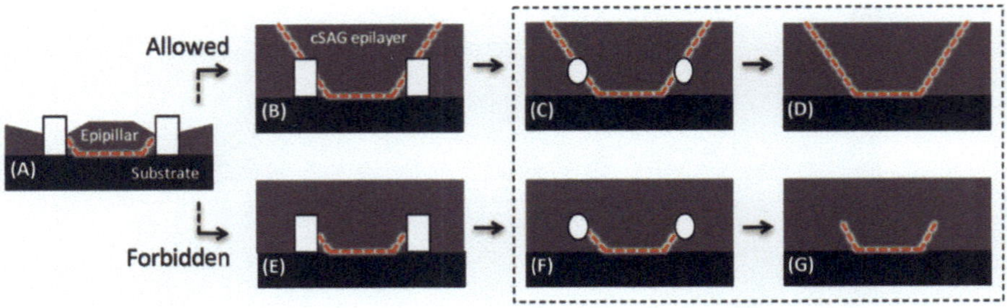

Figure 4.19 (A) A lattice-mismatched layer (gray) is grown on a substrate (black) in trenches defined by an oxide pillar (white). The dislocation in the crystal in the trench (epipillar) can be trapped by the oxide trenches. (B) However, continued growth of the crystal out of the trenches results in the re-emergence of the dislocation. (C) and (D) The mechanism of the process by replacing the oxide layer with a hypothetical crystalline material. (E) A forbidden situation, (F) and (G) tantamount to a dislocation terminating inside the crystal.

Source: Reprinted from W.E. McMahon, M. Vaisman, J.D. Zimmerman, A.C. Tamboli, and E.L. Warren, Perspective: fundamentals of coalescence-related dislocations, applied to selective-area growth and other epitaxial films, APL Mater. 6 (12) (2018) 120903, https://doi.org/10.1063/1.5047945, with the permission of AIP publishing.

interfaces. Fig. 4.19C, D, F, and G are "thought experiments" to understand why this is the case. Slowly minimizing the size of the oxide stripe while replacing it with crystalline material shows that the forbidden case is equivalent to a dislocation terminating in the interior of a crystal. That said, threading dislocation reduction can still occur in ELOG/coalesced films for altogether different reasons. One possible mechanism could be several closely spaced threading dislocations trapped at the lower oxide interface may emerge as a single threading dislocation (with a Burgers vector that is the sum of the individual burgers vectors of the trapped threading dislocation). In summary, dislocations are topological defects and any defect reduction mechanism needs to satisfy a simple set of rules or run the risk of being unphysical. Thankfully, there is now a greater appreciation of this in both the crystal growth and device design community.

4.5 Recombination-enhanced processes

We have discussed several thermally activated phenomena involving point defects and dislocations in Sections 4.2–4.4. We now move toward understanding what happens when nonradiative electron or hole recombination at a defect generates phonons or lattice-vibrations that enhance some of these otherwise thermally-activated processes via what is appropriately termed REDRs. REDRs include enhanced point defect reactions, as well as enhanced dislocation glide and climb. At the outset, much less is known about exactly which steps are enhanced and which remain unaffected even for simple multistep reactions such as point defect

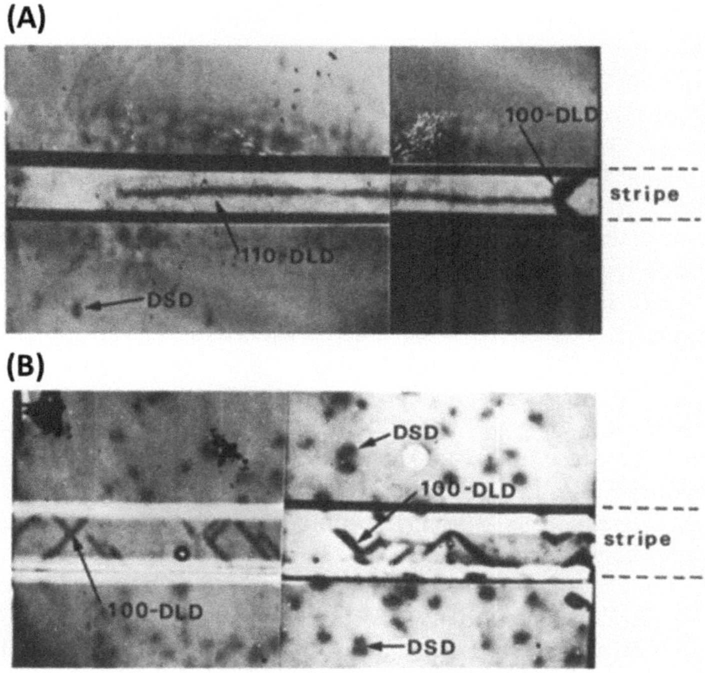

Figure 4.20 Electron beam-induced current images of dark spot defects (DSDs) and dark line defects (DLDs) corresponding to different dislocation structures in a stripe GaAs/AlGaAs double heterostructure laser. Note the two orientations of DLDs, (A) ⟨110⟩ and (B) ⟨100⟩, and how they only form on the laser stripe.

Source: Reprinted from K. Ishida and T. Kamejima, TEM study of dark line defect growth from dislocation clusters in (GaAl)As-GaAs double heterostructure lasers, J. Electron. Mater. 8 (1) (1979) 57–73, https://doi.org/10.1007/BF02655641 by permission of the Journal of Electronic Materials, copyright 1979.

diffusion. Yet, it is reasonable to expect that REDR should only enhance reactions for which there is an underpinning thermodynamic driving force. In other words, REDR might be thought of as a catalyst for a reaction. Fig. 4.20 shows a rapidly degraded GaAs/AlGaAs laser due to REDR [122]. Here, dark spot defects, corresponding to threading dislocations, undergo recombination-enhanced climb (REDC) and glide (REDG) processes to yield ⟨100⟩ and ⟨110⟩ oriented dark line defects (DLDs), respectively. This is evident by defect transformation only occurring under the laser stripe. This section focuses on the basics of this process and understanding what makes certain material systems more tolerant to REDR failure.

4.5.1 Background

Stoneham categorizes recombination-enhanced processes as those that are enhanced by increasing the carrier density [123]. These carriers may be created by light, ionizing radiation, or electrical injection, however, the net effect remains the same.

Light-induced reactions have been known in wider gap soft ionic solids for over a century (such as silver halides used in early photography). However, they are unexpected in the more covalently bonded semiconductors like GaAs. This is partly because the energy for typical defect creation and motion in semiconductors can be larger than the bandgap [124]. Nonradiative recombination of an electron and a hole does not yield enough excess energy to create many defects; for example, GaAs used in vertical cavity surface emitting lasers (VCSELs) has a 1.4 eV bandgap energy, but requires 2.0 eV for atomic motion, leaving 0.6 eV of additional thermal energy required for defect motion. Further, the electron–phonon coupling is much weaker in semiconductors and this does not present a way to localize energy for a reaction (such as via self-trapped excitons in ionic materials). Perhaps this is why we find that recombination-enhanced processes manifest via trap states at preexisting defects in semiconductors, where the energy required to excite defects to metastable states or enhance their diffusion is also lower.

Lang has reviewed several candidate mechanisms behind how nonradiative recombination activates the movement of atoms [124]. Stoneham categorizes them as (1) local heating models, (2) Bourgoin-Corbett (or saddle-point) mechanism, and (3) local electronic excitation models [123]. We only discuss the local heating model here, more correctly termed the phonon-kick model, as this is the most likely candidate for REDR of point defects and dislocations in GaAs [125]. Dean and Choyke specifically mention that a phonon kick is not the same as local heating or a thermal spike. In the case of local heating, the energy released upon carrier capture is equipartitioned over the local phonon modes of the defect. The phonon-kick model refers to energy specifically directed into select local phonon modes that more efficiently use the released energy toward atom motion.

4.5.1.1 The phonon kick model

The trapping or detrapping of charge at a defect leads to a change in the surrounding distribution of charge, and hence, a change in the position of the neighboring atoms. This is known as a spatial configuration change [126]. It is useful to visualize this process by plotting the total energy of the system (elastic + electronic) as a function of this spatial configuration change using a 1D representation of the atomic configuration. Of course, it is not obvious what the best method is to convey the complex 3D motion of all neighboring atoms to 1D. Sumi shows how to describe a configuration coordinate Q as a fixed linear combination of various normal mode coordinates (each with a coordinate q_i and angular frequency ω_i) [127]. As an example, the change in the nearest-neighbor distance for a defect with equally spaced atoms would represent a breathing mode [126]. The adiabatic approximation then allows for calculation of the electronic energies of this system with the ion cores remaining frozen.

Following Lang, we show a set of configuration coordinate diagrams for three states of a single defect D in Fig. 4.21 [124]. In the ground state, the defect is neutral D^0, the total energy curves resemble parabolas as bonds are treated like elastic springs for small deviations away from the lowest energy position Q_0. Vibrations in

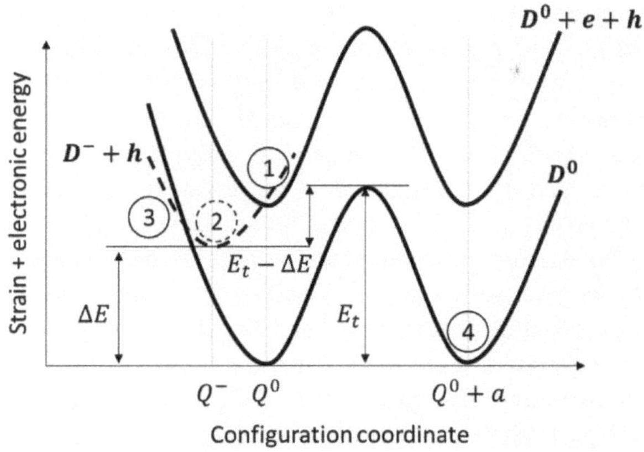

Figure 4.21 A configuration coordinate diagram illustrating the phonon kick model for recombination-enhanced defect migration of an electron trap. The neutral defect D^0 captures an electron, and then captures a hole to complete nonradiative recombination (sequence 1−2−3). The thermal barrier E_t is lowered by a value ΔE, allowing the defect to migrate to the next atomic position (sequence 4).

Source: Adapted from D.V. Lang, Recombination-enhanced reactions in semiconductors, Annu. Rev. Mater. Sci. 12 (1982) 377−398, https://doi.org/10.1146/annurev.ms.12.080182.002113.

the state lead to excursions away from Q_0, dictated by the shape of the parabola (the stiffness of the bonds). The energy of this defect state $D^0 + e + h$ with a delocalized electron and hole in the conduction and valence bands, respectively, looks similar except being shifted/raised by the band gap energy. Here, we assume that the electron and hole are at the band edges. Finally, the defect $D^- + h$ is shown having trapped an electron (localized), D^-, along with a delocalized hole (still at the valence band edge). This corresponds to a change in the atom configuration around the defect.

With this setup, we are now in a position to see how the phonon kick enhances atomic migration over a thermal barrier height E_t following nonradiative recombination at the defect. The points of intersection of these various curves are particularly important as these represent transition pathways when the system loses energy via vibronic modes (heat, no light). First, the $D^0 + e + h$ state decays into the $D^- + h$ pathway. Note that switching between the curves of these different states occurs at the same configuration coordinate, representing a sudden change in the electronic configuration but no change yet to the atom configuration. At this point, the system finds itself in an excited vibrational state of the $D^- + h$, slowly arriving at the low-energy configuration of this state by emitting a number of phonons. The defect then captures the hole, transitioning the system to a highly excited vibrational state of the D^0. Here, it is now possible that additional thermal vibrational energy will just push the system over the barrier to a neighboring lattice point in the crystal. The perceived barrier height is $E_t - \Delta E$. Typically, there is only a small probability that any particular recombination event would lead to the movement of the defect. This is characterized by an efficiency prefactor. Lang estimates this prefactor to be about 10^{-3}–10^{-5} for characteristic REDRs in semiconductors.

In Fig. 4.21, we assumed that the coordinate Q of lattice distortion due to carrier recombination (the phonon kick) is the same as that of the defect reaction (atom migration). This need not be the case in general. Sumi has presented an analysis for when the energy-accepting coordinate is at an angle to the reaction coordinate in phonon-coordinate space [127]. This sets to what extent the energy ΔE goes toward lowering the barrier E_t. We can even imagine an extreme situation where phonons emitted following recombination are all vibrating the defect atom orthogonal to the path of migration and no enhancement occurs due to recombination.

In summary, the electron–hole recombination energy supplements the available thermal energy to overcome the barrier to motion. It is not necessary that all of the recombination energy is channeled toward overcoming the barrier, hence, the difference between the thermally measured barrier and the recombination-enhanced barrier can be at most the recombination energy, but is usually smaller. Further, the prefactor of the reaction is very slow due to an efficiency factor associated with capture. This feature distinguishes it from the other models mentioned here. Maeda suggests that the phonon-kick model most likely explains how REDG (and possibly REDC) occurs [128].

4.5.2 Recombination-enhanced dislocation glide

4.5.2.1 Features of REDG

Unique to semiconductors, it is possible for minority carrier recombination at dislocations to lead to enhanced glide of dislocations in the presence of excess stress. We reiterate that REDG only enhances the tendency of the dislocation to glide in the presence of stress (the thermodynamic driving force), thus all the geometric rules of thermal dislocation glide (critical thickness, glide planes etc.) remain valid. This process has been studied extensively in bulk crystals of Si, Ge, GaAs, and InP, all showing a characteristic reduction in the activation energy for glide. The essential features of REDG behavior are summarized by an equation with two terms: [50]

$$v = v_0\left(\frac{\tau_{eff}}{\tau_0}\right)^m \exp\left(-\frac{E_t}{kT}\right) + v_e\left(\frac{\tau_{eff}}{\tau_0}\right)^m \left(\frac{I}{I_0}\right)^\gamma \exp\left(-\frac{E_t - \Delta E}{kT}\right). \tag{4.28}$$

The first term captures the pure thermally activated glide (in the absence of minority carriers) akin to Eq. 4.9, and the second term represents the enhancement by REDG. The prefactor of the second term is approximately linear with the minority carrier density $I(\gamma \approx 1)$. The reduction in activation energy ΔE is always less than the bandgap but the exact value depends on the core chemistry (α, β, or screw) of the dislocation. Fig. 4.22A shows the temperature dependence of thermal glide (under dark conditions) and REDG (electron-beam irradiation) velocities for the α, β, and screw dislocations in externally-stressed bulk n-GaAs determined using cathodoluminescence videos. A dramatic reduction in the activation energy and increase

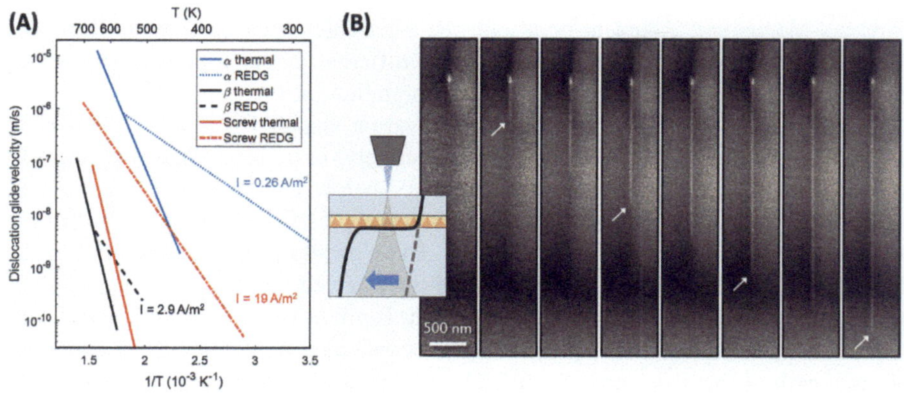

Figure 4.22 (A) Temperature dependence of the thermal (dark) and recombination-enhanced (electron beam) glide velocities of α, β, and screw dislocations in bulk n-GaAs crystals under an externally applied stress of 26 MPa. Note that without pumping through the beam current, dislocations only can grow at very high temperatures they would likely never see. (B) Electron channeling contrast images of recombination-enhanced dislocation glide of a threading dislocation at room temperature laying down a misfit dislocation in GaAs/InAs quantum dot/GaAs/Si layers in response to thermal-expansion mismatch-induced tensile strain (about 150–200 MPa). The inset shows a schematic of this process under e-beam excitation.

Source: (A) Adapted from K. Maeda and S. Takeuchi, Recombination enhanced mobility of dislocations in III-V compounds, J. Phys. Colloq. 44 (C4) (1983) C4-375-C4-385, https://doi.org/10.1051/jphyscol:1983445, by permission of EDP sciences. (B) Adapted from J. Selvidge et al., Defect filtering for thermal expansion induced dislocations in III-V lasers on silicon, Appl. Phys. Lett. 117 (2020) 122101, https://doi.org/10.1063/5.0023378.

in glide velocities at low temperatures is seen for all dislocations. We can directly visualize this process using electron channeling contrast imaging, another scanning electron microscope technique. In Fig. 4.22B, we see a plan-view image of a threading dislocation (dot) in a GaAs/InAs quantum dot/GaAs/Si structure lay down misfit dislocations at the InAs interface under the action of REDG [129]. However, as we have mentioned, REDG only enhances an otherwise sluggish reaction that has a thermodynamic driving force. Here, the driving force for dislocation glide is provided by thermal expansion mismatch induced stress between III-V and silicon.

4.5.2.2 Mechanism of REDG

Maeda suggests that the phonon-kick model likely explains how REDG occurs [128]. We have seen in Section 4.3.3 that double kink formation and migration over Peierls barriers are both thermally activated processes that control the glide velocity of dislocations. An enhancement of glide velocity following nonradiative recombination could arise from a lowering of the barrier for either or both of these processes. However, these are not identical processes. Enhancing double kink formation requires nonradiative recombination at a site along a straight segment of dislocation. On the other hand, enhancing kink migration requires nonradiative recombination at the kink. As Maeda and Takeuchi point out, the electronic states involved in these two recombination events are different and hence different amounts of barrier lowering can be expected for kink formation and kink migration [50].

In Fig. 4.22A, Maeda shows that the activation energy for glide is reduced by 0.7 and 1.1 eV for α and β dislocations, respectively, in n-GaAs. This is remarkably similar to the energies of the electron (0.68 eV below CB) and hole traps (1.1 eV below CB) measured in these dislocations and suggests an efficient utilization of the phonon-kick toward the defect reaction. Experiments consistently show a linear dependence of enhanced glide velocity and the injected current. This suggests that at least the double kink formation is enhanced. It might be possible to know if kink migration is also enhanced by performing glide velocity measurements as a function of the length of the dislocation. In n-GaAs, Maeda suggests that thermal dislocation glide of α-dislocations operate in the kink-collisionless regime and further suggests that only kink formation is enhanced by carrier injection. Enhanced dislocation glide at distances greater than the injected minority carrier diffusion length provides further proof for kink-migration not being enhanced. This is possible when double kink formation is enhanced at the site of nonradiative recombination close to injection, but then this kink migrates thermally deeper into the crystal [50].

4.5.2.3 Impact on lasers

Several researchers have proposed that $\langle 110 \rangle$ DLDs form via REDG of 60° dislocations in GaAs-based lasers. Remember that the $\langle 110 \rangle$ direction is simply a consequence of the FCC lattice where such glissile dislocations in GaAs have low energies, irrespective of whether glide occurs thermally or via REDG. A key requirement for REDG is the thermodynamic driving force in the form of stress in the film that exceeds the critically resolved shear stress. Without this, REDG would only wiggle the dislocation back and forth rather than extend it in one direction. Stress exceeding the critical limit can be attained even in lattice-matched systems due to device fabrication steps such as oxide stripe or apertures. Consequently, $\langle 110 \rangle$ DLDs can often be eliminated by simply minimizing strain from thin-film deposition steps such as dielectrics or metals. In other systems like III–V/Si, thermal expansion mismatch induces stresses significantly beyond the critical value, and REDG has been observed in such structures in the absence of fabrication-induced stresses (Fig. 4.22B) [107,130]. $\langle 110 \rangle$ DLDs are sometimes a source of the more extant $\langle 100 \rangle$ DLDs that form via recombination-enhanced dislocation climb [131]. We see this in Fig. 4.20A (right side).

4.5.3 Recombination-enhanced dislocation climb

4.5.3.1 Features of REDC

With our discussion of the energetics of point defects and dislocations, we can now appreciate the details of enhanced climb of dislocations upon minority carrier injection (and nonradiative recombination). In almost all cases, researchers have determined that the nature of climb is interstitial in nature, that is, atoms are attaching to the extra plane corresponding to the dislocation. This does not imply that the system has excess interstitials, as both interstitial absorption and vacancy emission would result in "interstitial climb." We also assume that the sluggish point defect migration toward or away from the dislocation needed for the climb is also enhanced by nonradiative recombination now occurring at the point defect.

DLDs due to dislocation climb occurs predominantly along the ⟨100⟩ direction in (001)-normal oriented lasers (i.e., 45 degrees-off of the cleaved facets), in contrast to those from REDG along the ⟨110⟩ directions (i.e., parallel to the cleaved facets). Fig. 4.20 shows this clearly. This DLD network tends to stay confined to the active region, likely due to being the region where most of the minority carriers recombine. Hutchinson and Dobson proposed a simple model to explain why DLD growth via dislocation climb leads to ⟨100⟩ growth [132]. We show this in Fig. 4.23A and B. Recall that pure climb lengthens the dislocation in the plane normal to the Burgers vector (climb plane shown in red). Threading dislocations generated during epitaxy of GaAs are mostly of 60° character, as an example with a Burgers vector $\vec{b} = a/2[011]$. Accordingly, the plane of pure climb is the (011) plane. As the threading dislocation climbs in this plane by expanding the half plane associated with the dislocation, portions of the dislocation line exit the active region and stop climbing, leaving behind horizontal segments of dislocation along both the top and bottom active region/clad interfaces (Fig. 4.23B). The orientation of these horizontal segments is the line intersection of the climb plane and the (001) confining active region/clad interfaces. This is the [100] direction. The only portion that continues to climb is the tiny threading segment that remains in the active region connecting the top and bottom ⟨100⟩ horizontal segments (Fig. 4.23B, blue arrow). Such a structure is known as a dislocation dipole as they are two very closely spaced horizontal dislocations with opposite line directions and yet have the same Burgers vector as they are a single dislocation (analogous to two dislocations with the same line direction and opposite Burgers vector).

According to Hutchinson and Dobson's model, a single threading dislocation should yield only a slender extended dipole along the ⟨100⟩, with minimal lateral expansion on the (001) plane. Experimentally, a single threading dislocation is seen to yield rather extensive dislocation structures with many dipole elements [133,134] (see Fig. 4.23C). Fig. 4.23D shows a TEM image of even more complex dislocation dipoles formed from a cluster of dislocations D1 − D4 that originate at an inclusion in an AlGaAs/GaAs laser. One can now appreciate the complexity of such a structure—the sum of the Burgers vector of this cluster must be zero (Frank's rule, since they originated together). This sets not only the respective climb planes but also the sense of helical twisting in the formation of the dipoles from these clusters (see Fig. 4.14C and D for comparison). Fig. 4.23E shows a sketch of this cluster, with the line direction sense and climb planes indicated.

While threading dislocations from substrate defects were significant in the early days of lasers in the 1970s, they are not an issue with the nearly defect-free native substrates available today. More likely sources of DLD nucleation are "punch out"

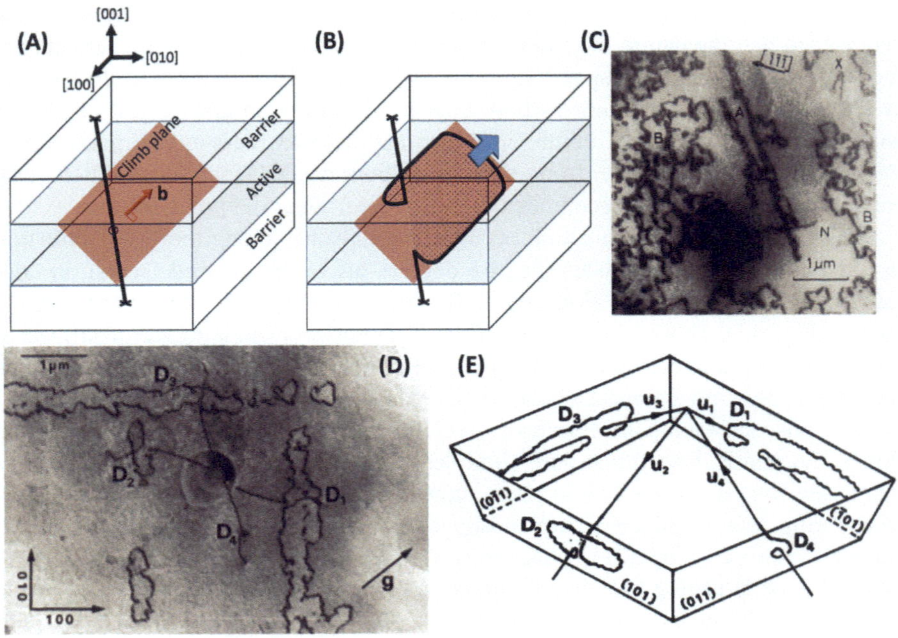

Figure 4.23 Process for recombination-enhanced dislocation climb leading to the formation of a dislocation dipole. (A) A threading dislocation is shown with a Burgers vector b = a/2 [011]. The climb plane for such a dislocation (orthogonal to $\vec{b}$) is shown in red. (B) As climb proceeds along this plane, the horizontal dislocation segments cross the active layer/barrier interface, following which there is a rapid reduction in the amount of minority carrier recombination. Hence, dislocation climb stops (or slows down significantly) beyond the barrier but continues elsewhere (blue arrow). This leads to ⟨100⟩ oriented dipoles (C) TEM image of a dislocation network "A" that shows several dipole arms that originate from a threading dislocation in a degraded GaAs/AlGaAs laser. (D) TEM image of a dislocation cluster (D1−D4) originating from an inclusion in a GaAs/AlGaAs laser, the sum of the Burgers vectors of this cluster has to equal zero. These dislocations form ⟨100⟩ dipoles. (E) Sketch of the growth of these dipoles along the climb planes for each of these dislocations.

Source: (A) and (B) Adapted from P.W. Hutchinson and P.S. Dobson, Defect structure of degraded GaAlAs-GaAs double heterojunction lasers, Philos. Mag. J. Theor. Exp. Appl. Phys. 32 (4) (1975) 745−754, https://doi.org/10.1080/14786437508221617. (C) Reprinted from P.W. Hutchinson and P.S. Dobson, Defect structure of degraded GaAlAs-GaAs double heterojunction lasers, Philos. Mag. J. Theor. Exp. Appl. Phys. 32 (4) (1975) 745−754, https://doi.org/10.1080/14786437508221617, with permission of Taylor & Francis Publishers. (D) and (E) Reprinted from K. Ishida and T. Kamejima, TEM study of dark line defect growth from dislocation clusters in (GaAl)As-GaAs double heterostructure lasers, J. Electron. Mater. 8 (1) (1979) 57−73, https://doi.org/10.1007/BF02655641, with permission of the Journal of Electronic Materials, copyright 1979.

dislocations that occur due to mechanical stress or cracking, such as occurs at the interface between the semiconductor and oxidized layers in oxide VCSELs or at sharp corners (stress concentrators). Punch out dislocations are not a new type of defect, they are still conventional glissile dislocations and the name only refers to their source of punched in segments of crystal surfaces that lead to prismatic dislocation loops that travel into the crystal along a glide cylinder. These dislocations yield extensive dipole networks that are connected with ⟨110⟩ segments lying along the n-cladding/active interface and each dipole has a tiny threading segment at its tip, which continues to climb to elongate the network. Single 60° threading dislocations in VCSELs result in rather different lateral dislocation loop networks, while still confined to the active region and now have a predominantly (001) orientation normal [135,136]. This would not be possible by continuous climb or pure glide alone as neither the glide planes nor the climb planes correspond to (001). A model to explain this proposes that if any segment of the dislocation dipole stuck at the confining active region ends up with a screw-character, they can cross-slip by glide onto another {111} planes and "reflect" back into the active region to continue to climb until it reaches the opposite active region interface [137]. Multiple cross-slip events of the dislocation [138], akin to bouncing of the dislocation between the top and bottom active region interfaces can yield an overall (001)-oriented dislocation loop. As recombination-enhanced glide (via cross-slip of screw segments) is invoked, stress must play the differentiating role in the formation of dislocation networks in VCSELs versus GaAs/AlGaAs broad area lasers. The high stress generated around the oxide aperture of a VCSEL might play a key role here.

We note one difference between observations of recombination-enhanced climb and thermal climb. In the former, the dislocations are considerably longer as a result of climb (see the dipoles for example in Figs. 4.19 and 4.22). Lengthening of dislocations comes with a significant energy penalty, and should reduce the driving force for climb if the osmotic force due to point defects being out of equilibrium drives climb. Due to the extraordinary length of dislocations in DLD networks, we suspect that such networks in a laser are not stable when the laser is not operating (and can be reduced by annealing); perhaps they are diffusion-limited aggregates that remain frozen much like self-organized or dendritic structures that form in nature [139]. This is an open area for research.

4.5.3.2 Origin of point defects

There is debate regarding the source of these extra atoms in REDC. Researchers have proposed extrinsic and intrinsic models, so called based on whether the point defects involved in climb preexist in the material or not. The intrinsic model first proposed by O'Hara suggests that there are no preexisting interstitials. Nonradiative recombination at the dislocation leads to dislodging of atoms in the immediate lattice to form interstitial atoms that go on to attach to the dislocation core [140]. This simultaneously leaves behind vacancies that diffuse outwards into the bulk. The tendency for interstitials to be more strongly attracted to the strain field of dislocations (known as the interstitial bias in metallurgy [141]) is invoked to explain why vacancies do not attach to the dislocation core instead. One of the main arguments in its favor is that the intrinsic model does not require a high density of preexisting interstitials that would have to be present in the competing extrinsic model. We demonstrate using a back of the envelope calculation. Fig. 4.23C shows an extensive interstitial network of dislocations, nearly covering the entire active layer plane. Such a network corresponds to roughly 10^{15} atoms/cm^2 that must have diffused to (or away from) the dislocation from the vicinity of the active region where minority carrier recombination occurs. If this source of point defects is taken to be 100 nm around the active region, we obtain a rather large concentration of point defects on the order of 10^{20}/cm^3 connected to the dislocation network, a rather large number that have not been detected using any other techniques.

On the other hand, researchers have questioned the validity of the intrinsic model from the standpoint of thermodynamics and kinetics [13,142]. The maximum available energy for an REDR process is the band gap energy and this is lower than the formation energies of a pair of vacancies and interstitials (a Frenkel defect). We must keep in mind that dislocation climb in the compound semiconductor GaAs requires both a gallium atom and an arsenic atom to maintain lattice polarity (see Fig. 4.13). This then requires the rate of generation of gallium and arsenic point defects to be equal. Even if the energy liberated following REDR is enough to create these point defects, no explanation is provided as to why these vacancies simply do not react with the interstitials instantly and annihilate. The fact that the interstitials attach to the dislocation and increase the total length and energy of the dislocation suggests that there must be a counterbalancing reduction in energy elsewhere as a result of liberating vacancies into the bulk. A complete intrinsic model that results in an overall lowering of the energy of the laser active region is missing.

The extrinsic model attempts to overcome these issues by assuming a preexisting out-of-equilibrium population of point defects, the removal of which provides the requisite thermodynamic driving force for REDC. One example of this driving force in LPE-grown GaAs is the removal of excess gallium interstitials from the bulk, proposed by Petroff and Kimerling [65]. We have seen that interstitial point defects are energetically expensive, and the system can use the dislocation as a sink for these interstitials. These point defects equilibrate at the dislocation via interstitial climb. However, the source of the arsenic atoms needed to complete climb comes via the dislocation expelling arsenic vacancies into the bulk. We have evaluated the driving force of this very process in Section 4.3.5 (see Eq. 4.23). The extrinsic model overcomes the need for requiring an excess of both gallium and arsenic point defects this way, still being an overall feasible reaction because of the rather large formation energies (hence low equilibrium concentrations) of interstitial atoms in semiconductors. One other point that led Petroff to the extrinsic model was the experimental observation that degradation was more rapid in group V-poor epitaxy, a result that was confirmed by Waters and Hill, when they saw that reliability was degraded for low V/III ratios in MOCVD-grown epitaxy [143].

The extrinsic model is certainly not yet validated as the preexisting excess point defects have not been found. Interstitial climb is seen in MBE- and MOCVD-grown AlGaAs/GaAs devices, and yet gallium interstitials are unlikely to be present at high concentrations in such growth techniques due to the arsenic-rich conditions used here (see Fig. 4.2C). Later, Petroff suggests that interfaces between the active region/barrier may serve as a source for point defects based finding dislocation loops there in aged lasers [42]. This may be possible as interfaces may be local variations in stoichiometry, but systematic studies linking interface preparation and reliability are not available. Indeed, the formation of dislocation loops first at interfaces may simply be a consequence of higher nonradiative recombination occurring at interfaces.

To reconcile these differences, Tan and Gosele propose a modified intrinsic model that provides the missing driving force in O'Hara's model for vacancies to be expelled into the bulk. They suggest that sample cool down after growth quenches in a nonequilibrium (deficit) population of gallium vacancies in n-type GaAs due to the Fermi level effect [13]. We have described this in Section 4.2.1. They propose that dislocations then serve to remove this deficit by absorbing gallium atoms from the lattice (as interstitials) and expelling gallium vacancies into the bulk. They do not discuss the arsenic atoms necessary for climb. However, as we saw in the extrinsic model, a net osmotic force for climb is still possible even if the arsenic vacancies are in equilibrium. Today, research clarifying the extrinsic and intrinsic models has all but halted and dislocations no longer plague GaAs-based lasers except for a very small minority of cases that can be rooted out with proper burn-in procedures in laser reliability. Perhaps, understanding REDC may become relevant again with the rise of III–V on silicon lasers where dislocations cannot be eliminated.

4.5.4 Tolerance to rapid degradation

The choice of "material system" (meaning substrate, cladding materials, and active region), is one of the most important decisions in building semiconductor lasers that are resistant to rapid degradation. There is a wealth of data suggesting indium-containing lasers have a greatly reduced likelihood of rapid degradation. InP, InGaAsP, and InGaAs (greater than 5–7% indium) lasers show improved tolerance to REDC compared to GaAs-rich lasers [144,145]. Additionally, Waters et al. show that compressively strained InGaAs QW grown below the critical thickness (pseudomorphic) are immune to rapid degradation compared to similarly tested GaAs QW devices [146]. Ueda has identified key material parameters that may play a role in setting the impact of REDR processes [147]. They are (1) bandgap energy, (2) electronic states associated with dislocations and point defects, (3) thermodynamics of defect formation, both dislocations and point defects, and (4) external and internal strain. We first compare unstrained GaAs and InP devices using these bulk material criteria that set REDR. We will then address the impact of strain and dilute indium discussing the InGaAs QW results.

4.5.4.1 Bandgap energy

This sets the upper bound on the amount of energy deposited into the lattice via nonradiative recombination. A narrow gap semiconductor would have less energy available to aid in point defect migration or dislocation growth. Ueda, Waters, and others mention this as one reason why lower bandgap materials would naturally be more tolerant. This fails to explain why InP (1.34 eV) or InGaAsP (1.5 eV) either show no or very slow REDC, despite almost the same bandgap as GaAs (1.42 eV).

4.5.4.2 Electronic states associated with defects

It is conceivable that if defects have reduced nonradiative recombination, the kinetic enabler of REDR is turned off. Dow and Allen, calculating that anion vacancies have no states in the gap for the InP-rich compositions of the InGaAsP, carried forward this vein of thought [148]. More recently, Park et al. also calculate that the energy level of recombination centers introduced by α dislocations finding them to depend sensitively on the indium fraction in InGaAs; harmful in GaAs, but benign in InAs [149]. They show that β dislocations remain harmful with recombination states in the gap even for InAs. The relevance and accuracy of such calculations to actual devices is not clear. Certainly both MBE- and MOCVD-grown InP show evidence for deep levels (electron traps) adjudged by DLTS. Likewise, the III–V photovoltaic literature has confirmed that threading dislocations are definitely nonradiative in both InP [150] and InGaAs [151]. At present, there is not enough experimental evidence to suspect that defects are benign in these degradation-tolerant materials, so we look to other possible aspects to explain why the 1.3 and 1.55 μm InGaAsP alloys are indeed resistant to DLDs via dislocation glide and climb.

4.5.4.3 Thermodynamics of defects

Ueda has highlighted a very interesting correlation between the existence of prismatic dislocation loops (loops that grow via climb from point defect coalescence, see Fig. 4.14A and B) in heavily doped bulk crystals of a particular semiconductor and rapid degradation of lasers of the same material [147]. GaAs, AlGaAs, GaAsP, etc. all show evidence for dislocation loops in quenched bulk crystals, whereas InP and InGaAsP do not. Two factors are necessary for prismatic loops to form—an excess of point defects underscoring a thermodynamic driving force to precipitate out and form a dislocation loop, and sufficient point defect mobility to enable the defects to come together to coalesce into this dislocation loop. Density functional theory calculations have revealed that the energy required to form indium vacancies in InP (n-type, phosphorus rich) is much greater than gallium vacancies in GaAs (n-type, arsenic rich) [152]. Thus, at any temperature, fewer indium vacancies are expected, significantly reducing the tendency for loop formation.

Annealing radiation-induced damage also reveals key differences between GaAs and InP regarding defect migration. Radiation can create nonequilibrium quantities of simple point defects such as individual vacancies and interstitials (as opposed to clusters). Yamaguchi showed that the radiation damage-annealing rate by thermal means has an activation energy of 1.75 eV in GaAs, but only 1 eV for InP [153]. This is ascribed to the lower barriers to migration of point defects in InP. To find no dislocations loops in InP bulk crystals despite the possibility of faster point defect migration, suggests that the number of point defects must be very low. Interestingly, the radiation-induced lattice damage can also be healed by minority carrier injection [153,154]. Here, the thermodynamic driving force for healing has origins in the removal of these nonequilibrium point defects. While the activation

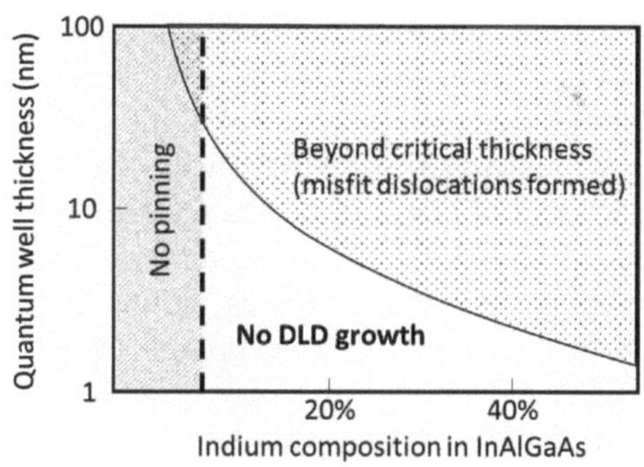

Figure 4.24 The effect of indium and strain on laser reliability are shown. DLDs have been successfully stopped in GaAs-based lasers by introducing 5% − 7% indium in the quantum wells, inducing compressive strain. The addition of indium below this threshold value (dotted) line does not prevent DLD growth. However, introducing too much indium causes the QW to cross the critical thickness for misfit dislocation formation (solid line).

Source: Adapted from J.A. Baumann, R.J. Dalby, R.G. Waters, S.L. Yellen, C. Harding, and A. Shepard, Growth of InAlGaAs strained quantum well structures for reliable 0.8 μm lasers, J. Electron. Mater. 23 (2) (1994) 207−216, https://doi.org/10.1007/BF02655271, with permission of the Journal of Electronic Material, copyright 1994.

energies were greatly reduced in both GaAs and InP due to REDR, GaAs once again had a higher activation energy of 0.98 eV compared to just 0.13 eV for InP when subject to carrier injection. This low migration energy has been credited for exceptional radiation hardness of indium-containing solar cells, where room temperature defect annealing is said to take place. Yamaguchi notes that the damage constant pertaining to radiation hardness of a solar cell trends with the "fraction of InP" in the crystal, GaAs > $In_{0.32}Ga_{0.68}As_{0.34}P_{0.66}$ > $In_{0.5}Ga_{0.5}P$ > InP [154]. These results are not unlike those of laser reliability.

4.5.4.4 External and internal strain

We have seen that strain has a profound influence on thermal glide and climb of dislocations. It can also modify the equilibrium point defect concentrations and their thermal migration energies. Therefore it is no surprise that strain is key to many reports of rapid degradation modes in lasers. We first discuss the clearest evidence of the importance of strain in the results of Waters et al. and Baumann et al. on compressively strained InGaAs and InAlGaAs QWs on GaAs that are immune to sudden failure, despite intentionally inducing damage [92,146]. Fig. 4.24 summarizes their observation. Curbing DLD growth requires at least 5% − 10% indium. Note that these quantum wells must be kept below critical thickness for misfit dislocation formation; the authors point out that exceeding the critical thickness leads to rapid failure [91]. We think the amount of indium is too small to significantly alter the bandgap, electronic states, or the bulk thermodynamics of point defects and we expect the GaAs component of the active region to set

REDR. The effect of indium and strain then likely acts directly to counter the growth of the dislocation despite REDR. We saw in Section 4.3.5 that biaxial compressive stress counteracts the osmotic force for interstitial climb. Hopgood et al. propose this as a mechanism to explain why these compressively strained InGaAs QWs do not rapidly degrade [64]. They also identify a critical value of indium content (~5% indium) in the QWs based on the magnitude of the osmotic force. This implicitly suggests that the role of indium is to merely induce compressive strain; experiments on indium-free compressive lasers such as GaAsP quantum wells on GaP substrates etc. can clarify this. As discussed in Section 4.3.4, Kirkby and others have explicitly suggested a role for indium in blocking climb and glide, even in the absence of compressive strain [51,130]. Small quantities of indium harden the lattice (increase the critically resolved shear stress) against dislocation formation in bulk crystals at high temperatures. Here, the larger atomic radius of indium serves as either a hardening agent in random alloys [155] or selectively incorporates near the dislocation to block kink migration [156]. This is verified in experiments where additional indium further reduces REDG in metamorphic InGaAs layers, all while keeping the strain constant and controlling the indium composition independently [130]. However, numerous attempts to make DLD-free 850 nm lasers or LEDs with just a few percent indium (below the 5% indium critical value) have all been unsuccessful. Even if they may have slowed the DLD growth, they fell short of "pinning the dislocations" the way a higher indium mole fraction does. More work is needed to verify whether indium is indeed segregating at the dislocation; based on atom probe measurements we suspect the indium alloy remains uniform [60]. Going beyond random alloys, Mahajan et al. propose that lateral phase separation in InGaAsP alloys leads to large internal strain fields and serves as a significant barrier to dislocation growth [157]. Lateral phase separation is almost ubiquitous in III–V semiconductors involving binary compounds with dissimilar lattice constants. Stringfellow has reviewed the thermodynamics of semiconductor alloys [158], but many of these microstructures are produced due to kinetic instabilities during growth that are not well-understood [159,160].

4.6 Outlook

Researchers today have both the experimental and theoretical tools to work with defects in materials. From the time of Hartmann first noting dislocation networks in GaAs lasers in 1974, advances in the accuracy of point defect and dislocation models have been nothing short of extraordinary. Progress in the reproducible synthesis of devices, electron microscopy, atom probe tomography, and scanning probe methods have more than kept up. Yet, one cannot shake the feeling that we do not know that much more about laser degradation than the researchers of that time. This might be tied to the extreme environment of the laser that remains difficult to probe and model. Additionally, laser degradation is tied to an economic force in the late stage of materials development, and problems have been very effectively solved by circumventing defects rather than getting to the heart of a mechanism. To this last point, the literature is replete with examples such as the comprehensive removal of factors that lead to dislocations in GaAs lasers in both edge-emitting and VCSEL designs. The fractal-like growth of dislocation dipoles in VCSELs represents perhaps the most beautiful arrangement of defects in any material, but the mechanism remains hidden as these issues no longer plague the billion or so devices sold annually.

Nevertheless, researchers in materials science can contribute significantly if they can keep up with the rapid progress in lasers made from new materials or designs where defects are yet to be controlled or eliminated. GaN, SiC, and III−V on Si provide avenues for the resurgence of the study of REDRs. We think it is vital to accurately probe and model the environment around defects in devices if useful suggestions are to be made. The electric, strain, and thermal fields need to be factored in along with the varying chemical potentials of constituent species. There are government and industrial research programs to understand the behavior of materials in extreme environments combining theory and experiments. Perhaps it is time to include semiconductor laser materials in this category.

References

[1] F. Tuomisto, I. Makkonen, Defect identification in semiconductors with positron annihilation: experiment and theory, Rev. Mod. Phys. 85 (4) (2013) 1583−1631. Available from: https://doi.org/10.1103/RevModPhys.85.1583.

[2] G.B. Stringfellow, Organometallic Vapor-Phase Epitaxy: Theory and Practice, Academic Press, 1998.

[3] M.D. McCluskey, E.E. Haller, Dopants and Defects in Semiconductors, CRC Press, 2012.

[4] F.A. Kröger, The Chemistry of Imperfect Crystals, North-Holland Publishing Company, 1964.

[5] D.T.J. Hurle, A comprehensive thermodynamic analysis of native point defect and dopant solubilities in gallium arsenide, J. Appl. Phys. 85 (10) (1999) 6957−7022. Available from: https://doi.org/10.1063/1.370506.

[6] S.Y. Chiang, G.L. Pearson, Properties of vacancy defects in GaAs single crystals, J. Appl. Phys. 46 (7) (1975) 2986−2991. Available from: https://doi.org/10.1063/1.321985.

[7] J. Gebauer, et al., Determination of the Gibbs free energy of formation of Ga vacancies in GaAs by positron annihilation, Phys. Rev. B 67 (23) (2003) 235207. Available from: https://doi.org/10.1103/PhysRevB.67.235207.

[8] G.A. Baraff, M. Schlüter, Electronic structure, total energies, and abundances of the elementary point defects in GaAs, Phys. Rev. Lett. 55 (12) (1985) 1327−1330. Available from: https://doi.org/10.1103/PhysRevLett.55.1327.

[9] W. Walukiewicz, Amphoteric native defects in semiconductors, Appl. Phys. Lett. 54 (21) (1989) 2094−2096. Available from: https://doi.org/10.1063/1.101174.

[10] W. Walukiewicz, Intrinsic limitations to the doping of wide-gap semiconductors, Phys. B Condens. Matter 302−303 (2001) 123−134. Available from: https://doi.org/10.1016/S0921-4526(01)00417-3.

[11] C.M. Wolfe, C.N. Wolfe, N. Holonyak, G.E. Stillman, Physical Properties of Semiconductors, Prentice Hall, 1989.
[12] H.-P. Komsa, A. Pasquarello, Comparison of vacancy and antisite defects in GaAs and InGaAs through hybrid functionals, J. Phys. Condens. Matter 24 (4) (2012) 045801. Available from: https://doi.org/10.1088/0953-8984/24/4/045801.
[13] T.Y. Tan, H.-M. You, U.M. Gösele, Thermal equilibrium concentrations and effects of negatively charged Ga vacancies in n-type GaAs, Appl. Phys. A 56 (3) (1993) 249–258. Available from: https://doi.org/10.1007/BF00539483.
[14] T.Y. Tan, Thermal equilibrium concentrations of point defects in gallium arsenide, J. Phys. Chem. Solids 55 (10) (1994) 917–929. Available from: https://doi.org/10.1016/0022-3697(94)90111-2.
[15] H. Seki, A. Koukitu, Thermodynamic analysis of molecular beam epitaxy of III–V semiconductors, J. Cryst. Growth 78 (2) (1986) 342–352. Available from: https://doi.org/10.1016/0022-0248(86)90070-9.
[16] G.B. Stringfellow, Thermodynamic aspects of OMVPE, Journal of Crystal Growth 70 (1984) 133–139. Available from: https://doi.org/10.1016/0022-0248(84)90258-6.
[17] A. Koukitu, T. Taki, N. Takahashi, H. Seki, Thermodynamic study on the role of hydrogen during the MOVPE growth of group III nitrides, Journal of Crystal Growth 197 (1999) 99–105. Available from: https://doi.org/10.1016/S0022-0248(98)00894-X.
[18] M. Ichimura, T. Wada, Calculation of native defect concentrations in GaAs grown by organometallic vapor-phase epitaxy, J. Cryst. Growth 115 (1) (1991) 479–483. Available from: https://doi.org/10.1016/0022-0248(91)90790-C.
[19] M. Ichimura, T. Wada, Calculation of point defect concentrations in GaAs grown by molecular beam epitaxy, J. Appl. Phys. 72 (3) (1992) 1200–1202. Available from: https://doi.org/10.1063/1.351806.
[20] Z. Bryan, et al., Fermi level control of compensating point defects during metalorganic chemical vapor deposition growth of Si-doped AlGaN, Appl. Phys. Lett. 105 (22) (2014) 222101. Available from: https://doi.org/10.1063/1.4903058.
[21] K. Alberi, M.A. Scarpulla, Effects of excess carriers on charged defect concentrations in wide bandgap semiconductors, J. Appl. Phys. 123 (18) (2018) 185702. Available from: https://doi.org/10.1063/1.5010945.
[22] D.V. Lang, R.A. Logan, M. Jaros, Trapping characteristics and a donor-complex (DX) model for the persistent-photoconductivity trapping center in Te-doped AlxGa1-xAs, Phys. Rev. B 19 (2) (1979) 1015–1030. Available from: https://doi.org/10.1103/PhysRevB.19.1015.
[23] P.M. Mooney, Deep donor levels (DX centers) in III-V semiconductors, J. Appl. Phys. 67 (3) (1990) R1–R26. Available from: https://doi.org/10.1063/1.345628.
[24] D.J. Chadi, K.J. Chang, Metastability of the isolated arsenic-antisite defect in GaAs, Phys. Rev. Lett. 60 (21) (1988) 2187–2190. Available from: https://doi.org/10.1103/PhysRevLett.60.2187.
[25] M.O. Manasreh, D.W. Fischer, W.C. Mitchel, The EL2 defect in GaAs: some recent developments, Phys. Status Solidi B 154 (1) (1989) 11–41. Available from: https://doi.org/10.1002/pssb.2221540102.
[26] D.J. Chadi, K.J. Chang, Theory of the atomic and electronic structure of DX centers in GaAs and AlxGa1-xAs alloys, Phys. Rev. Lett. 61 (7) (1988) 873–876. Available from: https://doi.org/10.1103/PhysRevLett.61.873.
[27] P.M. Mooney, Properties of DX centers in AlxGa1 – xAs and effects on heterojunction devices, MRS Online Proc. Library Archive, 184, ed 1990, doi: 10.1557/PROC-184-39.

[28] M.A. Zaidi, H. Maaref, J.C. Bourgoin, Minority carrier capture cross section of the EL2 defect in GaAs, Appl. Phys. Lett. 61 (20) (1992) 2452–2454. Available from: https://doi.org/10.1063/1.108149.

[29] J.C. Bourgoin, N.D. Angelis, The defect responsible for non-radiative recombination in GaAs materials, Semicond. Sci. Technol. 16 (6) (2001) 497–501. Available from: https://doi.org/10.1088/0268-1242/16/6/315.

[30] Y. Sin, Z. Lingley, N. Presser, M. Brodie, N. Ives, S.C. Moss, Catastrophic Optical Bulk Damage in High-Power InGaAs-AlGaAs Strained Quantum Well Lasers, IEEE Journal of Selected Topics in Quantum Electronics 23 (2017) 1–13. Available from: https://doi.org/10.1109/JSTQE.2017.2686336.

[31] T. Figielski, T. Wosinski, A. Makosa, Gettering or generation of the EL2 defects at dislocations in GaAs? Phys. Status Solidi A 131 (2) (1992) 369–375. Available from: https://doi.org/10.1002/pssa.2211310211.

[32] D. Vignaud, J.L. Farvacque, A quantitative study of the creation of EL2 defects in GaAs by plastic deformation, J. Appl. Phys. 65 (4) (1989) 1516–1520. Available from: https://doi.org/10.1063/1.343421.

[33] G.I. Taylor, The mechanism of plastic deformation of crystals. Part I.—Theoretical, Proceedings of the Royal Society of London. Series A, Containing Papers of a Mathematical and Physical Character, vol. 145, no. 855, pp. 362–387, Jul. 1934. Available from: https://doi.org/10.1098/rspa.1934.0106.

[34] E. Orowan, Zur Kristallplastizität. III, Z. Phys. 89 (9) (1934) 634–659. Available from: https://doi.org/10.1007/BF01341480.

[35] M. Polanyi, Über eine Art Gitterstörung, die einen Kristall plastisch machen könnte, Z. Phys. 89 (9) (1934) 660–664. Available from: https://doi.org/10.1007/BF01341481.

[36] J. Weertman, J.R. Weertman, Elementary Dislocation Theory, Macmillan, 1964.

[37] J.P. Hirth, J. Lothe, Theory of Dislocations, Krieger Publishing Company, 1982.

[38] D.B. Holt, B.G. Yacobi, Extended Defects in Semiconductors: Electronic Properties, Device Effects and Structures, Cambridge University Press, 2007.

[39] C.W. Zhao, Y.M. Xing, C.E. Zhou, P.C. Bai, Experimental examination of displacement and strain fields in an edge dislocation core, Acta Mater. 56 (11) (2008) 2570–2575. Available from: https://doi.org/10.1016/j.actamat.2008.01.045.

[40] W.K. Burton, N. Cabrera, F.C. Frank, Role of Dislocations in Crystal Growth, Nature 163 (1949) 398–399. Available from: https://doi.org/10.1038/163398a0.

[41] W.E. McMahon, M. Vaisman, J.D. Zimmerman, A.C. Tamboli, E.L. Warren, Perspective: fundamentals of coalescence-related dislocations, applied to selective-area growth and other epitaxial films, APL Mater. 6 (12) (2018) 120903. Available from: https://doi.org/10.1063/1.5047945.

[42] P.M. Petroff, Defects in III-V compound semiconductors, Semiconductors and Semimetals., Volume 22, Academic Press, 1985, pp. 379–403.

[43] D. Gerthsen, F.A. Ponce, G.B. Anderson, High-resolution transmission electron microscopy of 60[ddot] dislocations in si-GaAs, Philos. Mag. A 59 (5) (1989) 1045–1058. Available from: https://doi.org/10.1080/01418618908209836.

[44] H.R. Kolar, J.C.H. Spence, H. Alexander, Observation of moving dislocation kinks and unpinning, Phys. Rev. Lett. 77 (19) (1996) 4031–4034. Available from: https://doi.org/10.1103/PhysRevLett.77.4031.

[45] M. Peach, J.S. Koehler, The forces exerted on dislocations and the stress fields produced by them, Phys. Rev. 80 (3) (1950) 436–439. Available from: https://doi.org/10.1103/PhysRev.80.436.

[46] H. Alexander, P. Haasen, Dislocations and plastic flow in the diamond structure, in: F. Seitz, D. Turnbull, H. Ehrenreich (Eds.), Solid State Physics, 22, Academic Press, 1969, pp. 27–158.
[47] I. Yonenaga, Mechanical properties and dislocation dynamics in III-V compounds, J. Phys. III 7 (1997) 1435–1450. Available from: https://doi.org/10.1051/jp3:1997198.
[48] I. Yonenaga, K. Sumino, Effects of In impurity on the dynamic behavior of dislocations in GaAs, Journal of Applied Physics 62 (1987) 1212–1219. Available from: https://doi.org/10.1063/1.339672.
[49] G. Vanderschaeve, C. Levade, D. Caillard, Dislocation mobility and electronic effects in semiconductor compounds, J. Microsc. 203 (1) (2001) 72–83. Available from: https://doi.org/10.1046/j.1365-2818.2001.00901.x.
[50] K. Maeda, S. Takeuchi, Recombination enhanced mobility of dislocations in III-V compounds, J. Phys. Colloq. 44 (C4) (1983). Available from: https://doi.org/10.1051/jphyscol:1983445. C4-375-C4-385.
[51] P.A. Kirkby, Dislocation pinning in GaAs by the deliberate introduction of impurities, IEEE J. Quantum Electron. 11 (7) (1975) 562–568. Available from: https://doi.org/10.1109/JQE.1975.1068634.
[52] B.W. Dodson, The effect of cottrell locking on structural stability in strained-layer semiconductor alloy structures, J. Cryst. Growth 92 (1) (1988) 222–232. Available from: https://doi.org/10.1016/0022-0248(88)90453-8.
[53] J. Lothe, J.P. Hirth, Dislocation climb forces, J. Appl. Phys. 38 (2) (1967) 845–848. Available from: https://doi.org/10.1063/1.1709423.
[54] D.M. Barnett, W.C. Oliver, W.D. Nix, The binding force between an edge dislocation and a fermi-dirac solute atmosphere, Acta Metall. 30 (3) (1982) 673–678. Available from: https://doi.org/10.1016/0001-6160(82)90116-X.
[55] D. Blavette, E. Cadel, A. Fraczkiewicz, A. Menand, Three-dimensional atomic-scale imaging of impurity segregation to line defects, Science 286 (5448) (1999) 2317–2319. Available from: https://doi.org/10.1126/science.286.5448.2317.
[56] E. Cadel, A. Fraczkiewicz, D. Blavette, Atomic scale observation of Cottrell atmospheres in B-doped FeAl (B2) by 3D atom probe field ion microscopy, Mater. Sci. Eng. A 309–310 (2001) 32–37. Available from: https://doi.org/10.1016/S0921-5093(00)01688-9.
[57] E. Cadel, A. Fraczkiewicz, D. Blavette, Atomic scale investigation of impurity segregation to crystal defects, Annu. Rev. Mater. Res. 33 (1) (2003) 215–231. Available from: https://doi.org/10.1146/annurev.matsci.33.013102.095557.
[58] K. Thompson, P.L. Flaitz, P. Ronsheim, D.J. Larson, T.F. Kelly, Imaging of arsenic Cottrell atmospheres around silicon defects by three-dimensional atom probe tomography, Science 317 (5843) (2007) 1370–1374. Available from: https://doi.org/10.1126/science.1145428.
[59] Y. Yu, et al., Ag-segregation to dislocations in PbTe-based thermoelectric materials, ACS Appl. Mater. Interfaces 10 (4) (2018) 3609–3615. Available from: https://doi.org/10.1021/acsami.7b17142.
[60] B. Bonef, R.D. Shah, K. Mukherjee, Fast diffusion and segregation along threading dislocations in semiconductor heterostructures, Nano Lett. 19 (3) (2019) 1428–1436. Available from: https://doi.org/10.1021/acs.nanolett.8b03734.
[61] J.W. Matthews, The climb and glide of misfit dislocations, Philos. Mag. J. Theor. Exp. Appl. Phys. 8 (88) (1963) 711–713. Available from: https://doi.org/10.1080/14786436308211170.
[62] K. Takayanagi, K. Kobayashi, K. Yagi, G. Honjo, Heteroepitaxy of chalcogenite compounds I. Misfit dislocation formation in monolayer overgrowths, Thin Solid

Films 21 (2) (1974) 325–339. Available from: https://doi.org/10.1016/0040-6090(74)90119-9.
[63] G. Springholz, K. Wiesauer, Nanoscale dislocation patterning in PbTe/PbSe(001) lattice-mismatched heteroepitaxy, Phys. Rev. Lett. 88 (2002) 015507. Available from: https://doi.org/10.1103/PhysRevLett.88.015507.
[64] H. Wang, A.A. Hopgood, G.I. Ng, Analysis of dark-line defect growth suppression in InxGa1 − xAs/GaAs strained heterostructures, J. Appl. Phys. 81 (7) (1997) 3117–3123. Available from: https://doi.org/10.1063/1.364353.
[65] P.M. Petroff, L.C. Kimerling, Dislocation climb model in compound semiconductors with zinc blende structure, Appl. Phys. Lett. 29 (8) (1976) 461–463. Available from: https://doi.org/10.1063/1.89145.
[66] H. Strunk, U. Gösele, B.O. Kolbesen, Interstitial supersaturation and climb of misfit dislocations in phosphorus-diffused silicon, J. Microsc. 118 (1) (1980) 35–39. Available from: https://doi.org/10.1111/j.1365-2818.1980.tb00243.x.
[67] T. Figielski, Electron emission from extended defects: DLTS signal in case of dislocation traps, Phys. Status Solidi A 121 (1) (1990) 187–193. Available from: https://doi.org/10.1002/pssa.2211210122.
[68] J.-L. Farvacque, Theoretical aspects of the minority carrier recombination at dislocations in semiconductors, Mater. Sci. Eng. B 42 (1) (1996) 110–121. Available from: https://doi.org/10.1016/S0921-5107(96)01691-1.
[69] L. Pasemann, A contribution to the theory of beam-induced current characterization of dislocations, J. Appl. Phys. 69 (9) (1991) 6387–6393. Available from: https://doi.org/10.1063/1.348841.
[70] M. Yamaguchi, C. Amano, Efficiency calculations of thin-film GaAs solar cells on Si substrates, J. Appl. Phys. 58 (9) (1985) 3601–3606. Available from: https://doi.org/10.1063/1.335737.
[71] C. Donolato, Modeling the effect of dislocations on the minority carrier diffusion length of a semiconductor, J. Appl. Phys. 84 (5) (1998) 2656–2664. Available from: https://doi.org/10.1063/1.368378.
[72] K.K. Sabelfeld, V.M. Kaganer, C. Pfüller, O. Brandt, Dislocation contrast in cathodoluminescence and electron-beam induced current maps on GaN(0 0 0 1), J. Phys. D Appl. Phys. 50 (40) (2017) 405101. Available from: https://doi.org/10.1088/1361-6463/aa85c8.
[73] C.L. Andre, et al., Impact of dislocations on minority carrier electron and hole lifetimes in GaAs grown on metamorphic SiGe substrates, Appl. Phys. Lett. 84 (18) (2004) 3447–3449. Available from: https://doi.org/10.1063/1.1736318.
[74] J.S. Speck, S.J. Rosner, The role of threading dislocations in the physical properties of GaN and its alloys, Phys. B Condens. Matter 273–274 (1999) 24–32. Available from: https://doi.org/10.1016/S0921-4526(99)00399-3.
[75] J. Hornstra, Dislocations in the diamond lattice, J. Phys. Chem. Solids 5 (1) (1958) 129–141. Available from: https://doi.org/10.1016/0022-3697(58)90138-0.
[76] P. Sitch, R. Jones, S. Öberg, M.I. Heggie, Structures of dislocations in GaAs and their modification by impurities, Phys. Rev. B 50 (23) (1994) 17717–17720. Available from: https://doi.org/10.1103/PhysRevB.50.17717.
[77] S.P. Beckman, D.C. Chrzan, Dislocation cores and their electronic states: partial dislocations in GaAs, Phys. B Condens. Matter 340–342 (2003) 1001–1004. Available from: https://doi.org/10.1016/j.physb.2003.09.189.
[78] T. Wosiński, T. Figielski, Energy levels and electrical activity of dislocation electron states in GaAs, Acta Phys. Pol. A 83 (1) (1993) 51–58. Available from: https://doi.org/10.12693/APhysPolA.83.51.

[79] X. Xu, et al., Distortion and segregation in a dislocation core region at atomic resolution, Phys. Rev. Lett. 95 (14) (2005) 145501. Available from: https://doi.org/10.1103/PhysRevLett.95.145501.
[80] J.-S. Park, B. Huang, S.-H. Wei, J. Kang, W.E. McMahon, Period-doubling reconstructions of semiconductor partial dislocations, NPG Asia Mater. 7 (9) (2015). Available from: https://doi.org/10.1038/am.2015.102. Art. no. 9, Sep.
[81] P. Omling, E.R. Weber, L. Montelius, H. Alexander, J. Michel, Electrical properties of dislocations and point defects in plastically deformed silicon, Phys. Rev. B 32 (10) (1985) 6571–6581. Available from: https://doi.org/10.1103/PhysRevB.32.6571.
[82] G.P. Watson, D.G. Ast, T.J. Anderson, B. Pathangey, Y. Hayakawa, The measurement of deep level states caused by misfit dislocations in InGaAs/GaAs grown on patterned GaAs substrates, J. Appl. Phys. 71 (7) (1992) 3399–3407. Available from: https://doi.org/10.1063/1.350936.
[83] A. Djemel, J. Castaing, N. Visentin, M. Bonnet, Influence of thermal treatments on the electronic activity of dislocations in GaAs observed by cathodoluminescence, Semicond. Sci. Technol. 5 (12) (1990) 1221. Available from: https://doi.org/10.1088/0268-1242/5/12/013.
[84] P.B. Hirsch, A mechanism for the effect of doping on dislocation mobility, J. Phys. Colloq. 40 (C6) (1979). Available from: https://doi.org/10.1051/jphyscol:1979624. C6-117-C6-121, Jun.
[85] H. Iber, E. Peiner, A. Schlachetzki, The effect of dislocations on the optical absorption of heteroepitaxial InP and GaAs on Si, J. Appl. Phys. 79 (12) (1996) 9273–9277. Available from: https://doi.org/10.1063/1.362577.
[86] K.P. Homewood, M.A. Lourenço, Light from Si via dislocation loops, Mater. Today 8 (1) (2005) 34–39. Available from: https://doi.org/10.1016/S1369-7021(04)00677-7.
[87] V. Kveder, M. Badylevich, E. Steinman, A. Izotov, M. Seibt, W. Schröter, Room-temperature silicon light-emitting diodes based on dislocation luminescence, Appl. Phys. Lett. 84 (12) (2004) 2106–2108. Available from: https://doi.org/10.1063/1.1689402.
[88] J.W. Matthews, A.E. Blakeslee, Defects in epitaxial multilayers: I. Misfit dislocations, J. Cryst. Growth 27 (1974) 118–125. Available from: https://doi.org/10.1016/S0022-0248(74)80055-2.
[89] J. Zou, D.J.H. Cockayne, B.F. Usher, Misfit dislocations and critical thickness in InGaAs/GaAs heterostructure systems, J. Appl. Phys. 73 (2) (1993) 619–626. Available from: https://doi.org/10.1063/1.353372.
[90] W.D. Nix, Mechanical properties of thin films, Metall. Trans. A 20 (11) (1989) 2217. Available from: https://doi.org/10.1007/BF02666659.
[91] K.J. Beernink, P.K. York, J.J. Coleman, R.G. Waters, J. Kim, C.M. Wayman, Characterization of InGaAs-GaAs strained-layer lasers with quantum wells near the critical thickness, Appl. Phys. Lett. 55 (21) (1989) 2167–2169. Available from: https://doi.org/10.1063/1.102089.
[92] J.A. Baumann, R.J. Dalby, R.G. Waters, S.L. Yellen, C. Harding, A. Shepard, Growth of InAIGaAs strained quantum well structures for reliable 0.8 μm lasers, J. Electron. Mater. 23 (2) (1994) 207–216. Available from: https://doi.org/10.1007/BF02655271.
[93] L.B. Freund, The driving force for glide of a threading dislocation in a strained epitaxial layer on a substrate, J. Mech. Phys. Solids 38 (5) (1990) 657–679. Available from: https://doi.org/10.1016/0022-5096(90)90027-2.
[94] W.D. Nix, D.B. Noble, and J.F. Turlo, Mechanisms and kinetics of misfit dislocation formation in heteroepitaxial thin films, MRS Online Proceedings Library Archive, vol. 188, ed 1990, doi: 10.1557/PROC-188–315.

[95] R. Beanland, D.J. Dunstan, P.J. Goodhew, Plastic relaxation and relaxed buffer layers for semiconductor epitaxy, Adv. Phys. 45 (2) (1996) 87–146. Available from: https://doi.org/10.1080/00018739600101477.
[96] D.J. Dunstan, Strain and strain relaxation in semiconductors, J. Mater. Sci. Mater. Electron. 8 (6) (1997) 337–375. Available from: https://doi.org/10.1023/A:1018547625106.
[97] B.W. Dodson, J.Y. Tsao, Relaxation of strained-layer semiconductor structures via plastic flow, Appl. Phys. Lett. 51 (17) (1987) 1325–1327. Available from: https://doi.org/10.1063/1.98667.
[98] R. Beanland, Multiplication of misfit dislocations in epitaxial layers, J. Appl. Phys. 72 (9) (1992) 4031–4035. Available from: https://doi.org/10.1063/1.352257.
[99] R. Beanland, Dislocation multiplication mechanisms in low-misfit strained epitaxial layers, J. Appl. Phys. 77 (12) (1995) 6217–6222. Available from: https://doi.org/10.1063/1.359151.
[100] D.K. Biegelsen, F.A. Ponce, A.J. Smith, J.C. Tramontana, Initial stages of epitaxial growth of GaAs on (100) silicon, J. Appl. Phys. 61 (5) (1987) 1856–1859. Available from: https://doi.org/10.1063/1.338029.
[101] H.L. Tsai, Y.C. Kao, Generation and propagation of threading dislocations in GaAs grown on Si, J. Appl. Phys. 67 (6) (1990) 2862–2865. Available from: https://doi.org/10.1063/1.345457.
[102] M. Akiyama, Y. Kawarada, K. Kaminishi, Growth of GaAs on Si by MOVCD, J. Cryst. Growth 68 (1984) 21–26. Available from: https://doi.org/10.1016/0022-0248(84)90391-9.
[103] R. Alcotte, et al., Epitaxial growth of antiphase boundary free GaAs layer on 300 mm Si(001) substrate by metalorganic chemical vapour deposition with high mobility, APL Mater. 4 (4) (2016) 046101. Available from: https://doi.org/10.1063/1.4945586.
[104] S. Sharan, J. Narayan, Strain-relief mechanisms and nature of misfit dislocations in GaAs/Si heterostructures, Mater. Sci. Eng. A 113 (1989) 57–63. Available from: https://doi.org/10.1016/0921-5093(89)90293-1.
[105] P. Sheldon, K.M. Jones, M.M. Al-Jassim, B.G. Yacobi, Dislocation density reduction through annihilation in lattice-mismatched semiconductors grown by molecular-beam epitaxy, J. Appl. Phys. 63 (11) (1988) 5609–5611. Available from: https://doi.org/10.1063/1.340343.
[106] J.S. Speck, M.A. Brewer, G. Beltz, A.E. Romanov, W. Pompe, Scaling laws for the reduction of threading dislocation densities in homogeneous buffer layers, J. Appl. Phys. 80 (7) (1996) 3808–3816. Available from: https://doi.org/10.1063/1.363334.
[107] P.G. Callahan, B.B. Haidet, D. Jung, G.G.E. Seward, K. Mukherjee, Direct observation of recombination-enhanced dislocation glide in heteroepitaxial GaAs on silicon, Phys. Rev. Mater. 2 (8) (2018) 081601. Available from: https://doi.org/10.1103/PhysRevMaterials.2.081601.
[108] T. Ward, et al., Design rules for dislocation filters, J. Appl. Phys. 116 (6) (2014) 063508. Available from: https://doi.org/10.1063/1.4892162.
[109] M. Yamaguchi, M. Tachikawa, Y. Itoh, M. Sugo, S. Kondo, Thermal annealing effects of defect reduction in GaAs on Si substrates, J. Appl. Phys. 68 (9) (1990) 4518. Available from: https://doi.org/10.1063/1.346156.
[110] M. Yamaguchi, A. Yamamoto, M. Tachikawa, Y. Itoh, M. Sugo, Defect reduction effects in GaAs on Si substrates by thermal annealing, Appl. Phys. Lett. 53 (23) (1988) 2293–2295. Available from: https://doi.org/10.1063/1.100257.

[111] H. Okamoto, Y. Watanabe, Y. Kadota, Y. Ohmachi, Dislocation reduction in GaAs on Si by thermal cycles and InGaAs/GaAs strained-layer superlattices, Jpn. J. Appl. Phys. 26 (12A) (1987) L1950. Available from: https://doi.org/10.1143/JJAP.26.L1950.

[112] M. Yamaguchi, T. Nishioka, M. Sugo, Analysis of strained-layer superlattice effects on dislocation density reduction in GaAs on Si substrates, Appl. Phys. Lett. 54 (1) (1989) 24–26. Available from: https://doi.org/10.1063/1.100819.

[113] I. George, F. Becagli, H.Y. Liu, J. Wu, M. Tang, R. Beanland, Dislocation filters in GaAs on Si, Semicond. Sci. Technol. 30 (2015) 114004. Available from: https://doi.org/10.1088/0268-1242/30/11/114004.

[114] E.A. Fitzgerald, N. Chand, Epitaxial necking in GaAs grown on pre-pattemed Si substrates, J. Electron. Mater. 20 (7) (1991) 839–853. Available from: https://doi.org/10.1007/BF02665973.

[115] B. Kunert, W. Guo, Y. Mols, R. Langer, K. Barla, (Invited) integration of III/V hetero-structures by selective area growth on Si for nano- and optoelectronics, ECS Trans. 75 (8) (2016) 409. Available from: https://doi.org/10.1149/07508.0409ecst.

[116] B. Kunert, Y. Mols, M. Baryshniskova, N. Waldron, A. Schulze, R. Langer, How to control defect formation in monolithic III/V hetero-epitaxy on (100) Si? A critical review on current approaches, Semicond. Sci. Technol. 33 (9) (2018) 093002. Available from: https://doi.org/10.1088/1361-6641/aad655.

[117] Y. Ujiie, T. Nishinaga, Epitaxial lateral overgrowth of GaAs on a Si substrate, Jpn. J. Appl. Phys. 28 (3A) (1989) L337. Available from: https://doi.org/10.1143/JJAP.28.L337.

[118] Z.I. Kazi, P. Thilakan, T. Egawa, M. Umeno, T. Jimbo, Realization of GaAs/AlGaAs lasers on Si substrates using epitaxial lateral overgrowth by metalorganic chemical vapor deposition, Jpn. J. Appl. Phys. 40 (8R) (2001) 4903. Available from: https://doi.org/10.1143/JJAP.40.4903.

[119] A. Usui, H. Sunakawa, A. Sakai, A.A. Yamaguchi, Thick GaN epitaxial growth with low dislocation density by hydride vapor phase epitaxy, Jpn. J. Appl. Phys. 36 (7B) (1997) L899. Available from: https://doi.org/10.1143/JJAP.36.L899.

[120] Y. Wan, Q. Li, Y. Geng, B. Shi, K.M. Lau, InAs/GaAs quantum dots on GaAs-on-V-grooved-Si substrate with high optical quality in the 1.3 μm band, Appl. Phys. Lett. 107 (8) (2015) 081106. Available from: https://doi.org/10.1063/1.4929441.

[121] S. Zhu, B. Shi, Q. Li, K.M. Lau, 1.5 μm quantum-dot diode lasers directly grown on CMOS-standard (001) silicon, Appl. Phys. Lett. 113 (22) (2018) 221103. Available from: https://doi.org/10.1063/1.5055803.

[122] K. Ishida, T. Kamejima, TEM study of dark line defect growth from dislocation clusters in (GaAl)As-GaAs double heterostructure lasers, J. Electron. Mater. 8 (1) (1979) 57–73. Available from: https://doi.org/10.1007/BF02655641.

[123] A.M. Stoneham, Non-radiative transitions in semiconductors, Rep. Prog. Phys. 44 (12) (1981) 1251–1295. Available from: https://doi.org/10.1088/0034-4885/44/12/001.

[124] D.V. Lang, Recombination-enhanced reactions in semiconductors, Annu. Rev. Mater. Sci. 12 (1982) 377–398. Available from: https://doi.org/10.1146/annurev.ms.12.080182.002113.

[125] Y. Shinozuka, M. Wakita, K. Suzuki, Phonon-kick mechanism for defect reactions in semiconductors, Jpn. J. Appl. Phys. 51 (11S) (2012) 11PC03. Available from: https://doi.org/10.1143/JJAP.51.11PC03.

[126] D. Redfield, R.H. Bube, Photo-Induced Defects in Semiconductors, Cambridge University Press, 2006.

[127] H. Sumi, Phonon-kick mechanism for defect reactions enhanced by electronic excitation, J. Phys. C Solid. State Phys 17 (34) (1984) 6071–6086. Available from: https://doi.org/10.1088/0022-3719/17/34/009.

[128] K. Maeda, Recombination-enhanced dislocation glides—the current status of knowledge, MRS Online Proceedings Library Archivé, vol. 1195, ed 2009, doi: 10.1557/PROC-1195-B02-01.

[129] J. Selvidge, et al., Defect filtering for thermal expansion induced dislocations in III-V lasers on silicon, Appl. Phys. Lett. 117 (2020) 122101. Available from: https://doi.org/10.1063/5.0023378.

[130] E.T. Hughes, R.D. Shah, K. Mukherjee, Glide of threading dislocations in (In)AlGaAs on Si induced by carrier recombination: characteristics, mitigation, and filtering, J. Appl. Phys. 125 (16) (2019) 165702. Available from: https://doi.org/10.1063/1.5088844.

[131] T. Kamejima, K. Ishida, J. Matsui, Injection-enhanced dislocation glide under uniaxial stress in GaAs−(GaAl)As double heterostructure laser, Jpn. J. Appl. Phys. 16 (2) (1977) 233. Available from: https://doi.org/10.1143/JJAP.16.233.

[132] P.W. Hutchinson, P.S. Dobson, Defect structure of degraded GaAlAs-GaAs double heterojunction lasers, Philos. Mag. J. Theor. Exp. Appl. Phys. 32 (4) (1975) 745−754. Available from: https://doi.org/10.1080/14786437508221617.

[133] P. Petroff, R.L. Hartman, Defect structure introduced during operation of heterojunction GaAs lasers, Appl. Phys. Lett. 23 (8) (1973) 469−471. Available from: https://doi.org/10.1063/1.1654962.

[134] P. Petroff, R.L. Hartman, Rapid degradation phenomenon in heterojunction GaAlAs−GaAs lasers, J. Appl. Phys. 45 (9) (1974) 3899−3903. Available from: https://doi.org/10.1063/1.1663883.

[135] S. Xie, et al., Failure mode analysis of oxide VCSELs in high humidity and high temperature, J. Lightwave Technol. 21 (4) (2003) 1013.

[136] D.T. Mathes, *et al.*, Nanoscale materials characterization of degradation in VCSELs, in: Vertical-Cavity Surface-Emitting Lasers VII, 2003, 4994, pp. 67−82. Available from: https://doi.org/10.1117/12.482858.

[137] D.T. Mathes, Materials issues for VCSEL operation and reliability (Ph.D. thesis), 2002, Accessed: Nov. 13, 2019. [Online]. Available: https://ui.adsabs.harvard.edu/abs/2002PhDT155M/abstract.

[138] J. Matsui, K. Ishida, Y. Nannichi, Rapid degradation in double-heterostructure lasers. I. Proposal of a new model for the directional growth of dislocation networks, Jpn. J. Appl. Phys. 14 (10) (1975) 1555−1560. Available from: https://doi.org/10.1143/JJAP.14.1555.

[139] D.T. Mathes, Materials issues for VCSEL operation and reliability, ProQuest Dissertations And Theses; Thesis (Ph.D.)−University of Virginia, 2002.; Publication Number: AAI3065123; ISBN: 9780493839219; Source: Dissertation Abstracts International, Volume: 63-09, Section: B, page: 4322; 252 p.

[140] S. O'Hara, P.W. Hutchinson, P.S. Dobson, The origin of dislocation climb during laser operation, Appl. Phys. Lett. 30 (8) (1977) 368−371. Available from: https://doi.org/10.1063/1.89432.

[141] P.T. Heald, The preferential trapping of interstitials at dislocations, Philos. Mag. J. Theor. Exp. Appl. Phys. 31 (3) (1975) 551−558. Available from: https://doi.org/10.1080/14786437508226537.

[142] W. Frank, U. Gösele, The influence of intrisic defects on the degradation and luminescence of GaAs and other III−V compounds, Appl. Phys. 23 (3) (1980) 303−309. Available from: https://doi.org/10.1007/BF00914916.

[143] R.G. Waters, D.S. Hill, Influence of V/III variation on AlxGa1-xAs quantum well lasers grown by metalorganic chemical vapor deposition, J. Electron. Mater. 17 (3) (1988) 239−241. Available from: https://doi.org/10.1007/BF02652184.

[144] S.L. Yellen, et al., Reliability of GaAs-based semiconductor diode lasers: 0.6-1.1 mu m, IEEE J. Quantum Electron. 29 (6) (1993) 2058−2067. Available from: https://doi.org/10.1109/3.234469.
[145] O. Ueda, Materials-related reliability aspects of III−V optical devices, Mater. Sci. Eng. B 20 (1−2) (1993) 9−20. Available from: https://doi.org/10.1016/0921-5107(93)90387-3.
[146] R.G. Waters, D.P. Bour, S.L. Yellen, N.F. Ruggieri, Inhibited dark-line defect formation in strained InGaAs/AlGaAs quantum well lasers, IEEE Photon. Technol. Lett. 2 (8) (1990) 531−533. Available from: https://doi.org/10.1109/68.58039.
[147] O. Ueda, Degradation of III−V opto-electronic devices, J. Electrochem. Soc. 135 (1) (1988) 11C. Available from: https://doi.org/10.1149/1.2095535.
[148] J.D. Dow, R.E. Allen, Role of dangling bonds and antisite defects in rapid and gradual III-V laser degradation, Appl. Phys. Lett. 41 (7) (1982) 672−674. Available from: https://doi.org/10.1063/1.93609.
[149] J.-S. Park, J. Kang, J.-H. Yang, W.E. McMahon, S.-H. Wei, Polymerization of defect states at dislocation cores in InAs, J. Appl. Phys. 119 (4) (2016) 045706. Available from: https://doi.org/10.1063/1.4940743.
[150] R.K. Jain, D.J. Flood, Influence of the dislocation density on the performance of heteroepitaxial indium phosphide solar cells, IEEE Trans. Electron. Devices 40 (11) (1993) 1928−1934. Available from: https://doi.org/10.1109/16.239730.
[151] R.R. King, et al., 40% efficient metamorphic GaInP/GaInAs/Ge multijunction solar cells, Appl. Phys. Lett. 90 (18) (2007). Available from: https://doi.org/10.1063/1.2734507. 183516-183516−3, May.
[152] H.A. Tahini, A. Chroneos, S.T. Murphy, U. Schwingenschlögl, R.W. Grimes, Vacancies and defect levels in III−V semiconductors, J. Appl. Phys. 114 (6) (2013) 063517. Available from: https://doi.org/10.1063/1.4818484.
[153] M. Yamaguchi, K. Ando, Mechanism for radiation resistance of InP solar cells, J. Appl. Phys. 63 (11) (1988) 5555−5562. Available from: https://doi.org/10.1063/1.340332.
[154] M. Yamaguchi, Radiation-resistant solar cells for space use, Sol. Energy Mater. Sol. Cell 68 (1) (2001) 31−53. Available from: https://doi.org/10.1016/S0927-0248(00)00344-5.
[155] H. Ehrenreich, J.P. Hirth, Mechanism for dislocation density reduction in GaAs crystals by indium addition, Appl. Phys. Lett. 46 (7) (1985) 668−670. Available from: https://doi.org/10.1063/1.95523.
[156] F. Louchet, The reduction of dislocation density in GaAs by in doping: a specific interaction of in with the cores of 30° partial dislocations, J. Phys. Fr. 49 (7) (1988) 1219−1224. Available from: https://doi.org/10.1051/jphys:019880049070121900.
[157] S. Mahajan, B.V. Dutt, H. Temkin, R.J. Cava, W.A. Bonner, Spinodal decomposition in InGaAsP epitaxial layers, J. Cryst. Growth 68 (2) (1984) 589−595. Available from: https://doi.org/10.1016/0022-0248(84)90466-4.
[158] G.B. Stringfellow, Immiscibility and spinodal decomposition in III/V alloys, J. Cryst. Growth 65 (1−3) (1983) 454−462. Available from: https://doi.org/10.1016/0022-0248(83)90086-6.
[159] I.P. Ipatova, V.G. Malyshkin, A.A. Maradudin, V.A. Shchukin, R.F. Wallis, Kinetic instability of semiconductor alloy growth, Phys. Rev. B 57 (20) (1998) 12968−12993. Available from: https://doi.org/10.1103/PhysRevB.57.12968.
[160] K. Mukherjee, A.G. Norman, A.J. Akey, T. Buonassisi, E.A. Fitzgerald, Spontaneous lateral phase separation of AlInP during thin film growth and its effect on luminescence, J. Appl. Phys. 118 (11) (2015) 115306. Available from: https://doi.org/10.1063/1.4930990.

Chapter 5

Grown-in defects and thermal instability affecting the reliability of lasers: III–Vs versus III-nitrides

Osamu Ueda[1] *and Shigetaka Tomiya*[2,3]
[1]Meiji University, Tama-ku, Kawasaki, Japan, [2]Sony Corporation, Atsugi, Japan, [3]Tokyo Institute of Technology, Yokohama, Japan

5.1 Introduction

In the development of semiconductor lasers and related optical devices, it is crucial to improve the lifetime of the devices as well as achieving their high performance. The presence of defects in the epitaxial layers, especially in the active layer, is associated with the degradation of such devices. As will be described in some of the other chapters in this book, the so-called dark-line defects (DLDs) and dark-spot defects (DSDs) are observed in degraded lasers [1–5] and LEDs [6,7] fabricated by III–Vs. These dark defects correspond to interstitial-type dislocation dipoles generated as a result of climb motion at dislocations in the active layer [8–13]. Therefore it is extremely important to eliminate defects induced in the epitaxial layer during crystal growth processes to obtain highly reliable devices.

In this chapter, defects induced during crystal growth in various III–V compound semiconductors (III–Vs) and GaN-related semiconductors (III-nitrides) are described in detail and compared. We also discuss their generation mechanisms. Furthermore, we present two types of structures which are associated with thermal instability of crystals: composition-modulated structures (CMS) and ordered structures (OS). Based on these results, the influence of the defects and the thermally unstable structures on the reliability of lasers and methods for their suppression are presented.

5.2 Grown-in defects in III–Vs semiconductors for optoelectronics

5.2.1 *Classification of grown-in defects*

Generation of defects is one of most important issues in semiconductor crystal growth. Defects introduced during the growth procedure are classified into two types: interface defects and bulk defects.

Reliability of Semiconductor Lasers and Optoelectronic Devices. DOI: https://doi.org/10.1016/B978-0-12-819254-2.00004-7

The interface defects are also classified into two types, interface defects 1 and interface defects 2, as shown schematically in Figs. 5.1 and 5.2, respectively. The defects belongings to interface defects 1 are stacking faults, V-shaped dislocations, dislocation clusters, microtwins (MTs), and inclusions due to contamination and thermal damage of the substrate, and local segregation at the epi/sub and epi/epi interfaces. On the other hand, misfit dislocations and microtwins, interface defects 2, are generated by lattice mismatch between the epi-layer and the substrate during the growth and/or cooling processes after growth.

The stacking faults shown in Fig. 5.1A are generated from the heterointerface, extending on four equivalent {111} planes, forming square pyramid-shaped stacking faults [14,15]. In molecular beam epitaxy (MBE)-grown crystals, they often correspond to surface defects. Pairs of pits are often observed on the chemically etched surface of thin films. In the transmission electron microscope (TEM) image of the area including a pit pair, small dislocation segments are observed. From the shape of the dislocations, this dislocation pair is expected to be a V-shaped dislocation (Fig. 5.1B). Occasionally, preferential etching occurs at these V-shaped dislocations (see Fig. 5.1B) [16].

Microtwins are generated at the heterointerface through relaxation of stress, extending on the {111} planes as shown in Fig. 5.1C [15]. They occasionally terminate with {211}-type boundaries in the mid-layer. Inclusions are accidentally introduced during liquid phase epitaxy (LPE) growth by incorporation of the remaining droplets into the solid solution when the Ga- or In-rich melt is wiped from the substrate (Fig. 5.1D) [16]. In the cases of MBE and metalorganic vapor phase epitaxy

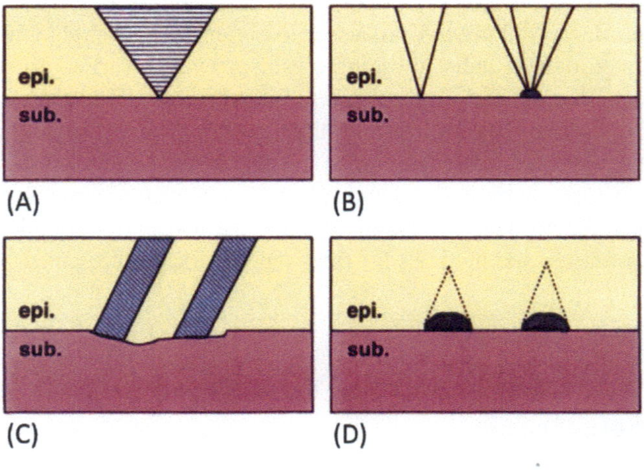

Figure 5.1 Schematic diagrams of interface defects 1 in III–Vs, which are caused by contamination, thermal damage (including evaporation of group V atoms from the surface) of the substrate or local segregation: (A) stacking faults; (B) V-shaped dislocations and dislocation clusters; (C) microtwins; and (D) "inclusions" for LPE and droplets for molecular beam epitaxy or metalorganic vapor phase epitaxy.

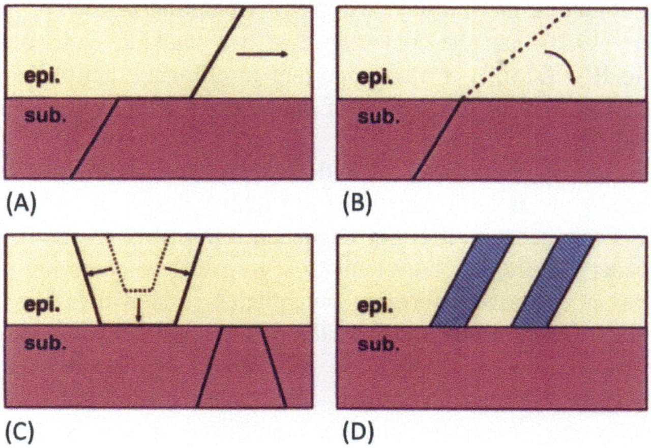

Figure 5.2 Schematic diagrams of interface defects 2 in III–Vs which are caused by lattice mismatch between the epi-layer and the substrate. (A) A misfit dislocation caused by a threading dislocation gliding out from the substrate; (B) a misfit dislocation caused by a threading dislocation bending out from the substrate; (C) misfit dislocations in the form of half loops, two features; (D) microtwins.

(MOVPE), droplets are generated by evaporation of group V element from the growth surface due to insufficient As overpressure (Fig. 5.1D).

Since the first three kinds of defects described above can be eliminated by cleaning of the substrate surface prior to the crystal growth and/or the protection of the substrate against evaporation of substrate atoms, they are expected to be generated by contamination and/or thermal damage of the substrate. Thus it should be recognized that careful control of the heterointerface is extremely important for the elimination of defects.

If there is a difference in lattice constants between the epi-layer and the substrate, misfit dislocations are generated when the thickness exceeds the critical thickness t_c, which depends on the structure and elastic properties of the material and the type of dislocations. There are several possible mechanisms for the generation of misfit dislocations as shown in Fig. 5.2A–C. First, threading dislocations (TD) propagating from the substrate may glide out (model 1, Fig. 5.2A) or bend out (model 2, Fig. 5.2B) to the edge of the wafer. We also expect that two dislocations with the same Burgers vectors in the substrate can terminate as one misfit dislocation at the hetero-interface, forming a half loop (right part of Fig. 5.2C). If the dislocation density in the substrate is low, another mechanism may occur as follows [17]: (1) nucleation of small half loops at the surface; (2) expansion of the half loops down to the interface; and (3) gliding out of the threading dislocation segments to the edge (left part of Fig. 5.2C).

In highly mismatched heteroepitaxy such as III–Vs/Si, microtwins (Fig. 5.2D) and/or stacking faults are very often generated, particularly during the initial stage of growth [18]. They propagate along the four equivalent {111} planes. However, they often annihilate in the mid-layer, that is, terminating in {211}-type boundaries for microtwins [19,20] and the coalescence of two stacking faults on the different planes [19]. In III–Vs/Si, it is also important to eliminate antiphase domains that are generated as a result of the growth of polar semiconductor GaAs on (001) non-polar semiconductor Si [21].

In Fig. 5.3, typical features of bulk defects are shown schematically. In most cases, they are generated in heavily (or highly) doped crystals as shown in Fig. 5.3A–C. Their generation is not dependent on growth methods. When super-saturated dopants are present, they tend to segregate locally during growth or the cooling process after growth, forming precipitates as shown in Fig. 5.3A. When there is a large lattice mismatch between the precipitates and the matrix, dislocation loops are often punched out from the precipitates by the relaxation of stress (Fig. 5.3B). In the formation of precipitates, excess native point defects can be introduced. They may condense at some nucleation centers to form dislocation loops (either faulted or unfaulted loops), as shown in Fig. 5.3C. Interstitial-type dislocation loops are often observed in heavily doped LPE-AlGaAs crystals with Ge [14].

Since the active layer of LDs and LEDs is undoped, they are not generated. However, in the case of vertical-cavity surface-emitting lasers (VCSELs), if they are generated in the p-distributed Bragg reflector (DBR) region they may grow larger, causing dark-line degradation. In heterojunction bipolar transistors (HBTs), in a heavily doped p^+-base layer, the existence of a high density of dislocation loops causes β (gain) degradation. On the other hand, dislocation loops are present even in undoped InGaP crystals (Fig. 5.3D) [22]. In this case, both vacancy-type and interstitial-type loops are found. In most III–V alloy semiconductor thin-film

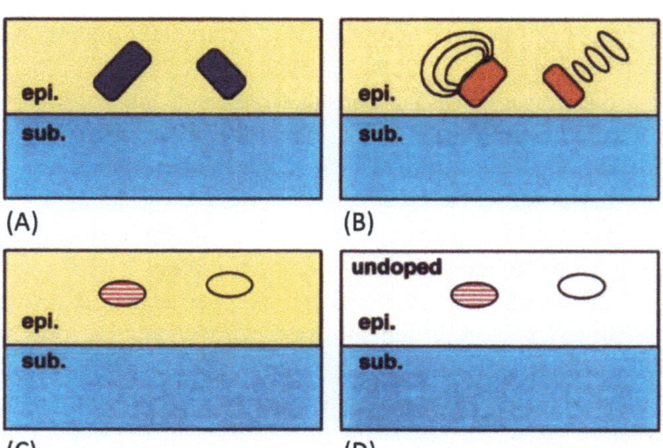

Figure 5.3 Schematic diagrams of bulk defects in III–Vs, which are often generated by heavy (high) doping: (A) precipitates; (B) precipitates accompanying punched-out dislocation loops; (C) faulted and/or unfaulted loops due to local segregation of dopant atoms; and (D) faulted and/or unfaulted loops due to generation of modulated structure.

growth, compositional fluctuation called CMS occurs (as described in Section 5.3.1). This phenomenon does not depend on the growth method. Therefore these loops are presumably generated by the condensation of excess native point defects induced as byproducts of the CMS.

In the following sections, the results on the structural evaluation of the interface defects 1, interface defects 2, and bulk defects are described, respectively.

5.2.2 Interface defects 1

Fig. 5.4A is a typical optical micrograph of the surface of a GaAsSb crystal grown at 590°C [15]. A high density of surface defects is observed. They are hillock-like defects lying along the <110> direction, and are similar to oval defects in the MBE-AlGaAs crystal [23]. The defect structure observed by TEM at these surface

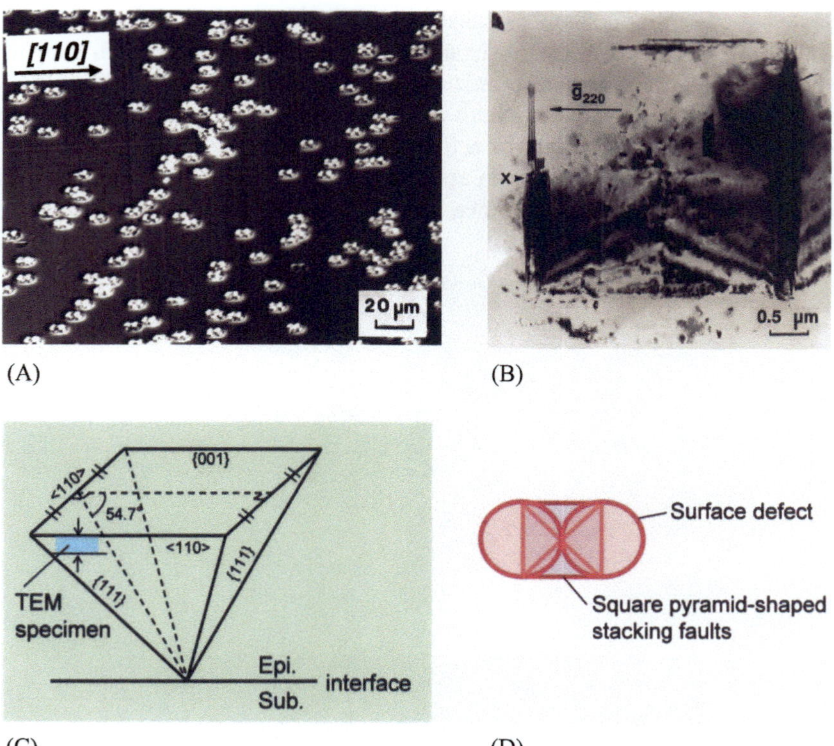

Figure 5.4 (A) An optical micrograph of defects revealed on the surface of an MBE-grown GaAsSb crystal on (001) InP substrate. (B) A (001) plan-view bright-field transmission electron microscope image of square pyramid-shaped stacking faults corresponding to the defect shown in (A). (C) A schematic diagram of the three-dimensional feature of the square pyramid-shaped stacking faults shown in (B). (D) A schematic diagram of the relationship between the square pyramid-shaped stacking faults and the hillock-like surface defects pair.

defects is shown in Fig. 5.4B. Stacking faults lying on the four equivalent {111} planes are observed. Since the thin TEM specimen includes only the surface region of the epi-layer, it is expected that the surface defects correspond to square pyramid-shaped stacking faults [24], as shown schematically in Fig. 5.4C. We have also found that the hillock-like defect pair is formed at the stacking fault as shown schematically in Fig. 5.4D. The stacking faults often multiplied, for example, at the region denoted by X in Fig. 5.4B. In addition, TD and small stacking faults are observed inside the square pyramid. By considering the length of the stacking faults at the surface region, their basal {111} planes, and the epi-layer thickness, it is expected that they originated at the heterointerface. We expect that dominant factors affecting their generation could be the evaporation of phosphorous atoms from the substrate and subsequent formation of droplets for group III elements, which act as the nucleation point at the tip of the pyramid. Therefore for the reduction of these defects we need appropriate control of As overpressure for MBE and V/III ratio for MOVPE during growth.

Next, we present an example of dislocation clusters generated from the epi-sub interface. In Fig. 5.5, we see an example where the transition from the InP to arsenide-based alloys was not optimized during development, and left an "indium droplet" that nucleated a stacking fault pyramid, which in turn started a climb dislocation network where it crossed the active region layers.

Microtwins are often observed in epi-layers grown at higher temperatures. They also originate from the heterointerface, extending to the surface region with {111}

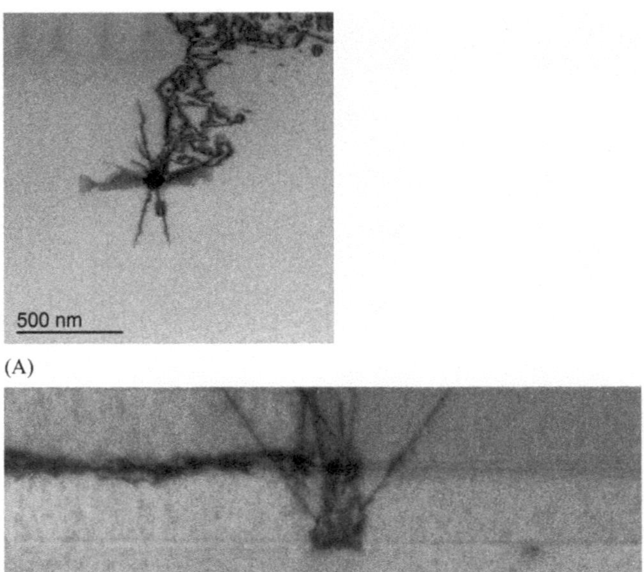

(A)

(B)

Figure 5.5 (A) A PV STEM image of a dislocation cluster and (B) an XS STEM image of extended dislocation networks from the dislocation cluster.

twin boundaries. Some terminate in the mid-layer with {211} boundaries. A cross-sectional high-resolution TEM (HRTEM) image of a GaAsSb crystal grown at 590°C from a region close to the heterointerface is shown in Fig. 5.6A [15]. In this image, amorphous-like area A and a shallow pit-like pattern just below area A are observed. A microtwin denoted by MT in Fig. 5.6A is observed close to defect A.

Based on these results, the following steps are suggested for the generation mechanism of the microtwins:

Step 1. Evaporation of phosphorous atoms from the substrate surface [formation of In-rich droplets (see A in Fig. 5.6A) on the growth surface];

Step 2. Generation of an amorphous region as a consequence of an In-rich melt;

Step 3. Growth of crystals with different lattice constants around the amorphous regions;

Step 4. Generation of microtwins (see MT in Fig. 5.6A) by relaxation of the stress resulting from the lattice mismatch between the normal GaAsSb region and those formed in Step 3.

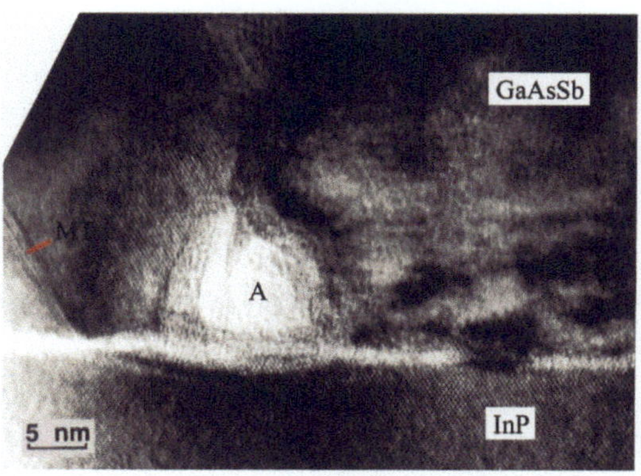

(A)

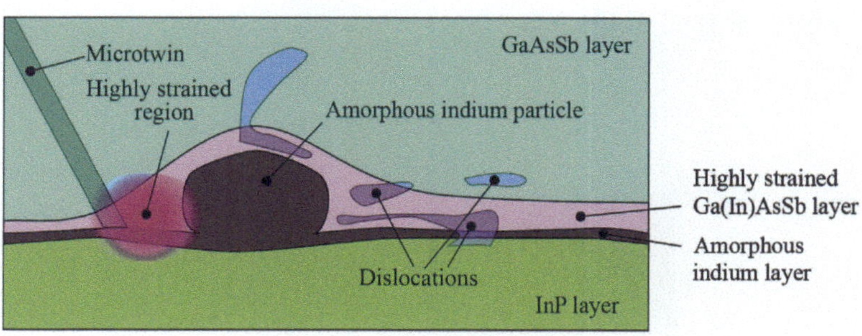

(B)

Figure 5.6 (A) A (110) cross-sectional high-resolution transmission electron microscope image of GaAsSb/(001)InP interface and (B) a schematic diagram for the high-resolution transmission electron microscope image in (A).

Therefore the elimination of microtwins can be achieved by methods similar to those described for the stacking faults.

Fig. 5.7 shows a TEM image of inclusions in an LPE-grown InGaAsP crystal on a (001) GaAs substrate [16]. They are lying in the <100> direction. From stereoscopic observation, they are found to originate from the interface between the InGaAsP-active layer and the n-InGaP cladding layer. These defects are assumed to be formed by the development of some inclusion from the interface to the direction of <01n>.

5.2.3 Interface defects 2

Typical TEM image of misfit dislocation in an InGaAsP/InP structure is shown in Fig. 5.8. They are lying in the [110] or the $[1\bar{1}0]$ direction [25]. The Burgers vectors of these misfit dislocations have been determined by contrast experiment using the 220, $2\bar{2}0$, 400, and 040 reflections. From the results, almost all of the dislocations were determined to be 60 degree dislocations whose Burgers vectors are of the type (a/2) <011> or (a/2) <101> inclined to 45 degree to the (001) plane, and pure edge and screw dislocations were rarely observed.

On the other hand, the interaction between misfit dislocations is observed as their density increases. This is thought to result from the interaction between dislocations, which is a common phenomenon in any epitaxial systems such as III–V mismatched heterostructures, III–Vs/Si, SiGe/Si, etc. Typical examples are described next.

5.2.3.1 Change in the nature of dislocations (formation of pure edge or screw dislocations)

Fig. 5.9 shows a TEM image of a pure edge misfit dislocation. In Fig. 5.9A, both dislocations lying in the [110] and the $[\bar{1}10]$ directions are visible. However, in Fig. 5.9B, the dislocation lying in the $[\bar{1}10]$ direction is completely out of contrast. Therefore, from the g · b invisibility criteria, the Burgers vector of the dislocation is determined to be (a/2) [110] (edge type). The edge dislocation is assumed to be

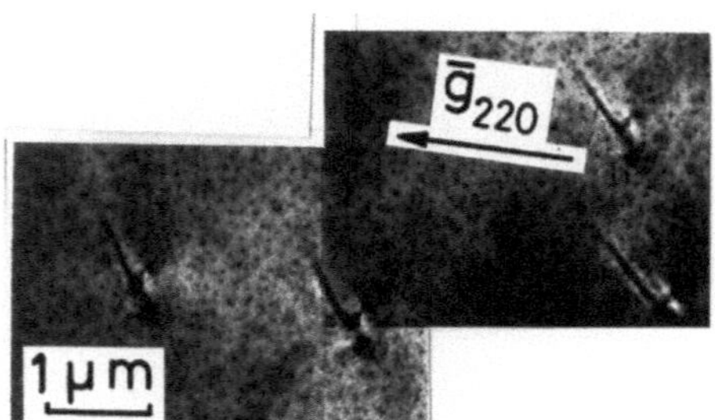

Figure 5.7 A plan-view transmission electron microscope image of inclusions in LPE-grown InGaAsP on the (001) GaAs substrate.

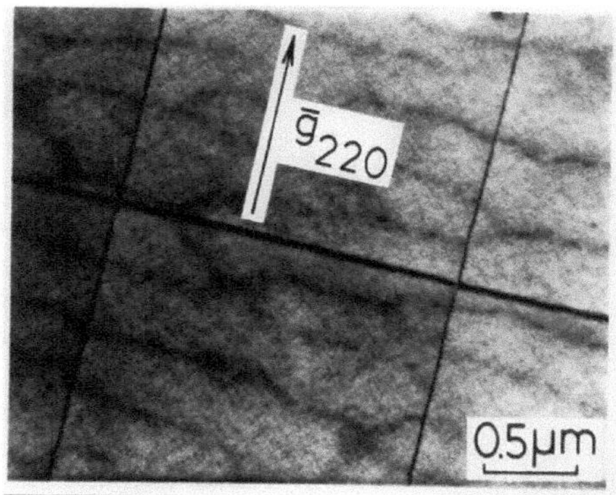

Figure 5.8 A bright-field transmission electron microscope image of misfit dislocations generated at the heterointerface of InGaAsP/InP [25].

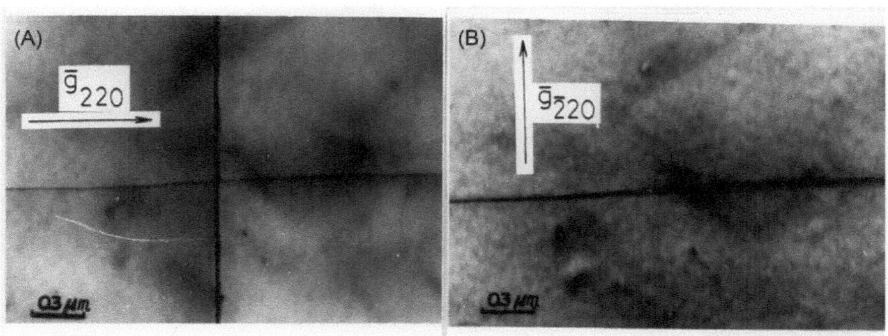

Figure 5.9 Bright-field transmission electron microscope images of a pure edge misfit dislocation.

generated by the reaction of two moving dislocations with different Burgers vectors, for example, by the reaction of two 60 degree dislocations as follows [26]:

$$\underset{(60^\circ)}{(a/2)[011]} \quad + \quad \underset{(60^\circ)}{(a/2)\left[10\bar{1}\right]} \quad = \quad \underset{(\text{edge or } 90^\circ)}{(a/2)[110]}$$

(screw for $\mathbf{u} = [110]$, edge for $\mathbf{u} = \left[1\bar{1}0\right]$, where $\mathbf{u}$ is a vector indicating the direction of the dislocation). Screw dislocation can also be generated by a similar reaction.

5.2.3.2 Annihilation reaction of misfit dislocations

A typical example of this phenomenon is shown in Fig. 5.10A–D. Fig. 5.10D is a bright-field TEM image of typical misfit dislocations caused by annihilation of two dislocations at their intersecting point. A contrast experiment shows that both

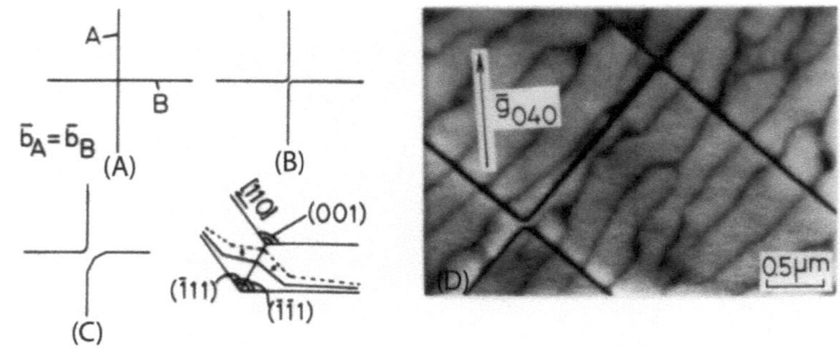

Figure 5.10 Annihilation reaction of misfit dislocation. (A)–(C) schematic diagrams for the mechanism of the annihilation and (D) a bright-field TEM image of this type of dislocations.

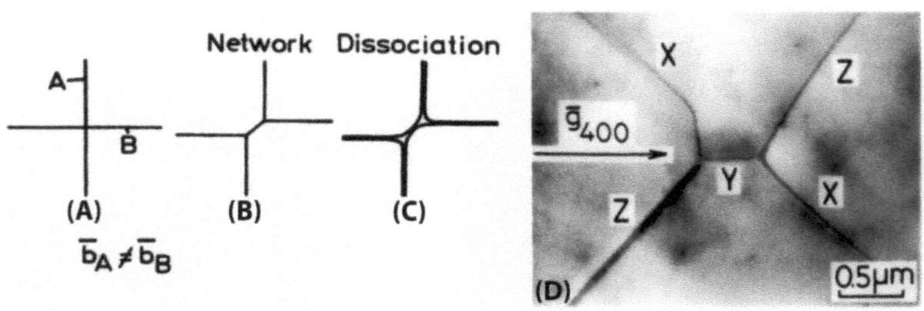

Figure 5.11 Formation of dislocation network at an intersecting point. (A)–(C) Schematic diagrams for the mechanism of the formation of dislocation network and (D) an example of a bright-field transmission electron microscope image of the dislocation network.

dislocations lying in the $<110>$ and the $\langle 1\bar{1}0\rangle$ directions are of 60 degree types with the same Burgers vectors. As schematically illustrated in Fig. 5.10A and B, the separated segment is L-shaped. These dislocations may be formed by dislocation glide motion at the edge of the L-shaped segment on the two different $<110>/\{111\}$glide systems after the annihilation reaction (see right-hand side of Fig. 5.10C) [27].

5.2.3.3 *Formation of the dislocation network at an intersecting point*

A typical example of this phenomenon is shown in Fig. 5.11A–D [25]. Fig. 5.11D shows a bright-field TEM image of an H-shaped dislocation network. In this case, the dislocation network consists of three segments denoted as X, Y, and Z. The segments X and Y are out of contrast under the $\bar{2}20$ and the 220 reflections, respectively. On the other hand, the segment Z shows residual contrast under the 040 reflection. These results indicate that the segments of the dislocation network differ from each other. This dislocation network could be generated as follows. First, two misfit dislocations with different Burgers vectors (A and B in Fig. 5.11A) arrive simultaneously at the intersecting point, and an H-shaped dislocation network with two nodes is generated (under the conservation of the total Burgers vector at each mode, as shown in Fig. 5.11B). After this reaction, each dislocation segment may dissociate into two partial dislocations if energetically favored (Fig. 5.11C).

5.2.4 Bulk defects

5.2.4.1 Defects in heavily Fe-doped InP crystals

First, we present results on precipitates generated in heavily Fe-doped InP crystals grown by MOVPE [28]. The InP crystals are grown on (001)InP substrates at 650°C. Vapor of ferrocene $Fe(C_5H_5)_2$ is used as an Fe-doping gas and its flow rate is varied in the range 0–100 mL/min. We evaluated first an Fe-doped InP crystal grown with the maximum gas flow rate (100 mL/min) which was examined. As previously reported by Nakahara and others [29], a large number of spherical FeP precipitates are observed in a dark-field image, particularly in the area close to the edge of the thin TEM specimen. They may be expected to stay on the surface of the specimen after etching since they are thought not to be dissolved by the Br_2/CH_3OH etchant. However, in the region slightly away from the edge of the specimen, the density of precipitates is found to be nearly constant, as shown in Fig. 5.12. Figs. 5.12A and B are bright-field and weak-beam dark-field TEM images of precipitates taken under the 220 and the 660 reflections.

From these images, one can differentiate between precipitates buried in the matrix and those lying on the surface in the following manner:

1. Precipitates exhibiting parallel Moiré fringes with constant spacing are buried in the matrix (we call these as type A precipitates).
2. Precipitates that do not exhibit Moiré fringes are lying on the surface (type B).

In the latter case, the probability of a direct double reflection between precipitates and matrix is considered to be much lower as a result of possible oxide presence and the small contact area between the two materials. Type B precipitates are

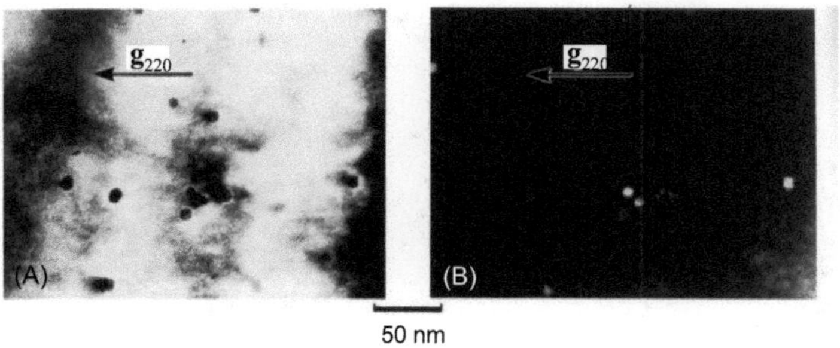

Figure 5.12 (A) A bright-field transmission electron microscope image of FeP precipitates in heavily Fe-doped InP crystal by metalorganic vapor phase epitaxy and (B) a weak-beam dark-field transmission electron microscope image of the region shown in (A).

generally out of contrast in the weak-beam image since there is no strain associated with them. However, they do occasionally exhibit bright contrast, presumably as a result of a sticking force between the precipitates and the matrix at the surface.

Fig. 5.13 is a plan-view high-resolution TEM image of a precipitate. The precipitate also exhibits Moiré fringes similar to those shown in Fig. 5.12. Lattice fringes corresponding to d_{200} for InP can be seen. Slight misalignments of lattice fringes are occasionally found at the precipitate–matrix interface (see arrows). This result suggests that there is little strain field at the interface, which is consistent with the fact that the lattice parameter of InP is very close to that of FeP. Thus the precipitates are coherent. Furthermore, from cross-sectional TEM observation, it was found that the precipitates are uniformly distributed in the crystal. On the other hand, precipitates attached to the surface were not found at all.

Next, the effect of the Fe-doping gas flow rate on the density and diameter of precipitates was also studied. Precipitates are observed in crystals grown with a gas flow rate higher than 20 mL/min. Their density increases as the gas flow rate increases. However, their diameter is almost constant. Precipitates are not observed in crystals grown with gas flow rates lower than 5 mL/min. These results are summarized in Table 5.1. From SIMS measurement, the total Fe concentration in sample 4 (gas flow rate is 5 mL/min) in Table 5.1 is determined to be $1 \times 10^{17}/cm^3$, which is the solubility limit of Fe in InP at 650°C.

Based on the results described above, a possible generation mechanism for FeP precipitates can be derived as follows:

Step 1: In the growing InP layer, when the doped Fe concentration exceeds the solubility limit, supersaturated Fe atoms tend to segregate at some nucleation centers via rapid diffusion.

Step 2: After Step 1, FeP precipitates start to grow.

Step 3: When the diameter of the precipitates reaches a critical value, they cannot grow any more against the strain field accumulated around them.

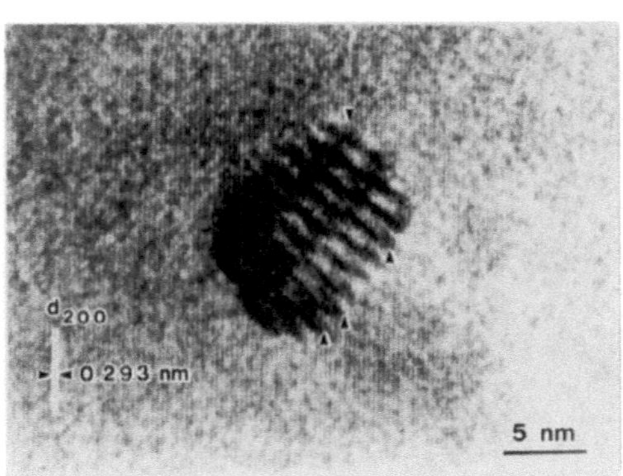

Figure 5.13 A high-resolution transmission electron microscope image of the FeP precipitates in a heavily Fe-doped InP crystal by metalorganic vapor phase epitaxy.

Table 5.1 Dependence of Fe-doping gas flow rate on the density and diameter of precipitates.

Sample no.	Gas flow rate (mL/min)	Precipitate density (cm^{-3})	Precipitate diameter (nm) average (max/min)	
1	100	1×10^{14}	8.3	(11.2/3.8)
2	30	3×10^{13}	5.8	(11.6/4.1)
3	20	8×10^{12}	12.9	(18.6/5.8)
4	5	$<5 \times 10^{10}$	–	–
5	1	$<5 \times 10^{10}$	–	–
6	0.5	$<5 \times 10^{10}$	–	–
7	0.2	$<5 \times 10^{10}$	–	–

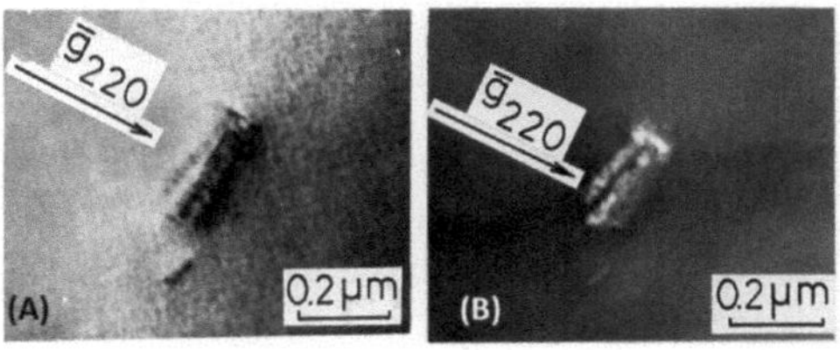

Figure 5.14 Transmission electron microscope images of rectangular-shaped precipitates obtained from heavily Zn-doped InGaAsP crystal: (A) a bright-field image and (B) a weak-beam dark-field image.

5.2.4.2 Defects in heavily (or highly) doped InGaAsP crystals

Next, we describe the results on the defects generated in heavily (or highly) doped LPE-grown InGaAsP crystals. The dopants are Zn, Cd (p-type), and Sn and Te (p-type) [30].

Zn-doped InGaAsP (1×10^{19}/cm^3): In InGaAsP (Zn) crystals, the observed defects were precipitate-like microdefects of 0.2–1.0 μm diameter. They were uniformly distributed, and their density was about 1×10^9/cm^3. Fig. 5.14A shows an example of a plan-view bright-field TEM image of typical precipitate-like microdefects. They are rectangular-shaped defects and are most frequently observed. They lie along the <110> direction and are about 0.25×0.1 μm^2 in size. A weak-beam dark-field TEM image of the defects was also taken as shown in Fig. 5.14B. This micrograph was obtained by a weak 220 beam under $s_{440} > 0$. As shown in this image, the defect consists of several microprecipitates (see those with dot-like bright contrast in Fig. 5.14B) and small dislocation tangles. These dislocation tangles may be caused by the relaxation of stress concentrated in the vicinity of the precipitates.

Cd-doped InGaAsP (1.5 × 10^{18}/cm^3*)*: In highly Cd-doped InGaAsP crystals, only particle-like microdefects or their clusters were found, and the precipitate accompanied by strain fields and/or small dislocation tangles, as observed in InGaAsP (Zn), were not observed. The density of these defects was very low compared with that in InGaAsP (Zn).

Sn-doped InGaAsP (2.5 × 10^{18}/cm3): In highly Sn-doped InGaAsP crystals, almost all of the observed defects were similar to those in heavily doped InGaAsP (Zn). These defects were typically 0.2–2.0 μm in diameter, uniformly distributed, and often accompanied by a large strain field. The estimated density of defects is nearly 1×10^{10}/cm^3. Bar-shaped precipitates lying in the <110> direction were very rarely observed.

Te-doped InGaAsP [(2.5–10) × 10^{18}/cm^3*]:* In heavily or highly Te-doped crystals, three kinds of defects were identified: (1) microdefect pairs with associated strain field (0.5 μm in diameter, 1×10^{10}–1×10^{11}/cm^3), (2) nonstructural precipitates (0.5 μm in diameter, very low density), and (3) dislocation dipole-shaped defects (0.2–0.3 μm long, very low density).

In all of the heavily or highly doped InGaAsP crystals, dislocation loops and faulted loops were not observed.

Generation mechanism of the defects in heavily or highly doped InGaAsP crystals: As described above, almost all of the defects observed in heavily or highly doped InGaAsP crystals are precipitates of certain kinds of compounds. Moreover, it was established by the weak-beam TEM technique that they were accompanied by large or small strain fields and small dislocation tangles (secondary defects caused by precipitate formation). The shape of the defects is quite similar to the microdefects observed in thermally annealed silicon crystals containing oxygen [31,32]. In this case, these structures may be a compound of silicon and oxygen (Si_xO_y), generated at nucleation centers during the thermal treatment. Furthermore, at high temperature (e.g., 1000°C), perfect dislocation loops are generated in the vicinity of the microdefects by prismatic punching as a consequence of the misfit stress field [32]. In the case of heavily or highly doped InGaAsP, if the defects are generated during crystal growth, a nonuniform distribution of dopants in abnormal growth conditions is expected to lead to their formation. A nonuniform dopant distribution can give rise to dopant-rich areas in the solid solution, causing defect generation. The possibility of defect formation during the cooling process after growth should also be considered. Immediately after crystal growth, supersaturated concentrations of dopant atoms exist interstitially in epitaxial layer. Microscopically, they take the form of localized solid solutions composed of the matrix elements and the dopant atoms. It is possible that these can segregate and form compounds in the epitaxial layer at temperatures below the growth temperature. In fact, considering the phase equilibrium data for binary compounds [33], it is easily seen that binary or ternary compounds consisting of the matrix elements In, Ga, As, and P, and the dopant atoms Zn, Cd, Sn, or Te, can exist as solid solutions in the epitaxial layers at these temperatures.

5.2.4.3 Defects in heavily or highly doped AlGaAs crystals

Next, defects in Si-doped and Ge-doped $Al_{0.1}Ga_{0.9}As$ epitaxial layers grown on (001) GaAs substrates by LPE are characterized by TEM [14]. The carrier concentration of the Si- or Ge-doped $Al_{0.1}Ga_{0.9}As$ layers varied in a range of $1 \times 10^{17}-9 \times 10^{18}/cm^3$. Faulted loops (that is, dislocation loops which consist of stacking faults) were observed locally in Ge-doped $Al_{0.1}Ga_{0.9}As$ layers. However, no faulted loops were found in Si-doped layers. Figs. 5.15A and B are TEM images of typical faulted loops in Ge-doped layers. They were 0.05–0.5 μm in diameter and existed in localized regions. Doubly faulted loops denoted by A and B in Fig. 5.15A and C in Fig. 5.15B were often observed. The density of the faulted loops tends to increase with increasing carrier concentration. These faulted loops are ellipsoidal-shaped, and the long axis is parallel to the $<110>$ direction indicating that they are lying on the {111} planes. Actually, their fault vectors are determined to be (a/3)$<111>$ by invisibility criteria in the TEM analysis. Furthermore, by an inside–outside contrast experiment, it was concluded that they are all determined to be interstitial type loops. Based on these results, one can conclude that they are Frank-type interstitial loops lying on the {111} planes.

It has been already reported that faulted loops and/or perfect dislocation loops are generated in bulk GaAs crystals heavily doped with Te [34,35], Se [35], and Si [36]. Narayanan and others reported that there was a high density of extrinsic faulted loops on the {110} plane and low density of intrinsic prismatic loops on the {110} plane in heavily Si-doped ($4 \times 10^{19}/cm^3$) GaAs crystals [36]. However, they were not observed in rather lightly Si-doped (approximately $1 \times 10^{18}/cm^3$) GaAs crystals. They concluded that the extrinsic faulted loops were generated by the segregation of Si atoms to the {111} planes. As for Si-doped $Al_{0.1}Ga_{0.9}As$ layers, the results of this work agree with those by Narayanan and others [36]. This is because the carrier concentration of Si-doped samples observed in this work varied in a range of $1 \times 10^{17}-3 \times 10^{18}/cm^3$. On the other hand, faulted loops were observed locally in comparatively lightly doped Ge-doped ($1 \times 10^{17}-9 \times 10^{18}/cm^3$) $Al_{0.1}Ga_{0.9}As$ layers. It is assumed that Ge atoms segregate more easily to the {111} planes than Si atoms in $Al_{0.1}Ga_{0.9}As$ layers. Therefore it is possible that the critical

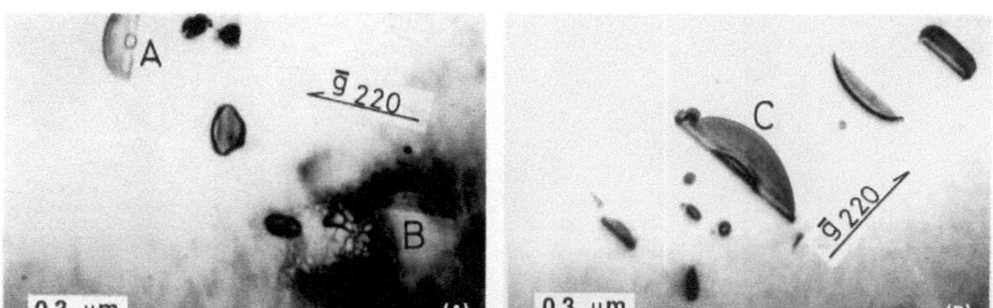

Figure 5.15 Transmission electron microscope images of faulted loops obtained from highly Ge-doped $Al_{0.1}Ga_{0.9}As$ crystals.

carrier concentration at which faulted loops are induced in the matrix depends on the dopants.

5.2.4.4 Defects in undoped InGaAsP crystals on GaAs substrates

The InGaAsP/InGaP material system is a promising one for visible and near-infrared "aluminum-free" lasers [37–39]. A typical plan-view bright-field TEM image of dislocation loops in undoped InGaAsP layer ($\lambda = 790$ nm) is shown in Fig. 5.16 [40]. These dislocation loops (which are indicative of gradual degradation) are 10–100 nm in diameter, and their long axis is in the <100> (loops denoted by L_1–L_4) or the <110> (loops denoted by L_5 and L_6) direction. By contrast experiment and inside–outside contrast experiment, the characteristics of these dislocation loops were determined as follows:

L_1 and L_2—fault vector: (a/2) $[10\overline{1}]$; character: intrinsic (vacancy type);

L_3 and L_4—fault vector: (a/2) $[\overline{1}0\overline{1}]$; character: intrinsic;

L_5 and L_6—fault vector: (a/3) $[\overline{1}\overline{1}1]$; character: extrinsic (interstitial type).

Thus both vacancy-type prismatic loops and interstitial-type Frank loops are shown to have been generated in the same crystal. As described previously, similar results have been observed in heavily Si-doped GaAs crystals. However, it is not yet understood why these different types of loops are generated on different planes.

Next, the depth profile of these dislocation loops was investigated. It is quite difficult to grow thick and flat LPE InGaAsP crystals (thicker than 0.5 μm) due to a strong liquid–solid interfacial reaction during growth and the subsequent local meltback of the substrate. Thus we focus only on a thick InGaP layer. Figs. 5.17A and B show typical plan-view bright-field TEM images of the region near the top surface of the InGaP layer. Dislocation loops are observed only in the region of the interface, implying that this region can act as a sink for excess point defects.

Concerning the generation of dislocation loops, several authors have shown that in GaAs and AlGaAs crystals, interstitial-type prismatic loops or faulted loops are

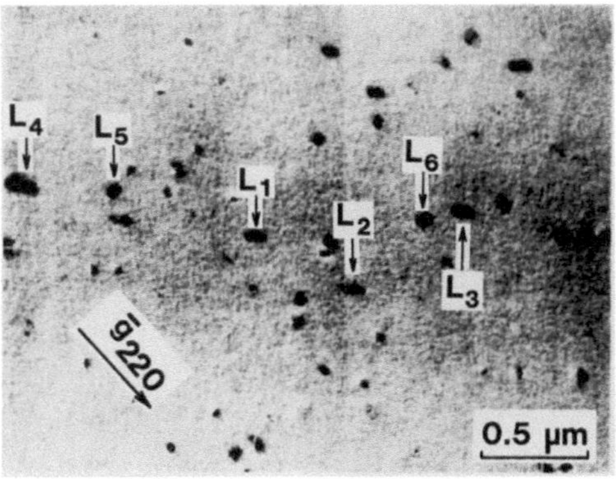

Figure 5.16 A plan-view bright-field transmission electron microscope image of two types of dislocation loops in undoped InGaAsP crystal grown on the (001) GaAs substrate.

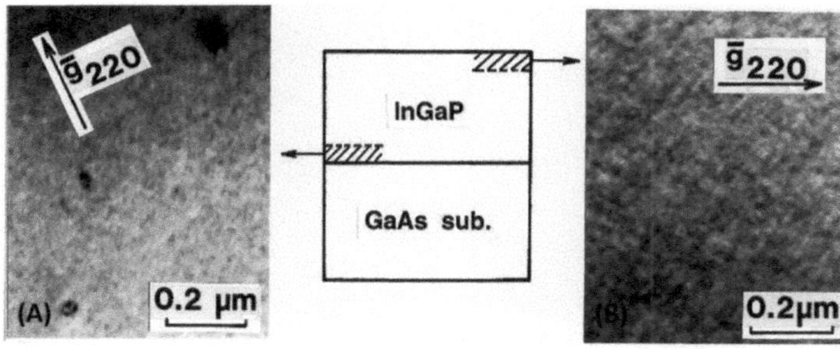

Figure 5.17 A depth profile of the dislocation loops in an undoped InGaP layer: (A) the region near the interface between the InGaP layer and the substrate and (B) the region near the top surface of the InGaP layer. Both plan-view transmission electron microscope images were taken by the 220 reflection.

generated by heavy doping or subsequent thermal annealing [14,34–36]. On the other hand, as previously described, no dislocation loops are generated in heavily doped InGaAsP crystals on InP substrates (only precipitate-like defects are present) [30]. In most cases, the generation mechanism for the dislocation loops is well explained by condensation of excess interstitial atoms at some nucleation centers [30,41].

However, in this case, both interstitial-type Frank loops and vacancy-type prismatic ones were generated in the same crystal, although the crystals were not intentionally doped. Therefore it is assumed that these dislocation loops are generated by excess interstitials and vacancies formed as byproducts of spinodal decomposition during growth and/or compositional fluctuation during cooling process after growth [40] (for detail, see Section 5.5.1).

5.3 Influence of grown-in defects on device reliability and degradation in III-V based optoelectronics

In this section we discuss the influence of four types of grown-in defects on the reliability of III–V optical devices: dislocations, dislocation loops, precipitates, and inclusions. Climb dislocations, also known as "dark line defects" or DLDs, are responsible for the majority of laser field failures observed. We discuss a number of causes of DLDs in the section that follows.

5.3.1 Influence of dislocations on reliability

Fig. 5.18A shows a bright-field TEM image of dislocation dipoles (the U-shaped section indicated by the arrow) corresponding to a $<100>$ DLD in a rapidly degraded AlGaAs DH LED [12]. Detailed analyses under various diffraction conditions reveals that these dislocation dipoles have Burgers vectors of $(a/2)<011>$ that are 45° inclined to the (001) plane and are of the interstitial type. These defects

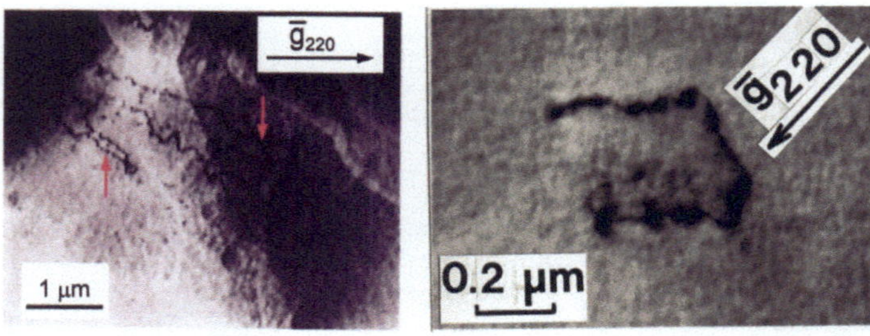

(A) (B)

Figure 5.18 Transmission electron microscope images of dislocation dipoles corresponding to <100> dark-line defects in rapidly degraded III–V optical devices: (A) a part of an extended dislocation network in a rapidly degraded AlGaAs DH LED and (B) a dislocation dipole at an initial stage of climb motion in a rapidly degraded visible InGaAsP/InGaP DH laser.

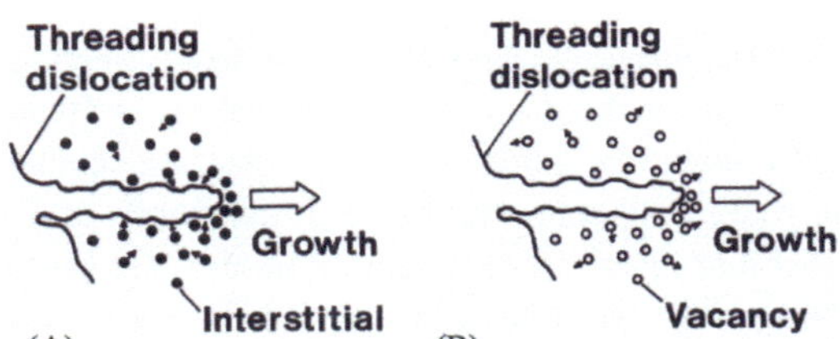

(A) (B)

Figure 5.19 Two proposed models for recombination-enhanced dislocation climb motion: (A) extrinsic defect model and (B) intrinsic defect model.

originate from TD that propagate from the substrate into the epitaxial layer during growth, and develop in the active region by recombination-enhanced dislocation climb (REDC) motion [42]. Prior to the above study, similar results also had been obtained for rapidly degraded AlGaAs DH lasers [8–11]. A dislocation dipole at an initial stage of dislocation climb motion in a rapidly degraded InGaAsP/InGaP visible DH laser is shown in Fig. 5.18B. Helical motion of the dislocation is clearly observed in this image. In addition, it is possible that dislocation dipoles are formed from (1) dislocation clusters originated from the interface between the epitaxial layer and the substrate as well as interface between epitaxial layers, (2) V-shaped dislocations, and (3) misfit dislocations.

Regarding the REDC, two typical models have been proposed. One is known as the "extrinsic defect model" [43] and is shown in Fig. 5.19A. In this model, only a single type of interstitial atom, for example, a Ga interstitial atom, is required to be absorbed into the dislocation core in order to enhance the dislocation climb motion, however it requires the emission of an arsenic vacancy. The second model, known as the "intrinsic defect model" [44], is shown in Fig. 5.19B. This requires emission of two types of vacancies (those of groups III and V atoms) from the dislocation core, for example, emission of Ga- and As-vacancies in GaAs.

It is currently unclear whether either of these two models is valid, but a number of physical parameters are believed to be candidates as factors contributing to REDC. First, since REDC is thought to involve the nonradiative recombination of minority carriers and to correspond to a local transfer of energy from a multiphonon mode to a lattice-vibrational mode excited by minority carrier nonradiative recombination at dislocations in the crystal [42,45], REDC is believed to depend essentially on the bandgap energy of the material. Table 5.2 summarizes the results on how readily REDC occurs in various compound semiconductors. In the III–V compound semiconductors, REDC occurs easily in most wide bandgap materials such as GaAs- and GaP-based ones, and InGaAsP materials lattice-matched to GaAs, while it does not in InGaAsP materials lattice-matched to InP. However, since InP, which has relatively wide bandgap energy, is also prone to suffer from REDC, bandgap energy alone is not a sufficient explanation. Other dominant factors for REDC are deep-levels which are related to such defects as dangling bonds and nonradiative recombination rates at the deep levels. In addition, since REDC is believed to be due to the absorption of interstitial atoms in the dislocation core or to the emission of vacancies, it is thought that the magnitudes of generation energy and migration energy for point defects may also have a major influence on REDC. In the case of the AlGaInAs/InP system which is a very popular material nowadays (the dominant material for most newer 1310 nm transceivers), it should be noted

Table 5.2 A summary of the easiness in recombination-enhanced dislocation climb under carrier injection or optical excitation in various compound semiconductors and alloy semiconductors.

Material	Easiness of recombination-enhanced dislocation climb
GaAs	Easy
AlGaAs	Easy
GaP	Easy
GaAsP	Easy
InP	Difficult
InGaAsP on InP	Difficult
AlInGaAs on InP	Moderate
InGaAsP on GaAs	Easy
InGaAsp on GaAs	Easy
ZnSe	Very easy
InGaN	Difficult
AlGaN	Difficult

that DLDs grow readily in it, although somewhat more slowly than GaAs quantum wells (QWs) (failure occurs over hundreds of hours, rather than <1 h in GaAs).

The choice of materials system has a strong impact on reliability, and in many cases, for a given wavelength, multiple material systems could be used, so selecting one that is resistant to REDC can significantly reduce the risk of random failures. If more flexibility is available, and the wavelength is not set by a standard, then it makes the job of choosing a reliable material system even easier.

5.3.2 Two-step degradation processes (REDG and REDC)

There are two major rapid degradation modes: REDC and recombination-enhanced dislocation glide (REDG). We have previously discussed REDC. On the other hand, REDG often takes place when the light-emitting region is highly strained. In this process, dislocations are generated by glide motion at a highly strained active region via optical excitation. Moreover, when REDG occurs during operation, the generated glide dislocations (or "punch out dislocations") may have the opportunity to climb immediately after they are generated or after a certain interval depending on the operating condition. Based on this assumption, we can expect with certain possibility to meet this degradation mode as follows:

5.3.2.1 Two-step degradation process in ridge-waveguide-type lasers (see Fig. 5.20):

The first step: REDG occurs at a region where a certain amount of stress is accumulated during the operation, for example, the bottom part of the ridge on both sides of the stripe (Fig. 5.20B).

The second step: dislocation networks are generated from these glide (or punch out) dislocations by REDC (Fig. 5.20C).

5.3.2.2 Two-step degradation process in AlGaAs VCSELs (see Fig. 5.21):

In VCSEL A, the MQW active layer is located below the bottom of the etched mesa, whereas in VCSEL B, the active layer is located in the middle of the etched mesa. In both cases, there is the possibility of generating the punch out dislocations during operation at four regions indicated by a–d in Fig. 5.21. In most commercially available high-quality VCSELs, the probabilities of causing cases a and b are very low during long-term operation in the field. On the other hand, elimination of the cases c is not easy because punch out dislocations are often originated from the crack at the oxidized sidewall of the etched mesa, so a deep proton implant is often used to electrically inactivate this part of the active region. Case d is also an issue for VCSEL A, and could be addressed with the same type of implant. In this case, the punch out dislocations are unintentionally introduced due to strain relaxation at several fabrication processes at high temperatures such as oxidation of the etched mesa and metallization of the p-electrode. Therefore it is suggested that we had

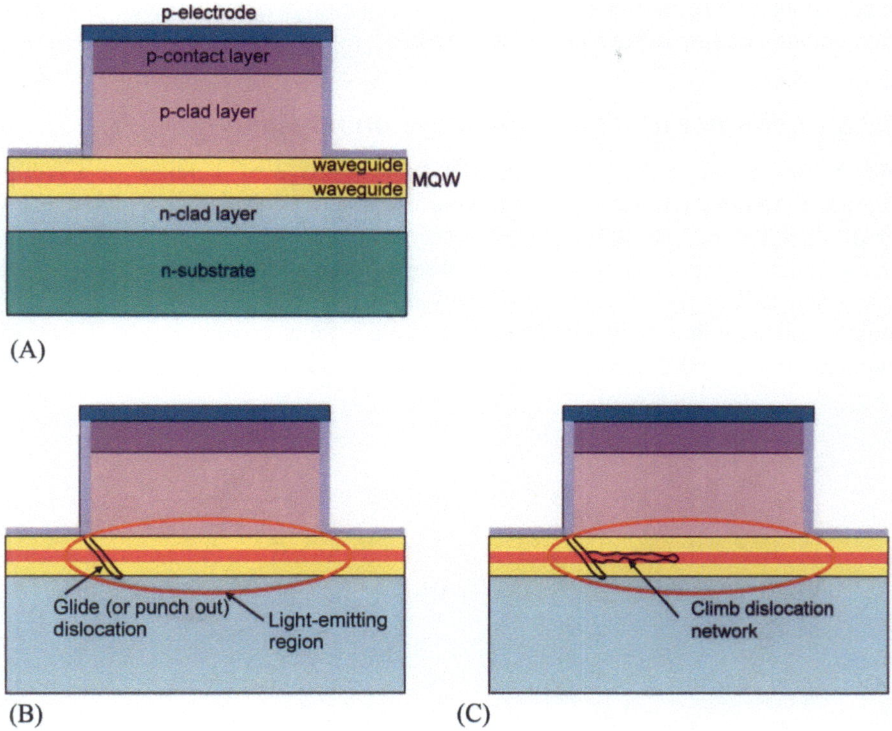

Figure 5.20 Schematic diagrams for a two-step degradation mode in a ridge waveguide-type edge emitter.

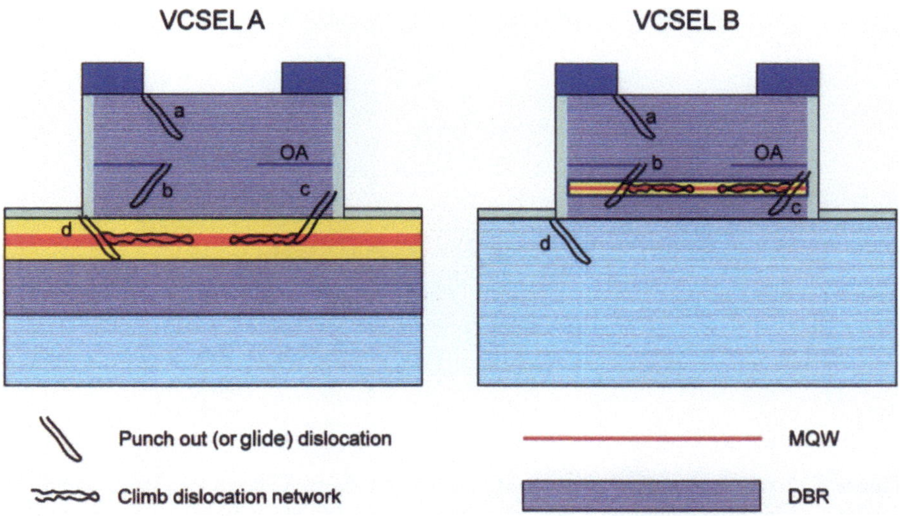

Figure 5.21 Schematic diagrams for two-step degradation mode in two types of VCSELs. In both diagrams, oxide aperture is indicated by the "oxide aperture" (abbreviated as OA in the figures).

better adopt VCSEL B from the viewpoint of reliability, which is much more common among commercial designs on the market.

5.3.3 Influence of dislocation loops on reliability

Fig. 5.22A shows a plan-view bright-field TEM image of a defect in a rapidly degraded AlGaAs DH LED [13]. It is an extended dislocation loop lying in the [010] direction on the (001) plane. It accompanies many jogs and has striking resemblance to the dislocation dipoles observed in rapidly degraded AlGaAs DH lasers and LEDs. By TEM contrast analysis, it was clarified that the dislocation loop is an interstitial-type prismatic loop with a fault vector of (a/2) [101] 45° inclined to the (001) plane. From this result, it is assumed that the extended

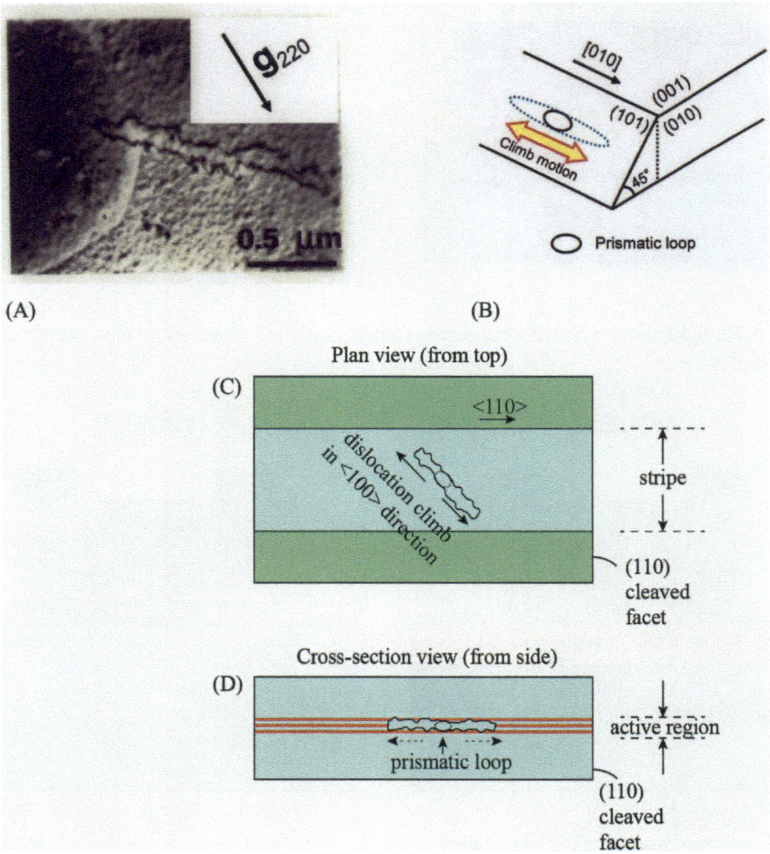

Figure 5.22 (A) A plan-view bright-field transmission electron microscope image of an extended dislocation loop from a prismatic loop on the (101) plane by REDC in a rapidly degraded AlGaAs DH LED and (B)–(D) schematic diagrams illustrating the generation mechanism for the extended dislocation loop in (A).

dislocation loop originated from a prismatic microloop on the (101) plane during operation by a mechanism similar to REDC (see Figs. 5.22B–D).

Another type of extended dislocation loop is shown in Fig. 5.23. Fig. 5.23A is a plan-view bright-field TEM image of an extended dislocation loop corresponding to a <110> DLD normal to the stripe in a rapidly degraded InGaAsP/InGaP DH laser [46]. The extended dislocation loop is a "faulted loop" which consists of a stacking fault and lies in the [110] direction on the (111) plane (whereas the prismatic loop shown in Fig. 5.22 is an unfaulted loop which does not consist of stacking fault). We call this a Frank loop. Similar to the defect shown in Fig. 5.22, this defect also has many jogs. The extended faulted loop was determined to be of an interstitial type with a fault vector of (a/3) [111] by TEM contrast analysis. The presence of both vacancy-type prismatic loops and interstitial-type Frank loops in undoped

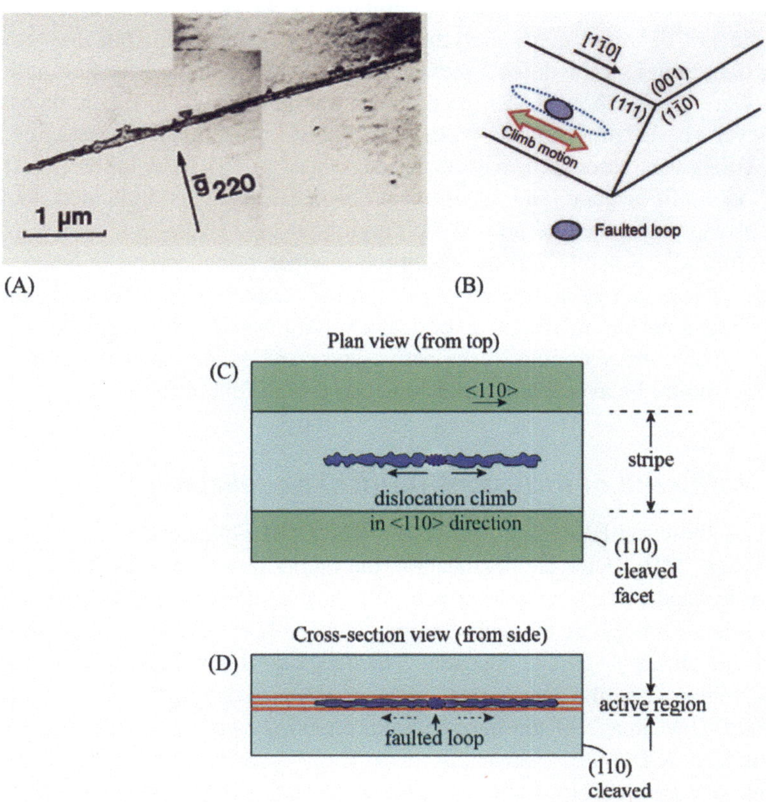

Figure 5.23 (A) A plan-view bright-field transmission electron microscope image of an extended dislocation loop originated from a faulted loop on the (111) plane by recombination enhanced dislocation climb motion in a rapidly degraded InGaAsP visible DH laser and (B)–(D) schematic diagrams illustrating generation mechanism for the extended dislocation loop in (A).

InGaAsP crystals grown on (001) GaAs substrates as byproducts of CMS formation is described in the previous Section 5.2.4. Thus it is expected that the faulted dipole was formed by the REDC of the interstitial-type Frank loop generated in the undoped InGaAsP active layer during growth (see Figs. 5.23B–D).

5.3.4 Influence of precipitates in heavily doped InGaAsP crystals on reliability

As described previously, most of the defects observed in the heavily doped InGaAsP crystals were precipitate-like microdefects about 0.1–2.0 μm in diameter, some of which were accompanied by small dislocation tangles and localized strain fields.

DSD that appeared in InGaAsP/InP DH LEDs aged at high temperature and have been proven to be bar-shaped precipitates by TEM [47]. However, the output power did not decrease even after the DSDs were generated. This may be due to the fact that, although the defects themselves act as nonradiative recombination centers, they are small (about 0.5 μm in diameter) in comparison to the thickness of the active layer (1–1.5 μm). On the other hand, in InGaAsP/InP DH lasers, the formation of DSDs was associated with rapid degradation [48]. A detailed investigation by cathodoluminescence and electron probe microanalysis indicated that these DSDs correspond to the Ga- and As-rich region.

Therefore it is possible that the precipitates which are observed in heavily doped InGaAsP crystals act as nonradiative recombination centers. Furthermore, they may develop under carrier injection or optical excitation and cause degradation of the materials. Thus heavy doping of the active layer and/or the cladding layer in lasers and LEDs should be avoided in order to achieve high reliability.

5.3.5 Influence of inclusions (In-rich) on reliability

Fig. 5.24A shows an EL image of InGaAsP/InP DH LED degraded during operation at 170°C for 100 h. After the degradation, the output power decreased by 30%–40% and large leakage currents were observed [49]. At the initial stage of operation, no dark defects were observed in the light-emitting region, but later, five <100>-oriented cross-shaped DLDs (C-DLDs) appeared. The darkness of the C-DLDs was strongest at the core region (i.e., the intersecting point of the cross). Fig. 5.24B is a plan-view bright-field TEM image of the light-emitting region shown in Fig. 5.24A. Five cross-shaped defects (CDs) are observed, corresponding to the C-DLDs in Fig. 5.24A. The CDs have one core and four arms, and the arms have several branches. The width of the arms is about 1 μm near the core region and becomes narrower away from the core. The core itself exhibits strong dark contrast. The arms and branches are not straight lines, but curve slightly from side to side. Contrast analysis was carried out on the core region and the edge of the arm. It was found that the CDs consist of two types of defect structures: (1) small dislocation clouds similar to cell structure in work-hardened metal, observed in the main body of the branches; (2) fibrous defects

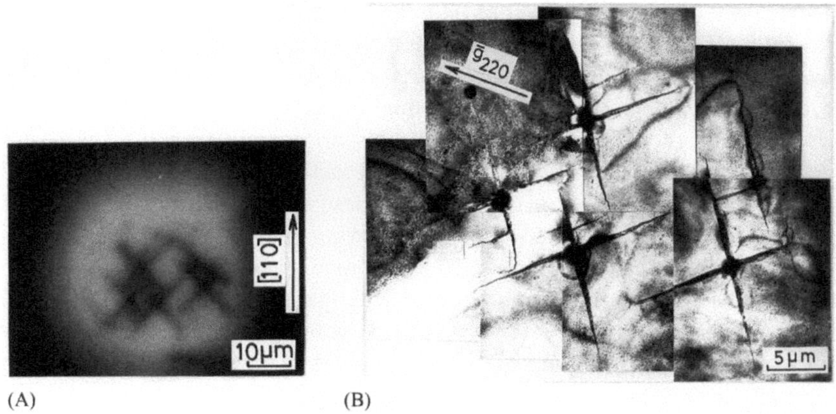

Figure 5.24 (A) An EL image of cross-shaped dark-line defects (C-DLDs) in a rapidly degraded InGaAsP/InP DH LED aged at higher than 170°C and (B) a plan-view bright-field transmission electron microscope image of cross-shaped defects corresponding to the C-DLDs shown in (A).

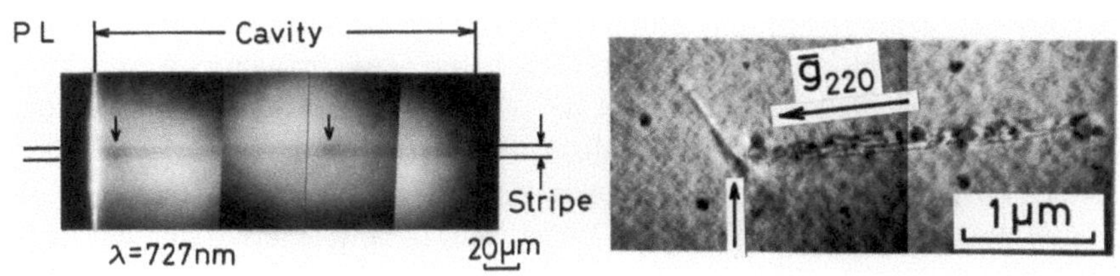

Figure 5.25 (A) Photoluminescence image of short <110> dark-line defects in a catastrophically degraded InGaAsP/InGaP visible DH laser and (B) a plan-view bright-field transmission electron microscope image of a defect structure corresponding to the short <110> dark-line defects shown in (A).

observed mainly around the ends of the CDs. Furthermore, by STEM/EDX analysis, it was clarified that the core regions of the CDs consist of an In-rich inclusion.

Based on these results, it was assumed that these defect structures were formed by the following mechanism: (1) heating and melting of the core region (In-rich inclusion) at the initial stage of operation; (2) subsequent propagation of molten regions from the core region to the four equivalent <100> directions; (3) formation of the two types of defect structures after the cooling of the molten regions. In the case of LDs and LEDs fabricated from MBE- or MOVPE-grown crystals consisting of In/Ga droplets, a similar degradation mode can be expected.

Fig. 5.25A shows a photoluminescence (PL) image of dark defects revealed in a catastrophically degraded InGaAsP/InGaP DH laser [16]. In this case, the dark defects exist inside the stripe region, and are in fact very short <110>DLDs. Fig. 5.25B is a TEM image of a defect corresponding to one of the short

<110>DLDs. The defect structure is quite similar to that observed in degraded diodes by COD, but develops from a defect lying in the direction of <100> (shown as an arrow in Fig. 5.25B). This type of defect was also observed in the InGaAsP active layer outside the stripe region, as shown in Fig. 5.7. As previously described, these are presumably defects originating from In-rich inclusion generated at the interface between the InGaAsP active layer and the n-InGaP cladding layer during the LPE growth. Based on these results, the mechanism for degradation in this case is assumed to be as follows: (1) at the initial stage of lasing, heating of the In-rich inclusion occurs; (2) immediately after this, melting of the inclusions takes place; (3) the molten region propagates in the <110> direction; (4) since the original inclusion is too small, the molten region terminates in a very short distance; (5) after cooling of the very short molten region, a defective region was formed there.

5.4 Grown-in defects in III-nitrides

5.4.1 Misfit dislocations

Since GaN substrates were not practically used in the early stage of development of III-nitrides-based light-emitting devices, GaN-based semiconductors were grown by performing heteroepitaxy on sapphire or SiC substrates. (Even now, sapphire or SiC substrates are still used for commercialized light-emitting diodes, while GaN substrates are only used for high-performance laser diodes.) A large number of structural defects are generated due to a difference in lattice constant and thermal expansion coefficient between GaN and the substrate. The crystal quality has been dramatically improved by the low-temperature buffer layer technology developed by Amano et al. [50], which has made it possible to reduce TD from $10^{11}/cm^2$ or more to $10^8-10^{10}/cm^2$.

Fig. 5.26 shows representative cross-sectional TEM images of a GaN film on the sapphire substrate grown by MOCVD. It can be observed that TD extending almost from the GaN film/sapphire interface to the c axis have occurred. These TDs are categorized into three types: pure edge with Burgers vector $1/3\langle\overline{2}110\rangle$ (**a**-type dislocation), a mixed character with Burgers vector $1/3\langle\overline{2}113\rangle$ ((**a** + **c**)-type dislocation), and pure screw with Burgers vector [0001] (**c**-type dislocation).

Fig. 5.27 shows magnified TEM images of the low-temperature buffer layer. The low-temperature buffer layer has many structural defects, including not only dislocations but also stacking faults, and has very poor crystallinity. However, the crystallinity of the upper part of the low-temperature buffer layer is improved. This is because many dislocations are inclined and do not propagate to the upper part due to a high probability of mutual interaction. Ponce et al. reported that in the early stage of growth of GaN on sapphire, three-dimensional growth occurred with many nucleation sites and tilted or twisted grains were formed [51]. Each grain boundary dislocation mitigates the tilt/twist between grains and misfit between GaN

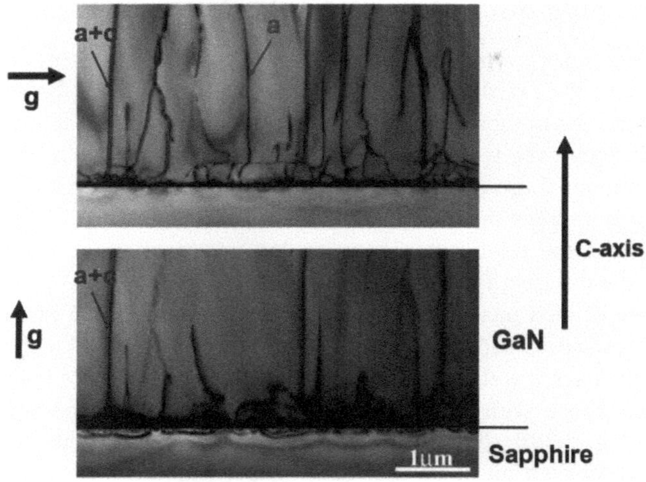

Figure 5.26 Representative microstructure of an MOCVD-grown GaN film on a sapphire substrate. The upper image can visualize all three types of dislocations, while the lower image can visualize mixed (**a** + **c**) and (pure screw) **c**-type dislocations.

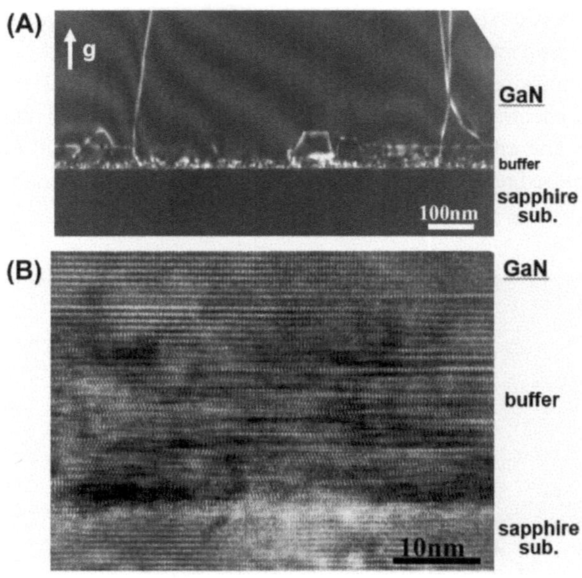

Figure 5.27 Magnified transmission electron microscope images of GaN film/sapphire substrate interface: (A) weak beam dark-field image; (B) high-resolution lattice image.

and sapphire. The dislocations are then inclined more by the three-dimensional growth, and the propagation of the dislocations to the upper part of the low-temperature buffer layer is suppressed.

5.4.2 Threading dislocation and optical properties [52,53]

It is known that when even a single threading dislocation is present in III–V light-emitting devices, the external quantum efficiency and reliability become poor, due to DLD growing out of those TD. By contrast, GaN-based LEDs show high external quantum efficiency even with TD as high as 10^8–10^{10} and have reliability that can withstand commercial use [54]. However, it was demonstrated that the reduction of dislocations can improve the lifetime of GaN-based laser diodes [55,56]. In this section, we describe the relationship between the nature of the TD and the optical properties of the different types of dislocations.

Three types of etch pits, labeled α, β, and γ, were observed in an HCl vapor-etched GaN surface as shown in the SEM image in Fig. 5.28A. The α-, β-, and γ-type etch pits have a well-defined hexagonal shape with a large core, an ambiguous hexagonal shape with a small core, and a polygonal shape without a core, respectively. Fig. 5.28B shows the depth profiles of the etch pits observed using AFM. The α-type etch pit has a triangular shape, which has a $\{1\bar{1}02\}$ facet. The γ-type etch pit has an arc shape. The β-type etch pit has a combination of a triangular and trapezoidal shape, which seems to be the result of a combination of α- and γ-types. Fig. 5.29 shows a cross-sectional TEM image of the three types of etch pits. The α-type etch pits originate mostly from a screw dislocation with Burgers vector of $\mathbf{b} = [0001]$, and slightly nanopipe. The β- and γ-type etch pits originate from a mixed dislocation with $\mathbf{b} = 1/3\langle 11\bar{2}3\rangle$ and an edge dislocation with $\mathbf{b} = 1/3\langle 11\bar{2}0\rangle$, respectively.

Band-edge emission was observed at an energy of 3.42 eV in the room-temperature PL spectra of Si-doped GaN. The Si concentration was maintained at $2.0 \times 10^{18}/\text{cm}^3$. The emission should be ascribed to donor-bound excitons. The PL intensity of band-edge emission is strongly related to the density of screw and

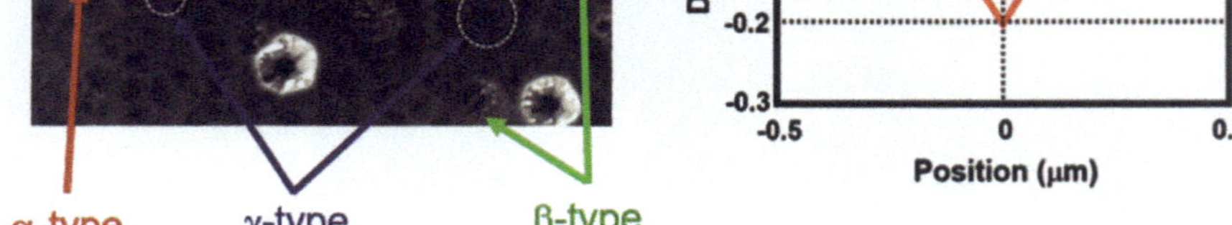

Figure 5.28 (A) Typical SEM image of an etched GaN surface. Three types of etch pits, labeled α, β, and γ, were observed. The contrast of the γ-type etch pit is faint, but its density is higher than the other two types. It is observed as a mottled or irregular "fish-scale-like" surface morphology with several dozen micron-sized domains observed in this image. (B) Schematic depth profiles of α-, β- and γ-type etch pits.

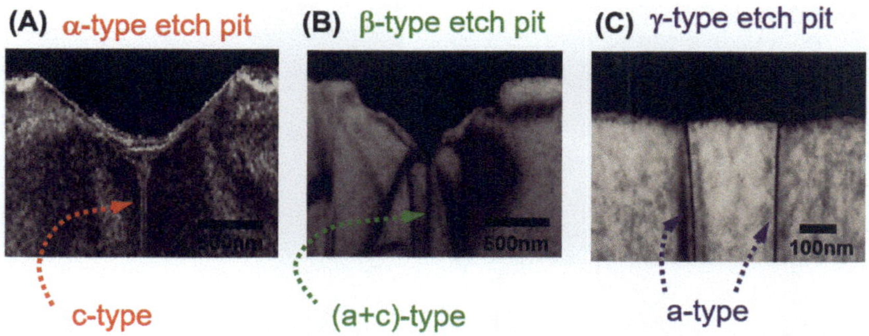

Figure 5.29 Cross-sectional transmission electron microscope images of α-, β- and γ-type etch pits.

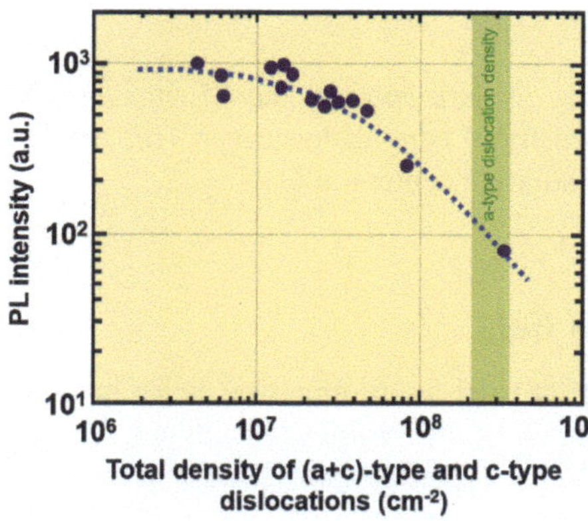

Figure 5.30 Room-temperature photoluminescence intensity dependence on the total density of (mixed) (a + c)-type and (screw) c-type dislocations for Si-doped GaN, where the density of (edge) a-type dislocations was nearly identical for all observations, as shown by the green band. The Si concentration was maintained at $2.0 \times 10^{18}/\mathrm{cm}^3$. The dislocation densities were estimated using etch-pit observation.

mixed dislocations, as shown in Fig. 5.30, although the edge dislocation dominated the total density of the three line defects and the density was relatively constant. The density of the edge dislocations was $2-3 \times 10^8/\mathrm{cm}^2$ for all of the experimental observations in the figure. This dependency indicates, therefore, that the screw and mixed dislocations should act as strong nonradiative centers in the GaN layer, and are responsible for most of the nonradiative recombination.

Fig. 5.31 shows an SEM image of etched GaN:Si surface and a monochromated CL image of as-grown GaN:Si surface. These images were taken using two $5 \times 5\ \mu\mathrm{m}^2$ samples which were cut from the next places of the same wafer. The CL image was taken at a wavelength of 362 nm, which corresponds to the band-edge emission of GaN:Si. Using etch-pit observation, the density of edge, screw, and mixed dislocations was estimated to be 1.0×10^9, 1.1×10^7, and $2.1 \times 10^7/\mathrm{cm}^2$,

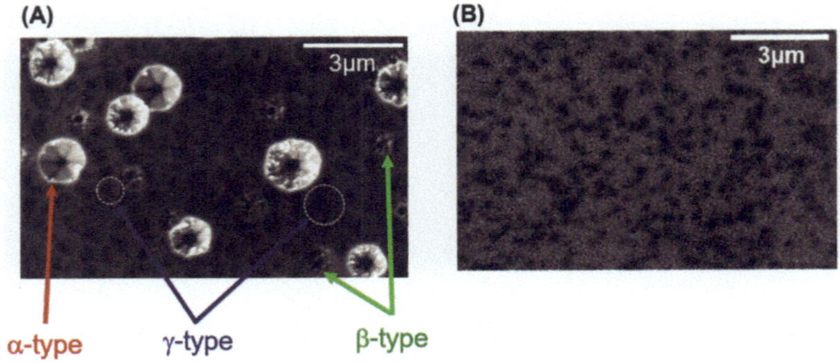

Figure 5.31 (A) SEM image of etched GaN:Si surface and (B) monochromatic CL image of as-grown GaN:Si surface at a wavelength of 362 nm. These images were taken using two 5×5 mm samples which were cut from the (neighboring) adjacent locations of the same wafer.

respectively. The density of dark spots in the CL images was around $5 \times 10^8/\text{cm}^2$, which is close to the density of edge dislocations. This result shows that edge dislocations also act as nonradiative centers in GaN.

5.4.3 V-defects in GaN

A GaInN ternary alloy is used as an emission layer in order to fabricate a GaN-based light-emitting device in the visible light region. The growth condition of the GaInN layer by MOCVD is typically 300°C lower than that of GaN since a lower growth temperature is required because indium is difficult to incorporate at higher growth temperatures. Low-temperature growth promotes the facet growth and formation of V defects, which consist of six $\{1\bar{1}00\}$ planes, and form an inverted hexagonal pyramid, as shown in Fig. 5.32. The V-defect mainly occurs from TD [57–59].

Since different facets form from the c-plane in the V-defect, the thickness/composition of GaInN layers vary, as shown in Fig. 5.33 [60]. Hanglieter et al. point out that the V-defect acts as a potential barrier for GaInN quantum well layers, which prevents injecting carriers into TD [61]. For example, when the indium composition of GaInN QW is 12% and the film thickness is 3.5 nm, the potential barrier is about 300–400 meV. Therefore the threading dislocation surrounded by the potential barriers can be prevented from acting as nonradiative centers. Additional scattering improves extraction efficiency in LEDs, but is detrimental to lasers, since it kicks photons out of the much narrower solid angle for most laser designs. Thus in LEDs, the external quantum efficiency is improved by V-defect scattering. Furthermore, the V-defects can be the source of reverse-bias leakage since V-defects accompanied by TD can act as a conduction path [62]. However, for the laser diodes, V-defects act as scattering centers and, thus the laser characteristics can be degraded.

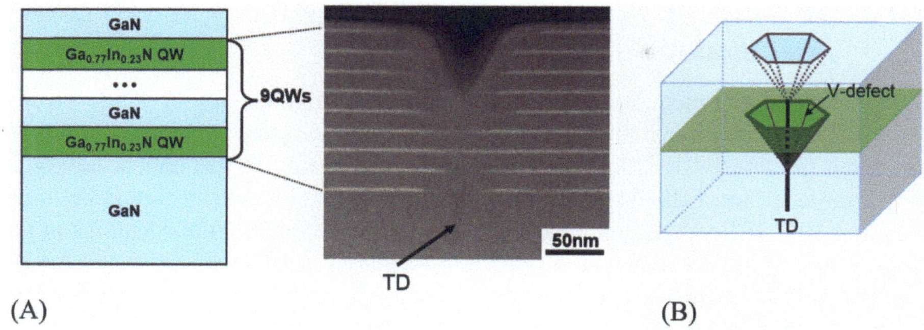

Figure 5.32 (A) Cross-sectional HAADF-STEM image of GaInN/GaN MQWs at and near a V-defect. The nine bright stripes are GaInN layers together with a schematic illustration of the sample structure. The V-shaped pit is initiated from the threading dislocation indicated by the black arrow. (B) Schematic illustration of V-defect.

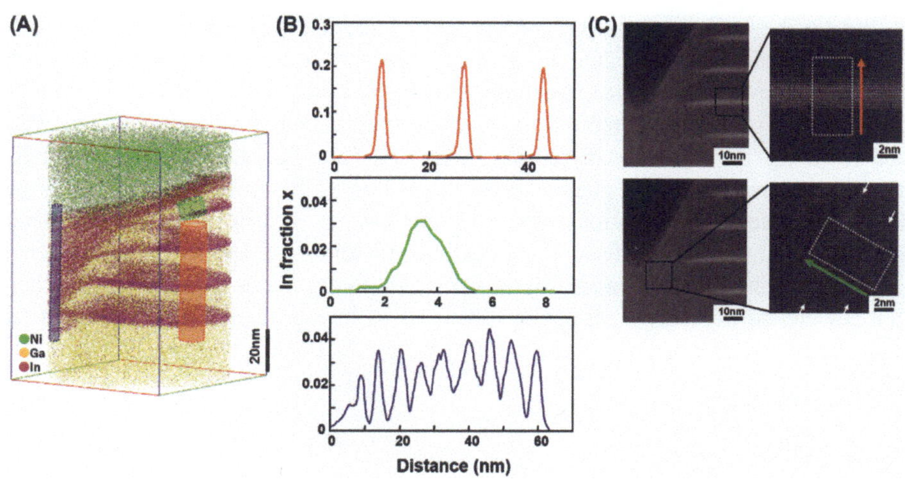

Figure 5.33 (A) Atom probe tomography of In (violet), Ga (yellow), and Ni (green) atoms at and near a defect (in this case, the nickel was used as a capping layer, and indicates where the sample surface would have been prior to sample preparation). We analyze the indium mole fraction vs. depth in three different-colored cylindrical data-sampling regions. The blue cylindrical region is in the vicinity of the center of the V-defect. The green cylindrical region is from the slope region of the fourth quantum well from the bottom, to give us a detailed profile for that one quantum wells. In the red cylindrical region, we look at the three flat quantum wells from the bottom. (B) One-dimensional In concentration profile from the red cylindrical region, the green cylindrical region, and the blue cylindrical region in Fig. 5.30A. (C) HAADF-STEM images taken from one of the quantum wells layers of the flat region and slope region. Faint inclined bright lines were observed along the V-defect slope.

5.4.4 Influence of dislocations on device reliability in GaN-based laser diodes

In the early development stage of GaN-based LDs, a high density of TD (typically $10^8-10^{10}/cm^2$) was inevitable, since samples were grown on sapphire substrates as described previously. It is known that a single dislocation in the optical device is very harmful in most III–Vs [63]. Therefore the dislocation density should be minimized to $10^3-10^4/cm^2$, which means that there are essentially no dislocations in the stripe region. Using a dislocation reduction technique such as the epitaxial lateral overgrowth (ELO) technique [64], the dislocation density can be minimized to $10^6/cm^2$, which is two orders of magnitude smaller than that in the stripe region. However, the TEM observations for the degraded GaInN LDs revealed that neither dislocation multiplication from the TD nor any structural changes to the TD were observed, despite the fact that there were 10 TD in the laser stripes of the LDs [65]. Therefore the dislocations in GaInN lasers do not seem to be harmful to device reliability, although dislocations act as nonradiative centers.

In the conventional III–V LDs, the existence of a single dislocation in the stripe region is fatal since dislocations can rapidly multiply (through growing DLD or "DLDs") from preexisting dislocations by carrier recombination [63]. There are two modes for the dislocation multiplication: REDG and REDC process.

We first discuss the REDG process. the crystal structure of conventional III–V LDs is a zinc-blende (cubic) structure, whose slip system is in the {111}<110> directions (Fig. 5.34A). Therefore dislocation multiplication by the REDG process occurs along the <110> directions, since the intersections between the (001) planes and the {111} slip planes are in the <110> directions. In these materials systems, 60-degree dislocations whose Burger's vector lies in the {111} slip plane are quite common. In order for dislocations to glide, an applied shear stress is required which is mainly caused by misfit strain due to lattice constant and/or thermal coefficient mismatch. An applied shear stress for 60-degree dislocations on the

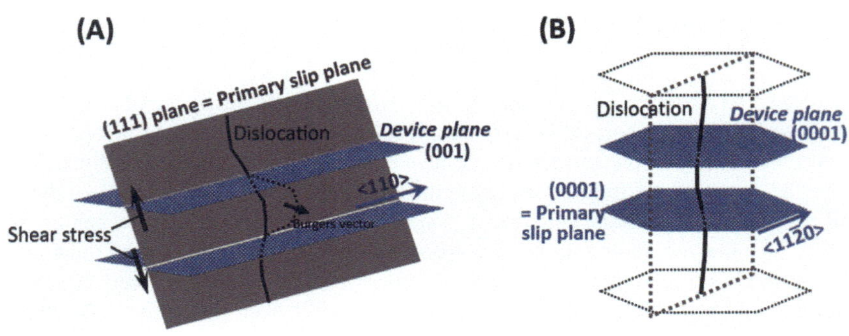

Figure 5.34 Schematic illustration of the relation between slip planes and threading dislocations. The case of (A) zinc-blende (cubic) structure and (B) wurtzite (hexagonal) structure. Dislocation in (A) easily glides, but dislocations in (B) hardly glide.

{111} plane is not zero, since the primary slip planes of the dislocations are inclined to the (001) plane of the device interface.

In the case of GaN-based LDs, their crystal structure is of the wurtzite (hexagonal) type and their primary slip system is the (0001) $\langle 11\bar{2}0 \rangle$ directions (Fig. 5.34B). The lattice misfit does not give rise to a shear stress on the primary slip systems, since the primary slip planes are parallel to the device interface (0001) plane. Thus the dislocations are thought to hardly glide.

On the other hand, the REDC process is accompanied with the absorption/emission process of point defects at dislocation cores during the electron–hole recombination process [42,66]. The REDC process is frequently observed in GaAs- and GaP-based optical devices, for example as DLD growth. The DLD growth is observed in the 808 nm InGaAsP LDs. However, DLD growth is not observed in 1.3 and 1.55 μm InGaAsP LDs. Therefore this process cannot be simply explained by a bandgap energy and materials system [42,66]. Several models for REDC process have been proposed [67]. The REDC process has not been observed in GaN-based LDs so far; however, this could be a critical issue for the development of higher power GaN-based LDs.

5.4.5 Mg-related pyramidal defects

P-type layers for GaN-related materials are usually obtained with doping Mg [67,68]. Since the Mg acceptor level is fairly deep, a high level of Mg doping is required to obtain a high concentration of free holes. However, the hole concentration saturates as the Mg concentration increases, so the doping is limited. In the course of our study of LDs, inverse pyramidal-shaped defects (hereafter referred to as "pyramidal defects") were often found in Mg-doped layers by TEM observations. Such defects also have been reported previously [69–71]. The pyramidal defects may increase the resistivity of the p-type layers, therefore these should be eliminated.

Fig. 5.35A shows a high-angle annular dark field scanning TEM (HAADF-STEM) image of the pyramidal defect. The structural model based on the previous results is shown in Fig. 5.35B [71]. The model shows that the inside of defects has inverse polarity to the matrix and the boundaries of the defects consist of Mg-rich layer such as Mg_3N_2 and $(Ga_xMg_{0.75(1-x)})N$ [71].

To investigate Mg-related defects having a three-dimensional structure in detail, a sample doped with about 1×10^{20} atoms/cm^3 of Mg (by secondary ion mass spectrometry (SIMS) analysis) was analyzed by three-dimensional atom probe analysis (3DAP). Fig. 5.36A shows a 3DAP reconstructed image. The purple dots in the figure represent Mg atoms, which are extracted from $^{24}Mg^{2+}$, $^{25}Mg^{2+}$, and $^{26}Mg^{2+}$ in the mass spectrum. The size of the defect was about 5–10 nm. Mg detected was at an average of 0.106 at%, which was about 9.3×10^{19} atoms/cm^3 and almost coincided with the SIMS result. Fig. 5.36B shows an individual defect, which is extracted and enlarged. Mg atoms are represented as spheres. Mg features appear to have an inverted pyramidal shape. The Mg concentration in the defects was analyzed by the solute analysis method [72,73] (Fig. 5.36C). The solute analysis is one

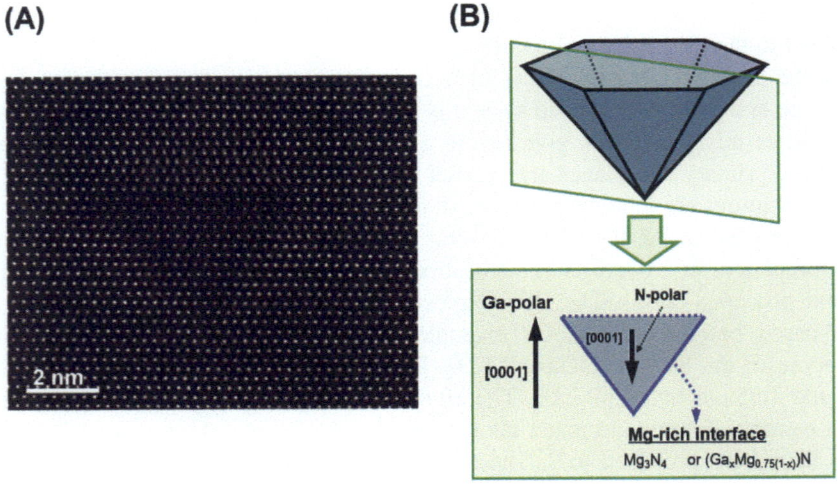

Figure 5.35 (A) High-resolution HAADF-STEM image of the pyramidal defects; (B) schematic illustration of the model of the pyramidal defects.

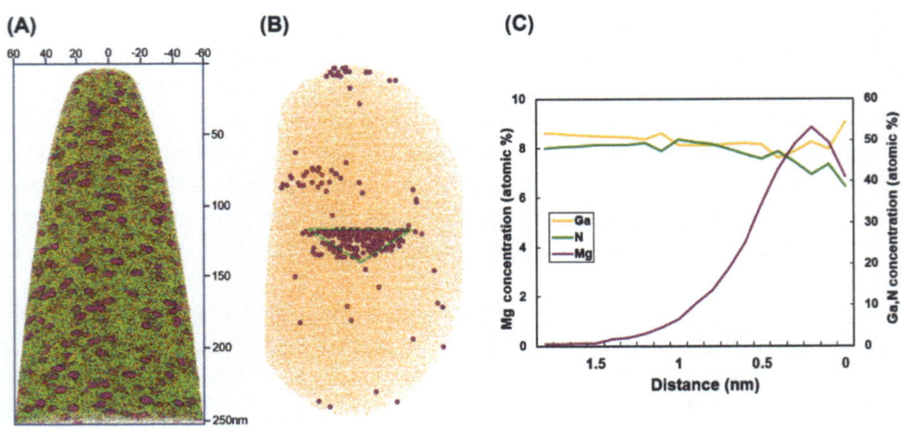

Figure 5.36 (A) Three-dimensional atom probe reconstructed image of Mg-doped GaN films. (B) Enlarged image of (A). (C) Solute analysis result of Mg, Ga, and N concentration of the Mg-related pyramidal defect.

of the most useful methods when the size of the precipitate is small, the shape is not spherical, and the density is high. It was been found that the maximum amount of defects was 9 at% (7.9×10^{21} atoms/cm^3), and Mg was contained at a concentration higher by about one order or more than the average concentration of the matrix.

We now briefly discuss the defect formation of the Mg-related pyramidal defects (Fig. 5.37). We adopt the following hypotheses: (1) surface segregation of Mg

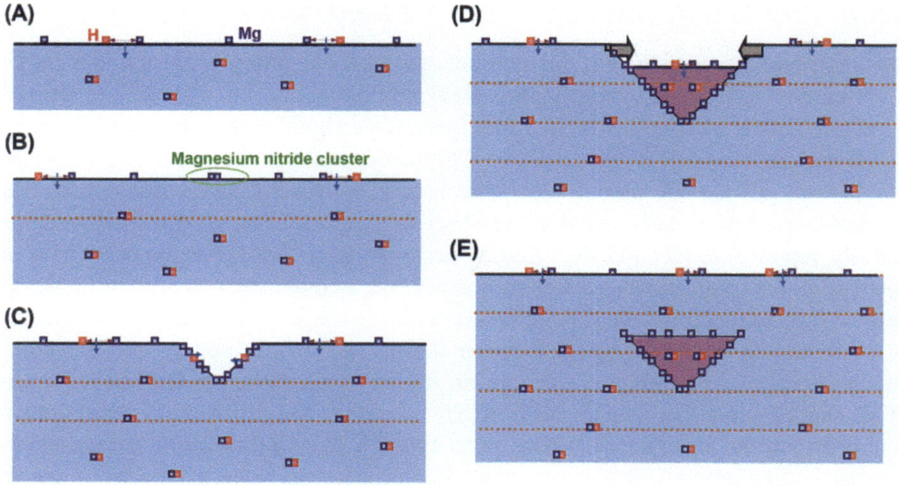

Figure 5.37 Schematic illustration of the formation model of an Mg-related pyramidal defect. (A) Surface segregation of Mg occurs during growth. Hydrogen enhances Mg incorporation in GaN. (B) Growth is suppressed at the point of the magnesium nitride clusters. (C) High exposure to Mg forms a faceted surface. (D) The polarity is changed inside the pits. The growth rate of an N-polar surface is slower than that of a Ga-polar surface. (E) The pits are soon buried by lateral growth from the surrounding Ga-polar matrix. A higher concentration of Mg in the vapor phase enhances the lateral growth.

occurs during growth, and (2) hydrogen enhances Mg incorporation in GaN as previously reported by Tomiya et al. [71]. These two phenomena were reported by Ptak et al. for the case of rf-plasma-assisted MBE [74], and the latter was predicted theoretically [75]. We accept these as being valid for MOCVD growth. As the amount of Mg precursor injected into the reactor increases to dope more heavily with Mg, Mg atoms start to segregate at the growth surface. When the concentration of hydrogen at the surface is sufficient, Mg will incorporate uniformly into the layer. However, when the level of hydrogen is insufficient, magnesium nitride clusters may be created at the growth surface.

High exposure to Mg forms a faceted surface, and the subsequent layer is of inverted polarity [76]. A similar phenomenon occurs in this case. Growth suppressed at the point of the magnesium nitride clusters, and consequently, small pits are formed, just like developing a faceted surface. The polarity also changed inside the pits. Since the growth rate of an N-polar surface is significantly slower than that of a Ga-polar surface in MOCVD growth, the pits are soon buried by lateral growth from the surrounding Ga-polar matrix. A higher [Mg] concentration inside the defects can be due to a slower growth rate. The formation of the pyramidal defects is then terminated. A higher concentration of Mg in the vapor phase enhances the ratio of the lateral growth rate to the vertical growth rate [77]. When [Mg] increases, the defects are consequently buried faster due to an enhancement of the lateral growth rate.

5.4.6 Gradual degradation of GaN-based laser diodes

As described previously, we did not observe any degradation phenomena in the GaN-based LDs due to rapid dislocation multiplication (i.e., "dark line defect" growth). This is in spite of the fact that there are a significant number of TD in the stripe regions.

We will address the gradual degradation of GaN-based LDs in this section. The gradual degradation sometimes is called "bulk degradation." Based on the author's previously reported results [65,78], gradual degradation is thought to proceed with an increase in the capture cross-section of nonradiative recombination centers at the active regions, due to the diffusion of point defects and/or impurities during laser operation.

A reduction of threading dislocation density is very important for extending device lifetime, and reducing the rate of gradual degradation in GaN-based lasers [65]. One possible explanation for how TD accelerate the diffusion of point defects and/or impurities is a pipe diffusion effect through TD. Optimization of magnesium doping is also crucial for improving the device lifetime since excessive or poorly incorporated magnesium will increase optical absorption and the threshold voltage of LDs. Therefore magnesium doping optimization can be one of the significant ways to reduce gradual degradation directly.

Meneghini et al. revealed by the deep-level transient spectroscopy (DLTS) study of degraded LDs that the increase in threshold current is strongly correlated with the increase in the concentration of a deep level (DL) which is located at 0.35–0.45 eV (generally labeled as the E2 level) below the conduction band [79]. One possible candidate for the origin of E2 is related residual Mg combined with hydrogen [80]. A second possible explanation is that E2 defects are caused by $Mg_{Ga}-V_N$ complexes [81].

Monemar et al. studied Mg-related acceptors in GaN by PL and reported that there were two types [82]. One, called A2, is stable for the thermal annealing process and can be Mg_{Ga}. Meanwhile, the other, called A1, is unstable for thermal annealing. Although A1 is not clearly identified so far, both H^+ and/or $V_N{}^+$-complexed Mg-related point defects are currently the two most likely candidates. Since degradation phenomena are unstable processes, A2 is the probable source for gradual degradation.

During device operation, these Mg-related point defects are decomposed into positively charged H^+ and/or $V_N{}^+$-complexed Mg-related point defects. These positively charged point defects may diffuse into active regions and thus drive device degradation.

It is known that nonabsorbing facets have been used in AlGaInP-based LDs as a countermeasure against facet-related degradation [83,84]. For this purpose, LDs with a current injection-free (CIF) region near the laser facet are a relatively simple and effective reliability-enhancing measure for high-power AlGaInP-based red LDs [85] and high-power GaN-based pure blue LDs [86,87]. Fig. 5.38A shows a schematic illustration of the structure of GaN-based pure blue LD CIF regions. Fig. 5.38B shows a fluorescence microscopy image of LDs with the CIF region after the aging test. The p-electrode metal was removed by chemical etching to observe the change of epitaxial luminescence intensity. It should be noted here that the sample was still "alive" and did not degrade by sudden failure due to facet

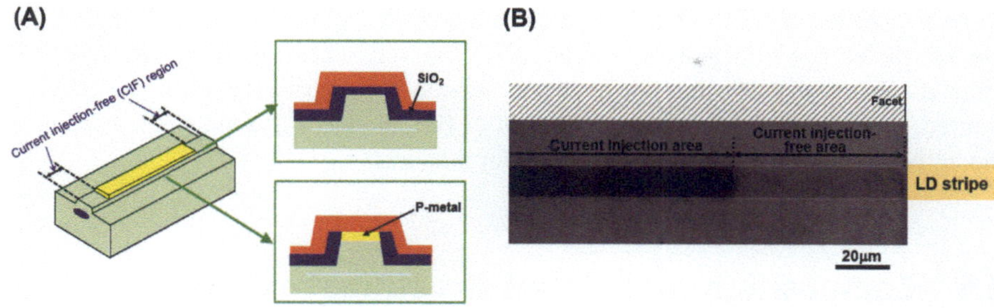

Figure 5.38 (A) Schematic illustration of the structure of GaN-based pure blue laser diode (LD) with current injection-free regions. (B) Fluorescence microscopy image of the aged LD with the current injection-free region.

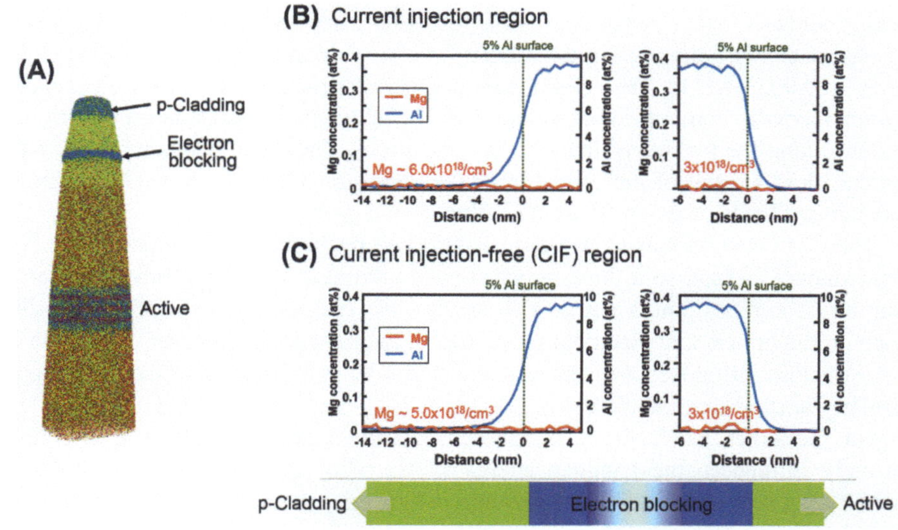

Figure 5.39 (A) Three-dimensional atom probe reconstructed image of aged high-power pure blue laser diode. The detailed vertical structure is shown in Reference [87]. (B) One-dimensional Mg and Al profile near the electron-blocking layer in the current injection region. (C) 1-D Mg and Al profile near the electron-blocking layer in the current injection-free region shows little difference from the injected region.

degradation. After the aging tests, no significant reduction in emission intensity was observed near the facet in the area of the CIF region. On the other hand, the emission intensity was significantly reduced in the current injection region in LDs with the CIF region.

To find out the physical entities of the diffusion process during device operation, we carried out a three-dimensional atom probe analysis. Fig. 5.39 shows 3DAP studies of the aged high-power GaN-based pure LDs observed in Fig. 5.38B. Mg concentrations were detected at about $3-6 \times 10^{18}$ atoms/cm^3 in the vicinity of the current-blocking layer, both in the current injection region and the CIF region. No notable difference in Mg atom profiles was observed. Therefore any significant Mg atom diffusion was not thought to occur. Although impurity detection of 3DAP is limited, this result may implicate diffusion of [H] and/or [V_N] species themselves to the active region as the cause of gradual degradation.

5.5 Composition-modulated structures and ordered structures in III-V based optoelectronics

In Section 5.2, the materials issues involved in the generation of various grown-in defects in III−V compound semiconductors and alloy semiconductors and III-nitrides are described in detail. Apart from the defect formation, in most of the III−V alloy semiconductors, other anomalous structures are generated during crystal growth. These are called "composition-modulated structures" [40,88−91] and "ordered structures" [89,90,92−131]. The composition modulated structures (CMSs) are generated by semiperiodic compositional variation of the alloy semiconductors on the growth surface along the two equivalent <100> directions. The ordered structures (OS) are generated as a result of the formation of "surface structures" [such as (2 × 4) structure] in two kinds of group III atoms during crystal growth.

The CMS corresponds to two-dimensional compositional modulation in the alloy semiconductors (and so a three-dimensionally columnar structure). Since the variation of the composition is not as high at 2%−3%, it does not have too much influence on the optical and electrical properties of the materials. However, formation of this structure introduces a certain amount of native point defects in the matrix particularly at the heterointerface.

On the other hand, the OSs, particularly the so-called CuPt-type structure, strongly influences the bandgap energy of the alloy semiconductor, such as an InGaP active layer for red lasers, and accelerates the gradual degradation. Therefore it is crucial to eliminate this structure for controlling the lasing wavelength and achieving high reliability.

In this section, we describe the results of a detailed investigation of the CMSs and OSs in various III−V alloy semiconductor TEM and related techniques. Furthermore, we also present recent results of these structures in III-nitrides.

5.5.1 Composition-modulated structures in III−V alloy semiconductors

5.5.1.1 Basic structure

Here we describe the basic structure of CMSs [40,88−91,132−134], where a vertical columnar structure forms, with alloy composition fluctuations, as we show below. Fig. 5.40 shows a TEM image of InGaAsP crystals in InGaAsP with different PL wavelengths. In all of the images, quasiperiodic diffraction contrast in the

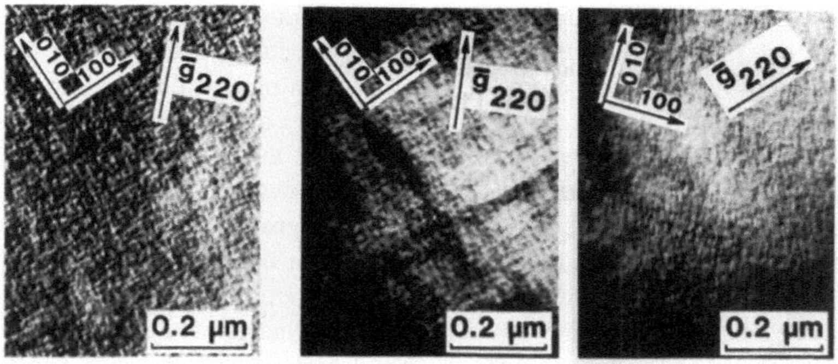

(A) (B) (C)

Figure 5.40 Bright-field plan-view transmission electron microscope images of composition modulated structures in InGaAsP alloy semiconductors with different photoluminescence wavelengths (i.e., different alloy compositions): (A) $\lambda_{PL} = 721$ nm, (B) $\lambda_{PL} = 730$ nm, and (C) $\lambda_{PL} = 800$ nm.

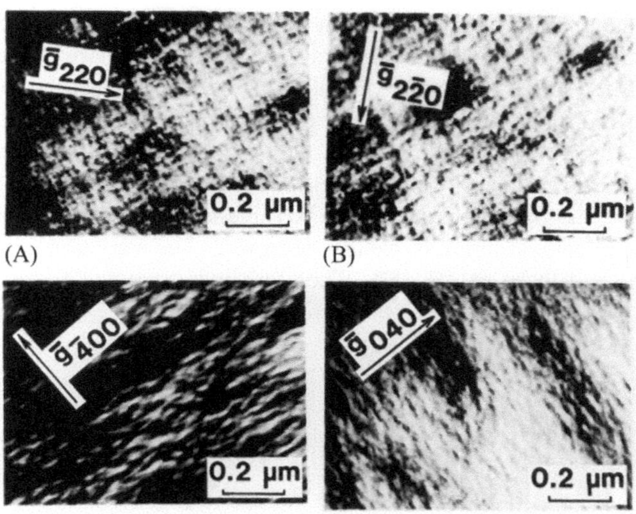

(A) (B) (C) (D)

Figure 5.41 Variation of diffraction contrast of the composition modulated structures: (A) $g = 220$, (B) $g = 2\bar{2}0$, (C) $g = \bar{4}00$, and (D) $g = 040$.

<100> and <010> directions is observed. Fig. 5.41 shows the variation of contrast of the modulated structure in InGaAsP crystal shown in Fig. 5.40B under four different reflections of 220, $2\bar{2}0$, $\bar{4}00$, and 040. In Figs. 5.41A, B, where $\mathbf{g} = 220$ and $2\bar{2}0$, respectively, periodic diffraction contrast is observed in the two equivalent directions of <100> and <010> similar to that in Fig. 5.40B. On the other hand, under the $\bar{4}00$ reflection, the [100]-oriented modulated structure is out of contrast,

and under the 040 reflection, the [010]-oriented modulated structure is out of contrast. Therefore these structures involve periodic strain in the two equivalent directions of <100> and <010>. However, no structural defects were caused by compositional variation in the crystal. It is clear that modulated structures with both long (50–200 nm) and short (5–20 nm) periodicities are formed.

In order to evaluate the cross-sectional shape of the modulated structure, cross-sectional TEMs have been performed. Fig. 5.42 shows $(1\bar{1}0)$ cross-sectional bright-field TEM images of modulated structure in LPE-grown InGaP crystal, under two different reflections. Figs. 5.42A and B were taken under the 220 and 004 reflections, respectively. It is noted that no periodic diffraction contrast is observed along the growth direction using the 004 reflection and that pronounced diffraction contrast along the <110> direction can be seen. These results lead us to conclude that the modulated structures are columnar-shaped. This is because the compositional modulation during crystal growth is two dimensional.

5.5.1.2 Degree of compositional variation in the modulated structure

To evaluate the compositional fluctuation in the modulated structure, STEM (scanning TEM)/EDX (energy dispersive X-ray spectrometry) analysis was performed [40]. A typical STEM image of the modulated structure obtained by the 220 reflection is shown in Fig. 5.43A. An EDX measurement was carried out along the scanning line denoted by A–B in this figure with a spatial resolution of about 15 nm. The results are shown in Fig. 5.43B. These are intensity ratios of the In Lα line to the Ga Kα line, the As Lα line to the Ga Kα line, and the P Kα line to the Ga Kα line. Compositional variation is clearly observed. The variation in the intensity ratios is estimated to be $\pm 4\%-5\%$. Therefore it is expected that the modulated

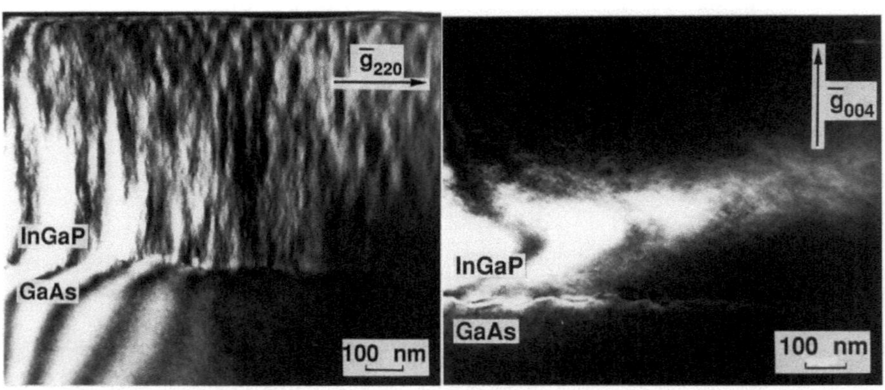

(A) (B)

Figure 5.42 A $(1\bar{1}0)$ cross-sectional bright-field transmission electron microscope image of composition modulated structure in LPE-grown InGaP crystal on a (001) GaAs substrate: (A) $\boldsymbol{g} = 220$, $s > 0$, (B) $\boldsymbol{g} = 004$, $s > 0$.

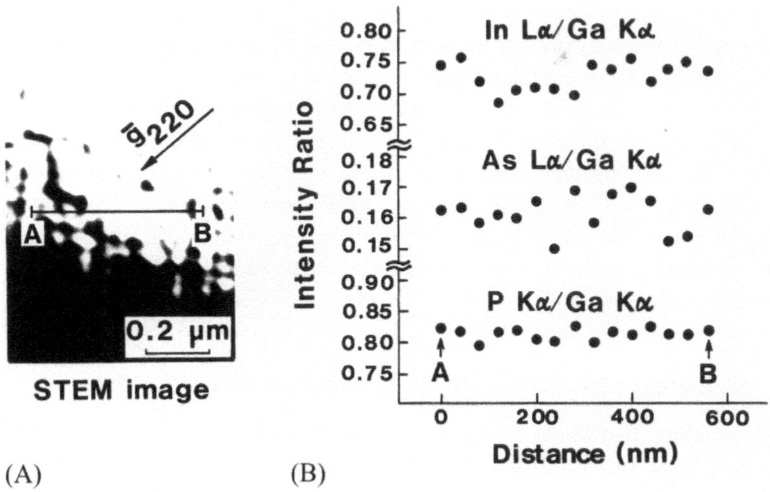

Figure 5.43 STEM/EDX analysis of modulated structure in InGaAsP crystal ($\lambda = 790$ nm). (A) A plan-view STEM image of InGaAsP crystal and (B) intensity ratios along the scanning line A–B in (A).

structure is associated with compositional modulation in the crystal during growth. However, from more quantitative EDX analyses on the modulated structure, the actual variation has been determined to be 2%–3%.

5.5.1.3 Origins of the two types of periodicities

We discuss here the origins of the two types of periodicities of the modulated structures. According to the calculation of the spinodal isotherms in InGaAsP quaternary system by De Cremoux and others [133], the compositions of InGaAsP shown in Figs. 5.40A and B are inside a region called the "spinodal region," in which the material is thermodynamically unstable, and tends to separate into two metastable phases during growth. Therefore the composition modulation can be associated with the spinodal decomposition in the crystal. However, that of crystal shown in Fig. 5.40C is outside the spinodal region. Therefore it is expected that the coarse structures are formed during growth by spinodal decomposition and that the fine structures are formed during the cooling process after growth.

5.5.1.4 Origin for the direction of modulation

In spinodally decomposed alloys which have asymmetry in elastic coefficients, when fluctuation of the concentration of solute atoms $(c - c_O)$ takes place, the variation of free energy ΔF in the crystal is given by [134]:

$$\Delta F = \int \left[(1/2)(f_0 + (2\eta^2 E)/(1-\nu))(c-c_0) + K(\nabla c)^2\right] dv \tag{5.1}$$

where f_0, η, ν, and $K(\nabla c)^2$ are the free energy per unit volume, the variation constant in length for variation of unit concentration, Poisson's ratio, and an additional energy term for composition gradients, respectively. Since the first and third terms are constant, the modulated waves extend so as to minimize the second term, which is associated with strain contribution to the free energy. In the case of cubic crystals, $E/(1-\nu)$ can be replaced by an elastic coefficient $Y_{[hkl]}$. The values $Y_{[100]}$, $Y_{[110]}$, and $Y_{[111]}$ for GaAs, GaP, InAs, and InP were calculated [40]. These results have shown that in all cases, $Y_{[100]} < Y_{[110]} < Y_{[111]}$. Therefore it is well explained that the modulated structures were generated in three equivalent directions of the [100], [010], and [001].

5.5.2 Ordered structures in III–V alloy semiconductors

The OSs in which two kinds of element atoms in column III or V lattice periodically arrange have been observed in various III–V alloy semiconductors [89,90,92–131]. In most cases, the OSs are generated locally in the crystal, and the degree of ordering is strongly affected by the growth method and growth conditions. The results for OS in III–V alloy semiconductors obtained are summarized as follows:

1. The most common structure in crystals grown on a (001) substrate is CuPt-type (see Fig. 5.44), and other structures are only occasionally observed.
2. They are generated very often in crystals grown by MOVPE, vapor phase epitaxy (VPE), and MBE, whereas no ordering takes place in LPE-grown crystals (except for one report for OSs in InGaAs by Nakayama and others [103]), which was not confirmed by others.
3. Regarding CuPt-type structures, one can observe only two variants in the (110) cross-section. In particular, atomic steps on the growth surface are believed to play an important role in the generation of the OSs since one of the two variants is preferentially enhanced when substrates tilted toward the <110> direction are used. To avoid ordering, growth is often performed on misoriented substrates (e.g., 15 degree-off-axis substrates for red InGaP lasers to maintain the desired random alloy).

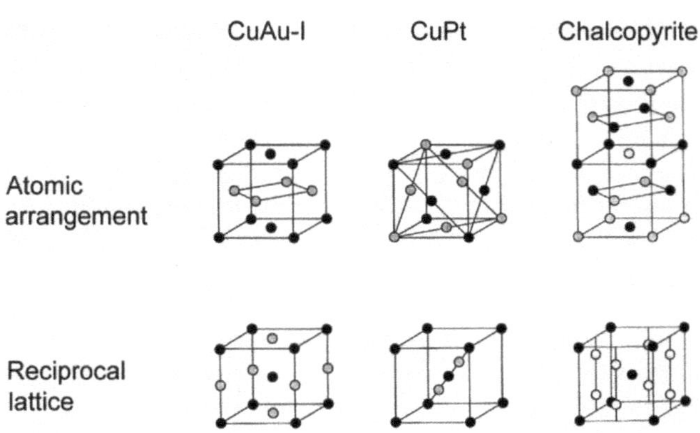

Figure 5.44 Schematic diagrams of atomic arrangement and corresponding reciprocal lattices of three types of ordered structures.

4. The ordering does not perfectly occur, and the ordered regions are plate-like microdomains lying on planes nearly parallel to the growth surface.
5. The degree of ordering depends on the growth temperature, V/III partial pressure ratio (in the case of MOVPE and MBE), and rotation velocity of the substrate.
6. Only strongly ordered InGaP crystals grown by MOVPE exhibit abnormal bandgap energies (up to 50 meV lower than the normal value).

As described in the last part of this section, this ordered structure tends to accelerate the gradual degradation due to its disordering, which is enhanced by diffusion of p-type dopants such as Zn atoms from the p-cladding layer to the active layer. Therefore it is crucial to eliminate the generation of this structure in the development of red LDs and VCSELs fabricated from In(Al)GaP DH structures.

In this section, we focus on the CuPt-type OSs in InGaP and related alloy semiconductors.

5.5.2.1 Basic structure of CuPt-type OSs in InGaP crystals

Figs. 5.45A and B are electron diffraction patterns obtained from the $(\overline{1}\overline{1}0)$ and $(1\overline{1}0)$ cross-sections of InGaP crystal grown by MOVPE at 630°C, respectively. Pronounced extra spots (here called "superstructure spots") and streaks are observed in the electron diffraction pattern from the $(\overline{1}\overline{1}0)$ cross-section but not in the one from the $(1\overline{1}0)$ cross-section. The characteristic features of the $(\overline{1}\overline{1}0)$ diffraction pattern are as follows:

1. The presence of superstructure spot pairs at positions indexed as [h ± ½, k − ½, l ± ½] and [h ± ½, k + ½, l ± ½] matrix spot (for example, the positions denoted by arrows in Fig. 5.45A);

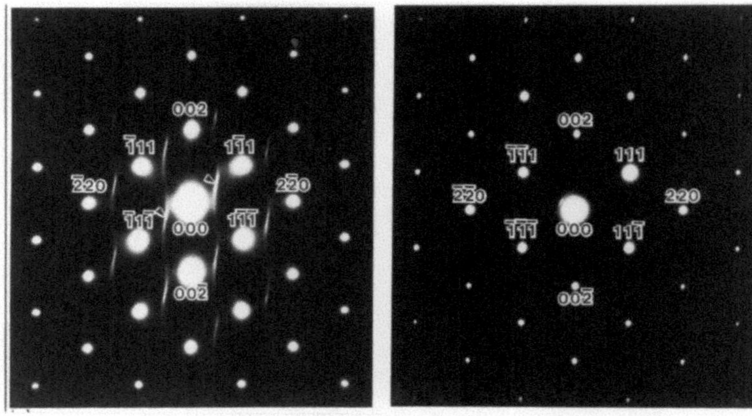

(A) (B)

Figure 5.45 Electron diffraction patterns obtained from ordered InGaP crystal: (A) the $(\overline{1}\overline{1}0)$ cross-section and (B) the $(1\overline{1}0)$ cross-section.

2. The appearance of streaks around the superstructure spots lying along both the [001] and $[00\bar{1}]$ directions, but slightly tilting (approximately 10°) toward the $[1\bar{1}0]$ and the $[\bar{1}10]$, respectively

These superstructure spots are associated with CuPt-type structure [89,102], which is one of the substitutional OSs in FCC crystals. The ordering occurs on the (111) planes of column III atoms with doubling in periodicity, that is, the OS corresponds to a monolayer superlattice of $(InP)_1(GaP)_1$ on the (111} plane. Since streaks are present around the superstructure spots, it is assumed that the ordering is not perfect, and that ordered regions correspond to microdomains. A typical $(\bar{1}\bar{1}0)$ cross-sectional high-resolution TEM image of OS in InGaP crystal is shown in Fig. 5.46A. In this image, doubling in periodicity of (111) lattice fringes, that is, the presence of arrays of brighter spots on each second $(1\bar{1}1)$ lattice fringe is clearly observed. Fig. 5.46B shows a schematic diagram of the atomic arrangement in a perfectly ordered InGaP crystal. In Fig. 5.46A, a brighter and a darker spot correspond to In-P and Ga-P atoms pairs, respectively. Fig. 5.47 shows a $(\bar{1}\bar{1}0)$

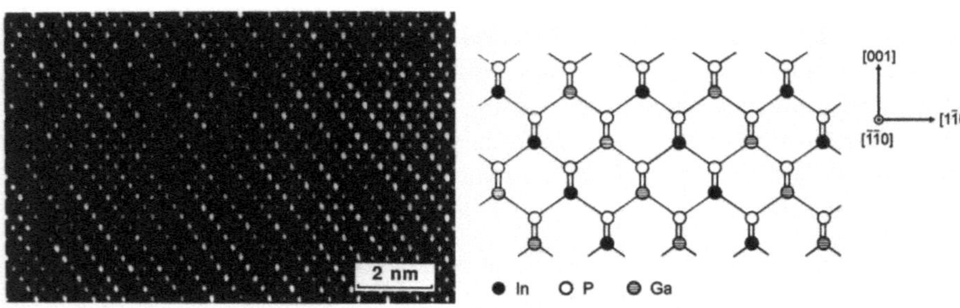

(A) (B)

Figure 5.46 (A) A cross-sectional high-resolution transmission electron microscope image of ordered regions in InGaP crystal and (B) a schematic diagram of atomic arrangement of ordered InGaP crystal (CuPt-type).

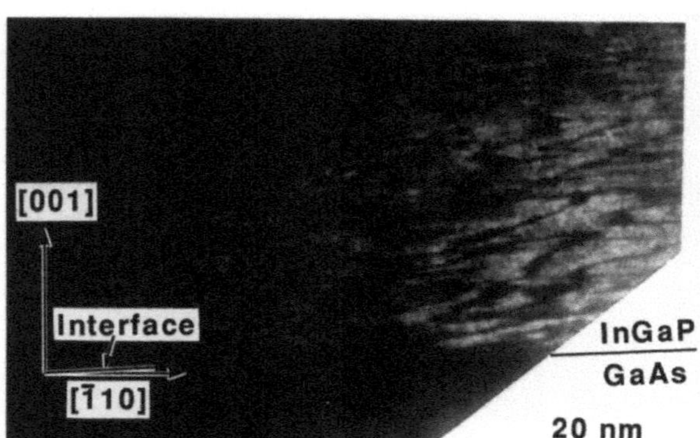

Figure 5.47 A $(\bar{1}\bar{1}0)$ cross-sectional dark-field transmission electron microscope image of ordered regions in InGaP crystal obtained from one of the superstructure spots.

cross-sectional dark-field TEM image of a strongly ordered InGaP crystal [different crystal from that shown in Fig. 5.45 (weaker ordering)] obtained from one of the superstructure spots. In this image, brighter regions correspond to ordered regions since the intensity is associated with only the superstructure spot. Moreover, the ordered regions are plate-like microdomains. In this case, approximately 80% of the total area is dominated by ordered domains [90].

5.5.2.2 Dependence on growth conditions

In the development of optical devices from various III–V alloy semiconductors, it is crucial to control the ordering since it strongly affects the bandgap energy and therefore the emitting wavelength of the active layer. In this section, the effects of various growth conditions on the degree of ordering are discussed.

Growth temperature: There have been several reports for various III–V alloy semiconductors for the influence of growth temperature on the degree of ordering which is based on the relative intensity of the superstructure spots in the electron diffraction pattern from the crystal [101,106,107,109–113]. In the case of CuPt-type ordering in InGaP crystals grown on the (001) GaAs substrates, there is a temperature range where the degree of ordering is maximal. The degree of ordering becomes lower when the growth temperature is lower or higher than this temperature range. It is assumed that this is due to the low migration velocity of deposited atoms on the growth surface at lower temperatures, suppressing the rearrangement of the deposited atoms, and to an enthalpy problem at higher temperatures, causing destruction of the OS.

Growth rate: There have been several works on the influence of growth rate on the degree of ordering [114–116]. In most cases, higher growth rates lower the degree of ordering. This may be explained by the fact that it is difficult for the deposited atoms on the growth surface to rearrange in the presence of rapid atomic diffusion. In an extreme case of low growth rate, McDermott and others reported the ordering in InGaP crystals by atomic layer epitaxy [115]. They found that the resulting crystal exhibits the lowest reported bandgap energy, indicating the highest degree of ordering.

Vapor pressure ratio: A detailed work on the effect of V/III ratio on the degree of ordering has been done by Suzuki and others [117]:

1. At V/III = 62, the ordering is strongest at 600°C–650°C (lower bandgap energy), and becomes weaker at 700°C–750°C (bandgap energy assumes normal value).
2. At V/III = 230, the ordering is strongest at 600–700°C, and weak ordering is observed at 750°C.
3. At V/III = 410, the ordering is strongest at 650°C–700°C, and becomes weaker when the growth temperature is lowered or raised outside this temperature range.

Therefore, in order to obtain random alloy of InGaP we need to grow the crystal at 700°C or higher with V/III lower than 100.

Doping: Effects of doping on the degree of ordering have been studied by several groups [118–120]. A detailed work on the evaluation of MOVPE-grown p-InGaP crystals doped with Zn or Mg has been done by Suzuki and others [118]. They found that the doped crystals with Zn or Mg exhibit a steep increase in bandgap energy for a carrier concentration higher than $1 \times 10^{18}/cm^3$. From an electron diffraction analysis, it was clarified that the undesired ordered structure is replaced by the desired random alloy for p-type doping higher than $1 \times 10^{18}/cm^3$. Possible mechanisms for this disordering by high p-doping are as follows:

1. During growth, the deposited dopant atoms interact with the matrix atoms on the growth surface, suppressing the ordering.
2. During growth, the dopant atoms diffuse through the InGaP layer and randomize the sublattice ordering.

They also studied the effect of n-type doping on the ordering [120]. They found a similar transition to the desired random alloy by doping with Si or Se. However, there was a difference in the threshold carrier concentration for the disordering between Si- and Se-doped InGaP crystals: n_{th} (disorder) = $3.8 \times 10^{17}/cm^3$ for Si-doping and n_{th} (disorder) = $2.0 \times 10^{18}/cm^3$ for Se-doping. They assumed that this difference was due to the difference in the sublattice occupation between Si and Se: Si and Se atoms favor the column III and V sublattices, respectively.

5.5.2.3 Dependence of substrate misorientation

There have been a number of reports on the effect of substrate misorientation on the degree of ordering in InGa(Al)P crystals. The results are summarized as follows:

1. In the case of crystal growth on a just (001) substrate, two variants of the four equivalent CuPt-type OS are equally generated, which are observed when viewed from the (110) cross-section.
2. When the crystal is grown on (001) substrates slightly (approximately 2 degrees) misoriented toward the <110> direction, only one variant is predominantly generated.
3. The degree of ordering decreases when the misorientation of the substrates toward the <110> direction increases.
4. Ordering does not occur on the (511)A, (311)B, and (111)B substrates [121,122] and (110) substrate [108,121].

For many applications such as red lasers, a wider bandgap for shorter wavelength is preferred, which is obtained with the random (nonordered) alloy. For this reason, most commercial red lasers are grown on the highly misoriented substrates such as the (311)B, or 15-degree-off (001) direction.

5.5.2.4 Effect of thermal annealing

We describe the thermal stability of the OSs for InGa(Al)P [123,124] and InGaAsP [125] grown on (001) GaAs substrates. It has been shown that strongly ordered InGaP crystals do not disorder during annealing at moderate temperatures (for example, 600°C). However, they tend to disorder by annealing at higher temperatures or for a long time. Gavrilovic and others have reported InGaP crystal disorder by annealing at 700°C for 96 h. Another study showed that ordered InGaAsP crystal disordered after annealing at 835°C for 4 h.

5.5.2.5 *Effect of impurity diffusion*

The effect of doping on the disordering of the ordered crystal was described previously. Since the dopant atoms can diffuse in the crystal at elevated temperatures, giving rise to disordering, intentional diffusion of impurities into the ordered crystal is expected to cause disordering of the crystal, similar to that observed in the case of Zn diffusion into AlAs/GaAs artificial superlattice [126]. Additional reports show that the OSs can be disordered by Zn diffusion [109,123,125,127].

5.5.2.6 *Generation mechanism of the CuPt-type ordered structures*

In this section, we discuss the generation mechanism of CuPt-type OSs in InGaP crystals. First, theoretical calculations based on bulk thermodynamics indicate that chalcopyrite and CuAu-I structures are energetically more favorable than a CuPt structure [128–130]. Thus the generation of CuPt-type structure is not associated with bulk ordering, but rather with surface kinetics during growth. Furthermore, the following steps are derived for the mechanism of CuPt-type ordering from the schematic diagram for the atomic arrangement of a perfectly ordered InGaP viewed from the (110) cross-section as shown in Fig. 5.46B:

Step 1: On the growth surface, In and Ga atoms are aligned in a row with doubling in periodicity in the $[1\overline{1}0]$ direction.

Step 2: During growth, each row of In and Ga atoms propagates in the same direction, forming an InP/GaP monolayer superlattice on the {111} plane.

Based on these considerations, several models for the ordering have been proposed. We describe here a possible model for the generation of OSs in InGaP crystals grown on (001) GaAs substrates by Suzuki and others [104] as follows:

Step 1: Interplane ordering of column III atoms occurs on a column V atom-stabilized surface where the In atoms stretch the lattice spacing, giving rise to an anisotropic site occupation affinity for Ga and In. This is considered to act as a driving force for the long-range ordering of alternate In and Ga line alignment within the (001) plane. Therefore this model is based on the fact that the size difference between Ga and In in InGaP enhances CuPt-type ordering on the growth surface due to size-mismatch-driven relaxations.

Step 2: For the evolution of the InP/GaP monolayer superlattice from step 1, atomic steps are required (see Fig. 5.48):

1. The configuration of the initial steps on the substrate surface is preserved during growth, where a layer-by-layer growth takes place.
2. The direction of step movement is normal to the [134] direction.
3. Only Ga atoms occupy the sites on the (111) microfacets at the step for a moderate growth condition range.
4. These preferentially settled Ga atoms on the microfacet act as phase lockers for the (001) ordered plane on the high side of the step.

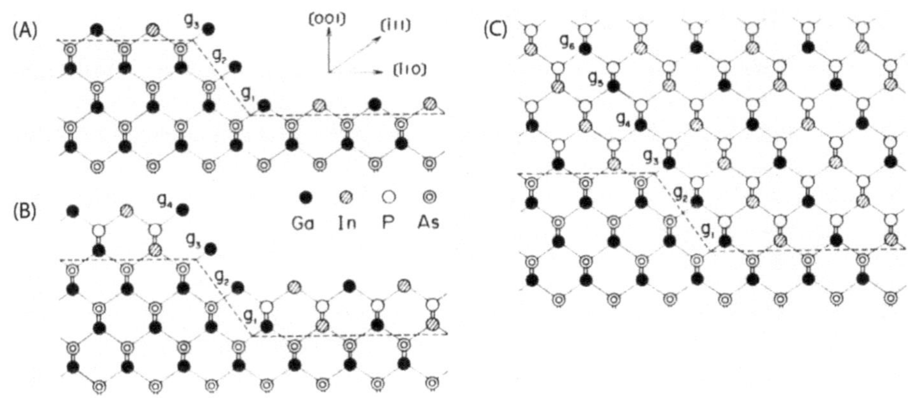

Figure 5.48 Schematic diagrams for generation of CuPt-type OSs in InGaP crystal grown on (001) GaAs substrate.

Source: Proposed by Suzuki and others, reprinted from T. Suzuki and A. Gomyo, J. Cryst. Growth, 93, 396 (1988), copyright 1988, with permission from Elsevier.

5.5.3 Influence of composition-modulated structures and ordered structures on the device reliability and degradation

As described previously, the CMSs are associated with quasiperiodic compositional variation presumably due to spinodal decomposition of the system. However, the compositional fluctuation estimated by EDX is 1%–2%, and the domains are columnar-shaped and quite stable. Therefore they do not greatly affect optical and electrical device properties. Moreover, in optical devices, the periodicity and amplitude of the compositional modulation do not greatly affect any of the three degradation modes: rapid degradation, gradual degradation, and catastrophic failure. However, as described previously, dislocation loops generated as byproducts of modulated structure generation can cause rapid degradation in the case of visible InGaAsP/InGaP DH lasers (see Figs. 5.16 and 5.23A for dislocation loops in an undoped InGaAsP active layer). This is also an issue in the future development of next-generation lasers fabricated from InGaAsP/InP, AlInGaAs/InP, AlGaInP/GaInP, and related materials.

OSs only affect the PL peak energies in strongly ordered InGa(Al)P [89,101,102,109] crystals by MOVPE. In other alloy semiconductors such as InGaAs, InAlAs, and GaAsSb, the degree of ordering is lower, and PL peak energies are not much different than in the random alloy.

Next, we describe a study on the degradation of an InGaP visible laser in which a strongly ordered InGaP layer is used as an active layer by Endo and others [131]. The LDs they used are red InGaP/InGaAlP LDs grown by MOVPE. The active layer is an undoped InGaP layer with a strongly ordered structure. The p- and n-type cladding layers are Zn-doped and Si-doped $In_{0.5}(Al_{0.6}Ga_{0.4})_{0.5}P$ layers, respectively. They carried out an accelerated aging test of these LDs under 50–150 mA and at 100°C–150°C. Fig. 5.49A shows the threshold current traces for the LDs operated at 75 mA at 150°C. The increase in pulsed threshold current was proportional to the square root of aging time. Fig. 5.49B shows lasing spectra measured at 25°C for one of the LDs. The lasing wavelength shifted from 676 to 663 nm, and the longitudinal mode changed from a single mode to a multimode. The result shown in Fig. 5.49A suggests that the degradation of these LDs is some impurity diffusion process. Further, the significant shift of the lasing wavelength shown in Fig. 5.49B is assumed to be associated with disordering of the OS in the InGaP active layer, shifting from an ordered alloy to a random alloy. Based on these considerations, one can conclude that the OS in the active layer disordered by

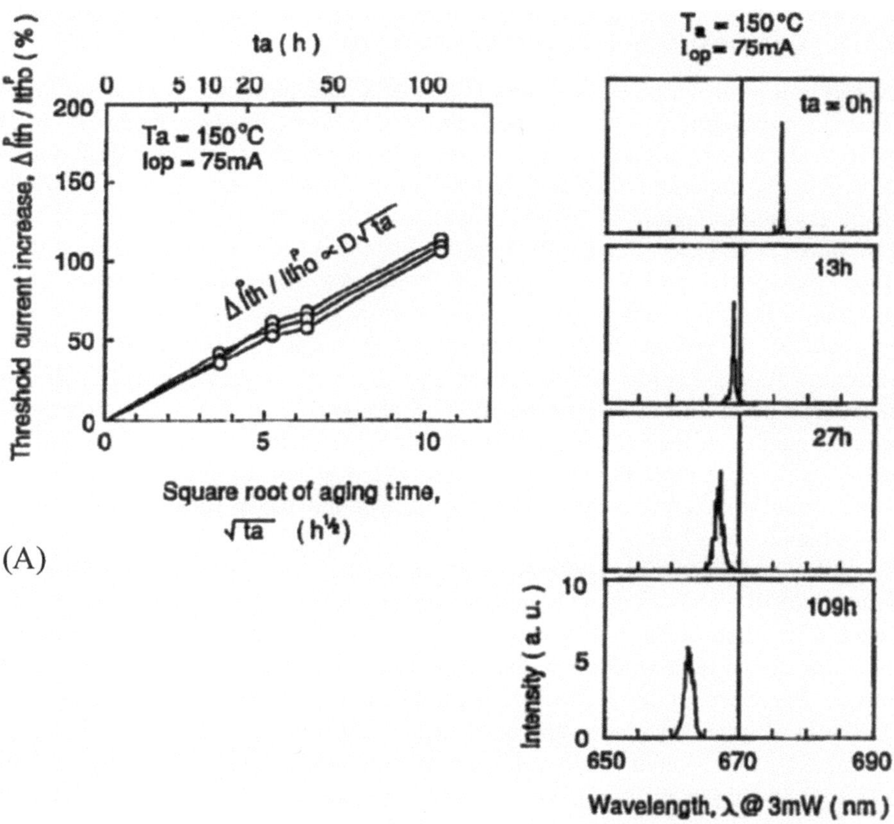

(B)

Figure 5.49 Changes in lasing characteristics at 25°C for laser diodes operated under 75 mA at 150°C. (A) Pulsed threshold current increase, approximately proportional to the square root of aging time and (B) lasing spectra at 0–109 h aging times for one of the laser diodes.

Source: Reprinted from K. Endo, K. Kobayashi, H. Fujii, and Y. Ueno, Appl. Phys. Lett. 64, 146 (1994), with permission from AIP Publishing.

Zn-diffusion from the p-cladding layer to the active layer by nonradiative recombination of minority carriers during operation, causing the blueshift of the lasing wavelength, and device degradation by the generation of point defects and/or point defect complexes. Therefore it is crucial to grow nonordered InGaP crystals by optimizing the growth condition (i.e., higher growth temperature, etc.) and to use substrates with high misorientation angles (i.e., >10°) in order to solve this problem, and avoid the very rapid degradation that occurs when aging is done on lasers with ordered alloy active regions. We may encounter this issue in the case of fabricating red VCSELs from InGa(Al)P/InGaAlP MQWs.

5.6 Thermal instability in III-nitrides

5.6.1 Composition-modulated structure

Ternary alloys of wurtzite III-nitrides (GaN, AlN, and InN) have been obtained by continuous changing of alloy compositions whose direct bandgap ranges from the value of 0.7 eV for InN to 6.0 eV for AlN. The GaInN alloy for the emission layer is particularly important in optical devices in the visible region. The first growth of single crystalline GaInN by MOCVD was realized by Nagatomo et al. [135] and Yoshimoto et al. [136]. Since then, considerable effort has been made for improving crystal quality and the amount of InN incorporation to realize high-efficiency LEDs and lasers. However, GaInN alloys are difficult to grow due to the large difference between Ga−N and In−N bond lengths and the high volatility of N on In/Ga. Matsuoka et al. [137] reported that lowering the growth temperature from 800°C to 500°C results in an increase in the In concentration range, but at the expense of reduced quality. A theoretical study by Ho et al. [138] shows that $Ga_{1-x}In_xN$ alloys are unstable over most of the composition range at normal growth temperatures which maintain crystal quality. This means that the phase separation of GaInN alloys inevitably occurs.

As described in the previous section, there are two types of deviation from randomness of III−V alloys: atomic ordering and phase separation. The latter is the state between two immiscible binary compounds such as GaN and InN. It is noted here that there are no reports so far on the observation of atomic ordering in wurtzite III-nitrides. Therefore, in this section, we focus on phase separation in GaInN alloys.

Rao et al. reported phase separation and composition modulation in $Ga_{1-x}In_xN$ layers more than 50 nm thick with various indium compositions x ranging from 0.12 to 0.34 (Fig. 5.50) [139]. The periodic composition modulation was observed. No particular direction is favored for the modulations since Young's modulus for hexagonal structures is isotropic in the (0001) plane. In addition, they also reported that the composition modulations were lying in the (0001) growth plane, implying that phase separation is two-dimensional in nature.

It is known that the composition of the GaInN ternary alloy can be shifted to reduce the deformation energy due to a lattice mismatch to the underlying GaN layer, and this phenomenon is known as the composition pulling effect [140].

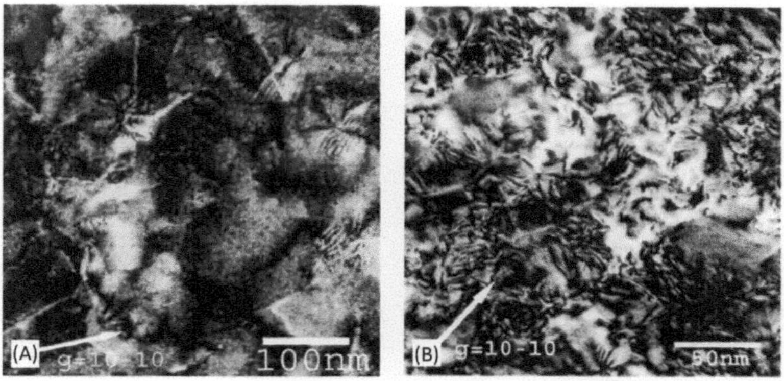

Figure 5.50 (A) Dark-field plan view transmission electron microscope images of $Ga_{0.88}In_{0.12}N$ showing speckled contrast and (B) $Ga_{0.78}In_{0.22}N$ showing domains with alternating light/dark contrast bands lying similarly oriented within a domain.

Source: Reprinted from M. Rao, D. Kim, and S. Mahajan, Appl. Phys. Lett., 85, 1961 (2004)., with permission from AIP Publishing.

This effect hinders the incorporation of indium in the early-stage growth of the GaInN layer because of an attempt to maintain coherent growth to lattice-matched to the underneath layer such as GaN. Thus the indium content is limited near the interface. This leads to the driving force for phase separation being lower at the growth interface. Since the GaInN layer in optical devices is usually used as a very thin quantum well (QW), it is essential to understand phase separation behavior for such thin layers.

TEM analysis is often applied to study nanoscale composition fluctuations. However, TEM for GaInN materials has several problems, such as sample preparation and e-beam damage. The 3DAP can provide the 3D composition information with subnanoscale resolution. Therefore 3DAP and TEM analyses were carried out complementarily for the study of GaInN composition fluctuation analysis [141,142]. Fig. 5.51 shows a cross-sectional HAADF-STEM image of the sample which consist of three pairs of the 3-nm-thick GaInN QW layers with different indium compositions [$x = 0.08$ (near-UV emission), 0.15 (blue emission) and 0.25 (green emission), and $T_{\mathrm{growth}} = 700°C$, 720°C, and 760°C from bottom to top]. Three different bright contrast bands are clearly observed corresponding to the three $Ga_{1-x}In_xN$ MQWs with their different indium compositions. Their contrasts are almost homogeneous. Each $Ga_{1-x}In_xN$ QW coherently grows on GaN.

Fig. 5.52A shows a 3D atom map of a 3DAP reconstructed image. Fig. 5.52B shows 1D indium composition profile extracted from Fig. 5.52A. Figs. 5.52C–E show the 2D indium composition map extracted from the green, blue, and violet plate regions in Fig. 5.52A. The colors correspond to the indium composition. The characteristic morphology patterns due to compositional variation in random directions are observed.

Fig. 5.53 shows the frequency distribution of In concentrations. Although the data were limited from each QW because of their thickness, each frequency

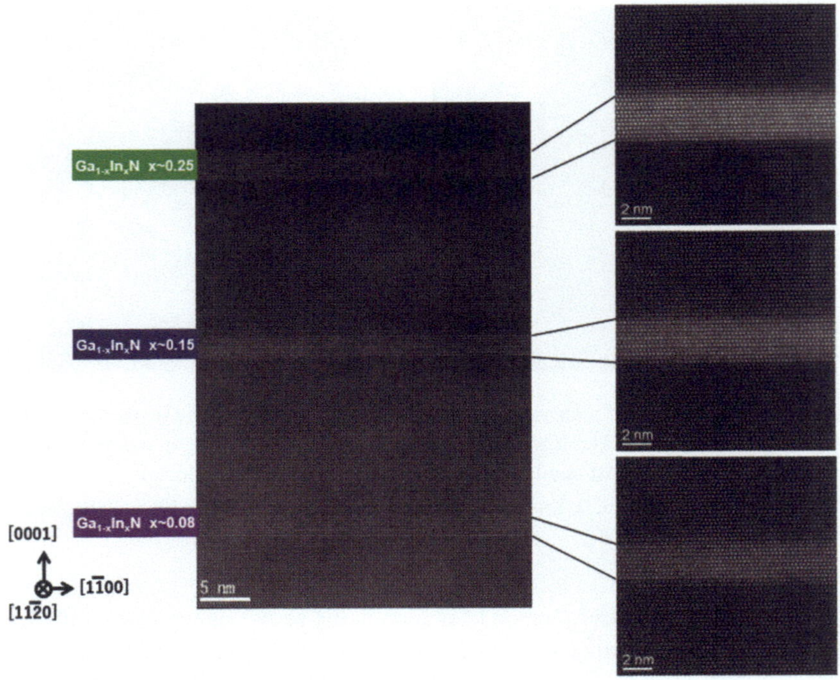

Figure 5.51 Cross-sectional HAADF-STEM image of the sample which consists of three pairs of the 3-nm-thick GaInN QW layers with different indium compositions.

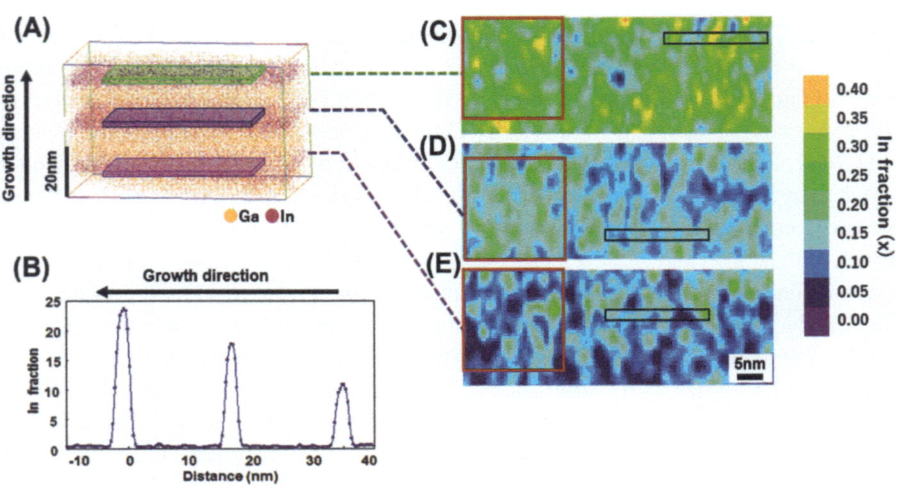

Figure 5.52 (A) Atom probe tomography of In (violet) and Ga (yellow). (B) 1D indium composition profile. 2D in-plane indium composition color maps of (C) the green plate region ($Ga_{0.75}In_{0.25}N$ QW), (D) the blue plate region ($Ga_{0.85}In_{0.15}N$ QW), and (E) the violet plate region ($Ga_{0.92}In_{0.08}N$ QW).

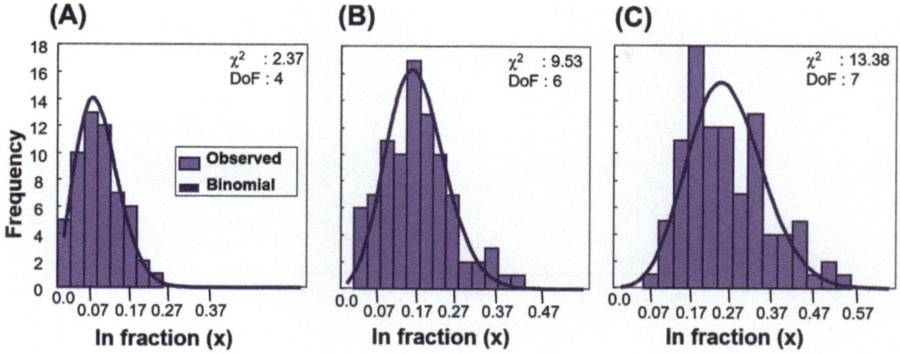

Figure 5.53 Frequency distributions of In concentrations of (A) $Ga_{0.92}In_{0.08}N$ QW, (B) $Ga_{0.85}In_{0.15}N$ QW, and (C) $Ga_{0.75}In_{0.25}N$ QW. Solid lines are binominal distributions.

distribution follows the binominal distribution. This indicates a homogeneous composition distribution. To compare the quantity of the homogeneity of three QWs, the χ^2 statistics enable significant testing of randomness. However, χ^2 values depend on analyzed size. Therefore it is useful to introduce the Pearson coefficient μ as follows:

$$\mu = \sqrt{\frac{\chi^2}{N + \chi^2}} \tag{5.2}$$

where N is the number of analyzed size [143]. $\mu = 0$ indicates a random distribution and $\mu = 1$ indicates a complete association with a nonrandom distribution of the solute atoms [143]. μ values of UV, blue and green QWs are 0.08, 0.149, and 0.169, respectively. This indicates that although μ values increase as In concentrations increase, μ values are relatively small and no significant deviation from random distributions occurred in all three QWs.

Fig. 5.54 shows the 1D indium composition profiles extracted from the black rectangular regions in Fig. 5.49C–E and the calculated profiles. Since the indium composition shows sinusoidal fluctuation, the estimated model function of the 1D indium composition profile can be written as

$$x = c\left\{ \text{sgn}(\sin t) \times \frac{1}{\sqrt{2\pi\sigma}} exp\left[\frac{-(d-m)^2}{2\sigma^2}\right]\right\} + x_{\text{mean}} \tag{5.3}$$

where c is the constant, σ is the standard deviation, d is the distance, m is the step function, and x_{mean} is the mean indium composition. The first term of the model function expresses a product of a square wave and a Gauss function. Since the composition modulation can be fitted by the Gauss function known as a solution of the diffusion equation, the composition modulation is governed by the diffusion process.

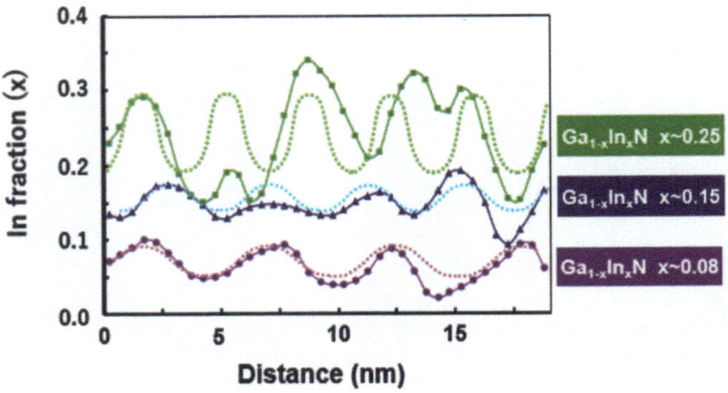

Figure 5.54 1D indium composition profiles obtained from black rectangular regions in Figs. 5.49C–E and calculated data (dotted lines).

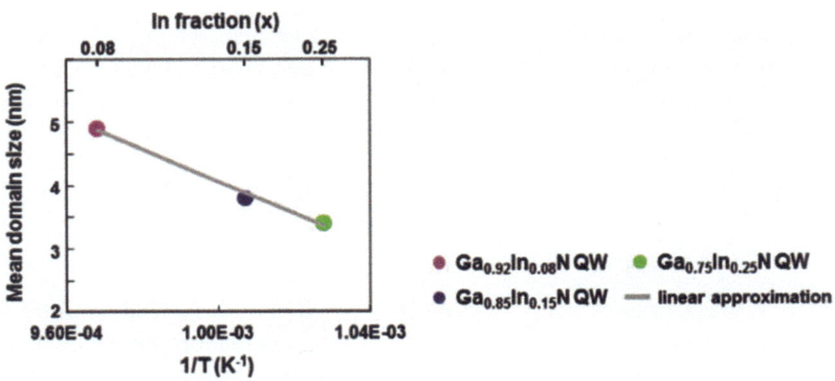

Figure 5.55 Correlations between indium composition (upper lateral axis) and reciprocal growth temperature (lower lateral axis) and estimated domain size from autocorrelation function. From the Arrhenius equation, activation energy $E = 25526 \times k_B = 2.2$ eV. k_B is the Boltzmann constant.

Fig. 5.55 shows the relationship between the mean domain size and indium composition or the reciprocal of T_{growth}. The mean domain size is derived using the autocorrelation function analysis which is calculated from the red rectangular region in Figs. 5.49C–E. The first minimum in the radial autocorrelation function indicates the mean domain size of the compositional variation, which is proportional to the period. The mean domain size becomes smaller with decreasing T_{growth} and also becomes smaller with increasing indium composition.

High-indium-content GaInN is readily induced to undergo phase separation, since composition fluctuations become more severe, and the degree of composition mixtures becomes more immiscible with increasing indium content [138]. Although higher-indium content GaInN QWs ($x = 0.25$) described above are located near the unstable region in the phase diagram [138], the indium composition distribution obeyed the binomial distribution, and no appreciable phase separation is observed. Since the GaInN MQWs were coherently grown on underlying GaN layers and thus strained, suppression of phase separation is partially due to strain, as Karpov theoretically predicted that elastic strain suppresses phase separation [144].

Nanoscale periodic compositional modulation on the (0001) growth plane is also observed in all $Ga_{1-x}In_xN$ QW layers. The occurrence of compositional modulations is due to the ability of the modulation to minimize the strain energy within GaInN QWs, as previously reported in the case of a metallic alloy [132].

The higher-indium-composition $Ga_{1-x}In_xN$ QWs used for green LEDs or LDs were grown at a low temperature to increase indium incorporation and to minimize indium re-evaporation during growth. The shortening of the period of compositional fluctuation with increasing indium composition is due to the decrease in the surface migration length upon lowering the growth temperature. The activation energy for this diffusion process in Fig. 5.52 is estimated to be 2.2 eV. This value is less than that for the bulk interdiffusion of Ga atoms and In atoms in GaInN alloy (3.4 eV in [145]) and is larger than that for the migration of Ga adatoms on GaN(0001) (0.4 to 1.8 eV in [146]). Therefore the compositional modulation in the GaInN MQWs is dominated by the surface migration process rather than the bulk interdiffusion process.

5.6.2 Influence of composition fluctuation on the device characteristics and degradation

In principle, the entire visible-light wavelength region can be covered by changing alloy composition in GaN-based optical devices with GaInN active layers. It is known that quantum efficiency of LEDs or LDs plummets rapidly when the emission is shifted from blue to green, in other words, when the indium content of GaInN QW increases. This is because growth of the GaInN layer with higher indium content causes phase separation and generates structural defects. For example, when the growth conditions are not optimized, indium-rich precipitates are frequently formed in the GaInN QW (see Fig. 5.56). These precipitates obviously reduced quantum efficiency and are detrimental to the device's lifetime. Even if no precipitates occur and there is no significant phase separation, the composition of

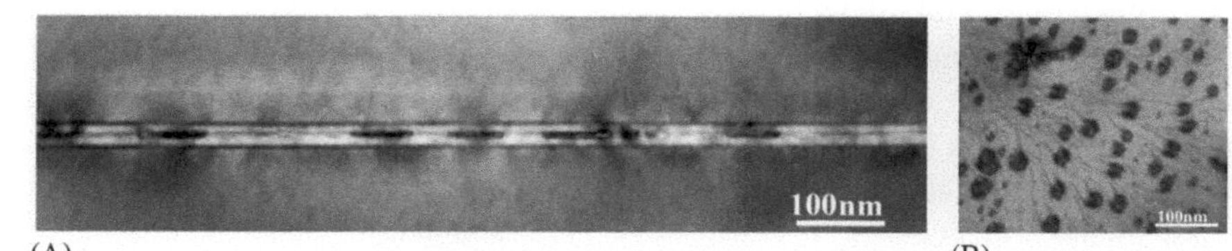

Figure 5.56 (A) Cross-sectional transmission electron microscope image of GaInN MQWs. Indium-rich precipitates and voids due to the formation of the precipitates are observed in the second quantum well layers. (B) Plan view transmission electron microscope image of GaInN MQWs.

the GaInN layer usually has spatial fluctuation in the GaInN layers [132,142–149]. This spatial fluctuation causes a large potential fluctuation in GaInN QWs because GaInN is an alloy material between the two materials with significantly different bandgap energies. In such systems, the potential fluctuation restricts the movement of carriers, and the carrier capture by nonradiative defects can be suppressed, which can improve the light emission efficiency [150–156]. On the other hand, the fluctuation also creates a problem that the optical gain peak becomes broadened, and the differential gain worsens in laser diodes [157,158]. In this way, the potential fluctuation in the GaInN active layers dramatically affects the characteristics of the GaN-based optical device.

As described above, the effect of GaInN composition fluctuation on the device characteristics seems to be different depending on the LED and the laser. However, the influence on the gradual degradation remains unknown. In general, compositional fluctuations modulate the Peierls potential in the crystal so that the movement of point defects is considered to be suppressed. For this reason, unless the composition fluctuation is not too large, gradual degradation occurs due to the point that defect diffusion process might be suppressed.

5.7 Conclusion

In this chapter, we have discussed the generation of various grown-in defects in III–V compound semiconductors (III–Vs) and GaN-related semiconductors (III-nitrides). The grown-in defects are classified into two types: interface defects and bulk defects.

In III–Vs, we have described six types of interface defects: stacking faults, V-shaped dislocations, dislocation clusters, misfit dislocations, microtwins, and inclusions, and three types of bulk defects: precipitates, precipitates with punched out dislocations, and dislocation loops. In particular, we have discussed the generation of climbed dislocation networks from punch out dislocations during operation of ridge-waveguide-type lasers and AlGaAs VCSELs.

In III-nitrides, we have also found interface defects: misfit dislocations and various types of TD derived from the misfit dislocations (screw-type, edge-type, mixed-type), and bulk defects: Mg-related pyramidal defects. Regarding these defects, generation mechanism, methods for suppression of the defects, and influence of them on the optical and electrical properties of the materials and devices reliability have been discussed. We have also suggested that the TD in the nitride LEDs and lasers enhance gradual degradation via pipe diffusion of various point defects (including dopants, residual impurities, and native point defects) and/or point defect complexes along them by recombination-enhanced point defect reaction and diffusion during the operation of lasers.

We have also described two types of structures due to thermal instability of the crystal during growth: CMS and OSs in III–Vs. The CMSs are observed as quasi-periodic diffraction contrast along the two equivalent directions of <100> and

<010>, and they are columnar-shaped in the <001> growth direction. They are observed in most III–V alloy semiconductors except AlGaAs. Since the compositional fluctuation along the CMSs is small, in the range 2%–3% in content, it is assumed that they have a minimal influence on the properties of the materials and device reliability. Three types of OSs are described in III–V alloy semiconductors: CuAu-I, CuPt, and chalcopyrite structures. Here we have focused on the CuPt structure since they are generated in the crystals such as InGa(Al)P crystals grown on the (001) substrates. We have discussed the generation mechanism for the OSs and the key parameters and processes controlling the degree of ordering: growth temperature, growth rate, vapor pressure ratio, doping, substrate misorientation, thermal annealing, and impurity diffusion. We have shown that the OSs strongly influence the bandgap energy of the materials and that the visible lasers with InGa(Al)P active layers are degraded rapidly due to recombination-enhanced diffusion of p-dopants such as Zn atoms during operation.

Furthermore, we have described the recent results on the CMSs in GaInN ternary alloy, which is particularly important in III-nitrides-based optical devices in the visible region. Although phase separation of GaInN is a significant concern because of its wide miscibility gap, frequency distributions of $Ga_{1-x}In_xN$ quantum well's composition follow the binominal distribution until at least indium composition up to $x < 0.25$, and no remarkable phase separation is observed. On the other hand, a very fine CMS of about 5 nm or less are formed in the (0001) growth surface. The period of the CMS becomes shorter as the indium composition increased. The correlation between the CMS and the degradation phenomenon of the optical semiconductor device due to the defect multiplication is unclear and is an area of future research.

References

[1] B.C. De Loach, B.W. Hakki, R.L. Hartman, L.A. D'Asaro, Proc. IEEE 61 (1973) 1042.

[2] R.L. Hartman, A.R. Hartman, Appl. Phys. Lett. 23 (1973) 147.

[3] N.R. Chignon, R. Ito, O. Nakada, IEEE J. Quantum Electron. QE-10 (1974) 81.

[4] R. Ito, H. Nakashima, O. Nakada, Jpn. J. Appl. Phys. 13 (1974) 1321.

[5] H. Yonezu, M. Ueno, T. Kamejima, I. Sakuma, Jpn. J. Appl. Phys. 13 (1974) 835.

[6] S. Yamakoshi, O. Hasegawa, H. Hamaguchi, M. Abe, T. Yamaoka, Appl. Phys. Lett. 31 (1977) 627.

[7] S. Yamakoshi, T. Sugahara, O. Hasegawa, Y. Toyama, H. Takanashi, IEDM Tech. Dig. (1978) 642.

[8] P.M. Petroff, R.L. Hartman, Appl. Phys. Let. 23 (1973) 469.

[9] P.M. Petroff, R.L. Hartman, J. Appl. Phys. 45 (1974) 3899.

[10] P.W. Hutchinson, P.S. Dobson, Philos. Mag. 32 (1975) 745.

[11] P.W. Hutchinson, P.S. Dobson, S. O'Hara, D.H. Newman, Appl. Phys. Lett. 26 (1975) 250.

[12] O. Ueda, S. Isozumi, T. Kotani, T. Yamaoka, J. Appl. Phys. 48 (1977) 3950.

[13] O. Ueda, S. Isozumi, S. Yamakoshi, T. Kotani, J. Appl. Phys. 50 (1979) 765.

[14] T. Kotani, O. Ueda, K. Akita, Y. Nishitani, T. Kusunoki, O. Ryuzan, J. Cryst. Growth 38 (1977) 85.

[15] O. Ueda, T. Fujii, and Y. Nakata, Proceedings of the International Conference on Science and Technology of Defect Control in Semiconductors, 1989, Yokohama, Japan 1990, 1091.
[16] O. Ueda, K. Wakao, S. Komiya, A. Yamaguchi, S. Isozumi, I. Umebu, J. Appl. Phys. 58 (1985) 3996.
[17] P.M.J. Marée, J.C. Barbour, J.F. van der Veen, J. Appl. Phys. 62 (1987) 4413.
[18] H. Takasugi, M. Kawabe, Y. Bando, Jpn. J. Appl. Phys. 26 (1987) L584.
[19] J.W. Lee, H.L. Tsai, J. Vac. Sci. Technol. B5 (1987) 819.
[20] K.W. Carey, F.A. Ponce, J. Amano, J. Appl. Phys. 54 (1983) 4414.
[21] O. Ueda, T. Soga, T. Jimbo, M. Umeno, Appl. Phys. Lett. 55 (1989) 445.
[22] O. Ueda, S. Komiya, S. Isozumi, Jpn. J. Appl. Phys. 23 (1984) L394.
[23] K. Akimoto, M. Dohsen, N. Watanabe, J. Cryst. Growth 73 (1985) 117.
[24] H. Kakibayashi, F. Nagata, Y. Katayama, Y. Shiraki, Jpn. J. Appl. Phys. 23 (1984) L846.
[25] O. Ueda, S. Komiya, S. Yamazaki, Y. Kishi, I. Umebu, T. Kotani, Jpn. J. Appl. Phys. 23 (1984) 836.
[26] J.F. Petroff, M. Sauvage, J. Cryst. Growth 43 (1978) 628.
[27] W. Hagen, H. Strunk, Appl. Phys. 17 (1978) 85.
[28] O. Ueda, K. Nakai, S. Yamakoshi, I. Umebu, Mat. Res. Soc. Symp. Proc. 138 (1989) 509.
[29] S. Nakahara, S.N.G. Chu, J.A. Long, V.G. Riggs, W.D. Johnston Jr., J. Crystal, Growth 72 (1985) 693.
[30] O. Ueda, I. Umebu, T. Kotani, J. Cryst. Growth 62 (1983) 329.
[31] T.Y. Tan, W.K. Tice, Philos. Mag. 34 (1976) 615.
[32] D.M. Maher, A. Staudinger, J.R. Patel, J. Appl. Phys. 47 (1976) 3813.
[33] M. Hansen, Constitution Binary Alloy, McGraw-Hill, New York, NY, 1958.
[34] D. Laister, G.M. Genkins, Philos. Mag. 23 (1971) 1077.
[35] P.W. Hutchinson, P.S. Dobson, Philos. Mag. 30 (1974) 65.
[36] G.H. Narayanan, A.H. Kachare, Phys. Status Solidi (a) 26 (1974) 657.
[37] B.W. Hakki, J. Electrochem. Soc. 118 (1971) 1469.
[38] G.B. Stringfellow, J. Appl. Phys. 43 (1972) 3455.
[39] S. Mukai, M. Matsuzaki, J. Shimada, Jpn. J. Appl. Phys. 20 (1981) 321.
[40] O. Ueda, S. Isozumi, S. Komiya, Jpn. J. Appl. Phys. 23 (1984) L241.
[41] O. Ueda, S. Komiya, S. Yamakoshi, I. Umebu, K. Akita, Jpn. J. Appl. Phys. Supplement 22-1 (1983) 243.
[42] P.M. Petroff, Semiconductors Insulators 5 (1983) 307.
[43] P.M. Petroff, L.C. Kimerling, Appl. Phys. Lett. 29 (1976) 461.
[44] S. O'Hara, P.W. Hutchinson, P.S. Dobson, Appl. Phys. Lett. 30 (1977) 368.
[45] R.D. Gold, L.R. Weisberg, Solid-State Electron. 7 (1964) 811.
[46] O. Ueda, K. Wakao, A. Yamaguchi, S. Isozumi, S. Komiya, J. Appl. Phys. 57 (1985) 1523.
[47] O. Ueda, S. Yamakoshi, S. Komiya, K. Akita, T. Yamaoka, Appl. Phys. Lett. 36 (1980) 300.
[48] M. Seki, M. Fukuda, K. Wakita, Appl. Phys. Lett. 40 (1982) 115.
[49] O. Ueda, S. Komiya, S. Yamakoshi, T. Kotani, Jpn. J. Appl. Phys. 20 (1981) 1201.
[50] H. Amano, N. Sawaki, I. Akasaki, Y. Toyoda, Appl. Phys. Lett. 48 (1986) 353.
[51] F.A. Ponce, B.S. Krusor, J.S. Major Jr., W.E. Plano, D.F. Welch, Appl. Phys. Lett. 67 (1995) 410.
[52] T. Hino, S. Tomiya, T. Miyajima, K. Yanashima, S. Hashimoto, M. Ikeda, Appl. Phys. Lett. 76 (2000) 3421.
[53] T. Miyajima, T. Hino, S. Tomiya, K. Yanashima, H. Nakajima, Y. Nanishi, et al., Phys. Stat. Zol. (b) 228 (2001) 395.
[54] S.D. Lester, F.A. Ponce, M.G. Craford, D.A. Steigerwald, Appl. Phys. Lett. 66 (1995) 1249.

[55] S. Nakamura, M. Senoh, S. Nagahama, N. Iwasa, T. Yamada, T. Matsushita, et al., Proceedings of the Second International Conference on Nitride Semiconductor, Tokushima (Japan), 1997, 444.
[56] T. Tojyo, T. Asano, M. Takeya, T. Hino, S. Kijima, S. Goto, et al., Jpn. J. Appl. Phys. 40 (2001) 3206.
[57] X.H. Wu, C.R. Elsass, A. Abare, M. Mack, S. Keller, P.M. Petroff, et al., Appl. Phys. Lett. 72 (1998) 692.
[58] I.-H. Kim, H.-S. Park, Y.-J. Park, T. Kim, Appl. Phys. Lett. 73 (1998) 1634.
[59] Y. Chen, T. Takeuchi, H. Amano, I. Akasaki, N. Yamada, Y. Kaneko, et al., Appl. Phys. Lett. 72 (1998) 710.
[60] S. Tomiya, Y. Kanitani, S. Tanaka, T. Ohkubo, K. Hono, Appl. Phys. Lett. 98 (2011) 181904.
[61] A. Hangleiter, F. Hitzel, C. Netzel, D. Fuhrmann, U. Rossow, G. Ade, et al., Phys. Rev. Lett. 95 (2005) 127402.
[62] P. Kozodoy, J.P. Ibbetson, H. Marchand, P.T. Fini, S. Keller, J.S. Speck, et al., Appl. Phys. Lett. 73 (1998) 975.
[63] O. Ueda, Reliability and Degradation of III-V Optical Devices, Artech House Publishers, Boston, MA/London, 1996.
[64] T.S. Zheleva, S. Scott, D.B. Thomson, K.J. Linthicum, P. Rajagopal, R.F. Davis, J. Electron. Mater. 28 (1999) L6.
[65] S. Tomiya, T. Hino, S. Goto, M. Takeya, M. Ikeda, IEEE J. Quantum Electron. 10 (2004) 1277.
[66] O. Ueda, Jpn. J. Appl. Phys. 49 (2010) 090001.
[67] H. Amano, M. Kilo, K. Hiramatsu, N. Sawaki, I. Akasaki, Jpn. J. Appl. Phys. 28 (1989) L2112.
[68] S. Nakamura, N. Iwasa, M. Senoh, T. Mukai, Jpn. J. Appl. Phys. 31 (1992) 1258.
[69] Z. Liliental-Weber, M. Benamara, W. Swider, J. Washburn, Appl. Phys. Lett. 75 (1999) 4159.
[70] P. Vennéguès, M. Benaissa, B. Beaumont, E. Feltin, P. De Mierry, S. Dalmasso, et al., Appl. Phys. Lett. 77 (2000) 880.
[71] S. Tomiya, S. Goto, M. Takeya, M. Ikeda, Mat. Res. Symp. Proc. 743 (2003) 839.
[72] D. Vaumousse, A. Cerezo, P.J. Warren, Ultramicroscopy 95 (2003) 215.
[73] L. Stephenson, M.P. Moody, P.V. Liddicoat, S.P. Ringer, Microscopy Microanal. 13 (2007) 448.
[74] A.J. Ptak, T.H. Myers, L.T. Romano, et al., Appl. Phys. Lett. 78 (2001) 285.
[75] J. Neugebauer, C.G. Van der Walle, Phys. Rev. Lett. 75 (1995) 4452.
[76] L.T. Romano, J.E. Northrup, A.J. Ptak, T.H. Myers, Appl. Phys. Lett. 77 (2000) 2479.
[77] B. Beaumont, S. Haffouz, P. Gilbart, Appl. Phys. Lett. 72 (1998) 921.
[78] S. Tomiya, O. Goto, M. Ikeda, Proc. IEEE 98 (2010) 1208.
[79] M. Meneghini, C. de Santi, N. Trivellin, K. Orita, S. Takigawa, T. Tanaka, et al., Appl. Phys. Lett. 99 (2011) 093506.
[80] A. Hierro, S.A. Ringel, M. Hansen, J.S. Speck, U.K. Mishra, S.P. DenBaars, Appl. Phys. Lett. 77 (2000) 1499.
[81] U. Kaufmann, M. Kunzer, M. Maier, H. Obloh, A. Ramakrishnan, B. Santic, et al., Appl. Phys. Lett. 72 (1998) 1326.
[82] B. Monemar, P.P. Paskov, G. Pozina, C. Hemmingsson, J.P. Bergman, T. Kawashima, et al., Phys. Rev. Lett. 102 (2009) 235501.
[83] Y. Ueno, H. Fujii, K. Kobayashi, K. Endo, A. Gomyo, K. Hara, et al., Jpn. J. Appl. Phys. 29 (1990) L1666.

[84] H. Hamada, M. Shono, S. Honda, R. Hiroyama, K. Matsukawa, K. Yodoshi, et al., Electron. Lett. 27 (1991) 661.
[85] Y. Xu, Y. Li, Q. Gan, Q. Cao, G. Song, L. Guo, et al., Opt. Eng. 45 (2006) 034205.
[86] O. Goto, Y. Ohizumi, M. Shoji, T. Tanaka, Y. Hoshina, M. Ohta, et al., IEICE Tech. Rep. ICICE 107 (207) (2007) 35.
[87] M. Ohta, Y. Ohizumi, Y. Hoshina, T. Tanaka, Y. Yabuki, K. Funato, et al., Phys. Stat. Sol. (a) 204 (2007) 2068.
[88] F. Glas, P. Henoc, H. Launois, Inst. Phys. Conf. Ser. 76 (1985) 251.
[89] O. Ueda, M. Takikawa, J. Komeno, I. Umebu, Jpn. J. Appl. Phys. 26 (1987) L1824.
[90] O. Ueda, M. Takikawa, M. Takechi, J. Komeno, I. Umebu, J. Cryst. Growth 93 (1988) 418.
[91] M.M. Treacy, J.M. Gibson, A. Howie, Philos. Mag. A51 (1985) 389.
[92] T.S. Kuan, T.F. Kuech, W.I. Wang, E.L. Wilkie, Phys. Rev. Lett. 54 (1985) 208.
[93] I.J. Murgatroyd, A.G. Norman, G.R. Booker, and et al., Proceedings of the Eleventh International Congress on Electron Microscopy, Kyoto, 1986, p. 1497.
[94] Y.B. Ihn, N. Otsuka, J. Klem, H. Morkoç, Appl. Phys. Lett. 51 (1987) 2013.
[95] T.S. Kuan, W.I. Wang, E.L. Wilkie, Appl. Phys. Lett. 51 (1987) 51.
[96] H.R. Jen, M.J. Cherng, G.B. Stringfellow, Inst. Phys. Conf. Ser. 83 (1987) 159.
[97] A. Gomyo, T. Suzuki, K. Kobayashi, S. Kawata, I. Hino, T. Yuasa, Appl. Phys. Lett. 50 (1987) 16.
[98] S. McKernan, B.C. de Cooman, C.B. Carter, J. Mat. Res. 3 (1988) 406.
[99] M.A. Shahid, S. Mahajan, D.E. Laughlin, H.M. Cox, Phys. Rev. Lett. 58 (1987) 2567.
[100] P. Bellon, J.P. Chevalier, G.P. Martin, E. Dupont-Nivet, C. Thiebaut, J. Andre, Appl. Phys. Lett. 52 (1988) 567.
[101] M. Kondow, H. Kakibayashi, S. Minagawa, Appl. Phys Lett. 53 (1988) 2053.
[102] A. Gomyo, T. Suzuki, S. Iijima, Phys. Rev. Lett. 60 (1988) 2645.
[103] H. Nakayama, H. Fujita, Inst. Phys. Conf. Ser. 79 (1986) 289.
[104] T. Suzuki, A. Gomyo, J. Cryst. Growth 93 (1988) 396.
[105] A. Gomyo, K. Kobayashi, S. Kawata, I. Hino, T. Suzuki, J. Cryst. Growth 77 (1986) 367.
[106] O. Ueda, T. Fujii, Y. Nakata, H. Yamada, I. Umebu, J. Cryst. Growth 95 (1989) 38.
[107] O. Ueda, M. Hoshino, Kodama, H. Yamada, M. Ozeki, J. Cryst. Growth 99 (1990) 560.
[108] O. Ueda, M. Takechi, J. Komeno, Appl. Phys. Lett. 54 (1989) 2312.
[109] A. Gomyo, T. Suzuki, K. Kobayashi, S. Kawata, I. Hino, Appl. Phys. Let. 50 (1987) 673.
[110] O. Ueda, M. Hoshino, M. Takechi, M. Ozeki, T. Kato, T. Matsumoto, J. Appl. Phys. 68 (1990) 4268.
[111] E. Morita, M. Ikeda, O. Kumagai, K. Kaneko, Appl. Phys. Lett. 53 (1988) 2164.
[112] C.S. Baxter, W.M. Stobbs, J.H. Wilkie, et al., Inst. Phys. Conf. Ser 117 (1991) 469.
[113] C. Nozaki, Y. Ohba, H. Sugawara, S. Yasuami, T. Nakanisi, J. Cryst. Growth 93 (1988) 406.
[114] D.S. Cao, E.H. Reihlen, G.S. Chen, A.W. Kimball, G.B. Stringfellow, J. Cryst. Growth 109 (1991) 279.
[115] B.T. McDermott, N.A. El-Masry, B.L. Jiang, F. Hyuga, S.M. Bedair, W.M. Duncan, J. Cryst. Growth 107 (1991) 96.
[116] S. Mahajan, M.A. Shahid, D.E. Laughlin, Inst. Phys. Conf. Ser. 100 (1989) 143.
[117] T. Suzuki, A. Gomyo, S. Iijima, K. Kobayashi, S. Kawata, I. Hino, et al., Jpn. J. Appl. Phys. 27 (1988) 2098.

[118] T. Suzuki, A. Gomyo, I. Hino, K. Kobayashi, S. Kawata, S. Iijima, Jpn. J. Appl. Phys. 27 (1988) L1549.
[119] S.R. Kurtz, J.M. Olson, J.P. Goral, A. Kibbler, E. Beck, J. Electron. Mat. 19 (1990) 825.
[120] A. Gomyo, H. Hotta, I. Hino, S. Kawata, K. Kobayashi, T. Suzuki, Jpn. J. Appl. Phys. 28 (1989) L1330.
[121] A. Gomyo, T. Suzuki, S. Iijima, H. Hotta, H. Fujii, S. Kawata, et al., Jpn. J. Appl. Phys. 27 (1988) L2370.
[122] S. Minagawa, M. Kondow, Electron. Lett. 25 (1989) 758.
[123] P. Gavrilovic, F.P. Dabkowski, K. Meehan, J.E. Williams, W. Stutius, K.C. Hsieh, et al., Appl. Phys. Lett. J. Cryst. Growth 93 (1988) 426.
[124] R.F. Bloom, W. Heuberger, P. Unger, P. Roentgen, Inst. Phys. Conf. Ser. 117 (1991) 473.
[125] W.E. Plano, D.W. Nam, J.S. Major, K.C. Hsieh, N. Holonyak Jr., Appl. Phys. Lett. 53 (1988) 2537.
[126] Y. Takahashi, Y. Horikoshi, Jpn. J. Appl. Phys. 19 (1980) 687.
[127] M.M. Al-Jassim, C.A. Warwick, G.R. Booker, Inst. Phys. Conf. Ser. 60 (1981) 357.
[128] J.E. Bernard, L.G. Ferreira, S.-H. Wei, A. Zunger, Phys. Rev. B38 (1988) 6338.
[129] S.-H. Wei, L.G. Ferreira, A. Zunger, Phys. Rev. B41 (1990) 8240.
[130] J.E. Bernard, R.G. Dandrea, L.G. Ferreira, S. Froyen, S.-H. Wei, A. Zunger, Appl. Phys. Lett. 56 (1990) 731.
[131] K. Endo, K. Kobayashi, H. Fujii, Y. Ueno, Appl. Phys. Lett. 64 (1994) 146.
[132] J.W. Cahn, Acta Met. 9 (1961) 795.
[133] B. de Cremoux, P. Hirth, J. Ricciardi, Inst. Phys. Conf. Ser. 56 (1981) 115.
[134] J.W. Cahn, J.E. Hilliard, J. Chem. Phys. 28 (1958) 258.
[135] T. Nagatomo, T. Kuboyama, H. Minamino, O. Omoto, Jpn. J. Appl. Phys. 28 (1989) L1334.
[136] N. Yoshimoto, T. Matsuoka, T. Sasaki, A. Katsui, Appl. Phys. Lett. 59 (1991) 2251.
[137] T. Matsuoka, N. Yoshimoto, T. Sasaki, A. Katsui, J. Electron. Mater. 21 (1992) 157.
[138] I.-H. Ho, G.B. Stringfellow, Appl. Phys. Lett. 69 (1996) 2701.
[139] M. Rao, D. Kim, S. Mahajan, Appl. Phys. Lett. 85 (2004) 1961.
[140] K. Hiramatsu, Y. Kawaguchi, M. Shimizu, N. Sawaki, T. Zheleva, R.F. Davis, et al., MRS Internet J. Nitride Semicond. Res. 2 (1997) 6.
[141] S. Tomiya, Mat. Res. Symp. Proc. 1195 (2009) 37.
[142] Y. Kanitani, S. Tanaka, S. Tomiya, T. Ohkubo, K. Hono, Jpn. J. Appl. Phys. 55 (2016) 05FM04.
[143] M.P. Moody, L.T. Stephenson, A.V. Ceguerra, S.P. Ronger, Micro. Res. Tech. 71 (2008) 542.
[144] S.Y. Karpov, MRS Internet J. Nitride Semicond. Res. 3 (1998) 16.
[145] C.-C. Chuo, C.-M. Lee, J.-I. Chyi, Appl. Phys. Lett. 78 (2001) 314.
[146] T. Zywietz, J. Neugebauer, M. Scheffler, Appl. Phys. Lett. 73 (1998) 487.
[147] Y. Narukawa, Y. Kawakami, M. Funato, S. Fujita, S. Fujita, S. Nakamura, Appl. Phys. Lett. 70 (1997) 981.
[148] S. Chichibu, K. Wada, S. Nakamura, Appl. Phys. Lett. 71 (1997) 2346.
[149] M.A. Maur, A. Pecchia, G. Penazzi, W. Rodrigues, A.D. Carlo, Phys. Rev. Lett. 116 (2016) 027401.
[150] J. Dalfors, J.P. Bergman, P.O. Holtz, B.E. Sernelius, B. Monemar, H. Amano, et al., Appl. Phys. Lett. 74 (1999) 3299.
[151] Y. Narukawa, Y. Kawakami, S. Fujita, S. Fujita, S. Nakamura, Phys. Rev. B 55 (1997) R1938.

[152] A. Kaneta, M. Funato, Y. Kawakami, Phys. Rev. B 78 (2008) 125317.
[153] A. Kaneta, K. Okamoto, Y. Kawakami, S. Fujita, G. Marutsuki, Y. Narukawa, et al., Appl. Phys. Lett. 81 (2002) 4353.
[154] S. Chichibu, T. Azuhata, T. Sota, S. Nakamura, Appl. Phys. Lett. 70 (1997) 2822.
[155] T.K. Uždavinys, D.L. Beverra, T. Ivanov, S.P. Denbaars, S. Nakamura, J.S. Speck, et al., Opt. Mater. Exp. 7 (2017) 3116.
[156] S.F. Chichibu, A.C. Abare, M.P. Mack, M.S. Minsky, T. Deguchi, D. Cohen, et al., Mater. Sci. Eng. B 59 (1999) 298.
[157] A.A. Yamaguchi, M. Kuramoto, A. Kimura, M. Nido, M. Mizuta, Jpn. J. Appl. Phys. 40 (2001) L548.
[158] A.A. Yamaguchi, M. Kuramoto, M. Nido, M. Mizuta, Semicond. Sci. Technol. 16 (2001) 763.

Chapter 6

Reliability of lasers on silicon substrates for silicon photonics

Justin C. Norman[1], Daehwan Jung[2], Alan Y. Liu[1], Jennifer Selvidge[3], Kunal Mukherjee[4], John E. Bowers[1,3] and Robert W. Herrick[5]
[1]Quintessent, Inc., Santa Barbara, CA, United States, [2]Korea Institute of Science and Technology, Seoul, South Korea, [3]Institute for Energy Efficiency, University of California, Santa Barbara, CA, United States, [4]Department of Materials Science and Engineering, Stanford University, Stanford, CA, United States, [5]Intel Corporation, Santa Clara, CA, United States

6.1 Introduction

The unparalleled manufacturing capabilities of the complementary metal-oxide-semiconductor industry offer an incredible opportunity for scalable production of devices incorporating integrated photonics. Globally, the silicon microelectronics industry produces millions of wafers per month at wafer diameters up to 300 mm. These same manufacturing techniques (e.g., dry and wet etching, dielectric growth/deposition, dopant implantation, and metallization techniques) and high-yield production controls can be directly extended to manufacturing photonic devices providing a shortcut to high-throughput manufacturing, high yield, and high reliability. Such scalability will enable the low-cost, widely deployable technologies necessary to meet growing demands for energy-efficient, high-bandwidth density I/O in data centers and high-performance computing, low-cost, size, weight, and power modules for supporting wireless 5G infrastructure, light detection and ranging (LIDAR) systems in autonomous vehicles, and sensors for environmental, biomedical, and process monitoring applications.

Indeed, the promise of scalable manufacturing of silicon photonics is already being realized. Several major silicon foundries have operating photonics lines manufacturing photonic integrated circuits on 200–300 mm wafers with high-performance waveguides, modulators, and detectors as building blocks. Included amongst these entities are IMEC, TSMC, STMicroelectronics, GlobalFoundries, TowerJazz, and the American Institute for Manufacturing (AIM) Integrated Photonics. Notably absent from each of these manufacturing platforms to date are lasers and amplifiers to serve as a light source. Hybrid integration of a separate III–V light source via external coupling or butt-coupling is typically used for light generation.

Alternative approaches to hybrid integration include heterogeneous integration where III–V wafers or dies are bonded directly to silicon photonic chips for

Reliability of Semiconductor Lasers and Optoelectronic Devices. DOI: https://doi.org/10.1016/B978-0-12-819254-2.00002-3

self-aligned processing with the laser light coupling evanescently to underlying waveguides [1] and heteroepitaxial integration where III–V material is grown directly on silicon or silicon-on-insulator substrates for various integration schemes [2]. Heterogeneous integration is the most mature process with active development underway at Juniper, Hewlett-Packard Enterprise, and Intel. Since 2017, Intel has been mass producing optical transceivers with demonstrated performance, yield, and reliability at a rate now exceeding 2 million transceivers/year [3]. Heteroepitaxial integration, on the other hand, offers a more economical means of production [4] but faces immense challenges to performance and reliability due to the presence of material defects such as dislocations, antiphase domains, and cracks that originate from the mismatch in crystal lattice constant and coefficient of thermal expansion between III–Vs and the silicon material. These challenges have relegated epitaxial integration to the laboratory setting for decades, but recent advancements in optimized material buffers and the use of quantum dot (QD) gain media have shown significant promise for future development [2].

In the following sections the unique reliability challenges posed by heterogeneous and heteroepitaxial integration approaches will be discussed. A brief summary of early-stage prospective technologies is presented first before shifting the focus to more immediate solutions and those with a longer history of investigation. As the techniques with the most short-term promise and historical investigation, heterogeneous integration of InP-based laser materials and direct epitaxial growth of InAs QD lasers (at the GaAs lattice constant) on silicon will be emphasized. In general, the challenges faced by heterogeneous InP devices and epitaxial QD lasers will be the same as those faced by other materials and integration schemes. Reliability aspects unique to integration on silicon (as opposed to those general to all semiconductor lasers) will be emphasized.

6.2 Early-stage prospective technologies

6.2.1 *Group IV lasers*

The indirect band gaps of group IV materials have largely precluded their use as a light source for silicon photonics. While a number of approaches to improve radiative efficiency have been investigated over several decades including the use of tensile strained and highly doped Ge [5], SiGe quantum wells [6], quantum cascade structures based on SiGe wells [7], and SiGe QDs [8], there is only one report of room-temperature, electrically injected lasing from a group IV material [9]. The device had a pulsed threshold current density of 280 kA/cm^2 and 1 mW of output power, making it impractical for commercial applications. Unfortunately, fundamental physics stands in the way of more efficient devices ever being achieved due to the indirect band gap and carrier-scattering effects in highly doped Ge that lead to a fourfold increase in homogeneous broadening and an order of magnitude reduction in carrier lifetime [10].

6.2.2 *Nanowire lasers*

Nanowire lasers represent an interesting possibility for epitaxial lasers on silicon. Due to their high aspect ratio and extremely narrow diameter, defect-free material can be grown close to the Si surface. Two general implementations of nanowire lasers have been demonstrated, in-plane and out-of-plane, referring to the orientation of the wire relative to the growth interface.

In-plane nanowire lasers, as developed by IMEC [11] and the Hong Kong University of Science and Technology [12,13], use v-grooved interfaces to suppress defect formation and confine the defects to the III–V/Si interface. These results have been successful in demonstrating optically pumped lasers and show obvious promise for direct waveguide coupling for silicon photonics due to the proximity of the gain medium to the silicon, but neither reliability studies nor electrically pumped lasing have been demonstrated to date. While challenging, electrical pumping could be achievable, but the active region must be designed to be truly dislocation free or to use a dislocation-tolerant gain medium such as QDs (as described below) to avoid the climb-induced formation of dark line defects (DLDs). Current demonstrations show a remarkable level of dislocation confinement to the interface, but some threading dislocations do still exist and could ultimately kill devices, limiting the potential yield of this technique. Also, if minority carriers are not kept away from the III–V/Si interface, the trapped dislocations could form climb segments that grow into the active region.

In out-of-plane implementations of nanowire lasers, the ability of free surfaces to relax stress is leveraged to provide defect-free gain media. The small diameter allows for the material to relax the lattice mismatch and residual thermal stress upon cooling from growth temperature through stretching/compressing the atomic bonds at the unconstrained free surfaces of the wire. Again, the challenge with nanowire lasers, even more so for out-of-plane designs, is how to implement electrical injection. While optically pumped nanowire lasers have been demonstrated on Si using GaAs–AlGaAs core–shell nanowires [14] and InP arrays in v-groove Si trenches [15], no electrically injected lasers have been demonstrated in either material platform. However, nitride materials, which already perform well in the presence of dislocations, have had much more success. Frost et al. demonstrated an InGaN/GaN broad area (1.5 mm $\times$ 10 μm) edge-emitting nanowire laser on (0 0 1) Si in 2014 [16]. The InGaN active layers formed disks in the wires essentially behaving like QDs. The device operated at room temperature under continuous-wave (CW) electrical injection with a threshold current density of 1.76 kA/cm^2 (I_{th} = 70 mA) and emission at 532 nm. The laser was aged at 150 mA at 300K for 1000 h, and an extrapolated time to a 50% reduction of the peak output power (6 mW) of 7000 h was determined.

6.2.3 Lattice-matched dilute nitrides

Dilute nitrides could be an excellent gain material for lasers on silicon. By alloying with appropriate amounts of the standard III–V materials, dilute nitrides with a wide range of band gaps can be formulated with lattice matching to Si or Ge. Band gaps in the material system have been theorized to span from the visible to the infrared. Notably, InGaNAsSb/GaAs quantum wells emitting in the O-band have been demonstrated on a Ge substrate with pulsed operation up to 80°C with a relatively low threshold of ~27 mA from a 350 μm cavity [17]. The output power was relatively low at 1–2 mW, but the laser operated with negligible degradation over 100 h of continuous aging. On GaAs substrates high-performance lasers spanning the O-band and C-band have been demonstrated with <1 kA/cm^2 threshold current densities and hundreds of milliwatts of output power, but the reliability was relatively poor at <100 h to double the bias for 10 mW [18,19]. Ga(N,As,P) lasers lattice-matched to Si have also been developed with electrical injection, but they have not yet reached room-temperature, CW operation [20].

6.2.4 Quantum cascade lasers on silicon

Quantum cascade lasers (QCLs) on silicon merit their own discussion with regards to reliability due to their unipolar nature. The primary degradation mechanism associated with dislocations in lasers on silicon, whether heterogeneous or epitaxial, is that of recombination-enhanced dislocation climb and glide. Since nonradiative recombination is a minority carrier limited process, QCLs should have extremely low levels of dislocation growth, and, thus, perhaps inherently long laser lifetimes relative to their quantum well or even QD laser counterparts. To date, no reliability studies have been conducted on QCLs on silicon, but lasers operating in the pulsed regime at room temperature, both through heterogeneous integration [21,22] and epitaxial growth on miscut and on-axis Si [23,24], have been demonstrated with performance very close to identical devices grown on native substrates.

6.3 Heterogeneous lasers on silicon

The technologies of the previous section are all under active investigation and will likely continue to progress, but to date, only one method of monolithic laser integration with silicon has led to commercial production: heterogeneous integration. The first heterogeneously integrated devices were lasers, as demonstrated by Fang et al. in 2006 [25], and since then, there have been substantial strides in improving laser performance and leveraging the newfound benefits of silicon integration. The primary benefit to integration with silicon waveguides is that the optical cavity of the device can be separated from the lossy III–V gain medium. Through careful design of the underlying Si waveguides and downstream resonant components, the confinement factor, cavity length, and mirror loss can be precisely controlled. Furthermore, cavity length and mirror loss can be tuned by taking advantage of thermo-optic or electro-optic effects in the resonator materials to change the refractive index. Together these properties can offer performance enhancements over native substrate devices for tunable, narrow linewidth, and dispersion-compensated applications [1]. The general form of a heterogeneously integrated device is shown in Fig. 6.1. The device consists of III–V laser epi bonded with or without a thin layer of material to a patterned silicon photonic wafer containing passive

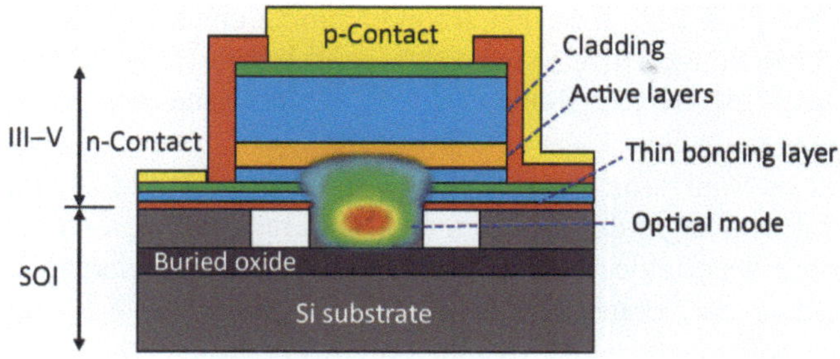

Figure 6.1 Schematic diagram of the cross-section of a heterogeneously integrated laser on a silicon photonic wafer illustrating the general epi structure and optical coupling to a silicon waveguide.

waveguides and other components. The III–V active region is kept close to the silicon to facilitate optical coupling.

6.3.1 Unique reliability considerations

From a reliability perspective, heterogeneous integration is excellent due to its inherently high-quality, native-substrate-grown epi material. For native substrate devices in the arsenide and phosphide material systems, dislocation densities are $<100\ \mathrm{cm}^{-2}$, and other defects such as antiphase domains and stacking faults are entirely absent.

The primary method for addressing defects in heterogeneously integrated lasers is the use of superlattice structures near the bonding interface. The trapping of point defects and dislocations at strained interfaces is a well-known phenomenon. In bonded laser material on silicon the thermal stress could cause dislocations to form at the bonded interface, but they may not remain there. For integration purposes the laser active region must be kept near the Si substrate to facilitate evanescent coupling. This proximity means the interfacial dislocations may undergo recombination-enhanced dislocation climb leading to their growth in the active region. If, instead, strained superlattice layers are incorporated below the laser active region, they could serve as mechanical barriers to prevent dislocations from climbing to the active quantum wells, and, if the band offsets are appropriately engineered, they can block minority carriers to limit recombination at the defective bonding interface. Such structures have been employed since the earliest days of wafer bonding showing substantial performance improvements [26] and even photoluminescence enhancement of the as-grown epi—presumably due to dislocations and/or point defects formed during growth gliding away from the active region to the strained superlattice.

6.3.2 Reliability results for heterogeneous lasers

Relatively little has been published to date on the reliability of heterogeneously integrated lasers for silicon photonics. The most notable studies are those of Srinivasan et al. [27] and Buffolo et al. [28,29], as well as a few conference presentations and papers from Intel [3,30,31]. These studies used similar epitaxial structures with identical bonding procedures. There has been successful deployment of commercial products from Intel [3], which show excellent performance and long lifetimes at accelerated-aging conditions.

The studies of Srinivasan et al. [27] and Buffolo et al. [28,29] looked at very similar material with slight deviations in the active region design and compressively strained bonding superlattices. Srinivasan et al. performed gradual aging experiments at constant accelerated conditions (50°C–85°C and 80–100 mA), while Buffolo et al. performed step-stress measurements, where the bias current was increased in 50 mA steps as a function of aging time from 200 to 950 mA over a period of 1000 min, and continuous aging tests at accelerated conditions (4.9–8.59 kA/cm^2 at 45°C) for heterogeneous distributed Bragg-reflector [29] and loop mirror lasers [28]. As the devices were aged, light–current–voltage (LIV) sweeps were performed to quantify laser degradation. The Buffolo et al. studies also performed capacitance deep-level transient spectroscopy (C-DLTS) to help identify the degradation mechanisms.

Both studies observed similar behavior and reached identical conclusions. In all cases, the lasers were observed to undergo gradual degradation of the threshold current with time. The threshold increase was found to correlate well with a decrease in subthreshold emission, strongly suggesting increased nonradiative recombination as the degradation mechanism. The C-DLTS results supported this conclusion by revealing the presence of four deep levels within the space charge region of the laser that increased in concentration over time. One of these traps matched closely with a commonly observed level associated with Al-containing materials which are present in the quaternary AlGaInAs throughout the laser active region. The others were not identified, but it is unlikely that they are associated with dislocations as none of the studies observed the onset of rapid degradation and reduction of slope efficiency associated with the DLDs that dislocations are known to form in this material system. This leaves point defects—possibly diffusing dopants, native defects, or contaminants during growth—as the degradation source, which are not enhanced by the heterogeneous embodiment of the laser and could have behaved similarly on native substrates.

As Robert Herrick showed in a 2019 OFC paper [31], three reliability advantages are believed to have helped heterogeneous bonded lasers in field deployment. The first has been lower current density relative to direct modulated lasers: these lasers are coupled with external modulators in existing applications, and thus do not have the restriction of requiring extremely high current density to get acceptable fall times. Therefore the lasers can be made larger and with fewer quantum wells, and the current densities can be nearly an order of magnitude lower. Consequentially, junction temperatures are also considerably lower. The second advantage comes from the lack of cleaved facets. In edge-emitting lasers, there are usually significant differences in both random failure rates and wearout failure lifetimes between simple cleaved facet lasers (which make up the large majority of commercially available lasers), and more expensive “window lasers” where additional processing has been used to form passive waveguides between the actively pumped laser section, and the passive section leading to the facet. However, the bonded lasers use tapers, which eliminate cleaved facets. The third and final advantage is the absence of significant lateral strain fields. In normal ridge lasers, the ridge redirects "⟨1 0 0⟩ DLDs," which normally want to travel at 45 degrees to the direction of light propagation bending the DLDs to travel along the ridge. In the bonded laser, the ridge of material does not define the electrically pumped area—that is restricted by an isolation implant—so there is no lateral strain field in the InP to contain and redirect DLDs, so the impact is far more limited.

The most convincing argument for the capability of reliable heterogeneous integration of lasers on Si is the 2019 deployment of >2,000,000 units per year of heterogeneous optical transceivers on Si by Intel [3] with over 3,000,000 units shipped by 2020 [32] and the first industry demonstration of a copackaged optical Ethernet switch [33]. Intel also recently disclosed its industry-leading field failure rate of 28 transceiver failures per million shipped, which has been approximately two orders of magnitude lower than that observed by competing transceivers [34]. The transceivers contain arrays of InP-based O-band distributed feedback lasers with near perfect uniformity in performance. The lasers operate up to 150°C CW and have less than 10% variation in the bias current needed for 10 mW of output power over 19,000 h of continuous aging at 80°C and 150 mA bias (Fig. 6.2).

Taken cumulatively, the results of the previously discussed studies seem to indicate that heterogeneously integrated lasers do not suffer significant additional reliability hurdles to their native III–V counterparts. Perhaps without bonding superlattices or with high-temperature bonding techniques, there could be additional degradation due to dislocations, but low-temperature plasma-assisted bonding and the inclusion of superlattices have obviated such concerns. While the previous results only discuss lasers based on InP substrates, lasers grown on GaAs and GaSb substrates have also been demonstrated to have high-performance levels when bonded to Si using similar low-temperature processes [35,36], but no reliability data have yet been reported for these structures.

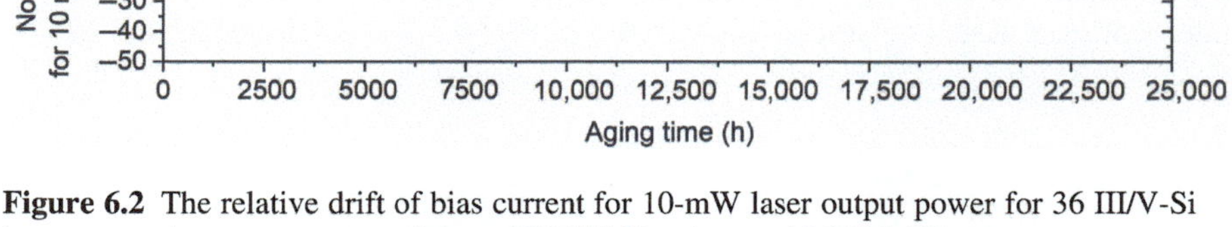

Figure 6.2 The relative drift of bias current for 10-mW laser output power for 36 III/V-Si lasers operating at a stress condition of 80°C/150 mA over 19,000 h [3].

6.4 Heteroepitaxial lasers grown on silicon substrates

The desire for direct heteroepitaxial integration of lasers on silicon is primarily economic. Heterogeneous integration requires additional manufacturing complexity and time for the bonding process. Through epitaxial growth, III–V integration can scale to the size of the silicon substrate and be performed in a single step rather than the series of steps associated with bonding: dicing, cleaning, dielectric deposition, plasma activation, die placement, and annealing. Production growth equipment is available with up to 450 mm platens for both metal–organic chemical vapor deposition and molecular beam epitaxy from multiple suppliers including AIXTRON, Veeco, and Riber, and they are already being widely used for multiwafer growth on much smaller III–V substrates. The barrier to commercial adoption of epitaxial III–V/Si integration is device reliability.

Decades of research have been devoted to developing high-performance III–V devices on silicon. The bulk of the effort has focused on minimizing dislocation densities through optimized buffer designs, generally utilizing some combination of graded buffers, strained superlattices, thermal annealing, and aspect ratio trapping. A detailed summary of the various techniques and their fundamental principles can be found in Ref. [37]. Despite ongoing efforts since the 1980s, the best achieved dislocation densities in III–Vs on Si are $\sim 2 \times 10^6\,\text{cm}^{-2}$, which were used for laser growth and achieved a (then) record device lifetime of 200 h [38]. Recent efforts have finally begun to push past this barrier showing promise for sub-$10^6\,\text{cm}^{-2}$ dislocation densities [39], but reliability results are not yet reported. It is important to note that most semiconductor lasers are thousands of μm^2 in area, and thus even the lowest dislocation densities demonstrated would still give hundreds or thousands of dislocations per laser. For most types of quantum well lasers sold today, even a single defect can nucleate a dislocation array that will grow until it kills the laser, which is why no one has previously made a reliable, high-performance laser grown on Si. The defect density of 2×10^6 is four orders of magnitude higher in dislocation density than currently available native III–V substrates (excluding nitrides). Better reliability has been achieved in InP/Si lasers, but these relied on commercially prohibitive buffers $>10\,\mu m$ in thickness and exhibited extremely low power conversion efficiencies [40–42]. Recently, the field of epitaxial III–V lasers on silicon has been revitalized due to the promising results achieved with QD active regions, which show an inherent tolerance to crystalline defects. In the last 4 years, the record device lifetime for O-band QD lasers on silicon has increased from $\sim 10^3$ to $>10^7$ h for continuous aging at 35°C [43], accompanied by strong improvements in laser threshold and efficiency. A summary of the development of heteroepitaxial lasers on silicon is presented in Fig. 6.3 and discussed in Ref. [2].

6.4.1 Unique reliability considerations

Silicon and III–V materials have significant differences in lattice constant, thermal expansion coefficient, and lattice polarity. A summary of these differences is given in

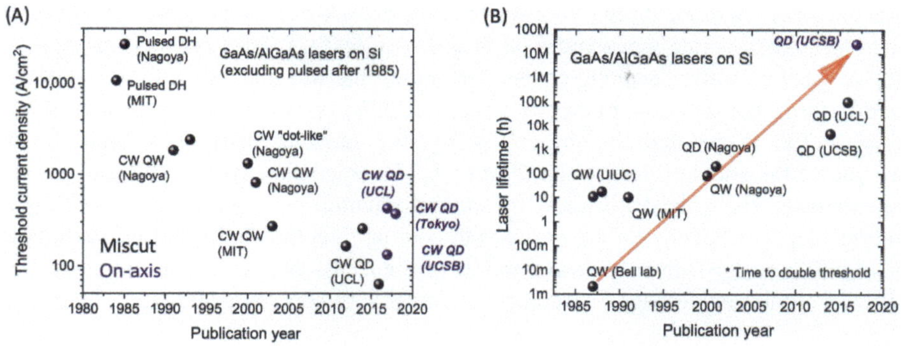

Figure 6.3 The historical trends in heteroepitaxial laser performance on silicon are presented for (A) threshold current density and (B) laser lifetime represented by the mean-time-to-failure defined according to a doubling of the initial threshold current density. Key institutions in the development of heteroepitaxial lasers on silicon have been labeled as well as a distinction being made for whether the silicon substrate utilized was miscut (black) or CMOS standard on-axis (0 0 1) Si (blue). The results are also labeled according to the active region of the device: DH, QW, and QD. *CMOS*, Complementary metal-oxide-semiconductor; *DH*, double heterostructure; *QD*, quantum dot; *QW*, quantum well.

Table 6.1 Summary of the mismatch in lattice constant and coefficient of thermal expansion of some III–V materials relative to silicon.

	GaAs	InP	GaSb
Lattice constant mismatch (%)	4.1	8.1	12.2
Coefficient of thermal expansion mismatch (%)	120	76.9	198

Table 6.1 (nitrides have been omitted due to varied mismatch depending on nucleation orientation [44]). When III–Vs are epitaxially deposited on Si, these mismatches lead to material relaxation in the form of dislocations and microcracks. The polarity mismatch can also lead to antiphase boundaries (APBs) where regions of III–III or V–V bonds form planar defects. Such boundaries generally occur across single-atomic surface steps due to the alternating sublattices of III and V in III–V requiring two monolayers of IV to maintain crystalline registry. From a reliability perspective the most influential of these defects are dislocations, which after decades of research development are still present in densities $\sim 10^6\ \text{cm}^{-2}$ in state-of-the-art epitaxial material on silicon. As discussed in Chapter 4, dislocations can undergo recombination-enhanced dislocation motion processes resulting in DLDs that grow and degrade laser performance in terms of both threshold current and slope efficiency. APBs are essentially a solved problem with multiple techniques including use of miscut Si substrates [45], surface pretreatment to form double atomic steps [46], specialized buffers [47], and growth on patterned substrates [48] having been demonstrated to eliminate APBs. Cracking in the III–V film comes about from the tensile

thermal stress induced during cool down from the growth temperature (250 MPa of stress or ~0.2% strain in arsenides on Si grown at 600°C). The cracks primarily limit device yield as wafer-spanning cracks penetrating the full thickness of the III–V epi readily form, but they also indirectly affect reliability through limiting the total thickness of III–V that can be grown on silicon, meaning there is a limit to the acceptable thickness of dislocation filtering buffer layers. For typical epitaxial growth temperatures, the critical thickness for crack formation is 4–6 μm for GaAs/Si [49] but this depends heavily on the postgrowth cooling rate, the ability of the substrate to bow during postgrowth cool down, and the presence of any stress concentrators such as nicks at the sample edge or macroscopic pregrowth contaminants.

6.4.2 Reliability results for quantum well lasers grown on silicon substrates

The aging results of quantum well lasers on silicon are best divided into those based on GaAs materials and those based on InP materials. The distinction between these two material systems is due to their stark differences in degradation mechanisms. Early results for lasers on Si indicated that GaAs-based lasers tended to undergo rapid degradation through the formation of DLDs in the laser active region [50]. In contrast, InP-based materials underwent much more gradual degradation despite similar dislocation densities, generally estimated via etch pit densities, of $\sim 10^6\ cm^{-2}$ [40,42].

For the InP-based lasers, no DLDs were observed; although, dark spots did appear in electroluminescence during the earlier stages of aging, they saturated in size and density as aging progressed [40]. This led to stable operation at elevated temperatures as high as 70°C for >9000 h of continuous aging at a bias current of 100 mA (11 kA/cm^2). Even double heterostructure lasers at the same dislocation densities were found to show saturated degradation when aged at 50°C. The results of this study are summarized in Fig. 6.4. In addition to significant initial degradation, we should note that the power conversion efficiency of the InP lasers was also quite low—presumably due to the defects creating a parallel path for recombination.

The best reported quantum well lasers from GaAs-based materials were reported by Kazi et al. [38] in 2001. They used epitaxial lateral overgrowth (ELO) to filter dislocations resulting in a film that showed an etch pit density of $2 \times 10^6\ cm^{-2}$. Lasers with single GaAs quantum wells were tested in constant output power mode for 1 and 2 mW emission, which corresponded to drive currents just above threshold, at room temperature. Despite the low dislocation density and extremely relaxed aging conditions, the lasers were observed to undergo rapid degradation in <200 h (see Fig. 6.5). Lasers were grouped into categories labeled type I and type II based on whether or not the cavity was located over one of the overgrowth windows, with type II being over the windows and type I being over a section of oxide mask. ELO is generally found to effectively trap dislocations as material grows through the mask and is limited by dislocations that are generated upon film coalescence above the mask; hence, the type II lasers would be expected to have higher dislocation densities than type I, possibly explaining the more rapid degradation rate. The performance of these devices is obviously far from commercial acceptability, and in

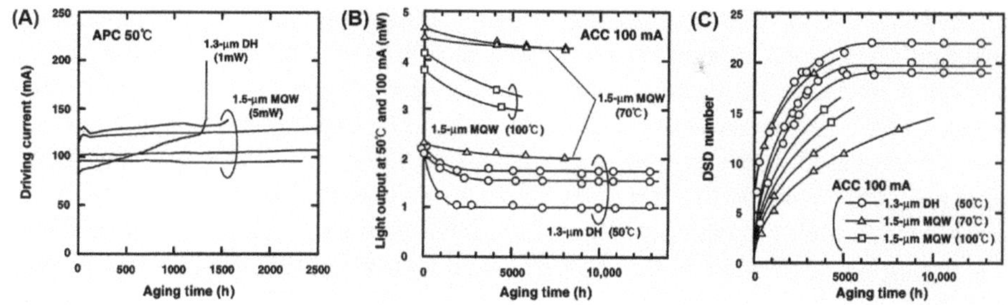

Figure 6.4 (A) The drive current required for 1 mW/facet (1.3 μm lasers) and 5 mW/facet (1.5 μm lasers) for InP-based lasers on silicon. (B) Laser output power during constant current aging at 100 mA for 1.3 μm DH lasers and 1.5 μm MQW lasers aged at elevated temperatures of 50°C–100°C. (C) The DSD as measured from electroluminescence as a function of aging time for constant 100 mA current stress at elevated temperatures from 50°C to 100°C. *DH*, double heterostructure; *DSD*, dark spot density; *MQW*, multiquantum well.

Source: Reprinted from T. Sasaki, et al., Aging tests of InP-based laser diodes heteroepitaxially grown on Si substrates, J. Appl. Phys. 84 (12) (1998) 6725–6728, with the permission of AIP Publishing.

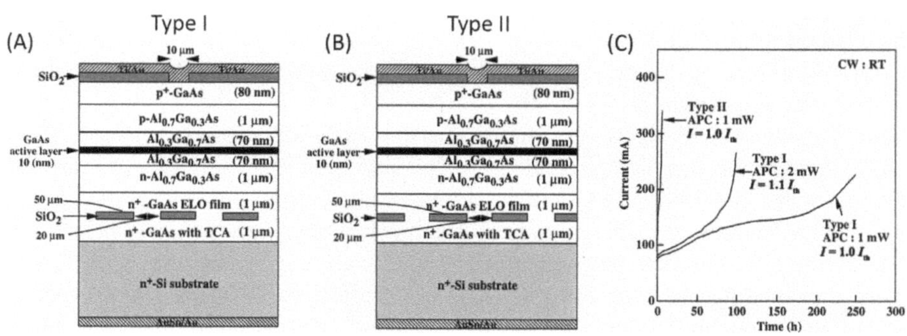

Figure 6.5 (A) Type I and (B) type II epi schematics for the two ELO schemes investigated. Type I places the current channel over a masked section from ELO growth, and type II places the current channel over a window in the ELO growth. (C) Aging results for APC aging of types I and II devices showing the necessary drive current for 1 mW or 2 mW of output power. *APC*, Automatic power control; *ELO*, epitaxial lateral overgrowth.

Source: Reproduced from Z.I. Kazi, et al., Realization of GaAs/AlGaAs lasers on Si substrates using epitaxial lateral overgrowth by metalorganic chemical vapor deposition, Jpn. J. Appl. Phys. 40 (8R) (2001) 4903, with permission. ©2001, The Japan Society of Applied Physics.

the nearly two decades since their publication, no GaAs-based quantum well laser grown on silicon has exceeded their reliability.

The proposed mechanism behind the enhanced performance of InP-based materials is that the defect states associated with dangling bonds lie closer to the band edge than in the arsenides [51], which reduces Shockley–Read–Hall recombination. With reduced nonradiative recombination, there should be reduced recombination-enhanced dislocation motion which is necessary to form DLDs. This conclusion is supported in native substrate devices where InP-based lasers are observed to have superior reliability characteristics. This trend would be expected to hold in other varieties of InP-based lasers as well (e.g., QDs). One drawback of the published InP/Si laser reliability work is that the lasers tend to have prohibitively thick buffers that would lead to high growth costs. Recent work has addressed this issue demonstrating CW lasing to 65°C with buffers $<4\,\mu m$ in thickness and dislocation densities of $1.15\times 10^8\,cm^{-2}$, but reliability results are not yet reported [52].

6.5 Reliability results for quantum dot lasers grown on silicon substrates

While impressive results have been achieved with InP-based quantum well lasers grown on silicon, none have managed to demonstrate the combination of commercially viable reliability, manufacturability, and high efficiency, and GaAs-based quantum well lasers lag even further in performance. One prominent reason for this is that quantum wells have little to impede lateral charge carrier diffusion granting dislocations a large radius of influence. This is not the case for QD active regions.

QDs represent quasi-zero-dimensional structures, with three-dimensional carrier confinement and an atom-like density of states [43]. The in-plane quantum confinement combined with relatively large energy level separations between states (>60 meV in the conduction band is typical) and high scattering rates to in-plane charge carrier mobility, give QD lasers a unique insensitivity to dislocations. In QD laser active regions, the charge carriers are much more likely to get captured and held by the dots and experience greater in-plane scattering by the dots as they move in the surrounding material. For this reason, the effective diffusion lengths tend to be $>4\times$ shorter in QD materials (down to 500 nm), depending on pumping densities. As a consequence, defects have a smaller capture radius in QD lasers, and thus the impact on performance is considerably smaller than in well-based lasers, as seen in Fig. 6.6.

Unfortunately, high-quality QDs are notoriously difficult to form in the InP materials system, and, to date, there is only one report of InAs/InP QD laser on Si under CW electrical injection at room temperature [53]. All of the successful laser results to date have been conducted using InAs/GaAs QD-in-a-well structures where the QDs are placed in InGaAs quantum wells (typically 10%–20% In content) to facilitate carrier capture and an emission wavelength near the technologically relevant O-band for optical communications.

Liu et al. performed a direct comparison of InGaAs/GaAs quantum well and InAs/GaAs QD laser material grown on silicon with identical dislocation densities ($2\times 10^8\,cm^{-2}$) [34]. Apart from the active region, the laser material was identical, and both sets of devices were processed simultaneously. The benefits of a QD active region were clearly visible. In the as-grown photoluminescence, the quantum

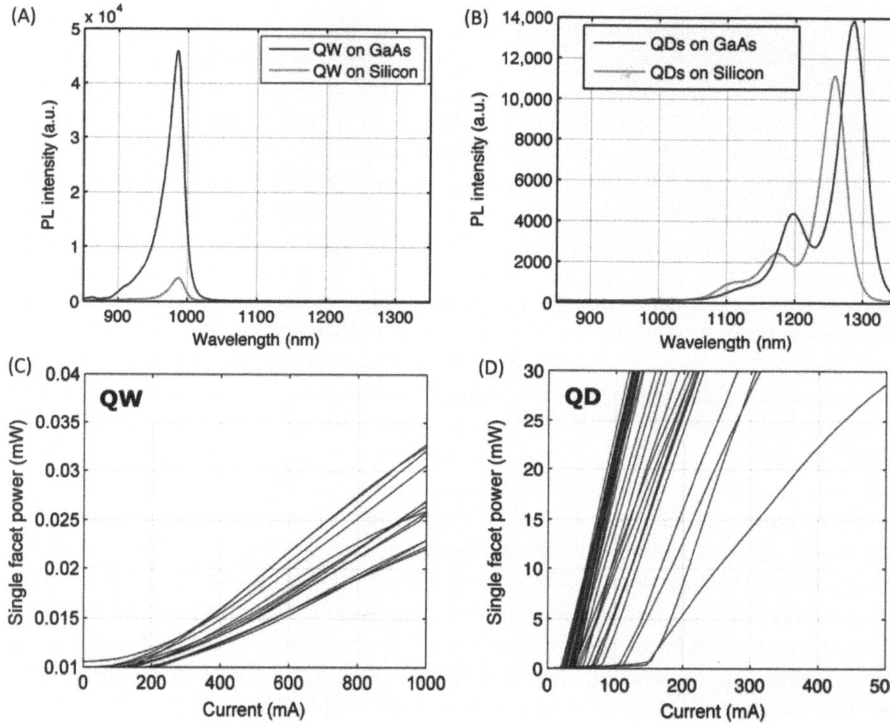

Figure 6.6 Photoluminescence of (A) QW and (B) QD material grown on silicon with identical dislocation densities of $2 \times 10^8\ cm^{-2}$ showing the enhanced tolerance of QDs to dislocations. The performance of ridge lasers with identical processing and device design grown on Si corresponding to (C) QWs and (D) QDs shows that the dot-based material lases with reasonable performance (i.e., low threshold current, high output power) while the well-based material fails to lase. *QD*, Quantum dot; *QW*, quantum well.

Source: Adapted from A.Y. Liu, et al., Quantum dot lasers for silicon photonics, Photonics Res. 3 (5) (2015) B1–B9 [55] with permission. ©2015, Chinese Laser Press.

well material showed a >90% reduction in intensity when comparing wells grown on silicon with those on native substrates while the QD material showed only a modest reduction in intensity of ~20% (see Fig. 6.6). At much lower dislocation densities of $7 \times 10^6\ cm^{-2}$, QD lasers on silicon show a photoluminescence intensity >95% of that obtained from native substrate material [43]. The lasers produced from the high dislocation density material showed relatively good performance for the QDs with threshold current densities of ~500 A/cm^2 and several 10s of milliwatts of single-facet output power. Meanwhile, the quantum well devices failed to lase. Both sets of devices showed essentially identical current–voltage characteristics, suggesting no difference in electrical properties. The time-zero performance results of this study were supported in the study by Liu et al. [54] which suggests the performance benefits of QDs on silicon persist to dislocation densities of $\sim 1 \times 10^6\ cm^{-2}$.

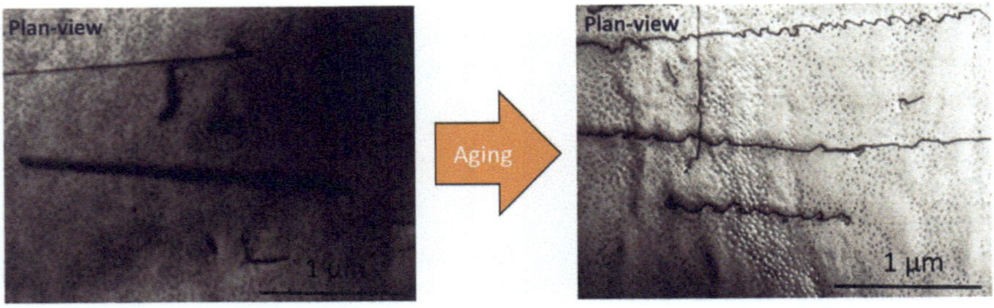

Figure 6.7 Threshold current density versus time for three generations of quantum dot laser material with dislocation densities of (A) 2.8×10^8, (B) 7.1×10^7, and (C) $7.3 \times 10^6\ \text{cm}^{-2}$. These lasers were aged at 35°C and twice their initial threshold current density.

Source: Reprinted from D. Jung, et al., Impact of threading dislocation density on the lifetime of InAs quantum dot lasers on Si, Appl. Phys. Lett. 112 (15) (2018) 153507, with the permission of AIP Publishing.

Returning to the lasers of Liu et al. [56], a few of the lasers were subjected to continuous aging at a constant current density of $1.1-2.5\ \text{kA/cm}^2$ at 30°C. The longest surviving devices were aged for more than 2700 h, with one device showing an extrapolated time to doubling its initial threshold current of 4600 h, exceeding by an order of magnitude the best previous result for an epitaxial GaAs-based laser on silicon including those with two orders of magnitude fewer dislocations [50]. A subset of these devices was evaluated postmortem using plan-view transmission electron microscope (TEM) to determine the degradation mechanism (as shown in Fig. 6.7). Comparing TEM images of the laser active region of aged and unaged samples revealed two interesting results: (1) both aged and unaged lasers contained high densities of misfit dislocations in the plane of the QD active region and (2) these misfit dislocations show the hallmark helical structure associated with dislocation climb after aging but not as extensively as is frequently seen in quantum well devices. The presence of misfits was surprising because the dot-in-a-well (DWELL) layers are well below the calculated critical thickness for the layer composition. Later studies have shown that these misfits result from the thermal expansion mismatch and form during postgrowth cool down. These results are described in later sections. The precise origins of reduced dislocation climb in QD active regions remain unknown, but it could be attributed to reduced nonradiative recombination due to the efficient carrier trapping by QDs, to a mechanical hardening of the crystal lattice by the dots themselves thus impeding climb, or a combination of the two effects.

In the 3 years following Liu et al.'s result [56], the dislocation density and laser performance of University of California, Santa Barbara (UCSB) QD material were improved through reductions in the dislocation density to $7 \times 10^6\ \text{cm}^{-2}$ [57], improvements in the QD density from $3-4 \times 10^{10}$ to $5-6 \times 10^{10}\ \text{cm}^{-2}$, and improvements in the QD size homogeneity from ~ 54 [58] to 28–30 meV [59] when grown on Si. These improvements combined to simultaneously decrease nonradiative recombination in the material while increasing the gain of the active region. Having this higher gain margin is highly beneficial to reliability as it allows the laser to tolerate more degradation before reaching its failure criterion. Another critical development was the addition of p-modulation doping (pMD) to the laser active region, which was first suggested to offer improved reliability by Liu et al. [56]. The benefits of pMD were reproduced in the higher quality material and shown to offer substantial improvements to laser reliability, most notably at elevated temperatures. These results are discussed below.

6.5.1 Dislocation density dependence of reliability in epitaxial quantum dot lasers on silicon

Following Liu et al.'s result, reliability testing was done on similar QD materials at a lower buffer dislocation density $\sim 10^6$ cm^{-2} by Chen et al. [60]. These lasers were broad area 50 μm × 3.2 mm devices and aged at relatively low-stress aging conditions of 26°C and 131 A/cm^2 ($1.75 \times I_{th}$) for 3100 h. The lasers showed relatively rapid initial degradation in the first 500 h that saturated and continued at a gradual rate to give an extrapolated time to threshold doubling of ~100,000 h. This was the first report of a QD laser on silicon with similar dislocation density to the best reported quantum well results and clearly demonstrated the benefits of a QD active region.

A systematic study was conducted by Jung et al. [61] investigating the impact of threading dislocation densities on the reliability of InAs/GaAs QD lasers on silicon. While QDs show remarkable tolerance to dislocations, they are far from immune, and a strong correlation between dislocation density and device lifetime was observed. Three generations of lasers were directly compared with dislocation densities of 2.8×10^8 cm^{-2} (Gen I), 7.1×10^7 cm^{-2} (Gen II), and 7.3×10^6 cm^{-2} (Gen III). The Gen I and Gen II lasers had seven dot layers and the Gen III laser had five dot layers. The laser epi design was otherwise identical, with dot layers spaced 37.5 nm from each other inside a 1.5 μm AlGaAs graded-index separate-confinement heterostructure. All ridges were etched through the laser active region with 30 nm Al_2O_3 passivation applied to the sidewalls and thick SiO_2 for electrical isolation. Initial studies were carried out at 35°C and twice the initial threshold current density of the device under test. Since, the study's publication, aging has continued, giving a total of ~10,000 h of aging data for the lowest dislocation density samples. A snapshot of aging results for lasers with similar geometry (ridge widths from 3–5 μm and cavity lengths 1000–1641 μm) at the three dislocation densities is given in Fig. 6.8.

The difference in degradation rate between the samples is significant in addition to significant improvements in the initial device performance at the lower dislocation densities. The initial threshold current density at the three dislocation levels was ~1700, ~1100, and ~250 A/cm^2. The depicted devices show extrapolated lifetimes (double initial threshold current) of ~1000, 20,000, and >10,000,000 h, respectively. The performance of the 7.3×10^6 cm^{-2} dislocation density devices clearly shows the potential for commercially viable performance near room temperature. These devices also have low absolute threshold currents of ~25 mA, and single-facet output powers of several 10s of milliwatts. The degradation of their

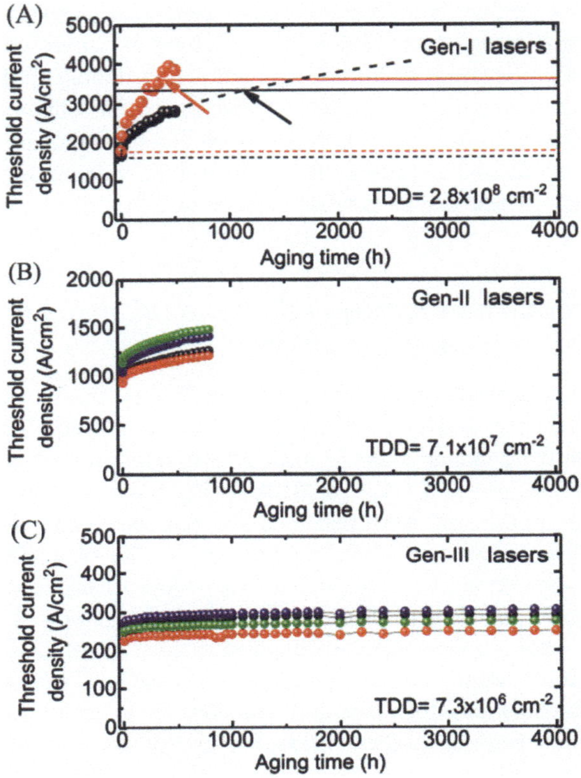

Figure 6.8 Relative changes in the slope efficiency with aging time of three generations of laser material with varying dislocation density: Gen I, $2.8 \times 10^8\ cm^{-2}$; Gen II, $7.1 \times 10^7\ cm^{-2}$; and Gen III, $7.3 \times 10^6\ cm^{-2}$. The inset shows light output versus bias current for the three lasers at various points during aging. Aging was performed at 35°C and twice the initial threshold current density for all samples.

Source: Reprinted from D. Jung, et al., Impact of threading dislocation density on the lifetime of InAs quantum dot lasers on Si, Appl. Phys. Lett. 112 (15) (2018) 153507, with the permission of AIP Publishing.

slope efficiency is shown in Fig. 6.9 with a mere 8% degradation over 4000 h in the lowest dislocation density devices.

A few devices from the Gen III laser material continued to be aged after the previous study for more than 1 year of continuous operation at the same conditions of 35°C and twice the initial threshold current density. The threshold current and bias current for 10 mW output power are shown in Fig. 6.10. The plots clearly show that extremely low rates of degradation are maintained throughout the period although a slight uptick in the curve is beginning to appear in the 10 mW bias.

From the initial aging data, it is notable that all three sets of samples show an initially higher degradation rate that saturates within a couple of hundred hours, as was observed in the previous studies by Liu et al. [56] and Chen et al. [60].

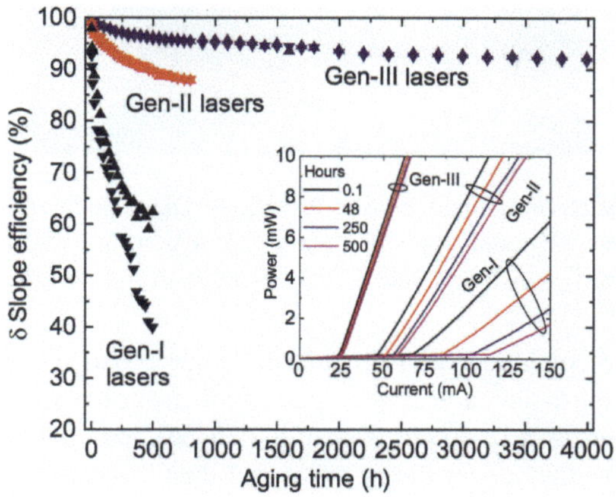

Figure 6.9 Extended aging results from the Gen III dataset described above are shown. (A) The threshold current and (B) the bias required for 10 mW of output power are shown for three devices over 10,000 h of continuous aging. Devices were aged at 35°C and twice their initial threshold current.

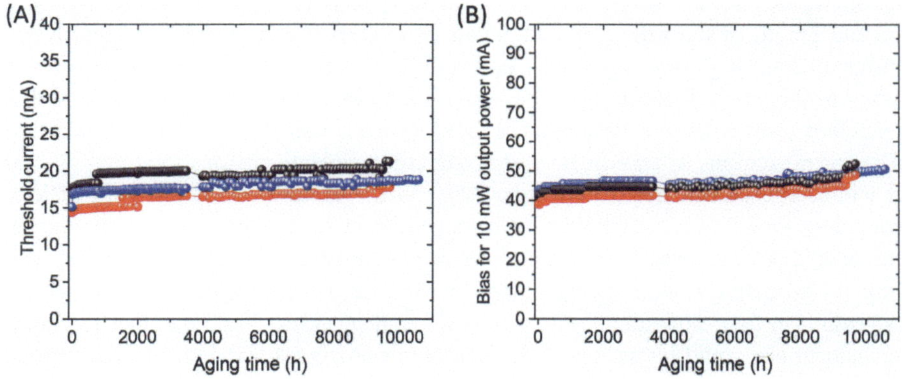

Figure 6.10 Transmission electron microscope image of the quantum dot active region of InAs/GaAs quantum dot lasers grown on silicon showing an unaged (A) and aged (B) device where the helical dislocation growth associated with climb processes is clearly visible.

Source: Adapted from A.Y. Liu, et al., Reliability of InAs/GaAs quantum dot lasers epitaxially grown on silicon, IEEE J. Sel. Top. Quantum Electron. 21 (6) (2015) 690–697.

The mechanism behind the early-stage degradation is unknown, but if it could be addressed, the mean-time-to-failure (MTTF) of the devices could be notably extended given the slower degradation rates thereafter. It is possible that the early-stage degradation could involve the generation of additional dislocation length in the active region through recombination-enhanced glide of dislocations from the buffer or laser cladding into the active region or the nucleation of new dislocations due to sidewall or facet damage from processing or other mechanisms

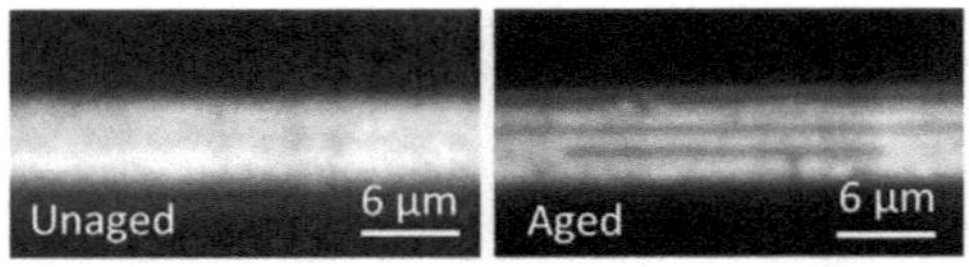

Figure 6.11 Backside electroluminescence of InAs/GaAs QD ridge lasers grown on silicon showing inhomogeneous degradation in the form of dark regions forming in sections of the laser cavity after aging at 60°C and $2 \times I_{th}$ until a 70% reduction in peak output power was observed.

standard to any semiconductor laser diode. Such glide-based sources would represent a greater incremental dislocation length than climb from existing dislocations given the relatively slow growth observed from Fig. 6.7 [56] and could support more rapid initial degradation as new dislocations arrive and then begin to slowly climb.

Backside electroluminescence has been performed on unaged and aged devices and showed nonuniform darkening of the laser active region. The results are shown in Fig. 6.11. At the scale of the laser cavity, electron channeling contrast imaging (ECCI) and plan-view TEM suggest relatively uniform distributions of dislocations in the active region (at $8 \times 10^6\ \mathrm{cm}^{-2}$ dislocation density, this would be one thread per 10 μm diameter). The climb from these dislocations should then appear as a local darkening of the electroluminescence with features at the appropriate density along the full length of the ridge. Instead, small regions of inhomogeneity and an overall uniform darkening were observed. As the uniform darkening is too dense to correspond to threading dislocations it may represent point defect clusters, which could be generated from recombination-enhanced dislocation climb [62]. There were additionally several regions of significant linear darkening several microns long and ~1 μm in width. These dimensions correspond well to the capture radius of misfit dislocations described in later sections. Given that these linear features are not apparent in the unaged device, these defects are either newly forming or migrating from the cladding to the active region during aging, or existing defects in the active region are increasing in their nonradiative impact as a result of aging. The nonradiative impact of dislocations could be increased through climb processes that increase the dislocation length per unit area or through the accumulation of point defects in the vicinity of misfit dislocations leading to elevated nonradiative recombination. A new length of dislocations could also be generated via recombination-enhanced glide of dislocations from outside the active region.

6.5.2 High-temperature reliability: the importance of p-modulation doping

To meet industrial standards for device reliability as demonstrated through accelerated-aging models and to meet basic performance requirements in many applications, lasers on silicon must demonstrate long lifetimes at elevated temperatures. For O-band InAs QD lasers, reliable operation at 80°C is necessary for data center and high-performance computing applications. When tested under continuous operation at twice the initial threshold current density at 60°C, even the highest performing lasers from Jung et al.'s study discussed above [61] were reduced to device lifetimes of 2500 h. This high level of acceleration is not unprecedented. Using a standard accelerated-aging model for MTTF:

$$\mathrm{MTTF} = AJ^{-n}\exp\left(\frac{E_a}{kT}\right) \tag{6.1}$$

where A is a fitting coefficient, n is the current acceleration factor, E_a is a characteristic activation energy for thermal acceleration, and kT is the thermal energy, gives, albeit with limited data points, an E_a of ~ 0.89 eV, which is similar to values obtained for InP-based semiconductor lasers. Fitting the current acceleration factor is also hindered by limited data currently. Due to the high variability in MTTF (e.g., see Fig. 6.13), an accurate value is difficult to extract, but the results suggest n is around 3–4.

Fortunately, improving the thermal performance of QD lasers is straightforward through the addition of p-type modulation doping to the laser active region. This technique was first employed by Shchekin et al. [63], and works by offsetting the rapid hole thermalization in InAs QDs through providing extra p-type carrier concentration. Due to the band offsets in the arsenide and phosphide material systems, InAs QDs always have much higher conduction band offsets than valence band offsets. The result is that the energy level separation between valence states is ~ 10 meV which is significantly less than kT. In contrast, the conduction band energy level separations can be >60 meV. These extra holes from p-doping help to dramatically increase the gain of the lasers [64,65], which allows them to tolerate the reduced injection efficiency at high temperature due to increased thermalization.

Returning to the lasers discussed in the previous section, when 10 nm of p-type doping was added to the 37.5 nm GaAs spacer between each of the five dot layers, the high-temperature lifetime improved significantly. The doping levels utilized were 5×10^{17}, 1×10^{18}, and $1.5\times10^{18}\ \mathrm{cm}^{-3}$, corresponding to ~ 10, 20, and 30 extra holes per QD, respectively. A schematic of one period of the laser active region is shown in Fig. 6.12. Fitting net modal gain data for these lasers to a first

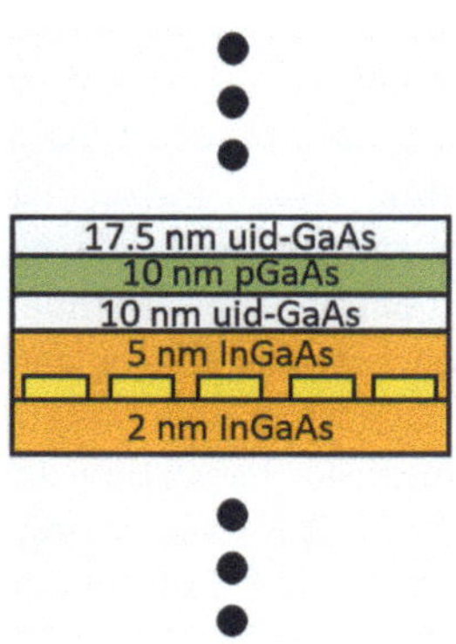

Figure 6.12 A schematic of one period of the p-modulation doped quantum dot laser active region. This structure is repeated 3–5 × as indicated in the various results presented in this chapter.

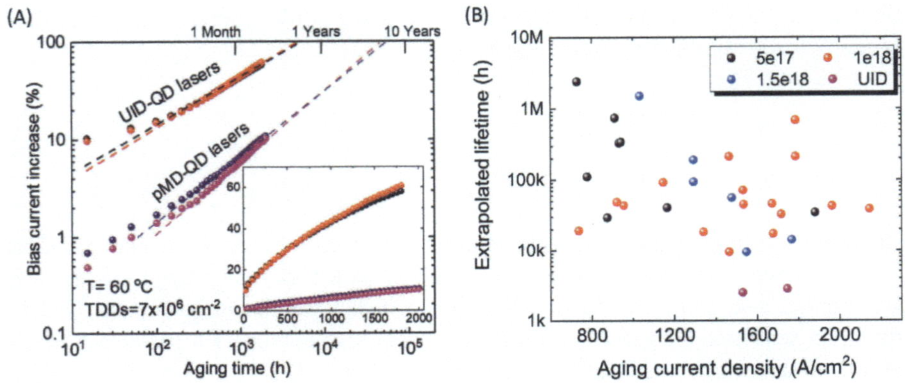

Figure 6.13 (A) Bias current increase for 10 mW of output power for UID and p-modulation doped (pMD) quantum dot lasers grown on silicon. (Inset) The same data on a linear scale. (B) Extrapolated lifetime defined as time to double initial threshold current density for different aging current densities at 60°C of devices with different p-modulation doping level. *UID*, Unintentionally doped.

Source: Adapted with permission from D. Jung, et al., Recent advances in InAs quantum dot lasers grown on on-axis (001) silicon by molecular beam epitaxy, Phys. Status Solidi A 216 (2018) 1800602. ©2018, John Wiley & Sons.

principles model suggests that the peak modal gain was increased from $\sim 20\ \mathrm{cm}^{-1}$ for the unintentionally doped active region lasers to $\sim 60\ \mathrm{cm}^{-1}$ in the p-modulation doped samples. The differential gain was also doubled in the p-doped samples from ~ 13 to $\sim 25\ \mathrm{cm}^{-1}$. Higher differential gain means that incremental degradation in optical loss or injection efficiency will result in smaller incremental degradation of threshold current and slope efficiency.

At the same dislocation density as the previous study, $\sim 7 \times 10^6\ \mathrm{cm}^{-2}$, the lasers showed a substantial improvement in 60°C reliability [43,61]. The extrapolated lifetime increased from ~ 2500 to $>50{,}000$ h based on the failure criteria of doubling the threshold current, and $>10{,}000$ h based on defining failure as a 50% increase in the bias required for 10 mW of output power. A compilation of aging data for several devices is shown in Fig. 6.13 as well as an extrapolated lifetime for the different doping densities and injection levels. Interestingly the degradation rate of the lasers did not correlate with the specific doping level. There was not a strong correlation between ridge width and extrapolated lifetime for ridges 6 μm and narrower, but there was some correlation between longer cavities, and hence lower mirror loss, and improved reliability, suggesting gain margin to be an important parameter, as would be expected if the intrinsic dislocation density were the limiting factor. However, the data are highly scattered. Considering that dislocations appear to be uniformly distributed, this suggests that there could be some extrinsic effect that has significant impact on reliability. Such effects could include processing defects or local variations in the distribution of misfit dislocations as suggested in the previous discussion of electroluminescence results on aged devices.

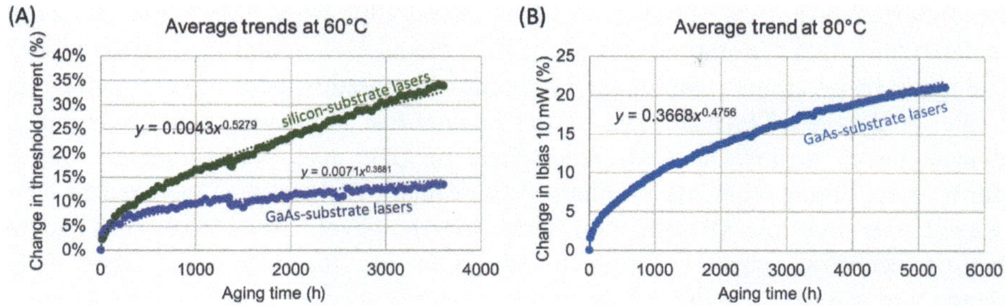

Figure 6.14 (A) Average trends in percent increase in threshold current for two lasers on GaAs and seven lasers on silicon subjected to continuous aging at 60°C. (B) Average percent change in bias current for 10 mW of output power for four GaAs lasers subjected to continuous aging at 80°C.

6.5.3 Reliability of quantum dot lasers on native substrates versus silicon

To confirm the hypothesis that defects related to growth on silicon were the primary source of degradation, identical epi structures were grown on GaAs substrates and aged at 60°C and even 80°C. At time zero, these lasers showed nearly identical threshold current and output power characteristics to the lasers on Si. After 5000 h of aging, the differences are apparent. The lasers on GaAs show very modest degradation rates as depicted in Fig. 6.14 with the lasers at 60°C extrapolating to an MTTF of 650,000 h (threshold doubling) and the lasers at 80°C showing extrapolated MTTF of 30,000 h (50% increase in bias for 10 mW output). The 80°C results are promising for commercial viability, and they indicate that the lasers on silicon have room for significant improvement before anything other than epitaxial material quality plays a dominant role in performance.

6.6 Degradation mechanisms in quantum dot lasers on silicon

In general, QD lasers appear to degrade in the presence of dislocations via similar mechanisms as bulk and quantum well devices in that recombination-enhanced climb has been observed, but there is one very notable distinction in that rapid failures due to DLDs have not been observed in QDs. A quantum well or bulk device containing dislocations will fail within hundreds of hours due to the formation of a dendritic network of nonradiative recombination (i.e., a DLD). DLDs result from recombination-enhanced climb and glide processes that create additional dislocation length in the device active region. While climb can be seen in the TEM images presented previously, extensive networks do not seem to form even after several thousand hours of continuous aging. One explanation could be that reduced nonradiative recombination rates at the dislocation due to efficient carrier capture by QDs prevent efficient motion of the dislocations laterally by dislocation climb. A second explanation is that the strain fields of the dots themselves inhibit dislocation motion analogous to precipitate hardening in metals. Or possibly both effects could play a role. A detailed TEM analysis of the geometrical structure of climb segments in aged lasers suggests that the strain field of QDs does not completely inhibit climb, since climb segments originating from misfit dislocations appeared to have crossed over multiple QDs in their formation; however, more work needs to be done to confirm carrier capture by the QDs as the sole mechanism [62]. Meanwhile, temperature-dependent cathodoluminescence (CL) results provide some clues as to the electrical interaction of QDs and dislocations.

The SEM CL study was performed on a single QD-in-a-well layer nominally identical to what was used in the lasers referenced above grown on a III–V/Si buffer with 7×10^6 cm^{-2} dislocation density. The QD layer was only capped with 100 nm of GaAs such that it could be imaged through the top surface with ECCI. Using ECCI, the locations of threading and misfit dislocations were determined and marked with platinum fiducials for subsequent spectrally resolved, temperature-dependent CL experiments. The correlated micrographs are shown in Fig. 6.15.

CL measurements were able to map the intensity of emission from the QD ground state, first excited state, and second excited state from 10K to room temperature (see Fig. 6.17).

These three states refer to the optical transitions resulting from the three-dimensional quantum confinement of the dots and are qualitatively analogous to the

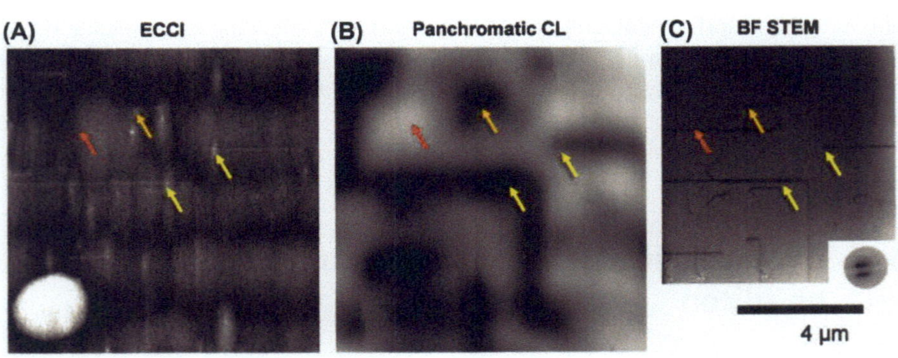

Figure 6.15 (A) Electron channeling contrast image (ECCI) of a sample containing a single quantum dot layer 100 nm below the surface, grown on silicon, with a threading dislocation density of 7×10^6 cm^{-2}. The straight line segments show misfit dislocations in the quantum dot layer. (B) Panchromatic CL image of the same region of material as in (A). (C) Plan-view BF STEM image of the same material as depicted in (A) and (B) with inset showing the (2 2 0) two beam diffraction condition. The arrows indicate points of commonality in the three images. *BF*, Bright field; *CL*, cathodoluminescence; *STEM*, scanning transmission electron microscope.

Source: Reprinted from J. Selvidge, et al., Non-radiative recombination at dislocations in InAs quantum dots grown on silicon, Appl. Phys. Lett. 115 (13) (2019) 131102 [66], with the permission of AIP Publishing.

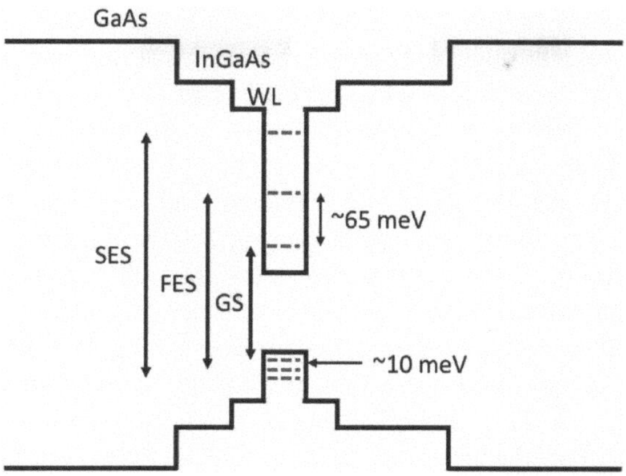

Figure 6.16 Schematic representation of the band structure of an InAs quantum dot in an In0.15−Ga0.85As well surrounded by GaAs. The schematic illustrates the three optical transitions termed the GS, FES, and SES of the quantum dot and surrounding band structure including the WL, well, and GaAs cladding. *FES*, First excited state; *GS*, ground state; *SES*, second excited state; *WL*, wetting layer.

textbook case of a particle-in-a-box or the hydrogen atom. An approximate schematic band structure is shown in Fig. 6.16.

In the plan-view CL images, dislocations appear as dark spots (threading dislocations) or lines (misfit dislocations) with radius corresponding to the diffusion length in the material and intensity reduction corresponding to the nonradiative recombination strength of the dislocation. The results showed that at the lowest temperatures, the dislocations would only darken emission from the second excited state of the dots, but as the temperature increased up to ~120−200K, the first excited state darkened, and then above 200K, the ground state darkened. The radius of the dark regions for all states remained constant with temperature and increased in emission reduction as temperature increased until saturating at a common value identical for each state. The result suggests that the carrier loss follows an identical pathway for all states within the dot (due to the identical radius of the dark regions for emission from each state), namely, a pathway of carrier capture by a dot, thermalization of the carrier back out of the dot, and then in-plane diffusion to a dislocation. Each state's emission saturates at a common relative reduction in intensity at high temperature, indicating that quantum confinement does not play as strong a role in limiting the impact of dislocations at elevated (>200K) temperatures. This points to the dislocation insensitivity of QDs resulting from reduced in-plane carrier diffusion lengths tied to high scattering rates in the wetting layer and surrounding quantum well. If carrier trapping by the dots were a significant factor, then the relative impact of the dislocation would be different for the different states, and this is only true at temperatures well below room temperature. This implies higher dot density could improve reliability through higher scattering rates (in addition to higher material gain).

An additional study on the degradation mechanisms in QD lasers was conducted by Buffolo et al. [67] through analysis of the electrical and optical properties of a laser throughout an accelerated-aging process. The authors performed step-stress testing where the lasers were continuously aged at incremental current levels that increased each hour. The devices correspond to the same material as the $5 \times 10^{17}\ cm^{-3}$ pMD samples described previously but with only three dot layers in the active region. The lasers were aged starting at 10 mA and aging concluded at 300 mA. This range of currents took the lasers through regimes of ground-state-only emission, dual-state emission, and excited-state-only emission. During aging, LIV curves and electroluminescence measurements were taken to monitor degradation. Key results of the paper were that gradual degradation was observed in the early stages of aging with an immediate uptick in the degradation rate when excited state lasing began. The change in degradation rate was observed in both the threshold current and slope efficiency, which were linearly (inversely) correlated with each other, signifying a decrease in injection efficiency presumably through nonradiative recombination. The degradation also showed a defined kink at the aging condition where excited state lasing became significant. The turn-on of the excited state stimulated emission quenches the ground state emission causing a relatively abrupt shift in the quasi-Fermi levels with higher carrier occupancy within the material outside the dots and increased nonradiative recombination rates to fuel dislocation growth.

6.6.1 Misfit dislocations in the laser active region

Returning to the CL images of Figs. 6.15 and 6.17, the outsized effect of misfit dislocations is clearly apparent. Misfits result in long segments of nonradiative recombination, potentially 10s of microns in length; whereas, threads result in single $\sim 1\ \mu m$ diameter dark spots. Intuitively, if the misfit dislocations could be removed, the overall laser performance should improve dramatically even if the thread density remains constant.

In pursuit of misfit-free device layers, several experiments were conducted to determine the origin of the misfit dislocations [68]. From both cross-sectional TEM and plan-view scanning TEM (STEM) tomography, Selvidge et al. observed that the misfits form only in the uppermost and lowermost dot layers. The misfits terminated with threading segments extending downward from the lower-layer misfits and upward from the upper-layer misfits, with an additional thread segment through the dot layers connecting them (see the schematic in Fig. 6.18). Through tomographic reconstructions from tilted PV-STEM images, the misfits were determined to originate from reverse glide of threading dislocations formed during the compressive relaxation of the III–V/Si buffer. Reverse glide of a threading dislocation occurs when the glide driving force is opposite in sign to the stress that was relaxed when the dislocation formed. Given that the threads formed during growth to relax the compressive lattice mismatch, the glide-triggering stress that formed the misfit must have been tensile (i.e., coming from the thermal expansion mismatch).

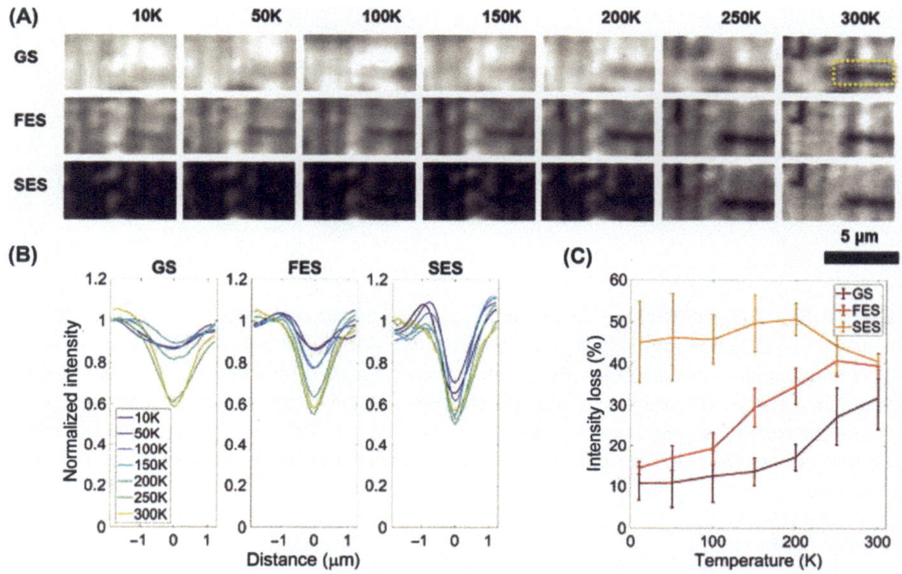

Figure 6.17 (A) Spectrally resolved CL images corresponding to emission from the quantum dot GS, FES, and SES transitions at varying temperatures from 10 to 300K. (B) CL line scans for the three dot transitions in the vicinity of a dislocation showing the relative loss in luminescence intensity for different temperatures. (C) Peak percentage reduction in luminescence intensity from (B) for the three transitions as a function of temperature. *CL*, Cathodoluminescence; *FES*, first excited state; *GS*, ground state; *SES*, second excited state.

Source: Reprinted from J. Selvidge, et al., Non-radiative recombination at dislocations in InAs quantum dots grown on silicon, Appl. Phys. Lett. 115 (13) (2019) 131102, with the permission of AIP Publishing.

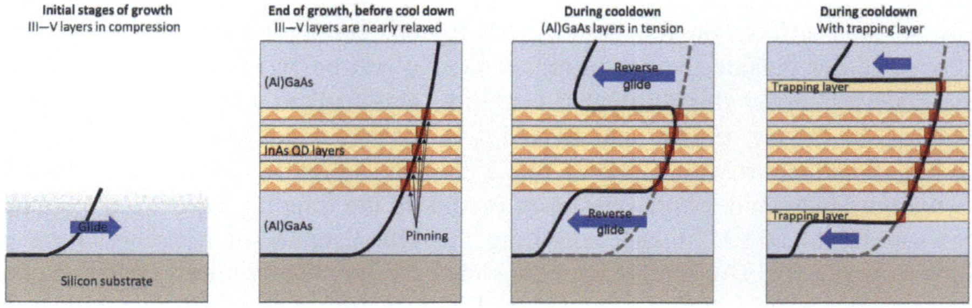

Figure 6.18 Schematic illustration of the process of misfit dislocation formation in the active region of heteroepitaxial lasers on silicon with the trapping layer concept illustrated in the last panel. Mechanistic details can be found in Ref. [68].

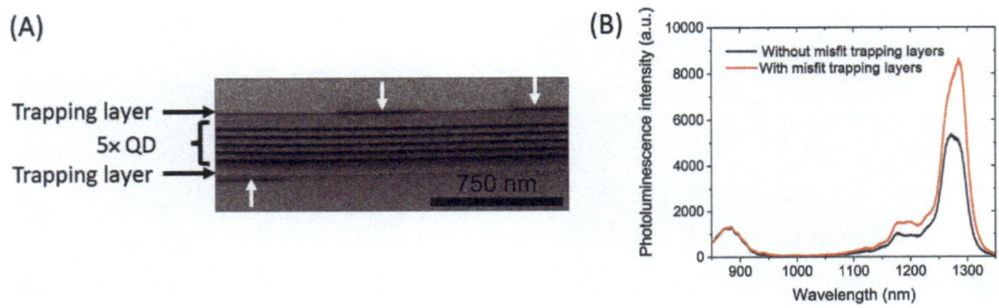

Figure 6.19 (A) Bright-field cross-sectional scanning transmission electron microscope image of a laser active region including five quantum dot layers with dislocation trapping layers above and below showing misfit dislocations (indicated by *white arrows*) confined to the trapping layers. (B) As-grown laser photoluminescence spectra for lasers with and without the misfit trapping layers. Differences in the shape of the emission profile are due to thin-film effects due to reflections at the silicon substrate and normal-incidence luminescence collection.

Reverse glide leads to the formation of misfits presumably due to pinning of threading segments by compositional fluctuations within the dot layers (this is analogous to alloy hardening in metallurgy) and precipitate hardening effects. Further analysis using energy-dispersive X-ray spectroscopy via STEM and atom probe tomography supported the conclusion that the misfits formed during cool down postgrowth. The dot layer compositions and thicknesses were not modified by the presence of misfits but were observed to be modified by the presence of threading dislocations (i.e., the threads were present during dot growth but the misfits were not). Localized growth modifications in the presence of dislocations are a well-established phenomenon.

Further elaboration on the mechanistic behavior of the misfit formation is beyond the scope of this chapter, but accepting the conclusion that misfits form during postgrowth cool down at the compressive interfaces provides a simple remedy to their effects on the laser active region: adding compressively strained alloyed layers outside the active region (above and below) to trap the misfits in the majority carrier (highly doped) cladding. This method was tested using 7 nm $In_{.5}Ga_{.85}As$ (above, p-side) and $In_{.15}Al_{.85}As$ (below, n-side) layers and found to be highly effective with CL and ECCI showing a >90% reduction of misfits in the dot layers and a high density of misfits in the trapping layers. An example cross-sectional STEM image exhibiting the effectiveness of trapping layers is shown in Fig. 6.19A. A corresponding increase in photoluminescence intensity from the material of >50% was observed (Fig. 6.19B) and laser performance was substantially improved. Lasers with trapping layers were found to exhibit multifold improvements in threshold, slope efficiency, and output power relative to control samples [68]. While these preliminary results show high promise, laser reliability testing is still pending.

6.7 The path forward and future improvements for quantum dot lasers grown on silicon substrates

Unfortunately, even the p-modulation doped laser results above are inadequate for commercial applications. To meet commercial reliability standards and close the gap with native substrate performance, improvements are likely needed to reduce the dislocation density, specifically in the laser active region, and improve the thermal performance of the devices. While buffer optimizations have yielded substantial improvements historically, the targeted removal of misfit dislocations is likely to have a much greater impact, perhaps even relaxing the requirements for the epitaxial buffer and allowing for thinner epi layers which lowers cost and moves closer to the goal of photonic integration with the Si underneath. From Jung et al.'s study on dislocation density [61], there appears to be clear evidence that further reductions could be beneficial. Considering that the best demonstrated III–V/Si dislocation densities are $\sim 2 \times 10^6\,\text{cm}^{-3}$, and that the data suggest that MTTF scales inversely with the cube of dislocation density, an extension of device lifetimes by a factor of ~ 8 could be achieved by just matching the best achieved dislocation densities. Intuitively, diminishing returns should set in as the average distance between dislocations becomes much larger than the in-plane diffusion lengths in the active region, but where that limit sits, remains to be seen.

Diving deeper into the materials science, reliability could be improved through improvements in the QDs themselves and in specifically addressing the misfit dislocations in the laser active region. As discussed above, even with their high quantum confinement, the QD ground states carrier densities are at thermal equilibrium with the surrounding well even significantly below room temperature. These results are consistent with previous fundamental investigations of InAs QDs in GaAs [69]. If materials could be developed with larger band offsets (particularly in the valence band) such that nonequilibrium carrier distributions were present at room temperature and above, then the insensitivity of the QDs to dislocations could improve dramatically since the temperature at which carriers thermalize out of the dot ground state could be increased. In currently demonstrated material designs, the dislocation tolerance of the dots comes about from the low in-plane diffusion length of carriers outside the dots due to in-plane scattering by the inhomogeneity of the surrounding well [66]. If the dot density could be increased, then the in-plane diffusion length should also decrease giving the dots further tolerance to dislocations (in addition to higher gain from a higher dot density). Intuitively, the greatest improvement in reliability could be obtained by eliminating the misfit dislocations in the laser active region. The misfit's interaction cross-section with the dot layers is much higher than threading segments that only intersect the active layers at a single point. With the knowledge that the misfits form from postgrowth cool down, trapping layers can be devised to avoid misfit formation in the dot layers, instead confining the dislocations to the majority carrier cladding regions where they could be relatively innocuous.

Other means of improvement are likely also possible, but further work is needed on failure analysis, and more statistically significant data are needed to determine any factors other than temperature, current density, and dislocation density that affect reliability. From the results above, there are clearly two regimes of laser degradation, one of rapid degradation in the first ~200 h and a much slower regime that follows. Understanding the early degradation mechanism(s) could provide insight to extending the laser reliability. These may be related to the inhomogeneous electroluminescence results above, and detailed failure analysis using TEM and through aging additional devices will likely shed light on the origin of the dark spots. Thus far, the only correlation with device reliability is gain margin, which is simply how much room a laser must degrade before failure. By selecting lasers with longer cavity lengths, and peak power $>5\times$ the use power at the operating temperature, significant degradation can be tolerated without a large impact on the bias current—rapid degradation is not seen until the peak power falls to $<2\times$ the target bias power. It seems likely that, analogously to native III–V lasers, there will be other factors that limit device performance as material improves, but larger data sets over broader ranges of temperatures, current densities, and device geometries are needed to reach statistically sound conclusions.

Another point of emphasis is that it is not clear that time-zero laser performance is a strong indicator of reliability. The gap in initial laser performance at elevated dislocation densities relative to native substrate devices is small, and diminishing returns have been observed even at dislocation densities as high as 7×10^7 cm^{-2}. This is true of the threshold current, slope efficiency, and output power, and is plotted in Fig. 6.20 for the slope efficiency as a function of the absolute dislocation density and the previously mentioned generations of laser material optimization, with Gen IV

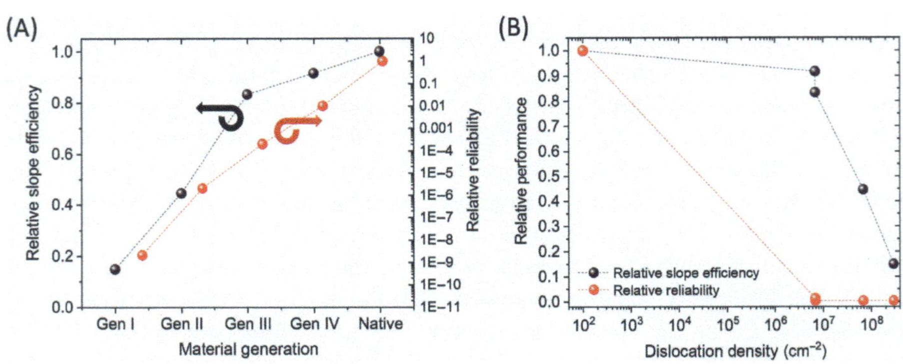

Figure 6.20 Performance relative to that of native substrate lasers (assumed 10^2 cm^{-2} dislocation density) is plotted for various embodiments of lasers on silicon. (A) The relative slope efficiency of the lasers (left axis) and relative reliability (right axis, referenced to mean-time-to-failure) is plotted for four generations of material on silicon: Gen I, 3×10^8 cm^{-2} dislocation density; Gen II, 7×10^7 cm^{-2} dislocation density; Gen III, 7×10^6 cm^{-2} dislocation density; and Gen IV, 7×10^6 cm^{-2} dislocation density with p-type doping in the active region. (B) The same data as in (A) are plotted as a function of dislocation density. These plots illustrate the lack of strong indicators in the preaging performance of device reliability.

representing the lasers with p-type modulation doping in the active region but still at $7 \times 10^6\ cm^{-2}$ dislocation density.

The lack of positive indicators of device reliability before aging complicates screening for commercial applications, necessitating a burn-in study of QD lasers on Si. Without a validated burn-in process, reliability testing must be performed before one can claim to have solved the discrepancy in performance between native substrate lasers and heteroepitaxial lasers on silicon; time-zero performance is not sufficient.

6.8 Conclusion

To summarize, the most technologically mature III–V on Si laser implementation is through heterogeneous integration via wafer or die bonding. Heterogeneous integration has been commercialized and is being produced at-scale for transceiver applications by Intel. Their demonstrated performance as well as published results shows that heterogeneous lasers are capable of similar reliability to their native substrate counterparts. They face challenges in the bonding step due to the elevated temperatures required, but through the incorporation of superlattices between the bonded interface and the gain region and through plasma-assisted low-temperature bonding processes, the formation of defects and their deleterious effects can be minimized.

The most economically desirable and promising contender for future development of III–V lasers on silicon is the heteroepitaxial integration of QD-based gain media grown on silicon substrates. Epitaxial integration poses significant economic benefits due to simplified processing and scaling to 300 mm and larger wafer sizes, and QD gain media have been demonstrated to show remarkable insensitivity in laser performance to dislocations in the laser active region. Long extrapolated lifetimes in excess of 10,000,000 h have been measured at 35°C, but more work needs to be done to reach similar performance at commercially relevant temperatures of >60°C, where extrapolated lifetimes are at best ~60,000 h with devices commonly undergoing rapid degradation at various points before 4000 h of continuous aging. One interesting observation is that lasers at both low and high temperatures show an initial relatively high degradation rate that saturates within a couple of hundred hours. If the initial degradation mechanism can be avoided, laser lifetimes would increase notably.

Going forward, significant work is needed to develop statistically valid accelerated-aging models for QD lasers on silicon. While a handful of studies have been conducted to date, no systematic study of the effects of gain margin, laser geometry, injection current, and temperature has been conducted. Furthermore, the failure mechanisms in QDs are not well understood. While some amount of dislocation climb in the laser active region has been observed, clear correlations between the extent of climb and the laser degradation have not been drawn. The observation of nonuniform darkening of the laser active region and rapid initial degradation also suggest multiple degradation mechanisms are likely in play.

Finally, we discussed many promising avenues to reach the necessary reliability targets, including reduced threading dislocation density, misfit trapping layers, and improvements in the dot design or growth to reduce carriers thermalizing out of the dots. These efforts pursue the complementary goals of defect density reduction in the active region and reduced activity of whatever defects may remain. Further optimization of these techniques individually and combining their use should lead to rapid progress in reliability improvement for heteroepitaxial lasers on silicon. While the discussion focuses on mismatched heteroepitaxy of GaAs- and InP-based materials specifically, these observations and goals are likely generalizable to many other materials systems that could be of interest for future technologies operating outside of the near infrared.

Acknowledgments

The authors thank Chen Shang, Eamonn Hughes, Aidan Taylor, Arthur Gossard, Pierre Petroff, M. Buffolo, M. Meneghini, and Michael Haney for helpful discussions. This research was supported by the Advanced Research Projects Agency-Energy (ARPA-E) No. DE-AR0001039 and ARPA-E No. DE-AR000067.

References

[1] T. Komljenovic, et al., Photonic integrated circuits using heterogeneous integration on silicon, Proc. IEEE 99 (2018) 1–12.

[2] J.C. Norman, et al., Perspective: the future of quantum dot photonic integrated circuits, APL Photonics 3 (3) (2018) 030901.

[3] R. Jones, et al., Heterogeneously integrated InP/silicon photonics: fabricating fully functional transceivers, IEEE Nanotechnol. Mag. 13 (2) (2019) 17–26.

[4] A.Y. Liu, J.E. Bowers, Photonic integration with epitaxial III–V on silicon, IEEE J. Sel. Top. Quantum Electron. 24 (2018).

[5] J. Liu, et al., Tensile-strained, n-type Ge as a gain medium for monolithic laser integration on Si, Opt. Express 15 (18) (2007) 11272–11277.

[6] G. Pizzi, M. Virgilio, G. Grosso, Tight-binding calculation of optical gain in tensile strained [001]-Ge/SiGe quantum wells, Nanotechnology 21 (5) (2009) 055202.

[7] G. Dehlinger, et al., Intersubband electroluminescence from silicon-based quantum cascade structures, Science 290 (5500) (2000) 2277–2280.

[8] R. Apetz, et al., Photoluminescence and electroluminescence of SiGe dots fabricated by island growth, Appl. Phys. Lett. 66 (4) (1995) 445–447.

[9] R.E. Camacho-Aguilera, et al., An electrically pumped germanium laser, Opt. Express 20 (10) (2012) 11316–11320.

[10] S.A. Srinivasan, et al., Carrier scattering induced linewidth broadening in in situ P-doped Ge layers on Si, Appl. Phys. Lett. 113 (16) (2018) 161101.

[11] B. Tian, et al., Room temperature O-band DFB laser array directly grown on (001) silicon, Nano Lett. 17 (1) (2016) 559–564.

[12] Y. Han, et al., Telecom InP/InGaAs nanolaser array directly grown on (001) silicon-on-insulator, Opt. Lett. 44 (4) (2019) 767–770.

[13] Y. Han, et al., Bufferless 1.5 μm III-V lasers grown on Si-photonics 220 nm silicon-on-insulator platforms, Optica 7 (2) (2020) 148–153.

[14] B. Mayer, et al., Monolithically integrated high-β nanowire lasers on silicon, Nano Lett. 16 (1) (2015) 152–156.
[15] Z. Wang, et al., Room-temperature InP distributed feedback laser array directly grown on silicon, Nat. Photonics 9 (12) (2015) 837.
[16] T. Frost, et al., Monolithic electrically injected nanowire array edge-emitting laser on (001) silicon, Nano Lett. 14 (8) (2014) 4535–4541.
[17] A.T. Aho, et al., High-power 1180-nm GaInNAs DBR laser diodes, IEEE Photonics Technol. Lett. 29 (23) (2017) 2023–2026.
[18] S.R. Bank, et al., Recent progress on 1.55-μm dilute-nitride lasers, IEEE J. Quantum Electron. 43 (9) (2007) 773–785.
[19] L.L. Goddard, et al., Recombination, gain, band structure, efficiency, and reliability of 1.5-μm GaInNAsSb/GaAs lasers, J. Appl. Phys. 97 (8) (2005) 083101.
[20] B. Kunert, et al., Near room temperature electrical injection lasing for dilute nitride Ga (NAsP)/GaP quantum-well structures grown by metal organic vapour phase epitaxy, Electron. Lett. 42 (10) (2006) 601–603.
[21] A. Spott, et al., Quantum cascade laser on silicon, Optica 3 (5) (2016) 545–551.
[22] G. Roelkens, et al., Silicon-based photonic integration beyond the telecommunication wavelength range, IEEE J. Sel. Top. Quantum Electron. 20 (4) (2014) 394–404.
[23] H. Nguyen-Van, et al., Quantum cascade lasers grown on silicon, Sci. Rep. 8 (1) (2018) 7206.
[24] Z. Loghmari, et al., InAs-based quantum cascade lasers grown on on-axis (001) silicon substrate, APL Photonics 5 (4) (2020) 041302.
[25] A.W. Fang, et al., Electrically pumped hybrid AlGaInAs-silicon evanescent laser, Opt. Express 14 (20) (2006) 9203–9210.
[26] A. Black, et al., Wafer fusion: materials issues and device results, IEEE J. Sel. Top. Quantum Electron. 3 (3) (1997) 943–951.
[27] S. Srinivasan, et al., Reliability of hybrid silicon distributed feedback lasers, IEEE J. Sel. Top. Quantum Electron. 19 (4) (2013) 1501305.
[28] M. Buffolo, et al., Degradation mechanisms of heterogeneous III-V/silicon loop-mirror laser diodes for photonic integrated circuits, Microelectron. Reliab. 88 (2018) 855–858.
[29] M. Buffolo, et al., Degradation mechanisms of heterogeneous III-V/silicon 1.55-μm DBR laser diodes, IEEE J. Quantum Electron. 53 (4) (2017) 1–8.
[30] H. Yu, et al. 100 Gbps CWDM4 silicon photonics transmitter for 5G applications, in: Optical Fiber Communication Conference, Optical Society of America, 2019.
[31] R. Herrick, et al., Reliable heterogeneous and monolithic integrated silicon photonics, in: Optical Fiber Communications Conference, San Diego, CA, 2019.
[32] P. Kennedy, Intel Silicon Photonics Update at Interconnect Day 2019, 2019. Available from: https://www.servethehome.com/intel-silicon-photonics-update-at-interconnect-day-2019/ (cited 2020 14 May 2019).
[33] Intel demonstrates industry-first co-packaged optics Ethernet switch. 2020. https://newsroom.intel.com/news/intel-demonstrates-industry-first-co-packaged-optics-ethernet-switch/#gs.2i2efz (cited 23 April 2020).
[34] Intel combines optics to its Tofino 2 switch chip. 2020. http://www.gazettabyte.com/home/2020/3/19/intel-combines-optics-to-its-tofino-2-switch-chip.html (cited 14 May 2020).
[35] G. Kurczveil, et al., Error-free operation in a hybrid-silicon quantum dot comb laser, IEEE Photonics Technol. Lett. 30 (1) (2018) 71–74.
[36] A. Spott, et al., Interband cascade laser on silicon, Optica 5 (8) (2018) 996–1005.
[37] J.E. Ayers, et al., Heteroepitaxy of Semiconductors: Theory, Growth, and Characterization, CRC Press, 2016.

[38] Z.I. Kazi, et al., Realization of GaAs/AlGaAs lasers on Si substrates using epitaxial lateral overgrowth by metalorganic chemical vapor deposition, Jpn. J. Appl. Phys. 40 (8R) (2001) 4903.
[39] C. Shang, et al., A pathway to thin GaAs virtual substrate on on-axis Si (001) with ultralow threading dislocation density, Physica Status Solidi A (2020) 2000402. Available from: https://doi.org/10.1002/pssa.202000402.
[40] T. Sasaki, et al., Aging tests of InP-based laser diodes heteroepitaxially grown on Si substrates, J. Appl. Phys. 84 (12) (1998) 6725–6728.
[41] M. Tachikawa, et al., Laser-diode-quality InP/Si grown by hydride vapor phase epitaxy, Jpn. J. Appl. Phys. 34 (6A) (1995) L657.
[42] M. Sugo, et al., Stable cw operation at room temperature of a 1.5-μm wavelength multiple quantum well laser on a Si substrate, Appl. Phys. Lett. 60 (4) (1992) 472–473.
[43] D. Jung, et al., Recent advances in InAs quantum dot lasers grown on on-axis (001) silicon by molecular beam epitaxy, Phys. Status Solidi A 216 (2019) 1800602.
[44] A. Dadgar, et al., Epitaxy of GaN on silicon—impact of symmetry and surface reconstruction, N. J. Phys. 9 (10) (2007) 389.
[45] H. Kroemer, Sublattice allocation and antiphase domain suppression in polar-on-nonpolar nucleation, J. Vac. Sci. Technol. B: Microelectron. Process. Phenom. 5 (4) (1987) 1150–1154.
[46] I. Németh, et al., Heteroepitaxy of GaP on Si: correlation of morphology, anti-phase-domain structure and MOVPE growth conditions, J. Cryst. Growth 310 (7–9) (2008) 1595–1601.
[47] J. Kwoen, et al., All MBE grown InAs/GaAs quantum dot lasers on on-axis Si (001), Opt. Express 26 (9) (2018) 11568–11576.
[48] Q. Li, K.W. Ng, K.M. Lau, Growing antiphase-domain-free GaAs thin films out of highly ordered planar nanowire arrays on exact (001) silicon, Appl. Phys. Lett. 106 (7) (2015) 072105.
[49] V. Yang, et al., Crack formation in GaAs heteroepitaxial films on Si and SiGe virtual substrates, J. Appl. Phys. 93 (7) (2003) 3859–3865.
[50] M.E. Groenert, et al., Improved room-temperature continuous wave GaAs/AlGaAs and InGaAs/GaAs/AlGaAs lasers fabricated on Si substrates via relaxed graded Ge_xSi_{1-x} buffer layers, J. Vac. Sci. Technol. B: Microelectron. Nanometer Struct. Process. Meas. Phenom. 21 (3) (2003) 1064–1069.
[51] D. Nolte, Surface recombination, free-carrier saturation, and dangling bonds in InP and GaAs, Solid-State Electron. 33 (2) (1990) 295–298.
[52] B. Shi, et al., Continuous-wave electrically pumped 1550 nm lasers epitaxially grown on on-axis (001) silicon, Optica 6 (12) (2019) 1507–1514.
[53] Y. Xue, et al., 1.55 μm electrically pumped continuous wave lasing of quantum dash lasers grown on silicon, Unpublished.
[54] Z. Liu, et al., Origin of defect tolerance in InAs/GaAs quantum dot lasers grown on silicon, J. Lightwave Technol. 38 (2019).
[55] A.Y. Liu, et al., Quantum dot lasers for silicon photonics, Photonics Res. 3 (5) (2015) B1–B9.
[56] A.Y. Liu, et al., Reliability of InAs/GaAs quantum dot lasers epitaxially grown on silicon, IEEE J. Sel. Top. Quantum Electron. 21 (6) (2015) 690–697.
[57] D. Jung, et al., Low threading dislocation density GaAs growth on on-axis GaP/Si (001), J. Appl. Phys. 122 (22) (2017) 225703.
[58] A.Y. Liu, et al., Electrically pumped continuous-wave 1.3 μm quantum-dot lasers epitaxially grown on on-axis (001) GaP/Si, Opt. Lett. 42 (2) (2017) 338–341.

[59] D. Jung, et al., Highly reliable low-threshold InAs quantum dot lasers on on-axis (001) Si with 87% injection efficiency, ACS Photonics 5 (3) (2017) 1094–1100.
[60] S. Chen, et al., Electrically pumped continuous-wave III–V quantum dot lasers on silicon, Nat. Photonics 10 (5) (2016) 307.
[61] D. Jung, et al., Impact of threading dislocation density on the lifetime of InAs quantum dot lasers on Si, Appl. Phys. Lett. 112 (15) (2018) 153507.
[62] K. Mukherjee, et al., Recombination-enhanced dislocation climb in InAs quantum dot lasers on silicon, J. Appl. Phys. 128 (2) (2020) 025703. Available from: https://doi.org/10.1063/1.5143606.
[63] O. Shchekin, J. Ahn, D. Deppe, High temperature performance of self-organised quantum dot laser with stacked p-doped active region, Electron. Lett. 38 (14) (2002) 712–713.
[64] Z. Zhang, et al., Effects of modulation p doping in InAs quantum dot lasers on silicon, Appl. Phys. Lett. 113 (6) (2018) 061105.
[65] J.C. Norman, et al., The importance of p-doping for quantum dot laser on silicon performance, IEEE J. Quantum Electron. 55 (6) (2019) 1–11.
[66] J. Selvidge, et al., Non-radiative recombination at dislocations in InAs quantum dots grown on silicon, Appl. Phys. Lett. 115 (13) (2019) 131102.
[67] M. Buffolo, et al., Physical origin of the optical degradation of InAs quantum dot lasers, IEEE J. Quantum Electron. 55 (3) (2019) 1–7.
[68] J. Selvidge, et al., Defect filtering for thermal expansion induced dislocations in III-V lasers on silicon, Appl. Phys. Lett. 117 (2020) 122101.
[69] M. Hutchings, et al., Fermi-Dirac and random carrier distributions in quantum dot lasers, Appl. Phys. Lett. 104 (3) (2014) 031103.

Chapter 7

Degradation mechanisms of InGaN visible LEDs and AlGaN UV LEDs

C. De Santi, A. Caria, F. Piva, G. Meneghesso, E. Zanoni and M. Meneghini
Department of Information Engineering, University of Padova, Padova, Italy

7.1 Introduction

GaN-based visible LEDs are now the standard choice for a wide range of general lighting applications and displays, whereas their UV versions are currently being developed to substitute in a cheaper and environmental-friendly way the traditional sources for curing, lithography, biological, medical, and detection applications. The degradation of a GaN-based optoelectronic system can be caused by various stress conditions and can affect different parts of the system [1]. Other than common operating bias stresses, additional degradation modes can be caused by electrostatic discharges [2–4], electrical overstress [5,6] and excessive current levels [7], and additional failure mechanisms can be present at the system level [8–10]. In this chapter, we review the most relevant findings involving point defect-related degradation in visible and UV LEDs in Sections 7.2 and 7.3, respectively. Each of the two parts discusses in detail the main processes present in state-of-the-art chips, such as the gradual generation of intrinsic and extrinsic defects and their diffusion in the structure.

7.2 Degradation of InGaN visible LEDs

Visible InGaN LEDs can be used to cover a wide range of possible applications. High-power white LEDs are the main source when a high luminous flux and color quality is needed, such as for general lighting applications in houses, shops, and cars, and the possibility of tuning the correlated color temperature and the color rendering index enables new special lighting applications for design and museums. Their lower-power counterparts are used as backlighting sources in a wide range of LCD displays. Monochromatic LEDs can be used for high-density and small-size displays, for signage and appliance lighting, and for detection, research, and telecommunication applications.

Reliability of Semiconductor Lasers and Optoelectronic Devices. DOI: https://doi.org/10.1016/B978-0-12-819254-2.00001-1

The epitaxial structure of a visible LED can significantly differ depending on the target wavelength and application, but typically is composed of a substrate (sapphire, silicon carbide, silicon, or gallium nitride) with the relevant nucleation layers and strain relief layers to ensure a good epitaxial quality, based on single or superlattice (SL) Al(Ga)N layers. A thick n-type region is grown on top of it, followed by an InGaN/GaN multiquantum well/quantum barrier structure, with typical periodicity values of three to five, with an indium molar fraction dependent on the target wavelength. Given the high electron mobility of gallium nitride, an AlGaN electron-blocking layer (EBL) follows to confine the electrons in the active regions, sometimes p-doped to reduce the penalty in turn-on voltage. The p-side of the device is completed by a p-GaN current-spreading layer and a thin highly doped region below the metal contact, used in order to improve the ohmic quality of the contact.

Defects in InGaN LEDs play a crucial role in device degradation, since they can act as nonradiative centers, lowering the quantum efficiency of the device and thus the overall efficiency of the LEDs [11–14]. During crystal growth, the formation of various type of defects is possible: point defects, that is, with size comparable to the atomic distance between lattice atoms, or extended defects, like dislocations, clusters, domain, voids, etc. The generation and diffusion of defects are the main causes of the gradual degradation of state-of-the-art LEDs. In order to reduce the degradation rate several approaches are possible. To limit the creation of new lattice defects a general increase in material, growth, and processing quality is important, as well as a possible derating of the operating conditions of the LEDs. More options are available in order to limit the diffusion of impurities, such as lowering their concentration gradient, ensuring a more stable position in the crystal lattice, or through the use of specific defect-trapping layers.

7.2.1 Point defects in LEDs

Several papers have analyzed the formation energy and corresponding energy level for major native defects in III-nitrides, both theoretically and experimentally (see Reference [15] and references therein). Native point defects can be interstitial impurities of nitrogen and gallium (N_i, Ga_i), nitrogen and gallium antisites (N_{Ga}, Ga_N), and vacancies (V_{Ga}, V_N, $V_{Ga}V_N$). The formation of these native point defects depends on the growth conditions (e.g., a nitrogen-rich or gallium-rich environment) and on the Fermi level position and doping [16]. Fig. 7.1 shows the formation energy as a function of the Fermi level for relevant native point defects in GaN calculated using density-functional theory (DFT) and assuming Ga-rich conditions.

From simulations, self-interstitial and antisite defects appear to be high-energy defects in GaN and therefore are unlikely to be generated during growth. Obviously, it is still possible to create them under nonequilibrium conditions, for instance by electron irradiation or ion implantation. Based on Fig. 7.1, however, only vacancies are expected to be present in significant concentrations in GaN, due to their low formation energies [15].

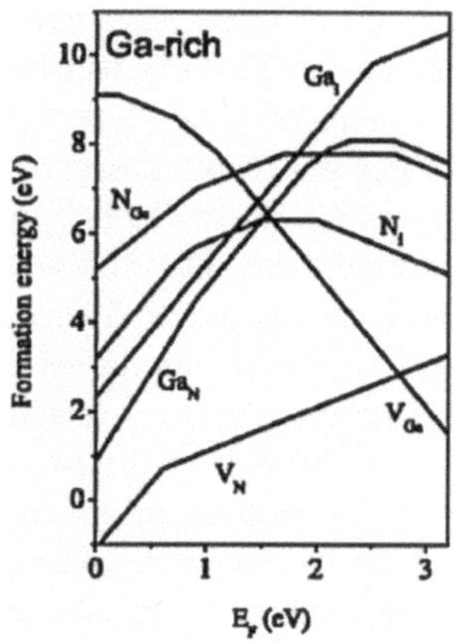

Figure 7.1 Calculated formation energies as a function of Fermi level for native point defects in GaN under Ga-rich conditions. The zero Fermi level corresponds to the top of the valence band. Only segments corresponding to the lowest-energy states are shown, their slope indicating the charge state. Kinks in the curves indicate transitions between different charge states.

Reproduced from [15], with the permission of AIP Publishing.

Nitrogen vacancies behave as shallow donors in GaN and have been considered responsible for the n-type "autodoping" for more than two decades [15]. However, their high formation energy in n-type GaN makes their spontaneous formation very unlikely during growth of nonintentionally doped GaN and hence they cannot be responsible for n-type conductivity. It is now well accepted that foreign impurities are responsible for such unintended conductivity, such as oxygen [15].

Gallium vacancies have been studied for years and their role in InGaN-based devices is still under investigation. In n-GaN they behave as acceptors, playing a role in donor compensation as well as in the frequently observed yellow luminescence (YL) [15]. The YL is a broad luminescence band centered around 2.2 eV, that has been observed in bulk GaN crystals as well as in epitaxial layers grown by different technologies [15]. Its intensity can vary significantly and good devices show almost no YL. Gallium vacancies (in isolated form or in complexes with an impurity) have been indicated as the most likely source of YL in the past and a number of experiments have confirmed that they play an important role in it [15]. The most direct evidence of a correlation between Ga vacancies and YL has emerged from positron annihilation spectroscopy measurements, which provide a direct probe of vacancies in the sample. It was found that the concentration of V_{Ga} correlates with the intensity of the YL, thus providing direct evidence for the involvement of the V_{Ga} acceptor levels in the YL [17]. Despite that, the details of the specific excitation and emission mechanisms responsible for the YL are still under discussion. The problem arises because standard DFT, used to calculate the formation energies and the transition levels of defects in Fig. 7.1, suffers from severe underestimation of band gaps [18]. Therefore the V_{Ga}

transition level of 1.1 eV above the valence band maximum, which was compatible with several experimental results on the YL [15], was moved deeper into the band gap by recent calculations based on more advanced techniques, so that the corresponding optical transitions would no longer match the observed YL band (see Fig. 7.2) [18]. Moreover, recent theoretical studies have revealed that the formation energy of isolated V_{Ga} is fairly high, making the formation of Ga vacancies more likely if they are complexed with other impurities such as oxygen or hydrogen. Calculated parameters for V_{Ga} complexes like V_{Ga}-3H and V_{Ga}-O_N-2H are in good agreement with experimental observations of YL, but are also close to those calculated for carbon in N-site, C_N (discussed below) [18]. It is therefore difficult to discriminate between YL due to V_{Ga} complexes and YL due to C_N, leaving its exact origin an open question.

In a recent work, Dreyer et al. indicated V_{Ga} complexes as efficient Shockley-Read-Hall (SRH) recombination centers in InGaN alloys with band gaps in the green and yellow region [19]. These authors stated that in bulk GaN the capture rates of electrons or holes is the bottleneck limiting the nonradiative recombination, due to the position of the energy levels associated with these defects that make the capture process inefficient. The situation changes significantly by moving from GaN to InGaN alloys, as shown in Fig. 7.3. In particular, as the In content in the alloy increases, the conduction band minimum (CMB) gets closer to the (+1/0) transition level of the complex. This results in an enhanced electron capture rate that eventually leads to higher SRH recombination.

Point defects also originate from impurities and relative complexes. The main foreign species used to obtain p-doped GaN is magnesium, which is an acceptor in III-nitrides. The Mg acceptor level is located above the valence band, with an ionization energy in the range 140–260 meV [18,20–22]. Recent studies have revealed

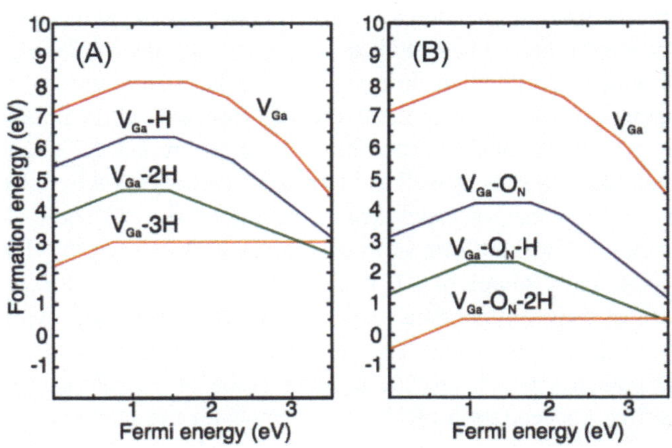

Figure 7.2 Calculated formation energies of point defects in GaN as a function of Fermi energy. (A) Gallium vacancy and complexes with H, (B) gallium vacancy and complexes with N-site oxygen.

Reproduced from [19], with the permission of AIP Publishing.

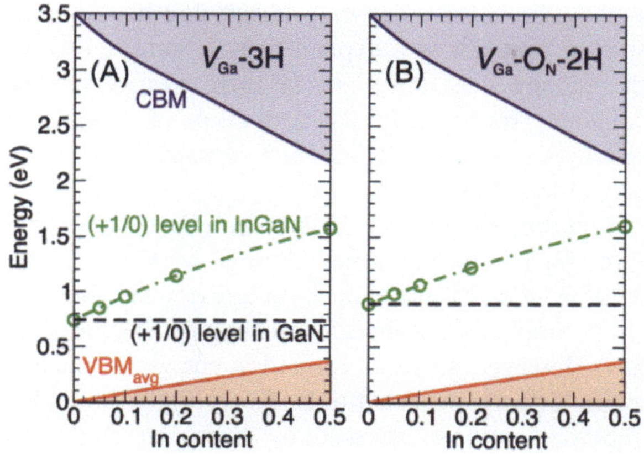

Figure 7.3 Position of the (+1/0) transition level within the band gap of InGaN as a function of In content for (A) the V_{Ga}-3H and (B) the V_{Ga}-O_N-2H complex. The black dashed line assumes the transition level is constant on an absolute energy scale and given by the value in GaN. The green dots (and the interpolating green dot-dashed line) take into account the varying In composition in the alloy.

Reproduced from [19], with the permission of AIP Publishing.

that Mg displays the features of a deep acceptor with a highly localized hole, in contrast to shallow acceptors where the hole is actually delocalized and its behavior is described by the effective mass theory [23].

Alternative acceptors such as Be, Ca, Zn, and C have been studied in the literature and are reviewed in Reference [15]. With the exception of carbon, none of them has been consistently and successfully used as a shallow dopant in GaN. In early studies [15] based on standard DFT calculations, which did not correctly describe hole localization at acceptors, it was argued that N-site carbon, C_N, should be a shallow acceptor located near $E_V + 0.2$ eV, C_{Ga} should be a shallow donor, and interstitial C (C_i) should be a deep center located near the midgap, behaving as a compensating acceptor in n-GaN and as a compensating donor in p-GaN. Subsequent calculations using the HSE hybrid functional reassessed the situation, showing that C_N in GaN is a deep acceptor having an ionization energy of about 0.9 eV [24]. This acceptor is considered to be responsible for compensation and high resistivity in C-doped GaN. Moreover, the calculated absorption and emission lines for C_N also indicate that it is a likely source of YL.

Calculations for simple substitutional donors, Si in Ga site (Si_{Ga}) and oxygen in N site (O_N) predicted a high solubility, low ionization energy and no strong lattice relaxation effects leading to metastabilities [15]. High levels of n-type conductivity have been reported many times in GaN crystals grown at high temperature and high pressure. Their characteristics were found to be very similar to other epitaxial films intentionally doped with oxygen, leading to the conclusion that unintentional oxygen incorporation results in n-type conductivity of bulk GaN [15]. High-pressure experiments also showed that, at a certain pressure threshold, a strong freeze-out of electrons occurs. This behavior was explained by assuming that oxygen donors in GaN under high pressure can move out of their normal substitutional position toward interstitial sites, resulting in the formation of DX-like acceptors. Such behavior was not found in the case of Si, which remains a shallow donor in GaN at similar pressure conditions [15].

An extremely important impurity in III-nitrides is hydrogen, since it can be easily introduced in high concentrations during MOCVD growth and forms stable complexes with dopants and defects, altering their electrical properties. In GaN, hydrogen is an amphoteric impurity behaving as a donor in p-GaN and as an acceptor in n-GaN. However, since its formation energy is much lower in p-GaN, the hydrogen concentration is expected to be much higher in p-GaN than in n-GaN [23]. A very important complex formed by hydrogen is the Mg−H complex. Hydrogen bonding with the Mg acceptors to form neutral complexes during growth is widely believed to be the main cause of the reduced p-type conductivity in GaN in early experiments, particularly in the case of layers prepared by MOVCD [25]. To break such Mg−H complexes and obtain high hole concentrations, high-temperature annealing in a hydrogen-free atmosphere is nowadays a common routine procedure for GaN films where a hydrogen carrier gas was used [25]. Hydrogen and oxygen may also form complexes with Ga vacancies, as discussed earlier, eventually limiting the optical performance of the devices by enhancing the SRH recombination [19].

Native defects can form complexes with other atoms: for example, V_{Ga} can form $V_{Ga}-O_N$ complex, whose formation is more energetically favorable than V_{Ga} itself [26]. The same is true for the $Mg_{Ga}-V_N$ complex: according to theoretical calculations, its formation energy is lower than the sum of the formation energies of single V_N and Mg_{Ga} [27]. Complexes with hydrogen are also possible, for example $V_{Ga}-H_n$, with n up to four hydrogen atoms, or V_N-H, although this complex can be formed only during high-temperature growth due to repulsion between V_N and H^+ [28].

7.2.2 Native defects: thermally activated and current-driven degradation

The impact of defects on device degradation depends strongly on their formation and migration barriers, since the energies reached during the normal device operation are limited. Reshchikov and Morcoç summarized the properties of these kind of defects, including their formation energies, their ionization energies, and their migration barriers [29].

Extended native defects are generated due to poor lattice matching between the substrate (generally sapphire or silicon) and the GaN buffer, or between AlGaN/InGaN and the GaN layers. In particular, the lattice mismatch between InGaN and GaN makes the fabrication of long-wavelength devices difficult, due to the high indium content required. High indium content in the quantum wells leads to the formation of dislocations and, for very high indium content, even phase separation [30].

Dislocations act as nonradiative centers in GaN devices, thus being detrimental to the device efficiency. Studies have shown that the radiative efficiency and luminescence of the devices have a direct correlation with the amount of threading dislocations [31,32].

The dislocations lead to the growth of V-shaped pits. Although this kind of structure is detrimental to the radiative efficiency of the LEDs, pit dimensions can have an inverse correlation with the leakage current [33]. Carriers from deep levels are thermally activated, and this process is enhanced by field-assisted emission barrier lowering (Poole-Frenkel effect; see Fig. 7.4) [33]. The effectiveness of this process depends on the Poole-Frenkel barrier height: since this barrier is higher in devices with large V-shaped pits than in devices with smaller pits, the first devices have less leakage current [34].

The degradation processes involving native defects are usually thermally activated or current-driven. Thermally activated processes usually involve diffusion, which is enhanced by the temperature. Low migration barriers, below 1 eV, allow the defects to diffuse not only during the high-temperature growth and annealing, but also in the relatively low-temperature device operative conditions, especially in the active region. On the other hand, current-driven processes involve the carrier flow, that can interact with the crystal lattice, transferring energy and thus enhancing the creation of new defects or their annihilation. Native defects can also ease impurity diffusion, driven by the previous processes [35].

Meneghini et al., by studying the degradation of InGaN laser diodes (LDs), found the signature of a deep level situated 0.35–0.45 eV below the conduction band, whose concentration increased with the device degradation [36]. This deep level has been correlated in previous studies to nitrogen antisites (N_{Ga}) [37]. However, the formation energy of these defects is high and thus its formation during device operation is less favorable than other types of defects.

Near extended defects, the diffusion and creation of point defects (native or non-native) is enhanced. Made et al. studied the generation of defects on InGaN-on-silicon LEDs during constant current stresses [38]. They analyzed InGaN-GaN LEDs with five quantum wells by means of electro-optical characterizations, deep-level transient spectroscopy (DLTS), scanning transmission electron microscopy with electron energy loss spectroscopy (STEM-EELS), and cathodoluminescence.

By DLTS analysis they found a deep level of 0.26 eV below the conduction band (E_c-0.26 eV) on the stressed sample, that was not present in the unstressed device. This deep level has a small capture cross-section of $\sim 10^{-15}$ cm^2, that is a

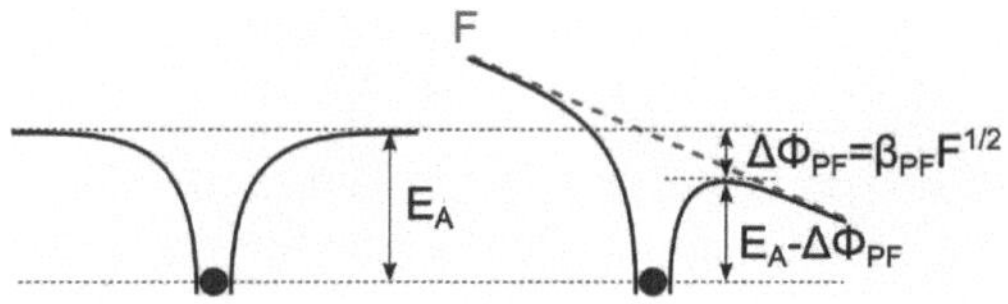

Figure 7.4 Sketch of the Poole-Frenkel barrier-lowering process.

characteristic of point defects. Defects with similar energy levels are associated with nitrogen vacancies [39,40].

Based on STEM-EELS analysis they found that, near dislocations like in Fig. 7.5A, the nitrogen concentration is lower in the stressed sample than in the unstressed one as in Fig. 7.5B, suggesting the formation of nitrogen vacancies during the stress. V_N has a relatively low formation energy of ~ 1 eV in p-GaN [16], but it is unlikely that the vacancies generated in p-GaN migrate toward the quantum wells because of their very low diffusion coefficient [41]. They associated the creation of the nitrogen vacancies with the presence of dislocations and magnesium: dislocations are a preferential path for Mg diffusion, that can lower the formation energy of nitrogen vacancies.

Further analysis showed that, even if the mobility of dislocation in GaN is very low with respect to other III–V compound semiconductors [42], reverse-bias and forward current stresses can induce a degradation [43]. The degradation due to the high-injection carrier flow is not related with threading dislocation, but the degradation during reverse-bias stress is related to material changes due to avalanche breakdown at the boundaries of the space-charge region near preexisting dislocations.

De Santi et al. [44,45] analyzed the time-dependent dielectric breakdown in low-power commercial LEDs submitted to reverse-bias stress. This condition is common

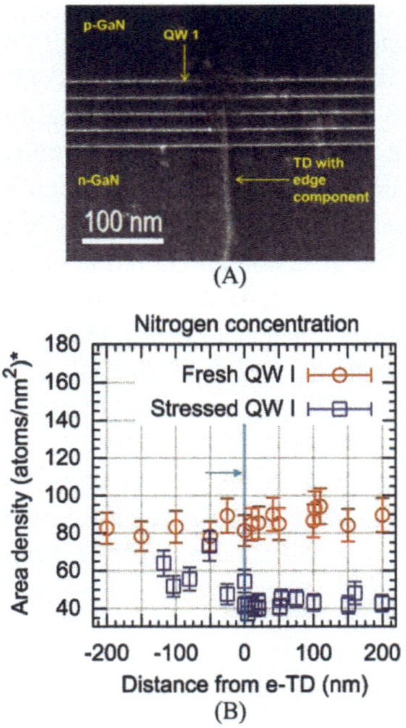

Figure 7.5 (A) STEM cross-section of the epi stack in the active area. A threading dislocation (TD) with an edge component is visible. (B) Nitrogen quantification along QW1 on unstressed and stressed sampled.

Reproduced from [38], with permission from Elsevier.

in driverless AC luminaries, where the devices are submitted to an intense reverse-bias stress due to the lack of current rectifiers. They first measured the breakdown characteristics of the devices, finding a value of about −65 V. They then submitted several devices to a constant reverse-bias stress at bias level smaller than breakdown voltage, employing electroluminescence (EL) measurements to evaluate the emission and the relation with leakage current, finding a direct proportionality between these quantities.

They also noticed emission hot-spots in the EL measurements, where the number of spots remains constant during stress, but emission intensity increases with stress time. The graph in Fig. 7.6 shows the increase in leakage current at −50 V, that is well correlated with the EL signal. On the inset it is possible to see that the

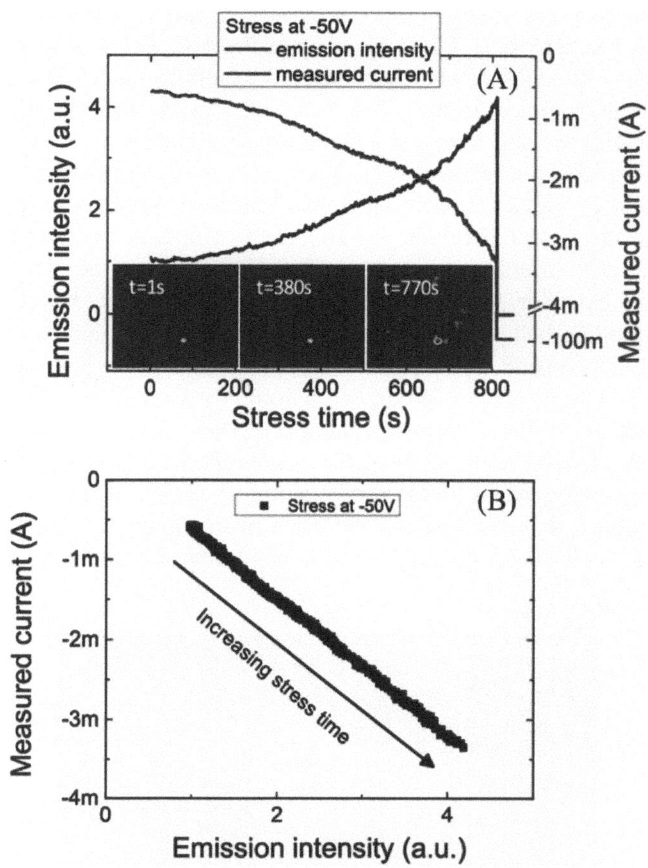

Figure 7.6 (A) Time dependence of the reverse current and the EL signal during stress at −50 V; (inset) EL map of the device under stress; (B) correlation between EL signal and leakage current.

emission hot-spots are present also on the unstressed device: the effect of the reverse-bias stress is the enhancement of the EL emission from these spots.

They investigated the failure mode of the devices by using electron microscopy. In Fig. 7.7 it is possible to see the EL signal before device failure (B) and the optical image of the failed device (C), in which it is clear that the failure is associated with one of the hot-spots. They concluded that hot-spots correspond to preferential leakage paths, due to tunneling or hopping through defects. When the device is submitted to high electric fields, close to the breakdown field, further defects form along these paths due to percolation processes. By increasing reverse voltage the very high current density flowing through these paths leads to thermal damage, which results in the catastrophic failure of the device.

Emission hot-spots were related to the presence of dislocation in the samples by Moseley et al. [46], who investigated the effects of threading dislocations on leakage current on deep UV LEDs. They fabricated two different LEDs, with a similar total dislocation density of 3×10^9 cm^{-2}, but different open-core threading dislocation (nanopipes) densities of approximately 10 nanopipes and 1100 nanopipes per device. The high nanopipe density device showed a lower turn-on voltage, a higher current at medium forward bias, and a higher leakage current.

To determine the role of nanopipes and their extension inside the devices, they performed local I−V characterizations on two different n-AlGaN and n-AlGaN−p-GaN templates by means of conductive atomic force microscopy, finding some hot-spots in which the current was increased. The n-AlGaN template hot-spots showed a nonrectifying behavior, whereas the n-AlGaN p-GaN template hot-spots showed a rectifying diode-like behavior. The authors hypothesized that conductive nanopipes were formed only inside the n-AlGaN layer, and created a p−n junction when reaching the p-GaN layer, thus producing a nonlinear leakage current effect on the whole LED structure. The structure without the p-GaN layer, instead, had conductive nanopipes exposed, thus showing the nonrectifying behavior. To confirm this hypothesis, they performed a selective etching of the surface with sulfuric acid on both templates, that exposed the presence of nanopipes on the surface of the structure. They observed hexagonal etch pits only on the n-AlGaN template

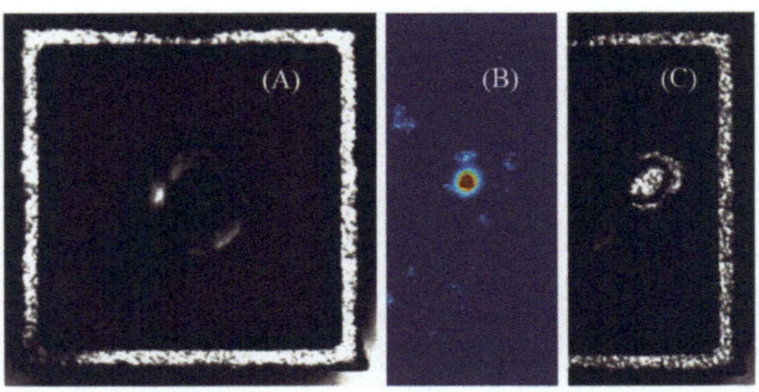

Figure 7.7 (A) Photo of a device before stress. (B) False color EL map of the device before catastrophic failure. (C) Image of the failed device after stress at −50 V.

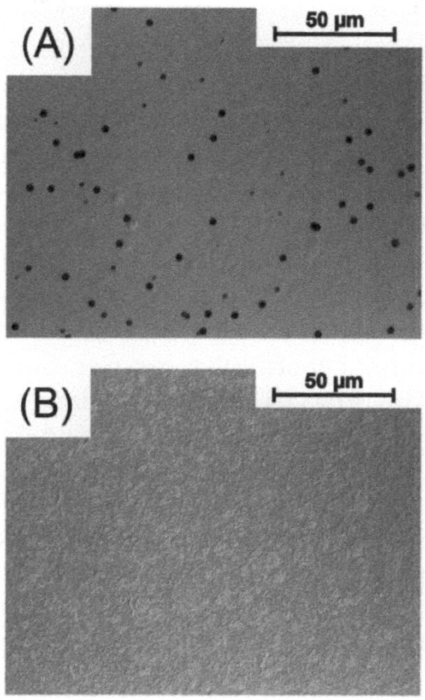

Figure 7.8 Optical images at 50× magnification of (A) n-AlGaN template and (B) n-AlGaN–p-GaN template after 90 s etching in hot sulfuric acid.
Reprinted from [46], with the permission of AIP Publishing.

(without p-GaN, Fig. 7.8), thus confirming that nanopipes are terminated in the n-AlGaN layer and thus produce a nonlinear current leakage effect when the p-GaN layer is grown above.

7.2.3 Impurities: degradation caused by diffusion

Impurity diffusion through a semiconductor follows the so-called Fick's second law, that in the one-dimension case can be expressed as:

$$\frac{\partial C}{\partial t} = D\frac{\partial^2 C}{\partial^2 x} \tag{7.1}$$

where C is the concentration of the impurity, x is the position along the considered dimension, and D is a parameter called diffusivity (or diffusion coefficient) that describes the ability of an impurity atom to move through the lattice. This formulation neglects many aspects of diffusion that become important at very high impurity concentrations. Therefore it must be used with caution when dealing with high impurity concentrations, because the diffusivity itself is a function of the concentration.

A relevant impurity in GaN-based LEDs is Mg that, as already discussed, is the preferred p-type dopant species. Controlling the diffusion of magnesium during the growth process is crucial for obtaining efficient devices. A moderate presence of Mg in quantum wells, detectable with secondary ion mass spectrometry (SIMS) analysis, can improve the optical performance of LEDs, due to increased overlapping of the electron and hole wavefunctions induced by partial screening of the internal fields [47]. Magnesium close to the active region may also enhance hole injection and thus increase the efficiency [48]. However, an excessive spread of Mg atoms into the active region leads to higher nonradiative recombination in the QW and therefore degrades the optical properties [48]. Another impurity that can significantly modify the electrical and optical properties of GaN LEDs is hydrogen. As discussed earlier, hydrogen can be easily introduced during growth and can form complexes with gallium vacancies that behave as efficient nonradiative recombination centers, thus reducing the emitted optical power of InGaN-based LEDs [19]. Hydrogen is also responsible for partial passivation of the Mg dopant, resulting in higher series resistance and operating voltage of the devices due to a lower hole concentration [25,49].

The diffusion coefficient D is expected to follow an Arrhenius law:

$$D = D_0 \exp\left(-\frac{E_a}{k_B T}\right) \tag{7.2}$$

where D_0 denotes a preexponential factor, E_a is the activation energy of the diffusion process, k_B is the Boltzmann constant, and T is the absolute temperature. Eq. (7.2) shows that the diffusivity has a strong dependence on the temperature, which is therefore a key parameter to control during growth.

Diffusion of impurities and/or defects has also been proposed as a degradation mechanism for GaN-based devices. Takeya et al. [50] performed aging experiments on AlGaInN-based LDs under 30 mW continuous-wave operation at 60°C, revealing that the lifetime of these devices is dependent on the dislocation density in the basal ELO-GaN layer and that the degradation rate is almost proportional to the square root of aging time. The authors proposed a degradation mechanism governed by the diffusion of defects toward dislocations. The accumulated point defects such as vacancies, Mg interstitials, or their complexes around the dislocations probably enhance the capture cross-section of nonradiative recombination centers [50]. Nam et al. [51] investigated the degradation of GaN-based LEDs under the same aging conditions and suggested the diffusion of Mg atoms and their accumulation in the active region as the main degradation mechanism. In particular, post-aging SIMS analysis revealed a much higher concentration of Mg in multiple quantum wells (MQW) of samples having a higher dislocation density, suggesting that dislocations are a possible diffusion path of point defects such as Mg, vacancies, as well as nonrecombination sources. However, other studies [52] indicated that degradation can take place even without changes in the SIMS profile, so the role of Mg diffusion is still under debate. In a more recent work, Orita et al. [41] showed that variation of the nonradiative recombination lifetime of stressed

GaN-based LDs, τ_{nr}, can be fitted with an analytical model based on a gradual diffusion process (see Fig. 7.9):

$$\Delta(1/\tau_{nr}) = \frac{KN_0}{w}\int_0^w erfc\left(\frac{z}{2\sqrt{Dt}}\right)dz \tag{7.3}$$

where w is the thickness of the separate-confinement heterostructure (SCH) including the InGaN MQW active layers, K is a coefficient related to the capture rate of carriers, and N_0 is the concentration of diffusants at the interface between the doped layer and the SCH layer.

Based on this model, the authors extracted a diffusion coefficient D of $1.9 \times 10^{-19}\ \mathrm{cm^2\ s^{-1}}$ and compared it to the reported values in the literature. They found that no impurity or point defect except hydrogen, H, in bulk diffusion explains such a high value of estimated D. Hydrogen acts as a deep donor and is theoretically predicted to enhance diffusion of defects and stabilize them. In addition, forward biasing allows H to drift through the p–n junction, which is consistent with the report that current injection causes a gradual degradation in InGaN-based LDs (see Reference [41] and references therein). Hydrogen was therefore suggested as a strong candidate for the diffusing species involved in the degradation process. On the other hand, the authors could not rule out diffusion through dislocations, which is more rapid than in bulk. Magnesium and oxygen are known to diffuse easily via threading dislocations in GaN, and the activation energy for diffusion through dislocation is typically lower than that for diffusion in bulk material [41]. The role of dislocations as the preferred path for diffusion was already confirmed by previous theoretical studies based on molecular dynamics simulations, showing that at 1000K the

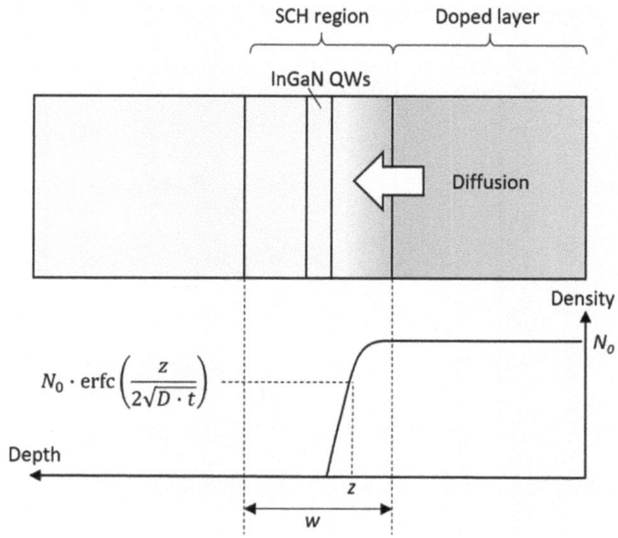

Figure 7.9 Proposed model of diffusion-involved gradual degradation.

diffusivity of interstitial Mg atoms inside the dislocation core is approximately three orders of magnitude larger than that of interstitial Mg atoms located outside the dislocation [53].

To better understand the generation of defects in the device caused by the stress, it is possible to follow the variation in the SRH nonradiative recombination coefficient (called A) [54]. The electron–hole recombination rate R in an LED is modeled by the so-called ABC model $R = An + Bn^2 + Cn^3$. In this model, n is the carrier density, A is the SRH coefficient, ascribed to the recombination through nonradiative center, B is the radiative coefficient, due to the radiative recombination (so the term Bn^2 is proportional to the light emitted L by the LED), and C is the Auger coefficient, ascribed to the recombination where the energy is given to a third carrier which is excited to a higher energy level. In steady-state and high injection conditions, the continuity equation can be written as

$$\frac{J}{qd} = An + Bn^2 + Cn^3 \tag{7.4}$$

where J is the current density, A is proportional to N_T (the defect density), and d is the thickness of the active region. If nonradiative recombination prevails (condition reached at low current levels), Bn^2 can be neglected, so n is proportional to the injected current and, for this reason, the emitted optical power (Bn^2) is proportional to the square of current $(L \propto I^2)$. Instead, if the radiative recombination dominates (higher current levels), the term An can be neglected, so the emitted optical power linearly depends on the current ($L \propto I$). Instead, if very high current levels are reached, Cn^3 prevails in Eq. (7.4), reducing the total power emitted by the device.

An analysis of the variation in the A coefficient and in the capacitance–voltage (C-V) characterization was carried out by La Grassa et al. [55] in the case of visible LEDs.

The authors found a square-root dependence between the A coefficient and the stress time (Fig. 7.10), that supports the hypothesis of a relation between the

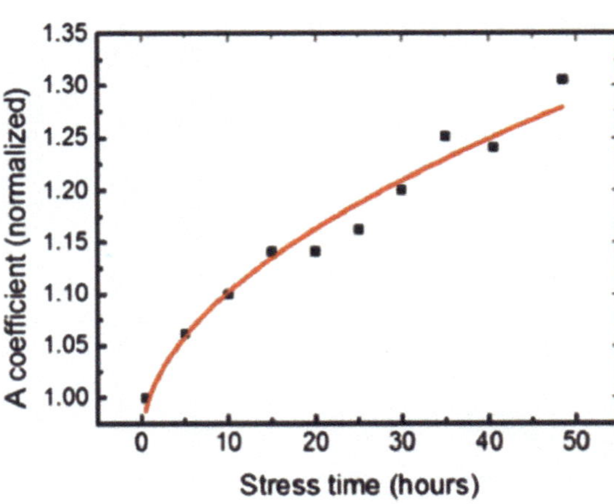

Figure 7.10 Variation of the A coefficient measured during the stress at 125°C.
Reprinted from [55], with the permission from Elsevier.

degradation and a diffusion process [41]. By analyzing the C-V measured during the stress (Fig. 7.11), they reported an increase in the capacitance; to better understand this behavior, the authors integrated the capacitance between 1.5 V and 2 V, in order to shrink the depletion region enough to bring the n-side space charge region edge close to the QW area.

The result (Fig. 7.12) indicates that the increase in free charge density in the active region follows the square-root of stress time. It is worth noting that this procedure could lead to unexpected artifacts if the internal electric field varies for some reason. In this case, the variation of C-V would be associated with different areas of the device, invalidating the analysis. For this reason, there are two possibilities: to plot the capacitance as a function of the depletion region width, under the assumption of a step p^{++}–n junction, or to calculate the apparent charge density

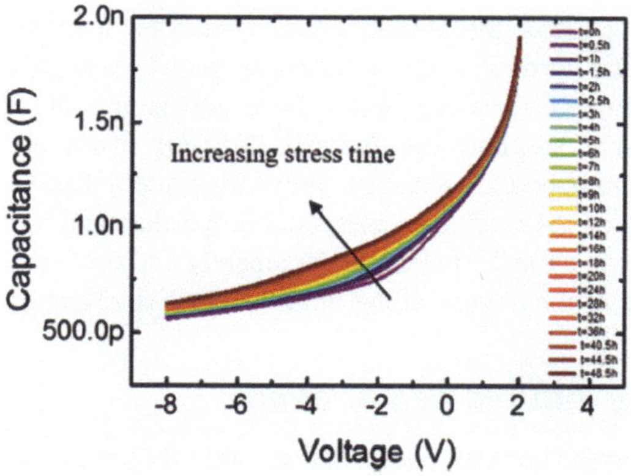

Figure 7.11 C-V measurement during the stress.
Reprinted from [55], with the permission from Elsevier.

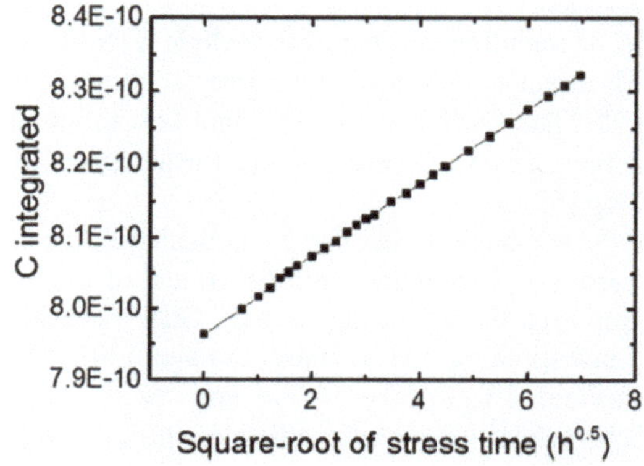

Figure 7.12 Integral of the capacitance between 1.5 V and 2 V.
Reprinted from [55], with the permission from Elsevier.

and plot it as a function of the depletion width. La Grassa et al. chose this last option, and they reported an increase in charge density that confirms the previous results. Moreover, this increment is linearly correlated to the increase in the SRH recombination coefficient, A. All these results confirm the hypothesis of a diffusion of defects toward the active region, that leads to optical degradation.

Limiting the diffusion can be a difficult task, given the influence the specific diffusing species and the device structure have on its kinetics, but some general ideas can apply to a wide range of situations. Since the diffusion process is driven by the concentration gradient, a graded impurity profile will significantly slow down its speed. Additionally, the diffusion is found to be faster in bulk material with a high dislocation density, therefore all the possible solutions to limit their number will have a positive impact on the diffusion-related degradation. The diffusion speed is also influenced by the bonding state of the diffusion impurity, and is typically faster for interstitial impurities and slower for substitutional impurities, due to the need for additional energy to break the atomic bond. Therefore, a more favorable substitutional/interstitial ratio through improvements in growth and processing quality can lead to an improvement. One last option is to add in the structure a defect getter layer, with the aim of stopping the impurity diffusion inside of it, for example, by means of a high number of vacancies, preventing them from reaching the active region. The design of the thickness, electrical properties, and vacancy concentration in this layer is a complex task, and can unfortunately end up having a negative impact on the performance of the device, along with the positive effect on its reliability.

7.2.4 Optically induced degradation

Thermal- and current-induced degradation are the most studied degradation mechanisms of optoelectronic devices, while degradation due to photons is often neglected in many models. However, the photons can induce degradation, and this needs to be taken into account. Most of the research into optical field-induced degradation has been carried out on LDs, because of the very high amount of photons in the optical active zone.

In LDs, common degradation mechanisms include degradation of the dielectric mirror coatings [56], direct damages of the mirror coatings [57], and catastrophic optical damage (COD). The COD is caused by light absorption at nonradiative centers, that causes an increase in temperature and further generation of defects in a positive-feedback loop [58].

However, LEDs do not have as intense an optical field as LDs, and not many studies have been carried out to understand the effects of the photons on degradation. Eliseev et al. reported the effect on the bare GaN material of subpicoseconds pulses at 400 nm, investigating the surface damage [59]. They found a laser-induced damage threshold (LIDT) of $\sim$36 TW cm^{-2} with a pulse width of 150 fs, that induces clean pits without signs of melting of the material. For longer pulses the thermal effects are the main cause of degradation. Elhadj et al. investigated the damage induced by 1064 nm pulsed laser with pulse width ranging from 9.5 to 0.3 ns, finding that the damage was thermally induced by localized defect-driven absorption [60,61]. Ščiuka et al. studied the damage on thin GaN and InGaN films caused by femtosecond pulses with various excitation wavelength (1030, 515, and 343 nm) [62]. They found a decrease in LIDT with increasing indium content and decreasing wavelength. Moreover, the optical pulses lead to the annealing of indium clusters present in the untreated material and, for higher pulse energies, to the formation of gallium clusters around the edges of the etched area.

De Santi et al. carried out a series of experimental tests on MQWs InGaN-GaN devices stressed by a continuous-wave 405 nm laser excitation [63]. The device is a structure developed for high-efficiency light collection with high periodicity $In_{0.15}Ga_{0.85}N/GaN$ MQW (25 pairs) grown on a sapphire substrate [64]. This structure can be used as a part of multijunction solar cell, as a high-power photodetector, or as a receiver in high-power laser-based wireless power transfer, thus being submitted to very high optical powers [65,66]. They used a stress optical power of 361 W cm^{-2}, that is comparable to the optical power reached by LEDs for projection applications (up to 461 W cm^{-2}).

They stressed the device for 486 h by illuminating it under open circuit condition (i.e., without any electrical connection) and characterized the device by means of dark I–V and photoluminescence maps. The I–V curves in Fig. 7.13 show an increase in the reverse and forward leakage current; a common signature of the increase in deep level concentration after the purely optical stress.

The PL maps in Fig. 7.14 show a decrease in luminescence efficiency of the stressed device (B) with respect to the unstressed device (A). The area where the PL decreases corresponds to the laser spot area during the stress (C). The PL decrease can be related to (1) an increase in the defects in the active zone, (2) an augmented escape of the carrier from the quantum wells, (3) surface reflectivity variation, or (4) thermal degradation due to the laser beam.

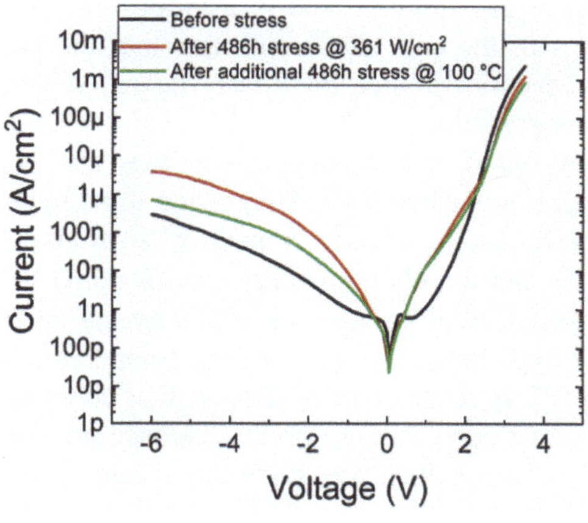

Figure 7.13 Current–voltage dark characterization of the unstressed device (black), optical-stressed device (red), and thermal stressed device (black).

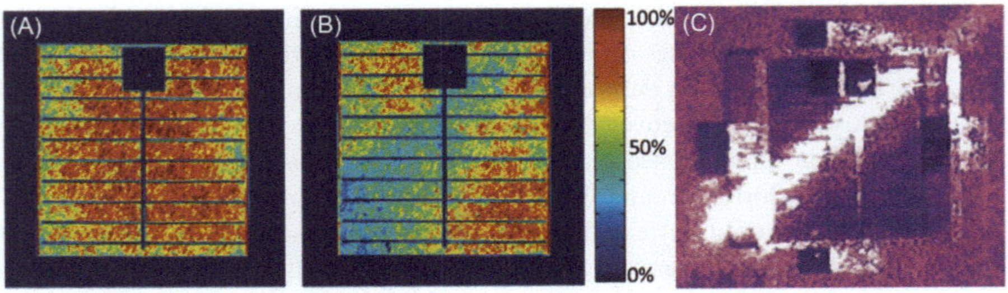

Figure 7.14 False color PL map before (A) and after (B) 486 h open-circuit optical stress at 361 W cm^{-2}. The PL maps are recorded at room temperature at 437 nm under 405 nm resonant excitation laser. (C) Laser spot position on the device during optical stress.

Since there is no current flowing in the device, due to the open-circuit condition, the escape is ruled out as a source of PL decrease. To exclude the variation of the surface reflectivity the authors mapped the surface reflectivity, whose pattern does not match the laser diode map, thus excluding this mechanism.

To exclude thermal degradation, they submitted an unstressed sample to a pure thermal stress. First, they estimated the temperature reached by the sample under optical stress by a thermal camera, finding about 30°C. Then, assuming a 60°C worst-case temperature error, they submitted the unstressed sample to a thermal stress at 100°C for the same amount of time of optical stress.

In Fig. 7.15 it is possible to compare the outcomes of the 486 h thermal stress in air at 100°C on an unstressed sample (A) and on a previously optically stressed sample (B). The photoluminescence map shows the difference in luminescence between after and before the thermal stress, normalized by the before value. The two degradation patterns are similar: there is a uniform degradation of the devices, enhanced in the edges of the cell. This confirmed that the degradation modes are different, because otherwise the area previously degraded would be affected in a different way by the degradation.

In conclusion, they found that the degradation was due to the photons, having excluded other degradation mechanisms. They made some hypotheses on the degradation mechanism: direct lattice damage is unlikely, since the GaN bond strength is evaluated to be 8.9 eV and the photon energy used is 3.06 eV [67]. Another possibility is the dehydrogenation of gallium vacancies by the incoming photons. Their results supported this hypothesis, as shown by the comparison of the YL before and after the stress: since YL is related to V_{Ga} [68] or their complexes with oxygen [26] or carbon [69], they concluded that the optical damage consists in the breaking of $V_{Ga}-H_n$ complexes and subsequent activation of gallium vacancies.

Other optically driven degradation mechanisms involve the LED package, rather than the chip itself [70]. Commercial LEDs need to be packaged in order to be used in lighting systems. Typically, an LED package includes an encapsulant, that protects the bare chip and increases light extraction efficiency, and a lens, that shapes

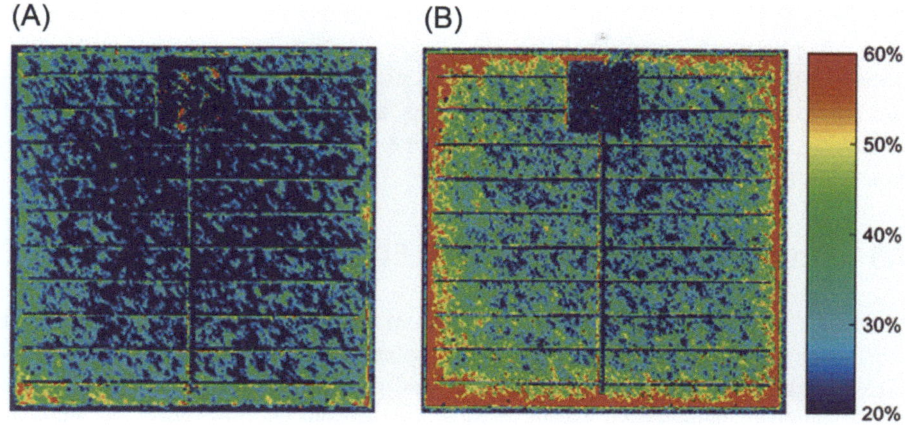

Figure 7.15 Photoluminescence map of the device at 437 nm under 405 nm resonant excitation after 486 h stress at 100°C in air on (A) an untreated device and (B) a previously stressed device for 486 h in open circuit configuration under 405 nm monochromatic illumination at 361 W cm^{-2}.

the light beam exiting from the LED. Different optical materials are used in LEDs, such as silicone or epoxy as encapsulant and bisphenol-A polycarbonate (BPA-PC) or polymethyl methacrylate (PMMA) for secondary optical elements. Moreover, in white LEDs phosphors are used to convert blue photons into longer wavelength photons, obtaining a broadband yellow emission superimposed on the nonabsorbed blue emission. Phosphors are usually combined with the encapsulants on the top of the LED chip. All these materials are subject to degradation due to heat transfer, environment, and optical absorption.

A typical outcome of the optical degradation of the LED package is the chromaticity shift, that is a change in the chromaticity coordinates (u',v') in the CIE1976 diagram. The direction of the shift can be related to the type of degradation: a change along the blue–yellow axis can be related to a change in the relative emission of blue and yellow peaks (for example, variation in the efficiency of the phosphors, discoloration of reflector, oxidation of secondary lens), whereas a green shift is observed when there is a variation in the phosphor emission wavelength.

Polymeric materials can undergo photodegradation, which leads to a decrease in the transparency of the material; a shorter wavelength typically causes higher degradation. Exposure to UV radiation causes discoloration and embrittlement of the silicone, whereas the photons cause photochemical degradation in PMMA; photodegradation of polycarbonate is observed under either UV or visible light.

The phosphors absorb high-energy photons, that allow their electrons to reach an excited state. The electrons then relax to the ground state by emitting a lower energy photon and heat. The phosphors themselves go through several degradation mechanism that lower the efficiency of the LED and also change the spectral characteristics of the emission, causing a chromaticity shift [71,72]. In particular, phosphors lower their efficiency with increasing temperature (thermal quenching), although phosphors used in LEDs have a relatively low thermal quenching. However, the temperature of the phosphors is usually higher than the chip temperature due to the losses in the conversion process, and this factor can enhance thermal quenching, thus causing a chromaticity shift due to the lowering of the yellow component in the LED spectrum.

7.3 Degradation of AlGaN UV LEDs

Nowadays, UV LEDs can be used for different purposes, from a fluorescent light source for lighting and displays, to light source for microscopes. UV-A LEDs (315–390 nm) are used to dry inks, resins, and paints, in security systems (such as identification of banknotes), and in medical applications such as blood gas analysis [73]. UV-B LEDs are used in phototherapy, especially for the treatment of diseases like psoriasis and vitiligo [74], and in lighting for plant growth [75]. Instead, UV-C LEDs allow carrying out of decontamination and sterilization of air and water, for example, for waste water treatment, recycling, and disinfection of medical equipment and food [76]. Finally, UV-B and UV-C LEDs are also used for detection and identification of biological molecules, because many gases, such as sulfur dioxide (SO_2) compounds, NO_x, ammonia (NH_3), and various biomolecules, like tryptophan, NADH, tyrosine, DNA, and RNA, have their own absorption bands in these spectral ranges [73–79].

The typical structure of an UV-LED (Fig. 7.16) is based on a sapphire substrate. The first layer grown is a thin layer of AlN. Above this layer, there is an AlGaN SL layer, used to reduce the number of defects. It is followed by a undoped AlGaN layer and by the active region, which consists of multiquantum wells (MQWs) of AlInGaN. This quaternary compound is used for the MQWs because it allows to

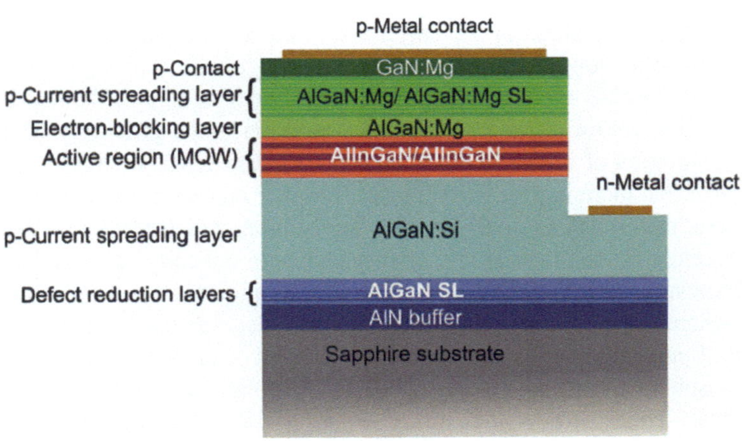

Figure 7.16 Typical structure of an AlGaN-based UV LED.
Reprinted from [80].

control the polarization fields inside the QW and to improve the morphology of the surface, thus reducing the density of defects. To prevent electron leakage current in the p-zone and to confine it to the active region, an AlGaN EBL doped with Mg is placed after the MQWs. The structure is completed by an AlGaN or AlGaN/GaN SL, doped with Mg, and a Mg-doped GaN layer, that is required in order to have a good ohmic contact.

7.3.1 Generation and diffusion of point defects

Many papers have reported studies about the degradation of AlGaN-based UV LEDs [41,51,81–83], where different aspects of the behavior of the LED during aging are addressed, including electrical and optical characteristics and analysis of the defect inside the active region. Moreover, the studies focus on different types of stress, electrical and thermal, in order to analyze and identify the causes of the degradation.

Monti et al. [84] studied the degradation of 308 nm UV-B LEDs stressed at a current density of 140 A/cm^2. They carried out optical and electrical characterization, deep-level transient spectroscopy (DLTS), and photocurrent spectroscopy (PC), in order to have a complete overview of the behavior of the device and to understand the physical origin of the defects. By the analysis of the optical power–current (OP-I) and power spectral density–current (PSD-I) measurements during the stress (Figs. 7.17 and 7.18), the authors noticed a degradation process that was more prominent at low current levels. This suggests an increase in the number of defects in the active region, with a subsequent increase in the SRH recombination rate. In fact, at low current levels, these centers can capture a larger part of the injected carriers and promote nonradiative events; at the same time, at higher current levels these centers are saturated by a high number of injected carriers, and so the influence on the total OP is, as a percentage, smaller.

To understand the physical mechanisms responsible for the OP degradations, Monti et al. investigated the electrical characteristics (I-V measurements) of the device.

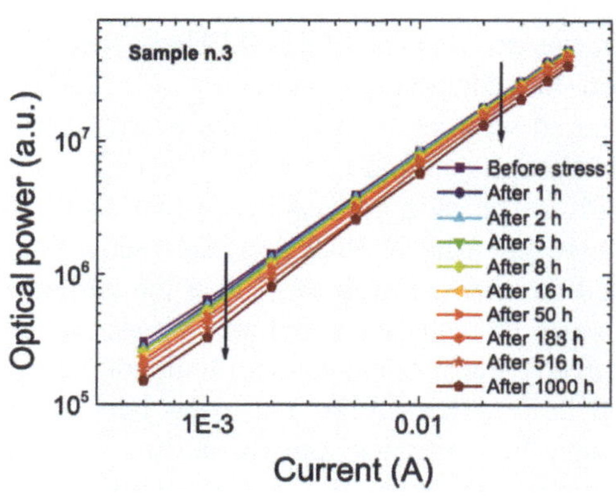

Figure 7.17 OP-I characteristics during the stress.

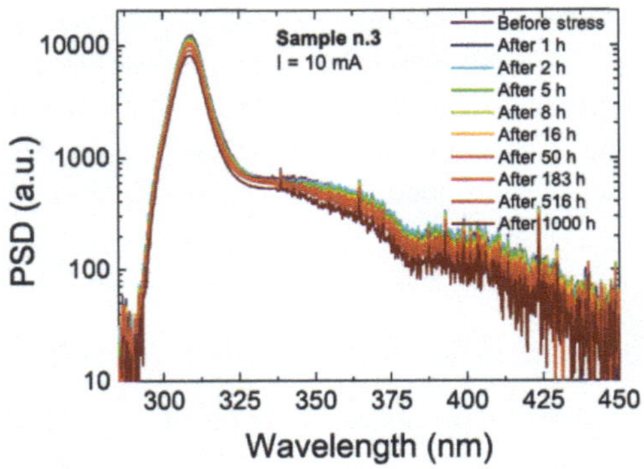

Figure 7.18 PSD at 10 mA during the stress.

In Fig. 7.19, they identified three regions: in the first region, the reverse bias region, they observed an increase in the leakage current, that has a logarithmic dependence on stress time; this suggests the generation of point defects within or around the active region. These induce the formation of parasitic current paths, that allow defect-assisted carrier transport along the active region [85]. In the second region, the bias region below the turn-on voltage (between 0 and 4 V), the authors observed an increase in the current governed by trap-assisted tunneling (TAT), related to an increase in the concentration of midgap defects. In fact, below the turn-on voltage, these new midgap defects (generated from the stress) allow the passage of electrons through the potential barrier by TAT, that leads to an increase in current [86,87]. In the third region, the region above the turn-on voltage (above 4 V), they noticed a decrease in the drive voltage during the first 2 hours of stress, related to the change in concentration in the active region of Mg traps, that leads to a lower resistivity of the p-type region. The same authors reported a direct confirmation of the generation of midgap states by photocurrent spectroscopy (PCS) in a comparable set of samples [88].

Another report on the degradation of UV-LEDs is that of Trivellin et al. [89], where they described the degradation processes of a 280 nm high-power DUV LED. The authors started with an analysis of the power spectral density (PSD) of the LED (Fig. 7.20) where they found two peaks, the main peak called PK1 and the parasitic one called PK2. The ratio between these two peaks changes with the current, in fact, the relative intensity of PK2 decreases when the current is increased. They also investigated the thermal drop of LED at the current of 1 mA (Fig. 7.21), finding a strong decrease of both peaks and an increase of the ratio between PK2 and PK1, that has a nearly linear behavior with temperature. The second peak PK2 is generally observed in deep UV-LEDs, and it has been associated with different causes in the literature, like nitrogen antisite defects [90], tunneling of carriers through the EBL in the p-GaN layer [91], and recombination in the last barrier of MQW region [92]. This YL in GaN is generally associated with gallium vacancies [26] or carbon impurities [24,69], and the temperature dependence of PK2/PK1

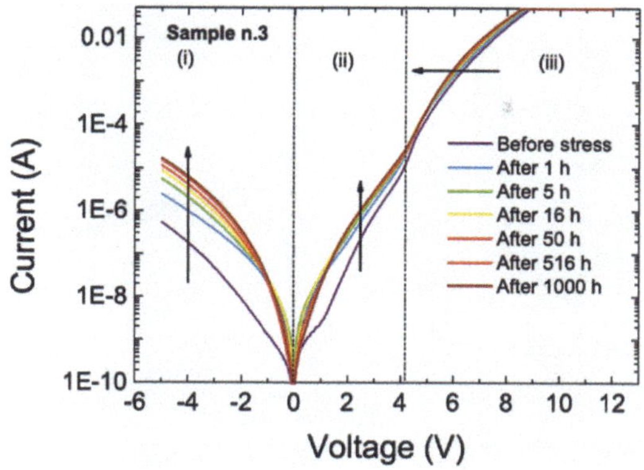

Figure 7.19 I-V characteristics during the stress, where there are three different regions identified.

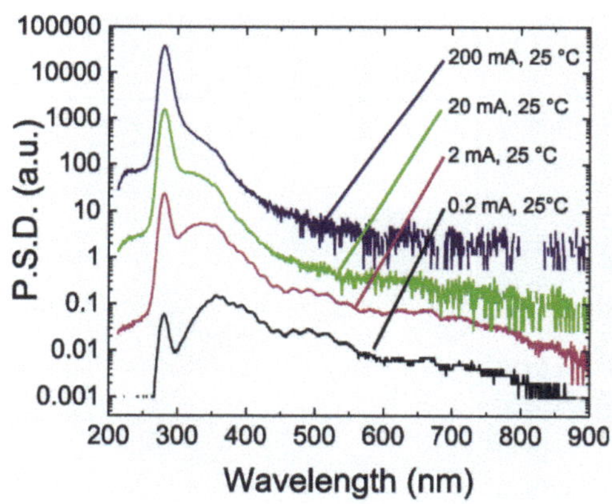

Figure 7.20 Effect of forward current on the spectrum emitted from DUV LED.

supports the hypothesis that the PK2 emission is related to a defect-assisted emission, because there is better transferring of carrier to the defect with higher temperatures. A decrease in the blocking properties of the quantum barriers is also reported in the literature [93].

In order to obtain a better understanding of the deep levels involved in the degradation, Monti et al. [84] used the capacitance deep-level transient spectroscopy (C-DLTS) technique on UV-B LED [94], that showed a spectra with three peaks (Fig. 7.22).

They extracted the Arrhenius plot of each peak and, after a comparison with the literature, they concluded that the deep level h_1 could be due to Mg-related acceptor traps or Mg–H complexes. For this reason, they correlated the change in this peak with the decrease in the drive voltage in the first hours of stress. In fact, the

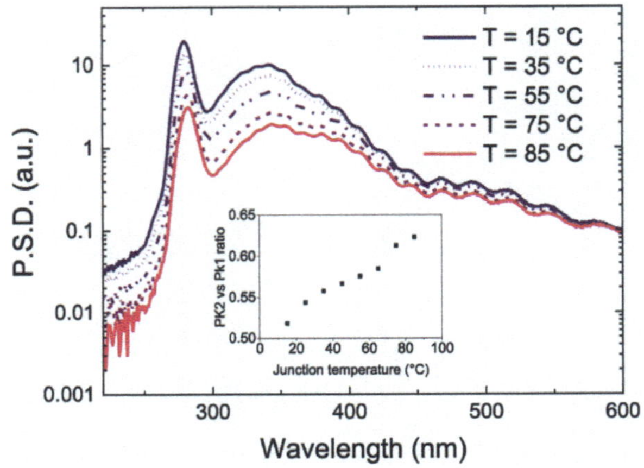

Figure 7.21 Effect of junction temperature on the spectrum of DUV LED. Inset: dependence of the ratio PK2/PK1 on junction temperature.

Reproduced from [89]. Copyright (2019) The Japan Society of Applied Physics.

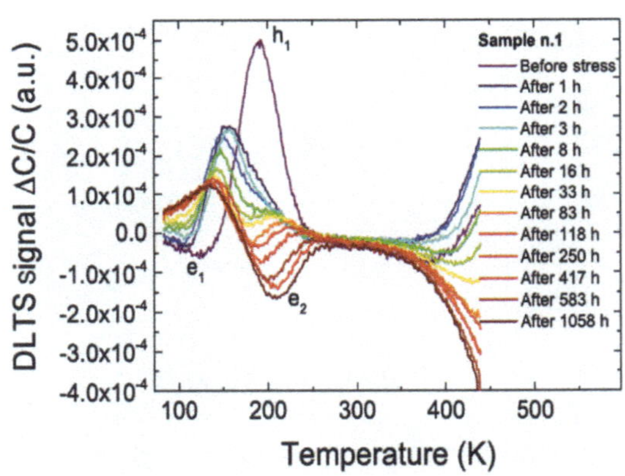

Figure 7.22 DLTS spectra during the stress ($V_{rev} = -5$ V, $V_{fill} = -3$ V, emission rate window 23.8 s^{-1}).

© 2017 IEEE. Reprinted, with permission, from [84].

reduction of Mg traps may involve a decrease in the resistivity of the p-type layers and the p-side contact, that leads to a decrement in the drive voltage. They also performed photocurrent spectroscopy (PCS) measurements, and found an increase in the concentration of a deep level located at E_c-2.25 eV, close to the midgap, that can behave as effective nonradiative recombination centers. This could explain the increase in the current below the turn-on voltage, that is associated with the increase in TAT processes.

Trivellin et al. used the same technique and reported (see Fig. 7.23) the presence of a defect for holes that is associated with nitrogen antisites. The authors, starting from this result, concluded the discussion saying that the increase in this defect should favor SRH nonradiative recombination, thus reducing the radiative efficiency at a lower current. However, after a current stress, they noticed an initial increase in the optical emission for PK3 that appeared during the aging (Fig. 7.24), that they ascribed to an increase in holes injected into the QW region, thus compensating for the effect of degradation at a lower current. This improvement could be the result of the activation of Mg-doping of the p-side [95], or the reduction of the energy barrier for the holes. The detection of PK3 only at higher current also suggests a possible relation to an electron leakage through EBL, that can be explained by the increase in defectiveness in the material, which generates TAT through

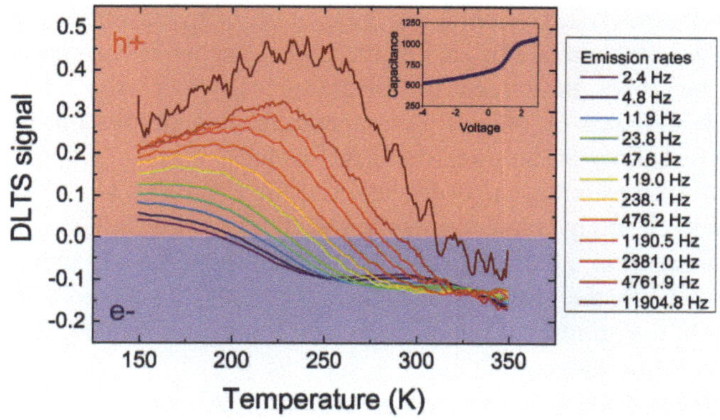

Figure 7.23 DLTS spectra at different emission rates, measured with $V_{fill} = 0$ V and $V_{meas} = -2V$. Inset: capacitance−voltage measurement.

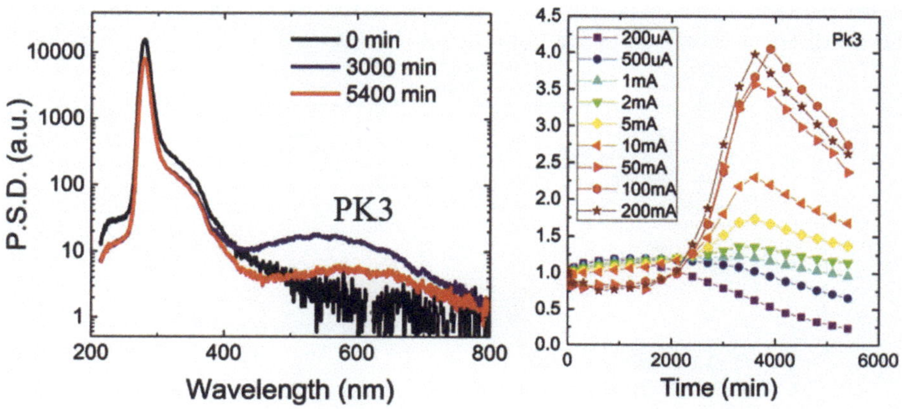

Figure 7.24 (Left) Spectral emission of the device measured at 100 mA, and (right) degradation kinetics of PK3.

EBL; this could be a possible explanation for the reduction in the energy barrier that reduces the confinement at higher current levels.

7.3.2 Role of magnesium

In the AlGaN-based UV LEDs, the resistivity of the material in the p-side region is an issue that is of critical importance. The high resistivity of this region, due to the low ionization of Mg used as p-dopant, results in an increased drive voltage of the diode, that leads to an increase in the dissipated power, especially at higher current levels. This power is dissipated as heat, and can favor various processes inside the LED, including thermal-activated degradation mechanisms [96–98]. Some of these are related to thermally activated defect generation or to current overflow and carrier escape from quantum wells, due to the higher carrier energy caused by higher temperature. Moreover, these mechanisms are typically thermally activated, with an exponential dependence on increasing temperature. For this reason, it is important to reduce the resistivity of the p-region of the device.

The role of magnesium in the degradation of the UV LEDs has been investigated by several research groups (see, for instance, References [99,100]).

Sung-Nam Lee et al. [101] reported that by increasing the Mg doping concentration in a UV-A LED, the photoluminescence of the MQWs decreases because there is an increase in the number of nonradiative recombination centers; these are probably caused by Mg-related point defects such as Mg-H, Mg interstitials, or Mg-related complexes. Moreover, they proposed that in AlInGaN-based LEDs the Mg dopant diffuses into the InGaN active layer by the applied voltage.

Nam et al. [102] reported a study on the effect of Mg concentration on the contact resistivity in p-GaN layers.

They compared different samples, with a p-GaN layer highly doped with different concentrations of Mg. If the highly doped p-GaN layer is not present, the resistivity is higher than any other cases [Fig. 7.25 (left)]. Moreover, if the concentration of Mg is increased, the resistivity decreased from 5.2E-2 Ωcm^2 to 9.4E-4 Ωcm^2. This is probably

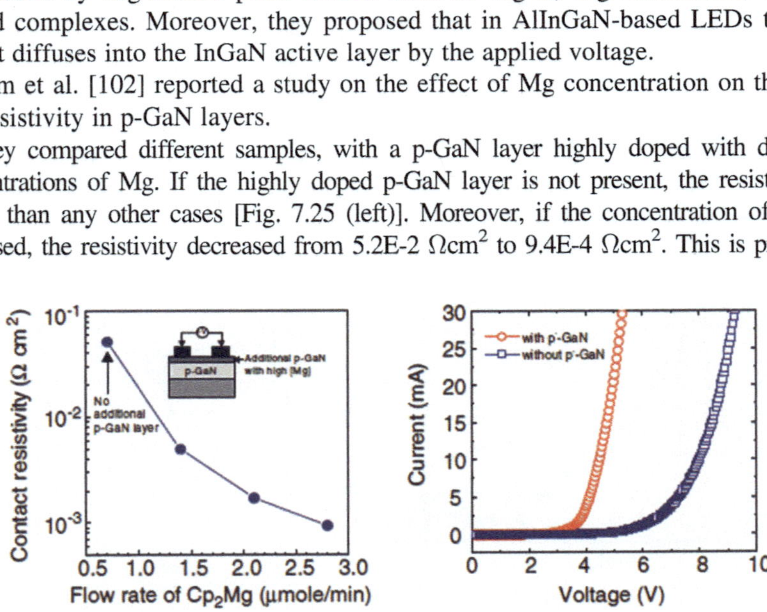

Figure 7.25 (Left) Contact resistivity plotted as a function of concentration of Mg. (Right) I–V characteristics of the two different sample.

Reprinted, with permission, from [102].

due to the fact that the Mg doping in the p-contact layer causes defect states, that help carrier injection due to TAT. Another consequence deriving from the introduction of this doped layer is the decrease in the drive voltage of the device [Fig. 7.25 (right)].

In a following work, the same group [51] extended the discussion, researching a correlation between the degradation of the device and the diffusion of Mg in the active region.

Fig. 7.26 reports the Mg concentration as obtained through SIMS; the plot shows a diffusion of Mg into the active layer after the aging test. This suggests that the diffusion and accumulation of Mg in the active layer is a possible degradation mechanism in AlInGaN LEDs. Moreover, the authors reported that the dislocations in the substrate (sapphire or free-standing GaN), that work like nonradiative centers, are a possible diffusion path for point defects, such as Mg and other vacancy.

This work was complemented by Marona et al. [103], who proposed a solution to the issue of the very narrow stripe width of a nitride laser diode, that makes having a reliable SIMS measurement impossible. The authors fabricated a laser diode with a very wide injection stripe and no mesa etching. The results of the test (repeated with two different SIMS systems) are reported in Figs. 7.27a and b, where it can be seen that there is no variation in the Mg concentration.

After this outcome, they decided to study the hydrogen profile after an aging test, and noticed no changes. Therefore they concluded that Mg, H, and Si impurity profiles are stable during the aging of the device. It is worth noting that in this paper the samples were different from the previous one, and this could be a possible reason for the differences in the results obtained in these two studies. For example, if we have a higher vacancy concentration in the last case, the magnesium which was

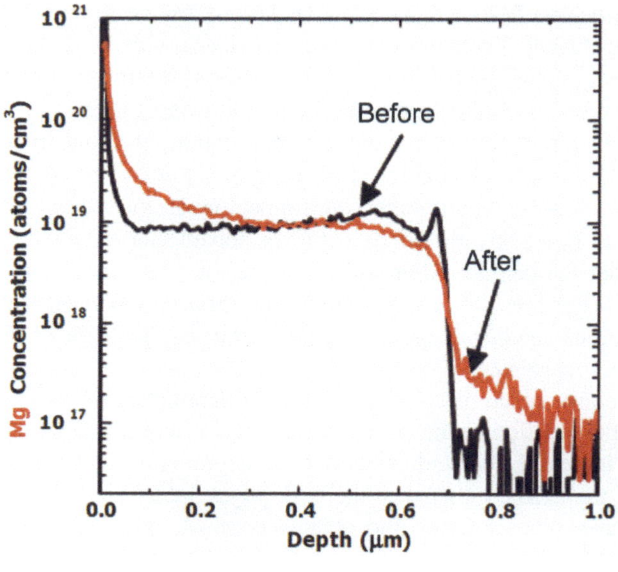

Figure 7.26 Mg-SIMS profile of LED samples.

Reprinted, with permission, from [51].

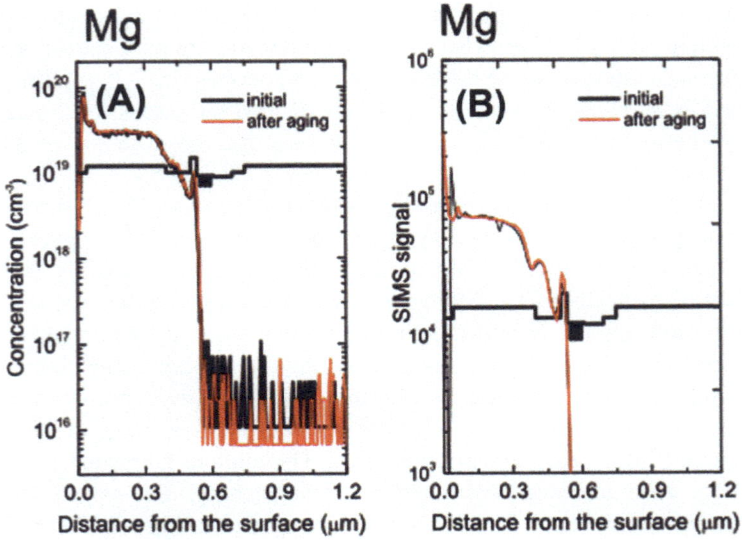

Figure 7.27 SIMS profile of Mg from different instruments: (A) Charles Evans Group Laboratory; (B) Institute of Physics Warsaw.

Reprinted from [103], with the permission of AIP Publishing.

detected as diffusing in the former one can be trapped by the vacancies, leading to a slower or not noticeable magnesium diffusion. Moreover, since the diffusing magnesium is located in an interstitial position, the difference could originate from better magnesium activation in the second case, leading to less or no interstitial magnesium available for diffusion. Therefore the last result does not exclude the possibility that both H^+ and Mg-related point defects may be responsible for the optical degradation of the devices based on GaN, as reported in the following papers.

Other papers report the interaction between magnesium and hydrogen, such as Reference [104], where Meneghini et al. supposed that the hydrogen incorporated in the passivation deposition process, during heating at 250°C, can interact with the LED surface, in order to form bonds with Mg dopant that reduce the acceptor concentration under the contact. Therefore if before the stress the behavior of the contact is ohmic, after the interactions Mg−H, the Schottky barrier at the metal/GaN interface increases, leading to an undesirable rectifying behavior of the contact, and a higher contact resistance.

Fig. 7.28 reports a summary of the electrical parameters of the LED, that shows an important degradation in the contact resistivity, whereas p-GaN resistance shows only a slight increase. This indicates that the degradation process mostly involves the surface of the LEDs, rather than the bulk p-type material.

Another paper that reported the influence of Mg in UV LEDs is Reference [105]. The authors report a comparison between the electrical characteristics (I-V) of two samples, one virgin and one stressed, in a very wide range of temperatures (Fig. 7.29). The stressed device presents lower drive voltage at low temperatures,

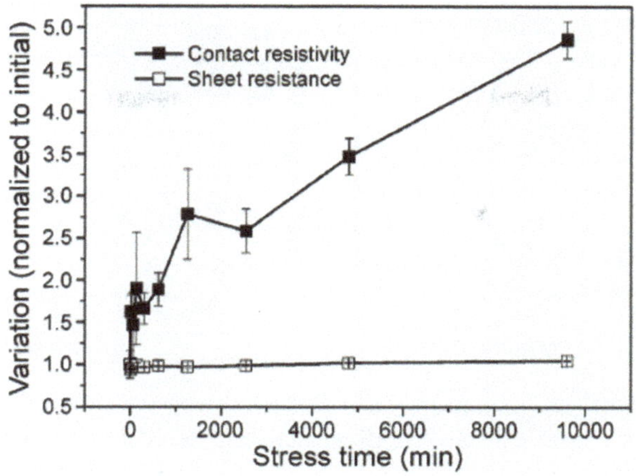

Figure 7.28 Variation of sheet resistance and contact resistivity before and during the stress at 250°C.

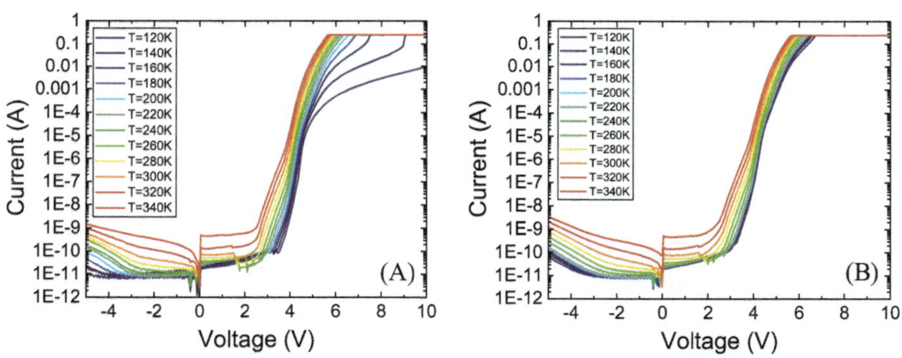

Figure 7.29 I-V characteristics in a wide temperature range of (A) an unstressed device and (B) a stressed device.

because aging induced the breaking of Mg−H bonds leading to a higher active hole concentration. This results in a higher number of free carriers, that causes a decrease in the resistivity of the p-type layers, which leads to a higher current at the same voltage and temperature.

In order to characterize the defects that appear during the stress, the authors performed capacitance DLTS (C-DLTS) measurements during the aging. In Fig. 7.30 they show the DLTS spectra with a reverse bias of −2 V and a filling pulse of 1.4 V. The plot shows two negative peaks that decrease in intensity as a consequence of stress. e_2 is located at the p-side and is related to majority-carriers trap levels, in this case holes. In Fig. 30B an additional peak e_4 with the same activation

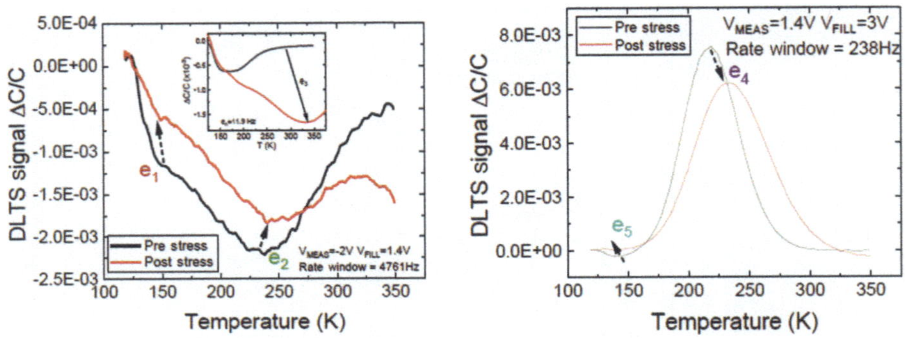

Figure 7.30 (Left) DLTS spectra before and after stress with $V_{MEAS} = -2$ V and $V_{FILL} = 1.4$ V. In the inset: extended DLTS measurement in order to detect peak e3. (Right) DLTS spectra before and after stress with $V_{MEAS} = 1.4$ V and $V_{FILL} = 3$ V.

energy of e_2 is shown. This suggest that, at a higher filling bias, it is possible to inject holes in the n-side, detecting the same level as e_2 but on the n-side. From the Arrhenius plots, the authors associated these peaks to Mg defects that decrease after stress, probably due to Mg dopant activation during aging.

They also found another peak, e_3, in the inset of Fig. 7.30 (left), that appeared after the stress; this peak is related to point defects or intrinsic defects.

7.3.3 Role of hydrogen

The use of hydrogen as a carrier gas during the growth of optoelectronic devices is a problem because hydrogen can interact with Mg; H is known to create Mg–H complexes, which may increase the resistivity of the material and behave as nonradiative recombination centers [15,19,106–108].

An extensive analysis of the role of hydrogen in the degradation of UV LEDs was done by Glaab et al. [109].

As reported in Fig. 7.31, the Mg concentration profile does not change significantly during 100 h of operation. The H profile follows roughly the Mg profile in the unaged sample, but after 10 h of stress it presents two peaks inside the active region, one near the EBL and one further towards the n-side. After 100 h of stress, the first peak migrates toward the second one, forming a single broad distribution. This indicates a shift of the hydrogen into the n-side during the stress, probably leaving behind charged point defects. These defects could act as nonradiative recombination centers and increase TAT and SRH phenomena, explaining the classic degradation mechanisms found in LEDs in previous works [84]. Moreover, this explains the increase in the p-conductivity and hole injection efficiency, due to the increase in Mg acceptors released by the dissociation of Mg–H complexes. Finally, the authors propose that H-containing defect complexes broke up by interaction with hot electrons generated by Auger recombination during operation of LEDs [110–112].

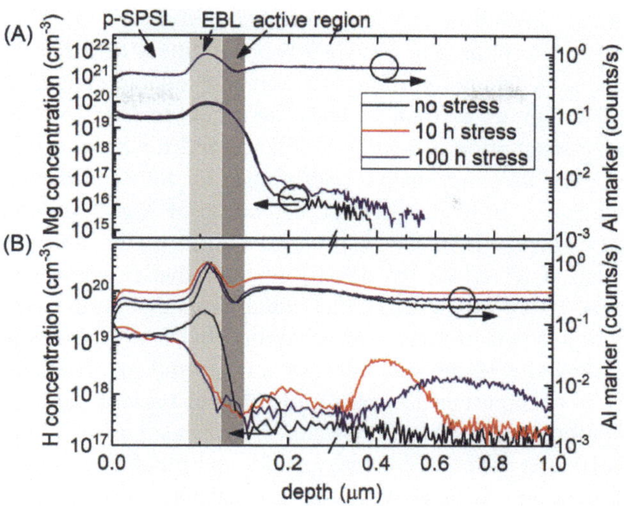

Figure 7.31 (A) Mg and (B) H concentration depth profiles at 0, 10, and 100 h as determined by SIMS.

The diffusion of hydrogen is confirmed also by Uedono et al. [113]. These authors carried out an extensive study of the role of impurities (such as Mg and H) and reported the presence of Ga-vacancies as major defects. The hydrogen atoms (for H-implanted samples) started to couple with vacancies above 800°C annealing, and the hydrogenation of vacancies enhanced vacancy clustering; instead, above 1000°C sub-nm-size voids were introduced. For Mg-implanted samples, vacancies start to agglomerate at 800°C annealing, introducing vacancy clusters like $(V_{Ga}V_N)_3$. Under illumination, these defects behave like electron trap centers. Introducing H, the clustering of vacancy type defects was suppressed, because there was formation of complexes between Mg atoms and H atoms. This technique shows an efficient way to hydrogenate the vacancies for optimizing the process of fabricating p-type GaN using Mg implantation.

7.4 Conclusions

In summary, we critically reviewed the main deep levels influencing the optical properties of LEDs. These defects have an impact not only on as-grown samples but also on the time-dependent degradation, due to thermally activated and current-driven defect generation processes or to photon-induced degradation. The root cause for the increase in defect concentration can be ascribed to several mechanisms, such as the local generation of point defects or the diffusion of impurities, such as magnesium and hydrogen. Moreover, all these concept are investigated in the case of UV emitters.

Improvements in crystal quality and dopant stability, as well as better control of foreign impurities, are important for GaN-based emitters to achieve better performance and reliability. Obtaining more efficient and reliable visible LEDs will help continue the LED lighting revolution, ensuring access to reliable light sources with reduced energy consumption to a wider audience and for more applications. In the case of UV emitters, there are many opportunities for future improvements in these structures. Finding a way to increase the activation of Mg as p-dopant, creating a better p-contact to favor the injection of carriers at the p-side, and reducing the diffusion of Mg and H to reduce the density of nonradiative centers are the biggest challenges for the future of UV LED, that could make them reliable and ready for a wider market. Better performance and reliability and lower cost could lead to the creation of a new series of devices, such as smaller and more effective UV sterilizers, which are very important for the sanitization of bacteria and viruses. Another relevant application, highly valued nowadays, is water purification, to permit the production of drinking water with cheap, small, and reliable systems in the regions of the world where an adequate supply is still an open issue. These are problems that we could resolve committing to the research and development of reliable UV LEDs.

References

[1] M.-H. Chang, D. Das, P.V. Varde, M. Pecht, Light emitting diodes reliability review, Microelectron. Reliab. 52 (5) (2012) 762–782. Available from: https://doi.org/10.1016/j.microrel.2011.07.063. May.

[2] M. Dal Lago, M. Meneghini, C. De Santi, M. Barbato, N. Trivellin, G. Meneghesso, et al., ESD on GaN-based LEDs: An analysis based on dynamic electroluminescence measurements and current waveforms, Microelectron. Reliab. 54 (9–10) (2014) 2138–2141. Available from: https://doi.org/10.1016/j.microrel.2014.07.122. Sep.

[3] C.M. Tsai, J.K. Sheu, P.T. Wang, W.C. Lai, S.C. Shei, S.J. Chang, et al., High efficiency and improved ESD characteristics of GaN-based LEDs with naturally textured surface grown by MOCVD, IEEE Photon. Technol. Lett. 18 (11) (2006) 1213–1215. Available from: https://doi.org/10.1109/LPT.2006.875063. Jun.

[4] S.-L. Chen, Enhanced electrostatic discharge reliability in GaN-based light-emitting diodes by the electrode engineering, J. Disp. Technol. 10 (10) (2014) 807–813. Available from: https://doi.org/10.1109/JDT.2014.2321460. Oct.

[5] M. Buffolo, M. Meneghini, A. Munaretto, C. De Santi, G. Meneghesso, E. Zanoni, Failure of high power LEDs submitted to EOS: dependence on device layout and pulse properties, IEEE Trans. Device Mater. Reliab. 17 (1) (2017) 191–196. Available from: https://doi.org/10.1109/TDMR.2016.2642167. Mar.

[6] N. Renso, M. Buffolo, C. De Santi, M. Ronzani, G. Meneghesso, E. Zanoni, et al., Failure limits and electro-optical characteristics of GaN-based LEDs under electrical overstress, Microelectron. Reliab. 88–90 (2018) 887–890. Available from: https://doi.org/10.1016/j.microrel.2018.06.054. Sep.

[7] N. Renso, M. Meneghini, M. Buffolo, C. De Santi, G. Meneghesso, E. Zanoni, Understanding the degradation processes of GaN based LEDs submitted to extremely

high current density, Microelectron. Reliab. 76–77 (2017) 556–560. Available from: https://doi.org/10.1016/j.microrel.2017.06.044. Sep.

[8] C. De Santi, M. Dal Lago, M. Buffolo, D. Monti, M. Meneghini, G. Meneghesso, et al., Failure causes and mechanisms of retrofit LED lamps, Microelectron. Reliab. 55 (9–10) (2015) 1765–1769. Available from: https://doi.org/10.1016/j.microrel.2015.06.080. Aug.

[9] W.D. van Driel, C.A. Yuan, S. Koh, and G.Q. Zhang, LED system reliability, in 2011 12th Intl. Conf. on Thermal, Mechanical & Multi-Physics Simulation and Experiments in Microelectronics and Microsystems, 2011, pp. 1/5-5/5. Available from: https://doi.org/10.1109/ESIME.2011.5765848.

[10] S. Lan, C.M. Tan, K. Wu, Reliability study of LED driver – a case study of black box testing, Microelectron. Reliab. 52 (9–10) (2012) 1940–1944. Available from: https://doi.org/10.1016/j.microrel.2012.06.023. Sep.

[11] S.J. Rosner, E.C. Carr, M.J. Ludowise, G. Girolami, H.I. Erikson, Correlation of cathodoluminescence inhomogeneity with microstructural defects in epitaxial GaN grown by metalorganic chemical-vapor deposition, Appl. Phys. Lett. 70 (4) (1997) 420–422. Available from: https://doi.org/10.1063/1.118322. Jan.

[12] T. Sugahara, H. Sato, M. Hao, Y. Naoi, S. Kurai, S. Tottori, et al., Direct evidence that dislocations are non-radiative recombination centers in GaN, Jpn. J. Appl. Phys. 37 (4A) (1998) L398–L400. Available from: https://doi.org/10.1143/JJAP.37.L398. Apr.

[13] D. Cherns, S.J. Henley, F.A. Ponce, Edge and screw dislocations as nonradiative centers in InGaN/GaN quantum well luminescence, Appl. Phys. Lett. 78 (18) (2001) 2691–2693. Available from: https://doi.org/10.1063/1.1369610. Apr.

[14] Q. Dai, M.F. Schubert, M.H. Kim, J.K. Kim, E.F. Schubert, D.D. Koleske, et al., Internal quantum efficiency and nonradiative recombination coefficient of GaInN/GaN multiple quantum wells with different dislocation densities, Appl. Phys. Lett. 94 (11) (2009) 111109. Available from: https://doi.org/10.1063/1.3100773. Mar.

[15] C.G. Van de Walle, J. Neugebauer, First-principles calculations for defects and impurities: applications to III-nitrides, J. Appl. Phys. 95 (8) (2004) 3851–3879. Available from: https://doi.org/10.1063/1.1682673. Apr.

[16] J. Neugebauer, C.G. Van de Walle, Atomic geometry and electronic structure of native defects in GaN, Phys. Rev. B 50 (11) (1994) 8067–8070. Available from: https://doi.org/10.1103/PhysRevB.50.8067. Sep.

[17] K. Saarinen, T. Laine, S. Kuisma, J. Nissilä, P. Hautojärvi, L. Dobrzynski, et al., Observation of native Ga vacancies in GaN by positron annihilation, Phys. Rev. Lett. 79 (16) (1997) 3030–3033. Available from: https://doi.org/10.1103/PhysRevLett.79.3030. Oct.

[18] J.L. Lyons, A. Alkauskas, A. Janotti, C.G. Van de Walle, First-principles theory of acceptors in nitride semiconductors, Phys. Status Solidi B 252 (5) (2015) 900–908. Available from: https://doi.org/10.1002/pssb.201552062. May.

[19] C.E. Dreyer, A. Alkauskas, J.L. Lyons, J.S. Speck, C.G. Van de Walle, Gallium vacancy complexes as a cause of Shockley-Read-Hall recombination in III-nitride light emitters, Appl. Phys. Lett. 108 (14) (2016) 141101. Available from: https://doi.org/10.1063/1.4942674. Apr.

[20] S. Brochen, J. Brault, S. Chenot, A. Dussaigne, M. Leroux, B. Damilano, Dependence of the Mg-related acceptor ionization energy with the acceptor concentration in p-type GaN layers grown by molecular beam epitaxy, Appl. Phys. Lett. 103 (3) (2013) 032102. Available from: https://doi.org/10.1063/1.4813598. Jul.

[21] G. Sabui, P.J. Parbrook, M. Arredondo-Arechavala, Z.J. Shen, Modeling and simulation of bulk gallium nitride power semiconductor devices, AIP Adv. 6 (5) (2016) 055006. Available from: https://doi.org/10.1063/1.4948794. May.

[22] B. Podor, Thermal ionization energy of Mg acceptors in GaN: effects of doping level and compensation, Proc. SPIE Int. Soc. Opt. Eng. (2001) 299–303. Available from: https://doi.org/10.1117/12.435848.

[23] A.Y. Polyakov, I. Lee, Deep traps in GaN-based structures as affecting the performance of GaN devices, Mater. Sci. Eng. R. 94 (2015) 1–56. Available from: https://doi.org/10.1016/j.mser.2015.05.001.

[24] J.L. Lyons, A. Janotti, C.G. Van de Walle, Carbon impurities and the yellow luminescence in GaN, Appl. Phys. Lett. 97 (15) (2010) 152108. Available from: https://doi.org/10.1063/1.3492841. Oct.

[25] J. Neugebauer, C.G. Van De Walle, Role of hydrogen in doping of GaN, Appl. Phys. Lett. 68 (13) (1996) 1829–1831. Available from: https://doi.org/10.1063/1.116027.

[26] J. Neugebauer, C.G. Van de Walle, Gallium vacancies and the yellow luminescence in GaN, Appl. Phys. Lett. 69 (4) (1996) 503–505. Available from: https://doi.org/10.1063/1.117767. Jul.

[27] I. Gorczyca, A. Svane, N.E. Christensen, Mg-O and Mg − V N defect complexes in cubic GaN, Phys. Rev. B 61 (11) (2000) 7494–7498. Available from: https://doi.org/10.1103/PhysRevB.61.7494. Mar.

[28] C.G. Van de Walle, Interactions of hydrogen with native defects in GaN, Phys. Rev. B 56 (16) (1997) R10020–R10023. Available from: https://doi.org/10.1103/PhysRevB.56.R10020. Oct.

[29] M.A. Reshchikov, H. Morkoç, Luminescence properties of defects in GaN, J. Appl. Phys. 97 (6) (2005) 061301. Available from: https://doi.org/10.1063/1.1868059. Mar.

[30] R. Singh, D. Doppalapudi, T.D. Moustakas, L.T. Romano, Phase separation in InGaN thick films and formation of InGaN/GaN double heterostructures in the entire alloy composition, Appl. Phys. Lett. 70 (9) (1997) 1089–1091. Available from: https://doi.org/10.1063/1.118493. Mar.

[31] T. Hino, S. Tomiya, T. Miyajima, K. Yanashima, S. Hashimoto, M. Ikeda, Characterization of threading dislocations in GaN epitaxial layers, Appl. Phys. Lett. 76 (23) (2000) 3421–3423. Available from: https://doi.org/10.1063/1.126666. Jun.

[32] A. Armstrong, T.A. Henry, D.D. Koleske, M.H. Crawford, K.R. Westlake, S.R. Lee, Dependence of radiative efficiency and deep level defect incorporation on threading dislocation density for InGaN/GaN light emitting diodes, Appl. Phys. Lett. 101 (16) (2012) 162102. Available from: https://doi.org/10.1063/1.4759003. Oct.

[33] M. Musolino, D. van Treeck, A. Tahraoui, L. Scarparo, C. De Santi, M. Meneghini, et al., A physical model for the reverse leakage current in (In, Ga)N/GaN light-emitting diodes based on nanowires, J. Appl. Phys. 119 (4) (2016) 044502. Available from: https://doi.org/10.1063/1.4940949. Jan.

[34] J. Kim, J. Kim, Y. Tak, S. Chae, J.-Y. Kim, Y. Park, Effect of V-Shaped Pit Size on the Reverse Leakage Current of InGaN/GaN Light-Emitting Diodes, IEEE Electron. Device Lett. 34 (11) (2013) 1409–1411. Available from: https://doi.org/10.1109/LED.2013.2280017. Nov.

[35] S. Limpijumnong, C. Van de Walle, Diffusivity of native defects in GaN, Phys. Rev. B 69 (3) (2004) 035207. Available from: https://doi.org/10.1103/PhysRevB.69.035207. Jan.

[36] M. Meneghini, C. de Santi, N. Trivellin, K. Orita, S. Takigawa, T. Tanaka, et al., Investigation of the deep level involved in InGaN laser degradation by deep level

transient spectroscopy, Appl. Phys. Lett. 99 (9) (2011) 093506. Available from: https://doi.org/10.1063/1.3626280. Aug.

[37] H.M. Chung, W.C. Chuang, Y.C. Pan, C.C. Tsai, M.C. Lee, W.H. Chen, et al., Electrical characterization of isoelectronic In-doping effects in GaN films grown by metalorganic vapor phase epitaxy, Appl. Phys. Lett. 76 (7) (2000) 897–899. Available from: https://doi.org/10.1063/1.125622. Feb.

[38] R. Made, Y. Gao, G.J. Syaranamual, W.A. Sasangka, L. Zhang, X.S. Nguyen, et al., Characterisation of defects generated during constant current InGaN-on-silicon LED operation, Microelectron. Reliab. 76–77 (2017) 561–565. Available from: https://doi.org/10.1016/j.microrel.2017.07.072. Sep.

[39] Z.-Q. Fang, D.C. Look, W. Kim, Z. Fan, A. Botchkarev, H. Morkoç, Deep centers in n-GaN grown by reactive molecular beam epitaxy, Appl. Phys. Lett. 72 (18) (1998) 2277–2279. Available from: https://doi.org/10.1063/1.121274. May.

[40] G.A. Umana-Membreno, J.M. Dell, T.P. Hessler, B.D. Nener, G. Parish, L. Faraone, et al., 60Co gamma-irradiation-induced defects in n-GaN, Appl. Phys. Lett. 80 (23) (2002) 4354–4356. Available from: https://doi.org/10.1063/1.1483390. Jun.

[41] K. Orita, M. Meneghini, H. Ohno, N. Trivellin, N. Ikedo, S. Takigawa, et al., Analysis of diffusion-related gradual degradation of InGaN-based laser diodes, IEEE J. Quantum Electron. 48 (9) (2012) 1169–1176. Available from: https://doi.org/10.1109/JQE.2012.2203795. Sep.

[42] L. Sugiura, Comparison of degradation caused by dislocation motion in compound semiconductor light-emitting devices, Appl. Phys. Lett. 70 (10) (1997) 1317–1319. Available from: https://doi.org/10.1063/1.118523. Mar.

[43] X.A. Cao, P.M. Sandvik, S.F. LeBoeuf, S.D. Arthur, Defect generation in InGaN/GaN light-emitting diodes under forward and reverse electrical stresses, Microelectron. Reliab. 43 (12) (2003) 1987–1991. Available from: https://doi.org/10.1016/J.MICROREL.2003.06.001. Dec.

[44] C. De Santi, M. Meneghini, M. Buffolo, G. Meneghesso, E. Zanoni, Experimental demonstration of time-dependent breakdown in GaN-based light emitting diodes, IEEE Electron. Device Lett. 37 (5) (2016) 611–614. Available from: https://doi.org/10.1109/LED.2016.2543805. May.

[45] C. De Santi, M. Meneghini, N. Renso, M. Buffolo, N. Trivellin, G. Mura, et al., Investigation of the time-dependent failure of InGaN-based LEDs submitted to reverse-bias stress, vol. 10124, 2017, p. 101240F. Available from: https://doi.org/10.1117/12.2256023

[46] M.W. Moseley, A.A. Allerman, M.H. Crawford, J.J. Wierer, M.L. Smith, A.M. Armstrong, Detection and modeling of leakage current in AlGaN-based deep ultraviolet light-emitting diodes, J. Appl. Phys. 117 (9) (2015) 095301. Available from: https://doi.org/10.1063/1.4908543. Mar.

[47] Z. Chen, N. Fichtenbaum, D. Brown, S. Keller, U.K. Mishra, S.P. Denbaars, et al., Influence of Mg doping on the morphological, optical, and structural properties of InGaN/GaN multiple quantum wells, J. Electron. Mater. 37 (5) (2008) 546–549. Available from: https://doi.org/10.1007/s11664-007-0332-0. May.

[48] K. Köhler, T. Stephan, A. Perona, J. Wiegert, M. Maier, M. Kunzer, et al., Control of the Mg doping profile in III-N light-emitting diodes and its effect on the electroluminescence efficiency, J. Appl. Phys. 97 (10) (2005). Available from: https://doi.org/10.1063/1.1901836.

[49] C. De Santi, M. Meneghini, M. Marioli, M. Buffolo, N. Trivellin, T. Weig, et al., Thermally-activated degradation of InGaN-based laser diodes: effect on threshold current and forward voltage, Microelectron. Reliab. 54 (9–10) (2014) 2147–2150. Available from: https://doi.org/10.1016/j.microrel.2014.07.073. Sep.

[50] M. Takeya, T. Mizuno, T. Sasaki, S. Ikeda, T. Fujimoto, Y. Ohfuji, et al., Degradation in AlGaInN lasers, Phys. Status Solidi C 0 (7) (2003) 2292–2295. Available from: https://doi.org/10.1002/pssc.200303324. Dec.
[51] O.H. Nam, K.H. Ha, J.S. Kwak, S.N. Lee, K.K. Choi, T.H. Chang, et al., Characteristics of GaN-based laser diodes for post-DVD applications, Phys. Status Solidi A 201 (12) (2004) 2717–2720. Available from: https://doi.org/10.1002/pssa.200405114.
[52] P. Perlin, Ł. Marona, P. Wisniewski, M. Leszczynski, P. Prystawko, M. Boćkowski, et al., InGaN laser diode degradation. Surface and bulk processes, MRS Proc. 1195 (2009). Available from: https://doi.org/10.1557/PROC-1195-B01-04. 1195-B01-04, Jan.
[53] K. Harafuji, K. Kawamura, Magnesium diffusion at dislocation in wurtzite-type GaN crystal, Jpn. J. Appl. Phys. Part 1 Regul. Pap. Short. Notes Rev. Pap. 44 (9) (2005) 6495–6504. Available from: https://doi.org/10.1143/JJAP.44.6495. A.
[54] M. La Grassa, M. Meneghini, C. De Santi, M. Mandurrino, M. Goano, F. Bertazzi, et al., Ageing of InGaN-based LEDs: effects on internal quantum efficiency and role of defects, Microelectron. Reliab. 55 (9–10) (2015) 1775–1778. Available from: https://doi.org/10.1016/j.microrel.2015.06.103. Aug.
[55] M. La Grassa, M. Meneghini, C. De Santi, E. Zanoni, G. Meneghesso, Degradation of InGaN-based LEDs related to charge diffusion and build-up, Microelectron. Reliab. 64 (2016) 614–616. Available from: https://doi.org/10.1016/j.microrel.2016.07.131.
[56] V. Kümmler, A. Lell, V. Härle, U.T. Schwarz, T. Schoedl, W. Wegscheider, Gradual facet degradation of (Al, In)GaN quantum well lasers, Appl. Phys. Lett. 84 (16) (2004) 2989–2991. Available from: https://doi.org/10.1063/1.1704861. Apr.
[57] L. Marona, M. Sarzynski, P. Wiśniewski, M. Leszczyński, P. Prystawko, I. Grzegory, et al., Comprehensive study of reliability of InGaN-based laser diodes, Proc. SPIE Int. Soc. Opt. Eng. 6485 (2007) 648504. Available from: https://doi.org/10.1117/12.699001.
[58] S. Tomiya, O. Goto, M. Ikeda, Structural defects and degradation of high-power pure-blue GaN-based laser diodes, Proc. SPIE Int. Soc. Opt. Eng. 6894 (2008) 68940N. Available from: https://doi.org/10.1117/12.767769.
[59] P.G. Eliseev, H.-B. Sun, S. Juodkazis, T. Sugahara, S. Sakai, H. Misawa, Laser-induced damage threshold and processing of GaN at 400 nm wavelength, Jpn. J. Appl. Phys. 38 (Part 2) (1999) L839–L841. Available from: https://doi.org/10.1143/JJAP.38.L839. Jul.
[60] S. Elhadj, J. Yoo, R.A. Negres, M.G. Menor, J.J. Adams, N. Shen, et al., Optical damage performance of conductive widegap semiconductors: spatial, temporal, and lifetime modeling, Opt. Mater. Express 7 (1) (2017) 202. Available from: https://doi.org/10.1364/OME.7.000202. Jan.
[61] J.-H. Yoo, M.G. Menor, J.J. Adams, R.N. Raman, J.R.I. Lee, T.Y. Olson, et al., Laser damage mechanisms in conductive widegap semiconductor films, Opt. Express 24 (16) (2016) 17616. Available from: https://doi.org/10.1364/oe.24.017616. Aug.
[62] M. Ščiuka, M. Dmukauskas, T. Grinys, A. Melninkaitis, Optical resistance of GaN and InGaN thin films, Proc. SPIE Int. Soc. Opt. Eng. 8530 (2012) 85300Y. Available from: https://doi.org/10.1117/12.977451.
[63] C. De Santi, A. Caria, N. Renso, E. Dogmus, M. Zegaoui, F. Medjdoub, et al., Evidence of optically induced degradation in gallium nitride optoelectronic devices, Appl. Phys. Express 11 (11) (2018) 111002. Available from: https://doi.org/10.7567/APEX.11.111002. Nov.
[64] E. Dogmus, M. Zegaoui, L. Largeau, M. Tchernycheva, V. Neplokh, S. Weiszer, et al., High structural quality InGaN/GaN multiple quantum well solar cells, Phys. Status

Solidi C 12 (12) (2015) 1412–1415. Available from: https://doi.org/10.1002/pssc.201510137. Dec.

[65] C. De Santi, M. Meneghini, A. Caria, E. Dogmus, M. Zegaoui, F. Medjdoub, et al., Degradation of InGaN-based MQW solar cells under 405 nm laser excitation, Microelectron. Reliab. 76–77 (2017) 575–578. Available from: https://doi.org/10.1016/J.MICROREL.2017.06.072. Sep.

[66] C. De Santi, M. Meneghini, A. Caria, E. Dogmus, M. Zegaoui, F. Medjdoub, et al., GaN-based laser wireless power transfer system, Materials 11 (1) (2018) 153. Available from: https://doi.org/10.3390/ma11010153. Jan.

[67] S. Porowski, Growth and properties of single crystalline GaN substrates and homoepitaxial layers, Mater. Sci. Eng. B 44 (1–3) (1997) 407–413. Available from: https://doi.org/10.1016/S0921-5107(96)01730-8. Feb.

[68] P. Kamyczek, E. Placzek-Popko, V. Kolkovsky, S. Grzanka, R. Czernecki, A deep acceptor defect responsible for the yellow luminescence in GaN and AlGaN, J. Appl. Phys. 111 (11) (2012) 113105. Available from: https://doi.org/10.1063/1.4725484. Jun.

[69] J.L. Lyons, A. Janotti, C.G. Van de Walle, Effects of carbon on the electrical and optical properties of InN, GaN, and AlN, Phys. Rev. B 89 (3) (2014) 035204. Available from: https://doi.org/10.1103/PhysRevB.89.035204. Jan.

[70] M. Yazdan Mehr, A. Bahrami, W.D. van Driel, X.J. Fan, J.L. Davis, G.Q. Zhang, Degradation of optical materials in solid-state lighting systems, Int. Mater. Rev. 65 (2) (2020) 102–128. Available from: https://doi.org/10.1080/09506608.2019.1565716. Feb.

[71] N. Trivellin, M. Yushchenko, M. Buffolo, C. De Santi, M. Meneghini, G. Meneghesso, et al., Laser-based lighting: experimental analysis and perspectives, Materials 10 (10) (2017) 1166. Available from: https://doi.org/10.3390/ma10101166. Oct.

[72] M. Buffolo, C. De Santi, M. Albertini, D. Carbonera, G. Rizzi, G. Granozzi, et al., Reliability of blue-emitting Eu2 + -doped phosphors for laser-lighting applications, Materials 11 (9) (2018) 1552. Available from: https://doi.org/10.3390/ma11091552. Aug.

[73] A. Endruweit, M.S. Johnson, A.C. Long, Curing of composite components by ultraviolet radiation: a review, Polym. Compos. 27 (2) (2006) 119–128. Available from: https://doi.org/10.1002/pc.20166. Apr.

[74] H. Hönigsmann, History of phototherapy in dermatology, Photochem. Photobiol. Sci. 12 (1) (2013) 16–21. Available from: https://doi.org/10.1039/C2PP25120E. Dec.

[75] M. Schreiner, J. Martínez-Abaigar, J. Glaab, M. Jansen, UV-B induced secondary plant metabolites, Opt. Photonik 9 (2) (2014) 34–37. Available from: https://doi.org/10.1002/opph.201400048. May.

[76] M.A. Würtele, T. Kolbe, M. Lipsz, A. Külberg, M. Weyers, M. Kneissl, et al., Application of GaN-based ultraviolet-C light emitting diodes – UV LEDs – for water disinfection, Water Res. 45 (3) (2011) 1481–1489. Available from: https://doi.org/10.1016/j.watres.2010.11.015. Jan.

[77] A. Žukauskas, N. Kurilčik, P. Vitta, S. Juršėnas, E. Bakienė, R. Gaska, Optimization of a UV light-emitting diode based fluorescence-phase sensor, Proc. SPIE Int. Soc. Opt. Eng. (2006) 63980Y. Available from: https://doi.org/10.1117/12.689907.

[78] M.S. Shur, R. Gaska, Deep-ultraviolet light-emitting diodes, IEEE Trans. Electron. Devices 57 (1) (2010) 12–25. Available from: https://doi.org/10.1109/TED.2009.2033768. Jan.

[79] M. Meneghini, D. Barbisan, L. Rodighiero, G. Meneghesso, E. Zanoni, Analysis of the physical processes responsible for the degradation of deep-ultraviolet light emitting

diodes, Appl. Phys. Lett. 97 (14) (2010) 143506. Available from: https://doi.org/10.1063/1.3497082. Oct.

[80] N.L. Ploch, Chip designs for high efficiency III-nitride based ultraviolet light emitting diodes with enhanced light extraction, Mater. Sci. (2015). Available from: https://doi.org/10.14279/depositonce-4678. no. July.

[81] A. Pinos, S. Marcinkevičius, J. Yang, R. Gaska, M. Shatalov, M.S. Shur, Optical studies of degradation of AlGaN quantum well based deep ultraviolet light emitting diodes, J. Appl. Phys. 108 (9) (2010) 093113. Available from: https://doi.org/10.1063/1.3506697. Nov.

[82] J. Glaab, J. Haefke, J. Ruschel, M. Brendel, J. Rass, T. Kolbe, et al., Degradation effects of the active region in UV-C light-emitting diodes, J. Appl. Phys. 123 (10) (2018). Available from: https://doi.org/10.1063/1.5012608.

[83] C. De Santi, M. Meneghini, G. Meneghesso, E. Zanoni, Degradation of InGaN laser diodes caused by temperature- and current-driven diffusion processes, Microelectron. Reliab. 64 (2016) 623–626. Available from: https://doi.org/10.1016/J.MICROREL.2016.07.118. Sep.

[84] D. Monti, M. Meneghini, C. De Santi, G. Meneghesso, E. Zanoni, J. Glaab, et al., Defect-related degradation of AlGaN-based UV-B LEDs, IEEE Trans. Electron. Devices 64 (1) (2017) 200–205. Available from: https://doi.org/10.1109/TED.2016.2631720. Jan.

[85] J. Glaab, C. Ploch, R. Kelz, C. Stölmacker, M. Lapeyrade, N.L. Ploch, et al., Temperature induced degradation of InAlGaN multiple-quantum well UV-B LEDs, MRS Proc. 1792 (2015). Available from: https://doi.org/10.1557/opl.2015.446. no. August.

[86] M. Mandurrino, G. Verzellesi, M. Goano, M. Vallone, F. Bertazzi, G. Ghione, et al., Physics-based modeling and experimental implications of trap-assisted tunneling in InGaN/GaN light-emitting diodes, Phys. Status Solidi A 212 (5) (2015) 947–953. Available from: https://doi.org/10.1002/pssa.201431743.

[87] M. Auf Der Maur, B. Galler, I. Pietzonka, M. Strassburg, H. Lugauer, A. Di Carlo, Trap-assisted tunneling in InGaN/GaN single-quantum-well light-emitting diodes, Appl. Phys. Lett. 105 (13) (2014). Available from: https://doi.org/10.1063/1.4896970.

[88] D. Monti, C. De Santi, S. Da Ruos, F. Piva, J. Glaab, J. Rass, et al., High-Current Stress of UV-B (In)AlGaN-based LEDs: defect-generation and diffusion processes, IEEE Trans. Electron. Devices 66 (8) (2019) 3387–3392. Available from: https://doi.org/10.1109/TED.2019.2920521. Aug.

[89] N. Trivellin, D. Monti, F. Piva, M. Buffolo, C. De Santi, E. Zanoni, et al., Degradation processes of 280 nm high power DUV LEDs: impact on parasitic luminescence, Jpn. J. Appl. Phys. 58 (2019) SCCC19. Available from: https://doi.org/10.7567/1347-4065/ab1393. no. SC.

[90] C. Stampfl, C.G. Van de Walle, Theoretical investigation of native defects, impurities, and complexes in aluminum nitride, Phys. Rev. B 65 (15) (2002) 155212. Available from: https://doi.org/10.1103/PhysRevB.65.155212. Apr.

[91] V. Adivarahan, S. Wu, W.H. Sun, V. Mandavilli, M.S. Shatalov, G. Simin, et al., High-power deep ultraviolet light-emitting diodes basedon a micro-pixel design, Appl. Phys. Lett. 85 (10) (2004) 1838–1840. Available from: https://doi.org/10.1063/1.1784882. Sep.

[92] M. Shatalov, A. Chitnis, V. Mandavilli, R. Pachipulusu, J.P. Zhang, V. Adivarahan, et al., Time-resolved electroluminescence of AlGaN-based light-emitting diodes with emission at 285 nm, Appl. Phys. Lett. 82 (2) (2003) 167–169. Available from: https://doi.org/10.1063/1.1536729. Jan.

[93] N. Trivellin, D. Monti, C. De Santi, M. Buffolo, G. Meneghesso, E. Zanoni, et al., Current induced degradation study on state of the art DUV LEDs, Microelectron.

Reliab. 88−90 (2018) 868−872. Available from: https://doi.org/10.1016/j.microrel.2018.07.145. Sep.

[94] D.V. Lang, Deep-level transient spectroscopy: a new method to characterize traps in semiconductors, J. Appl. Phys. 45 (7) (1974) 3023−3032. Available from: https://doi.org/10.1063/1.1663719. Jul.

[95] C. de Santi, M. Meneghini, S. Carraro, S. Vaccari, N. Trivellin, S. Marconi, et al., Variations in junction capacitance and doping activation associated with electrical stress of InGaN/GaN laser diodes, Microelectron. Reliab. 53 (9−11) (2013) 1534−1537. Available from: https://doi.org/10.1016/j.microrel.2013.07.053. Sep.

[96] M. Buffolo, C. De Santi, M. Meneghini, D. Rigon, G. Meneghesso, E. Zanoni, Long-term degradation mechanisms of mid-power LEDs for lighting applications, Microelectron. Reliab. 55 (9−10) (2015) 1754−1758. Available from: https://doi.org/10.1016/j.microrel.2015.06.098. Aug.

[97] D. Monti, M. Meneghini, C. De Santi, G. Meneghesso, E. Zanoni, Degradation of UV-A LEDs: physical origin and dependence on stress conditions, IEEE Trans. Device Mater. Reliab. 16 (2) (2016) 213−219. Available from: https://doi.org/10.1109/TDMR.2016.2558473. Jun.

[98] C.-Y. Hsu, W.-H. Lan, Y.S. Wu, Effect of thermal annealing of Ni/Au ohmic contact on the leakage current of GaN based light emitting diodes, Appl. Phys. Lett. 83 (12) (2003) 2447−2449. Available from: https://doi.org/10.1063/1.1601306. Sep.

[99] H. Amano, M. Kito, K. Hiramatsu, I. Akasaki, P-Type cconduction in Mg-doped GaN treated with low-energy electron beam irradiation (LEEBI), Jpn. J. Appl. Phys. 28 (12) (1989) L2112−L2114. Available from: https://doi.org/10.1143/JJAP.28.L2112. Dec.

[100] S. Nakamura, T. Mukai, M. Senoh, N. Iwasa, Thermal annealing effects on P-type Mg-doped GaN films, Jpn. J. Appl. Phys. 31 (Part 2) (1992) L139−L142. Available from: https://doi.org/10.1143/JJAP.31.L139. Feb.

[101] S.N. Lee, H.S. Paek, J.K. Son, H. Kim, K.K. Kim, K.H. Ha, et al., Effects of Mg dopant on the degradation of InGaN multiple quantum wells in AlInGaN-based light emitting devices, J. Electroceram. 23 (2−4) (2009) 406−409. Available from: https://doi.org/10.1007/s10832-008-9478-2.

[102] O.H. Nam, K.H. Ha, J.S. Kwak, S.N. Lee, K.K. Choi, T.H. Chang, et al., Recent progress of high power GaN-based violet laser diodes, Phys. Status Solidi C 2282 (7) (2003) 2278−2282. Available from: https://doi.org/10.1002/pssc.200303334.

[103] L. Marona, P. Perlin, R. Czernecki, M. Leszczyński, M. Boćkowski, R. Jakiela, et al., Secondary ions mass spectroscopy measurements of dopant impurities in highly stressed InGaN laser diodes, Appl. Phys. Lett. 98 (24) (2011) 1−4. Available from: https://doi.org/10.1063/1.3600338.

[104] M. Meneghini, L.-R. Trevisanello, U. Zehnder, G. Meneghesso, E. Zanoni, Reversible degradation of Ohmic contacts on p-GaN for application in high-brightness LEDs, IEEE Trans. Electron. Devices 54 (12) (2007) 3245−3251. Available from: https://doi.org/10.1109/TED.2007.908900.

[105] F. Piva, C. De Santi, M. Deki, M. Kushimoto, H. Amano, H. Tomozawa, et al., Stability and degradation of AlGaN-based UV-B LEDs: role of doping and semiconductor defects, Microelectron. Reliab. 100−101 (2019) 113418. Available from: https://doi.org/10.1016/j.microrel.2019.113418. Sep.

[106] S. Nakamura, N. Iwasa, M. Senoh, T. Mukai, Hole compensation mechanism of P-type GaN films, Jpn. J. Appl. Phys. 31 (Part 1) (1992) 1258−1266. Available from: https://doi.org/10.1143/JJAP.31.1258. May.

[107] S.F. Chichibu, A. Uedono, K. Kojima, H. Ikeda, K. Fujito, S. Takashima, et al., The origins and properties of intrinsic nonradiative recombination centers in wide bandgap GaN and AlGaN, J. Appl. Phys. 123 (16) (2018) 161413. Available from: https://doi.org/10.1063/1.5012994. Apr.

[108] A. Alkauskas, C.E. Dreyer, J.L. Lyons, C.G. Van de Walle, Role of excited states in Shockley-Read-Hall recombination in wide-band-gap semiconductors, Phys. Rev. B 93 (20) (2016) 201304. Available from: https://doi.org/10.1103/PhysRevB.93.201304. May.

[109] J. Glaab, J. Ruschel, T. Kolbe, A. Knauer, J. Rass, H.K. Cho, et al., Degradation of (In)AlGaN-based UVB LEDs and migration of hydrogen, IEEE Photon. Technol. Lett. 31 (7) (2019) 529–532. Available from: https://doi.org/10.1109/LPT.2019.2900156.

[110] K.K. Leung, W.K. Fong, P.K.L. Chan, C. Surya, Physical mechanisms for hot-electron degradation in GaN light-emitting diodes, J. Appl. Phys. 107 (7) (2010) 073103. Available from: https://doi.org/10.1063/1.3357312. Apr.

[111] J. Iveland, L. Martinelli, J. Peretti, J.S. Speck, C. Weisbuch, Direct Measurement of Auger electrons emitted from a semiconductor light-emitting diode under electrical injection: identification of the dominant mechanism for efficiency droop, Phys. Rev. Lett. 110 (17) (2013) 177406. Available from: https://doi.org/10.1103/PhysRevLett.110.177406. Apr.

[112] M. Meneghini, L.-R. Trevisanello, G. Meneghesso, E. Zanoni, A review on the reliability of GaN-based LEDs, IEEE Trans. Device Mater. Reliab. 8 (2) (2008) 323–331. Available from: https://doi.org/10.1109/TDMR.2008.921527. Jun.

[113] A. Uedono, H. Iguchi, T. Narita, K. Kataoka, W. Egger, T. Koschine, et al., Annealing behavior of vacancy-type defects in Mg- and H-implanted GaN studied using monoenergetic positron beams, Phys. Status Solidi B (2019) 1900104. Available from: https://doi.org/10.1002/pssb.201900104. May.

Index

Note: Page numbers followed by "*f*" and "*t*" refer to figures and tables, respectively.

A

Absorption coefficient, 14*f*
Accelerated aging, 53–54
  with temperature and bias, 55
Antiphase boundaries (APBs), 246–248
Apple Face ID, 3, 21
Aspect ratio trapping (ART), 150–152
Avalanche photodiodes (APDs), 24

B

Band gap, 13–14
Bathtub curve, 76
Beam divergence, 31–32
Blue laser reliability, 212–213
Bulk degradation, 212
Burgers vector, 123–128, 123*f*, 184–187, 202
Burn-in, 31, 54, 76–77, 96–98

C

Capacitance deep-level transient spectroscopy (C-DLTS), 244
Capacitance DLTS (C-DLTS) measurements, 301–302
Carrier confinement, 13–15
Case studies, 89
  case study where no formal reliability testing was done prior to product release, 101–102
  case study with dispute on whether 85/85 stress was excessive, 100–101
  case study of overstressed VCSELs, 90–92
  compact disk lasers in gigabaud link modules with inadequate wearout lifetime, 102–104
  dangers of copper contamination, 105–108
  high burn-in failure rate foretells early wearout failure, 109–111
  locked into a "no-burn-in" product definition, 96–98
  misunderstanding of true customer requirements, 94–96
  new integrated circuit packaging creates problems, 92–94
  reliability monitor incorrectly set up, 98–100
Catastrophic optical damage (COD), 288
Cathodoluminescence (CL), 259–260, 279
C-band, 5–7
Cd-doped InGaAsP, 190
Characteristic lifetime, 74
Climb dislocation, 134–139
Coarse wavelength division multiplexing (CWDM), 5–7
Coefficient of thermal expansion (CTE) mismatch, 93
Coffin Manson acceleration model, 59
Coherence length, 33
Composition-modulated structures (CMS), 177, 214–226
  on device reliability and degradation, 224–226
  in III–V alloy semiconductors, 214–218
Conduction band, 12–13
Conduction band minimum (CBM), 276
Continuous improvement, 52
Co-packaging, 37
Copper contamination, 105–108
Corrosion aging, 55–58
Corrosion in reliability testing, 55–58
Critical thickness, 144–146
Current injection-free (CIF) region, 212–213

Customer expectations, collecting information on, 49–51
Custom reliability test program for subcomponents, 64–66
  acceleration risk, 65–66
  device acceleration models, 64–65
  limitations in reliability testing, 66

D

Dark defects, 201–202
Dark line defects (DLDs), 96, 152–153, 159, 177, 241, 244–245, 248
  control through compressive strain, 165*f*
  detecting through failure analysis, 80–82
  ease of growth in different material systems, 63*t*, 195*t*
  eliminating through burn-in, 76–77
  failure due to DLDs originating from threading dislocations in non-native substrate lasers, 246–248, 259–260
  growth, 158–163
  two-step growth method in stripe lasers, 226–231
  two-step growth method in VCSELs, 231–232
Dark-spot defects (DSDs), 177
Data collection and analysis, 69–79
Data Communications (Datacom), 3–7
Deep-level transient spectroscopy (DLTS), 212, 279–280
Degradation mechanisms in InAs Quantum dot lasers on silicon, 259*f*
Degradation of AlGaN UV LEDs, 292–303
  generation and diffusion of point defects, 293–298
  hydrogen, role of, 302–303
  magnesium, role of, 298–302
Degradation of InGaN visible LEDs, 273–292
  degradation caused by diffusion, 283–288
  optically induced degradation, 288–292
  point defects in LEDs, 274–278
  thermally activated and current-driven degradation, 278–283
Delta specification, 31
Dense wavelength division multiplexing (DWDM), 5–7
Density-functional theory (DFT), 274
Designing for reliability (DfR), 49–53
  continuous improvement, 52
  customer expectations, collecting information on, 49–51
  designing in margin and product tradeoffs, 51–52
  failure mode and effects analysis (FMEA), 53
  ongoing reliability test (ORT), 52–53
Diffusion coefficient, 284
Diffusive glide, 129–130
Digital diagnostics monitoring (DDM), 90
Direct modulation of lasers, 22, 23*f*
Dislocation filter layers (DFLs), 151*f*
Dislocation loops, 180–181, 191–193, 198–200
Dislocations, electrical and optical properties of, 139–143
Dislocations of GaAs-based heteroepitaxy on silicon substrates, 146–152
DX center and EL2 defects, 122

E

Edge emitters, 19–21
Edge-emitting lasers, 21
EL2 defects, 122
Electroluminescence (EL) measurements, 280–281
Electron channeling contrast imaging (ECCI), 256, 260*f*
Electrostatic discharge (ESD) testing, 61
Energy dispersive X-ray spectrometry (EDX), 106
Energy gap, 13–14
Epitaxial growth, 25–27
Epitaxial lateral overgrowth technique, 150–152, 208, 248–249
Erbium-doped fiber amplifier (EDFA) technology, 5–7
Excited states, 11
Externally modulated lasers (EMLs), 81–82
Extrinsic defect model, 194

F

Fabrication, method of, 25–31
  burn-in, 31
  epitaxial growth, 25–27
  singulation and packaging, 29–31
  wafer fabrication, 27–28
  wafer test, 29

Face ID, 3, 21
Facial or gesture recognition, sensors for, 2–10
Failure analysis (FA), 80
Failure mode and effects analysis (FMEA), 53
Feedback grating, 27–28
Fiber optics, 36–37
Fiber optic transceivers, 1, 6*f*, 30–31, 34, 50, 62
Fick's second law, 283
Field failure method, 78
Filter layers, 151*f*
Frank's rule, 147–148
Future of optoelectronic devices, 37–41
  high-efficiency lighting and display markets, 41
  LiDAR, 38–39, 39*f*
  photonic band gap/plasmonics, 40–41
  photonic integrated circuits (PICs), 37–38
  quantum dot lasers, 39

G

GaAs, characteristics of simple point defects in, 114–122
  DX center and EL2 defects, 122
  negative temperature coefficient of gallium vacancies in GaAs, 119–120
  point defect thermodynamics and the Fermi level effect, 115–119
  quasi Fermi level effect, 120–121
  relevance to MOCVD and MBE growth, 120
GaAs, dislocations in, 122–143
  basic properties of dislocations, 123–124
  electrical and optical properties, 139–143
  optical properties of dislocations, 143
  point defect and solute atmospheres around dislocations, 130–134
  structure of dislocations, 124–126
  thermal climb of dislocations, 134–139
GaAs-based semiconductor lasers, 113
  recombination-enhanced processes, 152–166
Gallium phosphide (GaP), 13–14
Gallium vacancies, 275–276
Gas plasmas, light emission in, 11
Gigabaud link modules (GLM) system, 104
Glide dislocation, 126
Glide velocity, 129
GR-468 test, 62, 64
Gradual degradation, blue lasers, 212–213
Green gap, 18–19
Group IV lasers, 240
Grown-in defects in III-nitrides, 202–214
  gradual degradation of GaN-based laser diodes, 212–214
  influence of dislocations on device reliability in GaN-based laser diodes, 208–209
  Mg-related pyramidal defects, 209–211
  misfit dislocations, 202–203
  threading dislocation and optical properties, 204–206
  V-defects in GaN, 206–207
Grown-in defects in III–Vs semiconductors for optoelectronics, 177–193
  bulk defects, 180*f*, 187–193
  classification of grown-in defects, 177–181
  defects in heavily Fe-doped InP crystals, 187–188
  defects in heavily/highly doped AlGaAs crystals, 191–192
  defects in heavily/highly doped InGaAsP crystals, 189–190
  defects in undoped InGaAsP crystals on GaAs substrates, 192–193
  influence of grown-in defects on device reliability and degradation, 193–202
  interface defects, 178*f*, 181–187

H

Hazard rate, 72–73, 76, 94–96, 95*f*
Heat extraction, 16–17
Heteroepitaxial lasers grown on silicon substrates, 246–250
  reliability results for quantum dot lasers grown on silicon substrates, 250–259
  reliability results for quantum well lasers grown on silicon substrates, 248–250
  unique reliability considerations, 246–248
Heterogeneous lasers on silicon, 242–245
  reliability results for, 243–245
  unique reliability considerations, 243
Heterostructure, 14*f*, 19*f*, 27*f*
High-efficiency lighting, InGaN-based LEDs for, 2–3

High-efficiency lighting and display markets, 41
High temperature operating life (HTOL), 55
High-temperature storage (HTS), 55
Hydrogen passivation in AlGaN UV LEDs, 302–303

I

Impurity, 283–288
Indicator lamps, 13
Infrared wavelengths, 5–7
InGaN-based LEDs for high-efficiency lighting, 2–3
InGaN laser reliability, 208–209
In-plane lasers (IPLs), 19–21
Integrated circuits (ICs), 4–5

L

Laboratory test method, 78–79
Laser and light-emitting diode reliability, 34–35
  quality control, 35
  reliability qualification, 34–35
Laser-based range mapping (LIDAR), 1, 38–39, 39*f*
Laser-induced damage threshold (LIDT), 288–289
Lasers, 1, 11–12, 19–22
  for data- and telecommunications, 3–7
  direct modulation, 22
  history of, 7–10
  mirrors, adding, 20–21
  "population inversion", adding, 2–10
  for sensing arrays, 3
  vertical-cavity surface-emitting lasers (VCSELs), 21
  waveguiding, adding, 19–20
Lattice-matched dilute nitrides, 241–242
Lattice mismatch, 247*t*
L-band, 5–7
LiDAR, 38–39, 39*f*
Light–current–voltage (LIV) sweeps, 244
Light emission, 11–13
  in gas plasmas, 11
  spontaneous and stimulated emission from semiconductors, 12–13
  stimulated emission in gas lasers, 11–12
Light-emitting diode (LED), 1, 3*f*, 13–19
  carrier confinement, 13–15
  green gap, 18–19
  heat extraction, 16–17
  light extraction, 15–16
  "rollover" or "droop", 17–18
Light extraction, 15–16
Lognormal distribution, 73–74

M

Magnesium doping-driven aging, 298–302
Maverick versus wearout reliability failure modes, 66–67
Mean-time-to-failure (MTTF), 254–257
Mechanical aging, 58–61
Mega-scale cloud data centers, 1
Metal organic chemical vapor deposition (MOCVD), 25–26, 26*f*, 105
Metalorganic vapor phase epitaxy (MOVPE), 178–179
Microtwins, 178–179, 182–183
Misfit dislocations, 179, 202–203
  in laser active region, 262–264
  in GaAs on silicon, 146–147, 149, 263*f*
  in GaN on sapphire substrates, 202–203
Misfit strain between a film and a substrate, 144
Modulation rate, 34
Modulators, 22–23
Moisture sensitivity level (MSL), 61
Molecular beam epitaxy (MBE), 178–179
Multiple quantum wells (MQW), 284–285, 292–293
Multisource agreement (MSA) document, 4

N

Nanopipes, 282
Nanowire lasers, 240–241

O

O-band, 5–7
Ongoing reliability test (ORT), 52–53
Optical power–current (OP-I), 293, 293*f*
Optically-induced degradation of InGaN blue LEDs, 288–292
Optoelectronic applications, 2–10
  InGaN-based LEDs for high-efficiency lighting, 2–3
  laser, history of, 7–10
  lasers for data- and telecommunications, 3–7

lasers for sensing arrays, 3
Ordered structures (OS), 177
on device reliability and degradation, 224–226
in III–V alloy semiconductors, 218–223
Organo-metallic vapor phase epitaxy (OMVPE), 25–26
Osmotic force in dislocation growth, 135–137

P

Packaging, 29–31
Parts count method, 77–78
Peach–Koehler equation, 129
Peach–Koehler force, 136
Phonon kick model, 154–156
Photocurrent spectroscopy (PCS) measurements, 295–296
Photodetectors, 23–24
avalanche photodiodes (APDs), 24
photodiodes, 23–24
Photonic band gap/plasmonics, 40–41
Photonic integrated circuits (PICs), 37–38
P-modulation doping (pMD), 252–253, 256–258
Point defects in AlGaN UV LEDs, 293–298
Point defects in GaAs, 114–122
Point defects in LEDs, 274–278
Poole-Frenkel effect, 279, 279*f*
Power, 34
Power conversion efficiency (PCE), 14–15
Power spectral density–current (PSD-I), 293, 294*f*
Precipitates, 180*f*
Principles of operation for optoelectronic components, 11–24
carrier confinement, 13–15
lasers, 19–22
spontaneous and stimulated emission from semiconductors, 12–13
stimulated emission in gas lasers, 11–12
Probability density function (PDF), 72–73
Process change notice (PCN), 82–83
Proprietary information, 49–50

Q

Quadrature-amplitude modulation (QAM), 36–37
Quantum cascade lasers (QCLs) on silicon, 242
Quantum dot lasers, 39, 250–259
degradation mechanisms on silicon, 259–264
dislocation density dependence of reliability in, 253–256
high-temperature reliability, 256–258
misfit dislocations in the laser active region, 262–264
recommended future improvement paths for, 265–267
reliability of quantum dot lasers on native substrates versus silicon, 259
Quantum well lasers, 248–250
Quantum well lasers on silicon, 249*f*
Quasi Fermi level effect, 120–121

R

Random failure rates, calculating, 77–79
Rapid degradation, tolerance to, 163–166
Receiver, 4
Recombination-enhanced defect reactions (REDRs), 113, 152–153, 162
Recombination-enhanced dislocation climb (REDC), 152–153, 158–163, 193–196
features of, 158–161
point defects, origin of, 161–163
Recombination-enhanced dislocation glide (REDG), 152–153, 156–158, 196
features of, 156–157
impact on lasers, 158
mechanism of, 157–158
Recombination-enhanced processes, 152–166
background, 153–156
rapid degradation, tolerance to, 163–166
Reliability analysis, 72–73
Reliability engineering
building and running a reliability test lab, 68–69
data collection and analysis, 69–79
designing for reliability (DfR), 49–53
failure analysis (FA), 80
physics of failure and common failure mechanisms, 80–82
standards-based testing versus customized testing, 53–67
subcomponent-, component-, and system-level reliability testing, 67–68
Reliability models, 73–74

Reliability reports, 79
Risk priority number (RPN) score, 53

S
Scanning transmission electron microscopy (STEM) tomography, 262–264
Scanning transmission electron microscopy with electron energy loss spectroscopy (STEM-EELS), 279–280
Secondary ion mass spectrometry (SIMS) analysis, 106–107, 284
Semiconductor laser, 9–10, 18, 25–26, 31–32, 47
Semiconductor laser reliability testing, 65
Separate-confinement heterostructure (SCH), 284–285
Shape factor, 74
Shockley–Read–Hall (SRH) recombination, 249–250, 276
Silicon photonics, reliability of lasers on silicon substrates for, 239
  early-stage prospective technologies, 240–242
  heteroepitaxial lasers on silicon, 246–250
  heterogeneous lasers on silicon, 242–245
Single mode versus multimode, 32–33
Singulation and packaging, 29–31
Sn-doped InGaAsP, 190
Spinodal region, 217
Spontaneous emission, 11
SR-332 calculations, 77–78
Stacking faults, 178–182
Standards-based testing versus customized testing, 53–67
  accelerated aging, 53–54
  acceleration factors, 54–55
  categories of reliability testing, 55–61
  custom reliability test program for subcomponents, 64–66
  maverick versus wearout reliability failure modes, 66–67
  reliability test standards, 61–63
Stimulated emission, 11
Strained-layer superlattices (SLS), 149–150
Stress–strength model, 74–76
Super-luminescent LED (SLED), 20

T
Technology developments, 36–41
  fiber optics, 36–37
  future of optoelectronic devices, 37–41
Te-doped InGaAsP, 190
Telcordia GR-468 Standard, 34–35, 62
Telecommunications (Telecom), 3–7
Thermal instability in III-nitrides, 226–232
  composition-modulated structure, 226–231
  influence of composition fluctuation on the device characteristics and degradation, 231–232
Threading dislocations (TD), 179
Transceiver, 4, 6*f*
Transmitter, 4
Transmitter optical sub assembly (TOSA), 4–5
Trap-assisted tunneling (TAT), 294
Traps, origin of, 140–142
Trimethyl-aluminum (TMA), 25–26
Trimethyl-gallium (TMG), 25–26
Two-step degradation processes (REDG and REDC), 196–198

V
Valance band, 12–13
V-defects, 206–207
Vertical cavity surface emitting laser (VCSEL), 3, 9–10, 21, 90–92

W
Wafer fabrication, 27–28
Wafer test, 29
Waveguiding, 19–20
Wavelength-division multiplexing (WDM), 5–7
Wearout failure, 66–67
Weibull distribution, 74
Weibull plots, 59*f*

Z
Zn-doped InGaAsP, 189–190

FSC
www.fsc.org
MIX
Papier | Fördert gute Waldnutzung
FSC® C083411

Kolibri 360 GmbH
Ferdinand-Jühlke-Straße 7
99095 Erfurt, Deutschland
produktsicherheit@kolibri360.de

Theory and Technology of Multiscale Dispersed Particle Gel for In-Depth Profile Control

Theory and Technology of Multiscale Dispersed Particle Gel for In-Depth Profile Control

Caili Dai
China University of Petroleum (East China),
Qingdao, P.R. China

Guang Zhao
China University of Petroleum (East China),
Qingdao, P.R. China

Qing You
China University of Geosciences,
Beijing, P.R. China

Mingwei Zhao
China University of Petroleum (East China),
Qingdao, P.R. China

Yifei Liu
China University of Petroleum (East China),
Qingdao, P.R. China

Fulin Zhao
China University of Petroleum (East China),
Qingdao, P.R. China

ELSEVIER

Elsevier
Radarweg 29, PO Box 211, 1000 AE Amsterdam, Netherlands
The Boulevard, Langford Lane, Kidlington, Oxford OX5 1GB, United Kingdom
50 Hampshire Street, 5th Floor, Cambridge, MA 02139, United States

British Library Cataloguing-in-Publication Data
A catalogue record for this book is available from the British Library

Library of Congress Cataloging-in-Publication Data
A catalog record for this book is available from the Library of Congress

ISBN: 978-0-323-99849-9

Publisher: Glyn Jones
Editorial Project Manager: Naomi Robertson
Production Project Manager: Prem Kumar Kaliamoorthi
Designer: Greg Harris

Typeset by MPS Limited, Chennai, India

Contents

Preface *xi*

1. Introduction **1**

References 8

2. Preparation technology of bulk gel **11**

Section 2.1 Technical requirements for preparation of bulk gel 11

2.1.1 Requirement of reservoir type 11

2.1.2 Technical preparation requirements of bulk gel 11

Section 2.2 Bulk gel system for low-temperature and low-salinity reservoirs 13

2.2.1 Cross-linking mechanism of zirconium-based gel 13

2.2.2 Factors influencing the formation of zirconium bulk gel 14

Section 2.3 Bulk gel system for medium and high temperature reservoirs 22

2.3.1 Cross-linking mechanism of resin bulk gel 22

2.3.2 Influencing factors of resin bulk gel 23

2.3.3 Temperature resistance of resin bulk gel 29

2.3.4 Microstructure of resin bulk gel 33

Section 2.4 Bulk gel for ultra high temperature reservoirs 34

2.4.1 Cross-linking mechanism of strengthened bulk gel 34

2.4.2 Factors influencing gelation of strengthened bulk gel 36

2.4.3 Thermal stability of strengthened bulk gel 40

2.4.4 Microstructure of the strengthened bulk gel 44

References 45

3. The preparation technology of dispersed particle gel **47**

Section 3.1 Establishment of DPG preparation technique 47

Section 3.2 Optimization of process parameters for DPG preparation 52

3.2.1 Efficiency of DPG preparation 52

3.2.2 The influence of preparation parameters on the microscopic morphology of DPG 56

3.2.3 Effect of preparation parameters on DPG particle size 59

3.2.4 Effect of preparation parameters on viscosity of DPG particle 63

3.2.5 Effect of preparation parameters on zeta potential of DPG particles 67

Section 3.3 The preparation mechanism of multiscale DPG particle 70

3.3.1 Test principle 71
3.3.2 The bulk gel cross-linking reaction phase 72
3.3.3 The multiscale DPG particles shaping shearing phase 75
3.3.4 The preparation mechanism of DPG particles 78
Section 3.4 Development of multiscale DPG particles industrialized preparation software 81
3.4.1 The idea of establishment for multiscale DPG particles preparation model 81
3.4.2 Establishment of multiscale DPG particles preparation model 82
3.4.3 Development of multiscale DPG particles preparation software 85
Section 3.5 Design of the industrialized production equipment for DPG particles 89
3.5.1 Design industrialized workshop production equipment 89
3.5.2 Design skid-mounted production equipment of multiscale DPG particles 90
References 94

4. The profile control technology of multiscale DPG particles 97

Section 4.1 The characterization of DPG particles 97
4.1.1 The viscosity characteristics of DPG particles 97
4.1.2 Shear resistance 100
4.1.3 Thermal stability 101
4.1.4 Capability of self-growing and coalescence 102
4.1.5 Analysis of zeta potential 103
4.1.6 Macroscopic morphology 104
4.1.7 Microscopic morphology 105
Section 4.2 Matching relationship between DPG particles and pore throat 106
4.2.1 Establishing the mathematical model of matching relationship 106
4.2.2 Matching factor optimization 107
Section 4.3 The profile control performance of DPG particles 111
4.3.1 Capability of mobility control 111
4.3.2 Performance of profile improvement 113
4.3.3 Evaluation of the enhanced oil recovery potential 116
Section 4.4 The profile control mechanism of multiscale DPG particles 120
4.4.1 The profile control mode of multiscale DPG particles 120
4.4.2 Distribution behavior of DPG particles in porous media 126
Section 4.5 Oilfield application examples 127
4.5.1 Changqing Oilfield: ultralow-permeability microfractured well groups 128

4.5.2 Shengli Oilfield: typical well groups with high temperature and low permeability 134
4.5.3 Northwest Oilfield: high-temperature, high-salinity, medium- and high-permeability reservoir 139
References 153

5. DPG soft heterogeneous combination flooding technology 155

Section 5.1 Construction of DPG SHC flooding system 155
5.1.1 Compatibility research 155
5.1.2 Ability to reduce interfacial tension 157
5.1.3 Viscosity characteristics 160
Section 5.2 Characterization of the properties of the DPG SHC system 163
5.2.1 Microscopic morphology of DPG SHC flooding system 163
5.2.2 Viscosity stability 165
5.2.3 Ability to reduce interfacial tension 166
5.2.4 Expansion capacity 168
5.2.5 Wettability alteration 169
5.2.6 Shear resistance 172
5.2.7 Zeta potential 174
Section 5.3 Displacement performance of the DPG SHC flooding system 175
5.3.1 Evaluation of injectivity performance 176
5.3.2 Effect of permeability reduction at different positions of sand-pack 178
5.3.3 Profile improvement of SHC flooding system 179
5.3.4 Evaluation of oil displacement potential 182
Section 5.4 Interaction mechanism of SHC flooding system 188
5.4.1 The adsorption behavior of surfactants on the surface of DPG particles 188
5.4.2 Interface rheological behavior of the SHC flooding system 195
Section 5.5 Displacement mechanism of SHC flooding system 225
5.5.1 The chromatographic separation effect of SHC flooding system 225
5.5.2 Distribution of SHC flooding system in porous media 233
5.5.3 Sweep efficiency of SHC flooding system in porous media 239
5.5.4 Microscopic displacement mechanism of the SHC flooding system 244
Section 5.6 Field application case in zhejiang oilfield 251
5.6.1 General situation of the targeted well 251
5.6.2 Dynamic production parameters analysis of oil before operation 252
5.6.3 Tracking analysis of operation effect 254
References 256

6. DPG-strengthened polymer/surfactant combination flooding technology **259**

Section 6.1 Construction of DPS combination flooding system 259

6.1.1 Compatibility 260

6.1.2 Interfacial tension 260

6.1.3 Viscosity 264

Section 6.2 Characterization of DPS combination flooding system 268

6.2.1 Microscopic morphology of DPS combination flooding system 268

6.2.2 Thermal stability 269

6.2.3 Aggregation and expansion property 271

Section 6.3 Enhanced oil recovery performance of DPS combination flooding system 272

6.3.1 Mobility control performance 273

6.3.2 Profile improvement performance 276

6.3.3 Enhanced oil recovery performance 279

Section 6.4 Enhanced oil recovery mechanism of DPS combination flooding system 283

6.4.1 Synergistic thickening effect 285

6.4.2 Strengthened oil displacement efficiency 285

6.4.3 Improved mobility control performance 285

References 285

7. Profile control technology of the DPG particles three-phase foam system **287**

Section 7.1 Design of the DPG particles three-phase foam system 287

7.1.1 Modified Ross—Miles approach 288

7.1.2 Optimization of foaming agent concentration 289

7.1.3 Optimization of DPG particles concentration 289

7.1.4 Optimization of gas/liquid ratio 292

Section 7.2 Profile control ability of the DPG particles three-phase foam system 292

7.2.1 Stability factors of DPG particles three-phase foam system 293

7.2.2 Application performance of the DPG particles three-phase foam 300

Section 7.3 The stabilizing mechanism of the DPG particles three-phase foam system 303

7.3.1 Stabilizing mechanism of DPG particles three-phase foam 303

7.3.2 The generation and migration of DPG particles three-phase foam in porous media 308

Section 7.4 Injection parameters optimization of DPG particles three-phase foam 326
7.4.1 Optimization of injection pattern 327
7.4.2 Optimization of gas/liquid ratio 328
7.4.3 Optimization of gas injection velocity 328
7.4.4 Optimization of injection volume 329
Section 7.5 Application examples in oil fields 331
7.5.1 Geological survey of the test area 331
7.5.2 Development status of the pilot area 333
7.5.3 On-site construction 335
7.5.4 Construction effect analysis 336
References 338

8. Prospect of multiscale DPG in-depth profile control technology 339

8.1 Advancement of skid-mounted DPG online production/injection integration system 339
8.2 Further understanding on water control mechanisms of DPG chemical system 340
8.3 Development of multiscale DPG profile control technique in ultradeep reservoirs 340

Index 343

Preface

Water flooding is the main development method of China's oilfields. However, long-term water flooding aggravates reservoir heterogeneity, resulting in a high water-cut rate and low oil recovery. Polymer gel is an industry-recognized profile control agent that can decrease the formation heterogeneity and enhance oil recovery. The first-generation polymer gel that originated in 1988 is a chromium cross-linked polymer gel, whereas the second-generation polymer gel that originated in 2001 is a phenolic resin cross-linked polymer gel. These two kinds of gels have been successfully applied for water control in conventional oil fields. However, the oilfields have been gradually shifting from shallow to deep, low permeability or other complex reservoirs, while the traditional polymer gels have inherent drawbacks such as shear degradation due to injection device and seepage shearing in the formation, and dilution caused by contact with the reservoir minerals and fluids. The treatment risk has been increased due to uncontrolled gelation time and uncertainness of cross-linking due to shear degradation, changes in gellant compositions, and dilution effect. Besides, the traditional polymer gels have lower thermal stability, resulting in over cross-linking or syneresis in high temperature (>120°C) and high salinity (>200,000 mg/L) reservoirs. Therefore the polymer gels with the advantage of high temperature and high salinity resistance, controllable gelation, on-site online production, deep injection and migration, and high-strength water control ability are necessary. They bring a new direction for the development of water control technology.

Since 2004, Prof. Dai's scientific research team has focused on dispersed particle gel (DPG) research and has developed a novel DPG profile control technology. This technology is a complete innovation that overcomes the traditional polymer gel drawbacks. Specifically, DPG particles have been successfully prepared with bulk gel by a high shearing method on the ground. A series of original achievements have been made in preparation theories and methods, bulk gel, process equipment, and a "DPG + " system. After 17 years of scientific research, 32 invention patents were authorized, including 14 patents in the United States and Europe, and two computer software copyrights were authorized. Twenty industrial production lines with an annual output of 3000 tons and above have been set up, and the technical achievements have been industrialized in 17 oilfields at home and abroad, including Changqing, Tahe, Shengli, and Bohai, of China, and Kazakhstan. Water cutoff and oil have been enhanced significantly after applying these DPG treatments. Laboratory tests and field applications indicate that the in-depth profile control technology of multi-DPG particles is a new generation technology that provides a reference to water production control in high water-cut oilfields.

This book will help extend the influence of multiscale DPG in-depth profile control technology in the application of enhanced oil recovery, lead the advancement of enhanced oil recovery technology, and provide references for related petroleum technologies.

The authors would like to thank Yining Wu, Lin Li, Yongpeng Sun, Bin Yuan, Long He, Haien Yang, Renwei Qu, and Yingsheng Sun for their great help during the compiling of this book.

It is sincerely hoped that readers can put forward valuable opinions on the problems existing in this textbook, so that they can be revised for the next edition.

CHAPTER 1

Introduction

Enhanced oil recovery (EOR) is both an eternal theme of oilfield development and a crucial guarantee for China's energy strategic security. In China, the water-flooding oilfield accounts for a large proportion (>80%) at present. However, long-term water flooding during the development of the oilfield has resulted in aggravated heterogeneity of reservoirs, and the successive entrance of oilfields into the high water-cut stage which is "cancerous" for oil recovery. As a result, low recovery and poor economic performance occur, approximately two third of the oil remains underground, and those phenomena are even far worse in complex unconventional reservoirs, such as fractured reservoirs with low permeability and reservoirs with high temperature and high salinity.

Profile control and water plugging technology is recognized as an important method to improve the effectiveness of water flooding, with plugging agents as the key to ensuring its success. Up to now, bulk gels have been the most widely used plugging agents for profile control and water plugging across the world [1,2]. Bulk gels are a kind of viscoelastic material possessing a three-dimensional network structure formed by intermolecular cross-linking between polymers and cross-linking agents. The polymers are mostly partially hydrolyzed polyacrylamides. The cross-linking agents mainly include chromium ion, zirconium ion, phenolic resin, and polyethylenimine. Bulk gels have the advantages of adjustable gelation time and gelling strength, high plugging capacity, and good profile control effect. In practice, the first step is to inject the mixed solution of polymer and cross-linking agent into the target formation, and then shut the well and wait for gelling. Next, the bulk gel plugs the highly permeable zone to reduce the permeability of water. Finally, the production wells are open and the injection wells start to inject water, so that a balanced flooding is achieved and water-flooding effectiveness and oil recovery are improved. The profile control and water plugging mechanism of bulk gel is shown in Fig. 1.1. Although the gel treatments have been successfully conducted on many oilfields worldwide, the gelling solution is easily affected by the shear degradation due to the injection device and seepage shearing in the formation, and dilution caused by contact with the reservoir's minerals and fluids.

Theory and technology of multiscale dispersed particle gel for in-depth profile control
DOI: https://doi.org/10.1016/B978-0-323-99849-9.00001-1

Figure 1.1 The profile control and water plugging mechanism of bulk gel.

It usually leads to uncontrolled gelation time and uncertainness of the systems and increases the treatment risk. Therefore it is difficult to predict the gelation time, gel strength, and the depth of the gelatinization in the formation during the underground gelatinization process, which affects the effectiveness of profile control and water plugging process.

In addition, linked polymer solution (LPS) and colloidal dispersion gel (CDG) though different in names, have a similar adhesive nature. Both systems use low-concentration polymers and cross-linking agents to form a colloidal dispersion system with intramolecular cross-linking in the majority and intermolecular cross-linking in the minority. Thus the newly-formed system possesses low viscosity and both a colloid and a solution nature [3,4] (Fig. 1.2). Since there is no continuous network structure, it is able to enter the deep formations and control water production. With the deepening of research and the implementation of field testing, some scholars believe that LPS and CDG are a kind of partly cross-linking system, and that their performance is not only affected by the shear degradation and dilution effect but also by the contaminated water for preparation. Generally, LPS and CDG have poor resistance to high temperature and high salinity, and thus are not suitable for high temperature and high salinity reservoirs.

To solve the unpredictable situation during the kinetics reaction in the gelatinization process underground, the experts put forward the method of using prepreparation gel particles (PPG) to plug large pores or the highly permeable layer for profile control, at the ninth national water plugging conference, held in Hainan province in the 1990s. The PPG particles are mainly composed of monomer, cross-linking agent, auxiliary agent, and strength control agent (clay). Under certain conditions, the bulk gel with certain water absorption and swelling properties can be formed, and then a series of solid particles with different expansion abilities, different strengths, and different particle sizes are made through drying, crushing, granulation, and screening [5–7]. The PPG particles with the three-dimensional network structure contain a large number of

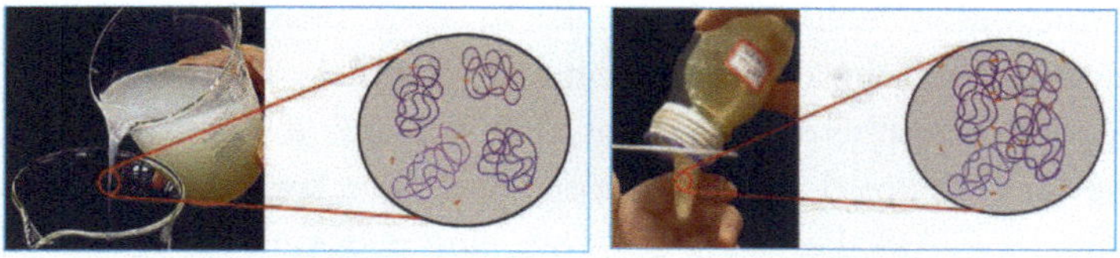

Figure 1.2 Difference between intramolecular cross-linking and intermolecular cross-linking.

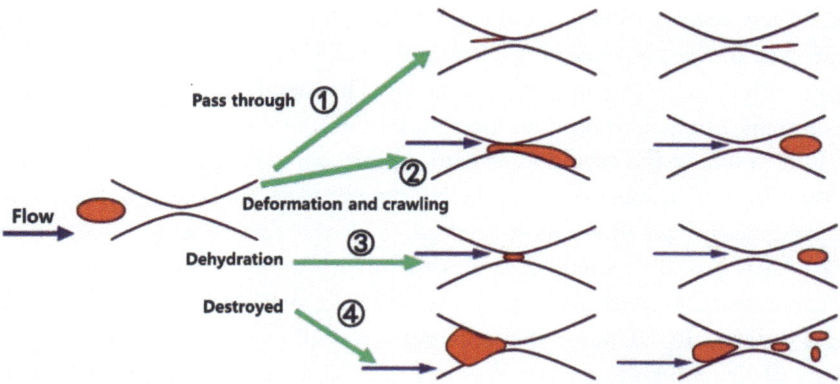

Figure 1.3 Mechanisms of particle passing through pore throat [8].

hydrophilic groups, which can adsorb a great amount of water and swell. The hydrophilic property makes the PPG significantly change its volume under different conditions. The three-dimensional skeleton structure generated by cross-linking gives it an acceptable strength, which results in plugging in the dominant channels and changing the flow direction of the fluid. Meanwhile, the viscoelastic material after water absorption and particle swelling can be reversibly deformed by external forces. When the external force is reduced, the deformation can be recovered to a certain extent of its original shape. In the profile control operation, the "amoeba" characteristics can be fully utilized to change the local pressure field of the reservoir, so as to realize the purpose of fluid diversion (Fig. 1.3). However, due to the limitation of the preparation conditions, the size of PPG particles often range from 1 to 5 mm, with swelling to 10–100 mm after adsorbing water. Since being affected by formation pressure, most swellable PPG particles are broken when they flow through the pore throat, which cannot make a complete use of PPG particles. In addition, the particle size cannot adapt to deep formations and their adaptability to low permeability reservoirs is weak, so it is unable to realize online continuous integration of production and injection as well.

Polymer microspheres, known as Bright Water overseas and originating from the idea proposed by BP Amoco, refer to a system of thermally activated particles (TAP). Featuring good swelling performance, temperature and salt resistance, size-controllability, and other characteristics, those microspheres can be applied to improve oil recovery. They are usually prepared by inverse (micro)emulsion polymerization, with a water-soluble monomer (acrylamide AM, acrylic acid, and salt tolerance of heat-resistant functional monomer AA) as the raw material. Using an emulsifier, the polymer microspheres are evenly dispersed in the oil (mineral oil, kerosene, diesel, etc.) and transformed into a microemulsion after mechanical stirring. The newly made microemulsion then starts a polymerization reaction under the action of the initiator and after that a water-in-oil emulsion system is formed. The particle sizes of the synthesized microspheres are in the nanometer scale, especially in the reverse microemulsion polymerization method, with the particle sizes ranging from 10 to 100 nm. The method requires a large number of emulsifiers and an accurate reaction temperature and reaction time, which is a complex preparation process. In addition, the polymer microspheres also can be prepared without an emulsifier by the method of inverse suspension polymerization and dispersed polymerization, but the particle sizes are relatively large.

Migration, plugging, elastic deformation, remigration, and replugging characteristics of polymer microspheres are the deep profile control and displacement mechanisms in the formation. Through these characteristics of polymer microspheres, the swept volume of the subsequent injected water is expanded [2,9–13]. In order to fully achieve the deep profile control capacity of polymer microspheres, accurate control of the particle size and particle size distribution of the microspheres is a key issue. When the particle size of the microsphere matches the pore throat of the reservoir rock, the plugging rate and elastic deformation migration pressure gradient of the microsphere reach their optima. If the particle size is too small, it cannot produce effective plugging at the pore throat. Comparatively, oversized microspheres cannot migrate to the full depth of the formation; they may even find difficulty in injection, which can bring about an undesired effect of profile control treatment. Field tests have shown that polymer microspheres are generally suitable for deep profile control in low permeability reservoirs. This method is often not suitable for formations with high permeability layers, large channels, and fractures. In recent years, researchers have designed large-sized core–shell structured microspheres to synthesize submicron and micron polymer microspheres, and endowed the microspheres with self-cross-linking function and formation adsorption capacity to solve the above problems. However, the cost is relatively high, and it is difficult for both the microsphere size and the reservoir to respond in time during the injection, which can result in a large matching uncertainty. Therefore their applications in oilfields are limited to some extent.

Due to the adaptability of polymer gels, PPG particles, and polymer microsphere plugging agents as well as the problems exposed in the field application, it is urgent to develop a new plugging agent to meet the requirements of high permeability layers or large pores with different permeabilities in actual reservoirs. That is the plugging agents should be easily injected, effectively plugged, and readily transported. In 2004 our research group initiated a new concept of dispersed particle gel (DPG) and the DPG particles were successfully prepared with a bulk gel by a mechanical shearing method. This new concept moved the chemical cross-linking reaction process from "underground to the ground's surface." The water control mechanism has been transformed from the "pouring sausage" type of polymer gel to the "plugging throat hole" type of DPG particles, realizing the "visibility" of the gel-forming process and the "controllability" of the profile control distance. It broke through the technical bottlenecks of uncontrollable underground polymer gel. Additionally, the oil phase as the dispersion medium, accurate polymerization conditions, uncontrollable reaction process, complicated process, and the high cost of traditional chemical emulsion polymerization method were also overcome.

Different from polymer microspheres prepared by a chemical method, multiscale DPG is a kind of water-based self-dispersive particle system (Fig. 1.4) [14–16]. In order to be adaptive to a long-term online injection for deep profile control in formations, the physical preparation method progresses from the coaxial cylinder shear cross-linking method and the pipe flow shear cross-linking method in basic indoor research, through the indoor small-scale high-speed mechanical shearing machine method, finally into the colloid mill shearing method available at industrial scale. Table 1.1 lists the comparison of several typical plugging agents.

The multiscale DPG particles prepared by this method have the following characteristics:

1. Wide distribution, at the scale from nm to mm.
2. Low viscosity (5–10 mPa · s), easy to inject into deep formations.
3. Shear resistance, good self-growth and coalescence ability, better deep profile control and displacement.

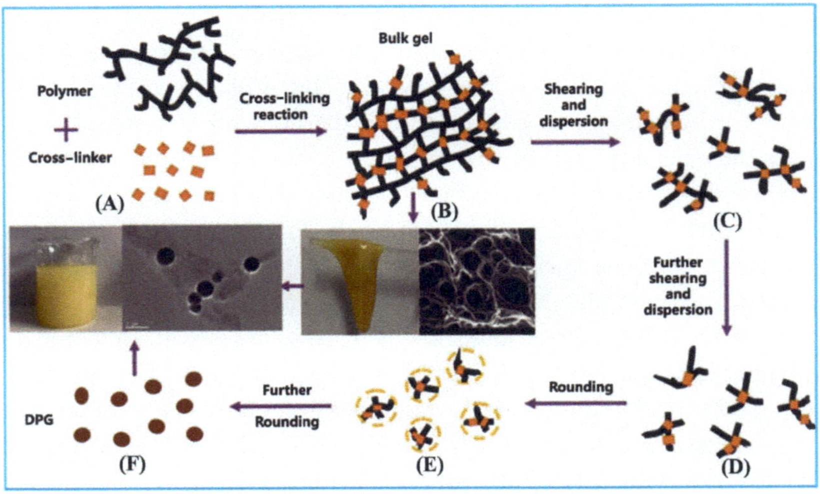

Figure 1.4 Flow chart of the multiscale DPG system: (A) polymers and cross-linkers for reaction; (B) formed bulk gel; (C) shearing and dispersion; (D) further shearing and dispersion; (E) rounding; (F) formed DPG.

4. Good temperature and salt resistance up to 150°C and 30×10^4 mg/L, respectively; suitable for wide reservoir conditions.
5. Simple preparation process and easy to adjust, available for online integration of production and injection, and adjusting the preparation parameters intelligently.
6. Easy to operate production equipment, and 2–8 tons/hour of production rate.
7. Environmentally friendly, fulfilling the requirements of national environmental protection.
8. Low cost, 20–40 yuan/ton (concentration used), satisfying the requirements of reducing oilfield costs and increasing efficiency at the low oil price.

According to the characteristics of multiscale DPG particles, the profile control mechanism can be described as follows (Fig. 1.5). When the DPG particles are injected into the formation, the particles preferentially enter high permeability zones. They can be retained and adsorbed on the rock surface when the channels are too large. Although bridging becomes more difficult in such high permeability zones, the particles could combine to form larger aggregates and then increase water flow resistance in these zones. With continuous injecting, the DPG particles could be deformed and enter the deep formation easily. When the DPG particle

Table 1.1 Comparison of several typical plugging agents.

Typical plugging agents	System characteristics	Characteristics of profile control and displacement	Oilfield applicability	Cost
Gel	Underground cross-linking, affected by shear, adsorption, formation water dilution, etc. uncontrollable gelation time and strength	Filling and sealing, and the usage is large	High concentration, short distance, small control radius, low concentration, poor cross-linking, short validity	Medium
PPG particles	Large particle size: 1–5 mm, swelling rapidly in contact with water, easy to pile up or break up near the well, and difficult to enter deep formation	Deformed and crawling, dehydrated or broken forms	Mostly used for pretreatment of large pores or high permeability layers	Low
Polymer microsphere	Emulsion polymerization of AM and other water-soluble monomers, precise preparation process, harsh synthesis conditions	Pore throat plugging, deformation through, replugging, remigration	Difficult to match the product parameters with the fractures (pores) dynamically, and weak profile control capacity	High
DPG particle	Cross-linked by HPAM ground, environmentally friendly; simple preparation, controllable particle size, and self-growth	Directly plugging, bridging across throat hole, deformation and in-depth migration	Online integration of production and injection, achieving a moderate adjustment of "water control-limited pressure boost -water injection linkage balance," available for dessert, hills, narrow offshore platforms, and other harsh environments	Low

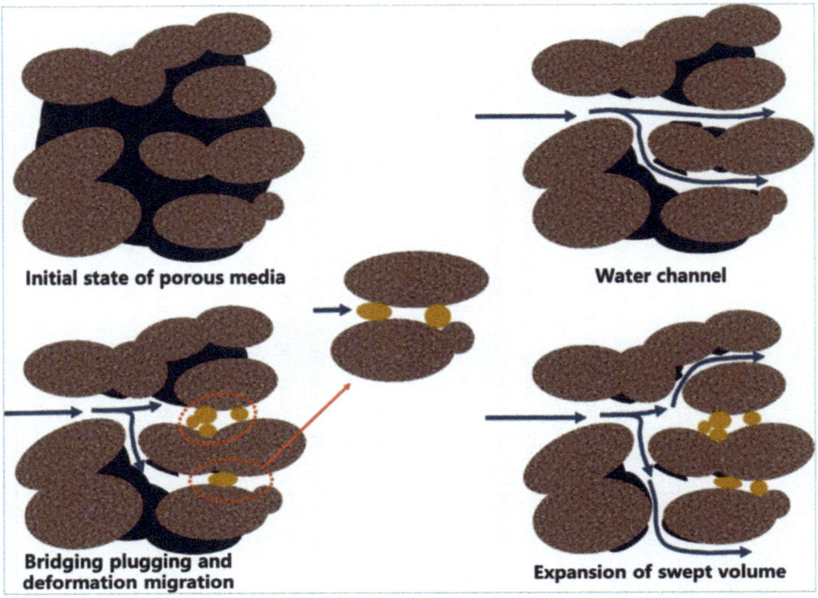

Figure 1.5 Schematic diagram of microscopic flooding mechanism of DPG particles.

radius is larger than the pore throat radius, the particles could be trapped and directly plug the entrance of the pore throat. When the DPG particle radius is a little smaller than the pore throat radius, two or three particles will be stranded in the pore throat space and bridge across the pore throats easily. However, when the pore throat radius is much larger than the DPG particle radius, many particles will combine together and form larger aggregates at the pore throat space. By trapping and bridging, the formed large aggregates can bridge across the high permeability channels and effectively reduce the formation permeability. Moreover, once the bridge is consolidated, the newly arriving particles will gather at the pore throat and form an effective plugging capacity, thus decreasing the subsequent water flow rate and yielding the following water into adjacent low permeability zones with high oil saturation, so as to expand the sweep area and enhance the oil recovery [17,18].

References

1. Zhao F. Oilfield chemistry(II). China University of Petroleum Press; 2010.
2. Sydansk R. A newly developed chromium (III) gel technology. SPE Reservoir Engineering 1990;5(3):346–52.

3. Mack J, Smith J. In-depth colloidal dispersion gels improve oil recovery efficiency. SPE 1994;27780.
4. Fielding R, Gibbons D, Legrand F. In-depth drive fluid diversion using an evolution of colloidal dispersion gels and new bulk gels: an operational case history of North Rainbow Ranch Unit. SPE 1994;27773.
5. Li Y, Liu Y, Bai B. Study on water plugging and profile control technology of bulk expanded particles. Oil Drilling and Production Technology 1999;21(3):65−8.
6. Bai B, Liu W, Li L, et al. Analysis of internal factors affecting the properties of pre-cross-linked gel particles. Petroleum Exploration and Development 2002;29 (2):103−5.
7. Bai B, Liu Y, Coste J, et al. Cause study on preformed particle gel for in-depth fluid diversion. SPE 2004;89468.
8. Coste J, Liu Y, Bai B, et al. In-depth fluid diversion by pre-gelled particles. Laboratory study and pilot testing. SPE 2000;59362.
9. Pritchett J, Frampton H, Brinkman J, et al. Field application of a new in-depth waterflood conformance improvement tool. SPE 2003;84897.
10. Frampton H, Morgan J, Cheung S, et al. Development of a novel waterflood conformance control system. SPE 2004;89391.
11. Mustoni J, Norman C, Denyer P. Deep conformance control by a novel thermally activated particle system to improve sweep efficiency in mature waterfloods of the San Jorge Basin. SPE 2010;129732.
12. Lei G. A new depth profile control technique for pore throat scale elastic microsphere. China University of Petroleum Press; 2011.
13. Yao C. Experimental and simulation study on percolation mechanism of pore throat elastic microsphere. China University of Petroleum; 2014.
14. You Q, Tang Y, Dai C, et al. Research on a new profile control agent: dispersed particle gel. SPE 2011;143514.
15. You Q, Tang Y, Dai C, et al. A study on the morphology of a dispersed particle gel used as a profile control agent for improved oil recovery. Journal of Chemistry 2014;33:1−9.
16. Zhao G, You Q, Gu C, et al. Preparation mechanism of multiscale dispersed particle gel system. Journal of Petroleum 2017;38(7):821−9.
17. Zhao G. Study on the construction and displacement mechanism of soft heterogeneous combination oil displacing system. China University of Petroleum; 2016.
18. Dai C, Zou C, Liu Y, et al. Matching law of elastic dispersed particle gel and pore throat and deep regulation mechanism. Journal of Petroleum 2018;39(4):427−34.

CHAPTER 2

Preparation technology of bulk gel

Multiscale DPG preparation technology overcomes the uncontrollability of polymer gel and the limitation of profile control distance in the formation. It transfers the chemical cross-linking reaction process from underground to the ground. This technological concept demands fast gelling of bulk gel system on the ground, controllable gelation time, and a gel viscosity that meets the requirements of the DPG particle preparation process. In this chapter, we carried out a systematic study on three types of typical bulk gels based on the requirements of the multiscale DPG system and the reservoir classification standard of Professor Zhao Fulin in China University of Petroleum (East China). These works lay the foundation for the industrial preparation of the concerned material.

Section 2.1 Technical requirements for preparation of bulk gel

2.1.1 Requirement of reservoir type

According to the temperature and formation water salinity of the reservoir classification standard from Professor Fulin Zhao, reservoirs can be divided into six types, which are given in Table 2.1. To satisfy the profile control treatment of these six types reservoir conditions, multiscale DPG should possess the corresponding bulk gel system. For class I—III reservoirs, temperature and salinity are relatively mild, and most of the existing bulk gel can meet the requirements of thermal stability under reservoir condition. For IV—VI reservoirs with high temperatures and high salinities, DPG particles present a relatively big step forward for the thermal stability properties of the bulk gel systems (Table 2.1).

2.1.2 Technical preparation requirements of bulk gel

The technology requires that the bulk gel system can be quickly gelled on the ground, that the gelation time is controllable, and that gel viscosity can satisfy the industrialized requirements of the DPG preparation technology. With reference to the standard of reservoir classification and the current situation of bulk gel, we specified the technical requirements in the bulk gel, as shown in Table 2.2.

Theory and technology of multiscale dispersed particle gel for in-depth profile control
DOI: https://doi.org/10.1016/B978-0-323-99849-9.00002-3

Table 2.1 Criteria for classifying reservoirs by temperature and formation water salinity.

Reservoir			High temperature and high salt			
Condition	**Low temperature low salinity (I)**	**Moderate temperature Moderate salinity (II)**	**Low–high temperature low–high salinity (III)**	**Medium–high temperature medium–high salinity (IV)**	**High–high temperature high–high salinity (V)**	**Extremely high temperature extremely –high salinity (VI)**
Temperature (°C)	<70	70–80	80–90	90–120	120–150	150–180
Formation salinity (mg/L)	$<1\times10^4$	$1\times10^4\sim2\times10^4$	$2\times10^4\sim4\times10^4$	$4\times10^4\sim10\times10^4$	$10\times10^4\sim16\times10^4$	$16\times10^4\sim22\times10^4$

Table 2.2 Technical requirements of bulk gel.

Reservoir type Entry	Type I reservoir	Type II–IV reservoir	Type V–VI reservoir
Salinity of solution water (mg/L)	≤ 30,000	≤ 30,000	≤ 30,000
Polymer dissolution time (min)	≤ 60	≤ 60	≤ 60
Gelation temperature (°C)	20–40	90–95	100–110
Gelation time (h)	≤ 4	≤ 10	≤ 4
Viscosity after gelation (mPa · s)	≤ 80,000	≤ 80,000	≤ 80,000
Thermal stability (days)	≥ 365	≥ 365	≥ 365
Source of raw materials	The raw materials have wide sources and low prices, which is convenient for large-scale field application		
Environmental requirements	Environment friendly, low pollution to the formation		

Section 2.2 Bulk gel system for low-temperature and low-salinity reservoirs

In the view of the characteristics of low temperature and low salinity reservoirs, the rapid formation of the zirconium bulk gel system was prepared by a functional polymer and environment-friendly zirconium cross-linking agent. This provides support for the preparation of multiscale DPGs suitable for class I reservoirs.

2.2.1 Cross-linking mechanism of zirconium-based gel

In the organic zirconium cross-linking agent, Zr^{4+} formed polynuclear hydroxyl bridged ions through complexation hydrolysis and hydroxyl bridged action, and the polynuclear hydroxyl bridged ions formed a cross-link reaction with the functional polymer with $-COO^-$ to form the zirconium bulk gel [1].

The specific reaction process is shown in Eqs. (2.1)–(2.5)

1. Complexation

$$Zr^{4+} + 8H_2O \rightarrow \left[(H_2O)_8Zr\right]^{4+} \tag{2.1}$$

2. Hydrolysis

The strong positive electric field of the multivalence metal ions repels the positively charged H^+ and makes it depleted from the complexed water, thus bringing about hydrolysis.

$$[(H_2O)_8Zr]^{4+} + H_2O \rightarrow [Zr(H_2O)_7OH]^{3+} + H_3^+O \quad (2.2)$$

3. Hydroxyl bridge action

In $[Zr\ (H_2O)_7\ (OH)]^{3+}$, there is a lone pair of electrons on the oxygen of OH, which can form a σ bond with the empty orbital of another metal ion, thus producing a bridging effect.

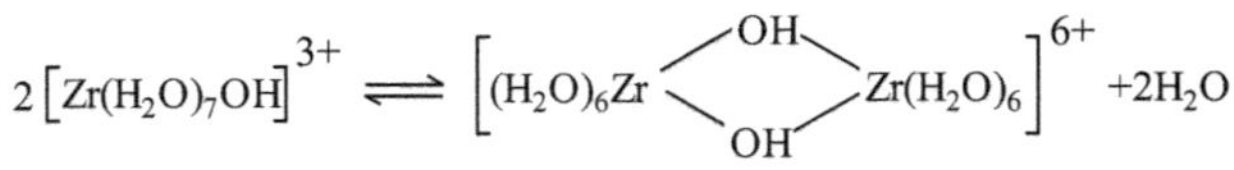

(2.3)

4. Further hydrolysis and hydroxyl bridging effect

Furthermore, the degree of polymerization increases with further hydrolysis and hydroxyl bridging to form polynuclear hydroxyl bridged complex ions. This reaction includes hydrolysis and polymerization. The equation is as follows:

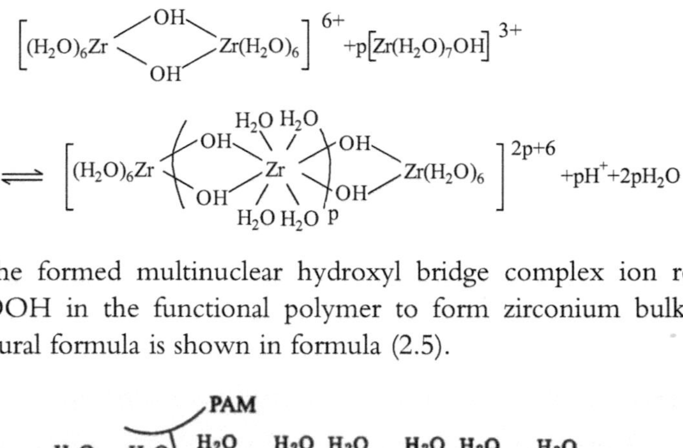

(2.4)

The formed multinuclear hydroxyl bridge complex ion reacts with —COOH in the functional polymer to form zirconium bulk gel. The structural formula is shown in formula (2.5).

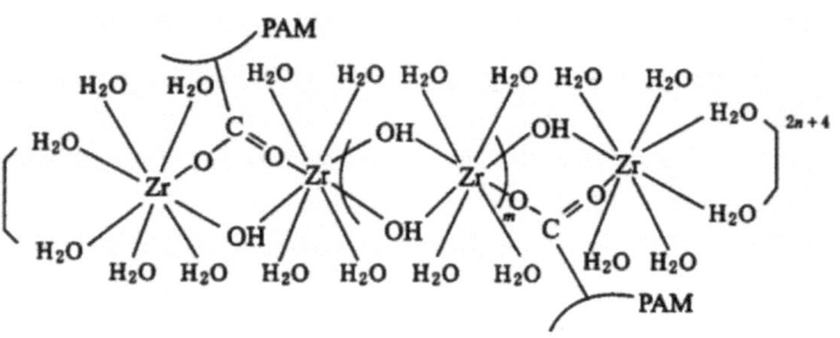

(2.5)

2.2.2 Factors influencing the formation of zirconium bulk gel

The gelation time was determined by the "hanging method." The strength of bulk gel was determined by the breakthrough vacuum method. The clear water salinity was 400 mg/L and the gelation temperature was 30°C.

2.2.2.1 Effect of concentration

The concentration of gelling solution affects the gelling performance of bulk gels. A 1.0% parent functional polymer solution was prepared by clear water. After diluting, the organic zirconium cross-linking agent with different mass fraction was added to obtain a series of gelling liquid systems with different concentrations of polymer and cross-linking agent. The effects of the concentration of functional polymer and organic zirconium cross-linker on the gelling properties of zirconium bulk gel were investigated at 30°C. The isograms of gelation strength and time are shown in Figs. 2.1 and 2.2.

It can be seen from Figs. 2.1 and 2.2 that the gelation time of the zirconium bulk gel system formed by the functional polymer and organic zirconium cross-linking agent is adjustable between 8 and 136 min, and the gelation strength is controllable between 0.04 and 0.071 MPa. The higher concentration the functional polymer and

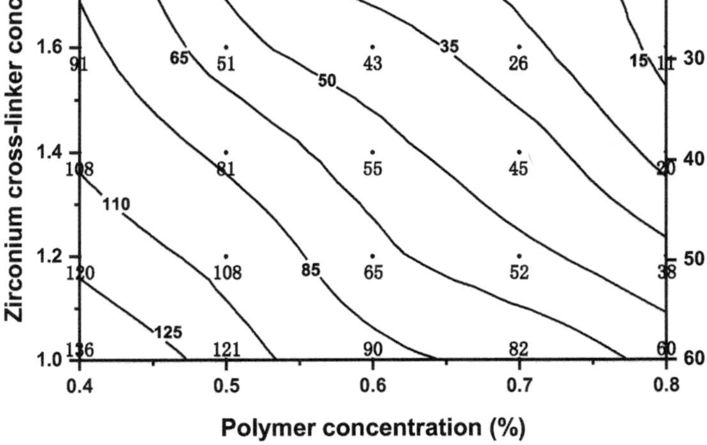

Figure 2.1 Isogram of gelation time (min) of zirconium bulk gel.

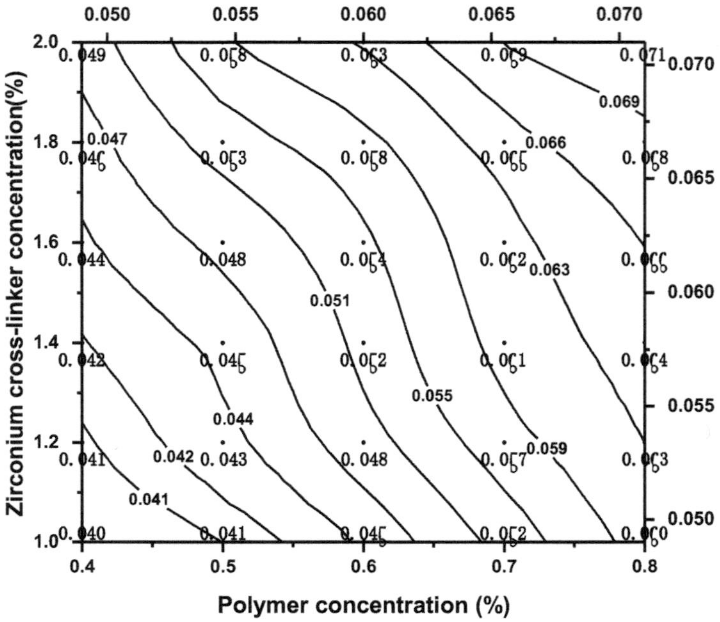

Figure 2.2 Isogram of gelation strength (MPa) of zirconium bulk gel.

the cross-linking agent, the faster the cross-linking reaction, and the stronger the zirconium gel. This is because the amount of $-COO^-$ provided to participate in the reaction increases as the concentration of the functional polymer increases. On the other hand, the higher concentration of the organic zirconium cross-linking agent leads to a higher concentration of the polynuclear hydroxyl bridge zirconium complex ions formed in the solution. This increases the bonding probability and cross-linking degree of $-COO^-$ and polynuclear hydroxyl bridge zirconium complex ions, so that a dense zirconium gel network structure is formed, which improves the strength of the bulk gel. Therefore in the later experiment, a zirconium bulk gel system with different strengths can be prepared by adjusting the concentration of functional polymer and cross-linking agent, so as to realize the preparation of multiscale DPG with different strengths.

Fig. 2.3 shows the macroscopic morphology of two typical zirconium gels. It can be seen that the higher the concentration of functional polymer, the higher the strength of the zirconium gel. This is consistent with the strength test results of the bulk gel measured by the above breakthrough vacuum degree method.

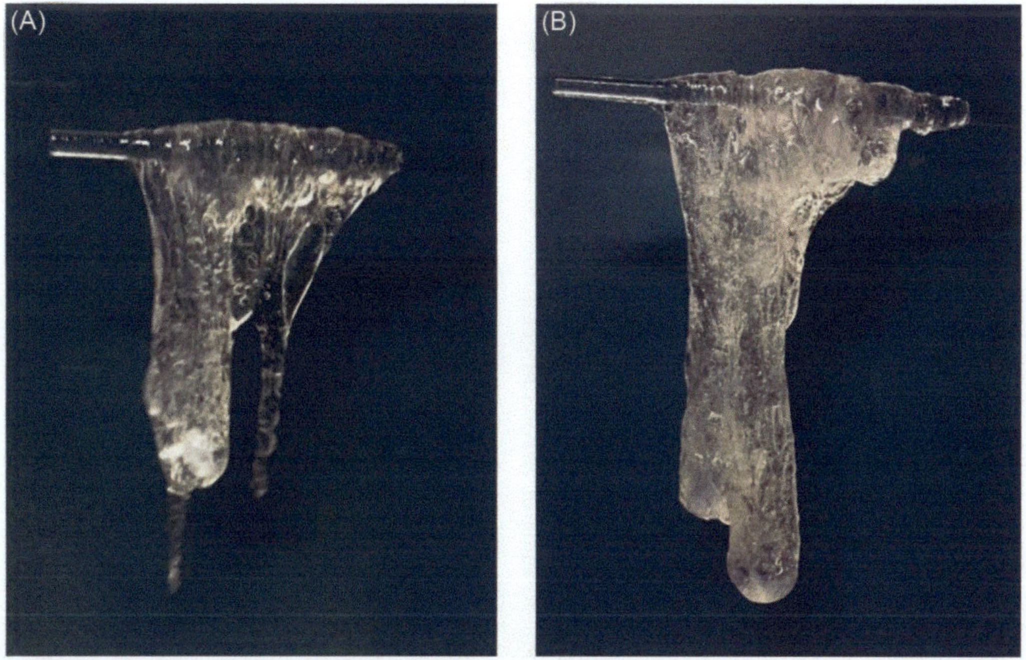

Figure 2.3 Zirconium gels formed by different formulations: (A) 0.6% functional polymer, 1.6% organic zirconium cross-linking agent; (B) 0.9% functional polymer, 1.6% organic zirconium cross-linking agent.

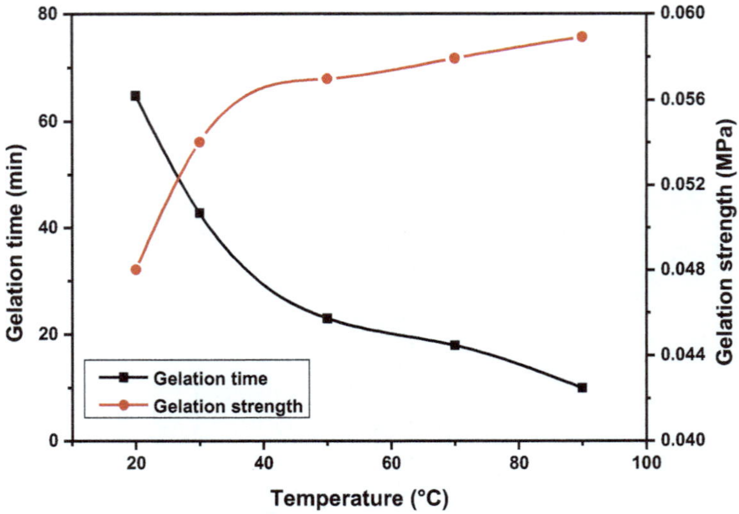

Figure 2.4 Influence of temperature on gelation performance.

2.2.2.2 Effect of temperature

The effect of temperature on the gelation performance of zirconium gel was investigated. The formula of zirconium gel in the test was 0.6% functional polymer and 1.6% organic zirconium cross-linking agent. The test results are shown in Fig. 2.4.

As can be seen from Fig. 2.4, with the increase of temperature, the gelation time of zirconium gel is shortened from 65 to 5 min, and the gelation strength increases from 0.054 to 0.06 MPa. The thermal motion of the molecule is also accelerated, increasing the probability of the binding of the $-COO^-$ group to the polynuclear zirconium ions and accelerating the reaction. Zirconium bulk gel can be gelled at 20°C, which indicates that the multiscale DPG particles can be prepared at room temperature. This simplifies the production, reduces energy consumption, and improves the efficiency of production.

2.2.2.3 Effect of salinity

The effect of inorganic salt ions on the gelation performance of zirconium gel was investigated at 30°C. The formula of zirconium gel was 0.6% functional polymer and 1.6% organic zirconium cross-linking agent. The test results are shown in Table 2.3.

It can be seen from Table 2.3 that the addition of three kinds of inorganic salt ions shortened the gelation time and increased the gelation

Table 2.3 Effects of inorganic salts on the gelation performance.

Type of inorganic salt	Concentration (%)	Gelation time (min)	Gelation strength (MPa)
NaCl	1.0	35	0.054
	5.0	31	0.057
	10.0	23	0.059
	15.0	18	0.060
	20.0	14	0.062
$CaCl_2$	0.1	25	0.054
	0.3	19	0.056
	0.5	15	0.060
$MgCl_2$	0.1	24	0.054
	0.3	22	0.055
	0.5	19	0.057

strength of the zirconium bulk gel system. The effects of three inorganic salt ions on the gelation performance of zirconium gel are as follows: $CaCl_2 > MgCl_2 > NaCl$. This is because the addition of inorganic salt ions compresses the electric double layer of the polymer. Besides, the hydration film becomes thinner, and the electric potential and the repulsion force between charged groups decreases. This facilitates the occurrence of the intermolecular cross-linking reaction, resulting in the shortening of gelation time and the increase of gelation strength. In addition, the stronger ability of divalent ion compression and diffusion renders the greater effect of the divalent ion on the gelation performance of zirconium bulk gel than that of the monovalent ion.

2.2.2.4 *Effect of pH*

At 30°C we investigated the effect of pH value on the gelation performance of the zirconium bulk gel. The zirconium gel was formulated into a 0.6% functional polymer plus 1.6% organic zirconium cross-linking agent. The pH value of the initial base solution was 7.49. The test results are shown in Table 2.4.

As can be seen from Table 2.4, either too high or too low pH value have an impact on the gelation performance of zirconium bulk gel—increasing the gelation time and weakening the gelation strength. Given overacidity or overalkalinity, the formation of polynucleate hydroxyl zirconium ions is inhibited, the concentration of polynucleate hydroxyl zirconium ions in the gel formation solution decreases, and the cross-linking

Table 2.4 Influence of pH value on the gelling performance of zirconium gel.

pH	Gelation time (min)	Gelation strength (MPa)
2.82	110	0.039
4.01	91	0.044
5.43	68	0.046
6.40	55	0.051
7.49	43	0.054
8.45	132	0.052
9.28	Not gelled	Not gelled

degree of the $-COO^{-}$ group and the zirconium complex ions in the polymer fall. As a result, the strength of the formed zirconium gel weakens. However, under the pH condition of the original base solution, the gelation effect is optimum. Therefore the adoption of neutral water is suggested for the preparation of gel solution in the production process of multiscale DPG.

2.2.2.5 Microstructure of zirconium gel

The microstructure of the zirconium gel was observed by using a Quanta 200 FEG scanning electron microscope in the field emission environment. A few samples of the zirconium gel were taken and placed on the sample table for direct observation. The test was set at an acceleration voltage of 15.0 kV, a gas pressure in the sample room of 313–455 Pa, and a temperature of 0°C. The test formula was a 0.6% functional polymer and 1.6% organic zirconium cross-linking agent. The results are shown in Fig. 2.5.

It can be seen from Fig. 2.5 that the developed zirconium bulk gel is a dense and ordered three-dimensional spatial network structure. It is composed of three typical parts: chain, mesh, and node. A possible reason for the formation of this spatial network structure is as follows. The $-COO-$in the functional polymer interacts with the zirconium polynuclear hydroxyl bridge complex ions to form the chains. Then, between the chains are formed regular mesh holes which are linked by nodes to form a dense spatial network structure. The dense spatial network structure confines the water in the meshes and prevents the water from escaping from the meshes. This is beneficial to the long-term thermal stability of zirconium bulk gel and lays a good foundation for the preparation of stable multiscale zirconium dispersed particle gel.

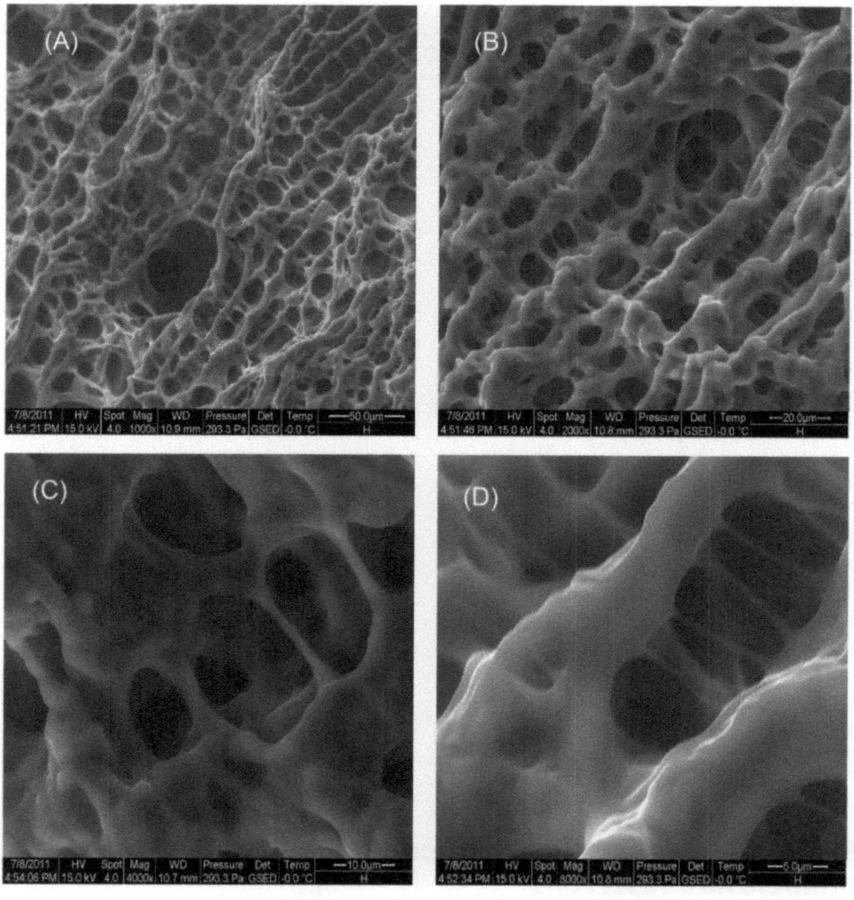

Figure 2.5 Microscopic morphology of zirconium gel at different magnification [2]: (A) 1000 times; (B) 2000 times; (C) 4000 times; (D) 8000 times.

2.2.2.6 Thermal stability of zirconium bulk gel

The gelled zirconium bulk gel was sealed within ampoules and placed in 30°C and 60°C incubators, respectively, for 1 year, with the intention of investigating the thermal stability of zirconium bulk gel. The test results are shown in Table 2.5. It can be seen that after the above process, the zirconium bulk gel shows no obvious dehydration, but its strength drops slightly. Nevertheless, the strength retention rate remains above 85%, indicating the good thermal stability of the zirconium bulk gel system. This phenomenon is mainly related to the dense network structure of the zirconium bulk gel. A denser network brings about a higher difficulty for water escaping from the mesh. In this sense, the zirconium bulk gel is

Table 2.5 Investigation on stability of zirconium gel.

Zirconium gel		Aging at 30°C for 1 year		Aging at 60°C for 1 year	
Functional polymer (%)	Organic zirconium cross-linking agent (%)	Dehydration	Strength retention (%)	Dehydration	Strength retention (%)
0.6	0.8	Nothing	87.85	Nothing	85.59
0.6	1.0	Nothing	95.04	Nothing	91.66
0.6	1.2	Nothing	95.64	Nothing	92.29
0.6	1.4	Nothing	96.71	Nothing	93.4
0.6	1.6	Nothing	96.78	Nothing	93.51
0.8	0.8	Nothing	88.33	Nothing	84.92
0.8	1.0	Nothing	94.64	Nothing	91.28
0.8	1.2	Nothing	95.87	Nothing	92.53
0.8	1.4	Nothing	97.73	Nothing	94.41
0.8	1.6	Nothing	97.78	Nothing	94.57
1.0	0.8	Nothing	98.24	Nothing	94.84
1.0	1.0	Nothing	97.14	Nothing	93.8
1.0	1.2	Nothing	97.94	Nothing	94.63
1.0	1.4	Nothing	98.13	Nothing	94.82
1.0	1.6	Nothing	98.43	Nothing	95.22

armed with good thermal stability. This outcome lays a good foundation for the preparation of zirconium dispersed particle gel.

Section 2.3 Bulk gel system for medium and high temperature reservoirs

With the increase of reservoir temperature and salinity, the requirements of DPG particles for bulk gel become higher. According to the current development status and the accumulation of our research group of the gel system, it is necessary to develop and to prepare multiscale DPG particles by resin-based bulk gel with rapid cross-linking on the ground to meet the needs of in-depth profile control treatments in II–IV reservoirs. Meanwhile, technical requirements and economic costs are also considered.

2.3.1 Cross-linking mechanism of resin bulk gel

The resin bulk gel was formed by a type II functional polymer, resin cross-linking agent, and coagulant. The resin cross-linking agent used in the test was formed by polycondensation of phenol and formaldehyde catalyzed by sodium hydroxide [3–6]. The reaction formula is shown in Fig. 2.6.

The toxicity of formaldehyde and phenol is reduced by a prepolycondensation reaction. The bulk gel formed by cross-linking the prepared resin cross-linking agent with a type II functional polymer and coagulant has good thermal stability. The cross-linking mechanism is shown in Fig. 2.7. The amide group ($-CONH_2$) in the type II functional polymer and the hydroxyl group ($-CH_2OH$) in the resin cross-linking agent undergo a dehydration condensation reaction to form bulk gel.

2.3.2 Influencing factors of resin bulk gel

The performance of the resin bulk gel directly determines the adaptability of production conditions of multiscale DPG. In this study, the basic properties of bulk gel were systematically investigated for the concentrations of type II functional polymers, cross-linking agent, and coagulant, along with the temperature, salinity, and shearing action, in an attempt to meet the technical requirements for the preparation of bulk gel and the production of multiscale DPG. The gelation time of bulk gel was determined by the GSC strength code method [7], and the gel strength of bulk gel was determined by the vacuum breakthrough method. Dehydration rate and strength retention rate characterize the thermal stability of the bulk gel.

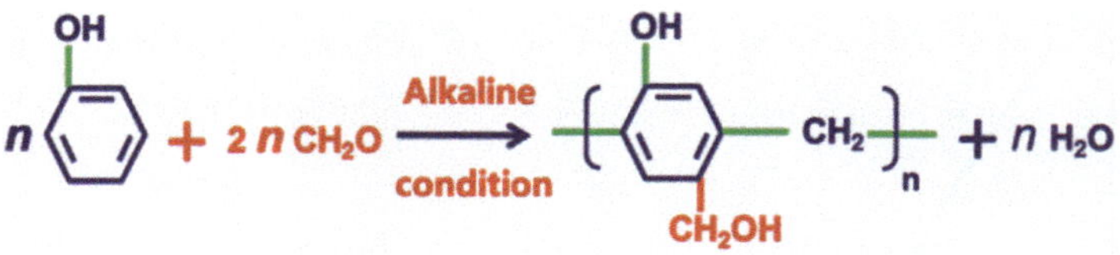

Figure 2.6 Formation mechanism of resin cross-linking agent.

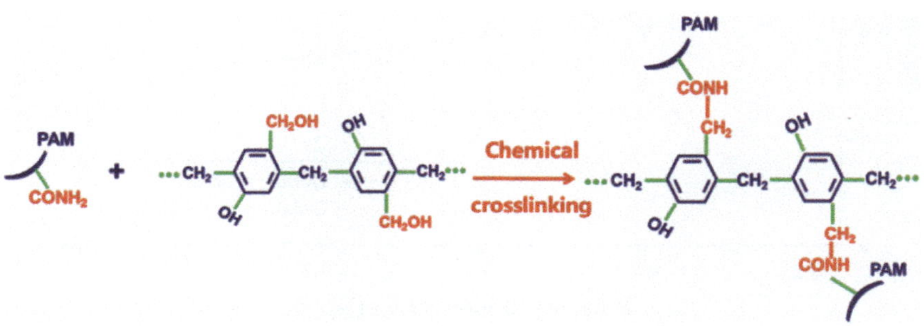

Figure 2.7 Formation mechanism of resin bulk gel [6].

The lower the dehydration rate of the bulk gel, the higher the strength of the bulk gel after aging. This indicates better thermal stability of the prepared bulk gel. The solution preparation was clear water with salinity of 400 mg/L, and the gelation temperature was 95°C.

2.3.2.1 The effect of type II functional polymer concentration

A type II functional polymer forms the skeleton of the bulk gel structure, which directly determines the gelling properties of the bulk gel. In the experiment, the concentration of resin cross-linking agent was fixed at 1.2%, and the concentration of coagulant at 0.1%. The influence of polymer concentration on the gelation time and strength were investigated. The test temperature was 95°C. The results are shown in Fig. 2.8.

As shown in Fig. 2.8, with the increase of the concentration of type II functional polymer, the gel strength enhances and the gelation time shortens. These findings are consistent with the regulation of cross-linking. The higher the polymer concentration is, the more acylamino ($-NH_2OH$) is provided to participate in the reaction, and the easier to start a dehydration condensation reaction with the oxhydryl, thus speeding up the reaction rate. Meanwhile, the tightly combined acylamino and oxhydryl can form bulk gel with high strength. The gelation time of the

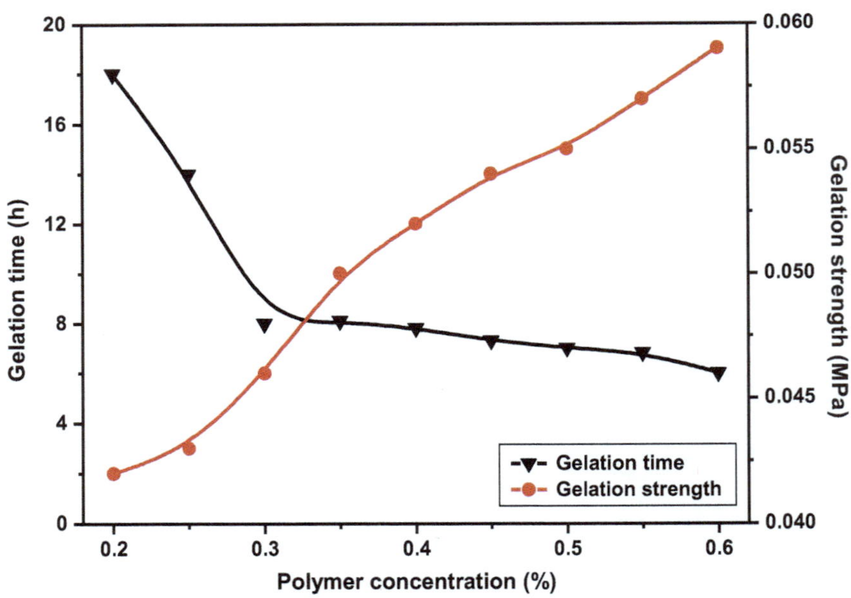

Figure 2.8 The effect of polymer concentration on the gelation time and strength.

self-developed resin bulk gel can be controlled between 6 and 18 hours, agreeing to the technical requirements for preparing multiscale DPG.

2.3.2.2 The effect of cross-linking agent concentration

By setting the concentration of type II functional polymer at 0.4% and the coagulant concentration at 0.1%, a series of bulk gel systems with different cross-linking agent concentrations were prepared. The effect of cross-linking agent concentration on the performance of resin bulk gel was tested at 95°C. The results are shown in Fig. 2.9.

As illustrated in Fig. 2.9, the gelation strength increases with the increase of cross-linking agent concentration, while the gelation time decreases. This is because the amount of acylamino ($-NH_2OH$) provided to participate in the reaction is large at the high cross-linking agent concentration. This increases both the probability of contact between the acylamino and the oxhydryl and the cross-linking density, and reduces the gelation time, thus increasing the strength of the bulk gel. During the preparation of the resin bulk gel on the ground, the fast cross-linking of the bulk gel can also be accomplished by adjusting the cross-linking agent concentration.

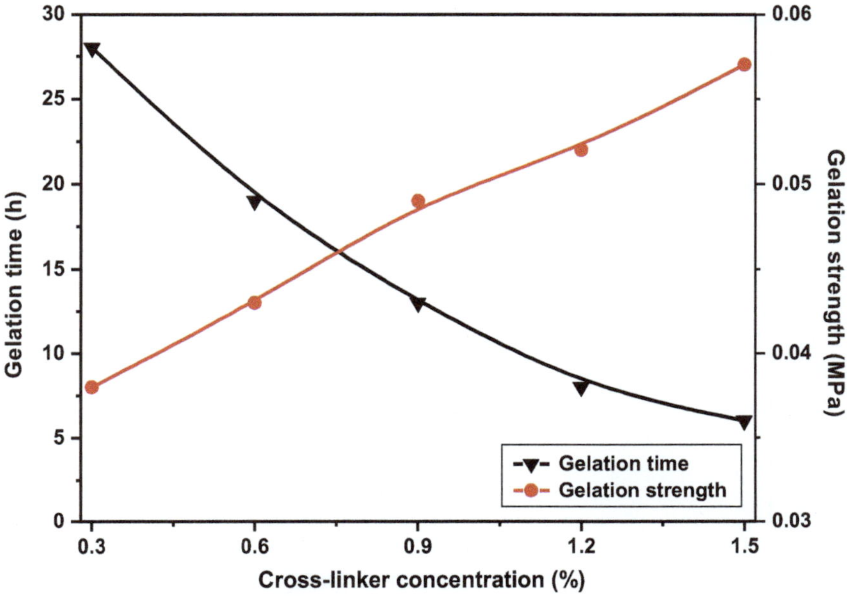

Figure 2.9 Effect of cross-linking agent concentration on the gelation time and gelation strength of bulk gel.

2.3.2.3 *The effect of cross-linking coagulant concentration*

The physical preparation method requires a short gelation time to prepare the bulk gel, because of the energy consumption and cost. When the polymer and cross-linking agent concentrations were 0.4% and 1.2%, respectively, the effect of the cross-linking coagulant concentration on the gelation time was tested at 95°C. As shown in Fig. 2.10, with the increase of the cross-linking coagulant concentration, the gelation time is shortened from the initial 20 hours to within 10 hours. This indicates that the addition of the coagulant can significantly reduce gelation time. After the addition coagulant, the network structure of bulk gel grows more intensive, which is favorable to combine the gel structure and shorten the gelation time.

2.3.2.4 *The effect of salt concentration*

Different mass concentrations of NaCl, $CaCl_2$, and $MgCl_2$ were added to the gel system. In this system, the concentrations of polymer, cross-linking agent for resin, and coagulant are 0.4%, 1.2%, and 0.1%, respectively. The gelation process is tested at 95°C, and the results are shown in Figs. 2.11 and 2.12.

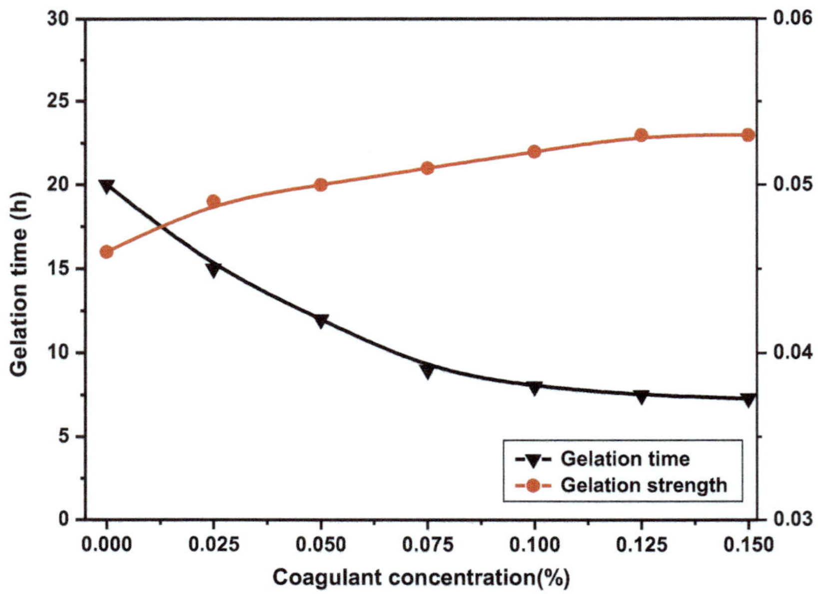

Figure 2.10 The effect of cross-linking coagulant concentration on the gelation time and strength.

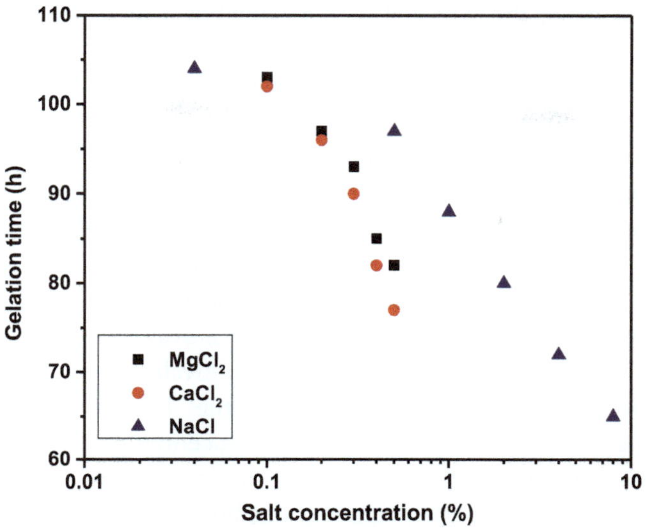

Figure 2.11 Effect of salt concentration on gelation time.

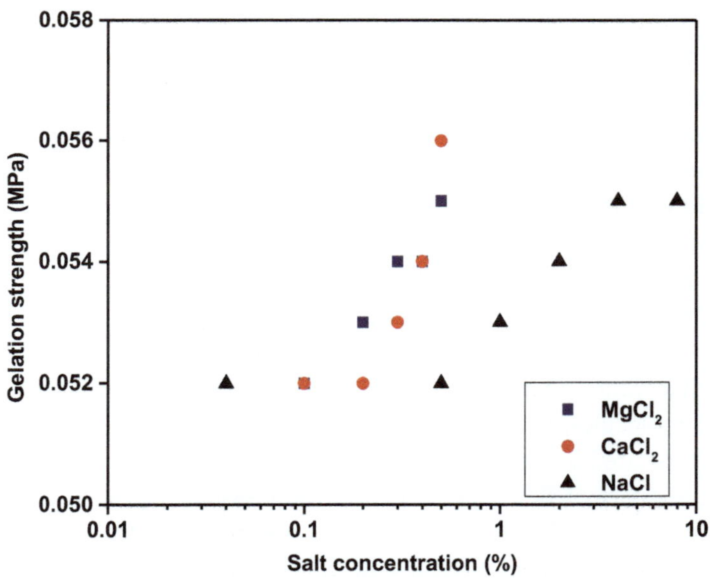

Figure 2.12 Effect of salt concentration on gelation strength.

It can be seen from Figs. 2.11 and 2.12 that the high salt concentration is able to shorten the gelation time and strengthen the gel strength. When the concentration of NaCl, $CaCl_2$, and $MgCl_2$ reaches 8%, 0.6%, and

0.6%, respectively, the bulk gel is maintained at a high concentration. It indicates a good salt toleration of the bulk gel. The effects of three kinds of salt on the bulk gel are ranked as $CaCl_2 > MgCl_2 > NaCl$. The addition of salt compresses the diffused double layer of polymer, leading to the thinning of the hydration shell. The decrease of Zeta potential and the reduction of repulsion force between charged groups makes the cross-linking reaction easy to carry out, shortens the gelation time, and increases the gelation strength. Therefore the addition of an appropriate amount of salt promotes the cross-linking reaction. However, the influence of divalent ions is greater than that of monovalent ions. Since Na^+ is a monovalent cation and Ca^{2+} and Mg^{2+} divalent cations, Ca^{2+} and Mg^{2+} have a stronger ability to compress the diffused double layer of polymer than that of Na^+. The higher salt concentration indicates a stronger compression ability which can make the cross-linking reaction easier to carry out [8]. The feasible preparation of bulk gel by different salt solutions with different concentrations indicates a good adaptability of the bulk gel system. This finding provides a solid foundation for multiscale DPG to adapt to industrial production in different regions.

2.3.2.5 The effect of temperature

Clean water was used to prepare a gel solution with 0.4% polymers, 1.2% resin cross-linking agent, and 0.1% coagulant. The prepared gel solution was sealed and placed at different temperatures, in order to investigate its gelation time. After gelation, the strength of the bulk gel was measured at room temperature.

It can be seen from Fig. 2.13 that when the temperature increases from 80°C to 105°C, the gelation time of the bulk gel is shortened from 38 to 6 hours, and the gelation strength increases from 0.048 to 0.052 MPa. Since increasing temperature speeds up the thermal movement of molecules, the binding probability of amide group and hydroxyl group is elevated and the cross-linking process of bulk gel is accelerated. Given the safety of field operation and the pressure resistance characteristics of the reactor, the bulk gel system is prepared at 90°C−95°C during the industrial production of multiscale DPG.

2.3.2.6 The effect of shearing

Shearing degradation takes place when a gelling solution is transported through injection equipment or flows during formation. Thus it is important to investigate the effect of shearing on the gelation time and strength

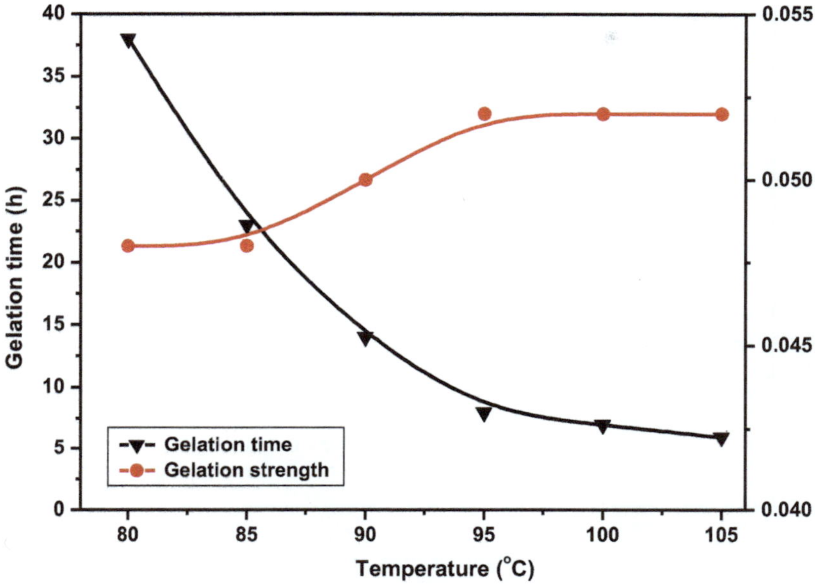

Figure 2.13 Effect of temperature on gelation time and strength.

of a gel system. The gelling solution with 0.4% polymers, 1.2% resin cross-linking agent, and 0.1% coagulant was prepared. To simulate shearing degradation under field conditions, the Waring agitator method was used at a shearing rate of 1000 rpm. The test temperature was 95°C and the results are shown in Fig. 2.14.

It can be seen from Fig. 2.14 that with the increase of shear time, the gelation time of the bulk gel increases and the strength decreases. The shear action shortens the polymer molecular chain and destroys the molecular structure, leading to the reduction of gelation strength. However, the reassembled polymer molecules will form bulk gel with cross-linker because the reaction rate of cross-linking will accelerate when the shear effect disappears. The gelation strength of the bulk gel still remains above 80%, which indicates that the bulk gel system has great shear resistance.

2.3.3 Temperature resistance of resin bulk gel

2.3.3.1 Stability investigation

The bulk gel was sealed in ampoule bottles, and then placed at 30°C and 95°C incubators for 1 year, respectively, to investigate the stability of the bulk gel. The results are shown in Table 2.6.

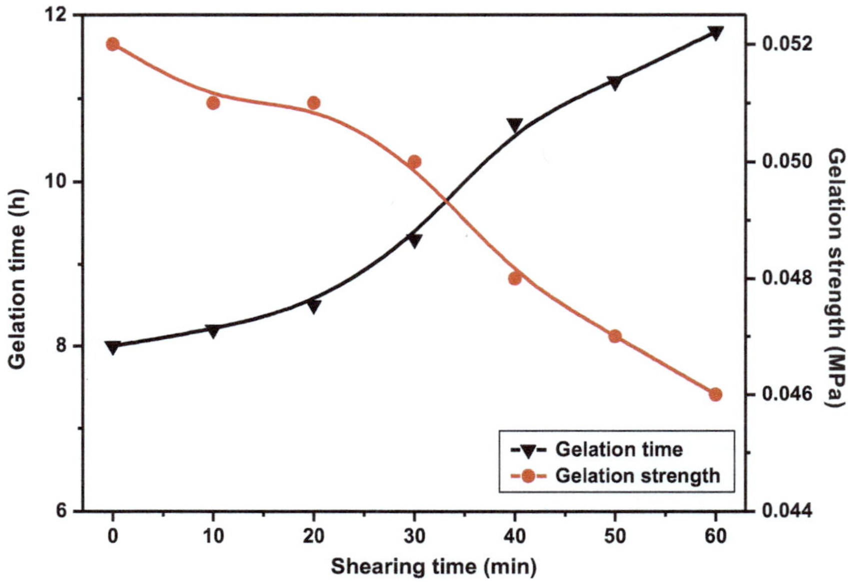

Figure 2.14 Influence of shearing on gelation time and gelation strength.

According to Table 2.6, the bulk gel is aging at 30°C and 95°C without obvious dehydration, and the strength of the gel slightly reduces. This indicates that the gel has good thermal stability. This is because the prepared gel has a dense network structure, which can prevent the water from separating from the gel structure and ensure its high strength and thermal stability.

2.3.3.2 Analysis of bulk gel temperature resistance performance

In order to further illustrate the thermal stability of the bulk gel, the bond broken temperature within or between the molecules of the gel was analyzed by differential scanning calorimeter (DSC), with a test temperature range of 50°C–150°C, a heating rate of 2.0°C/min, and a nitrogen flow rate of 20 mL/min. Bulk gel used in the test contained 0.4% type II functional polymer, 1.2% resin cross-linking agent, and 0.1% coagulant. The gel temperature was 95°C. The test results are shown in Fig. 2.15.

It can be seen from Fig. 2.15 that the DSC curve of the bulk gel shows a relatively wide gentle region before 120°C, and then the curve starts to rise sharply after 120°C. This indicates that the bulk gel possesses great structural stability and temperature resistance when below this

Table 2.6 Stability of resin bulk gel.

No.	Polymer (%)	Cross-linking agent (%)	Coagulant (%)	Aging at 30°C for 1 year		Aging at 95°C for 1 year	
				Dehydration (%)	Strength retention (%)	Dehydration (%)	Strength retention (%)
1	0.3	0.9	0.1	N/A	91.35	2	83.76
2	0.3	1.2	0.1	N/A	92.41	2	84.37
3	0.3	1.5	0.1	N/A	93.26	1	85.67
4	0.4	0.9	0.1	N/A	92.54	1	87.87
5	0.4	1.2	0.1	N/A	93.78	0	89.42
6	0.4	1.5	0.1	N/A	94.67	0	91.08
7	0.5	0.9	0.1	N/A	92.87	0	90.17
8	0.5	1.2	0.1	N/A	94.56	0	91.75
9	0.5	1.5	0.1	N/A	97.85	0	92.18

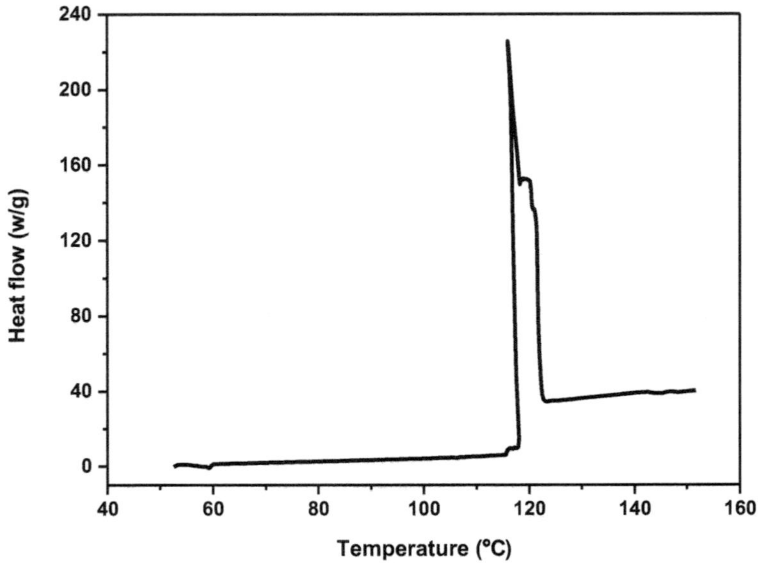

Figure 2.15 DSC curve of resin bulk gel.

temperature. As the temperature increases, the DSC curve drops sharply. It suggests that a small number of the chemical bonds with low bond energy in the structure of the gel begin to absorb heat and break. However, the number of these chemical bonds is too small to affect the strength of the bulk gel, so the curve changes in a relatively gentle manner. When the temperature reaches 120°C, the DSC curve rises sharply. This suggests that the structure of the gel has been destroyed over a wide range. The phase state of the gel transforms from a nonflowing state to a liquid state and loses its original strength, indicating the complete destruction of the structure of the gel [9]. Therefore the temperature of the reservoir should be lower than 120°C.

Based on the analysis of the factors affecting the formation of bulk gel, the concentrations of polymer and cross-linking agent determine the gelation time and strength. Bulk gels are equipped with temperature resistance, salt resistance, shear resistance, and thermal stability. Particularly, their temperature resistance up to 120°C enables them to meet the needs of preparing multiscale DPG. In the preparation process of the multiscale DPG system, the physical breaking does not change the chemical properties of the bulk gel. Therefore the multiscale DPG prepared from resin bulk gels are characterized with temperature resistance, salt resistance, and thermal stability.

2.3.4 Microstructure of resin bulk gel

To avoid the damage of the structure of the bulk gel, a Quanta 200 environmental scanning electron microscope (ESEM) was used to observe the microstructure of the resin bulk gel. During the test, a sample with a size of $\sim 0.3 \times 0.3$ cm was taken from the inside of the bulk gel and put into the copper grid. The initial pressure was set at 313–455 Pa, the temperature was set at 0°C, the acceleration voltage at 15 kV, and the working distance at 5–10 mm. The bulk gel used in the test was formulated with 0.4% type II functional polymer, 1.2% resin cross-linking agent, and 0.1% cross-linking coagulant, and gelled at 95°C. The test results are shown in Fig. 2.16.

Fig. 2.16 shows the microstructure of the resin bulk gel. According to Fig. 2.16, the bulk gel forms a compact and continuous three-dimensional structure after the gel formation, and there is no porosity in the three-dimensional compact structure. This indicates that the cross-linking density and strength between the polymer and the cross-linking agent molecules

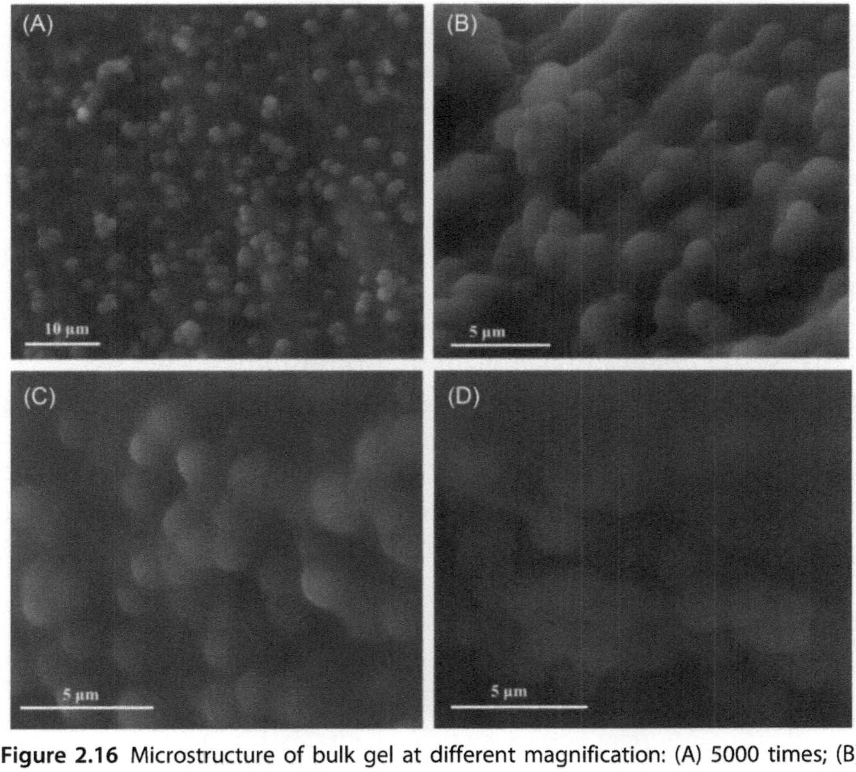

Figure 2.16 Microstructure of bulk gel at different magnification: (A) 5000 times; (B) 10000 times; (C) 10000 times; (D) 10000 times.

are very high. The environmental SEM results further show that many spheres with a particle size ranging from 1 to 3 μm are distributed on the surface of the gel, where they are closely combined and inset into the continuous gel surface. In the process of gelation formation, the acylamino of polymer and the hydroxyl of phenolic resin form an intensive spherical structure by dehydration condensation reaction. This structure, conducive to locked irreducible water inside the gel, makes the irreducible water hard to escape from spheroidal structure at high temperatures, thereby enhancing the long-term thermal stability of bulk gel [10]. During the preparation of DPG particles by the mechanical shear method, only physical shear is acting on the bulk gel, and the chemical structure of the bulk gel is not destroyed. Therefore DPG particles formed from bulk gel can still lock the irreducible water in the spherical structure, so that the prepared DPG particles has high thermal stability.

Section 2.4 Bulk gel for ultra high temperature reservoirs

For type V and type VI reservoirs, the temperature and salinity are comparatively high. The traditional bulk gels have poor tolerance to temperature and salinity because of their low cross-linking density and weak strength. By introducing inorganic nanoparticles into hybrid cross-links with polymer and cross-linker, a bulk gel strengthened with inorganic nanoparticles for high temperature (150°C) and high salinity (30×10^4 mg/L) was developed. It can be used to prepare multiscale DPG with good tolerance to high temperature and high salinity.

2.4.1 Cross-linking mechanism of strengthened bulk gel

The bulk gel strengthened with inorganic nanoparticles is composed of type III functional polymer, hexamethylenetetramine (HMTA), resorcinol, and inorganic nanoparticles. Through the decomposition of cross-linkers, cross-linking reaction, and nanoparticle adsorption, a kind of bulk gel system with tolerance to high temperature and high salinity is prepared [11–13].

1. The hexamethylenetetramine (HMTA) decomposes to form formaldehyde (CH_2O) and ammonia (NH_3) in an acidic environment or under high temperature. The chemical equation is shown as (2.6):

$$(CH_2)_6N_4 + 6\,H_2O \rightarrow 6\,CH_2O + 4NH_3 \tag{2.6}$$

2. The formaldehyde (CH_2O) reacts with resorcinol, forming 2,5-dihydroxymethyl hydroquinone. As is shown in (2.7):

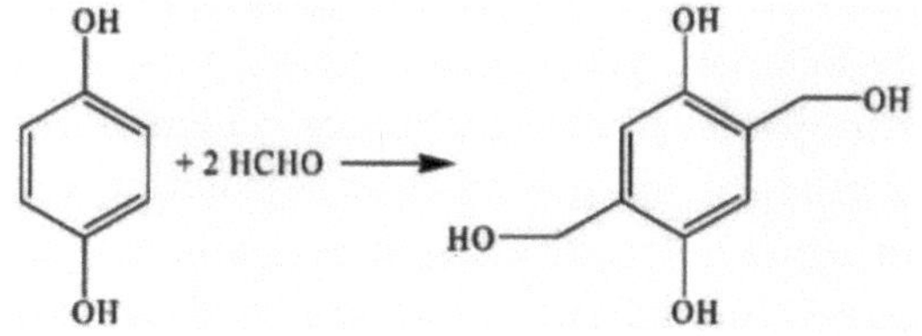

(2.7)

3. Then the phenolic resin precondensation molecules, which are the cross-linking agents cross-linked with polymer molecules, are generated by the condensation polymerization between 2,5-dihydroxymethyl and hydroquinone monomers, as shown in formula (2.8):

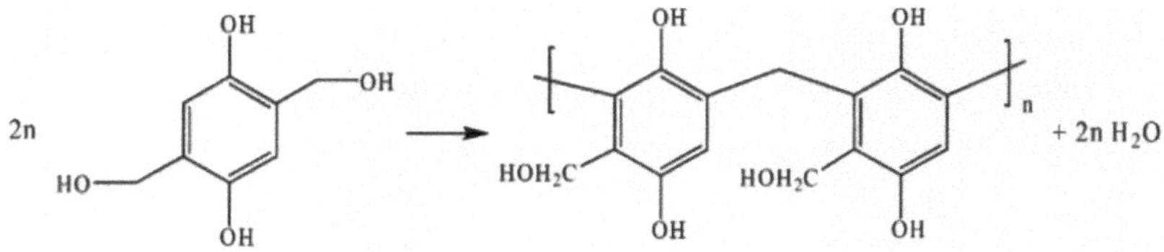

(2.8)

4. The amide group ($-CONH_2$) from the polymer molecules cross-links with the hydroxyl group ($-CH_2OH$) from the cross-linker, forming a stable uniformly distributed three-dimensional network structure. The cross-linking process is shown in Eqs. (2.9):

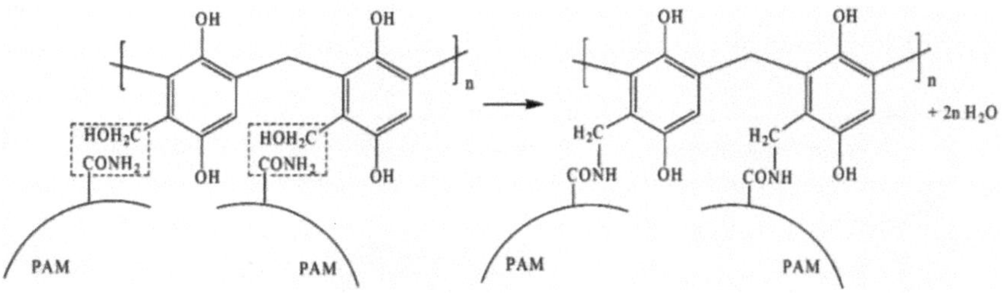

(2.9)

Through changing the cross-linking reaction conditions to adjust the above four steps, the gelation time and gelation strength of the bulk gel strengthened with inorganic nanoparticles can be controlled.

During the cross-linking reaction, the high temperature and high salinity resistance of bulk gel can be improved by adding inorganic nanoparticles. The strengthening effect of inorganic nanoparticles is mainly manifested in two aspects: the improved gelation strength and the strengthened thermal stability. Its strengthening mechanisms can be explained in the following two aspects.

1. Inorganic nanoparticles are adsorbed on polymer chain bunches and fill in the meshes of the gel structure, forming a certain microscopic arrangement structure. It provides strong support for the microstructure of the gel and improves the structural strength of the gel, as shown in Fig. 2.17.

Inorganic nanoparticles can increase the proportion of bound water in the bulk gel. There is a large number of hydroxyl groups (—OH) on the surface of inorganic nanoparticles. The hydrogen bonds formed between hydroxyl groups and water molecules turn some of the free water molecules into bound water. On the other hand, the inorganic nanoparticles have negative charges, which makes hydronium ions attracted by electrostatic force and further transforms part of the free water into bound water, as shown in Fig. 2.18. Hydrogen bonds and electrostatic attractions make a higher bound water ratio. The higher bound water ratio indicates stronger hydrophilicity of the gel, which makes it difficult for water molecules to separate out from the organic—inorganic hybrid cross-linked bulk gel. Thus the gel has better holding—water ability and heat resistance.

2.4.2 Factors influencing gelation of strengthened bulk gel

The gelation time is determined by the viscosity method. The gelation time (h) is defined as the time needed to reach the inflection point on the

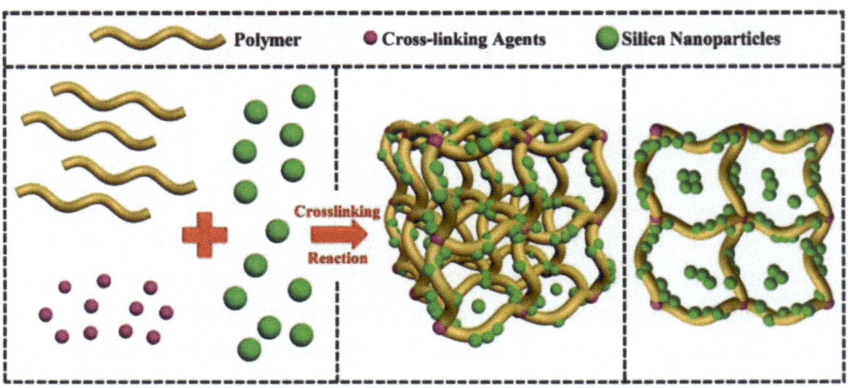

Figure 2.17 Schematic diagram of cross-linking process of bulk gel strengthened with inorganic nanoparticles.

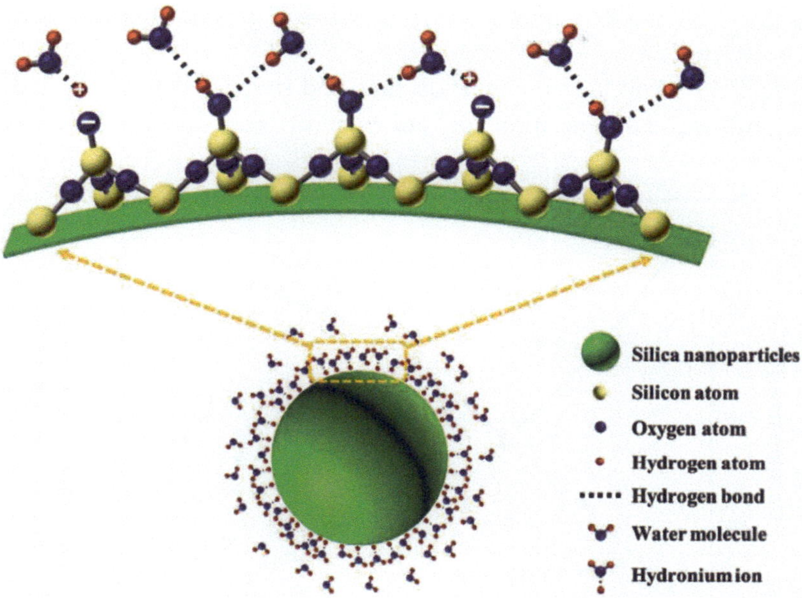

Figure 2.18 Schematic diagram of inorganic nanoparticles improves the proportion of bound water of bulk gel.

apparent viscosity versus time curve. The gelation strength is measured by the breakthrough vacuum method, and the dehydration rate and strength retention rate were used to characterize the thermal stability of the bulk gelation strengthened with inorganic nanoparticles.

In order to meet the technical requirements for the preparation and field application of the multiscale DPG, the influences of polymer, cross-linker, inorganic nanoparticles, temperature, and shearing to basic properties of the bulk gel strengthened with inorganic nanoparticles were investigated.

2.4.2.1 Influence of concentration on gelation performance

The concentrations of polymer and cross-linker are the key factors affecting the gelation performance. The gelation time and strength can be adjusted by changing the concentration of polymer or cross-linker. When the influence of polymer concentration was evaluated, the cross-linker concentration was fixed at 0.3%. When the influence of cross-linker concentration was evaluated, the polymer concentration was fixed at 0.6%. The gelation temperature was 120°C, and the salinity of simulated water was 2.1×10^5 mg/L (concentration of calcium and magnesium ions is 8.0×10^3 mg/L).

Fig. 2.19 shows that the gelation time and strength can be adjusted by changing the concentration of polymer or cross-linker. Under the

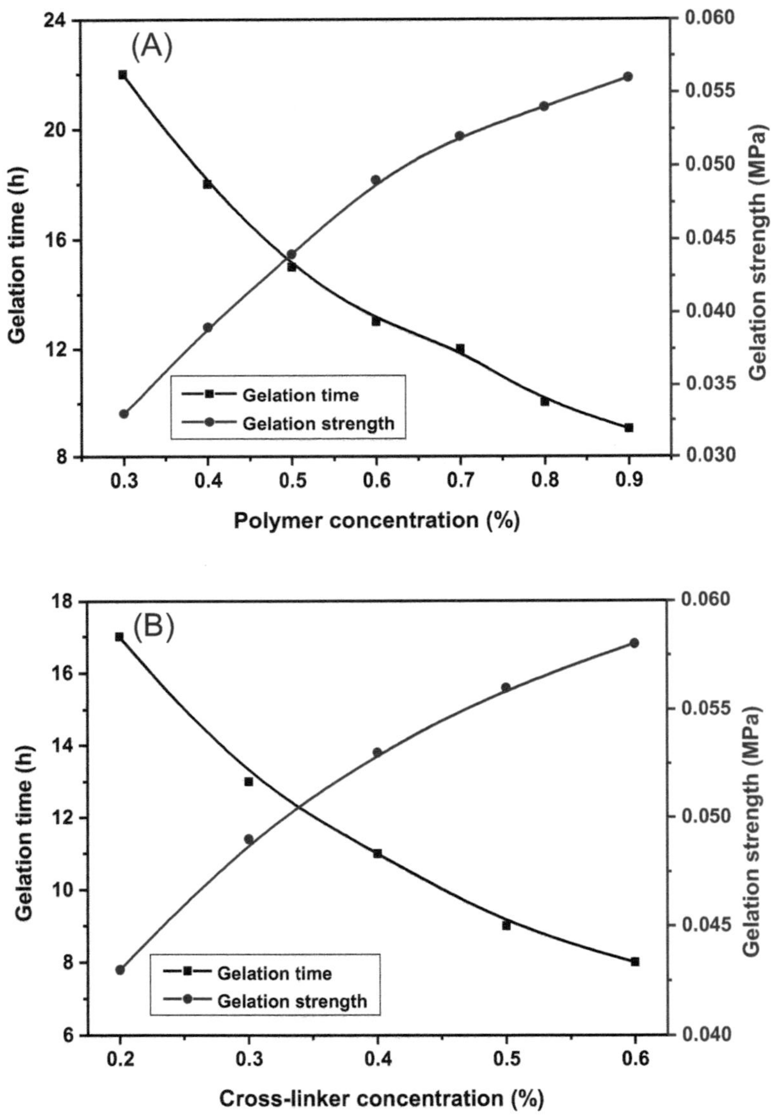

Figure 2.19 Influence of polymer and cross-linker concentrations on gelation time and strength: (A) influence of polymer concentration; (B) influence of cross-linker concentration.

conditions of target reservoir and concentrations of components in the test, the gelation time is controllable between 8 and 22 hours, and the gelation strength between 0.033 and 0.058 MPa. With the increasing of polymer or cross-linker concentration, the gelation time is shortened and the gelation strength is improved. This is because increasing the concentration of polymer or cross-linker provides more amide groups and hydroxyl groups that participate in the cross-linking reaction. This helps accelerate the gelation rate, shorten the gelation time, increase the cross-linking density, and improve the gelation strength of the bulk gel.

2.4.2.2 Influence of inorganic nanoparticles on the gelation performance

The influence of inorganic nanoparticles on the gelation performance was investigated, with 0.3% polymers, 0.3% cross-linker A, and 0.3% cross-linker B concentration, at 120°C. The test results are shown in Figs. 2.20 and 2.21. The gelation time was significantly shortened when the gelling solution is prepared by adding a certain number of inorganic nanoparticles. The gelation time decreased from about 22 hours to about 12 hours by addition of 0.3% nanomaterials. Nanoparticles could obviously improve the strength of bulk gel. When the concentration of nanomaterials increases to 0.3%, the gelation strength increases to 0.06 MPa. It can

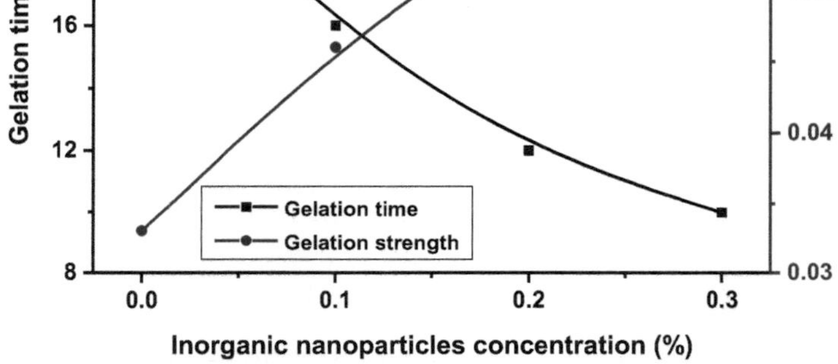

Figure 2.20 Influence of inorganic nanoparticles on the gelation time and strength.

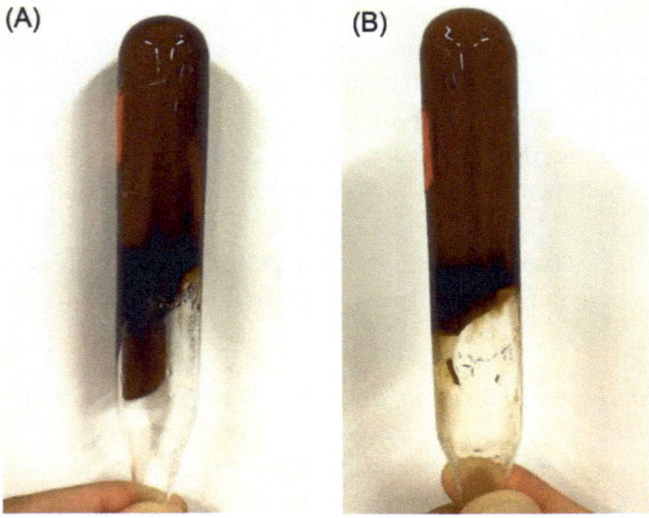

Figure 2.21 Influence of inorganic nanoparticles on the gelation strength: (A) not added; (B) 0.2% inorganic nanoparticles.

be seen from photos of the bulk gel shown in Fig. 2.21 that the strength of bulk gel is improved significantly after the addition of nanoparticles.

2.4.2.3 Influence of temperature on the gelation performance

Temperature is one of main factors that determine the cross-linking process of bulk gel. The effects of temperature on the typical gel and strengthened gel were compared. The tested gels were composed of 0.6% polymer, 0.3% cross-linker A, 0.3% cross-linker B, and 0.2% inorganic nanoparticles. The results are shown in Fig. 2.22. Temperature has a great influence on the gelation performance. With the increasing of temperature, the gelation strength is significantly increased and the gelation time is notably shortened. This is because the thermal motion of molecules in gelling solution increases, providing more energy for the cross-linking reaction. There are more polymer and cross-linker molecules involved in the reaction, improving the effective collision and contact probability between molecules. The reaction rate becomes faster and the cross-linking density is higher, which means shorter gelation time and higher gelation strength.

2.4.3 Thermal stability of strengthened bulk gel

The thermal stability of the bulk gel was investigated by inspecting the dehydration rate and strength retention rate. The test results are shown in

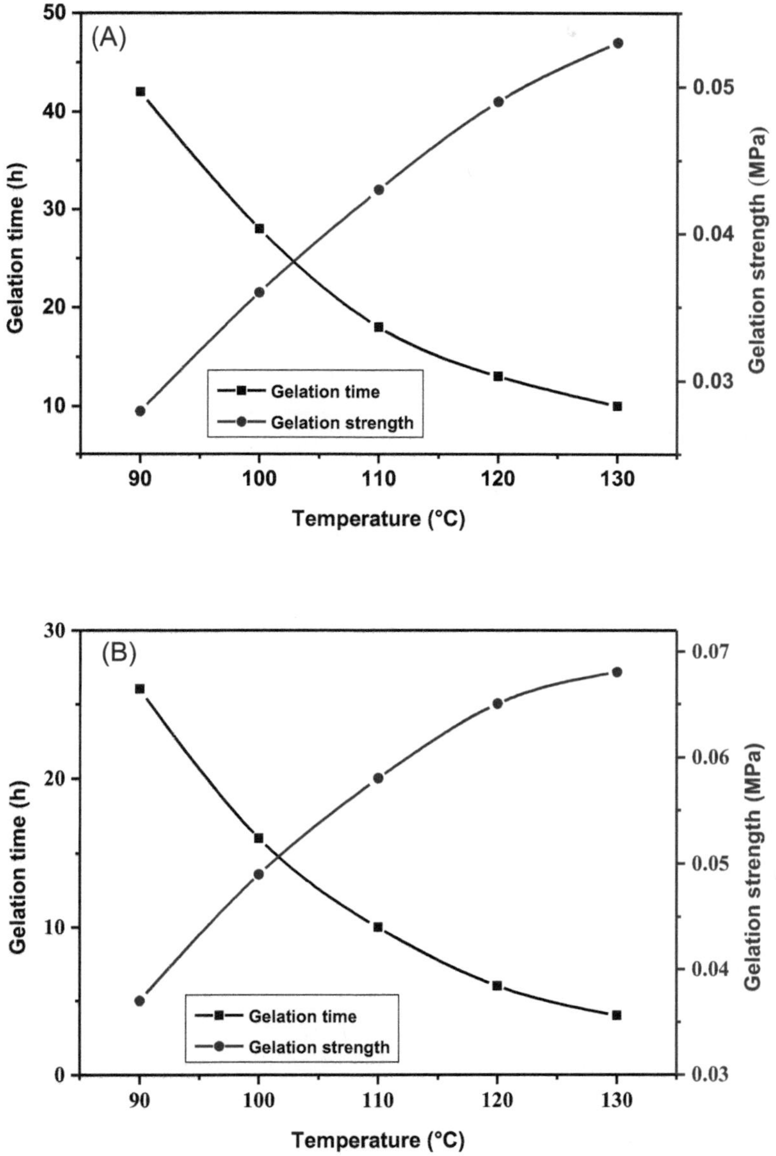

Figure 2.22 Influence of temperature on the gelation time and strength: (A) not added; (B) 0.2% inorganic nanoparticles.

Table 2.7. After 90 days of aging at 120°C, the two kinds of bulk gel samples don't engage in dehydration. The strength decreases so slightly as not to notice, indicating that the prepared bulk gels for ultrahigh

Table 2.7 Aging stability of bulk gel under high-temperature condition.

Plan	Aging 30 days		Aging 60 days		Aging 90 days	
	Dehydration	Strength retention rate (%)	Dehydration	Strength retention rate (%)	Dehydration	Strength retention rate (%)
0.6% polymer + 0.3% cross-linker A + 0.3% cross-linker B	No	94.3	No	89.4	No	88.5
0.6% polymer + 0.3% cross-linker A + 0.3% cross-linker B + 0.2% inorganic nanoparticle	No	97.7	No	95.2	No	94.6

temperature reservoirs had good thermal stability. In addition, the strength retention rate of gel with the addition of inorganic nanoparticles was higher than that of gel with no additions, indicating that the addition of inorganic nanoparticles could effectively improve the thermal stability.

The thermal stabilities of the bulk gels were further evaluated by DSC measurement. The DSC curves are shown in Fig. 2.23, the temperature at inflection point of bulk gel without inorganic nanoparticles is about 138.0°C. Before the inflection point, the gel samples absorb heat with the increase of temperature. When the temperature reaches 138°C, the bulk gel samples begin to release heat. This indicates that the energy absorbed by the sample has outgrown the energy barrier of chemical bonds breaking in the sample, thus leading to the breaking of chemical bonds and the damaging of the bulk gel's structure [14,15].

When the temperature exceeds 138°C, the structure of the bulk gel is destroyed, its phase state changes and its stability is reduced. Thus the critical stable temperature of the bulk gel prepared without inorganic

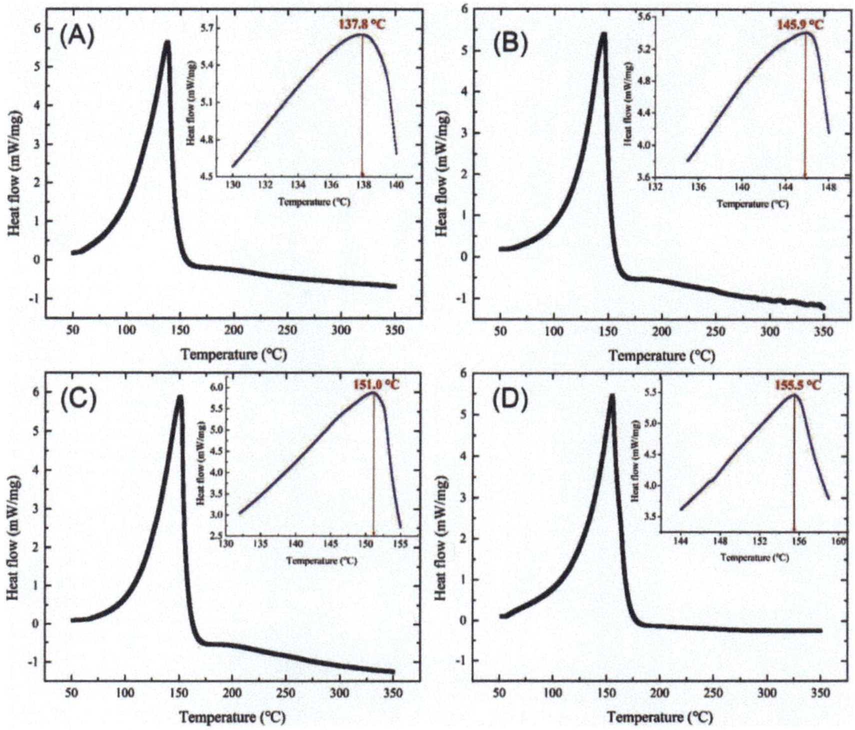

Figure 2.23 DSC curves of bulk gel: (A) not added; (B) 0.1% inorganic nanoparticle; (C) 0.2% inorganic nanoparticle; (D) 0.3% inorganic nanoparticle.

nanoparticles is 138°C. Similarly, the critical temperature to maintain thermal stability of the bulk gels prepared by adding inorganic nanoparticles can be raised from 138°C to about 146°C, 151°C, and 156°C, respectively. The prepared bulk gels can meet the thermal stability requirement of ultrahigh temperature reservoirs, and the addition of inorganic nanoparticles can significantly improve the thermal stability.

2.4.4 Microstructure of the strengthened bulk gel

The microstructure of the gels before and after the addition of inorganic nanoparticles was analyzed by ESEM measurement. The ESEM photos of the two kinds of bulk gel samples are shown in Fig. 2.24, the amide groups ($-CONH_2$) from the polymer cross-linked with the hydroxyl groups ($-CH_2OH$) from the cross-linkers, which contributes to a compact mesh-like

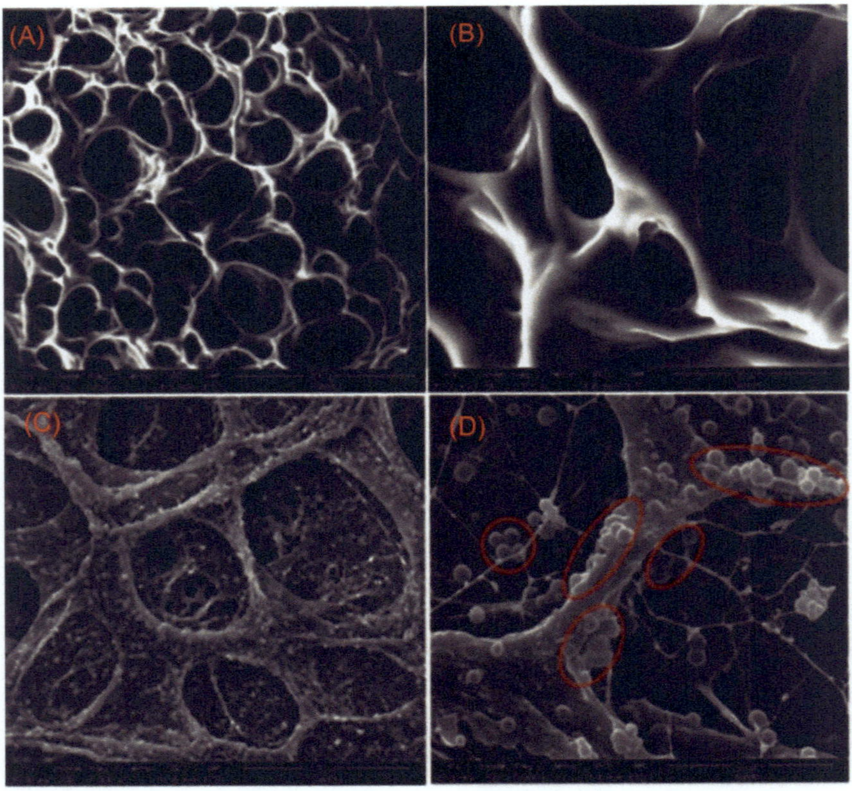

Figure 2.24 The ESEM photos of bulk gel: (A), (B) not added; (C), (D) added inorganic nanoparticles.

network structure. Nanoparticles were adsorbed on the skeleton and network structure of the bulk gel when the bulk gel was prepared with inorganic nanoparticles. The adsorption and arrangement of nanoparticles support the microscopic skeleton and improve the strength of the bulk gel.

On the other hand, due to hydroxyl groups and negative charges of the surface of inorganic nanoparticles, part of the free water in the bulk gel is transformed into bound water by hydrogen bonding and electrostatic attraction. The significant increase of bound water proportion in the gel makes it difficult for water molecules to separate from the bulk gel. Consequently, the water-holding capacity and stability of the gel are improved.

References

[1] Zhao F. Oilfield chemistry(II). China University of Petroleum Press; 2010.
[2] Ma L, Yang T, Li X, et al. Study on crosslinking reaction process, microstructure and influencing factors of gels formed by polymer and zirconium acetate. Oilfield Chemistry 2015;32(1):48–52.
[3] Zhao F. Oil recovery agent. China University of Petroleum Press; 1996.
[4] Dai C, Zhou H, Zhang G. Studies on influencing factors for formation of acidic polyacrylamide/chromium gel. Oilfield Chemistry 2002;19(1):29–32 46.
[5] Bryant L, Bartosek M, Lockhart T. Laboratory evaluation of phenol-formaldehyde polymer gelants for high-temperature applications. Journal of Petroleum Science and Engineering 1997;17:197–209.
[6] Zhao G, Dai C, Chen A, et al. Experimental study and application of gels formed by nonionic polyacrylamide and phenolic resin for in-depth profile control. Journal of Petroleum Science and Engineering 2015;135:552–60.
[7] Sydansk R. Conformance improvement in a subterranean hydrocarbon bearing formation using a polymer gel. United States Patent, No.4683949, 1987.
[8] Tang J, Tung M, Zeng Y. Compression strength and deformation of gellan gels formed with mono-and divalent cations. Carbohydrate Polymers 1996;29:11–16.
[9] Henry E, An L. Fall-off from extrapolated values of all chemical reactions at very high temperatures. Proceedings of the National Academy of Sciences 1975.
[10] Zhao G, Dai C, Zhao M, et al. The use of environmental scanning electron microscopy for imaging the microstructure of gels for profile control and water shutoff treatments. Journal of Applied Polymer Science 2014;131:2231–9.
[11] Ahmad M. A review of thermally stable gels for fluid diversion in petroleum production. Journal of Petroleum Science and Engineering 2000;26(1):1–10.
[12] Sengupta B, Sharma V, Udayabhanu G. Gelation studies of an organically cross-linked polyacrylamide water shut-off gel system at different temperatures and pH. Journal of Petroleum Science and Engineering 2012;81:145–50.
[13] Liu Y, Dai C, Wang K, et al. New insights into the hydroquinone (HQ)–hexamethylenetetramine (HMTA) gel system for water shut-off treatment in high temperature reservoirs. Journal of Industrial and Engineering Chemistry 2016;35:20–8.
[14] Liu Y, Dai C, Wang K, et al. Study on a novel cross-linked polymer gel strengthened with silica nanoparticles. Energy & Fuels 2017;31(9):9152–61.
[15] Zhao G, Dai C, You Q, et al. Study on formation of gels formed by polymer and zirconium acetate. Journal of Sol-Gel Science and Technology 2013;65(3):392–8.

CHAPTER 3

The preparation technology of dispersed particle gel

In this chapter, the feasibility of mechanical shearing technology to prepare multiscale DPG particles is discussed using a mathematics method, and the optimal technical parameters are studied in terms of bulk gel, mechanical shearing factors, and preparation procedures. Based on the change of viscosity, the preparation mechanisms in the course of shear cross-linking reaction of DPG particles are explained. Also, the relationship between DPG particle size and preparation technical parameters is established; industrialized software of DPG preparation is developed; production workshop and skid-mounted equipment line are designed; and simplistic and convenient DPG particles industrial production is achieved.

Section 3.1 Establishment of DPG preparation technique

Multiscale DPG particles are generated by using the mechanical shear method to apply a certain shear force on the bulk gel cross-linked on the ground. At present, there are four main methods to prepare DPG around the world: coaxial cylinder shear cross-linking method; pipe flow shear cross-linking method; high-speed mechanical shearing mechanical method; and colloid mill shearing method. The coaxial cylinder shear cross-linking method, proposed by Chauveteau [1,2], adopts a relatively small coaxial cylinder, which prevents it from adapting to the industrial production but confines it to laboratory study. The pipe flow shear cross-link method, mainly proposed by Fulin Zhao, Qing You et al. [3–5], applies a peristaltic pump as shear equipment to prepare DPG, which is, however, subject to the impaction of the peristaltic pump. Its displacement is too small—with a mere preparation of 5–20 mL per hour—to fulfill the demand of applications on the spot. In 2009 Caili Dai, Guang Zhao et al. [6–8] proposed the mechanical shear method which uses improved colloid mill as preparation equipment. This method is characterized with high production efficiency ($0 \sim 12$ tons/h), simple operation and continuous production, making it feasible in the practices. Therefore this section intends to employ

Theory and technology of multiscale dispersed particle gel for in-depth profile control
DOI: https://doi.org/10.1016/B978-0-323-99849-9.00003-5

mathematical theory and methods to analyze the technological feasibility of colloid milling to prepare multiscale DPG practices.

Colloid milling consists of stator and rotor (Fig. 3.1). When flowing through the gap between stator and rotor, the fluid passes through three grinding regions: rough grinding region, fine grinding region and ultrafine grinding region. The high-speed relative motion through the stator imposes on the fluid actions of shear, milling, and high-frequency vibration. According to the flow status between adjacent stator, a relevant mathematics model is built and the feasibility of the colloid mill shear method to prepare DPG is analyzed.

Since the same flow pattern exists in the gap among three grinding regions (rough grinding region, fine grinding region, and ultrafine grinding region), the analysis can be started with the monolayer flow pattern. In order to make easy the establishment of the model, in the course of theoretical analysis, we suppose the fluid flow to be laminar flow and the fluid to do circumferential flow between the stators, where δ is the gap distance between adjacent stator, r the radius, and ω the angular speed. A polar system is established as Fig. 3.2. Flow velocity is independent from the X-axis direction. Under the action of friction, the pressure gradient

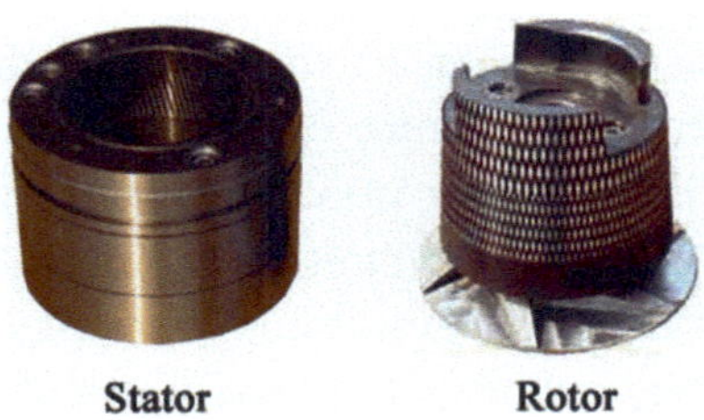

Figure 3.1 Stator and rotor of colloid mill.

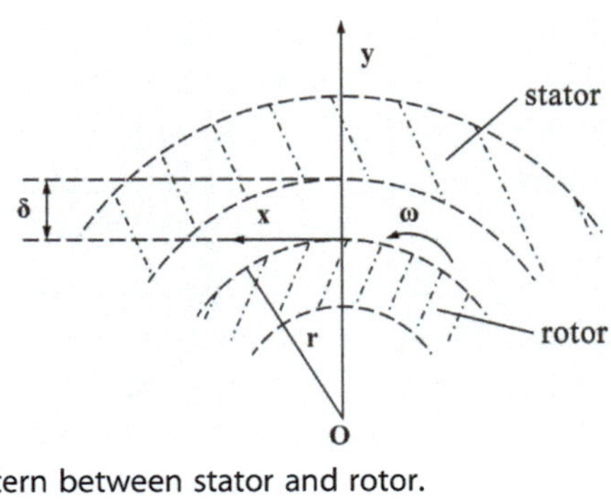

Figure 3.2 Flow pattern between stator and rotor.

decreases periodically, with a supposed pressure gradient of $\frac{dp}{dx}=c$ as constant.

The equation of fluid motion between stator is represented by the Couette flow vector different equation.

Continuity equation:

$$\nabla \vec{V} = 0 \quad (3.1)$$

Here: $\nabla \vec{V}$ is flow velocity gradient of fluid.

Momentum equation (N-S equation):

$$\rho\left(\frac{\partial \vec{V}}{\partial t} + \vec{V} \cdot \nabla \vec{V}\right) = -\nabla P + \mu \cdot \nabla^2 \vec{V} \quad (3.2)$$

Here: ρ is fluid density; t is shear time; $\vec{V}$ is flow velocity vector; ∇P is pressure gradient; μ is fluid dynamic viscosity; $\nabla^2 \vec{V}$ is flow velocity Laplace operator.

Boundary condition:

$$y = \delta, \quad V = 0; \quad (3.3)$$

$$y = 0, \quad V = r\omega; \quad (3.4)$$

Here: δ is gap distance between stator and rotor; r is radius of rotor; ω is angular velocity of rotor.

Substitute $\frac{dp}{dx} = c$, Eq. (3.1) into Eq. (3.2):

$$\mu \frac{d^2 V}{dy^2} - \frac{dp}{dx} = 0 \quad (3.5)$$

Here: $\frac{d^2 V}{dy^2}$ is second derivative of flow velocity to the y direction; $\frac{dp}{dx}$ is pressure gradient of x direction.

Therefore Eq. (3.5) is an ordinary differential equation to solve velocity. $\frac{dp}{dx}$ is independent with y, thus Eq. (3.5) can be expressed as:

$$V(y) = \frac{1}{2\mu}\frac{dp}{dx}y^2 + C_1 y + C_2 \quad (3.6)$$

Here: C_1 is integration constant; C_2 is integration constant.

Substitute boundary condition into Eq. (3.6):

$$V(y) = \frac{1}{2\mu}\frac{dp}{dx}y(y-\delta) + r\omega\left(1 - \frac{y}{\delta}\right) \quad (3.7)$$

From Eq. (3.7), $V(y)$ is the superposition of steady laminar flow and linear flow, that can be expressed as Eq. (3.8):

$$V = (y) = V_1(y) + V_2(y) \tag{3.8}$$

$$V_1(y) = \frac{1}{2\mu}\frac{dp}{dx}y(y - \delta) \tag{3.9}$$

$$V_2(y) = r\omega\left(1 - \frac{y}{\delta}\right) \tag{3.10}$$

Steady laminar flow $V_1(y) = \frac{1}{2\mu}\frac{dp}{dx}y(y - \delta)$ can be illustrated as shown in Fig. 3.3.

Linear flow $V_2(y) = r\omega\left(1 - \frac{y}{\delta}\right)$ can be illustrated as Fig. 3.4.

According to the law of Newton inner friction, the relationship between shear stress and strain can be expressed as Eq. (3.11):

$$\tau_{xy} = \tau_{yx} = \mu\left(\frac{\partial V_x}{\partial y} + \frac{\partial V_y}{\partial x}\right) \tag{3.11}$$

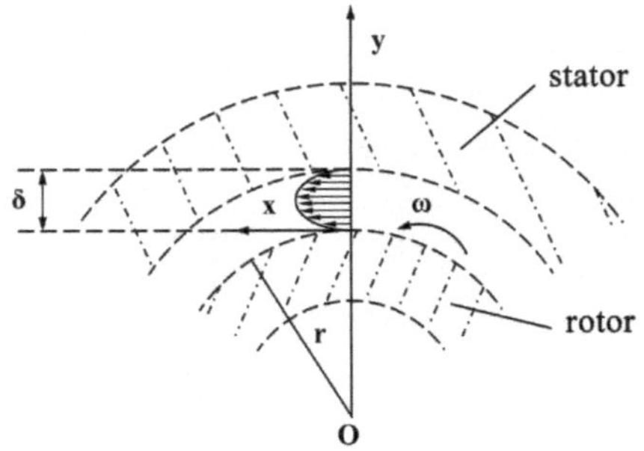

Figure 3.3 Steady laminar flow.

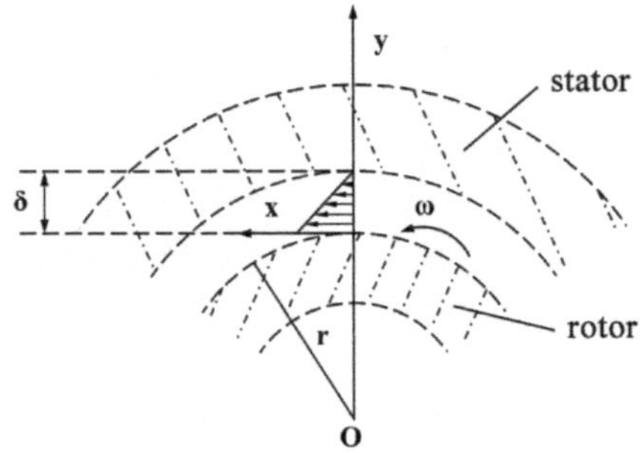

Figure 3.4 Linear flow.

Here: τ_{xy} is shear stress in y direction; τ_{yx} is shear stress in x direction; $\frac{\partial V_x}{\partial y}$ is gradient of x-direction velocity in y direction; $\frac{\partial V_y}{\partial x}$ is gradient of y-direction velocity in x direction.

Substitute Eqs. (3.8), (3.9), and (3.10) into Eq. (3.11):

$$\tau_{xy} = y\frac{dp}{dx} - \frac{\delta}{2}\frac{dp}{dx} - \frac{\mu r\omega}{\delta} \tag{3.12}$$

According to Eq. (3.12), it is obvious that $|\tau_{xy}| = \left|\frac{\delta}{2}\frac{dp}{dx} - \frac{\mu r\omega}{\delta}\right|$ reaches maximum when $y = \delta$ because $\frac{dp}{dx} = c < 0$. That is to say, the fluid is subjected to greatest shear stress at the wall of the stator. From Eq. (3.12), it can be found that the radius and the angular velocity of the stator are the same for specific colloid milling. The shear stress material subjected increases as gap distance of the stator decreases, and the particle size shrinks after shearing. Therefore the friction that the material receives in the gap can be managed by adjusting the gap between the radius of the stators through the positioning plate, so as to control the size of the particles prepared.

Flow rate could be expressed according to Poiseuille equation, the volume flow rate through dy is expressed as:

$$dq = vdA = 2\pi v(r + y)dy \tag{3.13}$$

Here: v is glow velocity; dA is unit cross section of fluid flowing into the gap.

Total flux through the gap:

$$Q = \int_0^q dq = \int_0^\delta 2\pi v(r + y)dy = \int_0^\delta 2\pi(r + y)\left[\frac{1}{2\mu}\frac{dp}{dx}y(y - \delta) + r\omega\left(1 - \frac{y}{\delta}\right)\right]dy \tag{3.14}$$

Or:

$$Q = \pi\left(-\frac{1}{12\mu}\frac{dp}{dx}\delta^4 - \frac{1}{6\mu}\frac{dp}{dx}\delta^3 + \frac{1}{3}\delta^2 r\omega + \delta r\omega\right) \tag{3.15}$$

Here $\frac{dp}{dx} < 0$, Eq. (3.15) indicates that if gaps between adjacent stators are bigger, the flux will be larger, therefore the flux of bulk gel can be controlled by the distance between adjacent stators to achieve the control of DPG production.

Considering Eqs. (3.12) and (3.15), the particle size and yield mainly depend on the shear distance of adjacent stator, the smaller shear distance results in bigger shear stress, smaller particle size, and lower yield.

Therefore according to the reality, a reasonable distance between adjacent stators can be selected to produce DPG particles. In light of the mathematical procedure, it is workable to produce DPG particles by colloid milling. Different DPG particle sizes can be prepared by adjusting the stator gap and shearing time. The size of DPG particles mainly depends on shearing time and shear stress, so a functional relationship between DPG particle size and shearing time can be established as Eq. (3.16) shows:

$$f(D) = f\left(t, \tau_{xy}\right) = f\left(t, \delta_y, \mu\right) \tag{3.16}$$

Here: D is DPG particle size; δ_y is gap distance between stator and rotor.

Combining Eq. (3.16) and the experiment results, a relevant mathematical model can be established, so as to lay the theoretical foundation for DPG preparation by colloid milling.

Section 3.2 Optimization of process parameters for DPG preparation

The industrialization production of DPG particles has the characteristic of simplicity and high efficiency. In this section, with the resin bulk gel as an example, the process of multiscale DPG preparation is optimized, and the influences of the bulk gel strength, the mechanical shear parameters (shear rate and shear time), and the bulk gel–water mixing ratio on the performances of DPG products are clarified.

3.2.1 Efficiency of DPG preparation

3.2.1.1 Procedure of DPG preparation

An improved industrial colloid mill was chosen as the shearing facility to prepare DPG particles. This facility is equipped with a condensing circulation system, which allows it to work for a long period of time and for its shearing rate to be controllable from 0 to 70 Hz, where the shearing rate is represented by frequency (Hz) in this study. The procedure of DPG preparation is as follows.

1. Prepare resin gel solution at room temperature, stir uniformly, and leave to cross-link in 95°C oven.
2. Put the bulk gel and water into colloid mill with a certain ratio, and start colloid milling. Then, adjust the shearing rate and shear different times until the dispersion system forms. The dispersed solution obtained is the product of multiscale DPG particles.
3. Pack the DPG production and keep it at room temperature.

3.2.1.2 *Method for testing DPG preparation efficiency*

It is hard to determine the degree of reaction from appearance state because of the even dispersing of DPG in the aqueous solution. The bulk gel still contains some acylamino groups after cross-linking between the cross-linker and polymer, so it will be efficient to use amylum–triodide to measure reaction efficiency. In this study, an amylum–cadmium iodide method was used to measure the reaction efficiency of DPG preparation.

3.2.1.2.1 Experimental principle

The amylum–cadmium iodide method utilizes the first-step reaction of Hofmann Rearrangement. The acylamino group from bulk gel or the polymer not involved in the reaction and bromine water react together to form N-bromoamide. The excessive bromine water can be removed by sodium formate. The N-bromoamide generated is further hydrolyzed to produce subbromic acid, a substance which can quantitatively oxidize iodine ion into blue triiodo–amylum complex. Therefore the degree of reaction can be determined by measuring its absorbance. This reaction continues in low acylamino group concentration, it can measure preparation efficiency rapidly in this way; the reaction is illustrated in Fig. 3.5.

3.2.1.2.2 Method of reagent preparation

Preparation method of buffer solution. 12.50 g sodium acetate trihydrate ($CH_3COONa \cdot 3H_2O$) was weighted accurately and dissolved in 400 mL

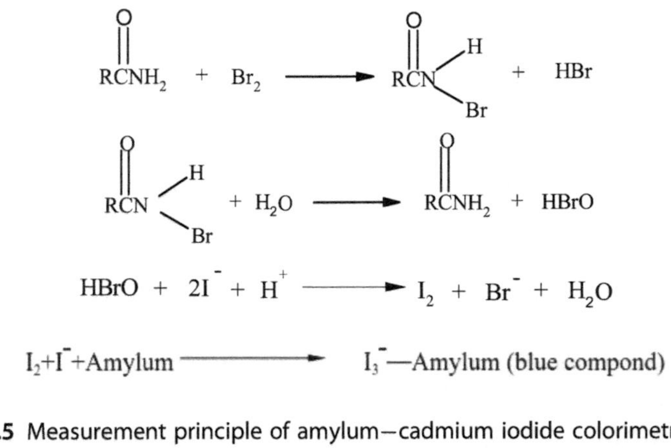

Figure 3.5 Measurement principle of amylum–cadmium iodide colorimetry.

deionized water. Then, 0.25 g of hydrated aluminum sulfate ($Al_2(SO_4)_3 \cdot 18H_2O$) was added. Acetic acid was used to adjust the pH of the solution to 4.0, and after that the solution was diluted to 500 mL for standby. Preparation of amylum–cadmium iodide reagent. First, 1.25 g amylum was weighted accurately and dissolved in 200 mL deionized water. The solution was then boiled for 10 minutes, cooled, and diluted to 400 mL. Next, 5.50 g of cadmium iodide was added to the solution. After the cadmium iodide was dissolved, the solution was filtered and diluted to 500 mL for standby.

3.2.1.2.3 Procedure of measurement

1. 5 mL of buffer solution was added into a 50 mL volumetric flask. The polymer content was controlled to 15–300 μg, and the test solution should not exceed 30 mL, which was then diluted to 35 mL with deionized water.
2. The solution obtained in the above step was mixed well, then into which 1 mL of saturated bromine water was added. After being shaken well, the solution was set to react for 10 minutes, and after that 3 mL sodium formate solution with 1% mass fraction was added.
3. After the reaction for 5 minutes, 5 mL amylum–cadmium iodide reagent was added immediately. Then, the new solution was diluted with deionized water to the particular scale, shaken, and stabilized for 10 minutes. The absorbance was determined by 722 spectrophotometer in 1 cm cuvette at 590 nm.

The standard absorbance curve of type II functional polymer was determined by the amylum–cadmium iodide method. The results are shown in Fig. 3.6.

3.2.1.2.4 Efficiency of DPG preparation

Bulk gel (0.12% functional polymer + 1.2% cross-linker + 0.1% coagulant) was sheared for 3–15 minutes in 20 and 40 Hz separately, and then centrifuged at 5000 rpm for 15 minutes in a high-speed centrifuge. The supernatant was taken to determine the preparation efficiency of DPG. The results are shown in Fig. 3.7.

It can be seen from Fig. 3.7 that the efficiency of DPG preparation by the mechanical shearing method is relatively high, mostly higher than 90%, and the influence of shear time and share rate on the reaction degree is small. Since the mechanical shearing method does not involve the chemical reaction of bulk gel in the preparation of DPG particles, it only

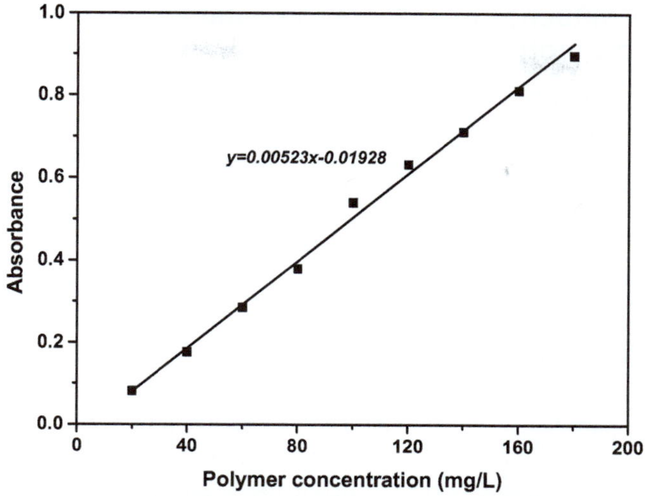

Figure 3.6 The standard absorbance curve of different polymer concentration.

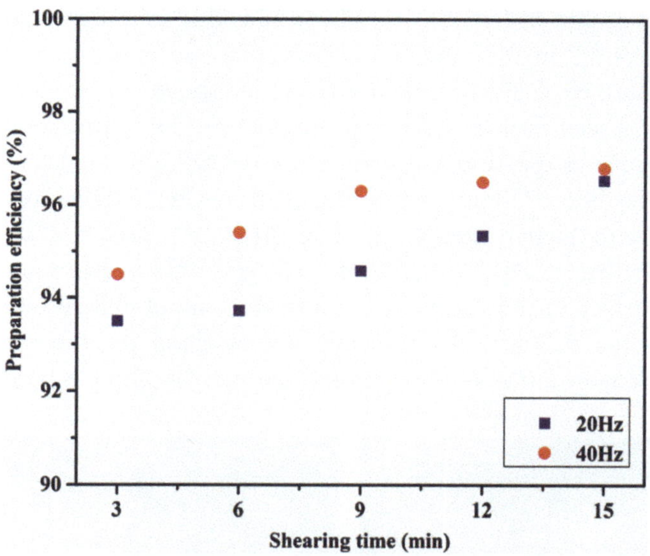

Figure 3.7 Preparation efficiency of multiscale DPG particles.

produces DPG particles by crushing and grinding the bulk gel with mechanical shearing force, without any degradation of raw materials. Therefore the mechanical shear method has the advantages of simple operation and high preparation efficiency.

3.2.2 The influence of preparation parameters on the microscopic morphology of DPG

Scanning electron microscopy (SEM) was used to observe the microscopic morphology of multiscale DPG samples. The operation method is as follows. Firstly, the newly prepared DPG sample was diluted to 300 mg/L. Then a small amount of the sample was placed on the copper mesh. Last, the copper mesh containing the sample was dried in a vacuum drying oven for later use.

3.2.2.1 Effect of bulk gel strength

The effect of bulk gel strength on the microscopic morphology of multiscale DPG particles was studied. Two kinds of bulk gels with different gelation time were selected: one with 0.25% type II functional polymer + 1.2% cross-linker + 0.1% coagulant, a gelation time of 14 hours, and a gel strength of 0.043 MPa; the other with 0.4% type II functional polymer + 1.2% cross-linker + 0.1% coagulant, a gelation time of 8 hours, and a gel strength of 0.052 MPa. Those bulk gels were mixed with water by a ratio of 3:2 in the modified colloid mill, and then sheared at 40 Hz for 6 minutes, respectively. The test results are shown in Fig. 3.8.

It can be seen from Fig. 3.8 that the prepared DPG particle is a sphere with a regular shape, and the particle size of DPG formed by the weak bulk gel is about 740 nm, while the particle size of DPG prepared by bulk gel with higher strength is about 1600 nm, which is significantly larger than that of DPG formed by weak gel. This is determined by the special structure of the stator and rotor in the colloid mill. When the bulk gel enters the colloid mill, it mainly moves along the direction of the rotor movement in the pore space, and the shearing force is perpendicular

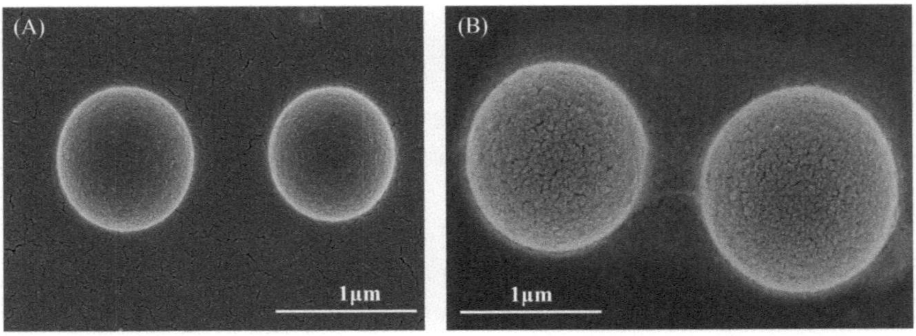

Figure 3.8 Effect of bulk gel strength on the microscopic morphology of DPG particles: (A) gelation strength, 0.043 MPa; (B) gelation strength, 0.052 MPa.

to the movement direction of the DPG. When the bulk gel enters the coarse grinding area of the colloid mill, the spherical structure of the bulk gel breaks due to the high-speed mechanical shearing force forming large and nonuniform particles. The particles formed at this stage then enter the fine grinding area, and the size of DPG particles is further decreased due to the influence of shearing force in this area. At the same time, under the influence of centrifugal force, the inner wall of the stator further rounds the DPG particles, making the particle shape more regular. The DPG formed at this stage finally enters the ultrafine grinding area, which receives balanced stress in the transverse and longitudinal directions, making the particle size and viscosity of the DPG further decrease. When the circulatory system is opened and the shearing time increased, the DPG enters the stator rotor gap and stator chute again, which further decreases the particle size and forms the DPG with uniform particle size distribution. When the bulk gel has higher strength, the cross-linking density grows, and the structure in the gel becomes harder to be broken. Therefore the particle size of the DPG formed is larger compared to that of DPG formed from bulk gel with lower strength at the same shearing parameters.

3.2.2.2 Effect of shearing rate

The bulk gel (formula: 0.4% type II functional polymer + 1.2% cross-linker + 0.1% coagulant) was mixed with water at a ratio of 3:2 and sheared for 6 minutes at a shearing rate of 10, 40, and 60 Hz. Samples were taken to observe the microscopic morphology of DPG particles and the results are shown in Fig. 3.9. It can be seen from the figure that DPG particles with regular shape can be formed under three different shearing rates, and the average particle size are 2900, 1580, and 940 nm. It can be seen from formula (3.12) that the shearing force exerted by the colloidal

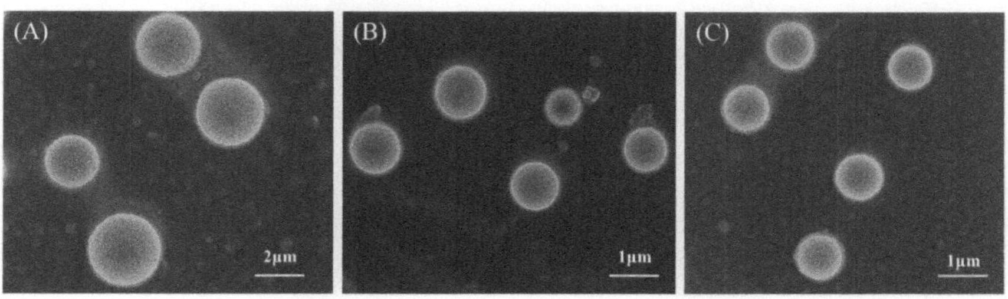

Figure 3.9 Effect of shear rate on the microscopic morphology of DPG particles: (A) 10 Hz; (B) 40 Hz; (C) 60 Hz.

mill on the bulk gel is bigger ($\tau_{60} > \tau_{40} > \tau_{10}$) when the shearing rate is higher, and the bulk gel is easier to be broken forming small particles. The shearing force only breaks the bulk gel at a low shearing rate, and the broken bulk gels can deform passing through the ultrafine grinding gap between the stator and the rotor as the broken gels are soft particles. However, if the shearing force is too small, these larger particles cannot pass through the stator chute and cannot be rounded, which makes the size of DPG particles prepared at a low shear rate larger.

3.2.2.3 Effect of shearing time

The bulk gel (formula: 0.4% type II functional polymer + 1.2% cross-linker + 0.1% coagulant) was mixed with water at a ratio of 3:2 and sheared at a shearing rate of 40 Hz for 2, 6, and 10 minutes, respectively. The microscopic morphology of the prepared DPG particles was observed as shown in Fig. 3.10.

It can be seen from Fig. 3.10 that the DPG particles prepared with different shearing time are spheres with regular shape. The average particle size distribution of DPG sheared for 2 minutes are 2200 nm, and the average size are 1200 nm after 10 minutes of shearing. This is because the bulk gel first breaks at the junction of balls under high shearing force, forming DPG particles with large sizes and irregular shapes. With the increase of shearing time, the larger particles shrink to smaller particles, and their shape turns regular after three-stage cyclic shearing. The DPG particles with smaller sizes rub against the stator wall under the action of centrifugal force and the particles are further rounded until they can pass through the chutes of the stator wall. After that, the shape and size of particles don't change with the increase of shearing time.

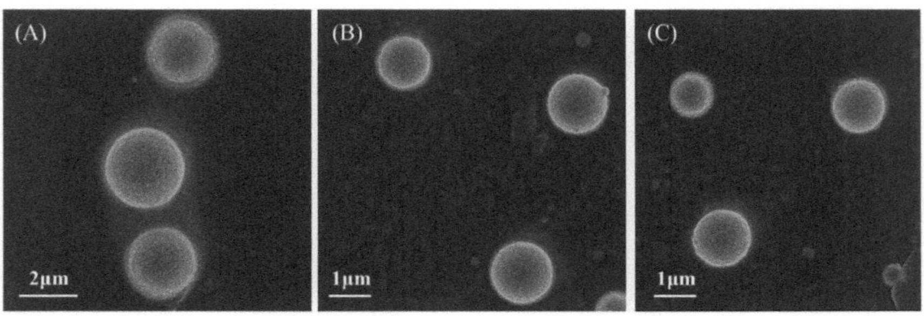

Figure 3.10 Effect of shear time on the microscopic morphology of DPG particles: (A) 2 min; (B) 6 min; (C) 10 min.

3.2.2.4 Effect of bulk gel–water mixing ratio

Water is not only a solvent but also plays the role of lubrication in the process of grinding to reduce the damage of high viscosity fluid to equipment. In the test, the bulk gel (formula with 0.4% type II functional polymer + 1.2% cross-linker + 0.1% coagulant) was mixed with water at ratios of 1:0, 3:2, and 1:4 and then sheared for 8 minutes at 40 Hz. Samples were taken to observe the microscopic morphology of the prepared DPG particles and the results are shown in Fig. 3.11.

It can be seen from Fig. 3.11 that the DPG prepared with different mixing ratios of bulk gel and water are spherical particles with regular shape, and the particle size distribution ranges from 1100 to 1400 nm. In the shearing process of bulk gel and water, the particles are more easily dispersed in water and the friction force between layers is smaller during flowing when the mixing ratio is smaller. The water is easy to carry the DPG particles pass through the stator chute, and the particles are easy to be ground round. However, this factor has little effect on the microscopic morphology of DPG particles under high shearing force.

3.2.3 Effect of preparation parameters on DPG particle size

Particle size is an important property parameter of the multiscale DPG particles. In this study, the effects of bulk gel strength, shearing rate, shearing time, and bulk gel–water mixing ratio on the particle size were investigated, and the Bettersize2000E laser particle size analyzer was used to measure the particle size.

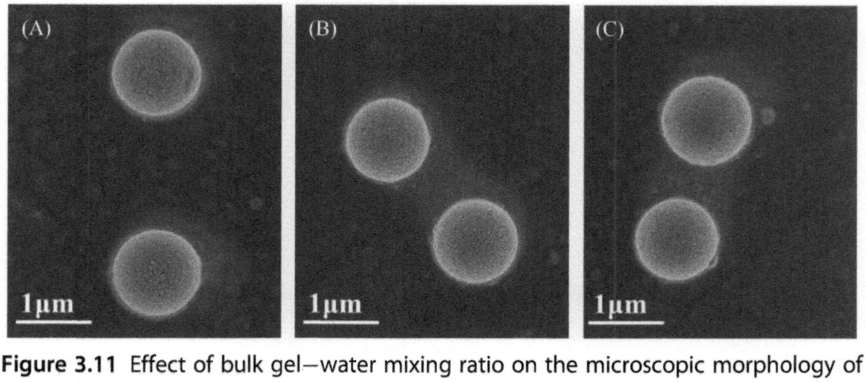

Figure 3.11 Effect of bulk gel–water mixing ratio on the microscopic morphology of DPG: (A) 1:0; (B) 3:2; (C) 1:4.

3.2.3.1 *Effect of bulk gel strength*

Two kinds of bulk gels with different gelation time were selected as follows. One bulk gel contained 0.25% type II functional polymer + 1.2% cross-linker + 0.1% coagulant, and had a gelation time of 14 hours and a gel strength of 0.043 MPa; the other with 0.4% type II functional polymer + 1.2% cross-linker + 0.1% coagulant, a gelation time of 8 hours, and a gel strength of 0.052 MPa. The bulk gel and water were mixed in a modified colloid mill at a ratio of 3:2 and sheared at 40 Hz for 6 minutes. The test results are shown in Fig. 3.12.

It can be seen from Fig. 3.12 that after shearing, the peak of the particle size distribution curve of the two bulk gel systems narrows. This indicates that the DPG formed by the mechanical shearing preparation method is uniform. After shearing for 6 minutes, the particle size of DPG formed from bulk gels with higher strength is 1160 nm, larger than that of those with lower strength which is 750 nm. When the polymer concentration is high, the cross-linking density between amide groups of polymer chain and hydroxyl groups of the cross-linking agent is high, and the structure of the formed gel is dense. This means a high strength of the bulk gel. Bulk gels with high strength are not easy to break under the same shearing conditions. Therefore the DPG formed grows in size.

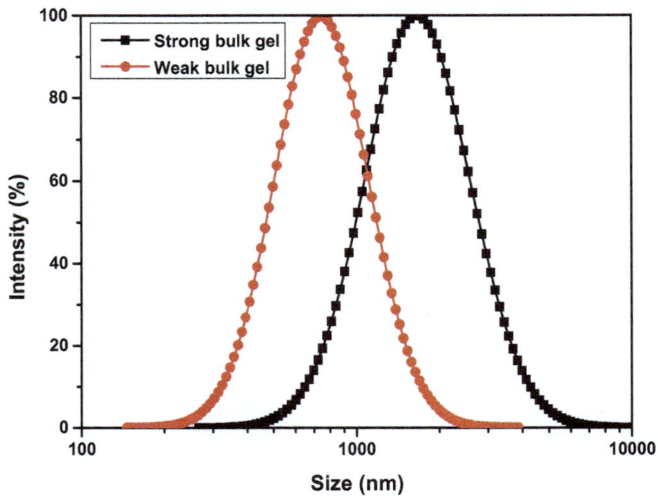

Figure 3.12 Effect of bulk gel strength on DPG particle size.

3.2.3.2 Effect of shearing rate

The bulk gel (formula: 0.4% type II functional polymer + 1.2% cross-linker + 0.1% coagulant) was mixed with water at a ratio of 3:2 and sheared for 6 minutes at a shearing rate of 20 to 60 Hz, respectively. Samples were taken to observe the particle size distributions of multiscale DPG and the results are shown in Fig. 3.13. It can be seen that the particle sizes of DPG prepared under different shearing rates are different, and the size distributions range from nanometer to micron. The average particle size of DPG prepared at a shearing rate of 20 Hz was 2800 nm, and that of DPG prepared at a shearing rate of 60 Hz was 950 nm. It can be seen from formula (3.12) that the shearing force exerted by the colloidal mill on the bulk gel is bigger when the shearing rate is higher, and the bulk gel is easier to break forming DPG particles with smaller sizes. Fig. 3.13 further shows that the particle size distribution curve of DPG formed at high shearing rate is narrow. It indicates that the DPG prepared by the mechanical shearing method is uniform.

3.2.3.3 Effect of shearing time

The bulk gel (formula: 0.4% type II functional polymer + 1.2% cross-linker + 0.1% coagulant) was mixed with water at a ratio of 3:2 and sheared at a shearing rate of 40 Hz for 2–10 minutes. Samples were taken to observe the particle size distributions of multiscale DPG and the results are shown in Fig. 3.14.

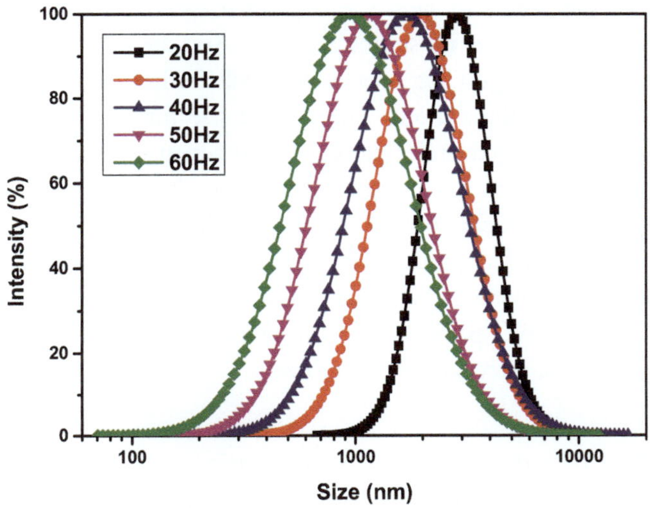

Figure 3.13 Effect of shear rate on particle size of DPG.

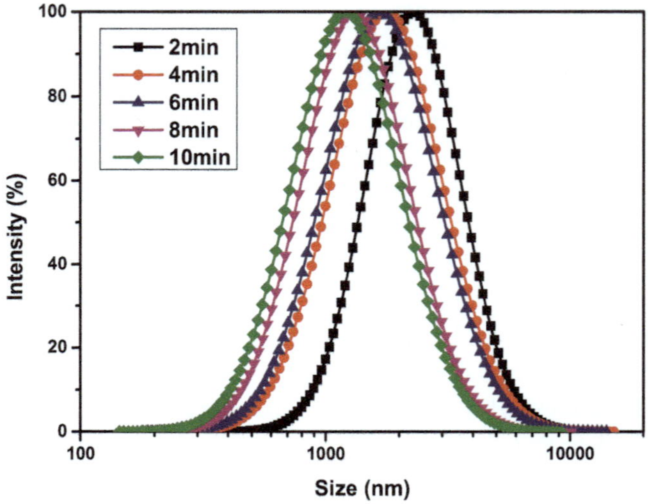

Figure 3.14 Effect of shear time on DPG particle size.

It can be seen from Fig. 3.14 that the particle size distribution of DPG becomes smaller and more uniform with the increase of shearing time. The particle size is distributed uniformly when the bulk gel is sheared for 2 minutes, which indicates that the mechanical shearing method can effectively prepare the multiscale DPG. The particle size does not change basically when the bulk gel is continuously sheared for 8 minutes, and the final average particle size is 1300 nm. The bulk gel is a viscoelastic fluid, it is broken to a coarse dispersion system under the action of high shearing force when it passes through the coarse shearing region (the first shearing region). And then the coarse dispersion system enters the fine grinding area, which will further decrease the particle size (the second shearing region). The particle size further decreases and the particle is further rounded under the combined action of shearing force, centrifugal force, and friction force when it enters the ultrafine grinding zone (the third shearing zone). It takes only 30 s for the bulk gel to complete the three-stage shearing cycle, which indicates that the mechanical shearing method is an efficient preparation method. When the shearing time increases, the shape and size distribution of DPG particles are more uniform under the action of shearing force, centrifugal force, and friction force.

3.2.3.4 Effect of bulk gel–water mixing ratio

In the test, the bulk gel (formula: 0.4% type II functional polymer + 1.2% cross-linker + 0.1% coagulant) was mixed with water at ratios of 1:0, 4:1,

3:2, 2:3, and 1:4, and then sheared for 6 minutes at 40 Hz. Samples were taken to observe the particle size distributions of DPG. The results are shown in Fig. 3.15.

It can be seen from Fig. 3.15 that when the bulk gel is mixed with water in different proportions, the particle size of DPG formed after shearing varies. The smaller the mixing ratio is, the larger the particle size is. This is because the viscosity of the DPG suspension is lower when the mixing ratio is smaller, and the DPG particles are easier to be carried by water to the stator chute. The DPG will move upward along the inclined slot of the first stage stator under the combined action of centrifugal force and friction force. A relatively homogeneous dispersed phase is formed due to the balanced force in the transverse and longitudinal directions of the DPG. Then the dispersed phase enters the second and third stator inclined slots, which makes the particle size further decrease. Owing to the high mechanical shearing force and cyclic shearing of DPG in the colloid mill, the preparation method has little influence on the particle size.

3.2.4 Effect of preparation parameters on viscosity of DPG particle

Bulk gels are highly viscoelastic fluids. After mechanical shearing, the network structure of bulk gels was broken and DPG particles with a certain viscosity was formed. The viscosity of DPG under different bulk gel

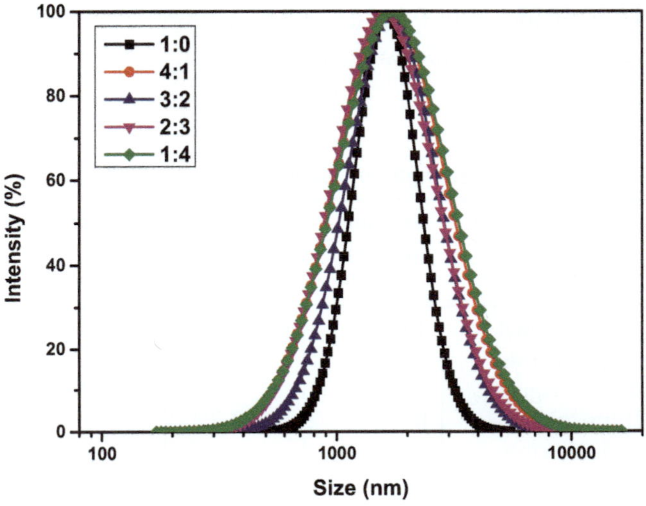

Figure 3.15 Effect of bulk gel–water mixing ratio on particle size.

strength, shearing rate, shearing time, and bulk gel—water mixing ratio were measured by Brookfield viscometer, for which the rotor speed was 6 rpm and the measurement temperature was 30°C.

3.2.4.1 Effect of bulk gel strength

The effect of bulk gel strength on the viscosity of DPG particles was investigated by selecting bulk gels with different gelation time. The bulk gel was placed in a colloid mill, and water of 40% the total volume was added, then the DPG particles were obtained by shearing for 6 minutes. The results are shown in Table 3.1.

It can be seen that the viscosity of multiscale DPG particles increases with the bulk gel strength. At the same concentration, the viscosity of DPG particles is much higher than that of type II functional polymer. Moreover, the lower the concentration, the more obvious this trend is. This is because the network structure formed by amide groups in polymer and hydroxyl groups in cross-linking agent is higher with a higher concentration of type II functional polymer, and the shearing to gel under the same shear condition is weak. The bulk gel with higher strength is not easy to break, and the particles adhere to each other. Therefore the viscosity of DPG particles formed after shearing is higher. For polymers with the same concentration, the increase of viscosity is mainly due to the intertwining of polymer molecules. When the concentration is too low, polymer molecules are only randomly distributed in the aqueous solution and cannot be entangled to form structural viscosity. Therefore the viscosity of polymer is lower at the same concentration.

Table 3.1 Effect of bulk gel strength on viscosity of DPG.

Formula	Gelation time (h)	Gelling strength (MPa)	Viscosity	
			500 mg/L	1000 mg/L
0.25% type II functional polymer + 1.2% cross-linker + 0.1% coagulant	14	0.043	5.8	7.1
0.4% type II functional polymer + 1.2% cross-linker +0.1% coagulant	8	0.052	7.6	9.7

3.2.4.2 Effect of shearing rate

The bulk gel (formula: 0.4% type II functional polymer + 1.2% cross-linker + 0.1% coagulant) was mixed with water at a ratio of 3:2, and sheared at shearing rate of 20 to 60 Hz for 6 minutes, respectively. The prepared DPG was diluted to 500 mg/L for viscosity measurement. The results are shown in Fig. 3.16.

It can be seen that with the increase of the shearing rate, the viscosity of DPG decreases sharply. When the shearing rate exceeds 50 Hz, the viscosity of DPG particles does not change at all, and the final viscosity of DPG particles decreases to 6.1 mPa · s. This is because the bulk gel used to prepare DPG particles has a high strength, and a low shearing rate could cut the bulk gel into larger particles. This results in the reduction of viscosity. However, a low shearing rate inevitably forms a low shearing force, which makes the particles adhere to each other and then results in higher viscosity of DPG particles formed.

3.2.4.3 Effect of shearing time

The bulk gel (formula: 0.4% type II functional polymer + 1.2% cross-linker + 0.1% coagulant) was mixed with water at a ratio of 3:2 and sheared at a shearing rate of 40 Hz for 2–10 minutes, respectively. The prepared DPG particles were diluted to 500 mg/L for viscosity measurement. The results are shown in Fig. 3.17. It can be seen that the viscosity

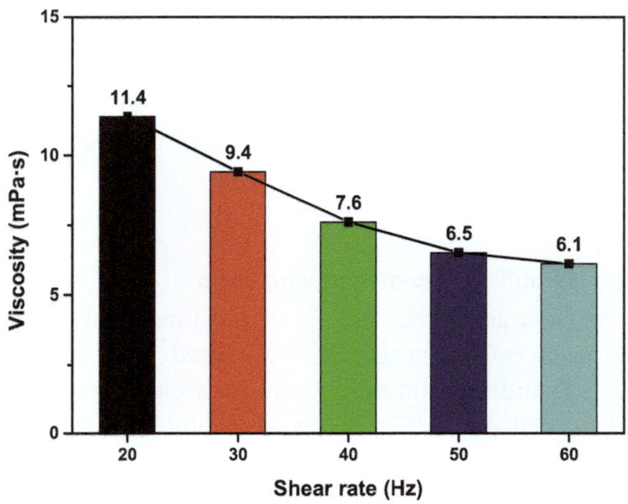

Figure 3.16 Effect of shear rate on viscosity of DPG particles.

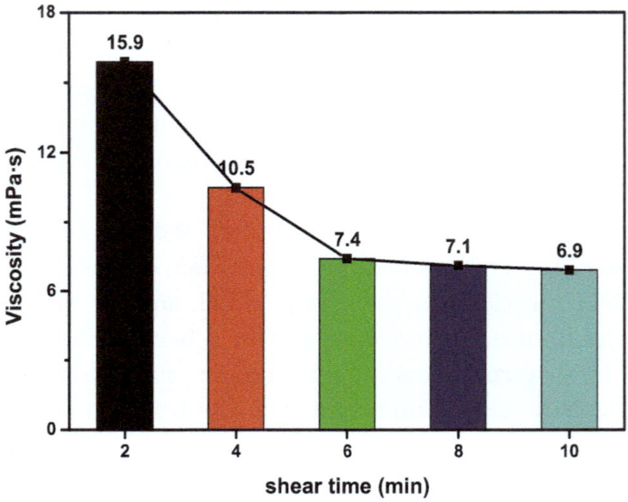

Figure 3.17 Effect of shearing time on viscosity of DPG particles.

of the prepared DPG decreases with the increase of shearing time under high-speed shearing. When the shearing time is more than 6 minutes, the viscosity of the DPG particles almost does not change any more. This is because, after the mechanical shearing of the high strength and dense network structure of the bulk gel, the network structure of the bulk gel is cut off to form smaller particles, so that the viscosity decreases. The formed DPG particles are smaller with longer shearing time, and the particles are harder to adhere to each other, making the viscosity lower. The particles formed after mechanical shearing are able to pass through the clearance between fixed rotors due to their small particle size. Then, friction generated by the mutual motion of fixed rotors basically no longer exerts shear action on the tiny particles formed, thus not triggering any further decrease in viscosity.

3.2.4.4 Effect of bulk gel—water mixing ratio

In the test, the bulk gel (formula: 0.4% type II functional polymer + 1.2% cross-linking agent + 0.1% coagulant) was mixed with water at ratios of 1:0, 4:1, 3:2, 2:3, and 1:4, and then sheared at a shearing rate of 40 Hz for 6 minutes. The prepared DPG particles were diluted to 500 mg/L for viscosity measurement, and the results are shown in Fig. 3.18.

It can be seen that the viscosity of DPG particles decreases rapidly with the increase of water proportion. When the ratio of bulk gel to

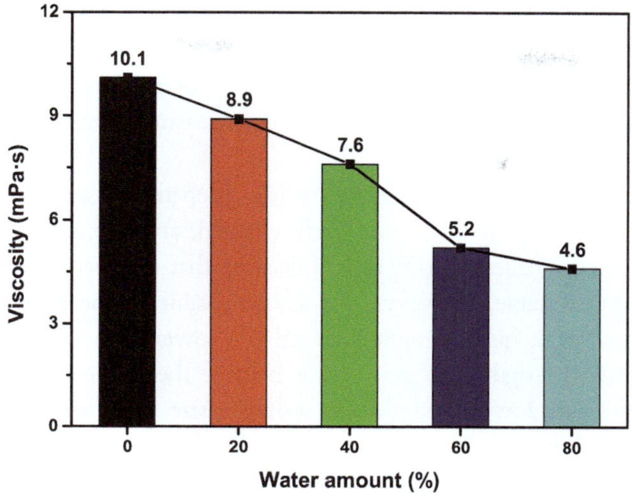

Figure 3.18 Effect of bulk gel–water mixing ratio on viscosity of DPG particles.

water is more than 2:3, the viscosity of DPG tends to be stable. This is because the formed multiscale DPG particles after shearing with low water proportion gather together, resulting in an increase of structural viscosity. However, after the shearing stops, it is difficult to separate the adhering particles by dilution with water. Therefore the viscosity of DPG formed with a low water proportion shear condition is higher.

3.2.5 Effect of preparation parameters on zeta potential of DPG particles

Zeta potential is an important parameter to characterize the stability of DPG particles. When the absolute value of zeta potential is higher than 30 mV, the particles are relatively stable, and when it is lower than 30 mV, particles tend to coalescence. The effects of bulk gel strength, shearing rate, shearing time, and bulk gel–water mixing ratio on the zeta potential of multiscale DPG particles were measured by NanoBrook Omni, Germany.

3.2.5.1 Effect of bulk gel strength

Two kinds of bulk gels with different gelation strength were selected, one with 0.25% type II functional polymer + 1.2% cross-linker + 0.1% coagulant, a gelation time of 14 hours, and a gel strength of 0.043 MPa; the other with 0.4% type II functional polymer + 1.2% cross-linker + 0.1%

coagulant, a gelation time of 8 hours, and a gel strength of 0.052 MPa. The bulk gel and water were mixed in a modified colloid mill at a ratio of 3:2 and sheared at 40 Hz for 6 minutes. The prepared DPG particles was diluted to 500 mg/L for zeta potential measurement, and the results are shown in Table 3.2.

It can be seen that the surfaces of particles prepared by two kinds of bulk gels with different strength are negatively charged, and the absolute values of zeta potential are greater than 30 mV, indicating that the prepared DPG particles are relatively stable. However, the absolute value of the zeta potential of DPG prepared with high-strength bulk gel was lower than that of DPG prepared with low-strength bulk gel. This is because the DPG particles prepared from bulk gel with lower strength have a smaller size. It means a larger specific surface area and stronger electrostatic repulsion, which makes the particles dispersed more stable. The negative charges on the surface of DPG particles are favorable for deep migration in the formation. Because most of the rock surface is negatively charged and the same charges repel each other, the adsorption of DPG particles in the near well zone can be avoided, so that the particles can migrate to depth and achieve the deep profile control.

3.2.5.2 Effect of shearing rate

The bulk gel (formula: 0.4% type II functional polymer + 1.2% cross-linker + 0.1% coagulant) was mixed with water at a ratio of 3:2, and sheared at a shearing rate from 20 to 60 Hz for 6 minutes, respectively. The prepared DPG was diluted to 500 mg/L for zeta potential measurement, and the results were shown in Fig. 3.19. The results show that the absolute value of the zeta potential of DPG particles increases with the shearing rate, but the zeta potential of DPG varies in a relatively small

Table 3.2 Effect of bulk gel strength on zeta potential of DPG particles.

Formula	Gelation time (h)	Gelation strength (MPa)	Zeta potential (mV)
0.25%II type II functional polymer + 1.2% cross-linker + 0.1% coagulant	14	0.043	−41.82
0.4%II type II functional polymer + 1.2% cross-linker + 0.1% coagulant	8	0.052	−31.22

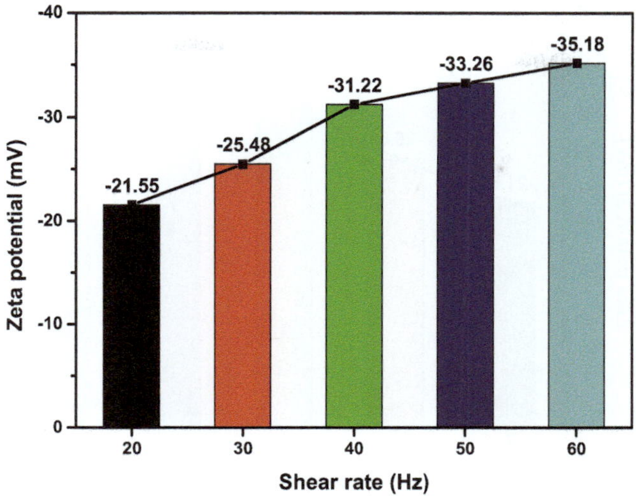

Figure 3.19 Effect of shear rate on zeta potential of DPG particles.

range. When the shearing rate is higher than 50 Hz, the zeta potential of DPG particles tends to be stable. This is ascribed to the fact that the higher the shearing rate, the smaller the particle size. The particles can be uniformly dispersed in the aqueous solution to form a stable colloidal system.

3.2.5.3 Effect of shearing time

The bulk gel (formula: 0.4% type II functional polymer + 1.2% cross-linker + 0.1% coagulant) was mixed with water at a ratio of 3:2 and sheared at a shearing rate of 40 Hz for 2–10 minutes. The prepared DPG was diluted to 500 mg/L for zeta potential measurement, and the results are shown in Fig. 3.20. Experimental results show that the average absolute value of the zeta potential of the DPG particles is higher than 30 mV after shearing for 6 minutes. The longer the shearing time is, the higher the average absolute value of zeta potential is. The electrostatic repulsion force between the particles is dominant, and is not easy to coalesce. These findings indicate that the DPG system is more stable. Since the DPG particles are formed by mechanical shearing to bulk gel, there are carboxylic acid groups on the surface of the bulk gel. The ionization of these groups leads to the negative charges on the surface of the DPG particles after shearing. The particles have smaller sizes when sheared longer, which means larger specific surface area and stronger electrostatic repulsion, giving the dispersed particles more stability. However, according to the average zeta potential distribution of the particles, the stability of DPG particles is

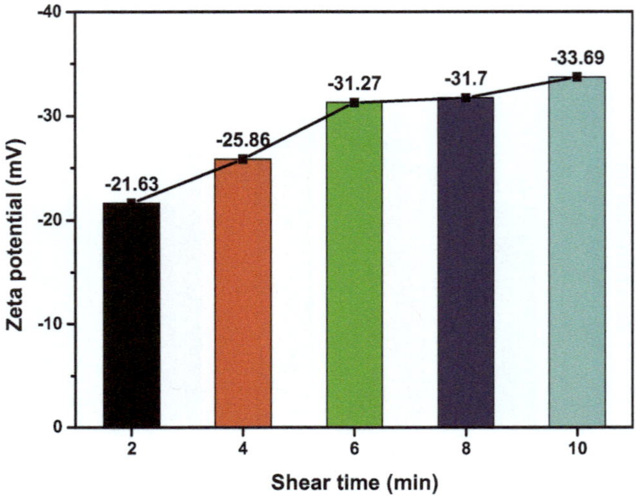

Figure 3.20 Effect of shear time on zeta potential of DPG particles.

relatively weak, and the weak stability is favorable for the DPG particles adjusting the seepage profile. When the particles migrate in porous media, the weak stability of the particles leads to the aggregation of particles, which is conducive to the microprofile control of a high permeability area, so as to achieve the purpose of seepage profile adjustment.

3.2.5.4 Effect of bulk gel–water mixing ratio

In the experiment, the bulk gel (formula: 0.4% type II functional polymer + 1.2% cross-linker + 0.1% coagulant) was mixed with water at ratios of 1:0, 4:1, 3:2, 2:3, and 1:4, and then sheared at a shearing rate of 40 Hz for 6 minutes. The prepared DPG particles were diluted to 500 mg/L for investigation on the influence of preparation method on the electric potential of DPG particles. The results are shown in Fig. 3.21.

It can be seen from Fig. 3.21 that with the increase of water proportion, the absolute value of the zeta potential of particles increases slightly, but within a limited range. This indicates that the DPG particles prepared with different water proportions possess good stability.

Section 3.3 The preparation mechanism of multiscale DPG particle

The technique of multiscale DPG particles has the characteristic of simplicity and high efficiency. This allows regulation of the size of multiscale

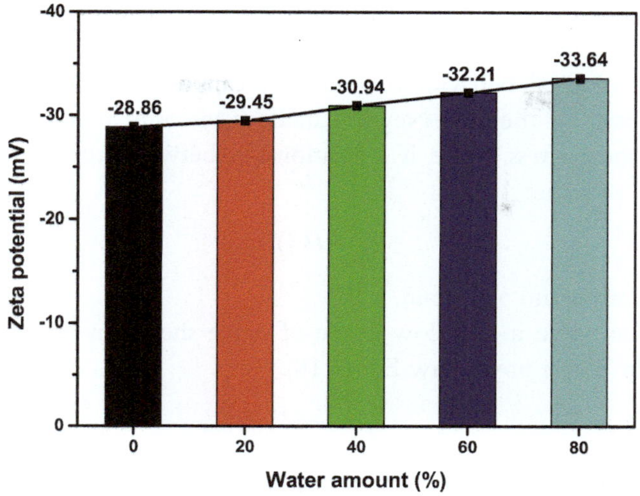

Figure 3.21 Effect of preparation method on the zeta potential of DPG particles.

DPG particles by controlling the mechanical shearing parameters and the strength of bulk gels. In the light of the conception and preparation process analysis, the whole preparation process can be divided into two major stages: the bulk gel cross-linking reaction phase and the DPG particle shaping shear phase.

3.3.1 Test principle

This study used HAAKE MARS 60 rheometer to determine the viscosity changes of the bulk gel cross-linking reaction phase and the DPG particle shaping shear phase. The rheometer employed in the test used CC41/Ti rotor (in the length of 55.03 mm and diameter of 41.42 mm). All tests were conducted at a temperature of 30°C.

In general, the flow curve of a solution covers two viscosity regions, corresponding to the zero-shear viscosity and shear viscosity limit, namely, η_0 and η_∞. In the regions of zero-shear viscosity, neither the structure nor the conformation of polymer molecules and bulk gels are subject to the shear force, so the viscosity of this region is determined to be zero-shear viscosity. When the shear is acting at a very high speed, the molecules of both polymers and bulk gels in a solution extend to the utmost. The viscosity at this point achieves the shear viscosity limit η_∞. It, however, is quite difficult to detect for most of the polymers or bulk gels. Given this fact, this study adopted zero-shear viscosity η_0 and flow index to characterize the rheological behavior of bulk gels and DPG particles in the process of their formation.

The rheological behavior of bulk gels can be characterized by their stable viscosity in the non-Newtonian flow viscosity region. Under the action of shear stress, there is a relationship between the stress and the flow curve of strain:

$$\tau = f(\gamma) \tag{3.17}$$

where τ is stress and γ is strain.

Thus the curve in the flow mode of stable shear can be fitted by the Ostwald de Waele power law Eq. (3.18):

$$\eta = k\gamma^{n-1} \tag{3.18}$$

where η is shear viscosity, k is consistency coefficient, and n is flow index.

3.3.2 The bulk gel cross-linking reaction phase

In this study, a bulk gel prepared by 0.4% II-type functional polymers, 1.2% cross-linking agent, and 0.1% coagulant was used as an example to expound the formation process of bulk gel. First, a gelling solution was placed at 95°C to react. Then, a sample was taken out at each regular interval and placed at room temperature for cooling. Next, those samples were analyzed by the rheometer at 30°C, with a view to an investigation on the rheological behavior of bulk gel during the process of gelling. The rheological curve of bulk gel at different gelation time are shown in Fig. 3.22, and the changes of zero-shear viscosity along with time in this gelation process in Fig. 3.23.

According to Figs. 3.22 and 3.23, in the initial phase, the viscosity of bulk gel increases in a slow manner with the time extending. After 2 hours of cross-linking reaction, the system viscosity starts to grow rapidly, and roughly stays unchanged and levels off when the cross-linking reaction is prolonged to 8 hours. This suggests the formation of bulk gel. With the viscosity change as a marker for this formation, bulk gels mainly undergo four phases. The first phase refers to the induction, the two continuous hours at the outset of the reaction. Following it is the rapid cross-linking phase for 2–4 hours. The third phase comes as the slow increase phase, lasting for 4–8 hours. After the 8-h formation, it enters into the viscosity stability phase, the fourth phase, where bulk gels are formed.

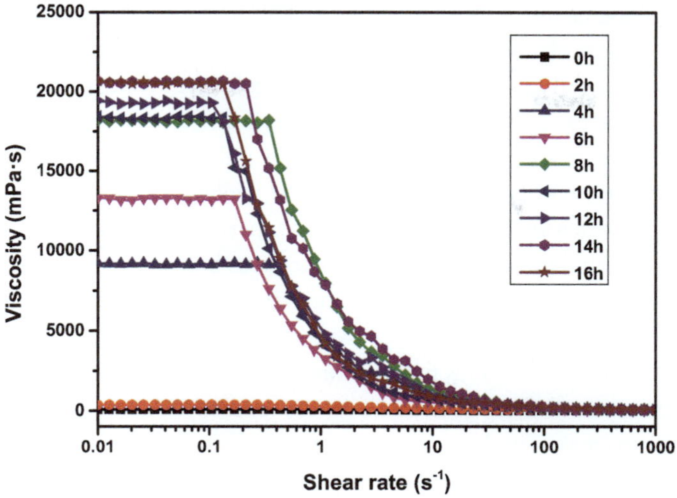

Figure 3.22 Rheological curves of bulk gels at different gelation time.

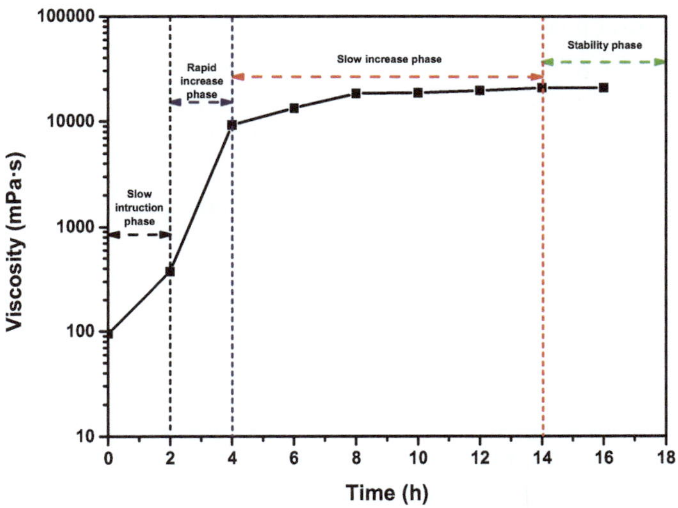

Figure 3.23 The viscosity changes of bulk gels with time in the process of gelation (at a shear rate of 0.1 s^{-1}).

1. The first phase: the slow induction phase

The rheological curves of bulk gel show that the zero-shear viscosity of both polymer solution and gelling solution is lower than the viscosity of gelling system undergoing 2 hours of reaction. This indicates an outset of the cross-linking reaction between the polymers and cross-linking agents in the gel solution. The molecular conformation of the gelling system changes, which further leads to a small change of its zero-shear viscosity (Fig. 3.23). The slow increase in the initial reaction phase suggests that the gelling system has not yet formed a spatial network structure. Therefore the viscosity increases in a mild and slow manner. This is because the cross-linking reaction of the gelling system takes place first inside the macromolecular chains of polymers. That is to say, the gelling system first undergoes an intramolecular cross-linking reaction that enables the formation of chemical cross-linking bonds inside the molecules. This intramolecular cross-linking of polymers, in turn, triggers contractions of the molecular chains. This furthermore results in a shrinking of these molecules in size and thus an increase in the viscosity. Additionally, the cross-linking progresses with being accompanied by dehydration condensation reaction between the polymers and cross-linking agents. However, the number of molecules undergoing intermolecular cross-linking is far smaller than that undergoing intramolecular cross-linking. This determines a slow increase in the viscosity in the initial phase.

2. The second phase: the rapid increase phase

   Rapidly growing viscosity is the most salient feature of this phase. At this point, the cross-linking reaction carries on in a relatively large reaction rate. At high temperature, the small phenolic resin molecules uniformly dispersing in the aqueous solution provide a host of hydroxymethyl groups. Those functional groups take place in a dehydration condensation reaction with the amide groups of the polymers, shaping a compact spatial structure, which thereby induces a dramatic increase in the viscosity of the gelling system.

3. The third phase: the slow increase phase

   Along with the reaction time, an increasing number of amide groups have reacted with hydroxymethyl groups in the solution. At this point, however, there are still a few of those groups joining in the reaction, making the viscosity of the bulk gel system increase slowly.

4. The fourth phase: the stability phase

In the fourth phase the bulk gel viscosity is no longer changing, basically. It suggests the completion of cross-linking reaction and that a stable, compact spatial structure has been formed between the polymers and cross-linking agents. Therefore the viscosity of bulk gel roughly does not change any more with time being prolonged.

To sum up, the formation of bulk gel is actually a course of chemically cross-linking reactions. By these reactions, a compact spatial structure is formed between the polymers and cross-linking agents through the induction phase, the rapid increase phase, the slow increase phase, and the stability phase. This new structure ensures the bulk gels possess certain strength and elasticity, which thus lays a foundation for the preparation of multiscale DPG particles by a mechanical shearing method.

3.3.3 The multiscale DPG particles shaping shearing phase

When the bulk gels were added into a colloid mill, the bulk gels were milled into particles with different sizes so that the system viscosity decreased under the action of shear force. Fig. 3.24 illustrates a change in the bulk gel viscosity with the time during the shearing phase, and Fig. 3.25 presents the relationship between the zero-shear viscosity changes of DPG particles and time throughout the shearing phase.

According to Figs. 3.24 and 3.25, the viscosity of bulk gel in the initial shear phase declines rapidly with the time. After being sheared continuously for 6 minutes, the system viscosity stays unchanged overall, which suggests the formation of DPG particles. With the viscosity change as the mark of DPG particles formation, the whole formation course can be divided into three typical phases: the crushing and rounding zone at the 0–2nd minutes in the initial shear phase; the further particle grinding zone during the 2nd–6th minutes; and the particle uniformity zone during the 6th–10th minutes.

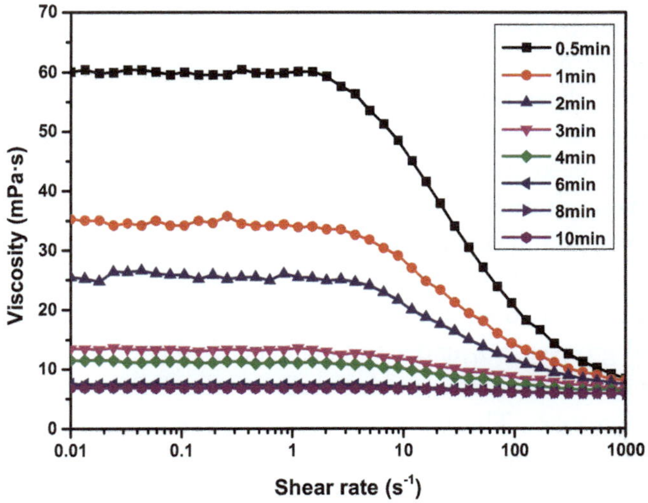

Figure 3.24 The rheological curves of DPG particles during the shearing phase.

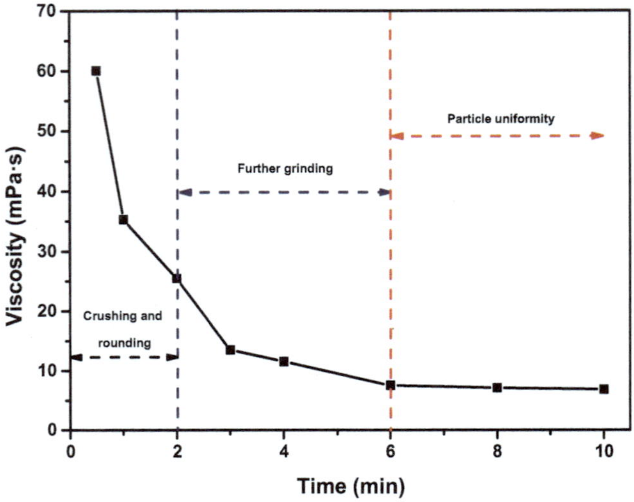

Figure 3.25 The changes of zero-shear viscosity of DPG particles with time during the shearing phase.

1. The crushing and rounding phase

   The sharp decline in system viscosity is an evident feature of this phase. In this phase, the system viscosity drops in a speedy manner, and the gel viscosity declines from 18,198.5 to 6.8 mPa · s. In the initial shear phase, the bulk gel enters into the colloid mill. Under the action of shear force, its continuous network structure is broken into discontinuous, coarse DPG particle systems with a large particle size, which thereby results in the dramatic decline of viscosity.
2. The further particle grinding phase

   As the shear time lengthens, the decline in the viscosity of the DPG particle system tends to slow down but by a moderate extent. In the cyclic shear mode, this coarse system further suffers shear action when passing through the gap between the stator and rotor. The DPG particles were subjected to the friction force perpendicular to the centrifugal force on the stator wall, which further shrank the size of those particles. In addition, the wall of the stator is made up of a host of inclined grooves. So, under the high-speed rotation of the rotor, friction is generated between the walls of the DPG particles and the inclined grooves. This friction brings the shape of those particles to the standard and decreases the mutual adhesion between the particles, such that a uniform dispersion phase is formed. Therefore the decline in the viscosity of the DPG particle system is limited in this phase.

3. The particle uniformity phase

In this particle uniformity phase, the viscosity of the DPG particle system remains unchanged overall with the time. Under the action of shear force and friction for a long period, the size of DPG particles shrinks, and their shape approximates to the standard. Thus the particles could smoothly pass through the gap between the stator and rotor and through the inclined grooves on the rotor wall. Moreover, a uniform dispersion phase is established. Therefore the viscosity does not reduce anymore and maintains at a balance state.

Fig. 3.26 further demonstrates the viscosity change of DPG particles throughout the cross-linking and shear phase. Before the shear, the amide groups on the polymer chain carry on a dehydration condensation reaction with the hydroxyl groups of the phenolic resin cross-linking agent to generate bulk gels. A high-viscosity system is thus formed by a low-concentration system. Involving chemical reactions between different molecules, this process forms a compact and continuous gel structure. After the formation, the bulk gel is given a certain mechanical shear force which breaks it into particles in different sizes. By controlling the strength of the mechanical shear force multiscale DPG particles are acquired. Only mechanical shear action is engaged in this phase, with no chemical reaction. The system viscosity presents a precipitous decline throughout the process. Nevertheless, the DPG particle system formed by high-speed

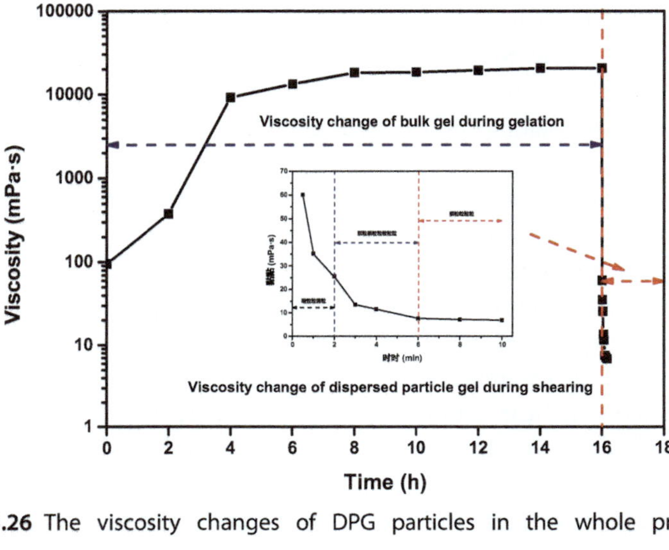

Figure 3.26 The viscosity changes of DPG particles in the whole preparation process.

shearing still holds a higher viscosity than that of the polymers at the same concentration. This indicates that DPG particles harbor high resistance to shearing. In other words, DPG particles can maintain a relatively high viscosity even after the shear by surface injection equipment, pipes, and underground seepage. They possess the characteristics of both particle plugging agents and polymer flooding.

3.3.4 The preparation mechanism of DPG particles

The viscosity of DPG particles changes throughout the two-phase process: the chemical cross-linking phase and the mechanical shear phase. The former phase involves chemical cross-linking reactions between different molecules, through which a compact bulk gel structure is formed. The latter, on the contrary, engages only mechanical shear action, which enables the realization of shaping shear for the DPG particles. Those particles, after going through both phases, are armed with both the stable nature of bulk gels and the characteristics of particle plugging agents.

3.3.4.1 The formation mechanism of bulk gels

In the initial cross-linking phase, the amide groups of the polymers undergo a dehydration condensation reaction with the hydroxymethyl groups of resin cross-linking agent, so as to form a compact, continuous gel network structure (Fig. 3.27). This process allows the transformation for the gel system to achieve a high viscosity from a lower one and to harvest a compact, ordered network molecular structure from irregular coils. The newly formed compact network structure envelopes water, thus it is able to ensure the stability of the bulk gel.

3.3.4.2 The shaping shear mechanism of DPG particles

The bulk gel was put under a strong shear force after its formation. With its compact network structure being broken, the bulk gel is made through grinding and rounding into DPG particles with a regular shape. DPG particles with different size distribution are obtained after regulation on the mechanical shear parameters (as Fig. 3.28). This process touches only upon mechanical shear action but no chemical reactions. Therefore the DPG particles through the physical shear action still hold the stability of bulk gel.

The preparation mechanism of DPG particles is further demonstrated in Fig. 3.29. The preparation process consists of the above ground bulk

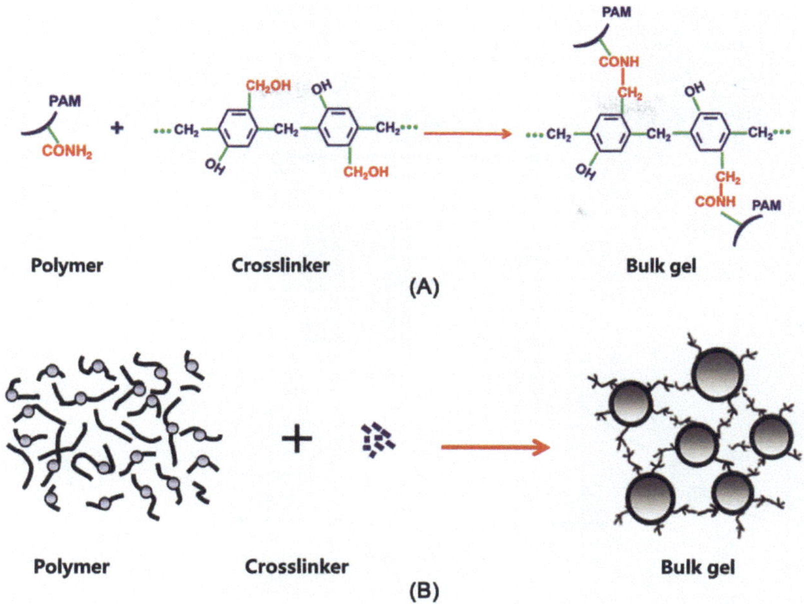

Figure 3.27 The mechanism of bulk gel formation: (A) bulk gel's cross-linking reaction; (B) the formation of bulk gels.

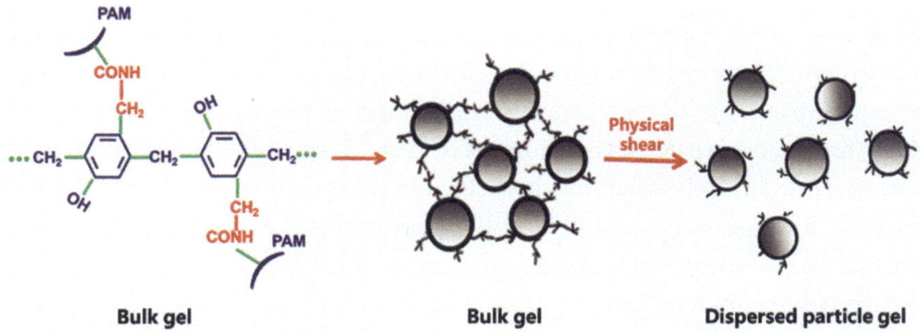

Figure 3.28 The shear formation of DPG particles.

gel gelation phase and the bulk gel shaping shear phase. In the first phase, the polymers and cross-linking agents reacts together into a bulk gel with a high viscosity up to 18,198.5 mPa · s. Through high-speed shear in the colloid mill, the viscosity of the bulk gel reduces to a drastic extent, because of which viscoelastic particles are formed. The coarse grinding of bulk gel and the further grinding of DPG particles brought about a uniform DPG system, which was formed by the special structure of the colloid mill. When entering into the colloid mill, the bulk gels move mainly in the direction of the rotor's rotation in the gap between pores. The

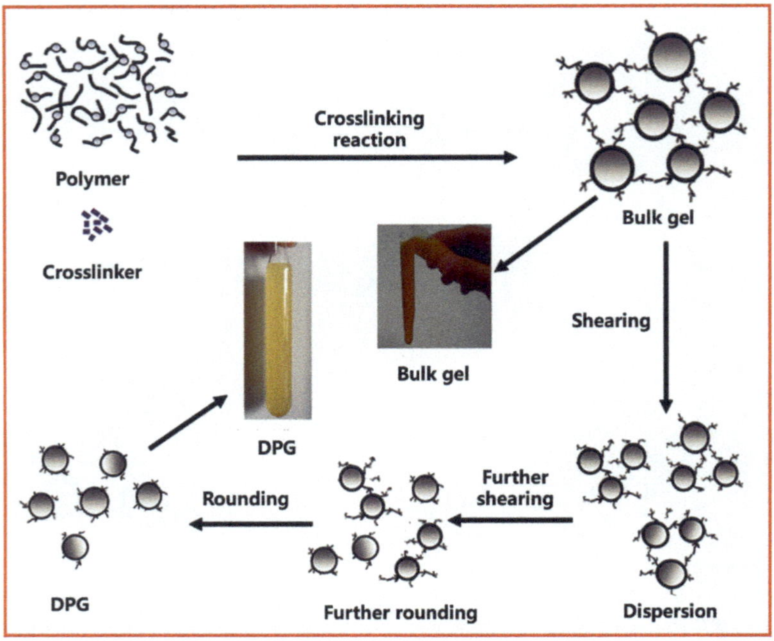

Figure 3.29 The formation mechanism of multiscale DPG particles.

shear force they bear is perpendicular to the direction of DPG particles' movement. Next, when the bulk gels enter into the coarse grinding zone, their compact network structure is crushed under the impact of high-speed shear force into particles with unevenly large sizes, and the viscosity drops sharply. Immediately after the particle formation, the DPG particles enter into the fine grinding zone. Here, they are narrowed down further in size because of the shear force. At the same time, these particles are subjected to the centrifugal force. They are rounded further by the inside wall of the stator such that their shapes are made standard. Then, these refined DPG particles go into the ultrafine grinding zone. Since they receive relatively balanced forces in the horizontal and vertical directions, their size is cut down even further, so is their viscosity. When the circulatory system is powered on and the shear time prolonged, the DPG particles enter again into the space between the stator and rotor and into the grooves of the former, which further shrinks their size. These new smaller DPG particles are thereby able to pass smoothly through the gap between the stator and rotor and through the inclined grooves on the wall of the rotor. Hence, neither the size nor the viscosity of the DPG particles would change with the shear time prolonging any more.

Section 3.4 Development of multiscale DPG particles industrialized preparation software

Particle size is an important parameter to characterize the properties of multiscale DPG particles. In the process of control profile treatments, only by injecting DPG particles in suitable size into the field formation will those particles regulate the profile control to a full play. Using a mechanical shearing device, this study sets up a relationship between the multiscale DPG particles and the preparation device process parameters (including shear rate, shear spacing, and shear time), and optimizes those parameters. Moreover, on this basis, a DPG particles scale preparation software has been developed, in order to actualize a simple and intelligent production of multiscale DPG particles.

3.4.1 The idea of establishment for multiscale DPG particles preparation model

In light of the derivation of mathematical equations, the size of multiscale DPG particles is dependent upon such equipment parameters as shear rate, shear time, and shear spacing. Thus it is feasible to establish a functional relation between the particle size and the shear rate, shear time, and shear spacing. Given the difficulty in precision regression of the concrete functional relation, an appropriate initial value must be given to the iteration adopted when a mathematical model is established; otherwise, a correct mathematical model would be hard to set up. However, the initial value could be not given if the 1stOpt method was adopted to establish the mathematical model, for it completes the mode through the global optimization algorithm. Therefore when establishing the mathematical model, we first used the 1stOpt software to regress the test data, build a corresponding mathematical model, and then set up parameters in accordance with the model so as to solve the variables required.

According to the analysis and research results of a single factor (shear time, or shear spacing, or shear rate) that affects the size of multiscale DPG particles, the particle size holds a good functional relation with a single factor of the three, under certain test conditions. Given the feasibility of the shear equipment, we fixed the shear spacing and altered the shear time and rate, so as to realize the preparation of DPG particles in different sizes. When shear time and shear rate came into consideration, however, the particle size showed a higher functional fitting relation with them. Since the distribution of prepared multiscale DPG particles is at the nanoscale, micron-level, and millimeter-level, three level zones were separated and mathematical models corresponding to each zone were established in an attempt to heighten the fitting precision. According to the test results, the Levenberg–Marquardt and universal global optimization algorithms were selected and the 1stOpt software adopted to perform continuous functional regression on the data. The idea of model establishment is illustrated in Fig. 3.30. First, the 1stOpt software was used to regress the data, set up the model, and optimize the parameters. On the basis of the model established, the python open source language was then utilized for programming to develop industrialized preparation software. Next, preparation parameters were selected in line with the software. This model establishment provides technical supports for the simplified industrialized production of multiscale DPG particles.

3.4.2 Establishment of multiscale DPG particles preparation model

The industrialized modified type-240 colloid mill was employed to regulate various factors including shear time, shear rate, and shear spaces. 185 groups of multiscale DPG particles size data were collected. According to the particle size distribution of DPG particles, most of the sizes were distributed in the nanoscale, micron-level, and millimeter-level zones,

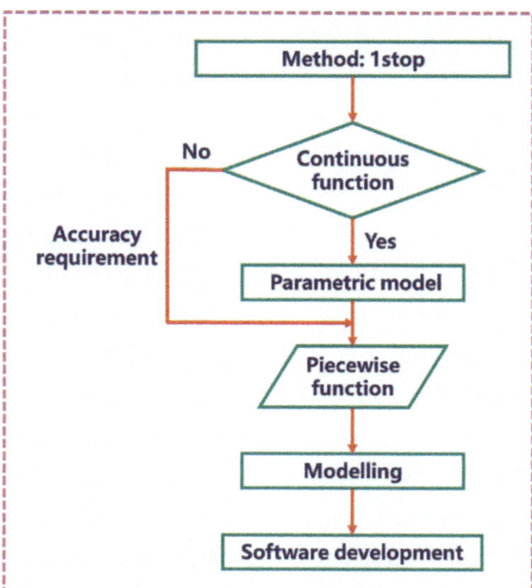

Figure 3.30 The steps of establishment of multiscale DPG particles model.

wherein the shear rate played as the dominant factor for this distribution. On this account, 1stOpt software was adopted to fit the particle size data, in order to establish nanoscale, micron-level, and millimeter-level DPG particles mathematical models.

For the nanoscale DPG particles that have been iterated for 21 times, the mathematical model after regression can be expressed as Eq. (3.19):

$$f(x_1, x_2, x_3) = a_1{}^*x_1 \wedge a_2 + a_3{}^*x_2 \wedge a_4 + a_5{}^*x_3 \wedge a_6 + a_7{}^*x_1 \wedge a_2{}^*x_2 \wedge a_4 + a_8{}^*x_2 \wedge a_4{}^*x_3 \wedge a_6 + a_9 \tag{3.19}$$

where f is particle size, nm; x_1 is shearing rate, rpm; x_2 is shearing time, min; x_3 is shearing space, μm; and a_i is a coefficient, as the scalar.

The correlation coefficient of the above mentioned fitting function model is $R = 0.9926$, indicating a relatively high fitting. The fitting function diagram and the test data and results are shown in Fig. 3.31. It can be seen that the mathematical model setup shares a high fitting precision with the DPG particles, thus it is capable of reflecting the practical production. Therefore this model is allowed to work as a process parameter model for the preparation of nanoscale DPG particles.

For the micron-sized DPG particles that have been iterated for 22 times, the mathematical model after regression can be expressed as Eq. (3.20):

$$f(x_1, x_2, x_3) = a_1{}^*x_1 + a_3{}^*x_2 \wedge a_4 + a_5{}^*x_3 + a_7{}^*\log(x_1)^*x_2 + a_8{}^*x_2{}^*x_3 + a_9 \tag{3.20}$$

where f is particle size, μm; x_1 is shearing rate, rpm; x_2 is shearing time, min; x_3 is shearing space, μm; and a_i is a coefficient, as the scalar.

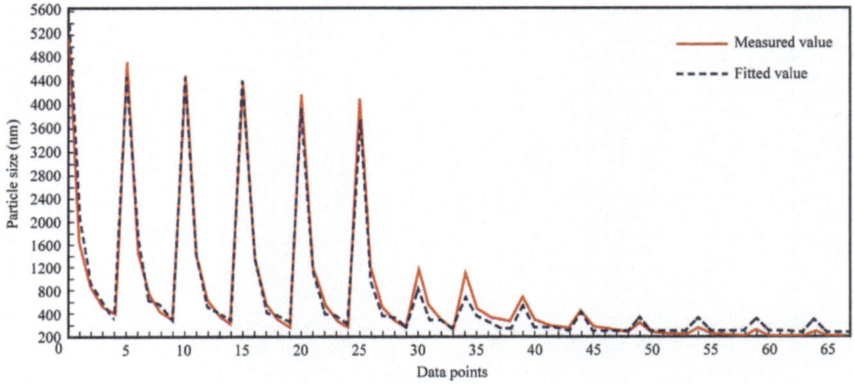

Figure 3.31 The function graph of numerical fitting of nanoscale DPG particles.

The correlation coefficient of the above mentioned fitting function model is $R = 0.9525$, indicating a relatively high fitting. The fitting function diagram and the test data and results are shown in Fig. 3.32. It can be seen that the mathematical model setup shares a high fitting precision with the DPG particles, which is capable of reflecting the practical production. Therefore this model is allowed to work as a process parameter model for the preparation of micron-level DPG particles.

For the millimeter-sized DPG particles that have been iterated for 22 times, the mathematical model after regression can be expressed as Eq. (3.21):

$$f(x_1, x_2, x_3) = a_1{*}\log(x_1) + a_3{*}x_2 \wedge a_4 + a_5 \wedge x_3 + a_6{*}\log(x_1){*}x_2 + a_7{*}x_2{*}\log(x_3) + a_8 \quad (3.21)$$

where f is particle size, μm; x_1 is shearing rate, rpm; x_2 is shearing time, min; x_3 is shearing space, μm; and a_i is a coefficient, as the scalar.

The correlation coefficient of the above mentioned fitting function model is $R = 0.9322$, indicating a relatively high fitting. The fitting function diagram and the test data and results are shown in Fig. 3.33. As can be seen, the mathematical model setup shares a high fitting precision with the DPG particles, which is capable of reflecting the practical production. Therefore this model is allowed to work as a process parameter model for the preparation of millimeter-level DPG particles.

It can be known from the mathematical models set up in different zones that in this study, the production process parameter model reaches a coincidence rate of above 90% with the industrialized production dynamic

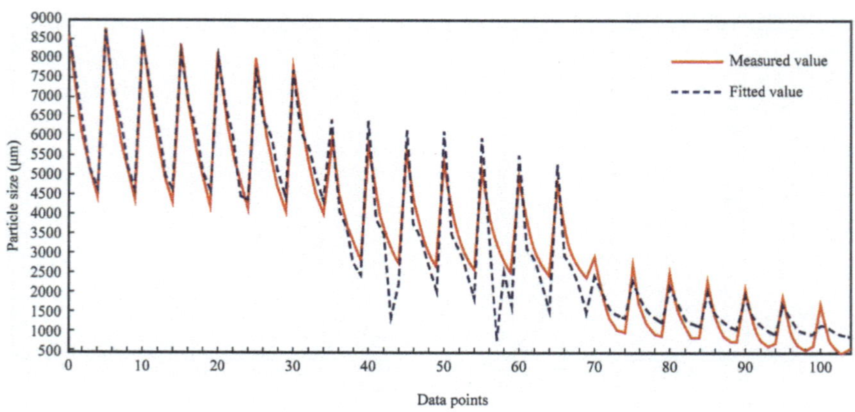

Figure 3.32 The function graph of numerical fitting of micron-sized DPG particles.

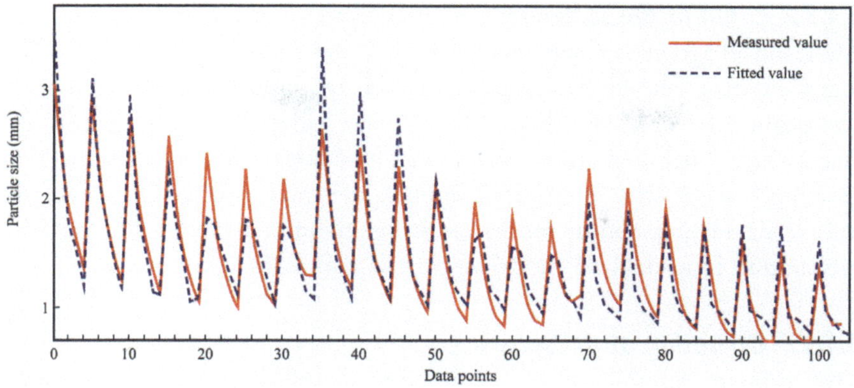

Figure 3.33 The function graph of numerical fitting of millimeter-sized DPG particles.

analysis. Therefore all these nanoscale, micron-level, and millimeter-level DPG particles mathematical models are viable to be the design model for industrialized production.

3.4.3 Development of multiscale DPG particles preparation software

On the basis of the established nanoscale, micron-level, and millimeter-level DPG particles mathematical models, the python open source language was used to program these models. Also, software for multiscale DPG particles large scale preparation was developed [9]. This software enabled the selection of suitable process parameters in line with production needs, so as to achieve fast preparation of DPG particles with different sizes.

3.4.3.1 Functions introduction of the software

Using this multiscale DPG industrialized preparation software, the third parameter for the preparation process can be acquired by inputting the required size of DPG particles and two preparation process parameters, in accordance with the particle size required in the actual production of oil storage application and in combination with the feasibility and convenience of the nanoscale/micron-level/millimeter-level DPG particles industrialized production.

Technical features: it adopted python open-source language and was characterized with good scalability, great fault tolerance, humanized interface design, and it was easy to operate and learn, etc.

3.4.3.2 Login of the software

After the program ran and was packaged, the software first presents the login interface (shown as Fig. 3.34), which required the authorized username and password. After login, the software entered into the main interface which included three functional modules: the nanoscale DPG particles industrialized preparation model, the micron-level DPG particles industrialized preparation model, and the millimeter-level DPG particles preparation industrialized model (shown as Fig. 3.35).

3.4.3.3 Operation of the software

The software running mode is illustrated as follows with the nanoscale DPG particle industrialized preparation module. First, click the "nanoscale DPG particle industrialized preparation module" button in Fig. 3.35 to enter the corresponding model interface (shown as Fig. 3.36). In the interface as Fig. 3.36,

1. To solve the shearing space of nanoscale DPG particles industrialized preparation.

   Input parameters: ① D_{DPG} represents the particle size for nanoscale DPG particles industrialized preparation, ranging from 100 to 1000, in nm; ② R represents the shearing rate of the mechanical shear device,

Figure 3.34 The login interface of the software used.

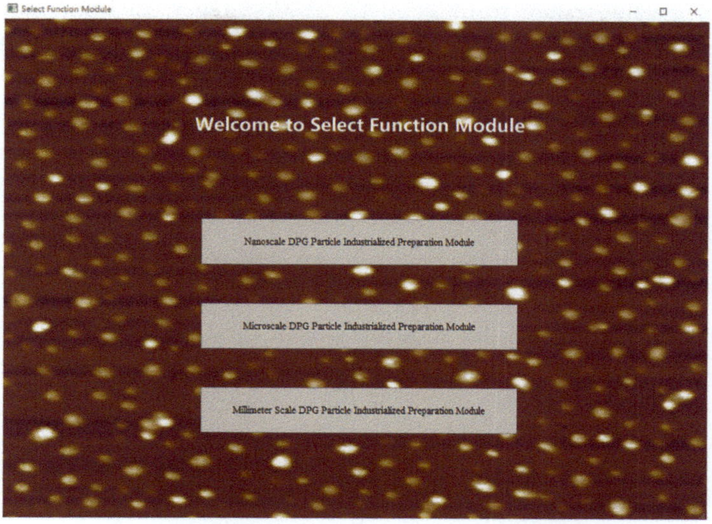

Figure 3.35 The main interface, and three major functional modules included.

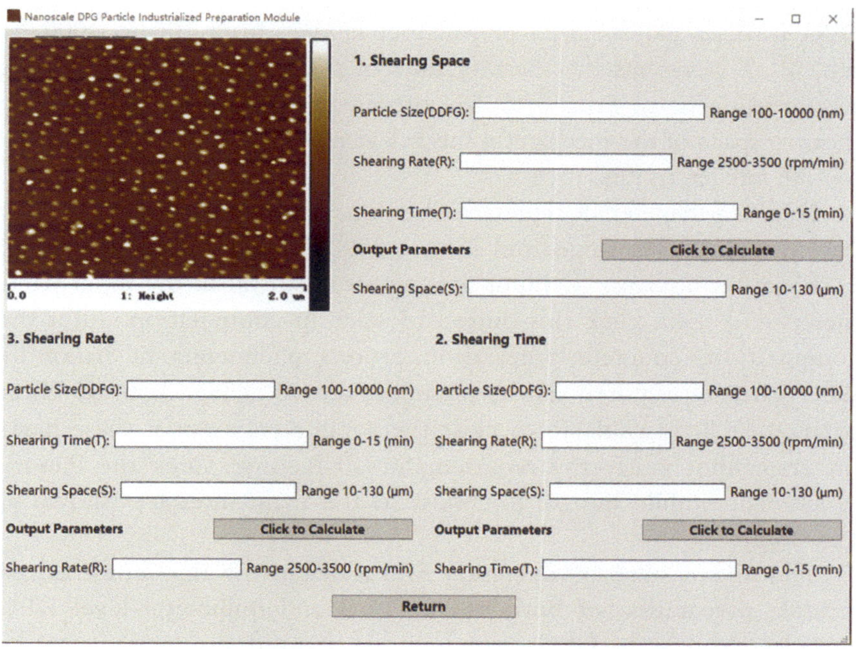

Figure 3.36 The module interface of industrialized preparation of nanoscale DPG particles.

with a value range from 2500 to 3500, integer, in rpm/min; ③ T represents the shearing time of the mechanical shear device, with a value range from 0 to 15, integer, in min.

Output parameter: S represents the shearing space of the mechanical shear device, with a value range from 10 to 130, integer, in μm.

2. To solve the shearing time of nanoscale DPG particles industrialized preparation.

   Input parameters: ① D_{DPG} represents the particle size for nanoscale DPG particles industrialized preparation, ranging from 100 to 1000, in nm; ② R represents the shearing rate of the mechanical shear device, with a value range from 2500 to 3500, integer, in rpm/min; ③ S represents the shearing space of the mechanical shear device, with a value range from 10 to 130, integer, in μm.

   Output parameter: T represents the shearing time of the mechanical shear device, with a value range from 0 to 15, integer, in min.

3. To solve the shear rate of nanoscale DPG particles industrialized preparation.

Input parameters: ① D_{DPG} represents the particle size for nanoscale DPG particles industrialized preparation, ranging from 100 to 1000, in nm; ② T represents the shearing time of the mechanical shear device, with a value range from 0 to 15, integer, in min; ③ S represents the shearing space of the mechanical shear device, with a value range from 10 to 130, integer, in μm.

Output parameter: R represents the shear rate of the mechanical shear device, with a value range from 2500 to 3500, integer, in rpm/min.

Function buttons: ① Input parameters: First, fill in the input parameters and then, click this button to start up computation. After the computation, concrete preparation process parameters of nanoscale DPG particles required can be obtained, shown as Fig. 3.36. ② Clear parameters: It is available to clear the parameters input if these parameters cannot satisfy the requirement. ③ Return: Click the Return functional module button to return to the main interface (shown as Fig. 3.35).

The above steps of operation are available for the industrialized module preparation of both micron-level and millimeter-level DPG particles. Multiscale DPG particles industrialized production can be achieved by this preparation software, where process preparation parameters are available for selection, in accordance with the need of production.

3.4.3.4 Quit of the software

After all the tasks are completed, return to the main software by clicking the close button at the upper right corner, and confirm that the software quits.

Section 3.5 Design of the industrialized production equipment for DPG particles

In the light of different working conditions, an industrialized workshop production device and skid-mounted production device for multiscale DPG particles were designed. The first device could meet the DPG particle production under routine working conditions, suitable for the operation of short-distance transportation in oilfields. The latter device not only works well with the operation under routine working conditions but also with the one under complicated working conditions such as deserts, tidal flats, hills, and offshore operating platforms. It overcomes the high-cost drawbacks including long-distance transportation and packaging. As of 2021, there have been 15 industrialized workshop production devices and seven skid-mounted production devices for multiscale DPG particles, allowing the realization of multiscale DPG particles oilfield application in the major blocks of 17 domestic oilfields.

3.5.1 Design industrialized workshop production equipment

The industrialized workshop production device can fulfill the production of multiscale DPG particles under routine working conditions and be suitable for short-distance transportation in oilfields. With Shengli Oilfield as an example, three factory production lines with an annual yield of over 5000 tons have been established around Dongying City, Shandong Province. They can meet the working needs of multiscale DPG particles in the Shengli Oilfield and its surrounding cities. For a transformation enterprise in the Shengli Oilfield, for example its industrialized workshop production device mainly consists of the chemical cross-linking reaction system, the shear grinding system, the high-viscosity fluid delivery system, the intelligent control system, and the storage system for DPG particles. It can meet such technical requirements for polymer dissolution, "above-ground instant gel formation" and "high-viscosity fluid delivery" of bulk gels, as well as "high-efficiency shearing and rounding" and "precise digital control" of multiscale DPG particles. Its physical design is shown in Fig. 3.37, and the real workshop in Fig. 3.38.

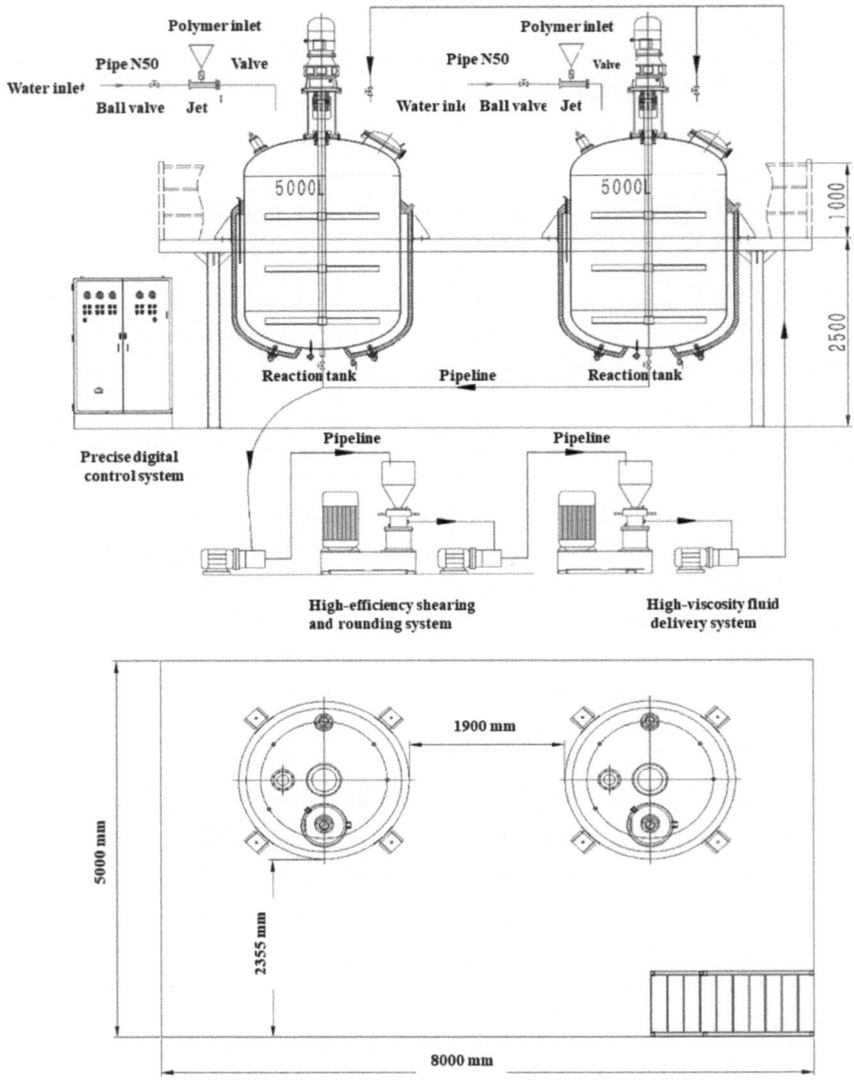

Figure 3.37 Flow chart design of industrialized production workshop.

3.5.2 Design skid-mounted production equipment of multiscale DPG particles

Not only is the multiscale DPG particles skid-mounted production device workable for the tasks under routine working conditions but also for the ones under sophisticated working conditions, including desserts, tidal flats, hills, and offshore operating platforms. It surmounts the high-cost

Figure 3.38 The real scene of the industrialized production workshop.

drawbacks such as long-distance transportation and packaging. To this end, the multiscale DPG particle skid-mounted production device was designed in this study. Its working conditions were extended from conventional areas to cover all-weather and 24 hours/7 days conditions such as deserts, hills, and small offshore platforms. A transformation enterprise of CNOOC Tianjin, for example has devised such a production device with a daily yield of 30 tons. This device is made up of a bulk gel chemical cross-linking reaction system, a high-viscosity fluid delivery system, high-efficiency shearing and rounding system, a convenient dispersion injection system and a precise digital control system. Its detailed design is shown in Fig. 3.39 and the actual device in Fig. 3.40. Thanks to its simple structure and the location of each system on the skid plate for the sake of the device movement, this production device can fulfill the skid-mounted online continuous production DPG particles as long as being connected with the electric cable through pipes. Meanwhile, the device is easy and convenient to operate. It thus reduces the costs of production and the labor intensity of workers, improving their work environment. With a precise digital control system, this device can also perform waterproof, overrun automatic alarm and power-off protection. This helps to improve the on-site safety of DPG particles production. This DPG particles skid-mounted production device we designed has successively been approved

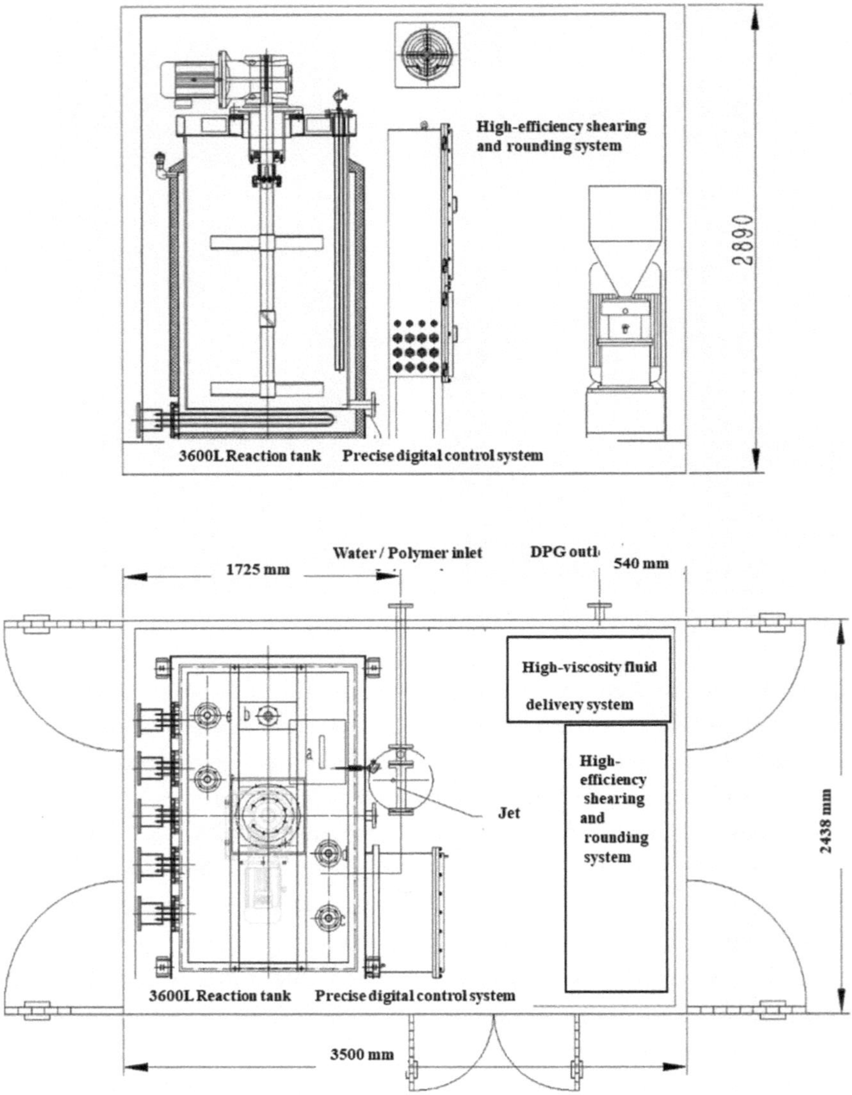

Figure 3.39 Flow chart design of DPG particles skid-mounted production equipment.

by the French Classification Society and the China Classification Society, contributing to promoting the international advancement of this technology.

The bulk gel chemical cross-linking reaction system consists of a jet, a material funnel, a water pipeline, a reaction tank, a heating device, a safety valve, and a fixed base. It allows the rapid dissolution of polymers and

Figure 3.40 Real DPG particles skid-mounted production equipment.

avoids the occurrence of the "fish-eye" phenomenon resulting from the conventional dispersion of polymers. This system can ensure that the gel-forming liquid reacts quickly under normal pressure and has a good heat preservation effect, avoiding energy waste. The high-efficiency shearing and rounding system is composed of an improved colloid mill, a water pipeline, and a safety valve, enabling it to realize the rapid shearing and rounding of bulk gels and the rapid adjustment of multiscale DPG particles production. The high-viscosity fluid delivery system is composed of a high-viscosity fluid pump, a water pipeline, and a safety valve. It can meet the pretreatment shear and fast transportation of fluids with a viscosity up to 10×10^4 mPa · s. The convenient dispersion injection system consists of a high-pressure plunger pump, flowmeter, a water pipeline, and a safety valve. The system can continuously inject the DPG particles into the water well in proportion, so as to save the injection cost. The precise digital control system is composed of a bulk gel chemical cross-linking reaction system controller, a high-efficiency shearing and rounding system controller, a high-viscosity fluid delivery system controller, a convenient dispersion injection system controller, an overcurrent protection device, an explosion-proof quick plug, and an explosion-proof cabinet. By being connected with these four systems through the explosion-proof quick plug, this system can function to waterproof, automatically alarm for overrunning, and provide power-off protection. It can not only ensure the safety of each system but also facilitate the quick connection of each system.

We take the multiscale DPG particle online skid-mounted production and injection device as an example to illustrate the specific operation steps:

1. A certain quantity of functional polymers was weighted. They were then placed through the bulk gel chemical cross-linking reaction

system into the reaction tank for bulk gel chemical cross-linking and stirred for 30–60 minutes.

2. Cross-linking agent, coagulant, and stabilizer were added through the jet mixing device into the functional polymer solution prepared by Step (1). Then, the mixed solution in the bulk gel chemical cross-linking reaction system was mixed and dispersed for 10–20 minutes, so as to obtain bulk gel solution.
3. The heating device of the bulk gel chemical cross-linking reaction system was powered on. The reaction temperature was controlled at 90°C–95°C, and the reaction time lasted for 5–8 hours. Then, a bulk gel system of DPG particles was prepared.
4. The high-viscosity fluid pump is turned on in the positive direction. Then, the bulk gel system prepared by Step (3) was transferred to the colloid mill of the high-efficiency shearing and rounding system, and the shear rate was regulated to the appropriate. Meanwhile, the high-viscosity fluid pump of the DPG particle storage system was positively turned on. The prepared DPG particles were next delivered to the stirring tank of this same system to complete the shearing and grinding once.
5. The high-viscosity fluid pump was turned on in reverse direction. Then, the DPG particles prepared in one cycle by Step (4) were transported to the colloid mill in the shear grinding system again, and the shear rate of the colloid mill was adjusted. At the same time, the high-viscosity fluid pump of the bulk gel chemical cross-linking reaction system was turned on. The prepared DPG particles were transported to the reaction tank of the bulk gel chemical cross-linking reaction system so as to complete the second shearing and grinding.
6. The 1st to 6th shearing and grinding were completed by Steps (4) and (5) in sequence, and then multiscale DPG particles were collected.
7. The multiscale DPG particles prepared by Step (6) were injected to the water well through the convenient dispersion injection system, and the multiscale DPG particle online skid-mounted production and injection treatment is finished.

References

[1] Chauveteau G, Tabary R, Renard M. Controlling in-situ gelation of polyacrylamide by zirconium for water shutoff. In: SPE international symposium on oilfield chemistry, Houston, Texas, February 16–19, 1999.

[2] Chauveteau G, Omari A, Tabary R, et al. New size-controlled microgels for oil production. In: SPE64988 presented at the 2001 SPE international symposium on oilfield chemistry conference, Houston, Texas, February 13–16, 2001.
[3] Zhu H, Zhu J, Sun S, et al. Application performance of a pre-crosslinking micro-gel dispersion for profile control and oil displacement. Petroleum Exploration and Development 2004;31(2):115–18.
[4] You Q, Tang Y, Dai C, et al. Research on a new profile control agent: dispersed particle gel. In: SPE enhanced oil recovery conference, Kuala Lumpur, Malaysia, July 19–21, 2011.
[5] You Q, Dai C, Tang Y, et al. Study on performance evaluation of dispersed particle gel for improved oil recovery. Transactions of the ASME: Journal of Energy Resources Technology 2013;135:1–7.
[6] Dai C, Zhao G, Zhao M, et al. Preparation of dispersed particle gel (DPG) through a simple high speed shearing method. Molecules 2012;17:14484–9.
[7] Zhao G, You Q, Gu C, et al. Preparation mechanism of multiscale dispersed particle gel. Acta Petrolei Sinica 2017;38(7):821–9.
[8] Zhao G, Dai C, You Q. Characteristics and displacement mechanisms of the dispersed particle gel soft heterogeneous combination flooding system. Petroleum Exploration and Development 2018;45(3):464–73.
[9] Dai C, Zhao G, Bo Q, et al. Industrial preparation software of multi-scale dispersed particle gel. Computer Software Copyright, 2020 SR0277181.

CHAPTER 4

The profile control technology of multiscale DPG particles

Deep profile control of multiscale DPG particles is a newly developed chemical water control technology, which harbors the potential to be applied in oilfields. To give full play to the role of DPG particles in deep profile control, it is necessary to study its application performance. In this chapter an investigation was carried out on their basic properties, matching the relationship with the pore throat of the reservoir, and the performance of the profile control. Its profile control mechanism was revealed from both macro- and microperspectives, so as to provide theoretical support for its application in oilfields.

Section 4.1 The characterization of DPG particles

In this section, modern analytical methods such as viscometer, laser particle size analyzer, scanning electron microscope, and zeta potential tester are used to characterize the viscosity, particle size, self-growth coalescence performance, micromorphology, and other properties of multiscale DPG [1]. The salinity of the simulated water used in the experiment was 5.0×10^4 mg/L, and the content of Ca^{2+} and Mg^{2+} was 3000 mg/L.

4.1.1 The viscosity characteristics of DPG particles

4.1.1.1 Viscosity—concentration relationship

The viscosity—concentration relationship of the multiscale DPG particles was studied and compared with the viscosity of type II functional polymers. The test results are shown in Fig. 4.1. It can be seen that the viscosity of the two systems increases with the increase of mass concentration. But the viscosity of type II functional polymer is significantly higher than that of DPG particles. This phenomenon becomes more obvious as the concentration increases. Because the bulk gel for DPG particles is a fluid with high viscoelasticity, although the high-shear effect cuts off the long chain bundles of the bulk gel, the bulk gel has certain viscoelasticity. The long chain bundle after shearing still maintains high viscoelasticity.

Theory and technology of multiscale dispersed particle gel for in-depth profile control
DOI: https://doi.org/10.1016/B978-0-323-99849-9.00004-7

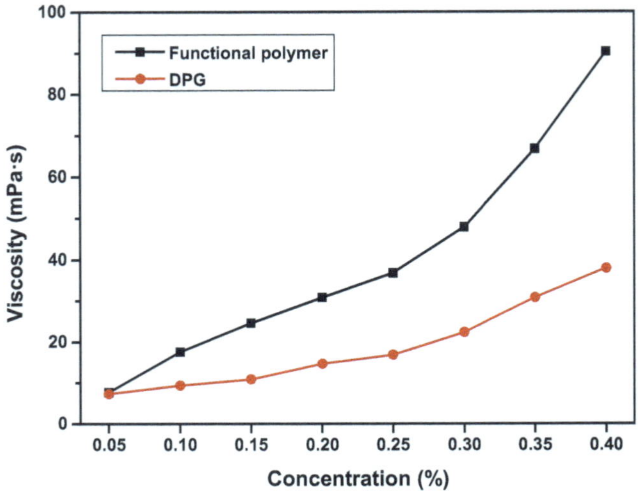

Figure 4.1 The relationship between viscosity and concentration of DPG particles and type II functional polymers.

Therefore the DPG particle liquid formed by the long-chain bundle still has high viscoelasticity. The viscosity of the multiscale DPG particle system increases with the mass fraction, which can be explained by the Krieger–Dougherty equation:

$$\frac{\eta}{\eta_{\text{medium}}} = \left(1 - \frac{\varphi}{\varphi_m}\right)^{-[\eta]\varphi_m} \tag{4.1}$$

where η is the viscosity of suspension; η_{medium} is the viscosity of base fluid; φ is the volume fraction of solid particles; φ_m is the maximum volume fraction of solid particles; and $[\eta]$ is the intrinsic viscosity index, which was 2.4 when the particles were spheres.

It can be known that the system viscosity increases as the concentration of DPG particles rises. When the content of DPG particles increases, the particles pack more tightly. At this time, the distance between the particles decreases and their free movement becomes more difficult, which thereby intensifies the interaction force between the particles. The flow resistance caused by this also increases, making the viscosity of the system increase abruptly.

4.1.1.2 Viscosity–temperature relationship

The effect of temperature on the viscosity of multiscale DPG particle system was investigated with the viscosity of type II functional polymer as a comparison, where the concentration of both was 0.25%. The test results are shown in Fig. 4.2. As the temperature increases, the viscosity of both systems decreases. Though the viscosity of polymers drops significantly, it is always lower than that of the same concentration of DPG particles. The increase in temperature accelerates the thermal movement of polymer molecules, reduces the radius of gyration between molecules, and weakens the interaction force. This, therefore leads to a decrease in the viscosity of the system. Moreover, the increase in temperature reduces the degree of solvation of polymer molecules, resulting in an increase in rotational resistance. As a result, the polymer molecular chain becomes curled, and its viscosity decreases. However, the increase in temperature will also accelerate the chance of collision between molecules and increase the viscosity. But the effect of this effect is relatively low, so the viscosity of the polymer decreases with the increase of temperature. For DPG particles, part of the system viscosity comes from the high viscoelastic bulk gel, because the bulk gel has higher temperature resistance. The other part comes from the internal friction generated by particle collision and contact. The increase in temperature accelerates the collisions between particles and brings about an increase in viscosity. Therefore the viscosity of multiscale DPG particles decreases slightly with the increase of temperature.

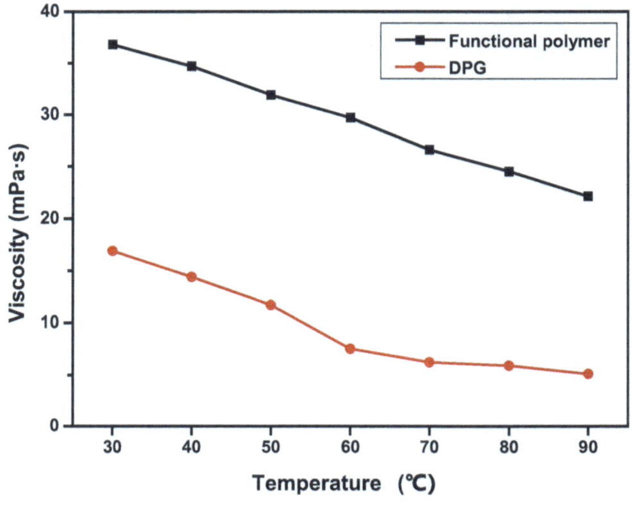

Figure 4.2 The relationship between viscosity and temperature of DPG particles and type II functional polymers.

4.1.2 Shear resistance

In this study, the functional polymer of bulk gel was used as a comparison to investigate the shear resistance of multiscale DPG particles. At room temperature, using a waring shearing machine, the corresponding viscosity was measured at a shear rate of 1000 rpm, different shear times, and 30°C. Both concentrations were 0.25%, and the test results are shown in Fig. 4.3. According to the results, as the shear time increases, the viscosity of the functional polymer drops to a sharp extent while that of the multiscale DPG particle system to drop a slight extent. This is because the functional polymer is dispersed in the solution in random coils. The high-speed shear cuts the random coils into numerous coils of smaller molecular weight. However, it is difficult for small molecule coils to form an effective winding structure. Therefore the viscosity of the polymer is greatly reduced. Multiscale DPG particles are a stable system formed by high-viscoelastic bulk gel through high-speed mechanical shearing. Under this shearing action, the particle size and the solid content of the particles in the solution are basically unchanged, but the mutual contact and collision between the particles can still produce effective viscosity.

The test further investigated the viscosity recovery performance of the two systems after shearing. The results are shown in Fig. 4.4. It can be seen that after 24 hours of standing at room temperature, the viscosity recovery rate of the functional polymer is 50%. This is because the

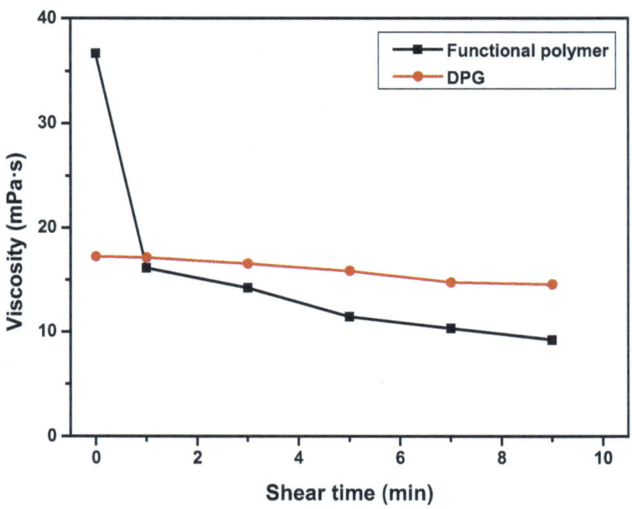

Figure 4.3 The effect of shear time on the viscosity of multiscale DPG particles and type II functional polymers (6 rpm).

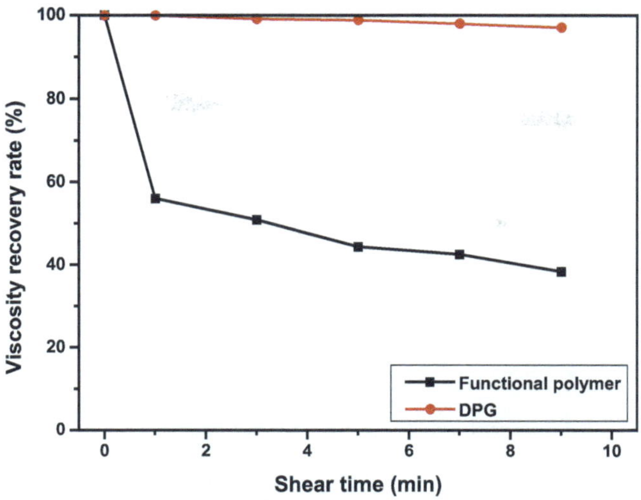

Figure 4.4 The viscosity restoration capability of DPG particles and type II functional polymers (still placement for 24 h).

sheared high-molecular-weight random coils repolymerize to form structural viscosity, but the sheared random coils cannot be completely recovered. The viscosity of multiscale DPG particles is basically unchanged, indicating their good shear resistance and stability.

4.1.3 Thermal stability

Simulated water was used to formulate type II functional polymer and DPG particles with a concentration of 0.25%. The viscosity stability of both systems at 90°C was investigated. The results are shown in Fig. 4.5. It can be seen that as the aging time increases, the viscosity of the two systems decreases. After 60 days of aging, the viscosity drop rate of type II functional polymer was higher than 76.8%, whereas that of the multiscale DPG particles dropped from 17 to 10.2 mPa · s, a decline rate of only 40%. High-temperature aging causes thermal oxidative degradation of polymers. This causes a hydrolysis reaction of the polymer side groups, which converts the amide groups into carboxyl groups. The main chain of the polymer is broken, reducing the molecular weight of the polymer. In addition, the high concentration of calcium and magnesium ions in simulated water can easily form complexes with polymers to form flocculation. Therefore the viscosity of type II functional polymer is greatly reduced after aging. According to the formation mechanism of multiscale DPG particles, bulk gel is a dense network structure formed by the

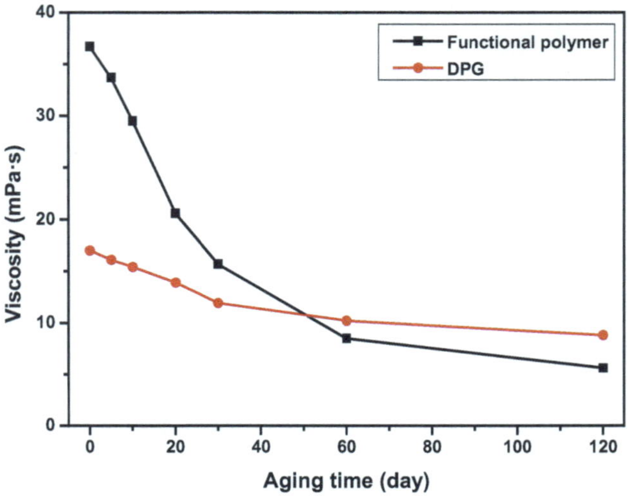

Figure 4.5 The thermal stability of DPG particles and polymers.

dehydration condensation between the amide groups of polymers and the hydroxymethyl groups of the cross-linking agent, thus possessing good viscosity stability. The mechanical shearing action only breaks the bulk gel and does not involve chemical reactions. Therefore the resulting multiscale DPG particles solution still has high thermal stability. DPG particles aggregate under high temperature conditions to form larger particles. The interaction force between these particles decreases, resulting in a decrease in viscosity. However, this decline is moderate, suggesting that DPG particles hold high thermal stability.

4.1.4 Capability of self-growing and coalescence

Multiscale DPG particles were prepared with high-salinity simulated water and aged in an oven at 90°C for different times to investigate the effect of aging time on the self-growth, coalescence, and expansion performance. In order to facilitate the test, samples to be tested were evenly shaken for 5 minutes every day, and the particle size change was measured by a laser particle size analyzer. Fig. 4.6 shows the change in the particle size of DPG particles before and after aging. After 20 days of aging, the average particle size of DPG particles increased from the initial 1.92 μm to 36.88 μm. The effect of high temperature increases the size of DPG particles. This not only involves the enlargement of the particles themselves but also mainly involves the self-growth and mutual aggregation of multiple particles. The larger agglomerates thus formed significantly increase the particle size. In addition, despite the high salinity of the

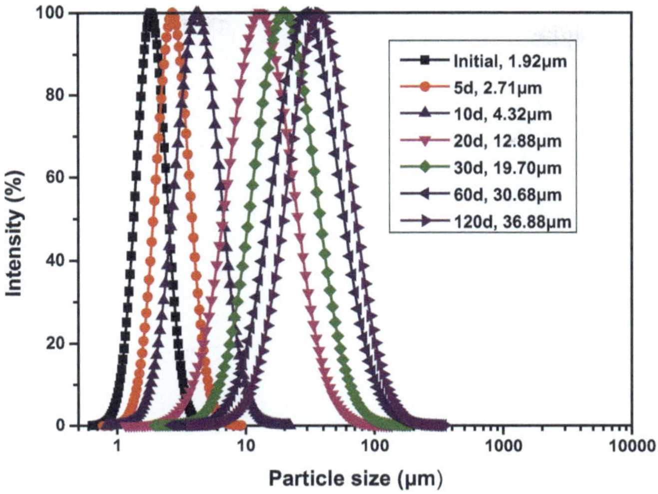

Figure 4.6 The particle size distribution of DPG particles before and after aging.

mixed solution, the particles still have the ability to self-grow and coalesce. This shows that multiscale DPG particles have relatively good resistance to high temperature and salt.

4.1.5 Analysis of zeta potential

Zeta potential is a measure of the strength of static electricity, charge repulsion or attraction between particles, thus a basic parameter that characterizes the stability of particles. The smaller the zeta potential, the more likely the particles are to coalesce. Since the prepared multiscale DPG particles are water-dispersible viscoelastic particles, their size is increased mainly through self-growth and coalescence at high temperature, so as to realize the deep profile control in the formation. Therefore it is necessary to study the surface potential of multiscale DPG particles. Considering the salt tolerance range of the Zeta potentiometer, the test only considers the potential changes of multiscale DPG particles in 10,000 mg/L simulated water (Na^+, 9100 mg/L; Ca^{2+}, 600 mg/L; Mg^{2+}, 300 mg/L). The results are shown in Fig. 4.7. The absolute value of the zeta potential of multiscale DPG particles after high-temperature aging decreases, indicating that the electrostatic repulsion between DPG particles decreases, and the particles self-grow to form large aggregates. When salt ions are added, the positively charged salt ions neutralize the negative charges on the surface of the particles, so that the electrostatic repulsion between the particles is reduced. Thus DPG particles tend to aggregate in simulated water.

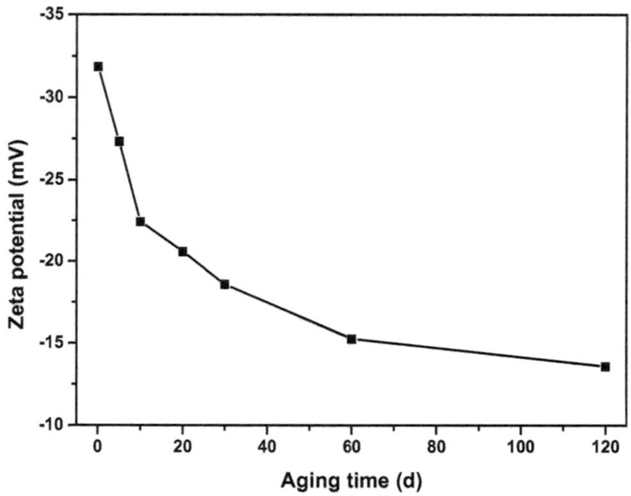

Figure 4.7 The zeta potential changes of DPG particles with the aging time.

In addition, calcium and magnesium ions have a higher ion valence than sodium ions, resulting in a stronger ability to neutralize the surface charge of particles. Hence, DPG particles are prone to agglomerate in the simulated water with high ion value, resulting in the increase of particles. This is confirmed by the particle size distribution curve of DPG particles after high-temperature aging. The self-growth and coalescence properties of multiscale DPG particles are beneficial for adjusting the formation profile. When DPG particles migrate in the porous media, they are affected by the physical and chemical properties of the formation. Therefore these particles self-grow and coalesce to grow up. This property is conducive to the formation of agglomerates with larger particle size, so as to achieve effective plug high-permeability area of the formation and reach the purpose of deep profile control.

4.1.6 Macroscopic morphology

The experiment examined the macroscopic morphology changes of multiscale DPG particles during the aging process (130°C, 220,000 mg/L), and the results are shown in Fig. 4.8. The aging process of multiscale DPG particles includes uniform dispersion of single particles, single-particle expansion, coalescence and self-growth of multiple particles, such that the particles have the characteristics of self-growth. The existence of DPG particle aggregates after 30 days aging suggests they have good resistance to high temperature and high salinity. The absence of degrading

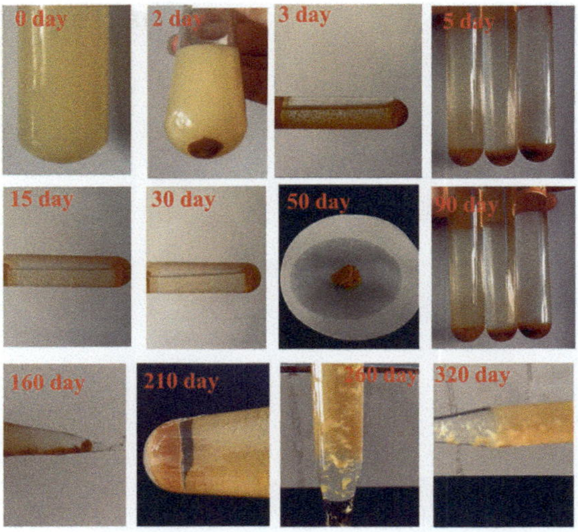

Figure 4.8 The macroscopic morphology of multiscale DPG particles before and after aging.

and dehydration of DPG particles throughout their aging process proves their good resistance to high temperature and high salinity.

4.1.7 Microscopic morphology

The initial microscopic morphology of multiscale DPG particles was spherical with regular shapes. DPG particles were prepared with simulated water. They were aged at 130°C for different time. Then, the microscopic morphology changes before and after aging were observed by scanning electron microscope. It can be seen from Fig. 4.9 that DPG particles are mainly dispersed uniformly in single particles before aging, and the particle size is mainly uniformly distributed between 1.5 and 2.1 μm. After 5 days of high-temperature aging, the particle size of a single particle becomes larger and the shape tends to be irregular. However, the expansion of a single DPG particle is limited. This feature enables the particles to maintain high strength. When DPG particles further age, multiple DPG particles begin to coalesce with each other. The degree of adhesion between particles increases, forming larger aggregates. After 30 days of aging, the self-growth of aggregates can reach over 30 times. This is because the diameter of DPG particles is smaller, the surface energy is larger, and there is a tendency for self-growth. Moreover, smaller particle size allows a more obvious tendency of coalescence. Before aging, DPG particles with the same charge repel each other, which prevents the particles from approaching each other.

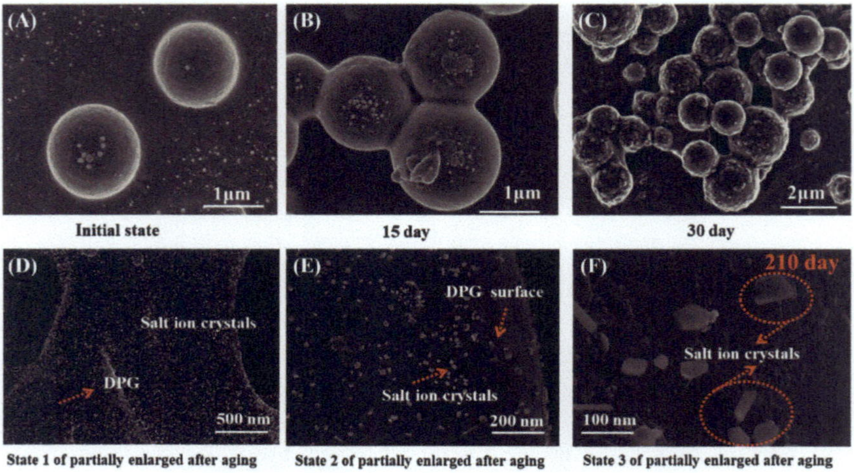

Figure 4.9 The micromorphologic changes of DPG particles before and after aging: (A) initial state; (B) 5 days; (C) 30 days; (D) partially after aging 5 days; (E) partially after aging 30 days; (F) partially after aging 210 days.

In addition, since the adsorbed ions and counter ions of DPG particles are hydrated (that is, water molecules are wrapped around the ions), they are particles with a hydration film. The existence of the hydration film makes a viscoelastic membrane between the particles, which prevents the mutual aggregation of DPG particles from becoming larger when they collide. Under these two actions, DPG particles can be evenly dispersed in the solution.

Section 4.2 Matching relationship between DPG particles and pore throat

The matching relationship between multiscale DPG and reservoir pore-throats is an important factor affecting the deep profile control effect of DPG particles. In order to ensure that DPG particles can be injected deep into the reservoir, the matching relationship based on the particle size of DPG and measurable reservoir parameters was studied by plenty of core flooding tests [2,3]. This may provide guidance for the application and implementation of deep profile control technology by multiscale DPG in the oilfield.

4.2.1 Establishing the mathematical model of matching relationship

The matching relationship between DPG and reservoir is represented by the matching factor (ψ), which is expressed as the ratio of the average particle size of DPG to the mean diameter of throats, as is shown in Eq. (4.2).

$$\psi = \frac{d}{D} \tag{4.2}$$

where ψ is the matching factor; d is the average particle size of DPG (μm); and D is the mean diameter of throats (μm).

As is shown in Eq. (4.2), the particle size of DPG is a measurable parameter, but the average diameter of pore-throats cannot be obtained directly. Thus reservoir permeability can be represented by the average diameter of pore-throats according to the Carman–Kozeny equation:

$$k = \frac{\phi D^2}{16f\mathrm{CK}\cdot\tau^2} \tag{4.3}$$

In this equation, k is the absolute permeability (μm^2); φ is the porosity; f_{CK} is the shape factor of Carman–Kozeny; and τ is the tortuosity.

For conventional reservoirs, the value of $f_{\mathrm{CK}} \cdot \tau^2$ is set to be 4.5. Substituting Eq. (4.3) into Eq. (4.2), the matching factor can be mathematically expressed using accessible parameters of reservoir, as given by Eq. (4.4).

$$\psi = d\left(\frac{72k}{\phi}\right)0.5 \tag{4.4}$$

It can be seen from Eq. (4.4) that the matching factor is determined by the average particle size of DPG, and the reservoir permeability and porosity.

4.2.2 Matching factor optimization

Based on core flooding tests, the matching factor was optimized by the injection capacity and plugging property of DPG particles. In order to ensure good injectivity, deep migration ability, and effective profile control of multiscale DPG particles, the optimization standard is as follows. Injection pressure won't begin to significantly rise until the DPG is injected by about 1.0 PV, and the plugging rate is not less than 80%.

Through changing the average particle size of DPG or the permeability of cores, the relationship between injection pressure or resistant coefficient and injection volume (PV) for different matching factors can be recorded, as shown in Figs. 4.10 and 4.11.

It can be seen from Figs. 4.10 and 4.11 that the injectivity and profile control characters of multiscale DPG at different matching factors are different. The results could be discussed in the following three cases.

1. When the matching factor was large ($\psi > 0.349$), the pressure began to rise sharply, and the output liquid decreased gradually and even disappeared after DPG particles were injected by 0.5 PV. This indicates that the oversized DPG particles rapidly accumulated and plugged the front of

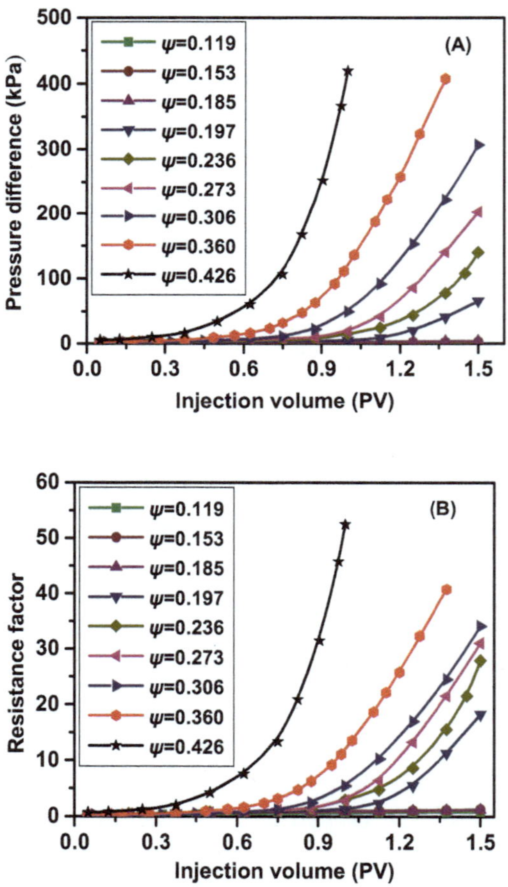

Figure 4.10 Injection of DPG particles at different matching factors (the average particle size of DPG used was fixed at 3.0 μm, and the permeability values of cores varied): (A) variation of injection pressure with injection volume. (B) variation of resistance coefficient with injection volume.

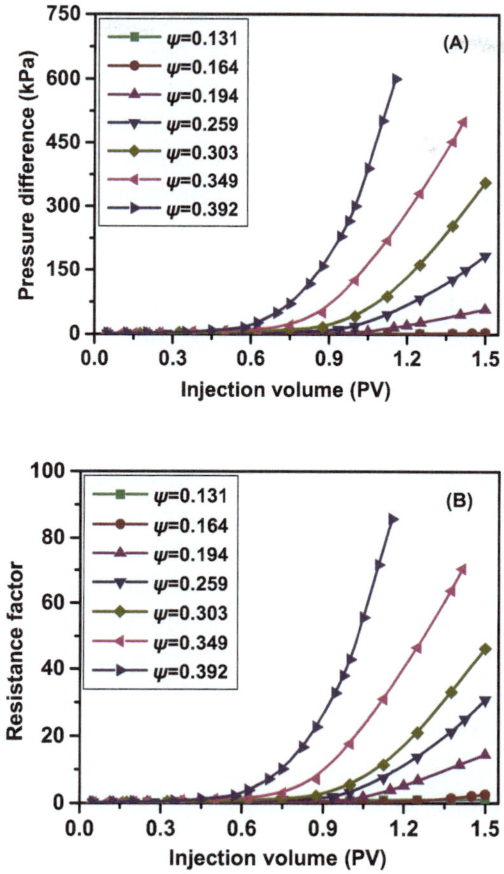

Figure 4.11 Injection of DPG particles for different matching factors (the permeability values of cores were fixed at about 2.0 μm^2, the average size of DPG particles varied): (A) variation of injection pressure with injection volume; (B) variation of resistance coefficient with injection volume.

the core, which led to a high DPG injection pressure and impeded a deep migration of DPG particles in porous media. Therefore the injection capacity of DPG is poor when the matching factor is large.

2. When the matching factor was moderate ($0.197 < \psi < 0.306$), the pressure began to rise obviously after the DPG particles were injected by 1.0 PV, the output liquid decreased but would not disappear. The reasonable increase of injection pressure indicates that the DPG particles performs a deep profile control effect, increasing the flow resistance, and improved the reservoir heterogeneity. Thus the DPG shows good injection capacity and plugging ability at the above matching factor range.

3. When the matching factor was small ($\psi < 0.185$), the injection pressures and output liquid flow almost remained steady until the end of injection. DPG could be obviously observed in the produced solution. This is because the DPG particles are smaller than the size of pore-throats, and the particles directly pass through the pore-throats to the end of the porous media. Though the DPG is with good injection capacity at small matching factors, no stable accumulation or bridging of particles is formed in pore-throats, which leads to poor plugging ability.

Therefore in order to ensure the DPG particles have good injectivity, deep migration ability, and effective profile control performance, the optimal matching factor range is about 0.20–0.30.

On the other hand, the profile control performance of multiscale DPG particles was studied to further verify the optimized matching factor range. Core flooding experiments were conducted to investigate the plugging properties of DPG particles for different matching factors. Firstly, saturate the core with brine. Then the core was flooded by brine until the injection pressure is stable. After that, 1.0 pore volume of DPG particles was injected. At last, subsequent brine flooding was carried out and the stable pressure was recorded. The plugging rate and residual resistance factor could be calculated based on the stable pressures, as shown in Fig. 4.12.

Fig. 4.12 shows that when the matching factors are higher than 0.20, the plugging rates of DPG are higher than 85%, which could meet the technical requirement of profile control treatment by DPG. With a comprehensive

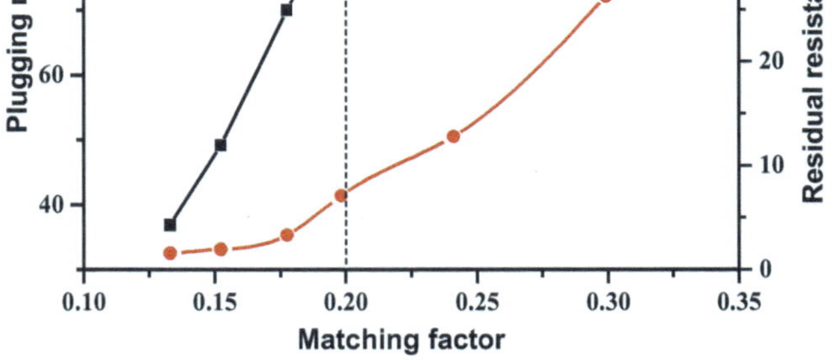

Figure 4.12 Profile control ability of DPG particles for different matching factors.

consideration of the injection and profile control ability of DPG in the core, the optimal range of matching factor is found to be determined as 0.20–0.30.

Section 4.3 The profile control performance of DPG particles

4.3.1 Capability of mobility control

With the help of the core flow test physical model, the plugging performance and erosion resistance of multiscale DPG particles under high temperature and high salt reservoir conditions were investigated. The simulated water salinity was 210,000 mg/L, (Na^+: 74,000 mg/L; Ca^{2+} and Mg^{2+}: 8000 mg/L), the test temperature 120°C, the average size of DPG particles 2.6 μm, and the concentration 0.1% (with a 1:4 ratio of bulk gel to simulated water). The flow chart of the test device is shown in Fig. 4.13, the test parameters in Table 4.1, and the test results in Fig. 4.14.

According to Fig. 4.14, the subsequent water flooding pressure difference rises dramatically after a long period of aging. The water flooding pressure difference tends to be stable and maintains a high level after multiple times of PV simulation water flushing. The plugging rate is maintained above 85%. Therefore it can be known that the prepared DPG particles possess a good capability of mobility control at high temperatures and salt.

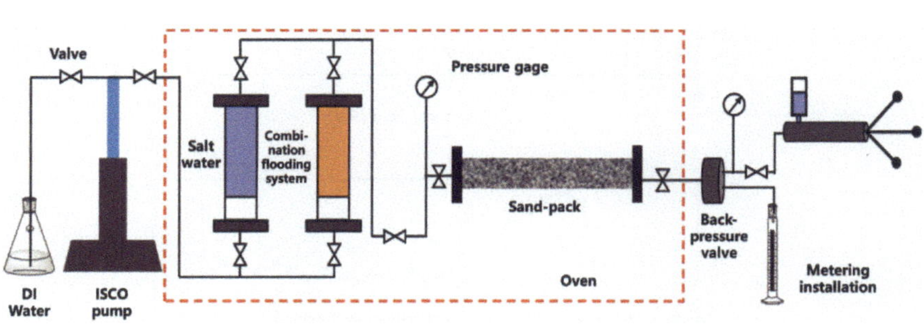

Figure 4.13 Schematic diagram of single sand-pack flow test device.

Table 4.1 Plugging performance test results.

Test No.	Permeability of sand-packs (μm^2)	Porosity of sand-packs (%)	Injection volume (PV)	Back pressure (MPa)	Days of aging (d)
1	0.70	29.6	1.0	20	10
2	0.67	28.7	1.0	20	30
3	0.72	30.2	1.0	20	60

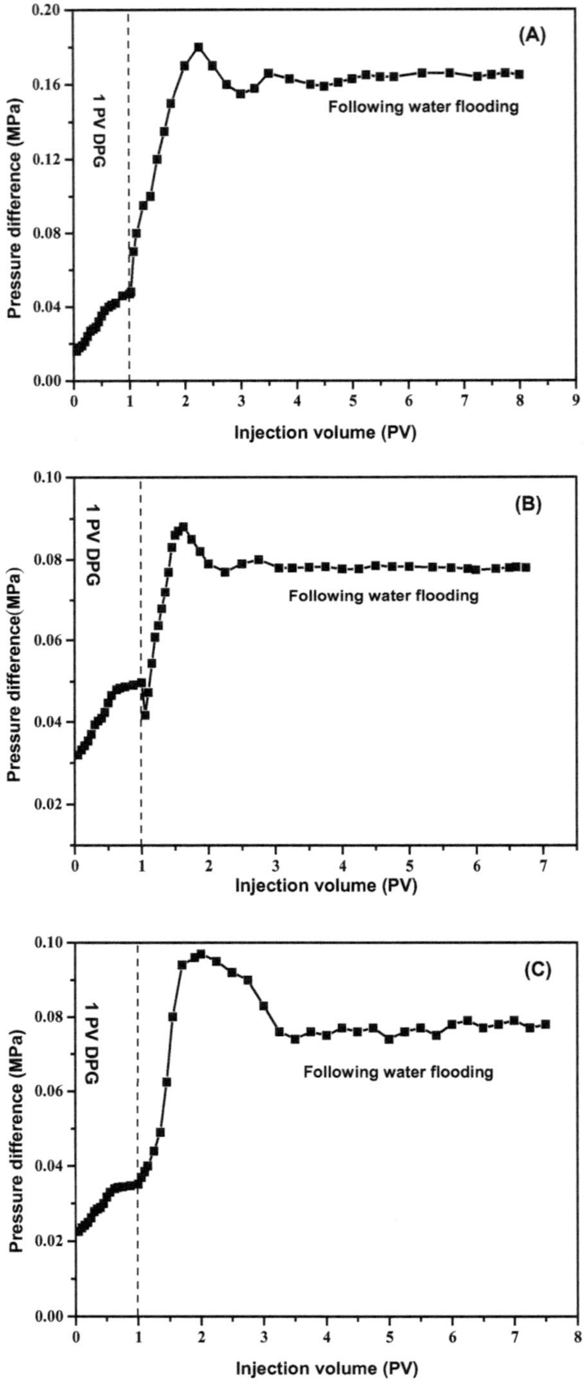

Figure 4.14 Results of plugging performance test at high-temperature and high-salinity condition: (A) aging for 10 days; (B) aging for 30 days; (C) aging for 60 days.

4.3.2 Performance of profile improvement

Multiscale DPG particles are flexible and can be deformed into deep formations through pore throats. The high-permeability layer was regulated by means of retention, adsorption, bridging, etc., so as to improve the formation profile. This study simulated the heterogeneity of the formation by establishing double sand-packs physical models with different levels of permeability rate. The high-permeability layer of the formation was simulated with a high-permeability sand-pack, and the low-permeability layer with a low permeability sand-pack. The reservoir control ability of multiscale DPG particles was also evaluated using shunt rate and profile improvement rate. The calculation of the profile improvement rate is shown in formula (4.5):

$$f = \frac{\frac{Q_{hb}}{Q_{lb}} - \frac{Q_{ha}}{Q_{la}}}{\frac{Q_{hb}}{Q_{lb}}} \tag{4.5}$$

where Q_{hb}, Q_{ha} is the water absorption before and after the profile control of high-permeability reservoir, respectively (%); and Q_{lb}, Q_{la} is the water absorption before and after the profile control of low-permeability reservoir, respectively (%).

The test flow chart is shown in Fig. 4.15. The specification of the sand-packs is 20 cm in length × 2.5 cm in diameter. Test steps: (1) cores with different permeability levels were filled; (2) water flooding was carried out at a pump speed of 0.5 mL/min until the pressure was stable; (3) 1 PV multiscale DPG particle (particle size 2.9 μm, injection concentration 0.1%) was injected at a pump speed of 0.5 mL/min; (4) the sand-packs were aged at 120°C for 7 days; (5) the following water flooding was performed at a pump speed of 0.5 mL/min; (6) the shunt rate and the

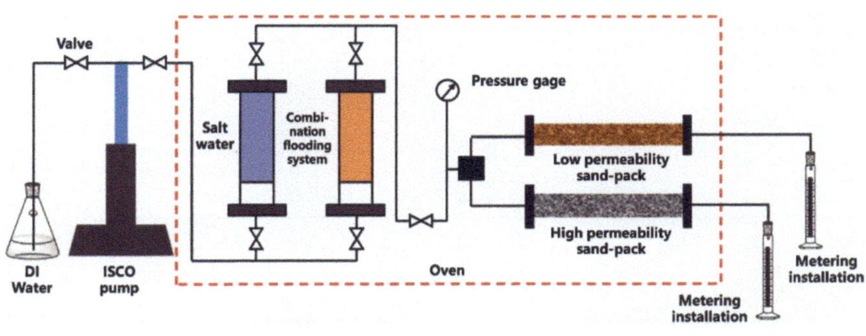

Figure 4.15 Schematic diagram of double sand-pack physical simulation device.

injection pressure of the high- and low-permeability pipes were measured separately; and (7) the improved performance of the water injection profile (as shown in Fig. 4.16) was under observation. In this test, the

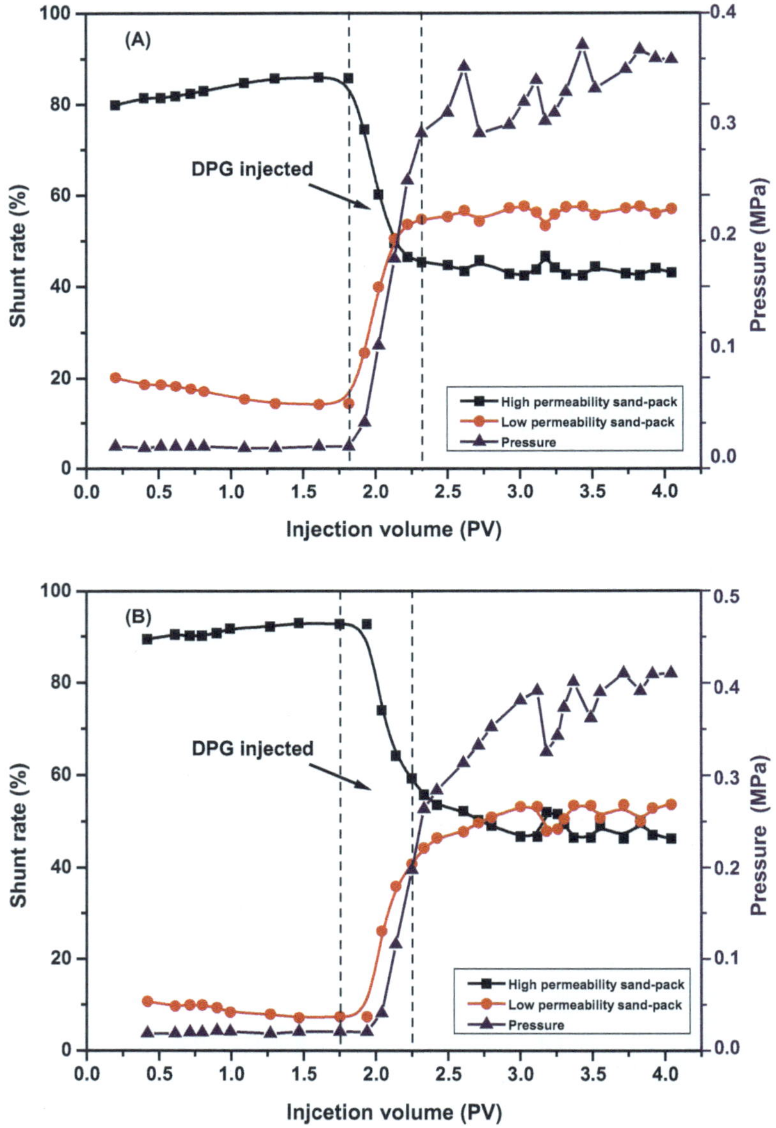

Figure 4.16 Results of profile improvement performance test: (A) permeability ratio of 2.8; (B) permeability ratio of 4.2.

Table 4.2 Parameters of profile improvement test.

Test No.	Permeability (μm^2)	Porosity (%)	Permeability ratio	Shunt rate (%)		Profile improvement rate (%)
				Before injection	After injection	
a	0.88	31.3	2.8	85.89	42.54	87.84
	0.31	28.2		14.11	57.46	
b	0.96	31.6	4.2	92.42	47.29	92.64
	0.23	27.6		7.58	52.71	

simulated water salinity was 210,000 mg/L (Na^+: 74,000 mg/L; $Ca^{2+}$and Mg^{2+}: 8000 mg/L) (Table 4.2).

According to Fig. 4.16, in the initial water flooding stage, the injected water starts to break through the high permeability sand-packs. At this time, the shunt rate of the high-permeability sand-packs exceeds 85%, while that of the low-permeability sand-packs is less than 10%. This is similar to what happens in heterogeneous reservoir development. When DPG particles are injected, the shunt rate of the high-permeability sand-packs decreases rapidly, while the shunt rate of the low-permeability sand-packs increases rapidly and exceeds that of the high-permeability sand-packs. The injection pressure rises rapidly. This indicates that during the process of injecting DPG particles, the particles preferably enter the high-permeability sand-packs, which effectively plugs them. This forces the following water to divert to the low-permeability sand-packs, thereby significantly increasing the swept volume. In the following water flooding process, the shunt rate of the low-permeability sand-packs continues to rise, the shunt rate of the high-permeability sand-packs continues to decrease, and the pressure continues to increase. After the following water flooding with multiple PVs, the shunt rate and pressure tend to balance. Finally, the profile improvement rate of test a (level difference 2.8) is 87.84%, and that of test b (level difference 4.2) 92.64% (Table 4.2). The final injection pressure of the latter is higher than that of the former. This shows that within a certain range, the greater the difference in permeability, the better the performance of improving the profile of DPG particles. Because the more serious the heterogeneity of the reservoir, the easier it is for DPG particles to enter the high-permeability area. Therefore both the direct plugging of a single particle or the aggregation of multiple particles can effectively plug the high-permeability reservoir, such that a good profile improvement is enabled.

4.3.3 Evaluation of the enhanced oil recovery potential

4.3.3.1 Evaluation of the enhanced oil recovery potential by sand-packs

With the aid of parallel sand-packs physical simulation experiments, the enhanced oil recovery potential of multiscale DPG particles was investigated. The simulated water salinity was 210,000 mg/L (Na^+: 74,000 mg/L; Ca^{2+} and Mg^{2+}: 8000 mg/L), the test temperature 120°C, the injection particle size 3.1 μm, and the concentration 0.1%. The sand-packs parameters are shown in Table 4.3, and the test results are shown in Fig. 4.17.

Fig. 4.17 shows the changes in the recovery factor and total recovery value of high-permeability and low-permeability sand-packs during water flooding, DPG particles injection, and following water flooding. It can be seen that in the initial water flooding stage, the recovery rate of high-permeability sand-packs is always higher than that of low-permeability sand-packs; and the increase in recovery of high-permeability sand-packs and low-permeability sand-packs grows with the difference in permeability. When the permeability gradient becomes larger, the total recovery rate drops (the total recovery rate of test a is 49.1%, and the total recovery rate of test b is 43.8%) after the initial water flooding. This suggests that the greater the permeability difference, the lower the extent of water flooding, and the higher the remaining oil content. With the injection of DPG particles slug, the recovery rate of high- and low-permeability sand-packs is improved. In the following water flooding stage, the recovery rate of high-permeability sand-packs increases slightly while that of low-permeability sand-packs increases significantly. With the following water flooding, the recovery rate of low-permeability sand-packed pipes even exceeds that of high-permeability sand-packs. This shows that DPG particles mainly enter the high-permeability sand-packs and effectively plug

Table 4.3 Experiment parameters of enhanced oil recovery test.

Test No.	Permeability (μm^2)	Porosity (%)	Initial oil saturation (%)	Permeability ratio	Injection volume (PV)	Aging time (d)
a	0.86	32.2	76.4	2.7	0.5	7
	0.32	29.3	75.2		0.5	7
b	1.02	32.7	78.6	4.3	0.5	7
	0.24	28.8	77.5		0.5	7

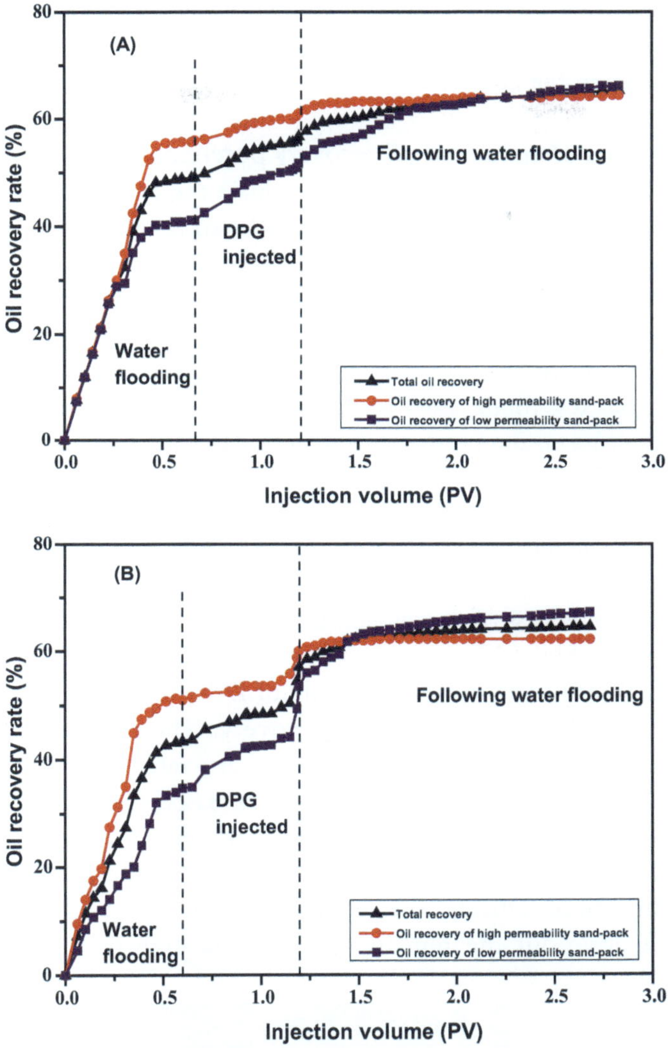

Figure 4.17 Potential results of enhanced oil recovery test: (A) permeability ratio: 2.7; (A) permeability ratio: 4.3.

them. Subsequent water flooding is transferred to a low-permeability area and greatly improves the subsequent water flooding sweep coefficient. A large amount of remaining oil in the low-permeability area is passively used, thereby greatly improving the total oil recovery. The final total recovery rate of test a is 65.3% while that of test b is 64.6%. In addition, the recovery rate of the two tests increases by 16.2% and 20.8%, respectively. The above results show that within a certain range, the increase of the permeability difference is beneficial to the performance of DPG particle profile control and enhanced oil recovery.

4.3.3.2 Results of enhanced oil recovery potential by visualized model

The plate sand test model was used to evaluate the enhanced oil recovery potential of multiscale DPG particles. In order to facilitate the observation of the migration state and displacement effect in the model, crude oil was dyed with methyl blue and the color was red, while DPG particles were dyed with tea tree phenol and the color was dark blue. The injected DPG particles were of a size of 5.6 μm and a concentration of 0.4%. The heterogeneous test model used is shown in Fig. 4.18. The entire test was carried out at room temperature. Test steps: (1) saturated water in model; (2) saturated oil; (3) water-flooding produced water-cut rate reached 98%; (4) DPG particles injected; (5) water flooding came into effect again until the produced liquid water-cut rate reached 98%, and the value-added recovery of each stage was calculated.

The whole process of the visualization displacement test of DPG particles is shown as Fig. 4.19.

Fig. 4.19 intuitively reflects the role of DPG particles in the process of profile control and enhanced oil recovery. The image analysis software calculated that the waterflood recovery rate was 27%, and the recovery rate of injected DPG particles increased to 46%. In the following water flooding process, the recovery rate increased by 3.3%. The increase in recovery is mainly concentrated in the process of DPG particle injection. During the water flooding process, the injected water will drive out the remaining oil in the high-permeability area,

Figure 4.18 Heterogeneous visualization test model.

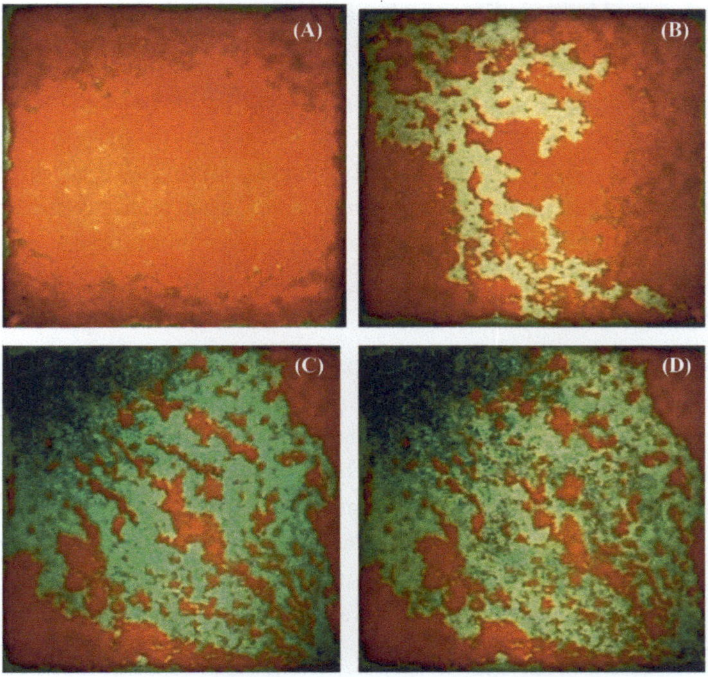

Figure 4.19 Visualization results of DPG particle displacement test: (A) saturated water and oil; (B) water flooding stage; (C) DPG particles injection stage; (D) subsequent water stage.

and gradually form a dominant channel between the water injection well and the oil well. The injected water breaks through into oil wells along the high-permeability channels, reducing oil production and limiting the recovery rate. When multiscale DPG particles are injected, the particles preferentially enter the high-permeability zones. Because DPG particles are flexible and can enter the deep zones through deformation, the plugging behaviors by single particles and multiple bridging can effectively reduce the permeability of high-permeability zones, and then divert the following water into the low-permeability zones. Furthermore, the medium- and low-permeability reservoirs are activated to effectively adjust the formation profile and drive out the remaining oil. The following water flooding stage can still improve the oil recovery, but the oil recovery increase is limited. Although DPG particles can effectively plug the high-permeability reservoir, the injected water breaks through again in the following water flooding process, which makes the oil recovery limited.

Section 4.4 The profile control mechanism of multiscale DPG particles

Using two types (plate sand-pack and glass etching) of visualized test models and core scanning electron microscopy, we studied the plugging and distribution behavior of DPG particles in porous media, as well as their driving out capability for the remaining oil, in order to reveal their profile control mechanism. This would be a support for the oilfield application of those substances.

4.4.1 The profile control mode of multiscale DPG particles

Two visualization models—plate sand-pack model and glass etching model—were used to study the reservoir profile control form of multiscale DPG particles. The test procedure is shown in Fig. 4.20. In the early stage of reservoir development, oil saturation is relatively high. However, in the middle and late stages, the formation gradually forms large pores with high water saturation. Two schemes were designed to study the deep profile control.

Scheme A: after visualization of saturated saturated water, DPG particles are directly injected for an observation of the role of DPG particles in the model.

Scheme B: after visualization of saturated crude oil and water flooding, DPG particles are directly injected for an observation of the role of DPG particles in the model.

4.4.1.1 Visualization results of plate sand-pack model test

4.4.1.1.1 The visualization results of the test in the saturated water model

The saturated water model was adopted to simulate the state of the late reservoir development, that is, there are high-permeability channels in the reservoir. The distribution of DPG particles under the saturated water

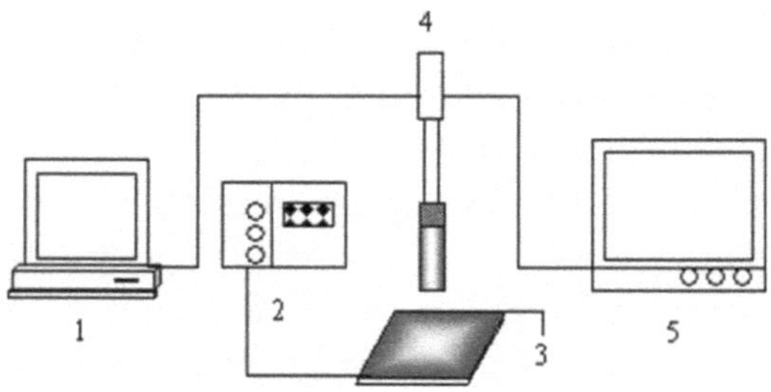

Figure 4.20 Flow chart of visualization simulation test: (1) computer; (2) micropump; (3) physical model; (4) camera; (5) television.

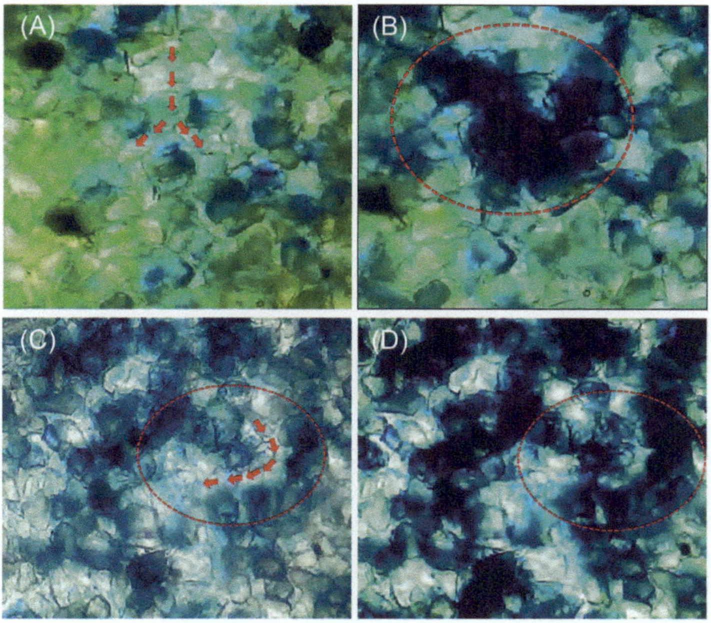

Figure 4.21 The distribution characteristics of DPG particle in the saturated water model: (A) the porosity throat distribution prior to the injection of DPG particles; (B) aggregation of DPG particles at the porosity throat; (C) large pores distribution prior to the injection of DPG particles; (D) the bridging role of DPG particles.

model is shown in Fig. 4.21. In the initial stage, DPG particles are mainly distributed along the high-permeability zones. With the injection of DPG particles, an effective plugging of the high-permeability channels is gradually formed. At this time, the subsequent fluid starts to turn to the low- and medium-permeability zones, and the pressure starts to rise. Because DPG particles are flexible and deformable to pass through the pore throat. After entering the pore throat, these particles restore to their original shape. Thus the formation profile is effectively achieved by directly plugging or bridging of DPG particles agglomerates with a larger particle size.

4.4.1.1.2 The visualization results of test in the oil model

The oil model was used to simulate the initial state of reservoir development. Fig. 4.22 shows that DPG particles are found mainly to enter into the water flow channel. When plugging the high-permeability water flow channel, they turn to the oil zones and drive out the remaining oil, so as to achieve the effect of deep profile control and oil displacement.

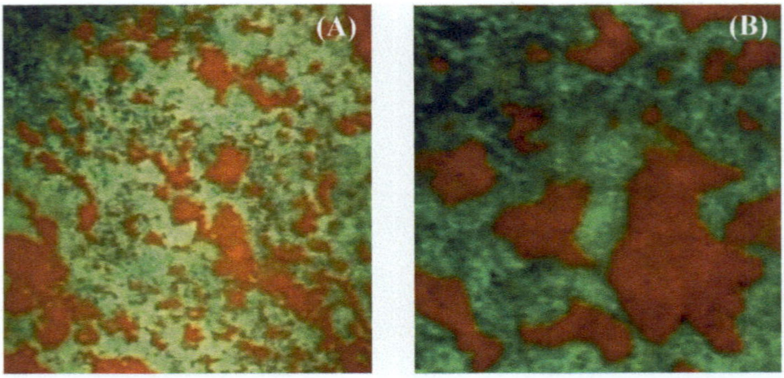

Figure 4.22 The distribution of DPG particles in the plate sand-pack oil model: (A) 5 × magnification; (B) 10 × magnification.

4.4.1.2 Visualization results of glass etching model test

4.4.1.2.1 The visualization results of test in the saturated water model

The glass etching model was adopted to observe the role of multiscale DPG particles in the model under saturated water conditions. According to Fig. 4.23, multiscale DPG particles can plug the high-permeability channels in four main ways in the model. Multiple particles stay in the high-permeability channel, making the subsequent injection fluid divert to the seepage channel (Fig. 4.23A); whereas the larger DPG particles directly occupy the seepage channel to achieve a plugging effect (Fig. 4.23B). When the pore throat is larger than the particle size of DPG particles, multiple particles bridge each other to achieve plugging. When the pore throat is smaller than the size of DPG particles, a single particle is directly plugged (Fig. 4.23C). Due to their large surface, DPG particles can be plugged by being adsorbed on the pore walls (Fig. 4.23D). Through these four methods, DPG particles can plug the high-permeability channel to the greatest extent, and then redirect the subsequent injected fluid to the low-permeability zones.

4.4.1.2.2 The visualization results of the test in the oil model

Under oil-saturated conditions, a glass etching model was used to determine the distribution of DPG particles in the model. The test results are shown in Fig. 4.24. It can be seen that the DPG particles can effectively plug the high-permeability channels mainly by directly plugging, self-growing aggregates of multiple particles bridging across throats, and adsorption on the walls of the channels. In these four ways, the

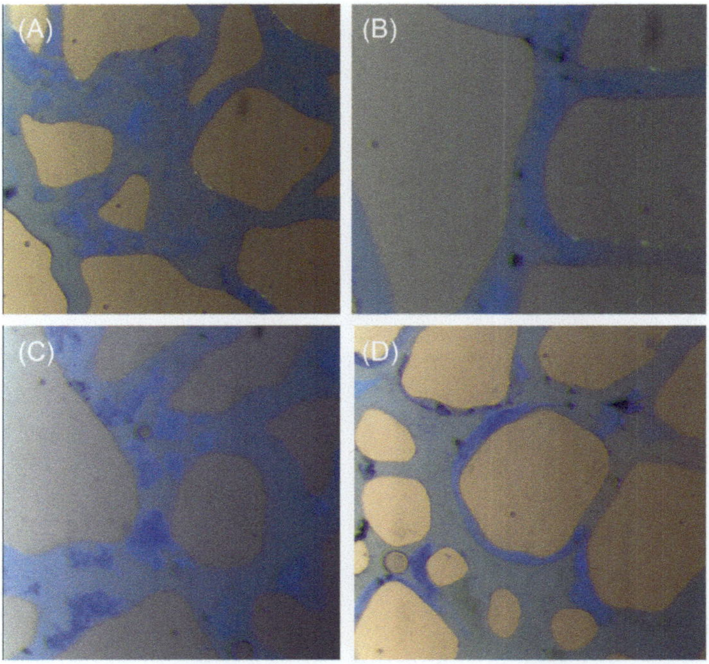

Figure 4.23 The distribution characteristics of DPG particles in the water-saturated glass etching model: (A) stagnation of DPG particles in large pore path; (B) occupation of large pore path by DPG particles; (C) bridging by multiple DPG particles or plugging by single DPG particle; (D) pore path being plugged by films that DPG particles make up.

subsequent injected water is forced to divert to the medium and low-permeability zones with higher oil saturation, and the swept volume of the injected water increases, thereby driving the remaining oil out and improving the oil recovery.

In order to further illustrate the distribution of DPG particles in the core, the distribution of multiscale DPG particles in the coexistence of oil and water phases was further observed under 40 × magnification. The test results are shown in Fig. 4.25. When entering the model, DPG particles directly plug the pore throats with a radius smaller than them (Fig. 4.25A); whereas the pore throats with a radius larger than the DPG particles are plugged by multiple particles through bridging or forming larger aggregates (Fig. 4.25B and C), such that subsequent injection water is forced to turn to the untouched area. DPG particles can also retain in the pore throats by adsorption. The subsequent injected water will flow

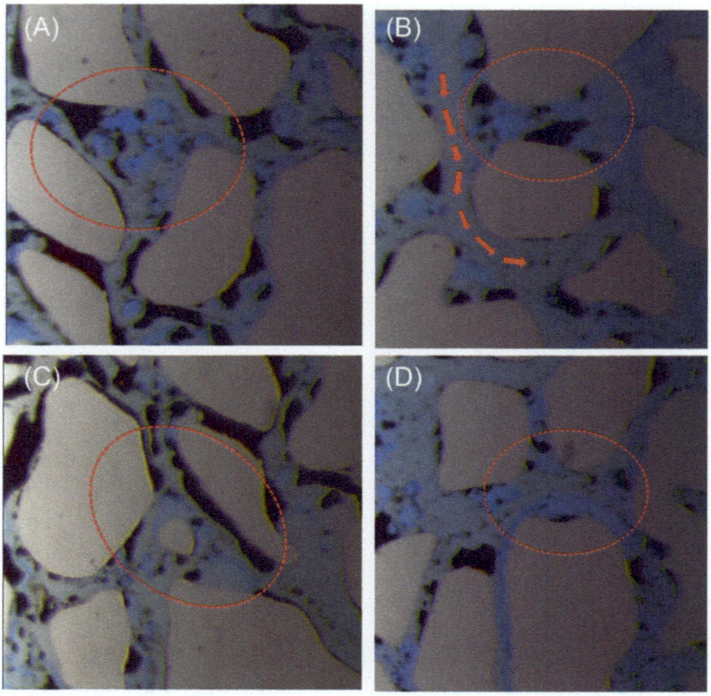

Figure 4.24 DPG particles distribution in the glass etching model: (A) plug by self-growing aggregates; (B) directly plug by a single particle; (C) and (D) plug by adsorption walls of the pore channels.

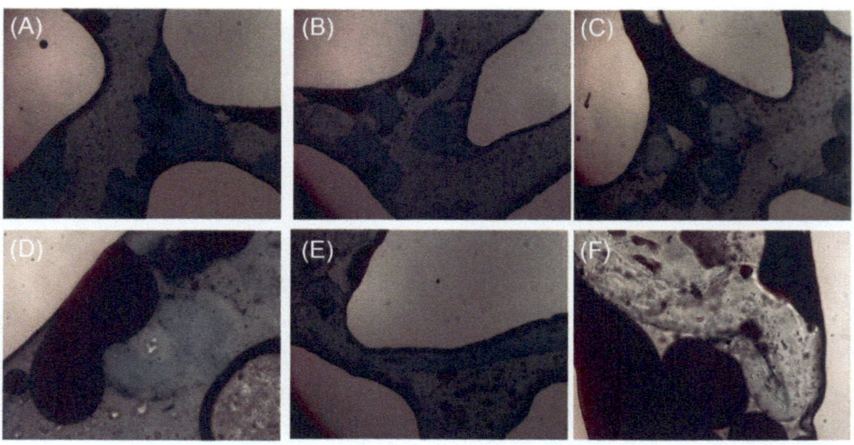

Figure 4.25 The distribution state of DPG particles in the model with magnification: (A–D) 40 × magnification; (E, F) 100 × magnification.

around to drive out the remaining oil attached to the surface of the hole wall (Fig. 4.25D). Since the prepared DPG particles are flexible and able to deform to pass through the pore throat, the negative pressure will generate and drive out the remaining oil when they pass the pore throat too (Fig. 4.25E).

In combination with the above test results of DPG particles in porous media, Fig. 4.26 depicts their mechanism of deep profile control and oil displacement. It can be seen from this Fig. 4.26 that in the early stage of reservoir development, with continuous water injection development, a water flooding channel is formed, leading to ineffective cycles of following water flooding (Fig. 4.26B) and reduction in oil recovery. At this stage, DPG particles are injected into the stratum, then migrate to the deeper stratum through deformation and breakthrough (Fig. 4.26D). By direct plugging, bridging plugging, retention, or adsorption effects, these

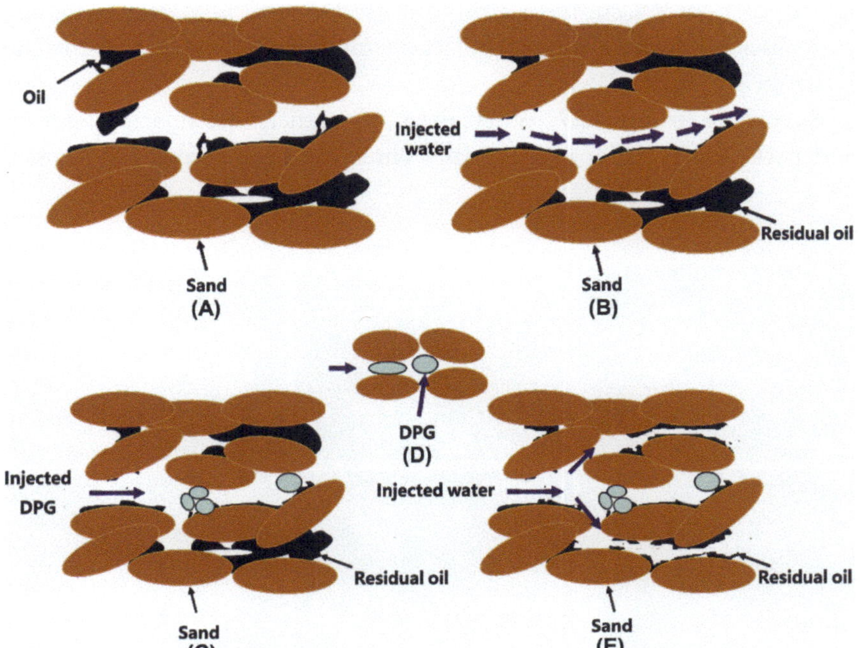

Figure 4.26 Schematic diagram of deep profile control and oil displacement mechanism of multiscale DPG particles: (A) initial oil state; (B) subsequent water displacement state; (C) DPG particles injection state; (D) state of DPG particles passing through the throat; (E) subsequent water displacement state.

particles achieve effective plugging of the high-permeability reservoir. Furthermore, the medium- and low-permeability reservoirs are activated to drive out the remaining oil (Fig. 4.26E). As a result, the recovery rate is improved.

4.4.2 Distribution behavior of DPG particles in porous media

Core scanning electron microscopy was used to observe the distribution behavior of multiscale DPG particles in porous media. The specific test steps are as follows: (1) the natural cores were extracted and dried, and their dried weight was measured; (2) those cores were saturated with water, and their wet weight was measured for the calculation of the pore volume and core permeability; (3) 1 PV of DPG particles (size of 2.2 μm) were injected at a pump rate of 0.2 mL/min, and then placed to stand at 90°C for aging for 5 days; (4) the abovementioned cores were taken out and placed in a freeze dryer at −80°C for gelation for 24 hours; (5) the frozen dried cores were crushed, from the middle of which fragments of the size of 1 cm × 1 cm were taken out, treated with gold sputtering, and then observed under a scanning electron microscope. The test results are shown in Fig. 4.27.

According to Fig. 4.27, When the DPG particles are injected into the sand-pack, the particles preferentially enter the high-permeability zones.

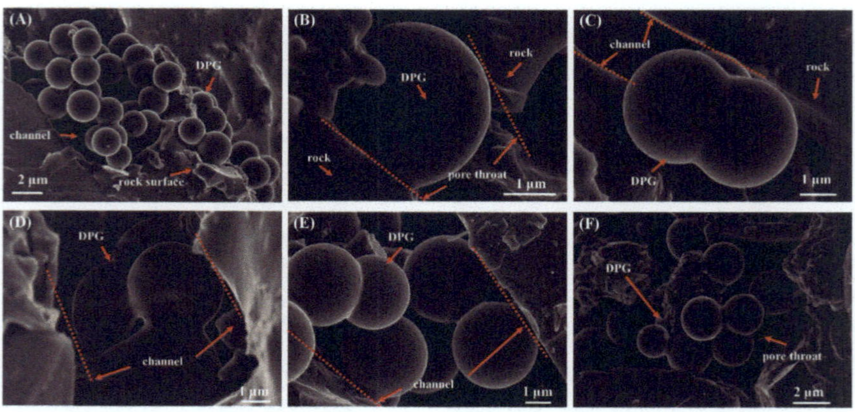

Figure 4.27 The distribution morphology of multiscale DPG particles in porous media [4]: (A) adsorption and retention in porous media; (B) directly plugging by one DPG particle; (C), (D) bridging by two or more DPG particles; (E), (F) plugging the big channel by bridging.

They can be retained and adsorbed on the rock surface when the channels are too large (Fig. 4.27A). Although bridging becomes more difficult in such high-permeability zones, the particles could combine to form larger aggregates and then increase water flow resistance in these zones. With continuous injecting, the DPG particles could be deformed and enter the deep formation easily. When the DPG particle radius is larger than the pore throat radius, the particles could be trapped and directly plug the entrance of the pore throat (Fig. 4.27B). When the DPG particle radius is a little smaller than the pore throat radius, two or three particles will be stranded in the pore throat space and bridge across the pore throats easily (Fig. 4.27C–E). However, when the pore throat radius is much larger than the DPG particle radius, many particles will combine together and form larger aggregates at the pore throat space. By trapping and bridging, the formed large aggregates can bridge across the high-permeability channels and effectively reduce the formation permeability. Moreover, once the bridge is consolidated, the newly arriving particles will gather at the pore throat and form an effective plugging capacity, thus decreasing the following water flow rate and yielding the following water into adjacent low-permeability zones. The above results show that retention, adsorption, trap, and bridge are four basic behaviors of DPG particles in the porous media. By these four basic behaviors, DPG particles can effectively reduce water production in high-permeability zones, and thus improvement occurs in the formation profile.

Section 4.5 Oilfield application examples

The oilfield application of multiscale DPG deep profile control technology was first carried out in the Xifeng block of Changqing Oilfield in September 2015, realizing a major shift from indoor research to oilfield application and opening a new chapter of multiscale DPG in field applications. As of March 2020, the multiscale DPG deep profile control and flooding technology has been used in Changqing, Tahe, Shengli, Bohai, Zhongyuan, Henan, Daqing, Jilin, Xinjiang, Zhejiang, Jidong, and another 17 domestic and overseas oilfields. The main oilfields have accumulated more than 1000 well-times and have achieved good water reduction and oil increase effects, laying a good foundation for the industrialization of this technology.

4.5.1 Changqing Oilfield: ultralow-permeability microfractured well groups

4.5.1.1 Geological profile of the test area

Baima Middle District is located in the southwest corner of the Ordos Basin and belongs to the main block of the Xifeng Oilfield. The development layer is Chang 8, the produced oil area is 66.60 km^2, the produced geological reserves are 3467.11×10^4 t, and the produced recoverable reserves are 728.09×10^4 t. The current daily oil production of Baima Middle District is 937 t, accounting for 36.1% the yield of Xifeng Oilfield. The main reservoir Chang 8 sedimentary phase is a delta front subface, with a buried depth of 1950–2300 m, an average of 2120 m, an average effective thickness of 15.8 m, average porosity of 10.5%, moderately developed reservoir porosity, original formation pressure of 18.1 MPa, and average permeability of 2.72×10^{-3} μm^2. It belongs to an ultralow-permeability reservoir. This block has the characteristics of a small maximum pore throat radius (1.63 μm), a small median radius (0.21 μm), and high drainage pressure (0.62 MPa).

4.5.1.2 Current development status of the experimental area

1. The degree of recovery increases, and the increase in water-cut accelerates.

   Baima Middle District has a high degree of recovery (18.29%) and a rapid increase in water-cut. The increase in water-cut in 2015 was mainly distributed to West 33–17, West 23, West 30–35, West 17, and West 13. Among them, the recovery degree of West 13, West 33–17, and West 23 unit are greater than 18.0%, as shown in Figs. 4.28 and 4.29.

2. The reservoir has strong heterogeneity with well-developed microfractures and shows the multidirectional water breakthrough on the plane.

   Baima Middle District belongs to the Triassic microfractured oil reservoir, with high unit fracture density, large reservoir penetration coefficient (see Table 4.4), short water breakthrough period (765 d), and rapid water-cut. At present, the comprehensive water-cut of a single well is greater than 55%.

3. The remaining oil saturation is relatively high, which has a certain potential for stimulation

   The comparative figures show that the current distribution of remaining oil reserves is highly correlated with oil saturation. Where the water flooding front reaches, the water saturation is higher and the

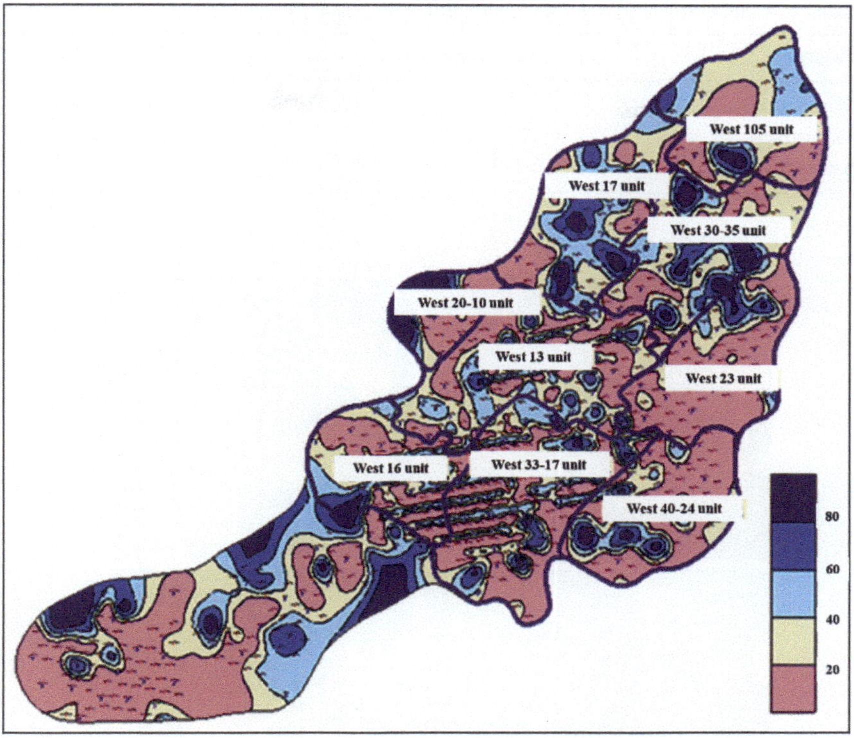

Figure 4.28 Water distribution map of Baima Central District.

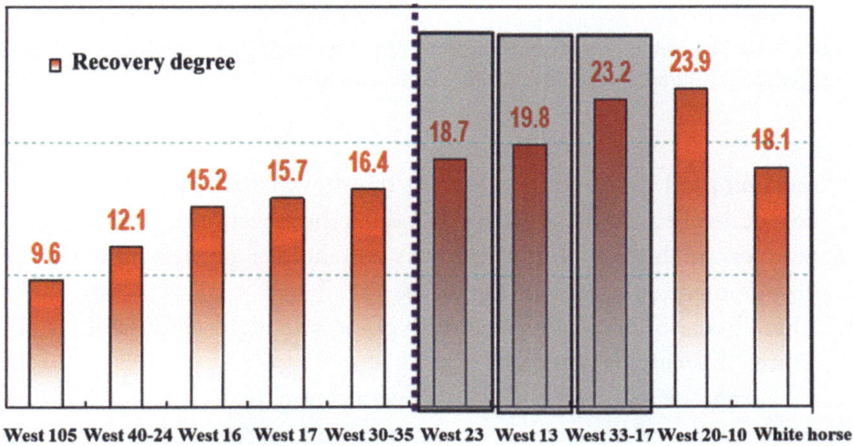

Figure 4.29 Histogram of extraction degree of Baima Central Subunit.

Table 4.4 Reservoir penetration coefficient and grade difference.

Value	Rush coefficient	Differential
Minimum	1.7	2.5
Maximum	21.1	1082.5
Average	6.8	190.3

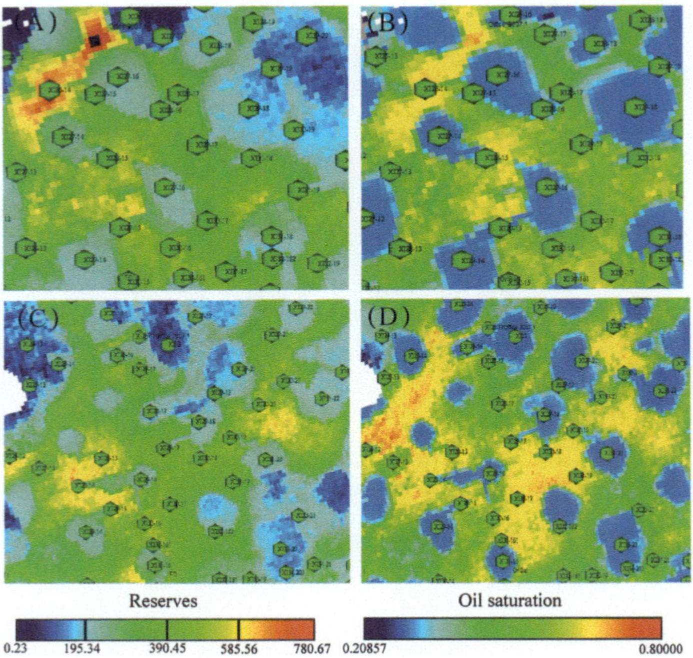

Figure 4.30 The remaining oil distribution in the block: (A) Part 1 reserves; (B) Part 1 oil saturation; (C) Part 2 reserves; (D) Part 2 oil saturation.

remaining oil distribution is less. At the sites neither being involved in nor controlled by local water flooding, however, the remaining oil reserves are higher; thus these sites still hold a greater potential for development, as shown in Fig. 4.30.

4. The direction of water breakthrough is relatively clear, and profile control has certain potential.

   Using the methods of mutual verification of tracer and dynamic analysis, the relationship between injection water well and oil well groups was analyzed. The results show relatively clear water

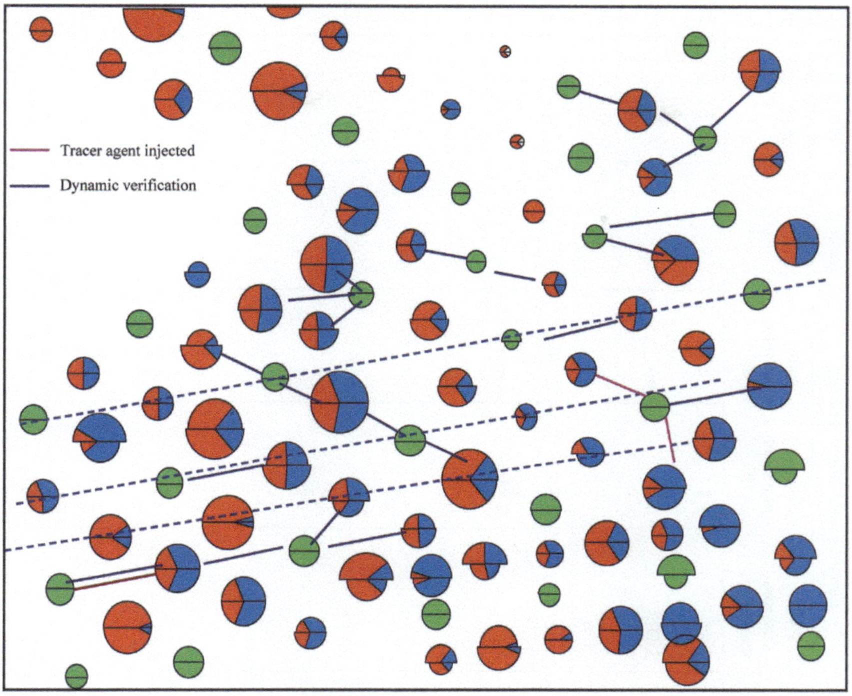

Figure 4.31 Schematic diagram of water breakthrough direction of oil wells in Baima Middle West 13 District.

breakthrough of well groups (see Fig. 4.31), which lays a good foundation for the next step of well group profile control treatment.

4.5.1.3 On-site construction of DPG profile control treatment

Based on the current development status of Baima Middle West 13 District in Xifeng Oilfield, the low-permeability microfractured well groups West 27–14 and West 27–16 were selected to carry out the pilot test of DPG profile control treatment. Before the operation, two well groups belonged to high-pressure water injection wells and the pressure-lifting space was small. Two well groups started to inject DPG particles in September 2015 and accumulated 3000 m^3 of DPG particles. As of July 2016 the cumulative oil increase had reached to 1439.7 t and had continuously been effective, as shown in Table 4.5. In 2016 three wells were further selected to expand the test shown in Table 4.6. In the five wells, the preadjustment water injection pressure was 21.4 MPa and the postadjustment pressure increased by 0.6 MPa on average. At present, all the wells can reach up to the allocation injection.

Table 4.5 Comparison of production data before and after the treatments in the West 27−14 and West 27−16 well groups.

Well number	Operation date	Well	Before the treatment			Current production treatment			Water reduction (%)	Daily increased oil (t)	Cumulative oil increase (t)
			Daily produced liquid (m^3)	Daily produced oil (t)	Water-cut rate (%)	Daily produced liquid (m^3)	Daily produced oil (t)	Water-cut rate (%)			
West 27−14	2015.9.8−10.27	West 27−15	3.85	1	69.3	3.84	1.17	64.3	−5	0.17	80.4
		West 28−15	13.09	5.4	51.5	12.02	4.06	60.3	8.8	0	19.76
		West 26−12	4.49	0.97	74.7	2.7	1.02	55.5	−19.2	0.05	18.27
		West 26−13	2.99	2.05	19.2	5.22	3.51	20.8	1.6	1.46	196.53
		West 26−14	6.21	0.99	81.3	10.54	2.81	68.6	−12.7	1.82	352.13
		West 27−13	11.05	6.71	28.6	10.56	6.93	22.8	−5.8	0.22	130.94
		West 28−14	4.9	1.4	66.2	10.22	2.78	68	1.8	1.38	101.83
West 27−16	2015.9.10−11.9	West 26−14	7.62	1.3	80	10.54	2.81	68.6	−11.4	1.51	263.73
		West 26−16	5.73	1.16	76.1	7.47	1.78	72	−4.1	0.62	48.85
		West 27−15	3.8	1.02	68.3	3.84	1.17	64.3	−4	0.15	72.34
		West 26−15	5.57	2.1	55.6	4.32	1.48	59.6	4	0	5.28
		West 27−17	4.77	1.78	56.1	5.02	1.95	54.2	−1.9	0.17	46.95
		West 28−17	6.57	3.8	32	6.35	3.98	26.2	−5.8	0.18	102.65
Total			80.6	29.7	56.2	92.6	35.5	54.4	−1.7	7.7	1439.7

Table 4.6 Comparison of production data of West 33–16, West 30–33, and West 32–35 well groups before and after the treatments.

Date	Well group	Operation date	Before treatment		After treatment		Oil wells (number)	Effective oil well (number)	Daily increased oil after operation (t)	Daily increase oil at the end of 2016 (t)	Cumulative oil increase (t)	Valid period (days)
			Wellhead pressure (MPa)	Injection volume (m^3)	Wellhead pressure (MPa)	Injection volume (m^3)						
2015	West 27–16	9.10	21.2	17	21.7	17	5	2	2.5	1.5	650	426
	West 27–14	9.02	22	15	22.5	15	6	2	2.2	1.1	658	431
2016	West 33–16	10.20	21.3	26	21.8	26	6	3	2	2.1	173	
	West 30–33	5.27	20.9	24	21.4	24	7	3	1.8	2	212	
	West 32–35	5.27	21.8	15	22.4	15	7	3	2.9	2.3	520	
Total/ Average	5		21.4	19	22.0	19	31	13	11.4	9.0	2213	428

This effectively alleviated the difficulty in high injection pressure and low water injection volume. Among the 31 wells of the construction well group, 13 wells were found to be effective, with an efficiency of 41.9%. At the end of 2016, the daily oil increase was 9.0 t, the average single well daily oil increase was 1.8 t, the cumulative oil increase was 2213 t, and the average validity period was more than 428 days, as shown in Table 4.6.

4.5.2 Shengli Oilfield: typical well groups with high temperature and low permeability

Since 2016 Shengli Oilfield Dongsheng Jinggong Petroleum Development Company, Petroleum Development Center, Linpan Oil Production Plant, Luming Oil Production Plant, Gudao Oil Production Plant, Xianhe Oil Production Plant, etc. have successively implemented multiscale DPG field tests. The good effect of water reduction and oil increase has been achieved. This section takes Gudao Oil Production Plant as an example to introduce the application of multiscale DPG in the high-temperature and low-permeability reservoirs of Shengli Oilfield.

4.5.2.1 Overview of the test area

1. *Reservoir geology*

   Block Ken 95, located in the east of the old area of Kenli Oilfield, is a narrow strip-shaped fault block reservoir located in the descending plate of the Kenli fault (Fig. 4.32). Its main oil-bearing strata are from Es_1 to Es_3Z, with a total of 33 oil-bearing layers over a large

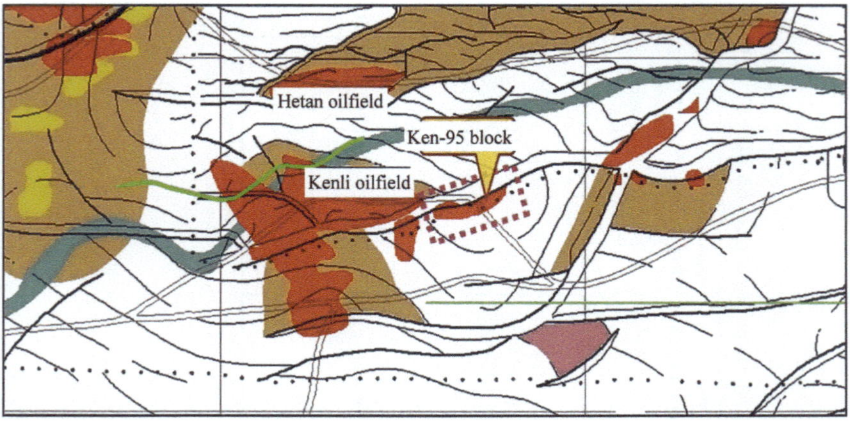

Figure 4.32 Exploration and deployment map of Block Ken 95.

longitudinal span. In 1993 and 2007, a total of 1.62 km^2 of superimposed oil-containing area and 4.54 million tons of geological reserves were reported. It has the following obvious characteristics:

a. The fault clamps the long and narrow fault block, with a stratum dip up to 2~31 degrees.

b. Semiopen fault blocks, long oil-containing well section (Es_1 section is 280 m), and insufficient natural energy.

c. A host of vertical small layers, with 33 oil-bearing layers.

d. $Es_1 \sim Es_2S$ are medium-permeability, and $Es_2X \sim Es_3Z$ are low-permeability.

2. *Reservoir characteristics*

The oil-producing layers in Ken 95 Oilfield are mostly isolated bodies, surrounded by faults or lithological boundaries. There is no edge water supply, giving them a closed or semiclosed state. Due to the large water-cut in the oil layer, the fluid flow in the production process is mostly oil and water two-phase flow, and the oil phase permeability is low. The reservoirs of Block Ken 95 are ones with medium-to-low porosity, medium-to-low permeability, low saturation pressure, and thin oil fault blocks. The specific reservoir characteristics are shown in Table 4.7).

3. *Development and production status*

Block Ken 95 was put into development in July 1991. In January 1993, it was switched to water injection development. Since the effect of water injection was not obvious, the injection was stopped in April 1999. In 2008, the reserves of Es_22 to Es_32 were discovered by rolling expansion in 2008. In October 2009, water injection was resumed and is currently in the fine injection−production adjustment stage. The results are shown in Fig. 4.33.

There are 14 oil wells in Block Ken 95, with a daily oil level of 31 t, a comprehensive water-cut of 89.3%, and five water injection wells with a daily injection level of 193 m^3 and an injection−production ratio of 0.7. Though the current recovery is only 14.5%, the oil wells still have a high exploitation potential.

4.5.2.2 Operated well group production situation

Based on the development status of Block Ken 95, the Ken K95-X26 well group was selected to carry out multiscale DPG deep profile control treatment to improve the heterogeneity of the reservoir and produce the

Table 4.7 Physical parameter table of Block Ken 95.

Item	Data	Item	Data	Item	Data
Production time (year · month)	1991.7	Bonding method	Base type, pore-base cement	Ground viscosity (mPa · s)	13
Water injection time (year · month)	1993.1			Underground viscosity (mPa · s)	2.96
Oil area (km^2)	1.62	Mud content (%)	9–12	Ground density (g/cm^3)	0.875
Effective thickness (m)	27.9	Permeability ($10^{-3} \mu m^2$)	14–66.3	Underground density (g/cm^3)	0.867
Geological reserves (10^4 t)	450.45	Porosity (%)	19.3	Water type	$NaHCO_3$
Mobilized reserves (10^4 t)	450.45	Oil saturation (%)	66.3	Total salinity (g/cm^3)	5945
Recoverable reserves (10^4 t)	61	Rock wettability	hydrophilic	Crude oil gelation point	28–35
Recovery (%)	13.5	Original formation pressure (MPa)	21.59	Sulfur content (%)	0.66
Stratum dip (°)	2–31	Saturated formation pressure (MPa)	2.9	Wax content (%)	20.7
Burial depth (m)	2309–3251	Original oil–gas ratio (m^3/t)	22.8	Deposition type	Delta to river facies
Formation temperature (°C)	95				

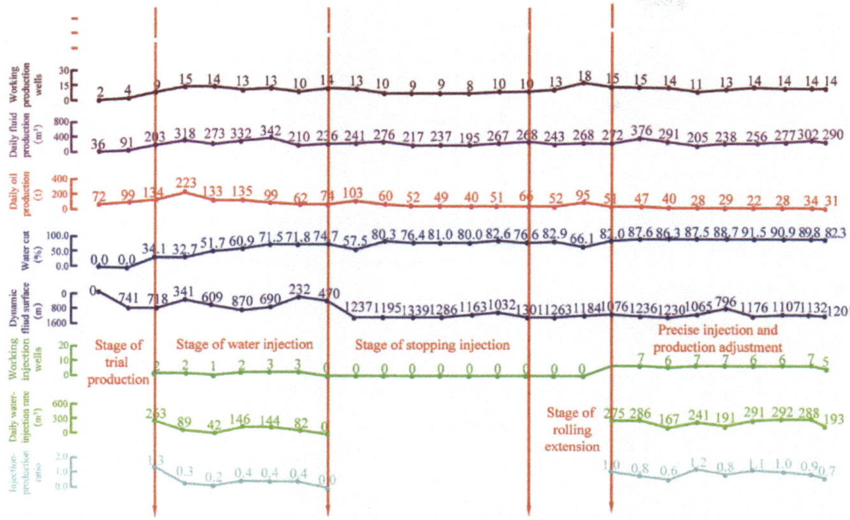

Figure 4.33 Historical development stage of Block Ken 95 in Kenli Oilfield.

remaining oil with poor physical properties, in order to further tap out the remaining oil from potential main layers.

1. *Oil layer and perforation of water injection well*

   The oil layer and perforation conditions of water injection well KLK95X26 are shown in Table 4.8.
2. *Water injection dynamics*

   Table 4.9 shows the current water injection performance of well KLK95X26, and Fig. 4.34 illustrates the current water injection performance of water injection well KLK95X26 of which the pressure increase space is limited.
3. *Corresponding oil well production*

   The K95-X26 wells group currently contains two oil wells in work, KLK95X25 and KLK95X28. Both wells have thick corresponding main layers, serious single-layer penetration, and difficulty in single-well fluid extraction. Their production status is shown in Table 4.10 and Figs. 4.35 and 4.36.

4.5.2.3 On-site construction of DPG particles treatment

Construction of the XI52-X435 well group started on August 20, 2019 and completed on October 2019. A total of 2200 m^3 of DPG was injected. After the completion of the construction, the maximum water reduction of the well group was 17.7%. As of August 10, 2020, the cumulative oil increase has been

Table 4.8 KLK95X26 oil layer and perforation data.

Oil (gas) layer	Layer number	Explanation number	Oil layer top depth (m)	Oil layer thickness (m)	Ø (%)	*K* (10^{-3} um^2)	Water saturation (%)	Log interpretation results
ES3Z	18	38	3496.6	2.2	14.58	6.49	54.17	Upper oil—water layer lower oil-bearing water layer
ES3Z	20	39	3540.1	16.9	12.75	65.836	65.83	Upper oil—water layer lower oil-bearing water layer

Table 4.9 Current water injection dynamics of Well KLK95X26 (2016/10/27).

Water injection well	Water injection method	Layer	Pump pressure (MPa)	Oil pressure (MPa)	Cover pressure (MPa)	Water injection (m^3/d)
KLK95X26	Atmospheric general injection	ES3Z(18)-ES3Z (20)	18.4	18.2	11	15/16

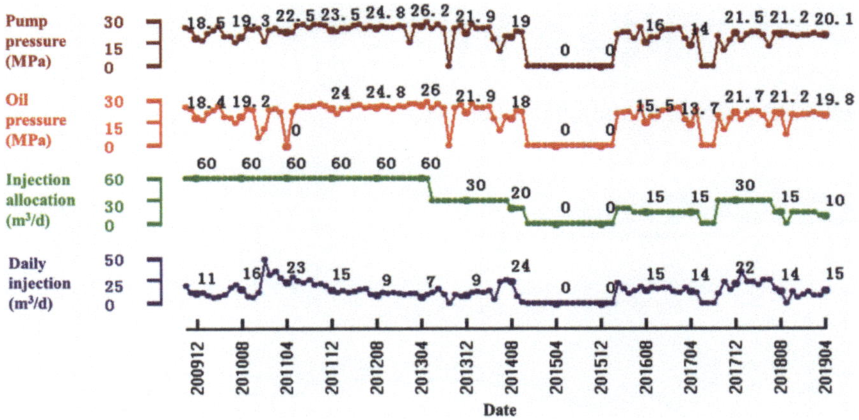

Figure 4.34 Current water injection performance of water injection well KLK95X26.

Table 4.10 KLK95X26 well group corresponding oil well production in October 2016.

Well number	Production layer	Daily fluid volume (t)	Oil volume (t)	Daily produced water (t)	Water-cut rate (%)	Well spacing (m)
KLK95X28	ES3Z	7.2	0.6	6.6	92.3	201.1
KLK95X25	ES3Z	3.2	1	2.2	67.7	226.4

up to 1100 tons, and the validity period exceeded 300 days (Figs. 4.37 and 4.38). It remains effective continuously.

4.5.3 Northwest Oilfield: high-temperature, high-salinity, medium- and high-permeability reservoir

Aiming to solve the problem of reduced water flooding efficiency caused by the breakthrough of early injected water in the YT2 block, injection of DPG was applied to achieve improved water injection efficiency and increased oil recovery, inhibiting the water channels and increasing the water flooding swept volume of underdeveloped reservoirs.

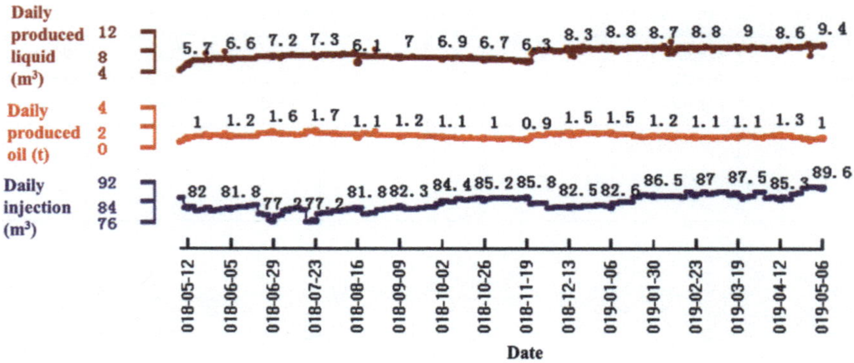

Figure 4.35 KLK95X25 daily production curve.

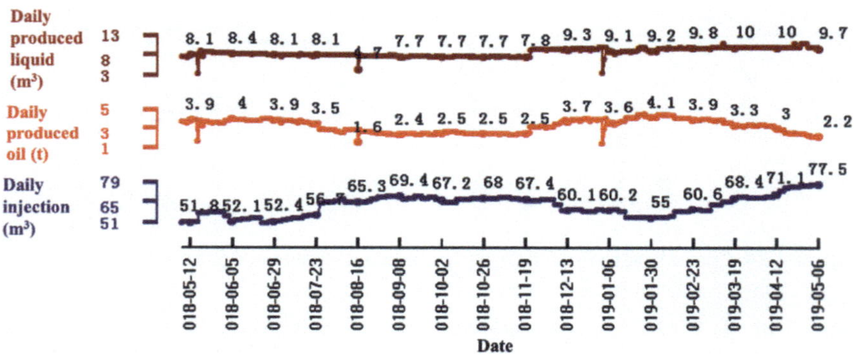

Figure 4.36 KLK95X28 daily production curve.

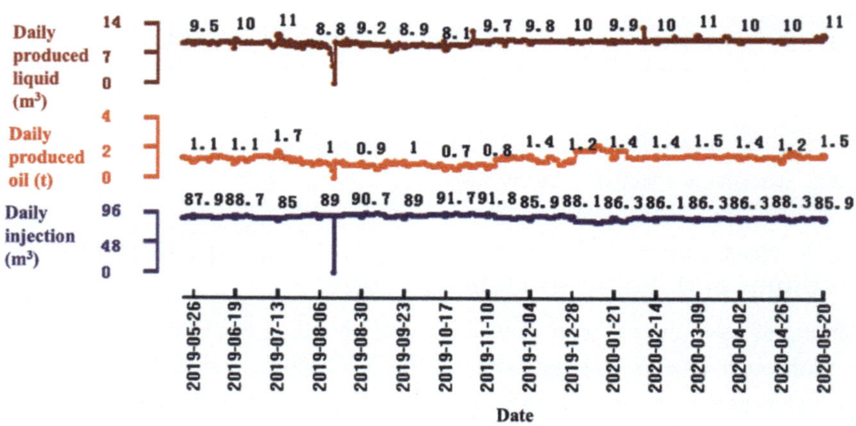

Figure 4.37 KLK95X25 production curve after construction.

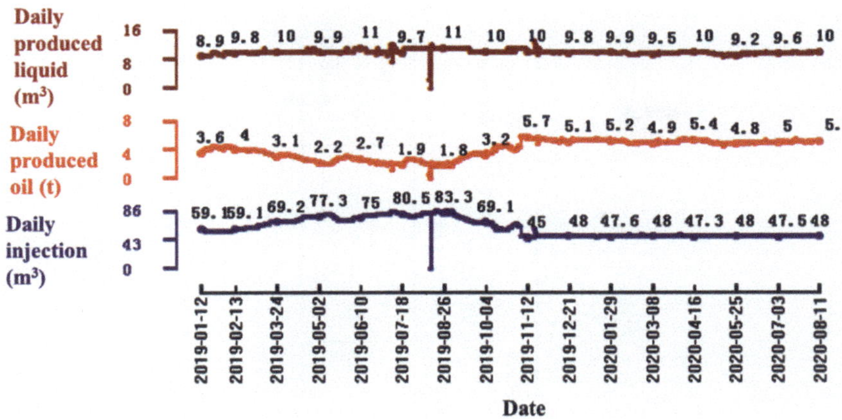

Figure 4.38 Daily production curve of KLK95X28 after construction.

The oil layer of the YT2 well group is a medium-porous and medium-permeable sandstone. It is a water-driving development reservoir. The effective thickness of the oil layer is thin (average 3–10 m), with obvious planar heterogeneity and interlayer and intralayer heterogeneity. The recovery is relatively low (13.3%) and a large amount of remaining oil is retained between wells. The oil layer is thus considered to harbor a clear potential for profile control treatment. Therefore the YT2 well group was selected to carry out the pilot test of DPG treatments.

4.5.3.1 Analysis of reservoir characteristics in YT2 well area

The reservoirs in the middle oil group of YT2 well block are characterized with positive rhythm. The upper part has fine lithology and the reservoir physical properties are relatively poor. The lower part has coarse grain size and the reservoir physical properties are good. It is considered that the vertical heterogeneity of YT2 well is relatively strong. The horizontal distribution changes rapidly. The sand in the middle of the river channel is thick and the physical properties are good. When the sand at the edge of the river channel is pinched, the shale content increases and the physical properties become worse. It can be seen from the plane permeability distribution and heterogeneity parameters of YT2 block that the vertical heterogeneity of the reservoir in the T_2a_3 member of the YT2 well group is relatively strong and the plane is relatively uniform. The large and small pore throats in the reservoir are relatively uniform, but the medium and small pore throats are the main ones, and the reservoir structure is relatively good. Considering the porosity and permeability parameters (Table 4.11), the

Table 4.11 Physical properties and heterogeneous parameters of T_2a_3 reservoir in YT2 well group.

Layer number	Layer position	Porosity (%)	Permeability ($10^{-3}\ \mu m^2$)	Heterogeneous parameters				Variation degree
				Direction	Differential	Rush coefficient	Variation coefficient	
T2	Middle	24.4	449.5	Vertical	3828	4.76	0.74	Serious
				Horizontal	7.8	2.4	0.65	More uniform

pore structure is a medium–high-porosity, high-permeability reservoir type with medium and small pore throats mainly, according to the pore structure classification and evaluation criteria of sandstone reservoirs.

4.5.3.2 Analysis of the relationship between injection and production

The YT2 well group was transferred to injection in May 2008 and water injection was resumed on April 7, 2011. The corresponding three oil wells are YT2–5H, YT2–9H, and YT2–25H, as shown in Fig. 4.39. The production process of YT2 well group is mainly divided into two stages:

In the oil production stage (July 1, 2006–February 28, 2008; November 8, 2009–January 3, 2010), the cumulative production of fluid was 21,434.2 t and the cumulative production of oil was 20,244.42 t.

In the transferring injection stage (May 8, 2008–June 8, 2009; April 7, 2011–present), YT2 well was transferred in May 2008 and the injection was stopped after accumulating 34,146 m^3. Water injection was resumed on April 7, 2011. The current (December 24, 2016) oil pressure is 5.5 MPa; the covering pressure is 5.5 MPa; the daily water injection volume is 35 m^3; the stage water injection is 67,045 m^3; the total well injection is 101,191 m^3, and the injection–production ratio is 0.37.

4.5.3.3 Analysis of fluid properties

Analysis of water injection samples were taken on February 28, 2009, with results as follows: Ca^{2+}: 12,195.06 mg/L; $K^+ + Na^+$: 69,965.17 mg/L;

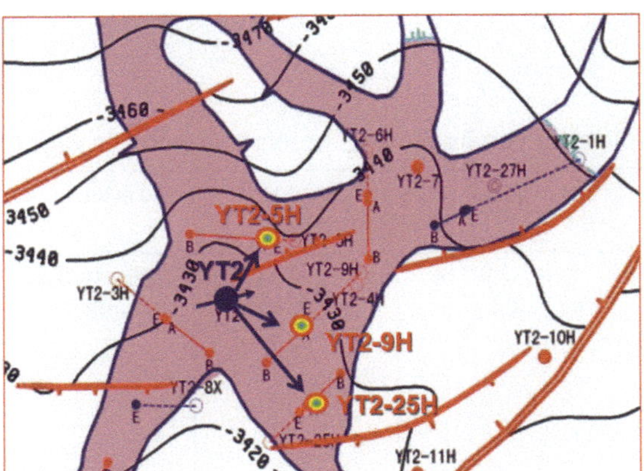

Figure 4.39 Correspondence between injection and production of YT2 injection–-production well group.

Cl^-: 133,280.66 mg/L; SO_4^{2-}: 150 mg/L; I^-: 4 mg/L; Br^-: 160 mg/L; total salinity: 217,156.61 mg/L; water type: calcium chloride; water density: 1.1460 g/cm^3; the specific composition is shown in Table 4.12.

4.5.3.4 Water absorption profile test

1. The logging interpretation results of the injection profile in July 2011 show that the current main suction zone of this well is still 4365–4367 m. The water absorption profile confirms the failed plugging of the layer and that the perforation zone located in 4372–4374 m has poor inhalation capacity (Table 4.13).
2. The logging interpretation results of the injection well profile in September 2012 show that currently, there are two inhalation sections of 4365–4367 m and 4372–4374 m, with the former as the main inhalation section; meanwhile, there may be casing leakage in the 4348.5–4351.5 m well section (Table 4.14).
3. The log interpretation result of the injection profile in May 2013 show that the well currently has two inhalation sections of 4365–4367 m and

Table 4.12 Reservoir fluid characteristics.

Ion type	$K^+ + Na^+$	Ca^{2+}	Cl^-	SO_4^{2-}
Ion content (mg/L)	69,964.17	12,194.06	133,280.66	150
Total salinity (mg/L)	217,156.61			
Water type	$CaCl_2$			

Table 4.13 Interpretation results of water absorption profile in July 2011.

Injection volume (m^3/h)	48 m^3/d		120 m^3/d	
Inhalation layer (m)	4365–4367 m	4372–4374 m	4365–4367 m	4372–4374 m
Injection layer temperature (°C)	97.5–98.5	98.5	97.7–98.5	98.5
Injection layer pressure (MPa)	39.7	39.8	46.8	46.9
Water absorption (m^3/d)	44.7	4.1	72.7	21.3
Apparent water absorption index	1.13	0.1	1.55	0.45

Table 4.14 Interpretation results of water absorption profile in September 2012.

Injection volume (m^3/h)	**80 m^3/d**		**40 m^3/d**	
Inhalation layer (m)	4365–4367 m	4372–4374 m	4365–4367 m	4372–4374 m
Injection layer temperature (°C)	96.11	97.10	97.10	97.15
Injection layer pressure (MPa)	56.21	56.30	52.27	52.32
Layered injection volume (10 m^3/h)	66	14	28	12
Apparent water absorption index (m^3/d/MPa)	1.17	0.25	0.53	0.21
Relative amount of layered injection (%)	82.5	17.5	70	30

Table 4.15 Interpretation results of water absorption profile in May 2013.

Injection volume (m^3/h)	**120 m^3/d**		**45 m^3/d**	
Inhalation layer (m)	4365–4367 m	4372–4374 m	4365–4367 m	4372–4374 m
Injection layer temperature (°C)	97.7	98.4	97.9	98.7
Flow pressure (MPa)	46.8	46.9	52.1	52.2
Layered injection volume (10 m^3/h)	85	24	29	16
Apparent water absorption index (m^3/d/MPa)	1.81	0.51	0.56	0.31
Relative amount of layered injection (%)	78	22	64.4	35.6

4372–4374 m, with the former 4365–4367 m as the main inhalation section. The bottom depth is 4374.0 m (Table 4.15).

4. The log interpretation results of the injection profile in November 2015 show that the well currently has two inhalation sections of 4365–4367 m and 4372–4374 m, of which the former is the main inhalation section (Table 4.16).

Table 4.16 Log interpretation results of water absorption profile in November 2015.

Number	Layer	Water-absorbing interval (m)	Effective thickness (m)	Injection area	Relative injection (%)	Absolute injection (m^3/d)	Injection strength ($m^3/d/m$)
1	T2a	4365.1–4367.1	2.0	1315.49	70.64	30.09	27.36
2		4371.7–4374	2.3	546.72	29.36	12.51	11.37
Total			4.3	1862.21	100	42.6	/

5. In October 2016, the YT2 well was designed to monitor the water absorption of the two perforated intervals of 4365.01–4367.01 m and 4372–4374 m. Due to the logging resistance at 4368.7 m, the flow data related to 4372–4374 m could not be obtained, and the flow rate of 4372–4374 m into the layer could not be calculated. This made it impossible to quantitatively analyze the specific water absorption of this layer.

4.5.3.5 Identification of dominant channels

According to the method of identifying and describing the dominant channel, the dominant channel of water channeling in the YT2 well group was determined, using logging curves, tracer response curves, and dimensionless chart judgments.

1. *Logging curve to judge the existence of dominant channels*

   The SP spontaneous potential curve in the YT2–5H and YT2–9H well logging curves is significantly shifted and the AC acoustic time difference curve increases. It is judged that there are water-flooded zones in these two wells, as shown in Fig. 4.40.

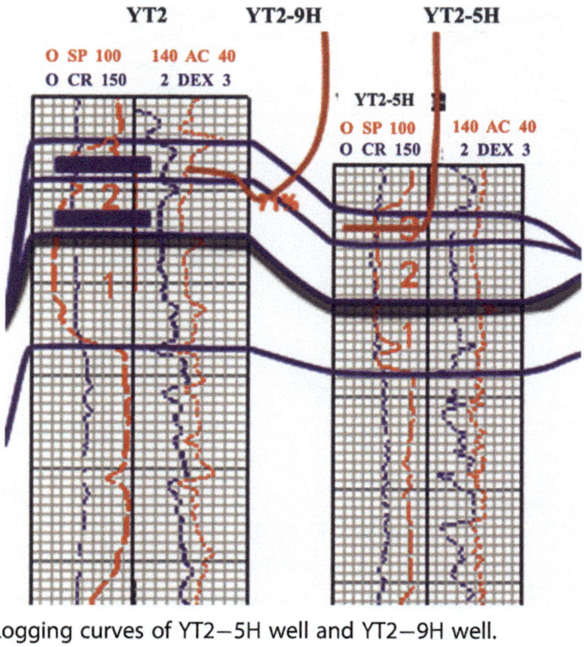

Figure 4.40 Logging curves of YT2–5H well and YT2–9H well.

2. *Tracer response curve to determine the lateral dominant channel*

Well YT2−5H monitored the tracer response (Tables 4.17 and 4.18; Fig. 4.41), indicating an explicit dynamic connection between wells YT2 and YT2−5H. No tracer output was detected in wells YT2−3 and YT2−6H; it is thus considered that these two wells have poor connectivity with well YT2. Well YT2−9H can neither obtain monitoring data due to long-term low water-cut, nor be judged. It is preliminarily determined that the lateral dominant channel exists between Well YT2 and Well YT2−5H.

Table 4.17 Tracer response table of YT2 well group.

Tracer injection well	Construction date	Name of injected tracer	Monitoring oil well	Monitor response
YT2	2011.11.2−2012.3.1	BY-3	YT2−3H YT2−5H YT2−9H YT2−6H	○ Δ ○ ○

Note: "Δ" means an obvious response of the tracer is detected, and "○" means no response of the tracer is detected.

Table 4.18 Basic situation of tracer output.

Water injection well	Oil well	Well spacing (m)	Peak concentration (cd)	Time to first sight of tracer (d)	tracer peak sight time (d)	Waterline forward speed (m/d)
YT2	YT2−5H	285	72.4	61	72	4.67

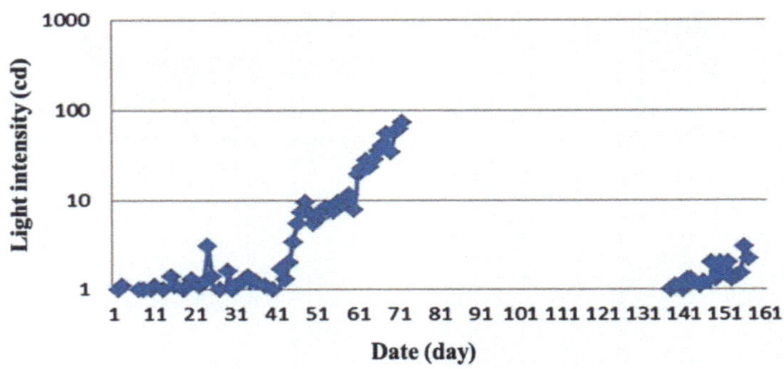

Figure 4.41 YT2−5H tracer monitoring diagram. Note: a total of 160 days of monitoring, 44 days of work over, and 21 days of shut-in resulted in incomplete curves.

According to the tracer curve shape—an incomplete waveform—the dominant flow channel between YT2 and YT2−5H is a fracture-type dominant channel.

4.5.3.6 Pilot test of DPG treatment

According to the concept of constant pressure drop stepwise in deep profile control, the injection of DPG with different particle sizes is dynamically adjusted. In June 2017, the first round of DPG treatment was implemented. A total of 7200 m^3 of DPG and 300 m^3 of weak gel protected slugs were injected. The first round began to take effect in October 2017, with cumulative oil of 3773 t, and the results are shown in Fig. 4.42. The second round of DPG treatment was carried out in June 2018, and a total of 7200 m^3 of DPG and 300 m^3 of weak gel protected slugs were injected. The second round of oil increase was 4778 t. The result is shown in Fig. 4.43. The cumulative oil increase in the two rounds exceeds 8551 tons, achieving a good effect of water reduction and oil increase.

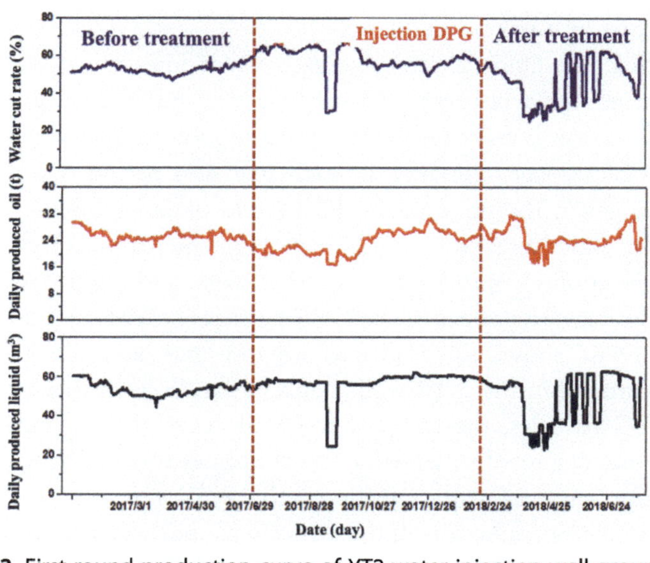

Figure 4.42 First round production curve of YT2 water injection well group.

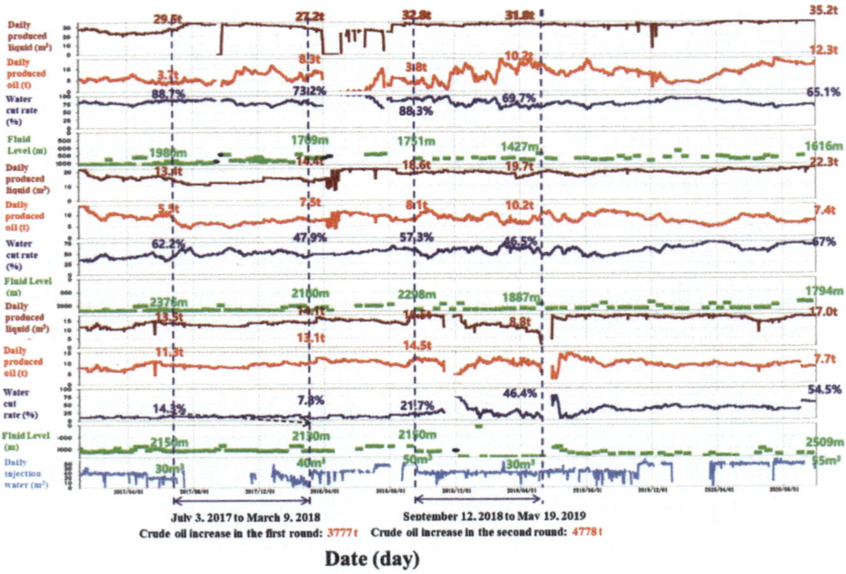

Figure 4.43 The second-round production curve of YT2 water injection well group.

4.5.3.7 The effect evaluation of water injection wells

The Hall curve evaluation, the water injection profile evaluation, and the indicator curve evaluation of water injection wells were used to further analyze the effect of injected DPG in the YT2 well group.

1. Hall curve of water injection well

   If different fluids are injected into water injection wells, different straight sections will be reflected on the Hall curve. The slope of each straight section is regressed by the curve section, in order to reflect the change of seepage resistance in each injection period. By comparing the changes in the slope of the Hall curve of water injection wells in different fluid injection stages, including the early water injection stage, the profile control agent injection stage, and the following water flooding stage, the effect of the DPG particles system can be evaluated.

   Fig. 4.44 shows the Hall curve before and after the injection of DPG particles in the YT2 well group. It can be seen that after injecting DPG particles, the slope of the Hall curve rises. It indicates that the oil-increasing effect is better and continues to be effective.

2. Indication curve of water injection well

   The water injection indicator curve reflects the relationship between the water injection volume and the injection pressure.

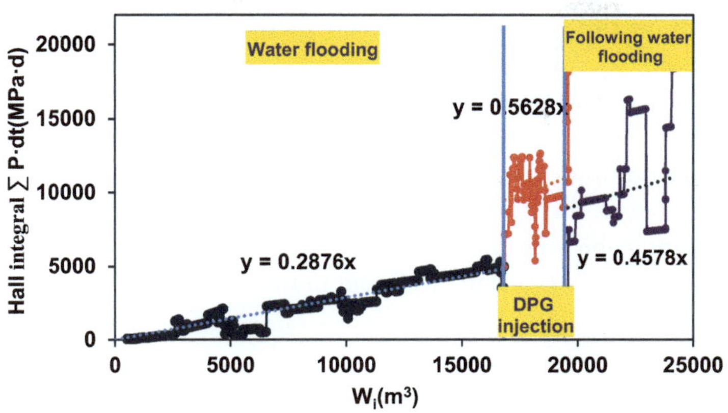

Figure 4.44 Hall curve of YT2 well group.

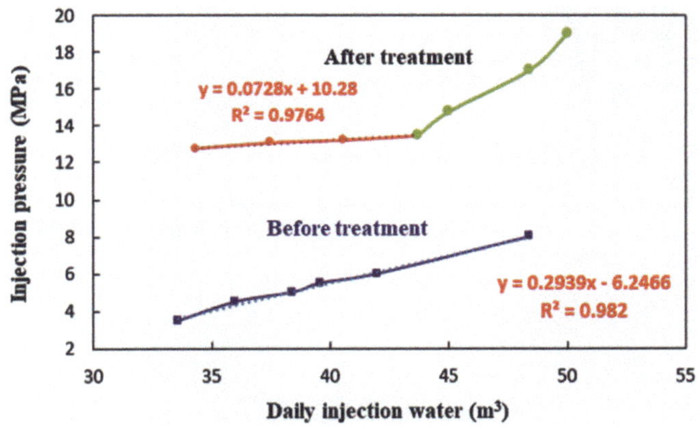

Figure 4.45 Indication curve of water injection well.

By judging the indicator curves before and after the measures, the effectiveness of the multiscale DPG profile control and flooding effect is determined. If the injection pressure rises to variable degrees, it is effective. In addition, there is a certain degree of increase in injection pressure, which can be considered from the increase in average pressure of a single well. The apparent water absorption index has different reduction degrees, which can be considered from the reduction range, as shown in Fig. 4.45. After the DPG operations, the curve moves to the upper right and the water absorption index becomes larger (increased from 3.4 to 13.7 $m^3/[d \cdot MPa]$). This suggests that the water absorption capacity of the formation has been increased and the measures are effective.

3. Water drive characteristic curve

Figs. 4.46–4.48 show the water drive characteristic curves of the YT2 well group corresponding to three oil wells after DPG injection. It can be seen that the slope of the curve of YT2–5H well decreases significantly before and after the operations while the slope of YT2–9H and YT2–25H wells continues to increase throughout the

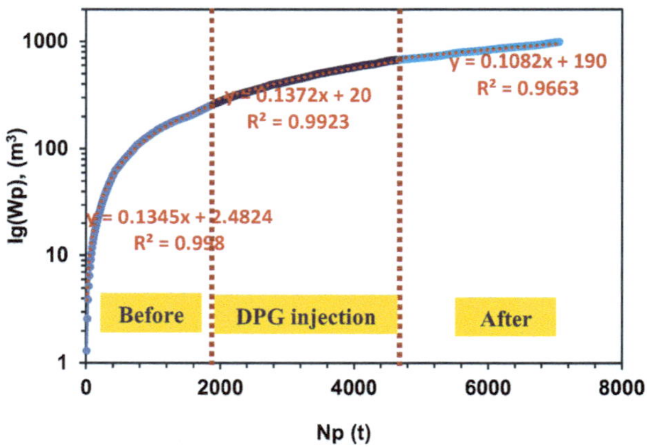

Figure 4.46 YT2–5H water drive characteristic curve.

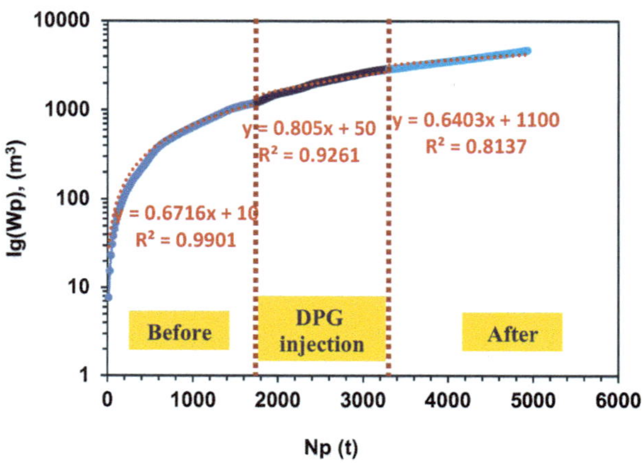

Figure 4.47 YT2–9H water drive characteristic curve.

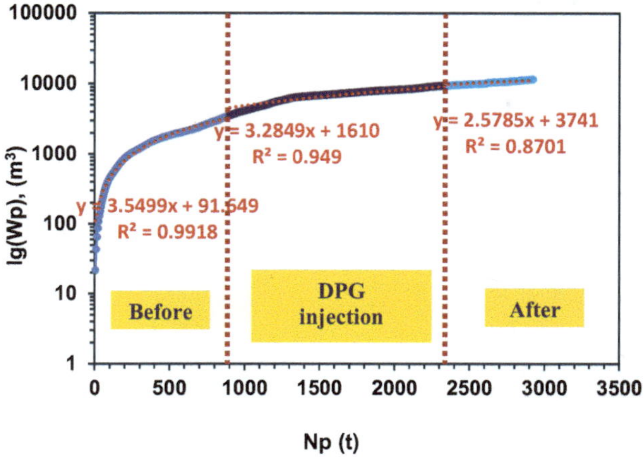

Figure 4.48 YT2–25H water drive characteristic curve.

plugging agent injection stage. This indicates that with the accumulation of oil production, the water production gradually decreases and the development effect changes. In particular, the YT2–5H well has the most significant effect.

References

[1] Zhao G. Research on the construction and displacement mechanism of soft heterogeneous combination flooding system. Qingdao: China University of Petroleum (East China); 2016. p. 40–6.

[2] Dai C, Zou C, Liu Y, et al. The matching rule of elastic dispersed particle gel with pore throat and the deep regulation mechanism. Acta Petrolei Sinica 2018;39 (4):427–34.

[3] Dai C, Liu Y, Zou C, et al. Investigation on matching relationship between dispersed particle gel (DPG) and reservoir pore-throats for in-depth profile control. Fuel 2017;207:109–20.

[4] Zhao G, You Q, Tao J, et al. Preparation and application of a novel phenolic resin dispersed particle gel for in-depth profile control in low permeability reservoirs. Journal of Petroleum Science and Engineering 2018;161:703–14.

CHAPTER 5

DPG soft heterogeneous combination flooding technology

In view of the complex conditions of medium—high temperature, medium—high salt oil reservoirs (high temperature, high salinity, and absolute heterogeneity), it is difficult for such oil recovery increase technologies as chemical flooding, polymer—surfactant binary combination flooding/polymer—surfactant—alkaline ternary combination flooding, gas flooding technology, which are derived from conventional EOR methods, to meet the needs of medium—high temperature and medium—high salt oil reservoirs. Based on self-developed multiscale DPG particles and temperature-resistant and salt-resistant surfactants, we constructed a DPG soft heterogeneous combination (SHC) flooding system and created a DPG soft heterogeneous combination flooding technology with microprofile control and oil displacement efficiency.

Section 5.1 Construction of DPG SHC flooding system

Taking the compatibility, viscosity, and interfacial tension of the flooding system as evaluation indicators, we constructed a DPG SHC flooding system suitable for medium—high temperature and medium—high salinity reservoirs. The salinity of the simulated water used in the experiment was 5.0×10^4 mg/L, the content of Ca^{2+} and Mg^{2+} was 3000 mg/L, and the simulated oil was crude oil from a block in Xinjiang Oilfield with a viscosity of 7.8 mPa · s at 90°C.

5.1.1 Compatibility research

On the basis of the previous research in the laboratory, considering the temperature and salt tolerance of sulfobetaine [1,2], this study uses Tetradecyl hydroxysulfaine surfactant (THSB) and multiscale DPG to construct the DPG soft heterogeneous combination flooding system. THSB surfactant is added to the DPG solution at room temperature and stirred evenly to prepare a soft heterogeneous combination flooding system. It was observed as to whether there was precipitation and flocculation. The test results are shown in Table 5.1 and Fig. 5.1. It can be seen

Theory and technology of multiscale dispersed particle gel for in-depth profile control
DOI: https://doi.org/10.1016/B978-0-323-99849-9.00005-9

Table 5.1 Compatibility investigation of DPG and surfactant.

Formula composition		30°C		90°C	
THSB (%)	DPG (%)	5 h	24 h	2 days	5 days
0.3	0.6	No precipitation, no flocculation	No precipitation, no flocculation	No precipitation, no flocculation	No precipitation, no flocculation
0.3	1.0	No precipitation, no flocculation	No precipitation, no flocculation	No precipitation, no flocculation	No precipitation, no flocculation
0.3	1.6	No precipitation, no flocculation	No precipitation, no flocculation	No precipitation, no flocculation	No precipitation, no flocculation

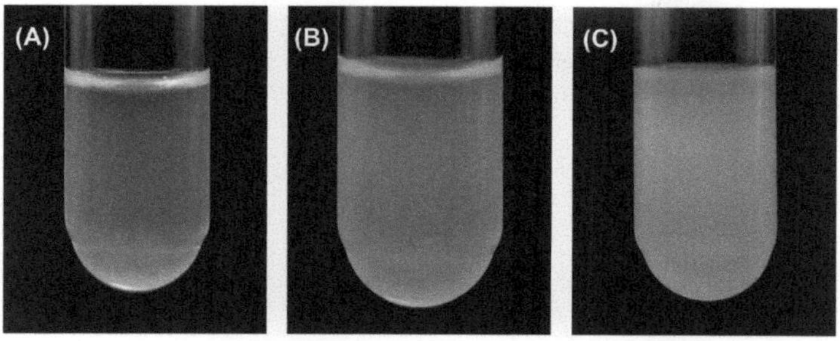

Figure 5.1 Compatibility of DPG SHC flooding system (0.3% THSB + 0.15% DPG): (A) 5 h; (B) 2 days; (C) 5 days.

that the DPG has good compatibility with THSB under the conditions of 30°C and 90°C. There is no precipitation and flocculation, indicating that the DPG has a good compatibility with surfactants.

5.1.2 Ability to reduce interfacial tension

5.1.2.1 The influence of surfactant concentration

The fixed concentration of DPG was 0.1% and the concentration of the surfactant was changed at a range of 0.01%–0.5%. The ability of a single surfactant and a soft heterogeneous combination flooding system to reduce the oil–water interfacial tension (IFT) was measured. The test results are shown in Fig. 5.2.

It can be seen from Fig. 5.2 that the changes in the reduction of oil–water interfacial tension between the single surfactant system and the soft heterogeneous combination flooding system are basically the same. In the low concentration range (0.01%–0.1%), the oil–water interfacial tension decreases with the increase of surfactant concentration. When the surfactant concentration exceeds 0.1%, the oil–water interfacial tension increases to 10^{-2} mN/m. This is due to the addition of surfactants. The surface energy of surfactant molecules in the interface phase is lower than the solution phase, forming a chemical potential difference and causing the hydrophilic groups of surfactant molecules to face the aqueous solution and turning the lipophilic groups toward the oil phase. The phase transfers to the oil–water interface and the active molecules are enriched at the oil–water interface to form a monomolecular adsorption film, forming a directional arrangement and reducing the interfacial tension. When the surfactant concentration further increases, the oil–water interface is filled with surfactant molecules and the surface

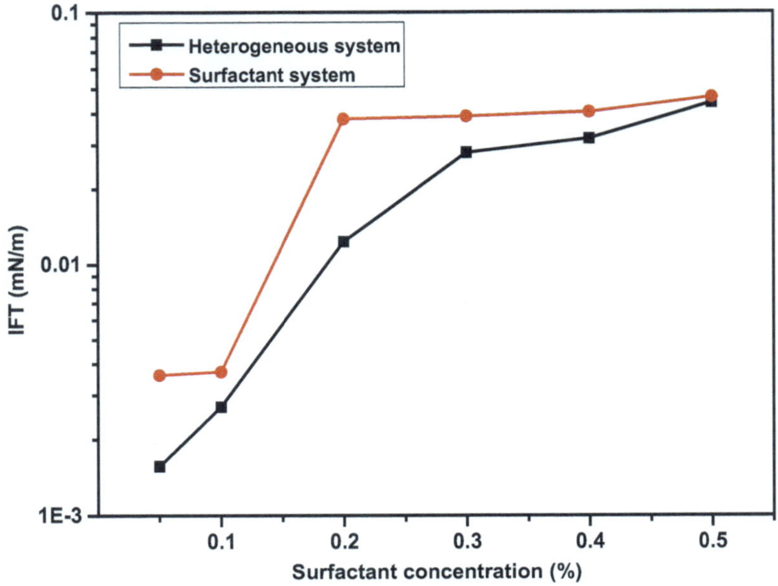

Figure 5.2 Interfacial tension reduces ability of single surfactant and soft heterogeneous combination flooding system.

can no longer accommodate extra molecules. At this time, the adsorption and desorption of surfactant molecules at the oil–water interface reaches a dynamic equilibrium [3,4]. Therefore the oil–water interfacial tension no longer changes, basically. When the surfactant concentration further increases, the surfactant molecules are difficult to adsorb on the interface, for the oil–water interface forms a stable adsorption layer. At the same time, the component, since being affected by the other substances of surfactants, is adsorbed on the oil–water interface. This helps reduce the adsorption sites of the surfactant adsorption molecules. This reduction, in turn, leads to an increase in interfacial tension. Fig. 5.2 further shows that when there exist DPG particles in the surfactant system, the oil–water interfacial tension slightly increases. Because the SHC flooding system is a viscoelastic system, it affects the transfer of surfactant molecules in the solution and the adsorption at the interface. This reduces the diffusion rate of surfactant molecules at the oil–water interface. In addition, the adsorption of the DPG at the oil–water interface occupies the adsorption sites of the oil–water interface. This adsorption behavior will also reduce the ability of the SHC flooding system to reduce the oil–water interfacial tension.

5.1.2.2 The influence of DPG particle concentration

The fixed concentration of the surfactant was 0.10%, and DPG particles of different concentrations were added, respectively. The dynamic interfacial tension of the SHC flooding system was measured at 90°C. The test results are shown in Fig. 5.3.

It can be seen from Fig. 5.3 that the ability of the SHC flooding system to reduce the oil−water interfacial tension decreases with the increase of the DPG concentration. Thus interface tension is stabilized between 3×10^{-3}–7×10^{-3} mN/m. In the experiment, the time for the SHC flooding system to reach a dynamic stable equilibrium interfacial tension is longer than the time for a single surfactant to reach the same tension. The higher the DPG concentration, the longer the heterogeneous combination flooding system takes to reach this tension. However, the final stable interfacial tension is basically the same. It shows that the addition of DPG extends the time for the combination flooding system to reach dynamic equilibrium interfacial tension, but has little effect on reducing the oil−water interfacial tension. When the DPG concentration is low, the surfactant molecules are easy to diffuse to the oil−water interface so that a strong adsorption layer structure is formed at the oil−water interface. At this time, the combination flooding system mainly exhibits

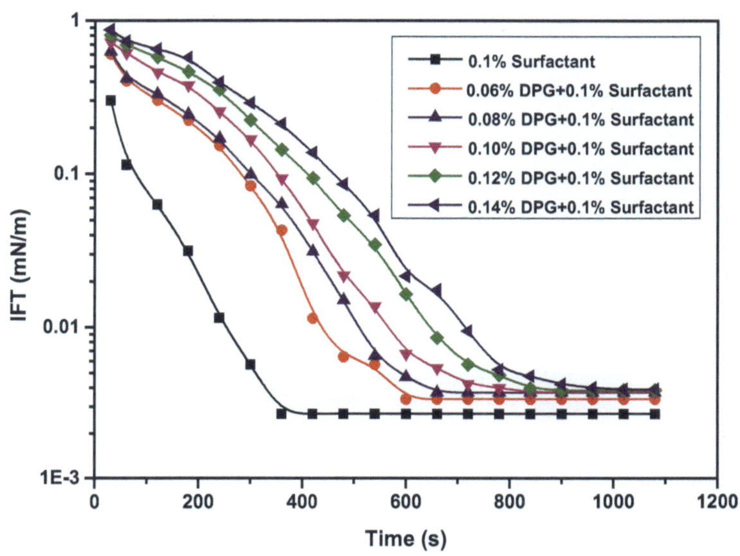

Figure 5.3 The effect of DPG concentration on the dynamic interfacial tension of SHC flooding system.

the properties of a polymer surfactant solution. As the concentration of DPG further increases, the viscosity of the combination flooding system increases accordingly. This reduces the transfer of surfactant molecules in the liquid phase and the interface diffusion speed, thereby prolonging the adsorption equilibrium time of the surfactant molecules at the interface. Since the SHC flooding system flows in a relatively slow process in porous media, however, it can be ensured that the surfactant molecules are adsorbed to the oil–water interface so that the interfacial tension reduces and the oil displacement efficiency improves.

5.1.3 Viscosity characteristics

5.1.3.1 The influence of surfactant concentration

Different concentrations of THSB were added to the DPG solution with a mass concentration of 0.1% to investigate the effect of surfactants on the viscosity of the SHC flooding system. The test temperature was 30°C. The results are shown in Fig. 5.4.

It can be seen from Fig. 5.4 that the addition of surfactants can slightly increase the viscosity of the SHC flooding system, but the change is not significant. This is because the surfactant is a dilute solution system. As the concentration increases, the surfactant may form micelles and increase the structural viscosity of the solution. In addition, under the influence of hydrogen bond and dispersion forces, the surfactant is adsorbed on the

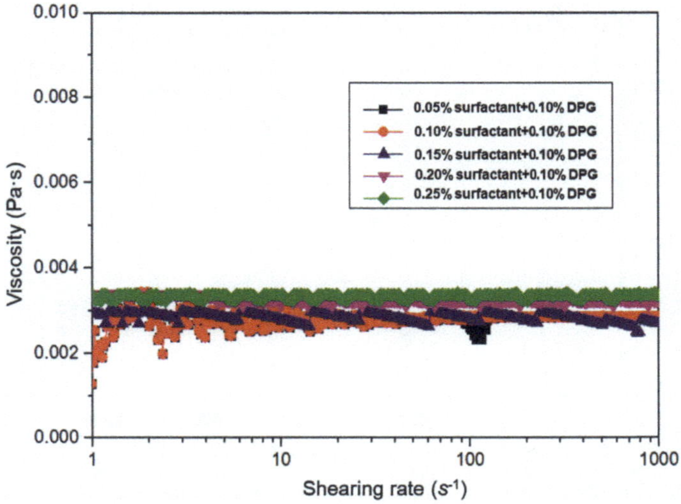

Figure 5.4 The effect of surfactant on the viscosity of SHC flooding system.

surface of DPG, increasing the radius of the particles and increasing the probability of collision between the particles. Therefore the viscosity of the heterogeneous combination flooding system increases slightly.

5.1.3.2 The influence of DPG concentration

The fixed concentration of the surfactant is 0.1% and DPG of different concentrations are added. The viscosity change of the SHC flooding system is measured at 90°C. The test results are shown in Fig. 5.5.

It can be seen from Fig. 5.5 that the viscosity of the SHC flooding system increases as the concentration of the DPG increases since the DPG is composed of highly viscous and elastic particles. When the concentration increases, the solid content of DPG in the solution rises and the distance between the particles narrows. The probability of intermolecular contact and collision increases, which increases the internal friction between molecules, resulting in an increase in the viscosity of the DPG solution. The increase in viscosity of the SHC flooding system is beneficial to the displacement. According to the water–oil mobility ratio formula, the greater the viscosity of the combination flooding system solution and the stronger the thickening ability of water, the smaller the water–oil mobility ratio and the larger the swept volume. As a result, an increased oil recovery is achieved. In addition, since the combination flooding system has viscoelasticity, it has the effect of similar polymer viscoelastic oil

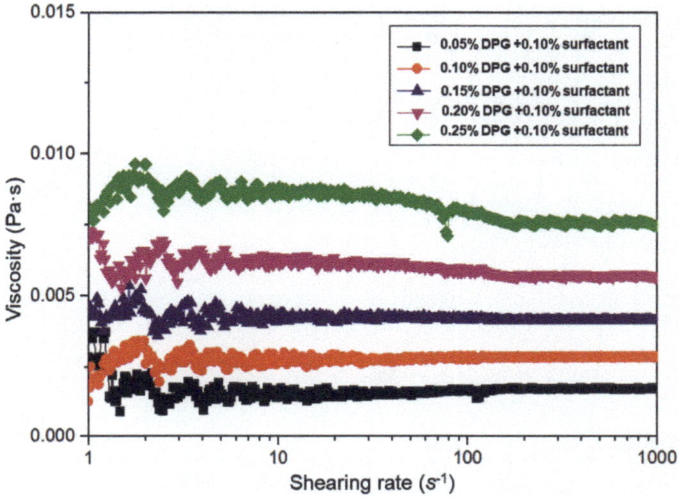

Figure 5.5 The influence of DPG concentration on the viscosity of SHC flooding system.

displacement. When the pore throat of the porous media is stretched, the combination flooding system adopts a straight conformation. After passing through the pore throat of the porous media, the stretching effect disappears and the curled conformation is adopted, so that the solution expands along the normal line of the flow direction. The negative pressure effect is used to drive out the remaining oil located in the pores.

5.1.3.3 Optimal concentration of the components of the SHC flooding system

Because the surfactant has little effect on the viscosity of the SHC flooding system, the experiment optimized the concentration of each component of the SHC flooding system from the interfacial tension. The results are shown in Fig. 5.6.

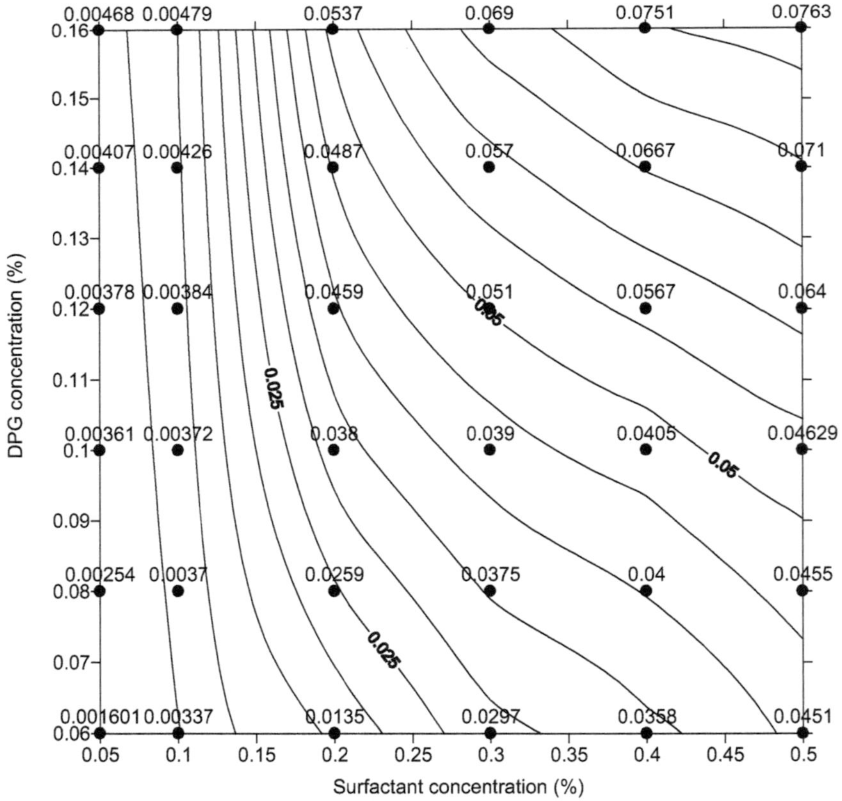

Figure 5.6 Optimal concentration of SHC flooding system.

It can be seen from Fig. 5.6 that as the surfactant concentration increases, the SHC flooding system has a stronger ability to reduce the oil–water interfacial tension. When the surfactant concentration exceeds a certain value, the interfacial tension increases instead. This indicates that there is an optimal concentration range for the surfactant concentration in the combination flooding system. Combining Figs. 5.4 and 5.5, it can be found that the viscosity of the combination flooding system increases with the increase of the DPG concentration. The low concentration range of surfactant has no effect on the viscosity of the combination flooding system while the high concentration of surfactant has the ability to synergistically increase the viscosity of the combination flooding system. The above studies show that the surfactant in the combination flooding system is the main controlling factor for reducing the interfacial tension between oil and water. Surfactant acts as a synergistic effect of increasing viscosity by adsorbing on the surface of DPG. The DPG is the main source of the increase in the viscosity of the combination flooding system. The increase in viscosity prolongs the time for dynamic equilibrium of the interfacial tension but has little effect on the interfacial tension. Considering the factors of low interfacial tension ($<10^{-2}$ mN/m), high viscosity, and low cost, the concentration range of surfactant used in the SHC flooding system is optimized to be 0.05%–0.12%. The used DPG concentration range is 0.06%–0.12%.

Section 5.2 Characterization of the properties of the DPG SHC system

The SHC flooding system has the characteristics not only of polymer/surfactant binary combination flooding but also of viscoelastic particles, due to the addition of DPG. This study characterized the nature of the SHC flooding system under the medium–high temperature and medium–high salinity reservoir conditions, from the aspects of its microscopic morphology, viscosity, ability to reduce interfacial tension, swelling performance, changes of wetting performance, shear stability, emulsification performance, and surface electrical properties.

5.2.1 Microscopic morphology of DPG SHC flooding system

The DPG SHC flooding system (0.1% DPG + 0.1% THSB surfactant) with simulated water was prepared and placed in a constant temperature oven at 90°C for 10, 15, and 30 days. Scanning electron microscopy

(SEM) was used to study the microscopic morphology changes. The test results are shown in Fig. 5.7.

As can be seen from Fig. 5.7, in the initial stage of aging, the DPG particles in the SHC flooding system are mainly single particles uniformly dispersed in the solution. After high-temperature aging, the individual DPG particles in the combination flooding system all begin to grow, but the expansion capacity is limited. This limited expansion ensures that DPG has high strength after high-temperature aging. With the further increase of the aging time, multiple DPG particles in the SHC flooding system collide with each other and the degree of adhesion between the particles increases, forming larger aggregates. The particle size of the DPG has increased from 2 μm to more than 30 μm. In addition, it can be seen from the micromorphology that the surface of the DPG particles is covered with a film after aging, which may be formed by the adsorption of surfactant in the combination flooding system on the surface of the particles during the aging process. The longer the aging time, the greater the aggregation degree of the DPG particles in the combination flooding system. This is because, before high-temperature aging, the SHC flooding system is a relatively stable system. The DPG in the combination flooding system is affected by electrostatic repulsion and can be uniformly dispersed in the solution. During high-temperature aging, the movement of DPG

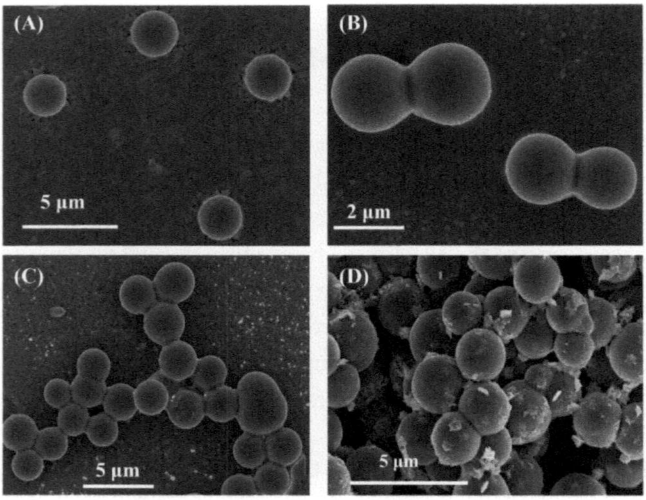

Figure 5.7 The microscopic morphology of the SHC flooding system before and after high temperature aging: (A) initial state; (B) aging for 10 days; (C) aging for 15 days; (D) aging for 30 days.

particles in the combination flooding system is intensified. The particles collide with each other to form large aggregates. The presence of salt ions in the simulated water with high salinity will neutralize the negative charge on the surface of the DPG particles and reduce the electrostatic repulsion between the particles, which accelerates the aggregation process of the heterogeneous combination flooding system. Therefore when the SHC flooding system migrates in the porous media, the surfactant in the combination flooding system can be better released to reduce the oil–water interfacial tension and improve the oil displacement efficiency while the aging and aggregation properties of DPG in the system are conducive to effective regulation of the formation of the hypertonic layer by the particles. When the pore throat is small, a single particle expands to realize the microscopic regulation of the pore. When the pore throat is large, multiple particles agglomerate to form larger aggregates and the pores are microscopically controlled. Then, the seepage profile is well adjusted and the swept volume is expanded. The oil recovery is maximized through the synergy of components of the SHC flooding system.

5.2.2 Viscosity stability

Simulated water was used to prepare SHC flooding system (0.1% DPG + 0.1% THSB surfactant), DPG, and polymer–surfactant binary combination flooding system (0.1% polymer + 0.1% surfactant). They were placed in a constant temperature oven at 90°C to age for different times. The viscosity change was measured with a Brookfield viscometer. The rotor was 0# and the speed was 6 rpm. The test results are shown in Fig. 5.8.

It can be seen from Fig. 5.8 that the viscosity of three systems decreases with the increase of aging time. The viscosity of the polymer–surfactant binary combination flooding system prepared with high-salinity simulated water decreases by 95% after 30 days of aging. While the viscosity retention rate of the SHC flooding system and the DPG system after 30 days of aging is higher than 60%, indicating that the SHC flooding system and the DPG systems have good viscosity stability under medium–high temperature and medium–high salinity conditions. This is because the viscosity stability characteristics of the SHC flooding system and DPG are mainly related to the prepared DPG. In light of the formation mechanism of the DPG, the bulk gel is a dense network structure formed by the dehydration condensation of the polymer amide group and the methylol of the cross-linking agent. This structure ensures that the bulk gel has high viscoelasticity and

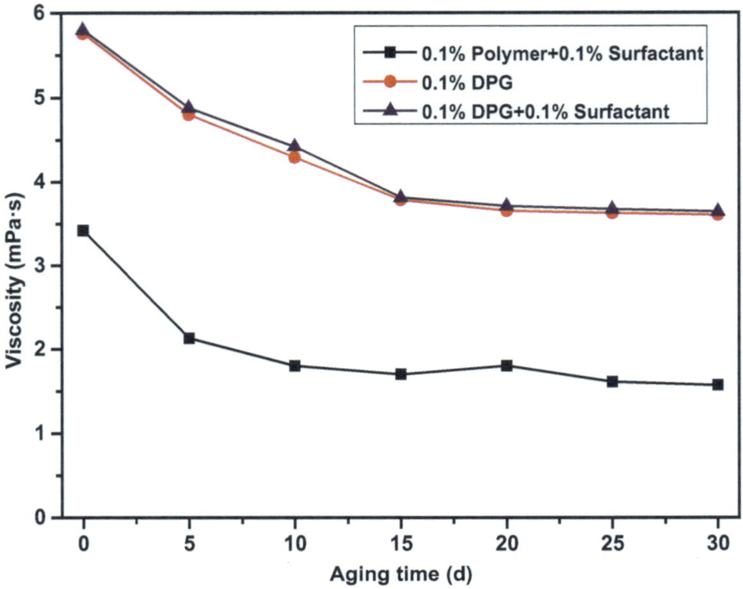

Figure 5.8 The effect of high temperature aging on the viscosity of polymer/surfactant, DPG, and SHC flooding system.

thermal stability. During the preparation of the DPG, the dense network structure of the bulk gel is only sheared by mechanical shearing and no chemical reaction is involved. Therefore the DPG formed from the bulk gel also has higher thermal stability. When DPG and surfactant are constructed into a SHC flooding system, the two components have good compatibility and the combination flooding system retains the characteristics of the thermal stability of DPG. Therefore after high-temperature aging, the SHC flooding system still has a high viscosity retention rate. When the SHC flooding system ages at high temperature, the DPG in the combination flooding system aggregates to form larger particles under the conditions of medium–high temperature and medium–high salinity conditions. The interaction between DPG particles is reduced, resulting in the reduction of the viscosity of the system to a certain extent. However, compared with polymers, the developed SHC flooding system still has higher resistance to high temperature and salinity.

5.2.3 Ability to reduce interfacial tension

The ability of the SHC flooding system to maintain the reduction of interfacial tension after aging is the key to ensure its improvement of the displacement efficiency. Therefore the experiment comparatively analyzed the interfacial tension changes of a single surfactant system (0.1%), a SHC flooding system (0.1% DPG + 0.1% THSB surfactant) aging for different time in a 90°C incubator. The results are shown in Fig. 5.9.

It can be seen from Fig. 5.9 that the interfacial tension of the SHC flooding system increases with the aging time. When the aging time increases to 20 days, the interfacial tension increases from 3.72×10^{-3} to 4.79×10^{-2} mN/m. After 15 days, the ability of the SHC flooding system to reduce the interfacial tension between oil and water remained basically unchanged, indicating that the combination flooding system has a better ability to reduce interfacial tension and good application potential. It is because the surfactant in the SHC flooding system is the sulfonate betaine surfactant. This surfactant has a chelating effect on the divalent salt ions in the simulated water, which enhances its salt tolerance. Under the condition of high concentration of salt ions, the salt ion shields the charge of the ion head, compressing the thickness of the surfactant ion atmosphere, destroying the hydration membrane of the hydrophilic group, reducing the electrostatic repulsion of the hydrophilic group, and promoting the adsorption of surfactant molecules at the oil–water interface. This makes surfactants form a tight interface layer arrangement structure.

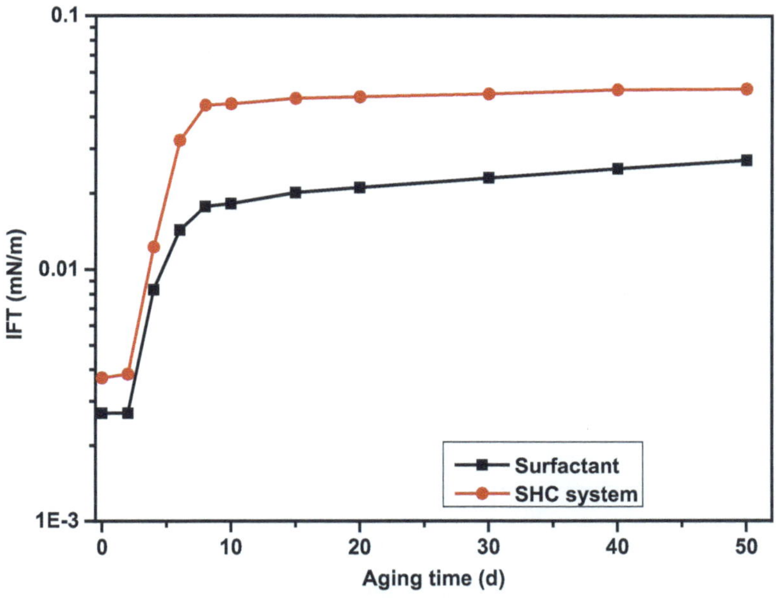

Figure 5.9 The interfacial tension of SHC flooding system changes with aging time.

Therefore the heterogeneous combination flooding system still has a better ability to reduce interfacial tension after aging. But at the same time, due to the effect of high temperature and high salinity, some surfactant molecules lose their effectiveness, reducing the adsorption capacity of surfactant molecules at the oil–water interface and making sparse the molecular layer of the interface layer. This results in an increase in interfacial tension. For the SHC flooding system, some surfactant molecules are adsorbed on the DPG surface due to the hydrophobic effect. This reduces the surfactant content in the combination flooding system. Therefore the ability of the SHC flooding system to reduce the oil–water interfacial tension is reduced.

5.2.4 Expansion capacity

This study employed a laser particle size analyzer to compare and analyze DPG (0.1%) and the SHC flooding system (0.1% DPG + 0.1% THSB surfactant) under the reservoir conditions. The result of coalescence expansion capacity under the condition is shown in Fig. 5.10.

It can be seen from Fig. 5.10 that the particle size of the two systems increases with the increase of aging time. In the early stage of aging, the

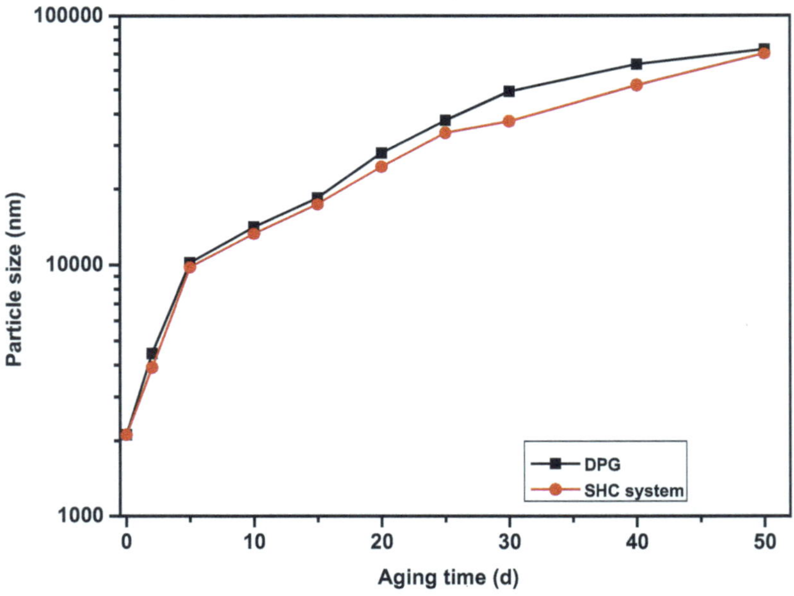

Figure 5.10 Expansion capacity of DPG and the SHC flooding system before and after high temperature aging.

particle size of the heterogeneous combination flooding system increases rapidly. When the combination flooding system is aged for 40 days, the particle size basically no longer increases. This phenomenon may be caused by the different strength charges of the heterogeneous combination flooding system. The surfactant in the combination flooding system has a strong negative charge, while the DPG particles have a weak negative charge. During the aging process, the strongly negatively charged surfactant is adsorbed on the DPG surface through hydrogen bonds and hydrophobic interactions. This increases the negative charge on the DPG surface and enhances the repulsive force between the particles, thus making the particles difficult to aggregate. Therefore the coalescence swelling capacity of the SHC flooding system is weaker than that of DPG alone. This feature is conducive to the smooth entry of the combination flooding system into the deep reservoir. When the combination flooding system was continuously aging for 50 days, the particle size increased from the initial 2.1 μm to 69.8 μm, indicating that the combination flooding system has better expansion capability. During the aging process, the DPG in the SHC flooding system realizes its own coalescence and expansion through three ways: a single particle achieves a limited expansion through water absorption, generally 1.5−2 times, to ensure DPG particles have some certain strength. Surfactant adsorbs on the surface of DPG particles to form a multilayer adsorption layer, which increases the effective radius of the particles. After continuous aging, the salt ions of the simulated water neutralize the negative charge on the surface of the particles, resulting in a decrease in the electrostatic repulsion between the particles and an increase in adsorption. The DPG particles grow up to form large aggregates, which increases the expansion capacity of the SHC flooding system.

5.2.5 Wettability alteration

In the process of fluid flow in reservoir, the long-term contact between crude oil and rock leads to the change of rock surface properties from hydrophilic to lipophilic, and the permeability of oil phase tends to decrease by 15%−85%. Both affect the oil recovery rate seriously. The surfactant in SHC flooding system can change the rock wettability by adsorption on the rock surface, which is beneficial to the stripping of crude oil from the rock surface. It is thus considered to play a positive role in enhancing oil recovery.

5.2.5.1 Experimental method

The wettability alteration of SHC flooding system was measured by optical projection method. The treated water-wet and oil-wet quartz plates were used to simulate formation rocks. Two kinds of quartz plates with different wettability were aged in surfactant solution and heterogeneous combination flooding system, and the wettability change on quartz plates was compared and analyzed. The specific measurement steps were as follows:

1. Quartz plates with length of 5 cm and width of 2 cm were soaked in 5% hydrochloric acid solution for 3 hours, then rinsed with deionized water repeatedly and dried at 90°C;
2. The oil-wet system was prepared by mixing 20% n-heptane with crude oil. The dried quartz plates were placed in the oil-wet system and aged at 90°C for 15 days to obtain lipophilic quartz plates, which were used to simulate the oil-wet sandstone surface;
3. Dry quartz plates were placed in simulated water and aged at 90°C for 15 days to obtain hydrophilic quartz plates, which were used to simulate the surface of water-wet sandstone;
4. The oil-wet and water-wet quartz plates obtained above were placed in surfactant solution (0.1%) and SHC flooding system (0.1% surfactant + 0.1% DPG) and aged for different days, respectively. Then the wettability alteration ability of the two systems at aged quartz plates was determined.

Among them, the initial contact angle of the modified water-wet quartz plate was 24.5 degrees, and the initial contact angle of the oil-wet quartz plate was 147.5 degrees. The results are shown in Fig. 5.11.

5.2.5.2 Experimental results and analysis

The results of SHC flooding system and surfactant changing the wettability of quartz plates are shown in Fig. 5.12. It can be seen that the aging of the two systems could reverse the wettability of the oil-wet quartz plates, transform the lipophilic quartz plates into hydrophilic quartz plates, and

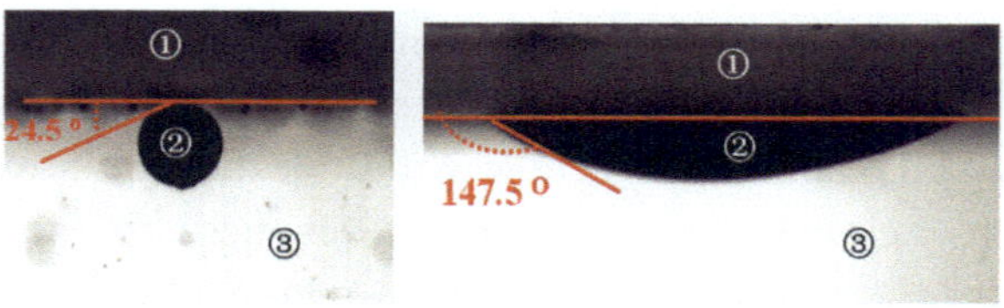

Figure 5.11 Wetting angle of modified quartz plate, ① quartz plate; ② oil drop; ③ water.

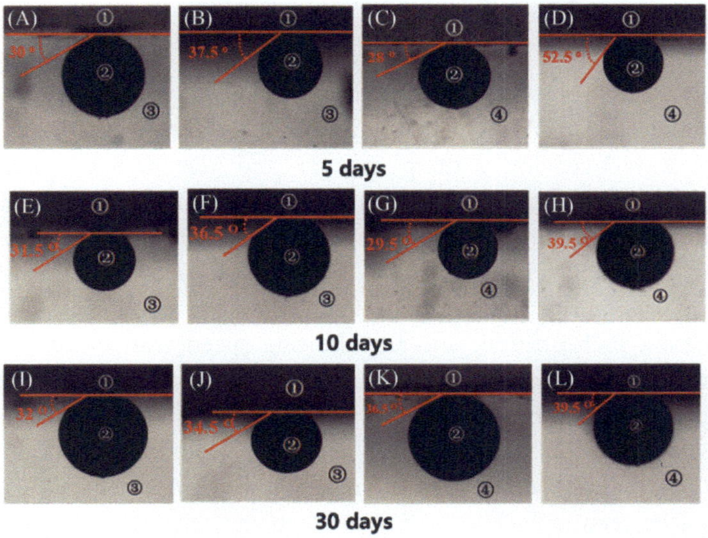

Figure 5.12 Wettability alteration of SHC flooding system and surfactant: ① Quartz plate; ② Oil drop; ③ Surfactant solution; ④ SHC flooding system, (A), (E), (I) Water-wet, surfactant; (B), (F), (J) oil-wet, surfactant; (C), (G), (K) water-wet, SHC flooding system; (D), (H), (L) oil-wet, SHC flooding system.

weaken the hydrophilicity of the hydrophilic quartz plates. Since the SHC flooding system has more surfactant active molecules before aging, the lipophilic groups face the quartz plate and are closely arranged on the surface of the quartz plate through hydrogen bonding to form a multilayer adsorption film, which significantly improves the wettability of the quartz plate surface. When the SHC flooding system ages, the effective content of surfactant active molecules decreases and reduces their adsorption on the surface of quartz. Also, the DPG in the heterogeneous combination flooding system is adsorbed on the surface of quartz by hydrogen bonding, decreasing the adsorption sites on the surface of quartz and weakening the adsorption of surfactants on its surface. Therefore the ability of a single surfactant to change the surface wettability of quartz plates is stronger than that of SHC flooding system. After continuous aging for 30 days, the system tends to be stable and the active molecules of surfactants are basically no longer reduced. So, the wetting properties of quartz plate are basically unchanged. This characteristic of the SHC flooding system ensures it changes the wettability of rock surface during the flowing process. This is beneficial to the stripping of crude oil from rock surface, and thus enhance oil recovery.

5.2.6 Shear resistance

The polymer–based oil flooding system is easily affected by injection equipment and flow shear in formation and the polymer viscosity can be greatly reduced. Especially in following water flooding stage, the injection pressure is reduced in a sharp manner and the subsequent mobility controllability gets limited. Thus the oil displacement effect is affected. Therefore shear stability is the long-term effective guarantee of combination flooding technology. In this part, a Waring high-speed shearing machine was used to simulate the shearing effect when the combination flooding system flows in formation, and the shearing speed was set at 1000 rpm, under the condition of 30°C and 6 rpm. In addition, the Brookfield viscometer was used to measure the polymer/surfactant flooding system (0.2% polymer + 0.1% THSB surfactant) and SHC flooding system (0.2% DPG + 0.1% THSB surfactant). The results are shown in Fig. 5.13.

It can be seen from Fig. 5.13 that when the shearing force is applied, the viscosity of the polymer/surfactant combination flooding system decreases rapidly with the increase of the shearing time. After 20 minutes of shearing, the viscosity drops by over 95%. This indicates a poor shear resistance of the combination flooding system. Since the viscosity of the

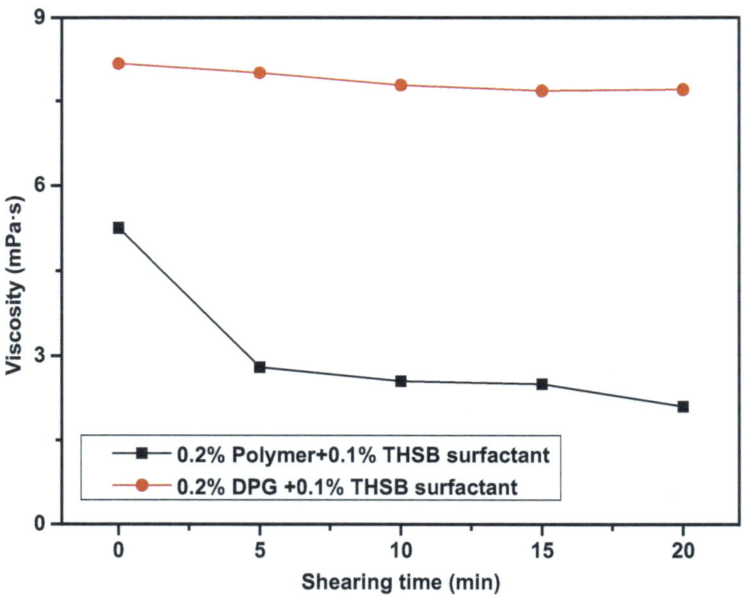

Figure 5.13 Shear resistance of SHC flooding system and polymer/surfactant combination flooding system.

polymer/surfactant combination flooding system mainly comes from the structural viscosity produced by the entanglement of polymer molecules, the long polymer chains are cut and the entangled structure is broken up after the application of shearing. Therefore the viscosity of the polymer/surfactant combination flooding system is reduced greatly. For the SHC flooding system, the viscosity retention rate of the system is still 98% after shearing for 20 minutes, which indicates that the system had good shear resistance. As the SHC flooding system is composed of small-molecular-weight surfactants and high-molecular-weight DPG particles, the viscosity mainly comes from the DPG particles. The results of previous studies demonstrate high shear stability of DPG. Therefore the SHC flooding system has high shear stability as well.

After shearing is ceased, the SHC flooding system and the polymer/surfactant combination flooding system are put to aging for 24 hours, to investigate the viscosity recovery ability of the two systems. It can be seen from Fig. 5.14 that the viscosity of the polymer/surfactant combination flooding system is basically not restored after being left for 24 hours. Under high-speed shearing, the long polymer chain is damaged. It is difficult for the chain to recover and the viscosity cannot be restored

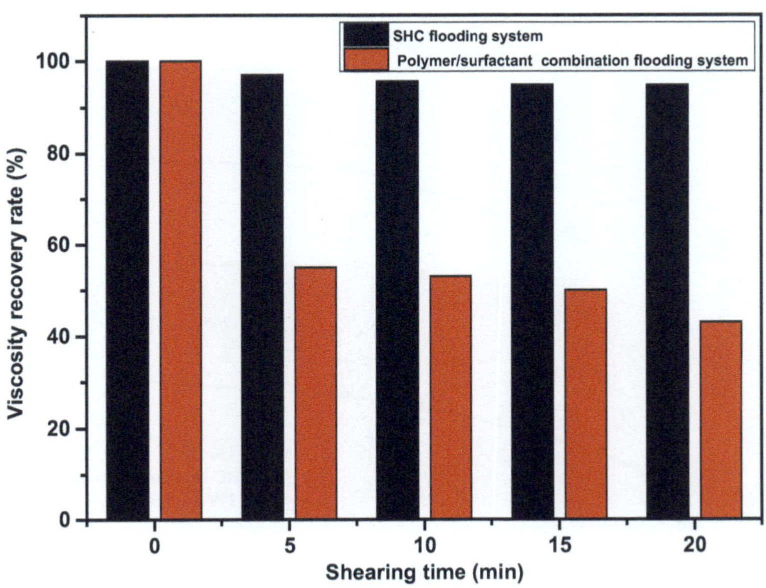

Figure 5.14 Viscosity recovery ability of SHC flooding system and polymer/surfactant combination flooding system (24 h).

completely. However, the viscosity recovery rate of the SHC flooding system reaches over 95% after being aged for 24 hours. Since the DPG in a heterogeneous combination system is a stable system, formed by high viscoelasticity bulk gel through high-speed mechanical shearing, the particle size of DPG and the solid content of the particles in the solution are basically unchanged when an external shearing force is in application. The effective viscosity can be maintained when the particles collide with each other.

5.2.7 Zeta potential

Zeta potential is an important parameter to characterize the stability of the SHC flooding system. If the components of the heterogeneous combination flooding system have different electrical properties, it is easy to form coagulations and reduce the displacement effect. The experiment analyzed the electrical properties of surfactant (0.1%), DPG (0.1%), and SHC flooding system (0.1% DPG + 0.1% surfactant) before and after aging at 90°C, respectively. Considering the salt tolerance range of the zeta potentiometer, the experiment only accounted for the electrical changes of the three systems in 10,000 mg/L simulated water (Na^{+}: 9100 mg/L; Ca^{2+}: 600 mg/L; Mg^{2+}: 300 mg/L). The experiment results are shown in Fig. 5.15.

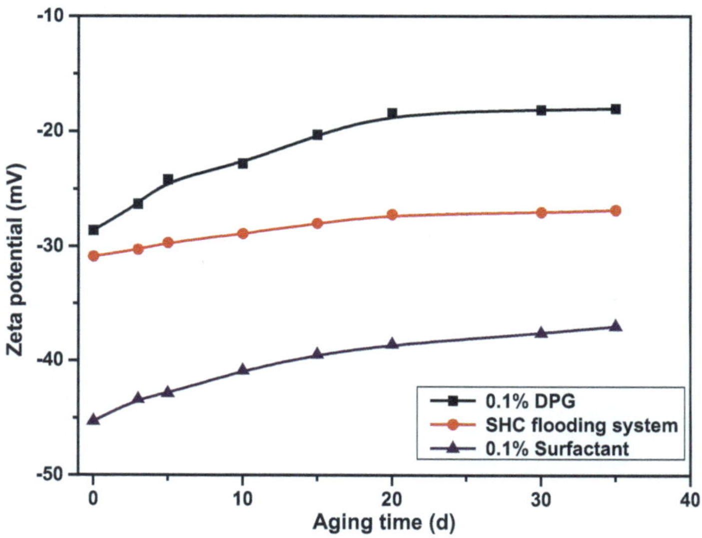

Figure 5.15 Influence of aging time on the zeta potential of the surfactant system, DPG, and SHC flooding system.

It can be seen from Fig. 5.15 that the surfactant system, DPG, and SHC flooding system are all negatively charged before aging, and the absolute zeta potential of the surfactant system is higher than that of the DPG system. When surfactant is added into DPG to form an SHC flooding system, surfactant is adsorbed on the surface of DPG particles. This increases the electronegativity of DPG, enhances the electrostatic repulsion between particles, and raises the absolute value of zeta potential of the combination flooding system. After 30 days of high-temperature aging, the absolute value of the surfactant zeta potential is still higher than 30 mV, indicating the relative stability of the system. However, the absolute value of the zeta potential of DPG and SHC flooding system decreases and changes from a stable system to an unstable system. Since the surfactant used is a sulfonate betaine type small molecule system, it has a chelating effect on the ions in the simulated water after high-temperature aging. This enhances its salt tolerance and makes it difficult for the small molecule system to aggregate. Therefore the surfactant system has a higher absolute value of zeta potential after aging, showing the characteristic of stability. For the SHC flooding system, the positively charged ions neutralize the negative charge on the particle surface, reducing the electrostatic repulsion between the particles. So, the SHC flooding system is easy to aggregate in salt water. In addition, the DPG particles in SHC flooding system have large specific surface area and high surface energy characteristics. The particles are easy to aggregate to reduce their surface energy. Macroscopically, the particles aggregate and become larger. And microscopically, the absolute value of the zeta potential decreases. When the SHC flooding system flows in the porous media, the two generate electrostatic repulsion since the surface of the formation rock particles is negatively charged. This helps ensure it enters into the deep formation smoothly without causing adsorption and retention in the near-wellbore zone, and to play a role deep displacement.

Section 5.3 Displacement performance of the DPG SHC flooding system

The displacement model of long sand-pack was established to study the in-depth injection ability of the SHC flooding system and its permeability reduction effect at different sand-pack positions. The reservoir conformance controllability and displacement potential of the SHC flooding system were investigated by a sand-pack flooding experiment to provide technical support for its oilfield application.

5.3.1 Evaluation of injectivity performance

To ensure the oil displacement effect of the SHC flooding system, the system must be equipped with the characteristics of "easy injection, easy movement, and in-depth working." The long sand-pack flooding instrument with multiple pressure tap was used to determine the injection performance SHC flooding system. There were five pressure measuring taps on the model, with the position distributions being ab = ef = 10 cm, bc = de = 20 cm, and cd = 40 cm. The core specification was 100 cm in length and 2.5 cm in diameter. The setup diagram is shown in Fig. 5.16.

Considering that the permeability near the water injection well in the formation is generally large due to the erosion by long-term water injection, in order to simulate the formation situation, gravel particles were used at the injection end when preparing sand-pack samples, and the specific measurement method was as follows:

1. Quartz sands of 60–80 mesh were used to fill the sand-pack, and the permeability and pore volume were calculated;
2. After the water flooding was stable, 1PV SHC flooding system (0.1% DPG + 0.1% THSB surfactant) was injected into the sand-pack at 0.5 mL/min at 90°C, and the pressure changes at five pressure measuring taps were recorded;
3. The abovementioned sand-pack was aged for 5 days in an oven at 90°C. Then, it was water flooded until the pressure was stable. The pressure changes during the water flooding process were recorded.

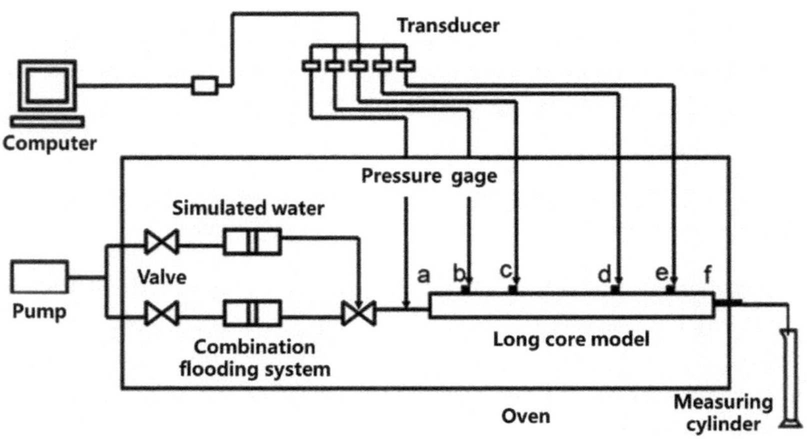

Figure 5.16 The long sand-pack flooding instrument with multiple pressure taps.

The permeability of the sand-pack was 1.07 μm^2 and the pore volume 154 mL. The pressure changed with the injected pore volume, as shown in Fig. 5.17.

It can be seen from Fig. 5.17 that the pressures at five pressure measuring taps decreases in the water flooding stage, but the decrease range was not large. When 1PV SHC flooding system is injected, the pressures of five pressure measuring taps all increase to variable degrees. Points 1, 2, and 3 increase first; when the injection volume is 0.6 PV, the pressures of points 4 and 5 began to increase. This indicates a movement of the SHC flooding system there. In the injection stage, when the core pore throat is smaller than the size of DPG particles in SHC flooding system, the DPG particles plug the core directly. When the pore throat of core is larger than that of the DPG particles size, multiple DPG particles bridge with each other to realize the microcontrol of high-permeability zones, thereby increasing the pressure at the five pressure measuring points. This indicates that the SHC flooding system has an active control effect on the high-permeability zones. However, the plugging ability of combination flooding system during injection process is relatively weak. Therefore the test further investigated the migration and plugging ability after

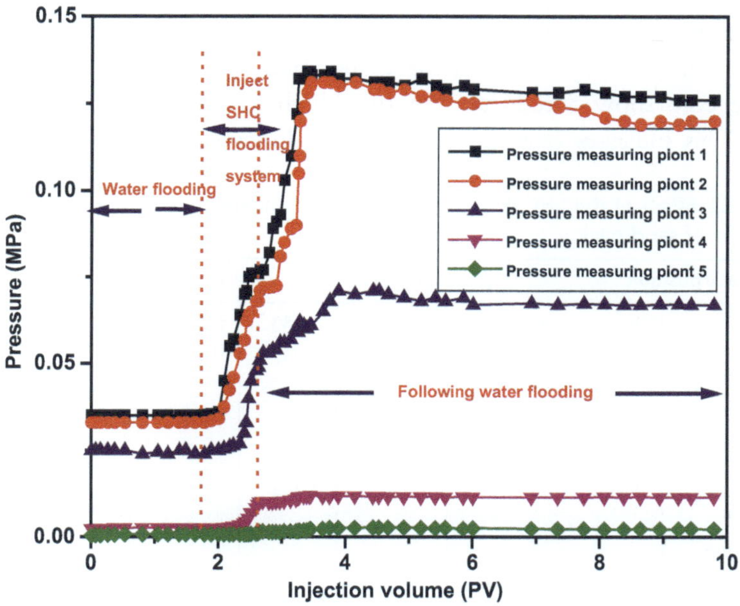

Figure 5.17 Injection pressure of the SHC flooding system with changes of pore volume.

high-temperature aging. The pressure at five pressure measuring taps all increases significantly. When at 1 PV, the water flooding reaches the maximum. After high-temperature aging, the DPG particles in the SHC flooding system can achieve reservoir control through an expansion of individual DPG particles or a coalescence of DPG particles so as to form larger aggregates. However, as the following water flooding progressed, the pressure begins to drop, indicating a reduction in the control of the combination flooding system. Since DPG is particles, the particles continue to migrate deep into the sand-pack after the breakthrough. Then, the pressure at the subsequent pressure measuring taps is maintained at a relatively high level, with the mobility characteristics of plugging, breakthrough, migration, and replugging. The heterogeneous combination flooding system is still kept at a relatively high pressure after the following water flooding for 7 PVs. This suggests a better reservoir conformance control effect of the system.

5.3.2 Effect of permeability reduction at different positions of sand-pack

In order to further illustrate the plugging ability of the SHC flooding system in the sand-pack, the permeability at each position was measured by the pressure drop at five pressure measuring points. According to Table 5.2, after injecting the SHC flooding system, different positions of the core show a reduction in their high permeability, reaching over 87%. This indicates that the combination flooding system can effectively

Table 5.2 Effect of the SHC flooding system on permeability reduction at different positions of sand-pack.

Sand-pack position		af	bf	cf	df	ef
Before injecting SHC flooding system	Pressure/MPa Permeability/μm^2	0.035 0.97	0.033 0.93	0.025 0.95	0.0027 3.77	0.0008 4.24
After injecting SHC flooding system	Pressure/MPa Permeability/μm^2	0.1260 0.269	0.120 0.254	0.0670 0.354	0.0116 0.878	0.0024 1.414
Permeability reduction/%		87.01	87.04	92.94	99.69	99.97

plug various positions of the sand-pack. The DPG particles in the SHC flooding system are the main controlling factor for reservoir conformance control. Since they are soft particles, the DPG particles can enter deep into the sand-pack through deformation by retention, adsorption, bridging, etc. The forms regulate the high-permeability parts effectively, thereby reducing the core permeability and achieving the purpose of microcontrol.

5.3.3 Profile improvement of SHC flooding system

Parallel sand-packs models with different permeability ratios were established to simulate the heterogeneity of formation. High-permeability sand-pack was used to simulate the high-permeability layers of formation while low-permeability sand-pack to simulate the low-permeability layers of formation. The shunt rate and profile improvement rate were used to evaluate the conformance controllability of the SHC flooding system. The measurement steps are follows. Cores with different permeability ratio were first prepared and then water flooded at a rate of 0.5 mL/min until the pressure was stable. Next, 1 PV SHC flooding system (0.1% DPG +0.1% surfactant) was injected into sand-pack at a rate of 0.5 mL/min, aged at 90°C for 5 days, and then flooded by water. The changes in pressure and fluid production during the test were recorded. The specification of sand-pack was 20 cm × in length × 2.5 cm in diameter. The results are shown in Table 5.3. The shunt rates and pressure changes of core models with different permeability ratio are shown in Figs. 5.18–5.21.

It can be seen from Table 5.3 that the SHC flooding system has good profile improvement capabilities for sand-packs with different permeability.

Table 5.3 Profile improvement ability of the SHC flooding system under different permeability.

Permeability ratio	Core type	Permeability (μm^2)	Shunt rate (%)		Profile improvement ability (%)
			Before injection	After injection	
1.32	Low permeability	0.295	62.71	56.31	53.87
	High permeability	0.388	37.29	43.69	
2.06	Low permeability	0.388	76.60	48.77	67.91
	High permeability	0.799	23.40	51.23	
4.07	Low permeability	0.334	85.77	40.98	76.11
	High permeability	1.358	14.23	59.02	
8.36	Low permeability	0.325	91.37	55.56	93.72
	High permeability	2.716	7.63	44.44	

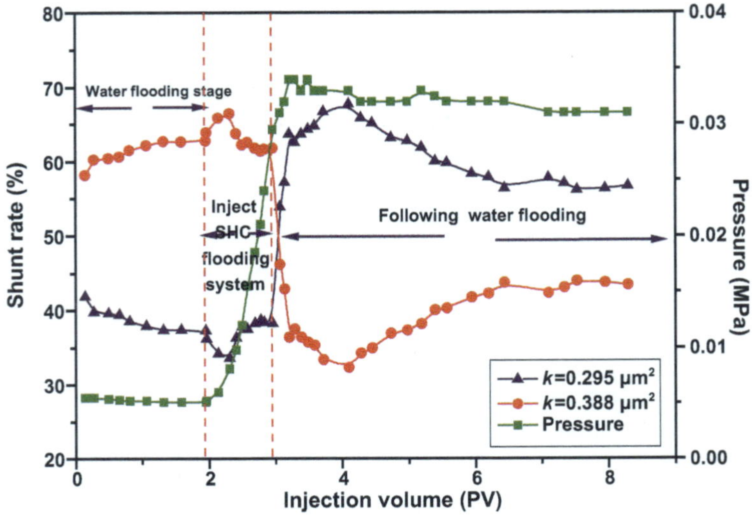

Figure 5.18 Shunt rate and pressure of sand pack models with different permeability ratio (permeability ratio:1.32).

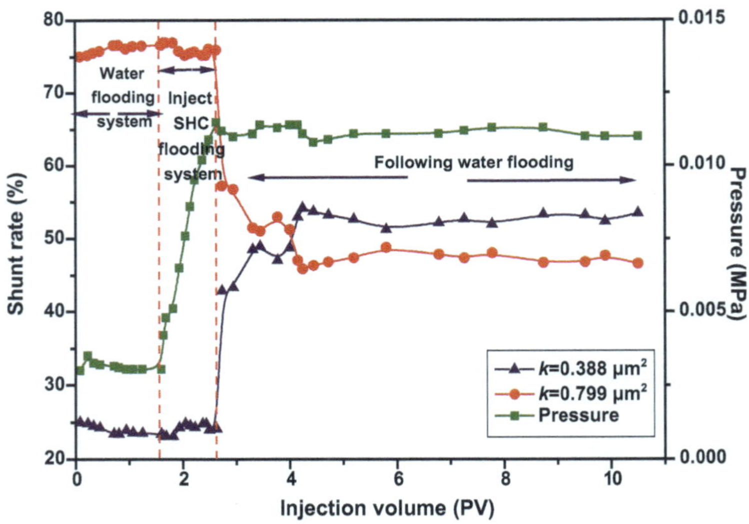

Figure 5.19 Shunt rate and pressure of sand pack models with different permeability ratio (permeability ratio:2.06).

At the same time, the larger the permeability difference, the stronger the profile improvement capability of the combination flooding system. During the injection process of the SHC flooding system, the combination flooding system preferentially enters the high-permeability layer which is affected

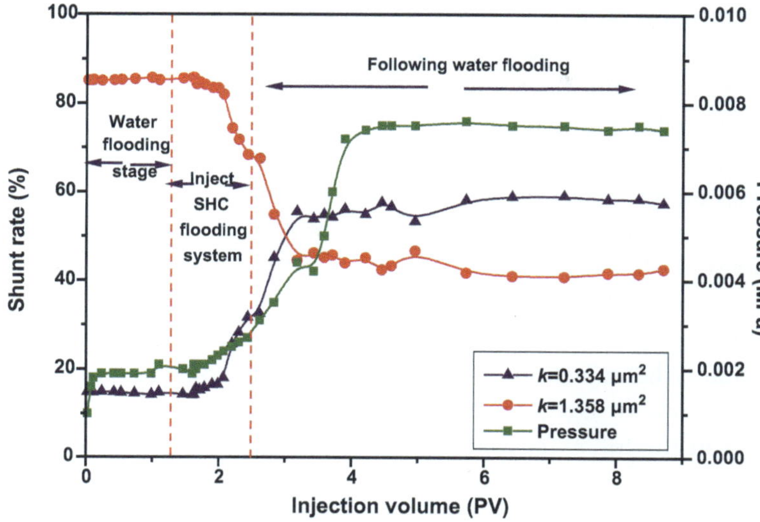

Figure 5.20 Shunt rate and pressure of sand pack models at permeability ratio of 4.07.

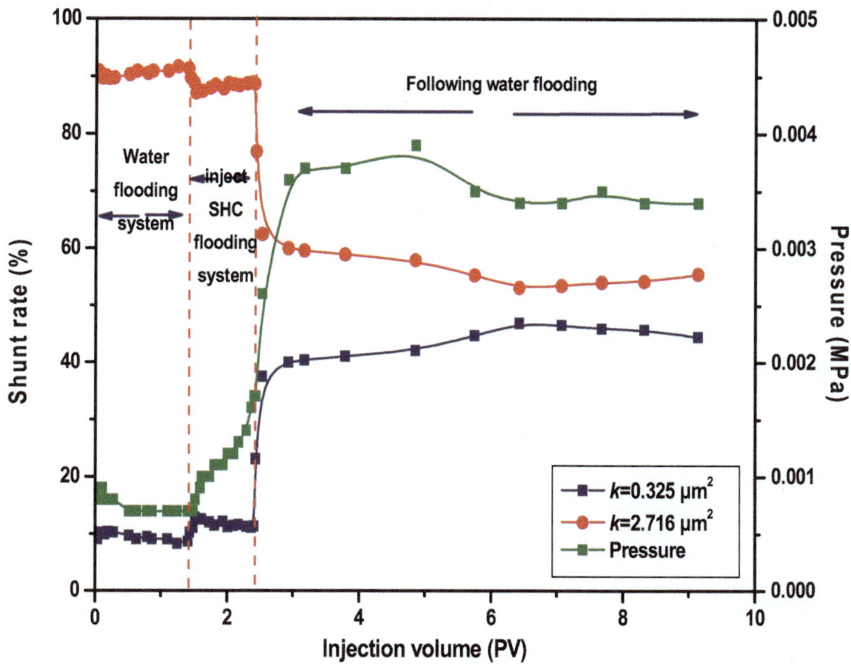

Figure 5.21 Shunt rate and pressure of sand pack models at permeability ratio of 8.36.

by the injection pressure. Then, effective plugging is formed in the high-permeability area by a means of retention, adsorption, bridging, etc. As a result, the following water is diverted into the medium–low-permeability layer. This helps improve the shunt capacity of the medium–low-permeability layer and then achieve microcontrol.

It can be seen from Figs. 5.18–5.21 that in the water flooding stage, the high-permeability pipe has larger shunt rate and lower water flooding pressure. When the SHC flooding system is injected, the fluid production of high-permeability sand-pack begins to decrease while the fluid production of the low-permeability sand-pack increases. At the same time, the injection pressure grows in an obvious way. When the sand-pack is aged, the flow rate of low-permeability sand-pack increases in the following water flooding process. This indicates that the SHC flooding system possesses a better profile improvement ability and can effectively control the high-permeability layer. It can be further known from the figures that the fluid profiles of different permeability models are improved after the SHC flooding system is injected. The greater the difference in permeability, the stronger the profile improvement capability of the SHC flooding system. Under the action of injection pressure, the greater the core permeability difference, the easier for the combination flooding system to transport in the high-permeability layer. This helps plug the high-permeability layers effectively and divert the subsequent water into the low-permeability layers. The permeability difference gets smaller as well. The SHC flooding system enters the low-permeability layers when entering the high-permeability layers, thus introducing certain pollution to the low-permeability layers and reducing the profile improvement ability. The good profile improvement capability of the SHC flooding system is beneficial to oil displacement. When the system is injected into formation, it can plug the high-permeability layers effectively, divert the subsequent fluid into the medium–low-permeability layers and displace the remaining oil. In addition, the surfactant in the SHC flooding system also displaces the residual oil in porous media through the effects of wettability reversal and emulsification, etc. The oil recovery then increases.

5.3.4 Evaluation of oil displacement potential

The single sand-pack flowing model was used to compare and analyze the displacement potential of surfactants, polymer/surfactant combination flooding systems, DPG and SHC flooding system. The specific measurement procedures are as follows:

1. Dry the core for 10 hours in an oven at 90°C, weigh it. Saturate the dried core with simulated water, weigh it again, and the pore volume can be calculated;
2. Water flooding until the pressure is stable, calculate the core permeability;
3. With the principle of "low pumping rate to high pumping rate," the core is saturated with simulated oil, and the core is placed in an oven at 90°C, and aged for 24 hours;
4. Water flooding at 0.5 mL/min until the water-cut of the produced liquid reaches 98%, and record the pressure, water production, and oil production during the process;
5. Inject 1 PV SHC flooding system (0.1% DPG + 0.1% THSB surfactant) at 0.5 mL/min, and record the pressure, water production, and oil production during the injection process, and then the core is placed at 90°C and aged for 5 days;
6. Water flooding until the water-cut of the produced liquid reaches 98%, record the pressure, water production, and oil production during the process, and calculate the recovery rate;

The SHC flooding system in step (5) above was replaced with 0.1% DPG particles, 0.1% surfactant, and a polymer/surfactant combination flooding system (0.3% polymer + 0.1% Surfactant), respectively. Then, the steps (1)−(6) above were repeated. The results are shown in Table 5.4. It can be seen that the injection of the four oil displacement fluids can effectively increase the recovery rate by more than 10%. The oil recovery increase of the SHC flooding system is significantly higher than that of surfactants, polymer/surfactant combination flooding systems, or DPG particles.

Figs. 5.22−5.25 show the curves of water-cut, oil production, and pressure during the displacement process of surfactant, polymer/surfactant combination flooding, DPG particles, and SHC flooding system. It can be seen that after long-term water flooding in the initial stage, the dominant flow channel is formed. This leads to the injection pressure dropping and the invalid water flooding cycle. At this time, when the SHC flooding system or DPG particles is injected, under the injection pressure, it can preferentially enter the high-permeability zones, and the high-permeability zones can be plugged by the retention, adsorption, or aggregation of DPG particles. Effective control diverts the following water to

Table 5.4 Displacement effects of different flooding systems.

System	Permeability (μm^2)	Length (mm)	Diameter (mm)	Pore volume (mL)	Oil saturation (%)	Water flooding recovery (%)	Final recovery (%)	Increased recovery (%)
Surfactant	0.778	934	50	9.34	76.02	49.30	60.56	11.26
Polymer/surfactant system	0.783	930	50	9.33	77.12	52.78	67.92	15.14
DPG particles	0.798	945	50	9.31	75.52	49.86	65.79	15.93
SHC flooding system	0.805	960	50	9.28	77.05	49.65	76.92	27.27

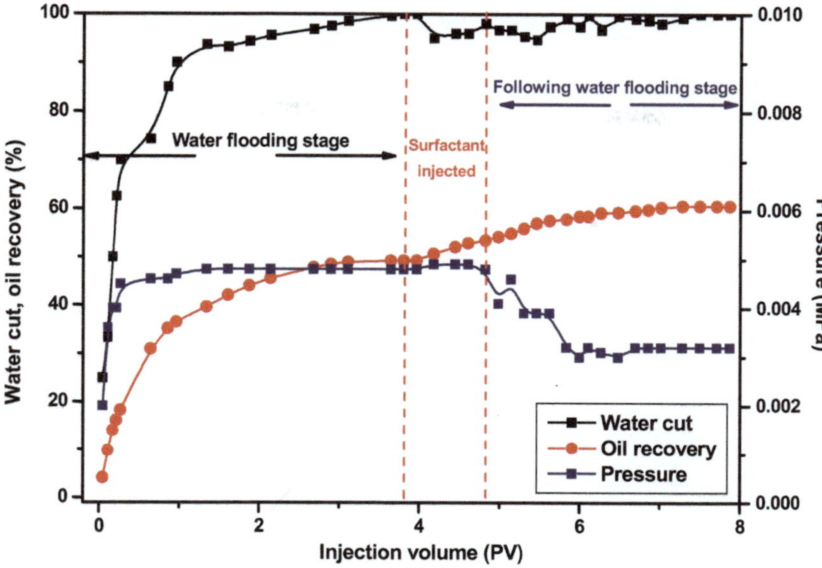

Figure 5.22 Displacement effect of surfactant system.

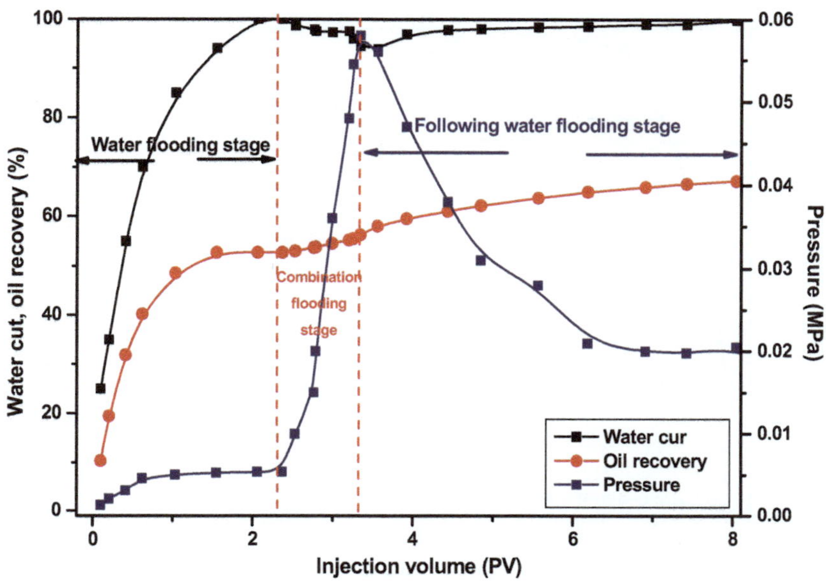

Figure 5.23 Displacement effect of polymer/surfactant combination flooding system.

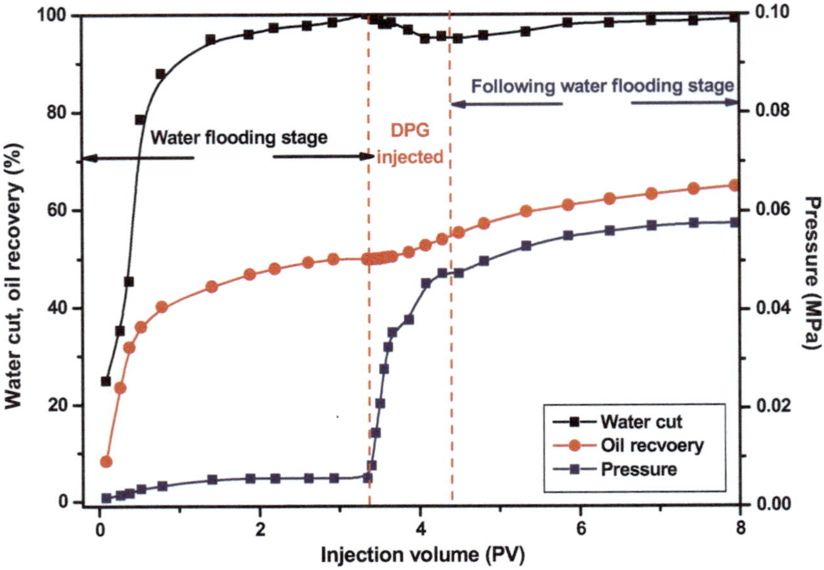

Figure 5.24 Displacement effect of DPG particles.

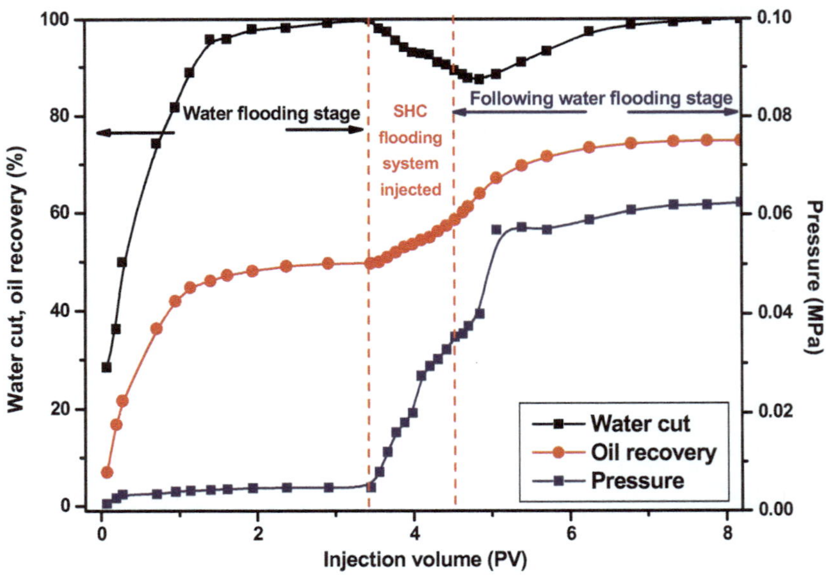

Figure 5.25 Displacement effect of SHC flooding system.

the upswept areas of the low- and medium-permeability zones, thereby mobilizing the low- and medium-permeability areas and displacing the remaining oil. For the surfactant flooding, since surfactant cannot change the water–oil mobility ratio of the following water flooding process, the injected surfactant protrudes in the core through the high-permeability part, which reduces the surfactant displacement effect and the recovery efficiency. The increase is mainly due to the contact with residual oil in the high-permeability zones, and the oil is stripped to increase the oil recovery. The surfactant injected in the following water flooding process cannot control the formation of the high-permeability zones effectively, so that the following injected water continues to intrude in the high-permeability zones, resulting in the injection pressure drop. For the polymer/surfactant combination flooding system, the pressure rises faster at the injection stage, and the surfactant/polymer combination flooding system has better mobility controllability. The pressure rises at this stage, mainly due to the viscosity of the polymer in the combination flooding system. During aging at high temperature and following water flooding stage, the injection pressure drops relatively fast. Due to the high temperature, the flow shear in formation and the polymer viscosity in the polymer/surfactant combination flooding system drop relatively large. The following mobility controllability is then limited. At the same time, the polymer of the polymer/surfactant combination flooding system will also be produced through high-permeability channels in the core. This helps reduce the viscosity of the combination flooding system. Therefore the following mobility controllability of the polymer/surfactant combination flooding system is weak. For the SHC flooding system and the DPG particles, the oil displacement effect is significantly higher than that of surfactant while the SHC flooding system has significantly higher oil displacement potential than the DPG particles. For the injection of DPG particles alone, the fluid profile can be effectively adjusted, but the remaining oil in the low-permeability zones can only be driven out in the following water flooding process. At this time there is still a considerable part of the remaining oil in the core. When injecting the SHC flooding system, the DPG particles in the system not only adjust the flow profile of conventional injection particles but also displace the remaining oil in the low-permeability zones. Besides, the surfactant in the SHC flooding system starts to contact with the remaining oil in the core. The remaining oil is stripped from the surface of the formation through emulsification,

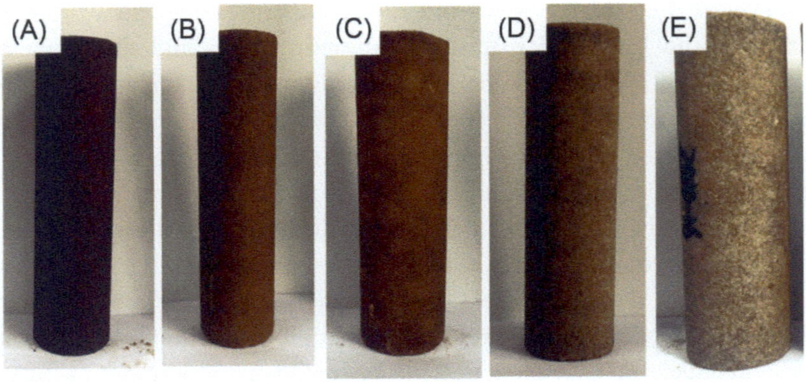

Figure 5.26 The core photos after injection of surfactant, polymer/surfactant combination flooding system, DPG particles, and SHC flooding system: (A) the initial state of core saturated with oil; (B) after surfactant injection; (C) after polymer/surfactant combination system; (D) after DPG particles; (E) after SHC flooding system.

wetting reversal, etc., so as to further improve the oil recovery. This is also confirmed by the experimental data of enhanced oil recovery in the following water flooding process of the SHC flooding system.

Fig. 5.26 shows the comparison of cores after displacement by surfactant, polymer/surfactant combination flooding system, DPG particles and SHC flooding system. The core color of the surfactant flooding is dark after the displacement, indicating that there is still more remaining oil in the core after the displacement. The color after the SHC flooding system is lighter, indicating that most of the remaining oil in the core has been displaced. This means good displacement potential of the SHC flooding system.

Section 5.4 Interaction mechanism of SHC flooding system

This section discusses the adsorption behavior of surfactants on the surface of DPG particles, the interfacial rheological behavior, and the force characteristics of the SHC flooding system. It also reveals the interaction mechanism, and provides a theoretical foundation for its application in oilfields.

5.4.1 The adsorption behavior of surfactants on the surface of DPG particles

The adsorption of surfactants on the surface of DPG particles can be regarded as a problem of the distribution of adsorbate molecules at the interface and continuous phase. Surfactant molecules can be adsorbed on the surface of DPG particles through electrical interaction, hydrogen bonding, hydrophobic interaction, etc., so as to form unique adsorption characteristics.

5.4.1.1 Principles, methods, and procedures of measurement

5.4.1.1.1 Measurement principle

The surfactant in the SHC flooding system is a sulfobetaine amphoteric surfactant. The adsorption behavior of the surfactant on the surface of the DPG particles in the combination flooding system can be determined by colorimetry. The principle of measurement [5,6] is that under the condition of pH = 1.0, betaine hydrochloride can react with Reinecke salt to form a red precipitate—a substance dissolvable in the 70% acetone solution—to form a pink solution. The reaction has a maximum absorption peak at 525 nm and it follows Beer's law when the betaine content is in the range of 0.1−12.5 mg/mL.

5.4.1.1.2 Preparation of chemicals

1. Preparation of betaine surfactant standard solution: add 0.15 g surfactant to 60 mL deionized water, stir evenly, and transfer to a 100 mL flask; use this solution to determine the standard absorbance curve of betaine surfactant;
2. Preparation of saturated Reinecke salt solution: weigh 1.5 g Reinecke salt into 90 mL of deionized water, adjust to pH = 1.0 with hydrochloric acid, stir at room temperature for 45 minutes, filter with suction to 100 mL. The solution must be prepared when needed;
3. Preparation of diethyl ether solution: add 1 mL of deionized water to 99 mL of anhydrous diethyl ether, stir well for later use;
4. Preparation of acetone solution: add 30 mL of deionized water to 70 mL of acetone and stir well for later use.

5.4.1.1.3 Measurement procedure

Determination of betaine surfactant standard curve: dilute the abovementioned standard solution of betaine surfactant to different mass concentrations, take 5 mL of the diluted standard solution, place it in the refrigerator (4°C) for 15 minutes, and add 1.5% 5 mL Reinecke salt in the refrigerator (4°C) for 3 hours, centrifuge at high speed for 15 minutes, take the precipitate, add 99% ether to rinse the precipitate, until the eluent is colorless, place the precipitate in a fume hood to evaporate until dry, and finally add 10 mL of 70% acetone to dissolve, and measure the absorbance of each solution at a wavelength of 525 nm. The reference solution is 70% acetone. Draw a standard curve of betaine surfactant with concentration and absorbance, as shown in Fig. 5.27. In the mass concentration range of 0%–0.12%, the absorbance has a linear relationship with the concentration, and the fitting accuracy reaches 87%. This result is consistent with Beer's law.

For the determination of surfactant adsorption on the surface of DPG particles in the SHC flooding system, a SHC flooding system (0.1% surfactant + 0.1% DPG particle) was prepared with high salinity simulated water and was aged at 90°C for different times. In order to avoid the influence of the DPG particles, the aged SHC flooding system solution was filtered with a 450-nm microporous membrane first, and then filtered with a 200-nm microporous membrane to obtain a clear solution. The clear solution is a sample to be tested, and it is determined according to colorimetry. When the surfactant concentration is too high, the SHC flooding system can be diluted appropriately and then measured. Then, the actual solution concentration can be calculated.

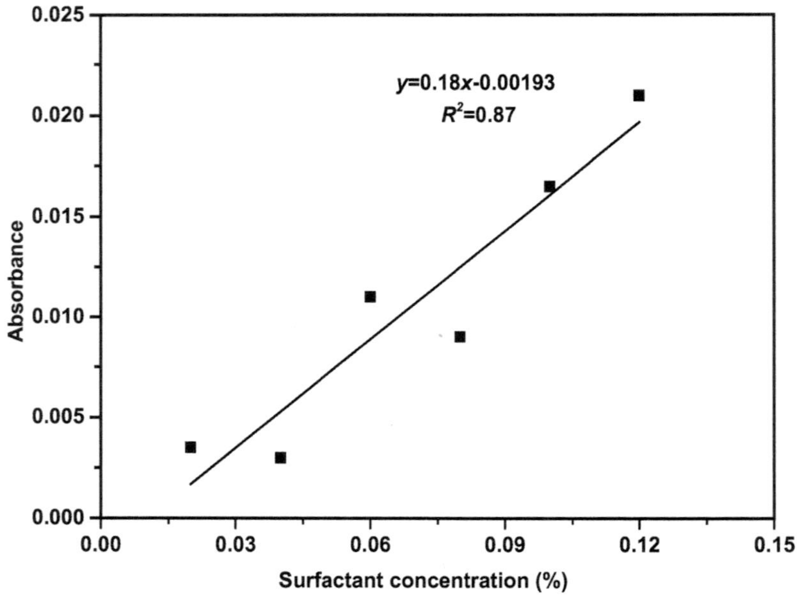

Figure 5.27 Standard curve of absorbance of betaine surfactant.

5.4.1.1.4 Calculation of static adsorption capacity

The adsorption behavior of surfactants on the surface of DPG particles under static conditions was introduced here. The calculation equation of adsorption capacity is shown in Eq. (5.1).

$$\tau = (C_0 - C) \times V/M \times 10^{-3} \tag{5.1}$$

where τ is static adsorption capacity (mg/g); V is volume of surfactant solution (mL); C_o is the initial concentration of the surfactant solution (mg/L); C is the equilibrium concentration of the surfactant solution (mg/L); and M is the mass of DPG particles in SHC flooding system (g).

5.4.1.2 The effect of adsorption time

The adsorption of surfactant on the surface of DPG particles is a dynamic equilibrium process. Fig. 5.28 shows the relationship between the adsorption amount of surfactant on the surface of DPG particles and the adsorption time.

It can be seen from Fig. 5.28 that the adsorption amount of surfactant on the surface of the DPG particles increases with the increase of adsorption time. When the adsorption time exceeds 2 days, the adsorption amount tends to balance and basically no longer changes. When the

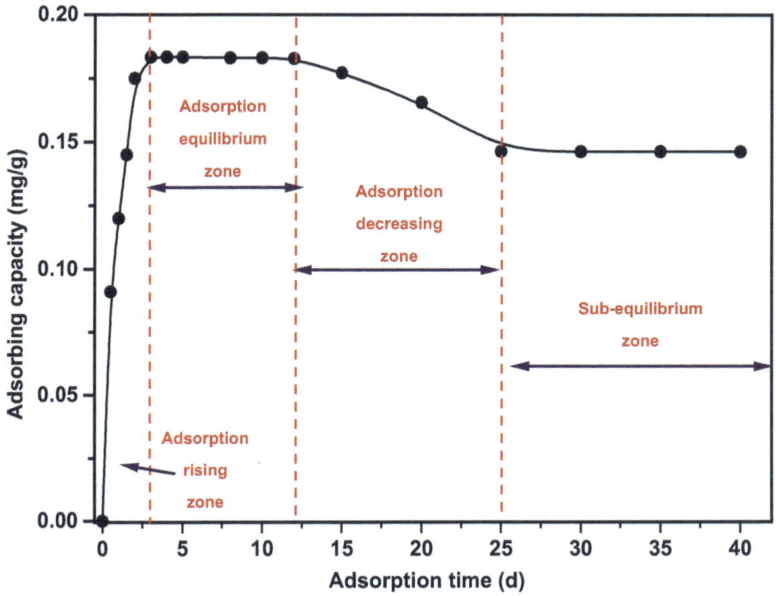

Figure 5.28 The effect of adsorption time on the adsorption behavior of surfactants.

adsorption time is increased further, the adsorption capacity of the surfactant decreases. After 30 days of aging, the adsorption capacity reaches equilibrium again and basically no longer changes. According to the trend of the adsorption curve, the adsorption behavior of the surfactant on the surface of the DPG particles can be divided into four areas: adsorption rising zone; adsorption equilibrium zone; adsorption decreasing zone, and subequilibrium zone.

5.4.1.2.1 Adsorption rising zone

The rapid increase in surfactant adsorption capacity is a distinctive feature of this zone. Since the surface of the surfactant and the DPG particles have the same charge, there are three main roles for adsorption: hydrogen bond adsorption formed between surfactants and particles; adsorption caused by dispersion force; and adsorption caused by hydrophobic association. In the initial stage, the first two forms of adsorption dominate. At this time, surfactant active molecules quickly occupy the adsorption sites on the surface of the DPG particles, resulting in a sharp increase in adsorption capacity. When the surfactant is further adsorbed on the surface of the DPG particles, the later interaction between the hydrophobic chains of the surfactant forms a surface aggregate, which further increases the adsorption of the surfactant molecules on the surface of the DPG particles. Under high-temperature conditions, the thermal movement of surfactant molecules intensifies, and this process will also be accompanied by the escape of adsorbed surfactant molecules from the surface of the DPG particles, that is desorption, but the adsorption dominates rather than desorption. Therefore the adsorption amount of surfactant in this interval increases sharply.

5.4.1.2.2 Adsorption equilibrium zone

The tangent of the slope in the adsorption curve in this region is approximately a horizontal straight line. This indicates that the adsorption and desorption of the surfactant on the particle surface reaches a dynamic equilibrium state. As the aging time increases, the particles of the DPG particles aggregate each other to reduce their specific surface area and reduce the effective adsorption area of the surfactant. This behavior leads to a decrease in the adsorption capacity; but at the same time, the surface of the DPG particles is electrically charged during the aging process. The decrease in performance reduces the electrostatic repulsion between the DPG particles and the surfactant, which is conducive to the adsorption of the surfactant on the surface of the DPG particles. This behavior further increases the adsorption amount of the surfactant on the surface of the particles. The adsorbed surfactant molecules make up for the decrease in the amount of surfactant active molecules adsorbed by the adsorption. Desorption and adsorption reach a dynamic equilibrium, and the adsorption capacity remains unchanged.

5.4.1.2.3 Adsorption decreasing zone

When the adsorption dynamic balance is maintained for a period of time, the adsorption amount of the surfactant on the surface of the DPG particles begins to decrease and the desorption dominates the process. Under the oil reservoir conditions, the positively charged salt ions neutralize the negatively charged DPG particles on the surface, so that the electrostatic repulsion between the DPG particles is reduced, which causes the DPG particles to aggregate with each other and reduce the effective adsorption area of surfactant molecules. Although betaine surfactants have a high salinity resistance, the chelation of some surfactant molecules with calcium ions reduces both the content of surfactant active molecules and the adsorption on the surface of DPG particles in the long-term aging process.

5.4.1.2.4 Subequilibrium zone

The adsorption capacity of surfactants in this area gradually stabilized, but still declined slowly, showing a state of subequilibrium. The adsorption behavior of the surfactant on the surface of the DPG particles in the SHC flooding system is a dynamic equilibrium process of adsorption and desorption. The adsorption behavior of the surfactant is not only affected by its own inherent properties, but also by the behavior of the solid particles and the properties of the dispersed phase in the SHC flooding system. The aggregation of solid particles, the thermal movement of surface-active molecules and the chelation of very few surfactant molecules with calcium and magnesium ions will reduce their adsorption on the surface of the DPG particles, causing the surfactant molecules to be released and re-dissolved in the dispersed phase. The desorption behavior of surfactant molecules on the surface of DPG particles is beneficial to SHC flooding. The slowly released surfactant molecules are readsorbed to the oil—water interface, reducing the oil—water interfacial tension, and then improving the displacement efficiency.

5.4.1.3 The effect of adsorption concentration

In order to further explore the adsorption behavior of surfactants on the surface of DPG particles, the adsorption type of surfactants was studied in the adsorption equilibrium zone. The mass concentration of DPG particles in the fixed SHC flooding system is 0.1%. The adsorption behavior of different concentrations of surfactants on the surface of the particles is studied. The adsorption equilibrium time is 3 days at 90°C. The results are shown in Fig. 5.29.

It can be seen from Fig. 5.29 that the adsorption amount of surfactant on the surface of the DPG particles increases with the increase of the surfactant concentration. When the surfactant concentration reaches 0.3%, the adsorption capacity basically no longer increases, and saturated adsorption is achieved. From the change of isotherm surfactant adsorption curve, the adsorption of surfactant on the surface of DPG particles is a typical Langmuir adsorption. When the surfactant concentration is low, the surfactant active molecules are adsorbed on the surface of the DPG particles through hydrogen bonding, hydrophobic association, dispersion force, etc., forming a single-layer adsorption film on the surface of the particles. At this time, there are still more adsorption sites on the surface of the DPG particles, and the adsorption amount increases sharply with the increase of the surfactant concentration. When the surfactant increases

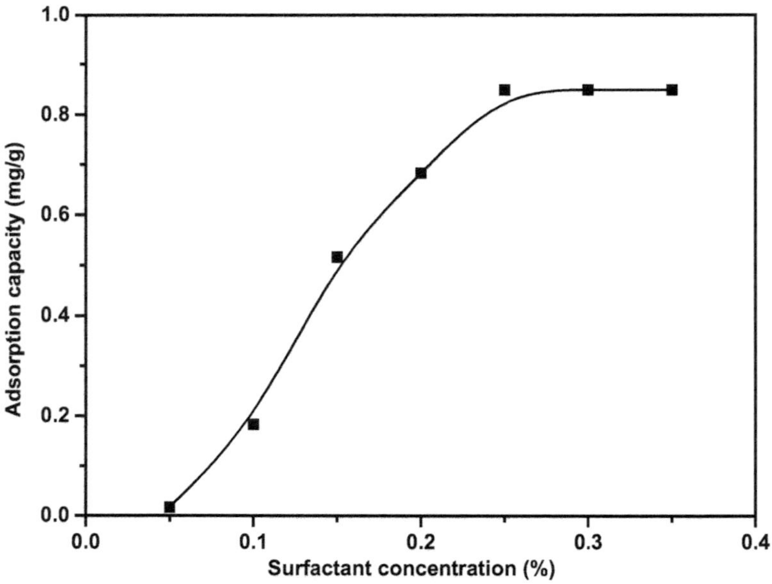

Figure 5.29 Isotherm adsorption curve of surfactant.

to a certain concentration, on the one hand, the active molecules in the surfactant occupy the unoccupied adsorption sites on the surface of the DPG particles. On the other hand, the active molecules in the surfactant are adsorbed on the surface of the DPG particles through the interaction between the hydrophobic chains to form a multilayer adsorption, which increases the adsorption capacity, and the adsorption capacity increases slowly at this time. However, the thermal motion of the surfactant molecules will also result in escaping from the adsorption surface of the DPG particles, forming a dynamic adsorption—desorption process. When the surfactant concentration exceeds 0.3%, the adsorption of the surfactant on the surface of the DPG particles reaches a dynamic equilibrium, and the adsorption amount no longer increases. The concentration at this time is the equilibrium adsorption concentration, and the corresponding adsorption amount is the saturated adsorption amount. At this concentration, the adsorption and desorption of surfactant molecules on the surface of the DPG particles reach a dynamic equilibrium, and the amount of adsorption basically no longer increases. The adsorption behavior of surfactants on the surface of DPG particles improves the ability of the SHC flooding system to reduce the oil—water interfacial tension, and makes the DPG particles have the ability to displace oil. In addition, the adsorption of surfactants on the surface of the DPG particles will also enhance the negative charge of the SHC flooding system. Since most of the reservoir surface is negatively charged, when the negatively charged SHC flooding system flows in the porous media, it is not easy to adsorb near the well zone due to the electrostatic repulsion. This is beneficial to the SHC flooding system transport deep into the formation.

5.4.2 Interface rheological behavior of the SHC flooding system

The interface expansion rheological method [7—9] is an important method to study the molecular interface behavior of the adsorption layer. The molecules adsorbed on the interface layer form an adsorption film that can change the microscopic properties of the interface layer. When the interface layer is disturbed, it will produce a tension response. By analyzing the process of restoring the balance of the interface layer, the interaction between the molecules of the interface layer and the aggregation morphology of molecules in the interface layer, etc. are studied. In this section, the interfacial rheological behavior of the SHC flooding system was discussed with an interfacial rheometer.

5.4.2.1 Theoretical foundation of interface rheological behavior

When the interface is periodically compressed and expanded, the interfacial tension will also change periodically. That is the expansion modulus will be produced. The expansion modulus can be used to characterize the strength of the interface film. Generally speaking, the larger the expansion modulus, the stronger the role of molecules adsorbed on the interface layer, and the higher the strength of the formed interface film. The expansion modulus is defined as the ratio of the change in interfacial tension to the change in the relative interface area, as in Eq. (5.2),

$$\varepsilon = (d\,\gamma)/(d\ln A) \tag{5.2}$$

where ε is expansion modulus; γ is interfacial tension; and A is interface area.

For a viscoelastic interface, there is a certain phase θ between the periodic change of the interfacial tension and the periodic change of the interface area, that is the phase angle of the expansion modulus. Therefore the expansion modulus can also be expressed in a plural form, as in Eqs. (5.3)–(5.5),

$$\varepsilon = \varepsilon d + i\,\omega\,\eta d \tag{5.3}$$

$$\varepsilon' = |\varepsilon|\cos\theta \tag{5.4}$$

$$\varepsilon'' = |\varepsilon|\sin\theta \tag{5.5}$$

where ε' is storage modulus, representing the contribution of the elastic part of the viscoelastic interface; and ε'' is loss modulus, characterizing the contribution of the viscous part to the expansion modulus.

The attention expansion interface rheometer produced by Bioolin Technology Co., Ltd. of Sweden was used to determine the interface rheology of the SHC flooding system. Through the periodic disturbance of the suspended droplets, the droplet size and interface area were changed. The interfacial tension response was measured by the drop shape analysis, and the interface properties of the combination flooding system were analyzed. The specific measurement steps are as follows. First, a syringe was used to add the sample to be tested into the sample tube and installed in the measurement chamber. Then, the crude oil was added to the U-shaped needle tube microinjector and immersed in the sample tube to form 2.5 μL oil droplets at the bottom of the U-shaped needle tube. Third, the parameters were adjusted, the measurement started, and the changes of expansion modulus recorded. In order to maintain the accuracy, a Sine disturbance of 0.5 Hz is applied to the droplet after its formation 20 seconds later. The expansion deformation ($\Delta A/A$, where ΔA is the deformation area and A is the interface area) was 10%, and all tests were conducted under the condition of 30.0°C ± 0.1°C.

5.4.2.2 Influencing factors of dynamic interfacial tension

5.4.2.2.1 The effect of surfactant concentration on the dynamic surface tension

The concentration of the DPG particles was fixed at 0.1%. The influence of the surfactant concentration on the dynamic interfacial tension of the SHC flooding system was investigated. The oscillation frequency was 0.5 Hz. The results are shown in Fig. 5.30.

It can be seen from Fig. 5.30 that as time increases, the interfacial tension of the SHC flooding system decreases gradually until it reaches a dynamic equilibrium. In the concentration range tested, the higher the concentration of surfactant in the SHC flooding system, the lower the interfacial tension when the dynamic equilibrium is reached, and the shorter the time required. The adsorption of the surfactant at the interface in the SHC flooding system involves two processes: diffusion control and energy barrier control. The diffusion control process makes the surfactant molecules in the combination

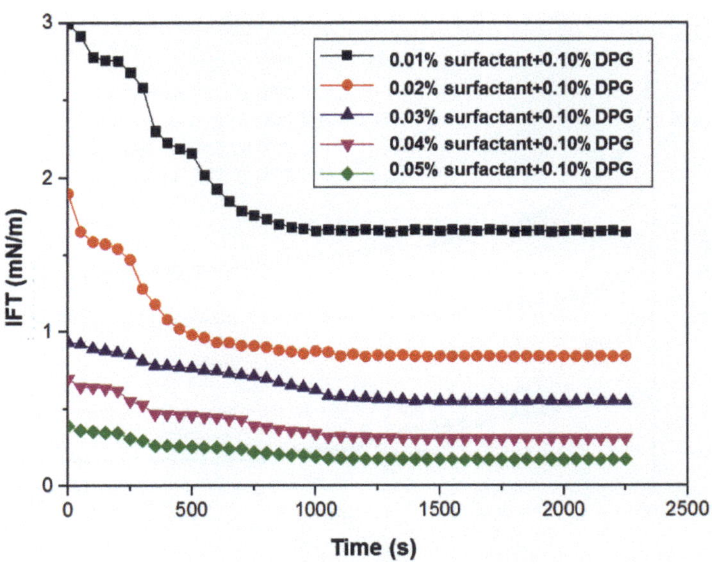

Figure 5.30 The change of surfactant concentration on the dynamic interfacial tension of the SHC flooding system.

flooding system diffuse from the bulk to the interface and adsorb on the interface layer, resulting in a continuous decrease in the interfacial tension, and finally reaching the dynamic equilibrium value of adsorption and desorption, and the dynamic interfacial tension no longer changes. The energy barrier control process is related to the behavior of active molecules adsorbed on the interface layer, involving changes in orientation, rearrangement, and aggregation of active molecules, but the number of molecules has not changed. Therefore the dynamic interfacial tension is maintained at a stable value.

5.4.2.2.2 The influence of DPG particles concentration on the dynamic interfacial tension

The surfactant concentration was fixed at 0.02%. The influence of the concentration of the DPG particles on the dynamic interfacial tension of the SHC flooding system was investigated. The oscillation frequency was fixed at 0.5 Hz. The results are shown in Fig. 5.31.

It can be seen from Fig. 5.31 that the stable interfacial tension of the SHC flooding system formed by the high-concentration DPG particles is significantly lower than that of the same kind of system formed by the low-concentration DPG particles. Since the SHC flooding system is a viscoelastic system, the viscoelasticity of the combination flooding system increases as

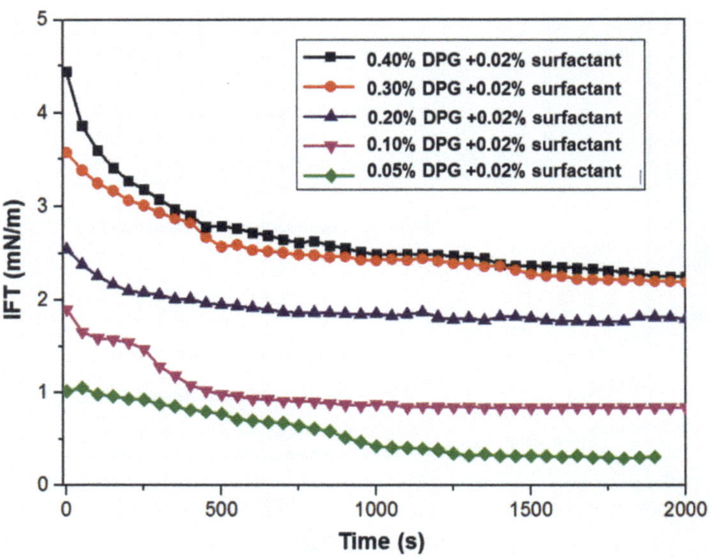

Figure 5.31 The effect of DPG particle gels concentration on the dynamic interfacial tension.

the concentration increases. This reduces the diffusion rate of surfactant molecules to the interface layer. Also, a higher concentration of DPG particles brings about a slower diffusion rate, resulting in higher interfacial tension of the combination flooding system. In addition, some surfactant molecules are affected by the hydrophobic interaction and adsorbed on the surface of the DPG particles. This not only reduces the number of active molecules in the interfacial layer but also leads to higher interfacial tension of the SHC flooding system formed by the high-concentration DPG particles.

5.4.2.2.3 The effect of aging time on the dynamic interfacial tension

Since the SHC flooding system is a long-term flooding process in the formation, the reservoir conditions have a certain impact on its interface properties. Therefore the dynamic interfacial tension of the SHC flooding system (0.1% DPG + 0.02% surfactant) under different aging times was investigated, at an oscillation frequency of 0.5 Hz. The results are shown in Fig. 5.32.

It can be seen from Fig. 5.32 that as the aging time increases, the equilibrium interfacial tension of the SHC flooding system decreases gradually. After 25 days of aging, the dynamic interfacial tension does not change basically. With continuous aging comes a decrease in the number of surfactant molecules, which leads to a decrease in surfactant molecules

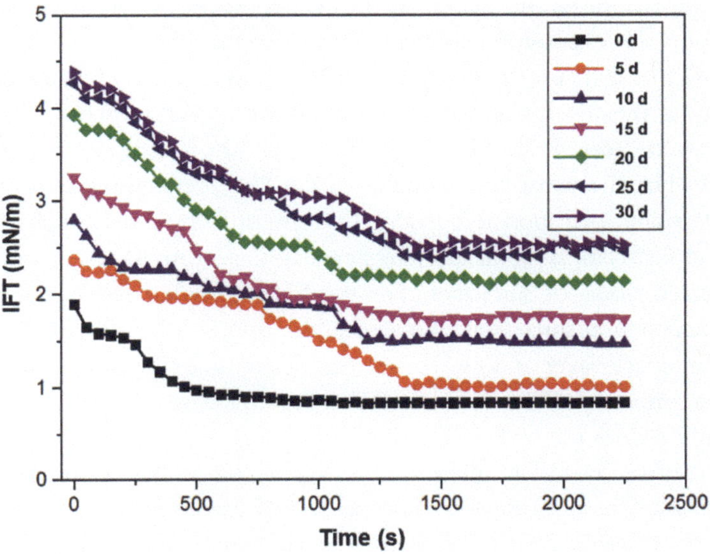

Figure 5.32 The effect of aging time on the dynamic interfacial tension.

adsorbed on the interface layer and to an increase in the interfacial tension of the system. When the aggregates exceed a certain size, affected by the difference in gravity, the aggregates gather at the bottom of the solution. This reduces the concentration of the internal system components in the liquid phase and thus triggers the desorption of molecules adsorbed on the interface layer. Under the impact of the concentration difference of the system substances, the molecules of the interface layer migrate gradually from the interface layer to the inside of the liquid phase, weakening the interaction between the molecules of the interface layer and reduces the expansion modulus. The results show that the aging is a process throughout which the molecules in the interface layer migrate from the surface to the interior of the liquid phase, and the interaction between the molecules in the interface layer changes from strong to weak.

5.4.2.3 Influencing factors of expansion modulus of SHC flooding system

5.4.2.3.1 The effect of surfactant concentration on the expansion modulus

The adsorption film formed on the interface layer of the SHC flooding system is a viscoelastic film. Thus there is always diffusion and exchange of adsorbed molecules between the interface and the bulk phase. The weaker the molecular diffusion exchange effect, the stronger the elasticity of the surface film, and the smaller the influence of the oscillation frequency on the expansion modulus. Therefore this section uses the amplitude of the expansion modulus with frequency to characterize the elasticity of the adsorption film. The results are shown in Fig. 5.33.

It can be seen from Fig. 5.33 that when the oscillation frequency is 0.01–0.1 Hz, the expansion modulus of the SHC flooding system increases with the rising oscillation frequency. The system formed by a high concentration of surfactant is significantly higher than that of the system formed by a low concentration of surfactant. For the SHC flooding system formed by low-concentration surfactants, the adsorption amount of surfactant molecules in the interface layer is low, so are the adsorption film strength and the expansion modulus. When the adsorption time increases, the surfactant molecules diffuse from the bulk to the surface interface and adsorb a large amount in the interface layer. The intermolecular interaction increases and the expansion modulus increases rapidly. The process of expansion modulus is also a process from weak to strong molecular interaction in the surface adsorption layer. For the SHC flooding system formed by a high concentration of

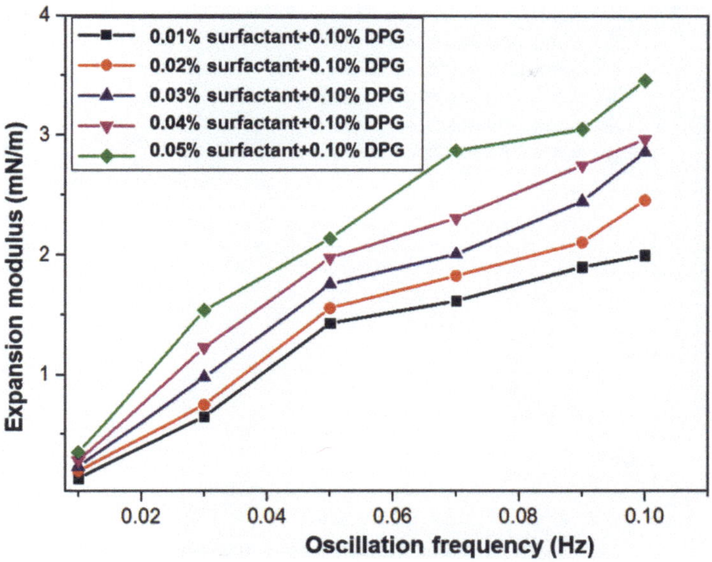

Figure 5.33 Effect of surfactant concentration on expansion modulus.

surfactant, the surfactant molecules form aggregates after the interface layer reaches saturation adsorption. The interaction force between the aggregates is weak, and the expansion modulus is relatively low. When the oscillation continues for a period of time, there may be three kinds of aggregates in the interface layer—aggregates between surfactant molecules, aggregates formed by hydrophobic interaction between surfactant molecules and DPG particles, and aggregates formed between DPG particles. The interaction between these three types of aggregates is higher than that between the aggregates formed by a SHC flooding system with a low concentration of surfactant. An increase is thus present in the expansion modulus.

5.4.2.3.2 The influence of DPG particles concentration on the expansion modulus

Fig. 5.34 shows the trend of the expansion modulus of the SHC flooding system with the oscillation frequency under different DPG particles concentrations. As the oscillation frequency increases, the DPG particles concentration rises, and the expansion modulus of the SHC flooding system drops down. For the SHC flooding system formed by low-concentration DPG particles, most of the surfactant molecules are adsorbed on the interface layer, enhancing the interface activity and increasing the expansion modulus of the interface layer. However, when the concentration of the

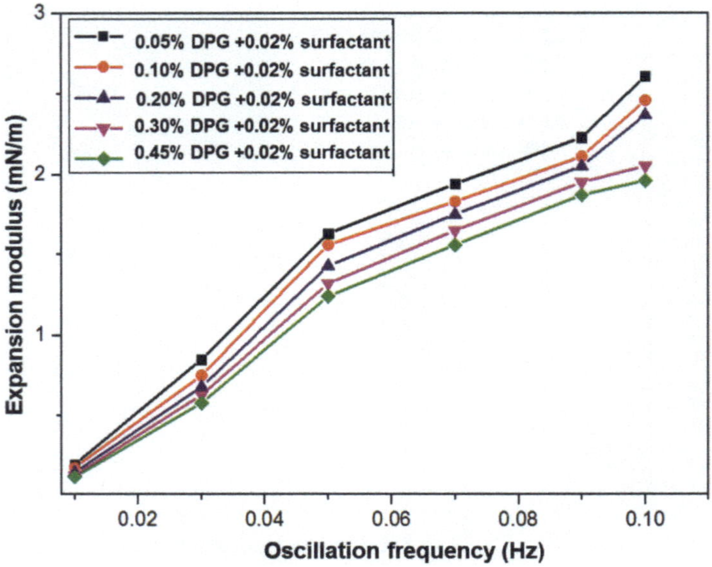

Figure 5.34 The effect of DPG particles concentration on the expansion modulus.

DPG particles increases, the surfactant is adsorbed on the surface of the particles by hydrophobic interaction, reducing adsorption on the interface layer. In addition, the small number of hydrophobic groups in the DPG particles in the combination flooding system makes it adsorbed on the oil–water interface. As the concentration of the DPG particles in the bulk phase increases, the exchange of particles between the interface and the bulk phase increases. The particle aggregates at the interface occupy the adsorption sites of the interface layer, further reducing the adsorption of surfactant molecules on the interface layer. Therefore the expansion modulus of the SHC flooding system decreases as the concentration of the DPG particles increases.

5.4.2.3.3 The effect of aging time on the dynamic surface expansion modulus

When the SHC flooding system is aged, the molecular orientation, rearrangement, and aggregation of the interface adsorption layer may affect the properties of the interface layer, and then affect the behavior of the molecules in the adsorption layer and the state of the interface film. Therefore the effect of aging time on the expansion modulus of the SHC flooding system (0.1% DPG + 0.02% THSB surfactant) can be measured. The results are shown in Fig. 5.35.

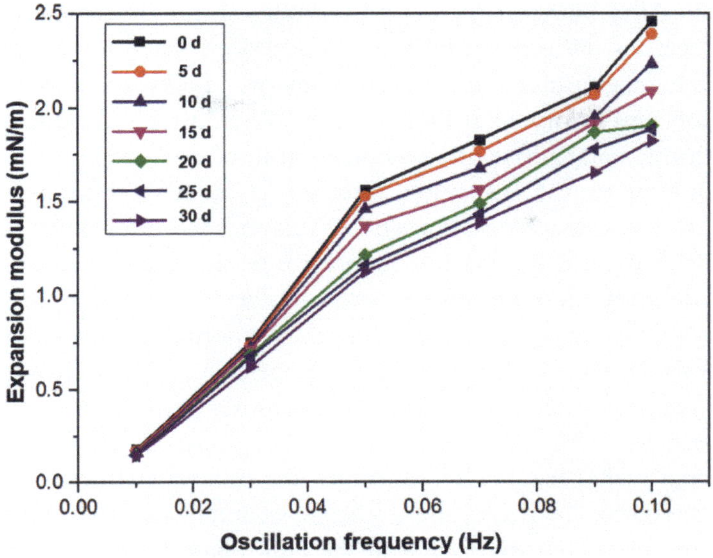

Figure 5.35 The effect of aging time on the expansion modulus.

It can be seen from Fig. 5.35 that as the aging time increases, the expansion modulus of the SHC flooding system decreases gradually. After 25 days of aging, the expansion modulus basically does not change. The SHC flooding system is uniformly dispersed in the solution before aging and forms an adsorption film on the interface layer. Hence, the expansion modulus is higher. After continuous aging, the activity of surfactant molecules in the combination flooding system decreases and desorbs from the interface layer. This helps reduce the molecular activity of the interface layer and results in a decrease in expansion modulus. In addition, the intensification of the thermal movement of DPG adsorbing surfactant molecules brings about an aggregation of particles with each other and a decrease in the concentration of the internal system components in the liquid phase. At the same time, the molecules of the interface layer migrate gradually from the interface layer to the inside of the liquid phase; and the molecules adsorption on interface layer is reduced. Also, the interaction between the molecules of the interface layer reduces and the expansion modulus decreases.

5.4.2.4 Discussion on the interface behavior of the SHC flooding system

Based on the results from the interface expansion rheometer, the interface adsorption behavior of the SHC flooding system is shown in

Fig. 5.36. After being in contact with crude oil, the SHC flooding system undergoes diffusion, preferential adsorption, macromolecular rearrangement, aggregation, and other behaviors. Together, a viscoelastic adsorption film is formed at the interface. When the oil phase is in contact with the water phase, the combination flooding system in the water phase diffuses to the oil–water interface layer. The diffusion rate is related to viscosity, substance concentration, and molecular weight. From the analysis of the dynamic interfacial tension and expansion modulus of the SHC flooding system, small-molecule surfactants are preferentially enriched at the oil–water interface to form an active adsorption film. Affected by hydrogen bonds, dispersion forces, hydrophobic groups, etc., surfactant molecules are also adsorbed on the surface of the DPG particles. This, however, changes the properties of the particles' surface and makes it easier to adsorb on the oil–water interface. When the concentration increases, the exchange of surfactants and DPG particles in the oil–water interface with the bulk phase increases. When the exchange reaches a dynamic equilibrium at the interface, a viscoelastic interface film is formed as well. In this equilibrium state, the SHC flooding system rearranges in the interface layer to form three kinds of aggregates: aggregates between surfactant molecules, which are formed by hydrophobic interaction between surfactant molecules and DPG particles, and aggregates formed between DPG particles. The interaction between these three kinds of aggregates is higher than that of a single surfactant molecule or a DPG particle aggregate. This phenomenon is expressed as an increase in the expansion modulus of the interface layer. The dynamic interfacial tension and expansion modulus of the SHC flooding system reflect the microscopic changes of the interaction force between the combination flooding system from weak to strong. When the interface concentration of the combination flooding system is low,

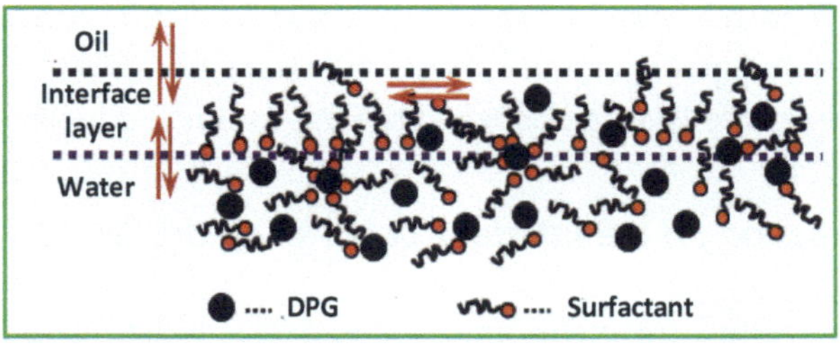

Figure 5.36 Schematic diagram of interface adsorption behavior.

the diffusion process between the bulk and the interface layer is in a dominant position. When the interface concentration is high, on the contrary, the rearrangement of macromolecules dominates. The adsorption film formed in the interfacial layer of the SHC flooding system helps reduce the oil–water interfacial tension and enhances the stability of the emulsion formed. This increases the sweep efficiency and the oil displacement efficiency of the combination flooding system.

5.4.2.5 Research on the force characteristics of the SHC flooding system

The force characteristics are mainly analyzed from aspects including van der Waals force (dispersion force, induction force, orientation force), hydrophobic interaction, electrostatic force, and hydrogen bond, to analyze the interaction characteristics between the DPG particles/SHC flooding system and core surfaces with different wettability, and between the DPG particles and the SHC flooding systems. Researchers mainly use rheometer, atomic force microscope, and surface force apparatuses to study the characteristics of the force. By measuring the stress curve by rheometer, the coexistence state of particles in the solution can be inferred. However, this is generally applicable to non-Newtonian fluids. Atomic force microscopy is generally suitable for measuring the characteristics of the force of particles with a particle size of 10–50 μm, and the surface of the particles is required to be smooth. The surface force apparatus [10–13] is a device for measuring the force with relatively high accuracy. The roughness of the sample on the mica is required to be less than 5 nm. In our research, a self-built surface force apparatus (SFA) was used to directly measure the interaction force of the above system. With the help of an analysis of the functional group of the SHC flooding system, we clarified the types of force between the SHC flooding system. As the surface of the formation rock is negatively charged and the surface of the formation is rough, mica was used to replace the surface of the formation rock to study the types of interaction forces between SHC flooding systems.

5.4.2.5.1 Principle of force measurement

SFA is mainly composed by a spectrometer, microscope objective lens, piezoelectric crystal, a spring adjustment rod, a force spring, an upper and lower adjustment rod, the main support frame, rigid double cantilever spring, light source, and a motor. The principle of SFA measurement is shown in Fig. 5.37. The specific measurement principle is as follows. When two sample surfaces approach to a certain range, a force comes into being between them.

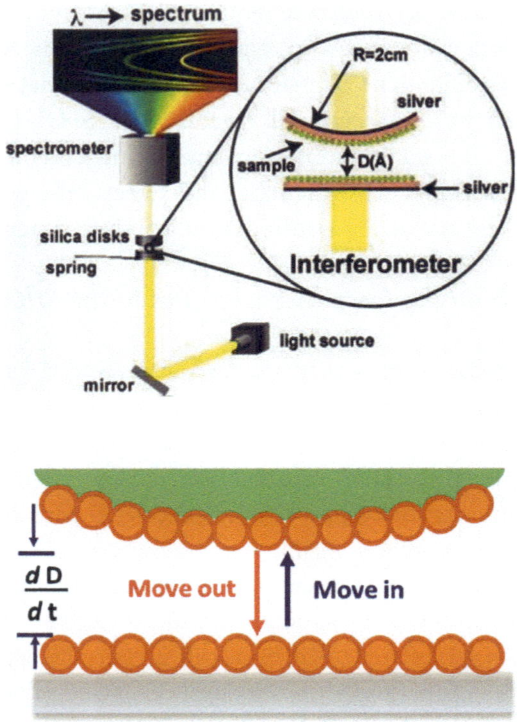

Figure 5.37 Schematic of measuring principle of surface force apparatus [14].

When two sample surfaces separate, there also comes a force between them, resulting in the deformation of the force measuring spring. The degree of deformation is a reflection of the force. During the measurement, the light source passes vertically through the sample surface, and light interference occurs on the sample surface between the silver-plated plates. The distance between the two sample surfaces can be obtained by analyzing the interference fringes. The difference between the position change of the sample caused by piezoelectric crystals and the actual change is the measurement of the spring deformation. The molecular force between the sample surface can be obtained by calibrating the spring elastic coefficient. The resolution of SFA to sample distance is 0.1 nm, and the force detection sensitivity can reach up to 10^{-8} N.

5.4.2.5.2 Design of force measurement scheme

Since the DPG particles are prepared by mechanical shearing of the bulk gel and this does not involve any chemical reaction, the chemical composition of the DPG particles is consistent with that of the bulk gel. In view of the requirements of SFA force measurement, in our study, interactions between the bulk gel are measured to analyze the force characteristics of DPG particles. By adding surfactant into the bulk gel precursor solution, the SHC flooding system is prepared and the force characteristics of the system are analyzed. Two schemes of experimental design are used to measure the interaction force between DPG particles and SHC flooding system. The specific test plans are as follows.

Scheme 1: for the interaction between the DPG particle system, on the one hand, the interaction between mica surfaces adsorbed with gel film and mica surfaces with different wettability are measured. On the other hand, the interactions between mica surfaces adsorbed with gel film are measured.

Scheme 2: for the interaction between SHC flooding system, on the one hand, the interaction between mica surfaces adsorbed with the SHC flooding system and the mica with different wettability are measured. On the other hand, the interactions between mica surfaces adsorbed with the SHC flooding system are measured.

Through the above two test schemes, the structural properties of the bulk gel and SHC flooding system were combined to clarify the force characteristics. Mica surfaces were modified with octadecyltrichlorosilane (OTS) to obtain three kinds of wettability: strong hydrophilic, medium hydrophilic, and weak oleophilic. The surface state of the core under different oil flooding stages was thus simulated. The contact angles of the three kinds of mica are shown in Fig. 5.38.

5.4.2.5.3 Molecular structural analysis of SHC flooding system

1. *Analysis of molecular structural characteristic of mica sheets*

   Since both the bulk gel and the SHC flooding system can adsorb on the mica surface, the properties of the mica surface directly determine the adsorption capacity of the bulk gel and the SHC flooding system. Mica is the general name of the hydrated aluminosilicate group minerals

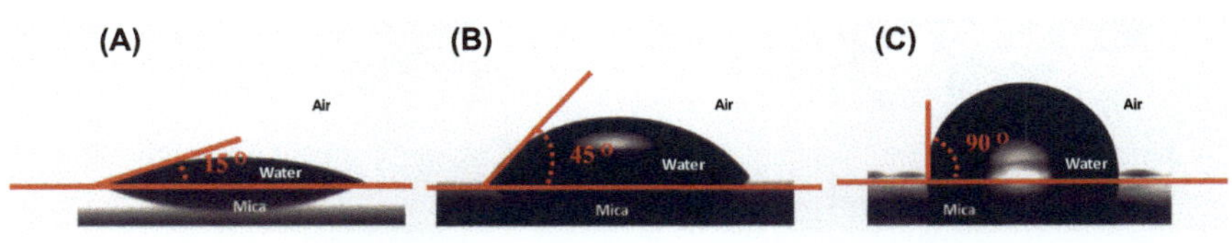

Figure 5.38 The wettability of mica surface: (A) strong hydrophilic, 15 degrees; (B) medium hydrophilic, 45 degrees; (C) weak oleophilic, 90 degrees.

containing lithium, sodium, potassium, magnesium, and other metallic elements with a layered structure. It is a complex silicate with a formula of $Al_2K_2O_6Si$. Mica refers to a combination silica layer composed of two layers of silicon–oxygen tetrahedra sandwiching a layer of aluminum oxide octahedra. The crystal structure of mica is shown in Fig. 5.39. Mica, besides, is a lamellar structure, with potassium ions bridging the adjacent hydroxyl layer and siloxane ion layer. When being peeled off, the mica layer breaks off at the point of the potassium ion. The newly stripped mica is negatively charged due to the excess hydroxyl groups.

2. *Analysis of molecular structural characteristics of bulk gel and SHC flooding system*

In order to analyze the characteristics of the forces between the SHC flooding systems, the structural characteristics of the systems were first investigated by infrared analysis. The chemical composition of the DPG particles was found consistent with the bulk gel in that the DPG particles were prepared by physical shear of the bulk gel, with no chemical reaction involved. Therefore its structure is characterized by measuring the structure of the bulk gel. In order to clarify the characteristics of the interaction between the SHC flooding systems, experiments were carried out to determine the force between the bulk gel and the bulk gel after adding surfactant.

In terms of the bulk gel system, the finished bulk gel (0.3% polymer + 0.6% cross-linking agent) was vacuum-dried at −20°C for 96 hours. Then it was ground and crushed and stored in a desiccator for later use.

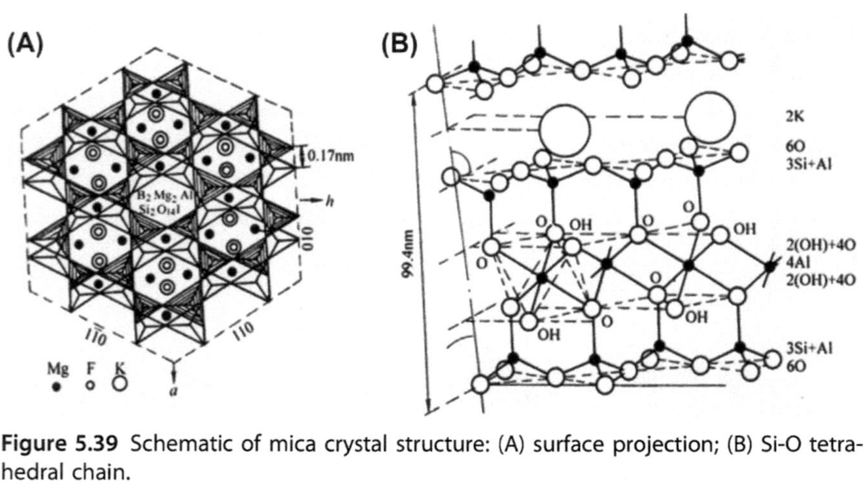

Figure 5.39 Schematic of mica crystal structure: (A) surface projection; (B) Si-O tetrahedral chain.

In terms of the SHC flooding system, 0.1% surfactant was added to the gelation solution (0.3% polymer + 0.6% cross-linker). After gelation, the system was dried in vacuum at −20°C for 96 hours. Then, the system was stored in a desiccator, after being ground and crushed, for later use.

The infrared tests of the bulk gel and the SHC flooding system were carried out. The results are shown in Fig. 5.40. It can be seen that the vibration absorption peak of the two systems concerned at 3348 cm^{-1} is the characteristic −OH infrared absorption peak. Comparison implies a shift of the infrared characteristic absorption peak of the bulk gel from the wave number of 1006−1020 cm^{-1}. This is the characteristic absorption band of the sulfonic acid group. The characteristic absorption peak at the wave number of 2944 cm^{-1} moves to 2922 cm^{-1}, which is attributed to the vibration absorption peak of the carbon chain −CH. Since it contains hydroxyl groups, the bulk gel surface is negatively charged. When the bulk gel is added with a surfactant to form a SHC flooding system, the structure is basically unchanged. The surfactant is a negatively charged system, which strengthens the electronegativity of the SHC flooding system.

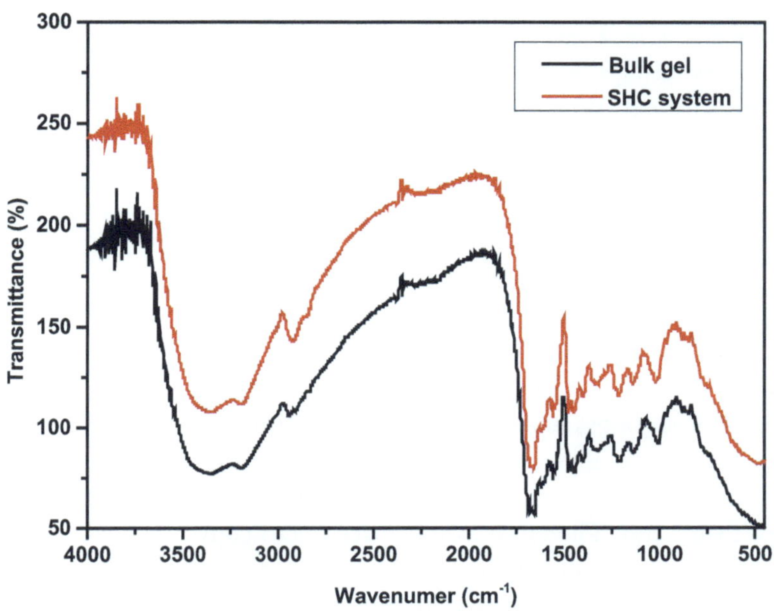

Figure 5.40 Infrared spectrum of bulk gel and SHC flooding system.

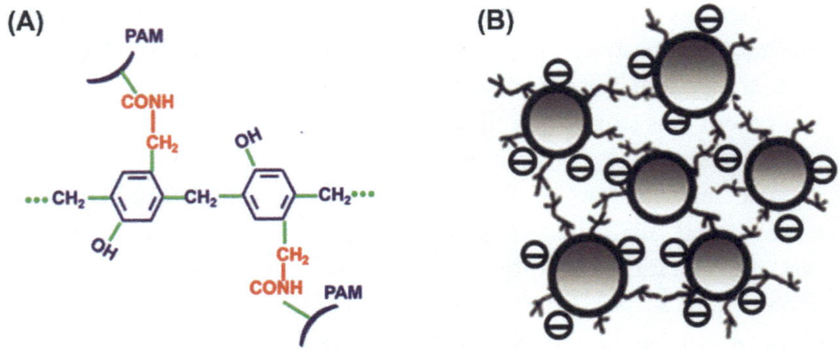

Figure 5.41 Structure analysis of bulk gel: (A) molecular structure of bulk gel; (B) structure schematic diagram of bulk gel.

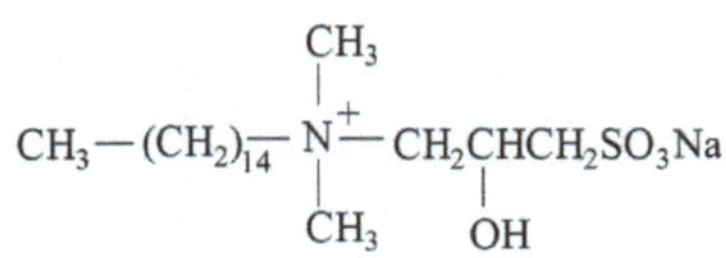

Figure 5.42 Molecular structure formula of sulfobetaine surfactant.

The structure schematic diagram of bulk gel is shown in Fig. 5.41. It can be seen that during the formation of bulk gel, the amide group of polymers reacts with the carboxyl group of phenolic resin cross-linker to form a bulk gel with a dense structure. There were redundant hydroxyl groups on the surface of the bulk gel, leaving the surface negatively charged. When betaine surfactant (Fig. 5.42) is added, the surfactant molecules with strong negative charge are adsorbed on the surface of bulk gel, thereby enhancing the surface electronegativity. Zeta potential test results also confirm the negative charge of both the DPG particles and the SHC flooding system. The addition of surfactants possessing strong electronegativity enhances the electronegativity of the SHC flooding system.

5.4.2.5.4 Roughness analysis of adsorption film in SHC flooding system

SFA surface force apparatus was used to analyze the force characteristics. The roughness of the film adsorbed on the silver-coated mica sheet should be less than 5 nm. Therefore atomic force microscopy was used to analyze the surface roughness of the bulk gel before and after the addition of surfactants. At 60°C, mica sheets were immersed in the bulk gelling solution (0.3% polymer + 0.6% cross-linker) and the SHC flooding system

(0.3% polymer + 0.6% cross-linker + 0.1% surfactant), respectively, for even adsorption. After being aged for 12 and 24 hours, respectively, the mica sheets were taken out and dried with nitrogen for testing. The test results are shown in Table 5.5 and Fig. 5.43.

Table 5.5 Roughness of the adsorption film formed by the SHC flooding system and DPG particles.

No.	System	Temperature (°C)	Reaction time (h)	Scanning range	Roughness (nm)
A	DPG	25	10	500 × 500 nm	0.43
B	DPG	60	10	500 × 500 nm	0.19
C	SHC flooding system	25	10	1 × 1 μm	0.61
D	SHC flooding system	60	10	1 × 1 μm	0.22

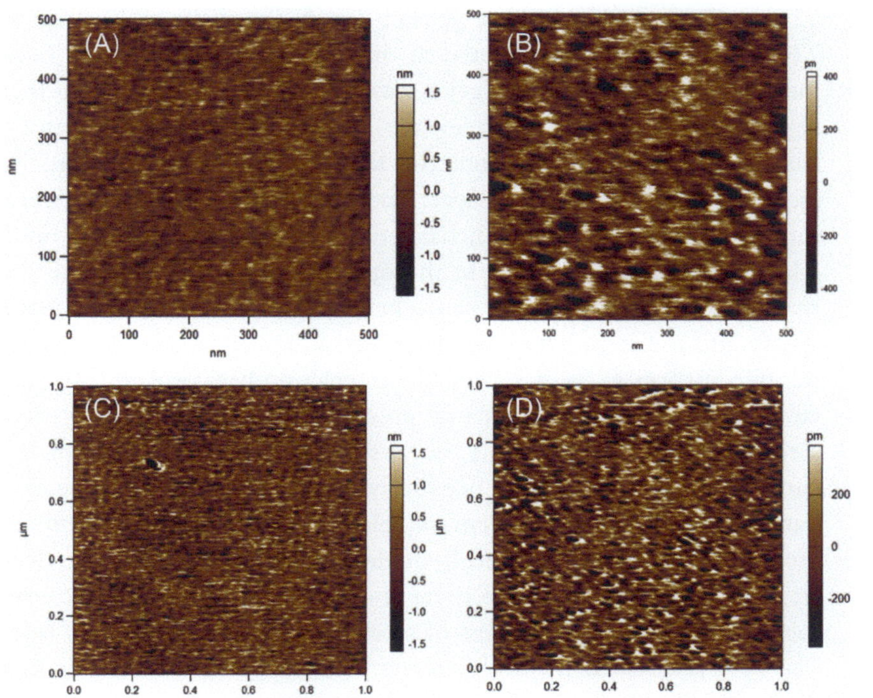

Figure 5.43 Microstructure of adsorption film of bulk gel and SHC flooding system: (A) DPG particles, 25°C; (B) DPG particles, 60°C; (C) SHC flooding system, 25°C; (D) SHC flooding system, 60°C.

It can be seen from Table 5.5 and Fig. 5.43 that both the DPG particles and the SHC flooding system can be well adsorbed on the mica surfaces. The longer the adsorption time, the thicker and rougher the adsorption films. However, the roughness of the adsorption films formed by the two systems is less than 5 nm. This indicates the availability of the adsorption films prepared by this method to measure the force. Considering such factors as easy oxidation of silver-coated mica sheet, and thickness and roughness of adsorption films, the adsorption time of the two systems was set as 24 hours and the reaction temperature at 60°C.

5.4.2.5.5 Analysis of interaction force characteristics of DPG particles

1. *Interaction force analysis between DPG and strong hydrophilic surface*

   The interaction force between DPG and the strong hydrophilic mica surface was measured under three different salt concentrations. The results are shown in Fig. 5.44.

   It can be seen from Fig. 5.44A that when the mica sheet adsorbed with DPG particles slowly approaches the strong hydrophilic mica surface driven by the motor, there is no force at a distance larger than 72 nm. The repulsion effect appears when the distance is close to 72 nm. This could be attributed to the steric repulsion resulting from the stretching chain of the gel network structure. With the further approaching of the two surfaces, the repulsion force increases until a hard wall is reached. This indicates that the adsorbed DPG particles are compressed to a most compact state and that the distance between the two surfaces remains unchanged (about 12 nm). At this time, when the motors are driven in the opposite direction to separate the two surfaces (red line), the distance between the two surfaces increases slowly and the steric repulsion decreases. When the distance between the two surfaces exceeds 72 nm, the interaction between the two surfaces disappears. In the separating process, the attraction effect generated is relatively weak. The mica surfaces before and after DPG particles adsorption are both negatively charged, leading to an electrostatic repulsion between them. While both surfaces have hydroxyl groups, the attraction induced by hydrogen bond outweighs the electrostatic repulsion. Therefore there is only a slight attraction between the two surfaces in the process of separation. It can also be seen from Fig. 5.44 that when the salt concentration increases, the compression of the DPG particles is enhanced during the approaching of two surfaces. This could be an explanation that the addition of salt ions

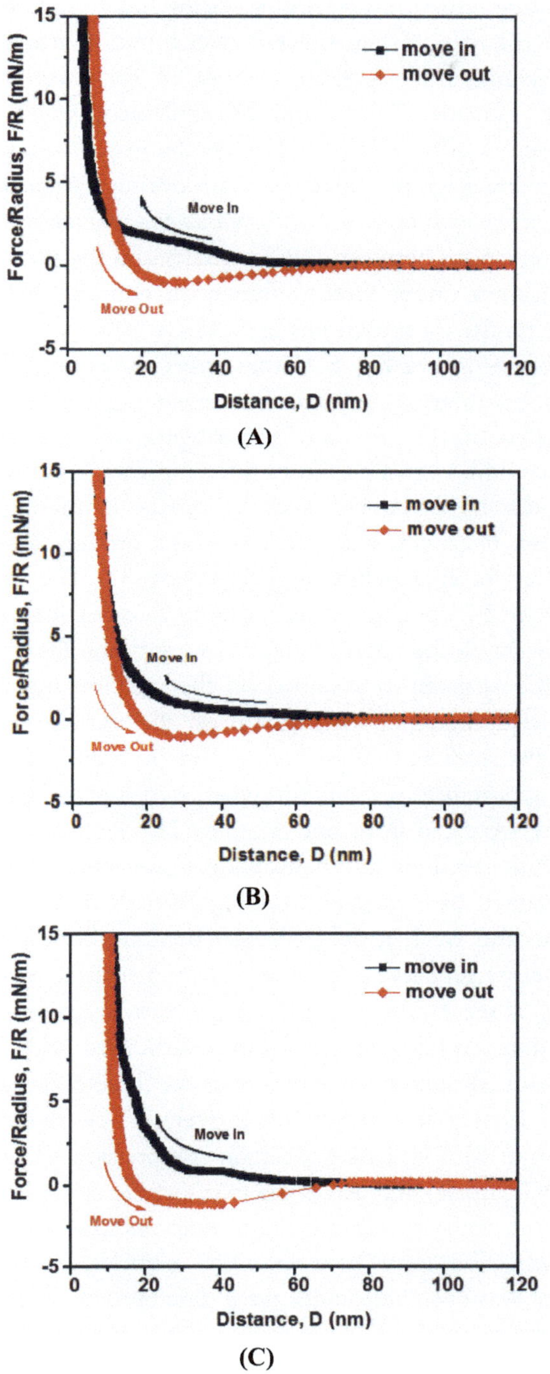

Figure 5.44 Interaction force between DPG particles and strong hydrophilic mica surface: (A) deionized water; (B) 1 mmol NaCl; (C) 100 mmol NaCl.

compresses the structure of DPG particles and further narrows down the surface distance. When the salt concentration reaches 100 mmol, the approaching and separating curves of the two surfaces fail to completely coincide. This could be explained where some of the DPG molecules adhere to the strongly hydrophilic surface and thus result in the breakage of molecular chain during separation process.

2. *Interaction force analysis between DPG particles and medium hydrophilic surface*

   The interactions between DPG particles and the surface of medium hydrophilic mica under three different salt concentrations were measured. The results are shown in Fig. 5.45.

   It can be seen from Fig. 5.45 that under the conditions of three salt concentrations, there is no attraction between the two surfaces, but a repulsion effect, in the process of approaching and separating. This phenomenon could be explained from three aspects: (1) the surface of mica adsorbed with DPG particles and the surface of medium hydrophilic mica are both negatively charged. This could produce a repulsion effect in the process of approaching and separating. (2) The steric hindrance effect is caused by stretching chain of DPG particles, for the steric repulsion increases with the approaching of two surfaces. (3) OTS containing methyl ($-CH_3$) group is adsorbed on the medium hydrophilic surface to form a typical hydrophobic group on the mica surface. The DPG system on the mica surface contains a host of hydrophilic groups, leading to the appearance of repulsion, when two surfaces approach to each other or are separated from one another. The repulsive force produced by electrostatic repulsion and hydrophobic interaction is greater than the attraction caused by hydrogen bonding. What's more, the addition of salt ions can also weaken the hydrogen bonding between the two surfaces. Therefore repulsive force between two surfaces are manifested in the process of separation. It could also be seen from Fig. 5.45 that the hard wall distance between mica surface adsorbed with DPG particles and medium hydrophilic mica surface under three different salt concentrations are 19, 13, and 11 nm, respectively. These results indicate that the addition of salt ions makes a certain compression effect on the structure of DPG particles and leaves the two surfaces closer.

3. *Interaction force analysis between DPG particles and weakly lipophilic surface*

   The interaction force between DPG particles and weak lipophilic mica surface was measured under three different salt concentrations. The results are shown in Fig. 5.46. It could be seen that under the conditions of three salt concentrations, the two surfaces have no attraction in the

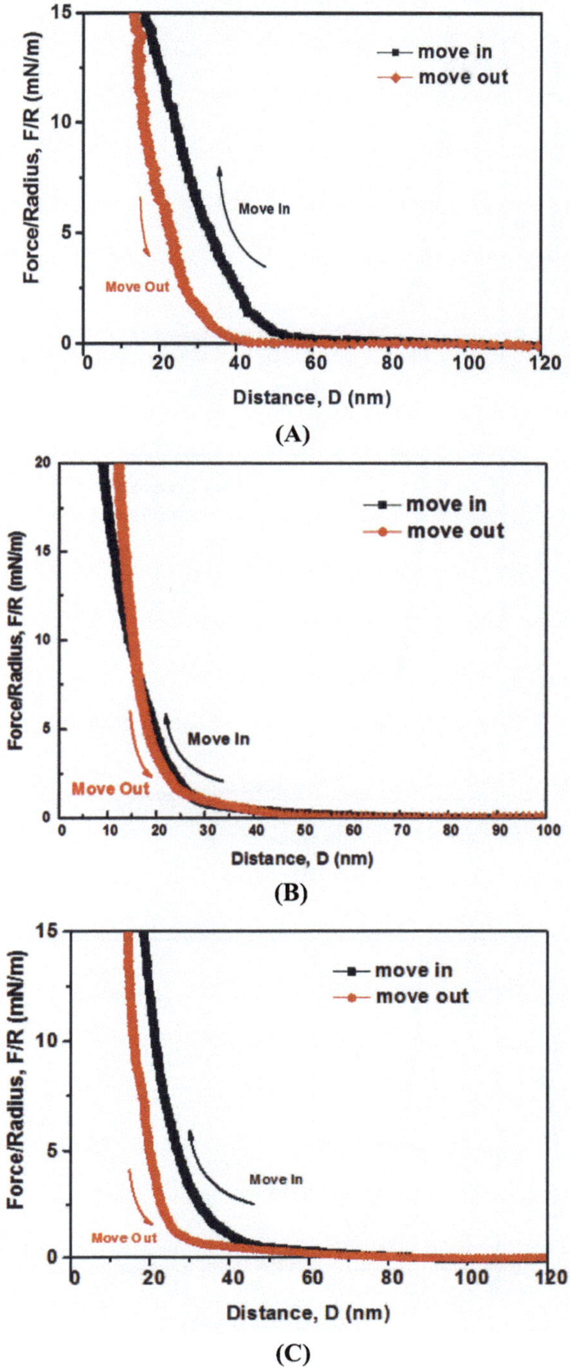

Figure 5.45 Interaction force between DPG particles and surface of medium hydrophilic mica: (A) deionized water; (B) 1 mmol NaCl; (C) 100 mmol NaCl.

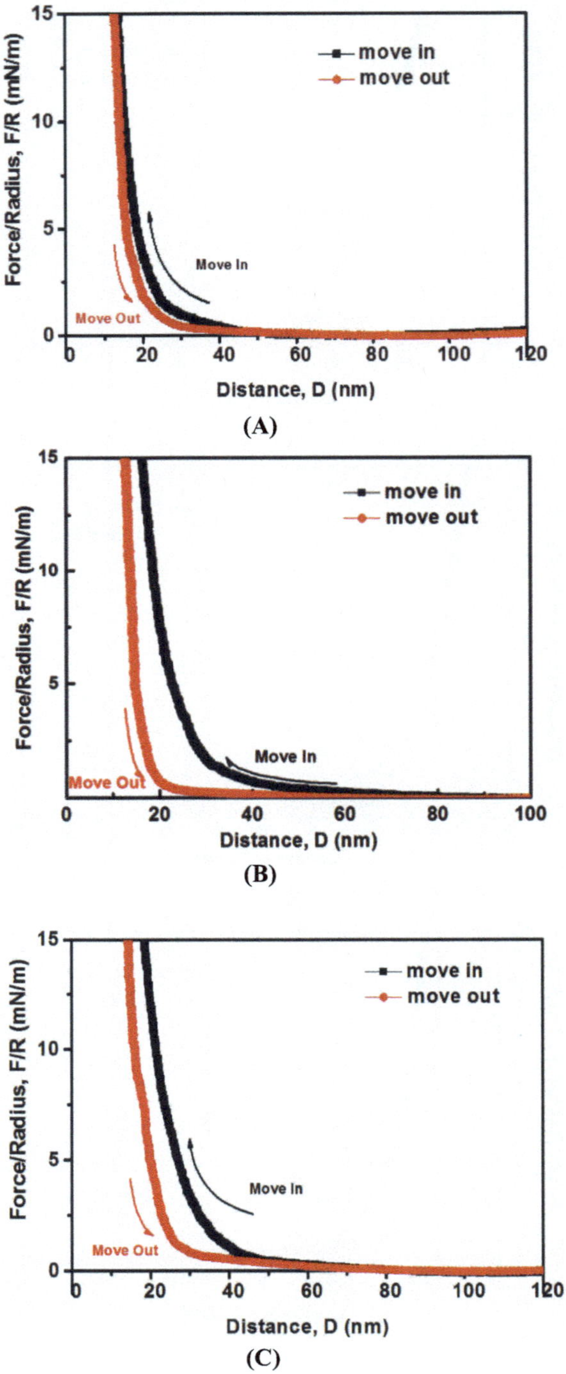

Figure 5.46 The interaction force between DPG particles and weak lipophilic mica surface: (A) deionized water; (B) 1 mmol NaCl; (C) 100 mmol NaCl.

process of approaching and separating. Due to the influence of steric repulsion and electrostatic repulsion, the two surfaces always show a repulsion effect. When the surface of mica changes from medium hydrophilic to lipophilic, the OTS is adsorbed on the mica surface and forms a directional arrangement. This endows the mica surface with more hydrophobic groups, further enhancing the repulsion between the two surfaces.

4. *Interaction force analysis between DPG particles and DPG particles*

   The interaction forces between mica surfaces adsorbed with DPG particles were measured under three different salt concentrations. The results are shown in Fig. 5.47. It can be seen that the two mica surfaces adsorbed with DPG particles exhibit a repulsive force during the process of approaching and separating. The adsorption of DPG particles on both surfaces generates a mutually attractive hydrogen bond between the two surfaces. However, the negatively charged DPG particle system strengthens the electronegativity of the mica surface. The repulsive force generated by electrostatic interaction is stronger than the attractive force generated by hydrogen bonding. This thus leads to an overall repulsion force, as demonstrated in Fig. 5.47. It can be further concluded from Fig. 5.47 that under three different salt concentrations, the curves do not overlap when the two surfaces are approaching and separating. This indicates the existence of the steric hindrance effect.

   To sum up, the interaction force between the DPG particles and the strong hydrophilic surface is dominated by hydrogen bonding, which is manifested as an attractive force. As the mica surface changes from strongly hydrophilic to weakly lipophilic, the electrostatic effect dominates and is manifested as a repulsive force. When the DPG particles are adsorbed on the mica surface, the repulsion effect during the separation of the two surfaces is significantly higher than that between the DPG particles and mica surfaces with different wettability. This indicates that electrostatic repulsion plays a leading role in the surface forces between DPG particle-adsorbed mica surfaces. The steric hindrance effect exists under different test conditions, which further enhances the repulsive force of the two surfaces.

5.4.2.5.6 The interaction force characteristic analysis of the SHC flooding system

1. *Interaction force analysis between the SHC flooding system and the strong hydrophilic surface*

   Fig. 5.48 shows the interaction force between the SHC flooding system and the strong hydrophilic surface. It can be seen that when

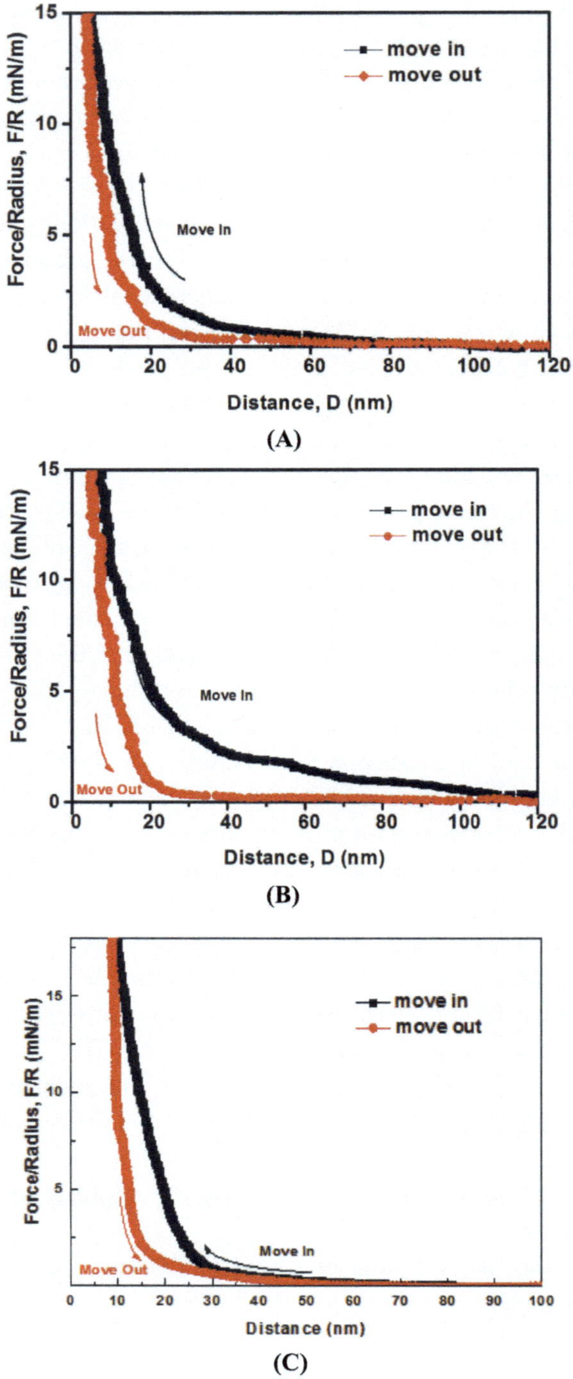

Figure 5.47 The interaction force between DPG particles and DPG particles: (A) deionized water; (B) 1 mmol NaCl; (C) 100 mmol NaCl.

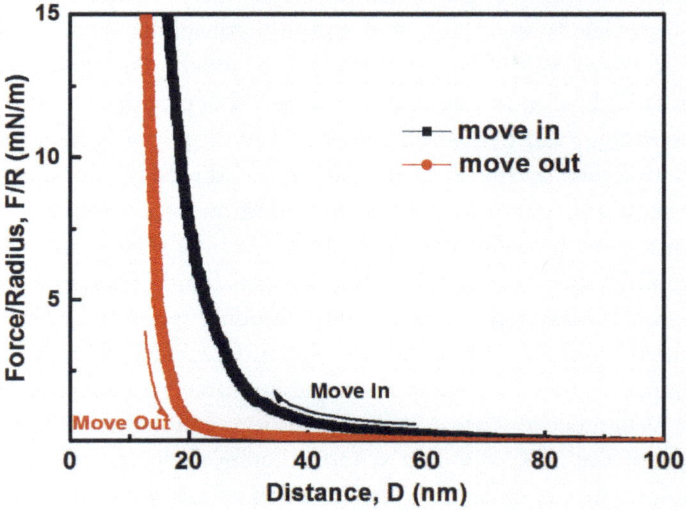

(A)

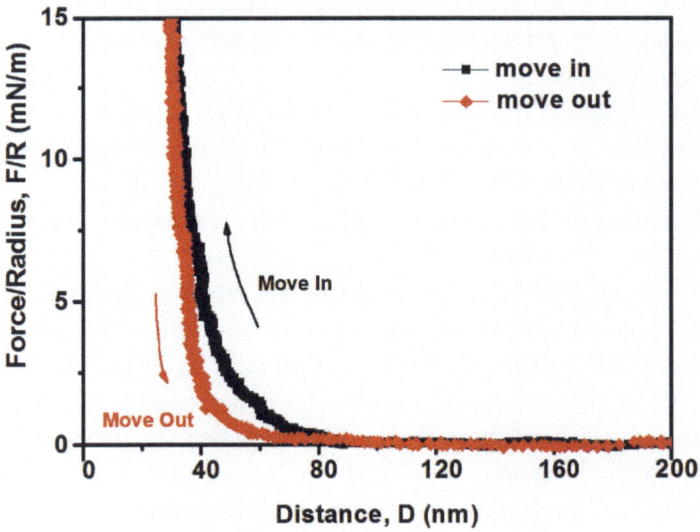

(B)

Figure 5.48 The interaction force between the SHC flooding system and the strong hydrophilic mica surface: (A) deionized water; (B) 100 mmol NaCl.

the SHC flooding system is approaching or separated from the surface of the strong hydrophilic mica, only the repulsive force is measured, with no attractive force existing. Based on the analysis in Fig. 5.44, hydrogen bonding is the main reason for the attraction between the DPG particle system and the strong hydrophilic surface. When a strongly negatively charged surfactant is added, the surfactant molecules can adsorb onto the surface of the DPG particles through hydrogen bonding, van der Waals force, and hydrophobic interaction. This helps to enhance the electronegativity of the SHC flooding system. The repulsion effect caused by the addition of surfactant outweighs the attraction between the single DPG particle system and the strong hydrophilic surface caused by the hydrogen bond. Therefore an overall repulsion is measured when the SHC flooding system is approaching or separated from the strong hydrophilic surface. The existence of steric hindrance further intensifies the repulsive force between the two surfaces. When salt ions are added, the salt ions can neutralize the negative charge of the SHC flooding system, thereby reducing the electrostatic repulsion. At the same time, the addition of salt ions also weakens the hydrogen bonding between the two surfaces. The synergistic effects result in the appearance of repulsive force at a longer distance.

2. *Interaction force analysis between the SHC flooding system and the medium hydrophilic surface*

   By changing the wettability of the mica surface from a strong hydrophilic to a medium hydrophilic state, the characteristics of the interaction force between the SHC flooding system and the medium hydrophilic surface were analyzed. The test results are shown in Fig. 5.49.

   It can be seen from Fig. 5.49 that the repulsive forces between the SHC flooding system and the medium hydrophilic mica surface are measured during the process of approaching and separating. The repulsive force measured in deionized water is significantly higher than that measured in salt solution. This is because when the strong hydrophilic mica surface changes to a medium hydrophilic state, the OTS is adsorbed on the surface and results in a large number of hydrophobic groups. When the two surfaces approach each other, their repulsion effect further increases. At the same time, the repulsion caused by the electrostatic effect and the steric hindrance effect is superimposed so that the repulsion effect between the SHC flooding system and the medium hydrophilic surface is significantly higher than that between the SHC flooding system and the strong hydrophilic surface.

3. *Interaction force analysis between the SHC flooding system and the weak lipophilic surface*

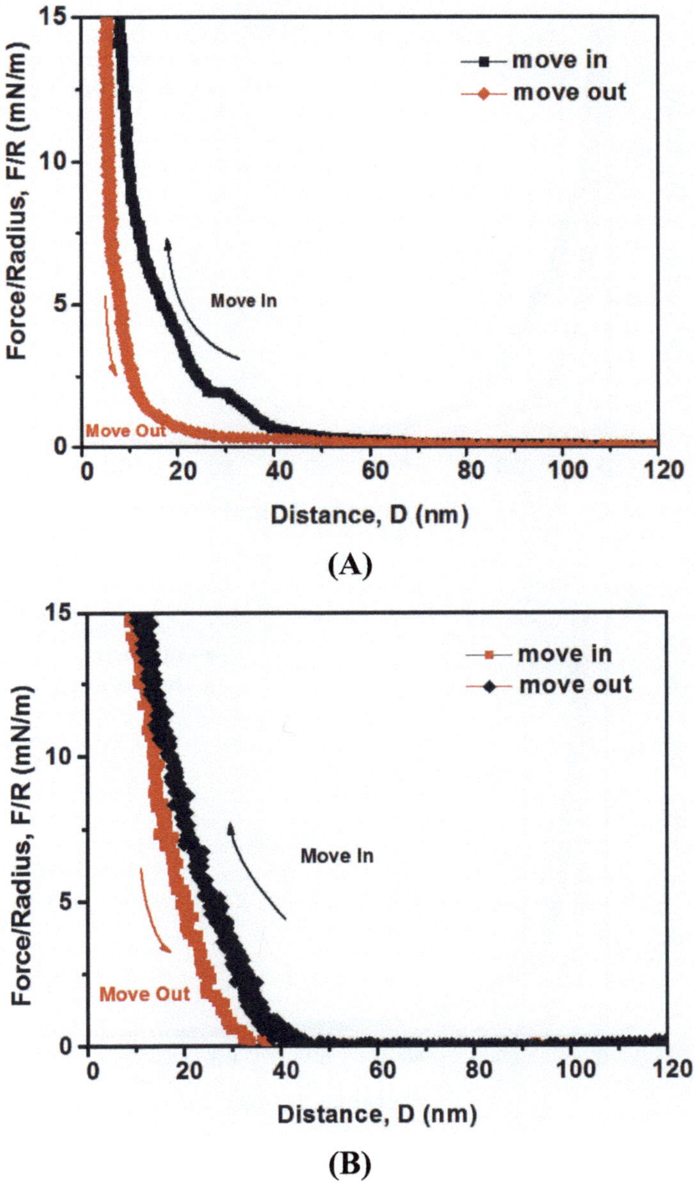

Figure 5.49 The interaction force between the SHC flooding system and the medium hydrophilic mica surface: (A) deionized water; (B) 100 mmol NaCl.

The characteristics of the interaction force between the SHC flooding system and the weakly lipophilic surface were further analyzed, as shown in Fig. 5.50. It can be seen that both the surfaces

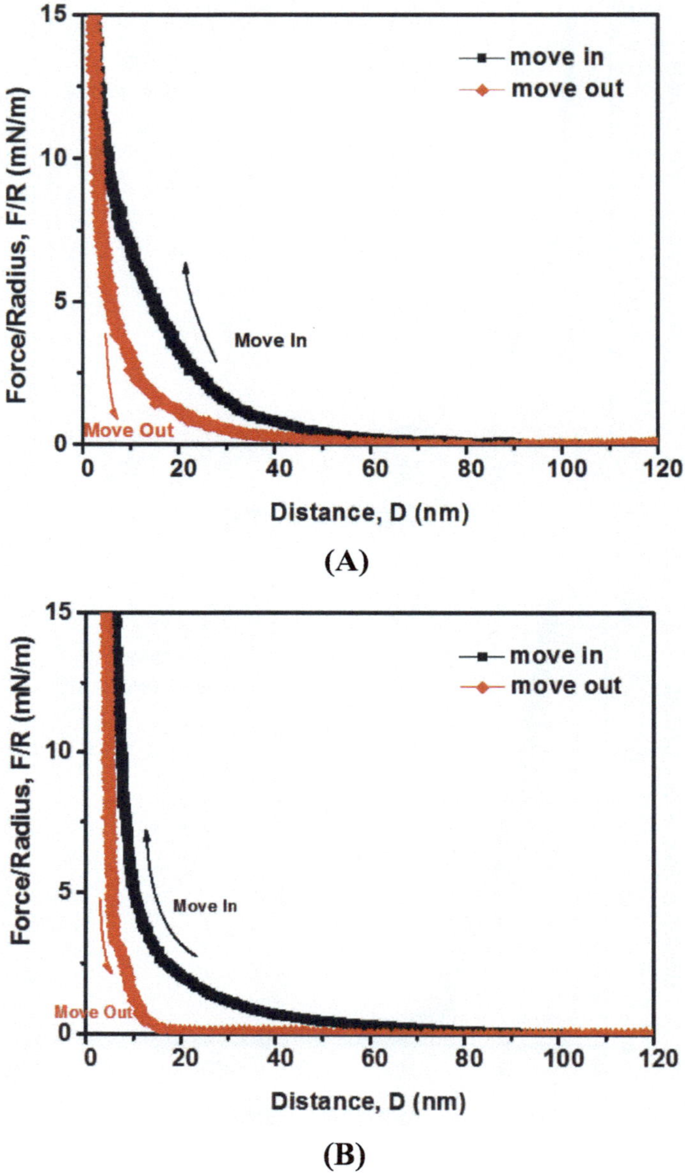

Figure 5.50 The interaction force between the SHC flooding system and the surface of weakly lipophilic mica: (A) deionized water; (B) 100 mmol NaCl.

exhibit a repulsive force during the process of approaching and separating, and that the repulsive force in deionized water is significantly higher than that in salt solution. The repulsion effect between the

SHC flooding system and the weakly lipophilic surface is significantly higher than that between the SHC flooding system and the strong and medium hydrophilic surfaces. The stronger the lipophilicity of the mica surface, the stronger the hydrophobic effect formed by the OTS adsorbed on the mica surface, and the more intensified the repulsion effect between the SHC flooding system and the surface. According to Fig. 5.50, under salt conditions, there is a steric hindrance effect during the approaching of two surfaces, which further enhances the repulsion.

4. *Interaction force analysis between the SHC flooding systems*

   The characteristics of the interaction force between the SHC flooding systems under two salt concentrations are analyzed. The results are shown in Fig. 5.51. It can be seen that the repulsive forces are measured during both the approaching and separating of two mica sheets adsorbed with the SHC flooding system. The adsorption of SHC flooding system on mica surfaces induces the mutually attractive hydrogen bonding between the two surfaces. However, the adsorption of negatively charged SHC flooding system strengthens the electronegativity of the mica surface. As a result, the repulsive force generated by electrostatic interaction outweighs the attractive force generated by hydrogen bonding. The critical-point distance of the steric hindrance effect generated during the approaching of two surfaces is about 2 nm. This may be caused by the compression of the stretched polymer chains of the dispersed particle gel of the SHC flooding system adsorbed on the mica surfaces. Therefore when the two surfaces are separated, a repulsive force at a shorter distance is measured. Fig. 5.51 further shows that under both salt concentrations, the curves measured during the approaching and separating of two surfaces do not overlap. This indicates the existence of the steric hindrance effect.

   The above results show that repulsion is the main interaction force between the SHC flooding system and the mica surface with different wettability. The steric hindrance effect exists under different test conditions, which further enhances the repulsion between the two surfaces. The addition of surfactants strengthens the negative charge of the SHC flooding system. Therefore the repulsive force between the SHC flooding system and the mica surfaces with different wettability is significantly stronger than that between the DPG particles and the mica surfaces with different wettability. The addition of inorganic salt neutralizes the negative charge of the two surfaces, so that there is reduction in the repulsive effect between the SHC flooding system and the mica surface

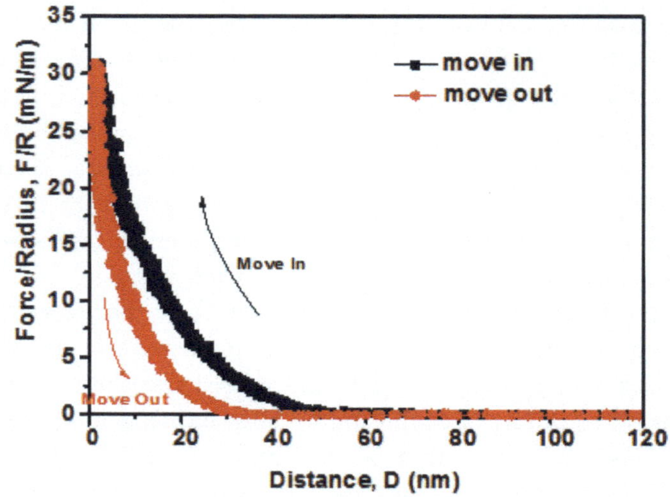

(A)

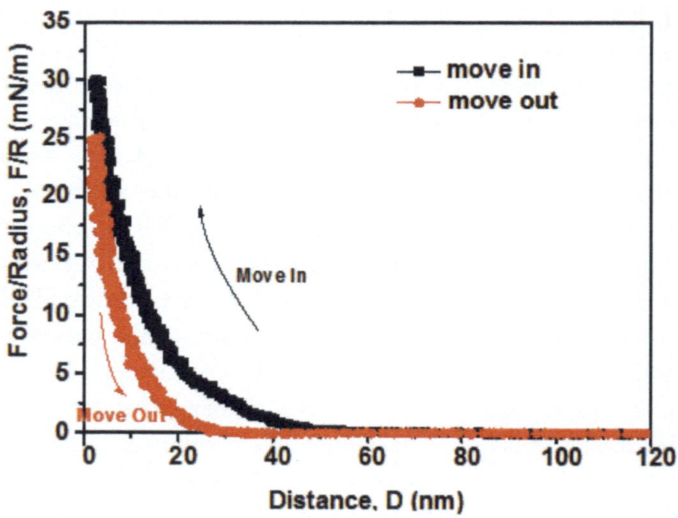

(B)

Figure 5.51 The interaction force between the SHC flooding systems: (A) deionized water; (B) 100 mmol NaCl.

with different wettability. As the surface of mica changes from strongly hydrophilic to weakly lipophilic, the repulsive force between the SHC flooding system and the mica surface gradually increases. Understanding the characteristics of interaction forces between SHC flooding system and mica surfaces is beneficial to the regulation of the reservoir by the combination flooding system. In the early stage of reservoir development, the oil saturation is high and the reservoirs are mostly oil-wet surfaces. With the continuous development of oil reservoirs, crude oil located in high-permeability areas is gradually driven out and the reservoirs tend to be heterogeneous. At this time, the surface of the high-permeability reservoir changes from lipophilic to hydrophilic while the surface of the medium−low-permeability reservoir still contains high oil saturation. Mostly, the surface of the reservoir is lipophilic. When the SHC flooding system is injected, the repulsion effect of the combination flooding system on the hydrophilic surface is weaker than that on the lipophilic surface. For this reason, the adsorption and retention capacity of the combination flooding system in the high-permeability reservoirs is stronger than in low-permeability reservoirs. This helps to improve the microcontrollability of the SHC flooding system.

Section 5.5 Displacement mechanism of SHC flooding system

An SHC flooding system bears such characteristics as high temperature and salinity resistance, aggregation expansion, efficient reduction of interfacial tension, etc. It also exhibits an excellent application potential in oilfields. In order to make full use of the SHC flooding system, it is necessary to explore its displacement mechanism. This section explores the chromatographic separation effect, the controllability for reservoirs and the displacement mechanism of SHC flooding system in porous media, so as to reveal its displacement mechanism.

5.5.1 The chromatographic separation effect of SHC flooding system

When the SHC flooding system migrates in porous media, the chromatographic separation phenomenon occurs to various degrees, with a possibility of affecting its displacement effect. When the SHC flooding system passes through the core model, each component has a different production concentration law at the outlet of the core if chromatographic separation takes place.

5.5.1.1 Test method and procedure

In this study, the dimensionless breakthrough time and dimensionless distance of equal concentration were used to characterize the chromatographic separation degree of the combination flooding system [15−17].

Dimensionless breakthrough time is defined as the injected pore volumes when the first drop of each component of the SHC flooding system is detected at the outlet of the core flooding model. Differences in dimensionless breakthrough time indicate the presence of a chromatographic separation effect between each component of the combination flooding system.

Dimensionless distance of equal concentration is defined as the pore volume difference between two components when two chemicals produced at the outlet of core flooding model achieve the relative same dimensionless concentrations. If the initial injection concentration of each component in the SHC flooding system is C_o, and the concentration at the outlet end at certain moment is C_i, then the relative concentration can be defined as C_i/C_o. In general, the relative concentration of each component of the combination flooding system is 0.5. The value can be reduced if less than 0.5. The larger the value of dimensionless distance of equal concentration, the higher the degree of chromatographic separation of the combination flooding system.

In this study, the columnar core was used as a model to investigate the chromatographic separation phenomenon in the SHC flooding system by ultraviolet spectrum method. The experiment mainly focuses on the influence of the concentration of DPG particles and formation permeability on the chromatographic separation of SHC flooding system. The specification of the columnar core is 8 cm in length and 2.5 cm in diameter. More specifically, the test steps are as follows:

1. Extract the columnar core, dry it at 90°C for 10 hours, and weigh the dry core.
2. Saturate the core with water, weigh the wet core, and then calculate the pore volume and core permeability.
3. Inject water into the core until the pressure is stable, and then inject the SHC flooding system (0.1% DPG + 0.1% THSB surfactant) into the core. During the whole process, the injection rate shall be kept constant at 0.2 mL/min. Collect the produced fluid of constant volume, and detect the concentration of each component of the combination flooding system at the outlet end until the concentration of each component at the outlet of the core is stable.
4. Age the core for 5 days and inject water into the core again. Collect the produced fluid of constant volume, and detect the concentration of each component of the combination flooding system at the outlet of the core until the concentration of produced fluid is stable.

5.5.1.2 The influence of DPG particle concentration

The influence of the concentration of DPG particles on the chromatographic separation of the SHC flooding system was investigated using 0.1% surfactant. The experimental results are shown in Table 5.6 and Figs. 5.52–5.54. It can be seen that when the SHC flooding system migrates in the porous media, the dimensionless breakthrough times of the DPG particles and the surfactant are different. The dimensionless breakthrough time of the DPG particles is less than that of the surfactant. This indicates the presence of chromatographic separation between the components of the SHC flooding system. When the relative

Table 5.6 Influence of DPG particle concentration on the chromatographic separation of the SHC flooding system.

Scheme for injection	Core permeability (μm^2)	Dimensionless breakthrough time		Injection pore volume when $c/c_0 = 0.5$		Dimensionless distance of equal concentration
		t_{DPG}	$t_{surfactant}$	DPG	surfactant	
500	1.104	0.365	0.669	1.346	1.709	0.363
1000	1.030	0.418	0.596	1.403	1.668	0.265
2000	1.134	0.313	0.518	1.225	1.487	0.264

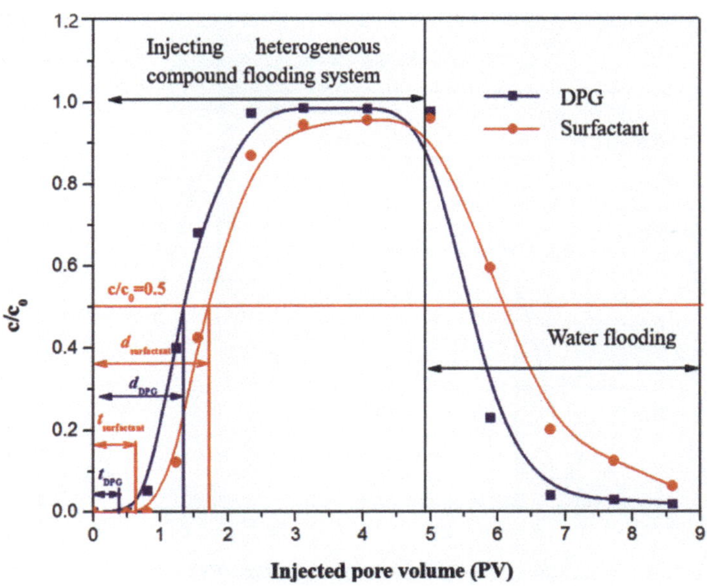

Figure 5.52 Effect of DPG concentration on the chromatographic separation effect of SHC flooding system (500 mg/L).

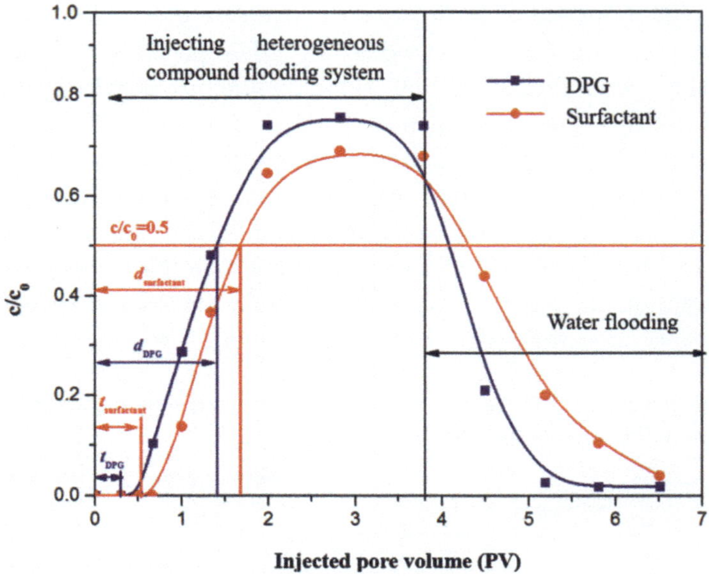

Figure 5.53 Effect of DPG concentration on the chromatographic separation effect of SHC flooding system (1000 mg/L).

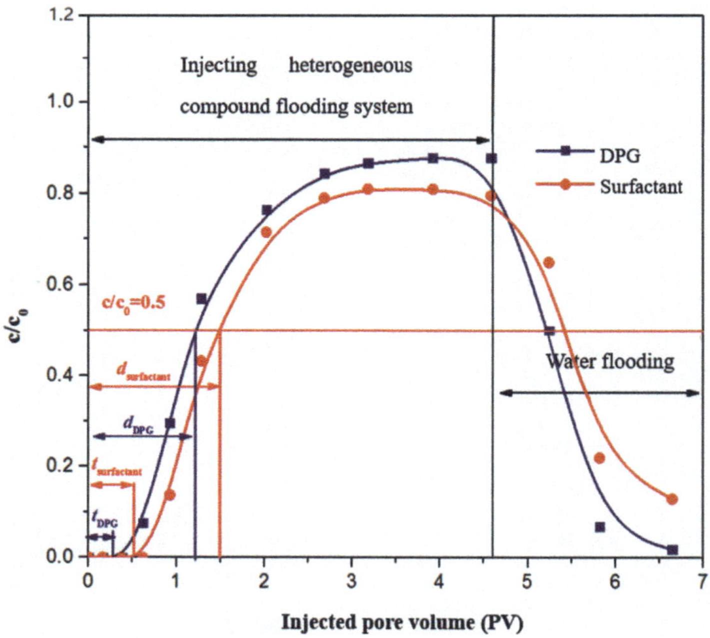

Figure 5.54 Effect of DPG concentration on chromatographic separation effect of SHC flooding system (2000 mg/L).

concentration of each component of the heterogeneous combination flooding system reaches 0.5, the SHC flooding systems formed by different concentrations of DPG particles have differences in the distance of equal concentration. The lower the concentration of DPG particles and the bigger the dimensionless distance of equal concentration, the more obvious the chromatographic separation effect. This could be explained by the higher the concentration of the DPG particles, the more surfactant molecules adsorbed on the DPG particles, and the more difficult it is to produce the surfactant.

Figs. 5.52–5.54 show the changes in the outlet concentration of each component in the following water flooding stage after aging of the SHC flooding system. It can be seen that due to the chromatographic separation phenomenon of the SHC flooding system, the concentration of the DPG particles at the outlet end increases rapidly with the increase of the injection volume until it becomes stable and balanced. While the pore volume of surfactant has reached a dimensionless breakthrough time, the concentration at the outlet end will also become stable and balanced rapidly. It can also be observed that the time needed for the concentration of the surfactant at the outlet end to reach a stable value is longer than that of the DPG particles. When the surfactant of the SHC flooding system migrates in the porous media, the pore volume and distance it flows through are longer than that of the DPG particles. Therefore the time needed for a concentration equilibrium at the outlet end is relatively longer in the following water flooding process. This phenomenon is beneficial to SHC flooding. Surfactant molecules can enter the low-permeability parts or pores which DPG particles cannot enter. After entrance, these molecules are adsorbed on the surface of rock particles to change the wettability and then exfoliate the crude oil, so as to enhance oil recovery.

5.5.1.3 The effect of permeability on chromatographic separation effect

The migration behavior of the SHC flooding system in different permeability cores is different. If the core permeability is too high, it is difficult for the SHC flooding system to control the high-permeability part of the reservoir. If the core permeability is too low, on the contrary, it is difficult to inject the combination flooding system. Therefore the chromatographic separation effect of the SHC flooding system under three different core permeability conditions was investigated. The formula of injected SHC flooding system is 0.1% DPG + 0.1% surfactant. The test results are shown in Table 5.7, Figs. 5.53, 5.55, and 5.56.

It can be seen from Table 5.7 that the dimensionless breakthrough time of DPG and surfactant is different under different core permeability conditions. This indicates the existence of chromatographic separation effect in the migration of SHC flooding system in porous media. The higher the core permeability, the longer the dimensionless breakthrough time of DPG, and the more obvious the chromatographic separation effect. This is ascribed to the existence of numerous multiscale pores of which quite a few are inaccessible. When the pore throat radius is much smaller than the DPG radius, DPG can

Table 5.7 Effect of core permeability on chromatographic separation effect.

K (μm²)	Dimensionless breakthrough time		Injection pore volume when $c/c_0 = 0.5$		Dimensionless distance of equal concentration
	t_{DPG}	$t_{surfactant}$	DPG	Surfactant	
0.42	0.300	0.542	1.293	1.504	0.211
1.03	0.418	0.596	1.403	1.668	0.265
1.65	0.538	0.908	1.597	1.982	0.385

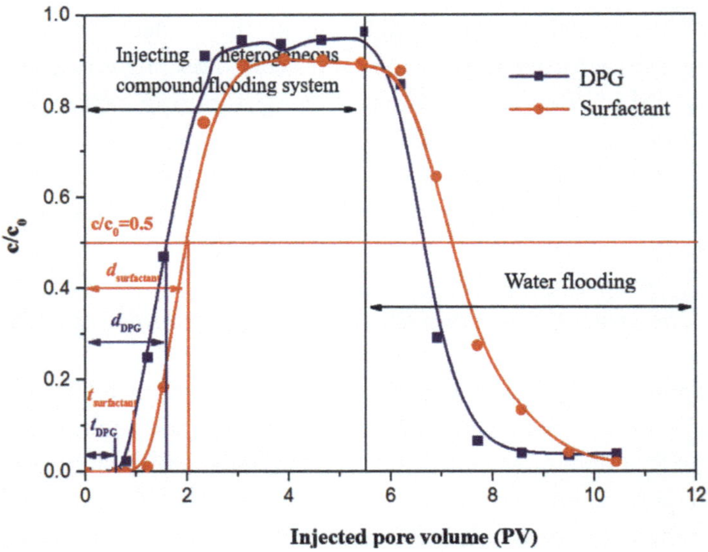

Figure 5.55 Effect of core permeability on chromatographic separation effect of SHC flooding system ($k = 0.42$ μm²).

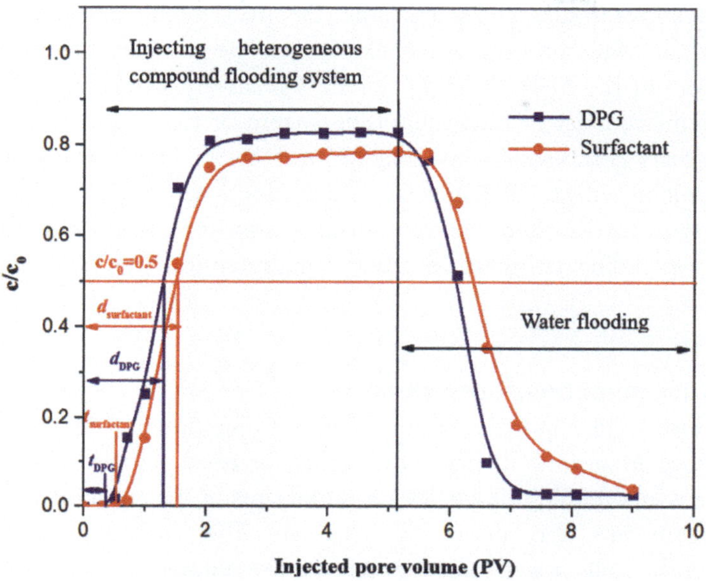

Figure 5.56 Effect of core permeability on chromatographic separation effect of SHC flooding system ($k = 1.65\ \mu m^2$).

only flow in relatively large pores and throat. Hence, the volume pores through which the DPG flows are the least, and their distance to the outlet is the shortest. However, surfactants possessing smaller diameters can enter almost all pores. Thus they need to pass through the longest distance to the outlet. The difference of path and pore volumes flowing through leads to the differential migration of components in the combination flooding system. As a result, a difference comes into being in dimensionless breakthrough time of the two components at the core output end.

Figs. 5.55 and 5.56 show the variation of the output concentration of each component of the SHC flooding system at the following water flooding stage. In the combination flooding system, the time needed for surfactant concentration to reach a stable value is longer than that for the DPG particles. Also, lower core permeability brings about a longer time required for the surfactant concentration to reach stable value. This is mainly due to the need for surfactants, with lower permeability, to pass through higher pore volume and longer distance. Therefore the concentration equilibrium time of produced liquid would be prolonged during the following water flooding process.

The above results show that the chromatographic separation effect exists in the SHC flooding system. However, the chromatographic separation effect of the SHC flooding system is relatively slight compared with the polymer/surfactant binary flooding system or the polymer/surfactant/alkali ternary combination flooding system [18,19]. This is favorable for its displacement effect. When the SHC flooding system migrates in core porous media, there are three main reasons for chromatographic separation, including competitive adsorption, retention loss, and multipath migration.

5.5.1.3.1 Competitive adsorption

Adsorption is an important physicochemical phenomenon of SHC flooding system in porous media. The surface of rock particles is negatively charged while the DPG particles and surfactants in the combination flooding system are also negatively charged. The electrostatic repulsion between the SHC system and the rock surface reduces the adhesion ability of the combination flooding system on the particle surface. Due to the existence of hydrogen bond and hydrophobic forces, however, both the DPG and surfactants are adsorbed on the surface of rock particles. This may influence the migration rate of each component in the SHC flooding system. Also, the differences in electric property, viscosity, and particle size of each component in SHC flooding system make the adsorption capacity of DPG and surfactant different on the surface of rock particles. Therefore competitive adsorption occurs and results in the differential migration velocity in porous media. Macroscopically, the dimensionless breakthrough time of each component in the combination flooding system at the output end of the core is different, resulting in chromatographic separation.

5.5.1.3.2 Retention loss

Retention loss is one of the important reasons for chromatographic separation in SHC flooding system. When the SHC flooding system migrates in porous media, the particles directly plug the pore throat if the pore throat radius is smaller than the DPG radius in the SHC flooding system; otherwise, the pore throat is plugged by multiple DPG particles bridging. Both types of plugging lead to the retention of DPG in the pore throat and thus result in the separation between the DPG and surfactant in the migration process, as well as their differential migration.

5.5.1.3.3 Multipath migration

Multipath migration is the most important factor causing chromatographic separation of SHC flooding system. There are quite a few inaccessible pores among the numerous pores distributed in different scales in the formation. When the pore throat radius is much smaller than the DPG radius, the particles cannot enter. They can only flow in relatively large pores and throats, sweeping the least pore volume and the shortest distance to the outlet. However, small molecular surfactants can enter most pores and travel the longest distance to the outlet. Different flow paths lead to different dimensionless breakthrough times of the two systems at the core production end.

5.5.2 Distribution of SHC flooding system in porous media

5.5.2.1 Experimental principle

Under the premise of leaving the external shape and internal structure of the core unchanged, the core micropores were dynamically scanned by *CT* scanning technology at different flooding stages to observe the porous flow states and the distribution characteristics of the flooding fluid. The fluid saturation distribution was obtained during the flooding process, for it directly reflects the mobilizing capability for the remaining oil and reveals the flooding mechanism. Based on the ray linear attenuation, the core porosity (φ), oil saturation (S_o), and water saturation (S_w) can be calculated with respect to the *CT* values, as follows:

Porosity:

$$\varphi = \frac{CT_2 - CT_1}{CT_w - CT_a} \tag{5.6}$$

Oil saturation:

$$S_o = \frac{CT_1 - CT_i}{CT_w - CT_o} \times \frac{CT_w - CT_a}{CT_2 - CT_1} \times 100\% \tag{5.7}$$

Water saturation:

$$S_w = \left(1 - \frac{CT_1 - CT_i}{CT_w - CT_o} \times \frac{CT_w - CT_a}{CT_2 - CT_1}\right) \times 100\% \tag{5.8}$$

where CT_a is air *CT* value; CT_1 is dry core *CT* value; CT_2 is water saturated core *CT* value; CT_w is water phase *CT* value; CT_o is oil saturated *CT* value; CT_i is core *CT* value at a time instance during the displacement process.

5.5.2.2 Experiment method and procedures

GE LIGHTSPEED 8-Slice Spiral *CT* Scanner was applied to scan the core, with an experiment temperature of 30°C, scanning voltage of 120 kV, current of 130 mA, minimum scanning layer thickness of 1.25 mm, and a resolution scale of 0.18 mm. To avoid the interference of conventional core holders with the X-ray of *CT* scanning equipment, a specially manufactured core holder using PEEK carbon fiber material is used for the flooding experiment. This core holder can enhance the passing strength of X-rays through the core and reduce measurement error due to radiation stress effect. The experiment apparatus is illustrated in Fig. 5.57. The specific procedures are as follows:

1. Extract and dry a natural core.
2. Fully saturate the core using water.
3. Displace the water saturated core to its irreducible water saturation by simulated oil using stepwise pressurization method. Scan to determine the irreducible water distribution and average irreducible water saturation.
4. Scan the core during flooding process to obtain oil and water saturation distributions as well as their heterogeneous characteristics until the water-cut rate reaches 98% and above.
5. Next, scan the core during injection of SHC flooding system (0.1% DPG + 0.1% THSB surfactant) to obtain the oil and water saturation distributions as well as their heterogeneous characteristics.
6. Subsequently, scan the core during water flooding to obtain the oil and water saturation distributions as well as their heterogeneous characteristics, until the water-cut rate reaches 98% and above.

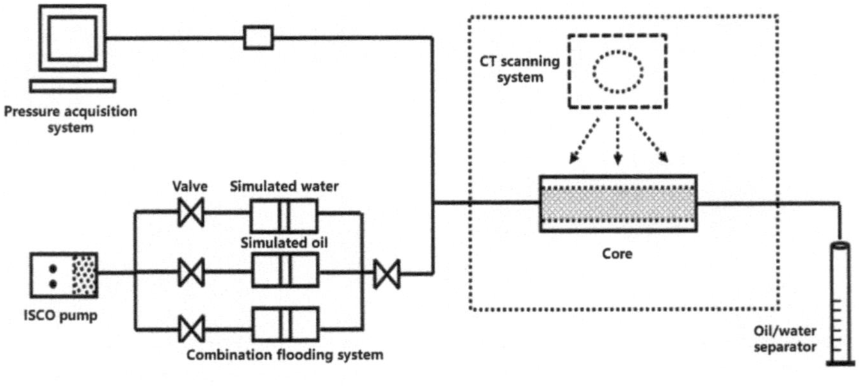

Figure 5.57 CT scanning experiment equipment diagram.

5.5.2.3 Analysis of the displacement process of the SHC flooding system

5.5.2.3.1 Core oil saturation analysis

Fig. 5.58 provides the original oil saturation distribution of different core sections. In the reconstructed 2D image, the color purple represents the rock, the color blue is the water, and the color red is the oil. Once the core is fully saturated by simulated oil, the original oil saturation of each section is between 60.1% and 79.9%. This indicates a quite good oil saturation of the core.

5.5.2.3.2 Dynamic changes of oil and water phases during water flooding

Fig. 5.59 shows a 2D reconstruction of the dynamic changes of the oil and water phases at a certain instance during water flooding. With the increase of the flooding time, the oil saturation continues to decrease. When the water flooding continues for 10 minutes, the water phase breaks through. At this time, due to the enhanced internal heterogeneity of the core, the subsequently injected water along the high-permeability

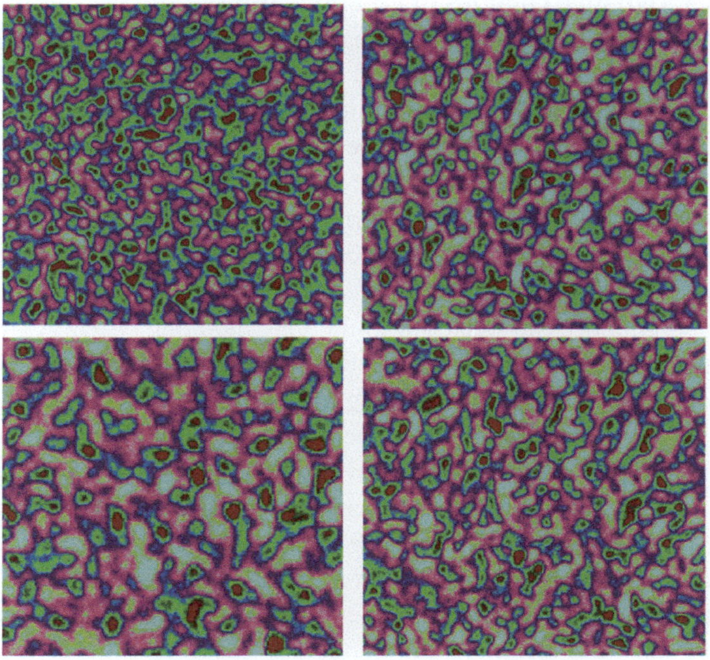

Figure 5.58 Original oil saturation distribution of each section.

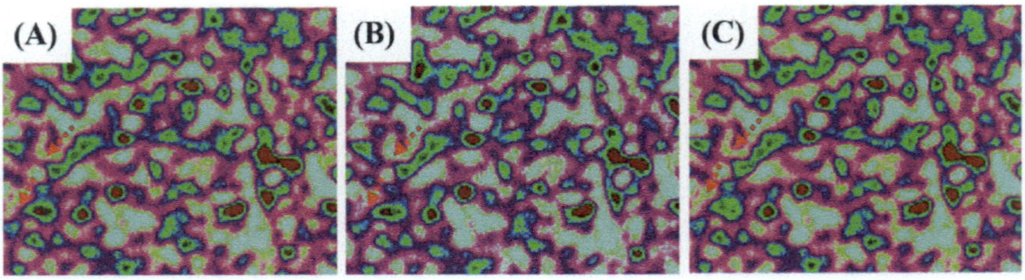

Figure 5.59 Dynamic changes of oil and water phases during water flooding: (A) initial distribution; (B) 10 min; (C) 20 min.

channels forms an ineffective circulation, and the oil saturation basically does not change with the increase of the flooding time. During the water flooding process, the crude oil in the highly permeable zones is driven out. After a certain period of water flooding, the heterogeneity of the core is strengthened, forming some dominant channels and resulting in effective circulation. Therefore as the water flooding time prolongs, the oil saturation of the core basically no longer changes, and the oil recovery factor does not increase any more.

5.5.2.3.3 Dynamic changes of oil and water during SHC flooding

Fig. 5.60 shows the dynamic changes of oil and water during SHC flooding. After the water flooding, the heterogeneity of the core is strengthened. This stimulates the injected water to breakthrough along the high-permeability channel. There is, however, a lot of remaining oil in the low-permeability zones of the core. When the SHC flooding system is injected, the DPG particles form an effective plugging of the high-permeability zones through retention, adsorption, direct plugging, or bridging plugging. But at this time, some remaining oil still stays in the smaller pores. Since the surfactant in the SHC flooding system is a molecular system small enough to cross through smaller pores, the residual oil is stripped from the rock surface through the wetting reversal and emulsification of the surfactant, thereby increasing the oil recovery rate. The DPG particles in the SHC flooding system can control the seepage profile of the heterogeneous core at the microscope level. This ensures the entry of the surfactant into the inaccessible pores of the DPG particles. At the same time, the viscosity increasing effect of the DPG particles in the SHC flooding system can raise the mobility ratio of the surfactant, reduce its diffusion rate, and promote the surfactant, so as to turn to the low-permeability part of the core and drive out the remaining oil. In this way,

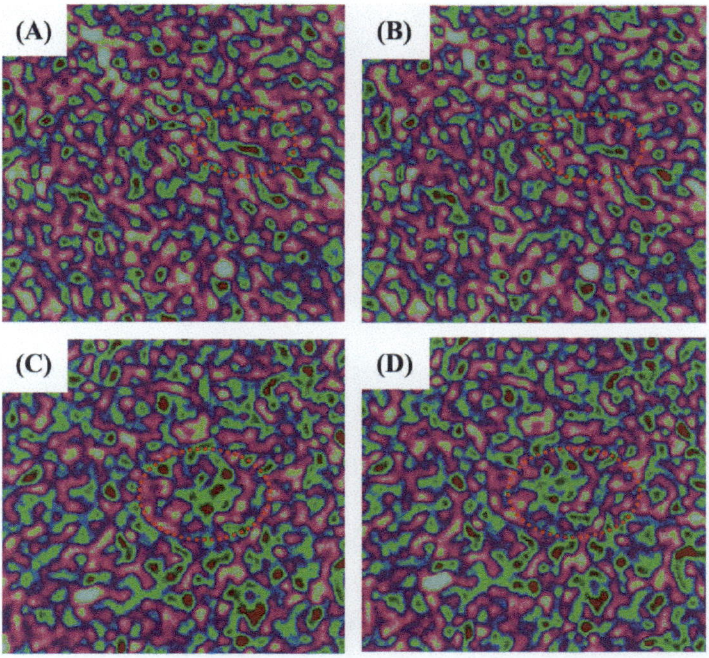

Figure 5.60 Dynamic changes of oil and water phases during SHC flooding: (A) 5 min; (B) 10 min; (C) 15 min; (D) 20 min.

the viscosity can enhance the microwashing efficiency of surfactants. The surfactant in the SHC flooding system can be adsorbed on the surface of the DPG particles so that the DPG particles has a certain degree of activity. When in contact with the surface of the rock formation, the DPG particles strengthen the ability of the SHC flooding system to change the wettability of the rock, and facilitate the stripping of oil from the rock surface. In addition, the adsorption of surfactants on the surface of the DPG particles can enhance the negative charge of the SHC flooding system. It can also promote the migration of the SHC flooding system to the deep part of the rock formation. The synergistic effect of the two components greatly improves the oil recovery.

5.5.2.3.4 Dynamic changes of oil and water during following water flooding

Fig. 5.61 shows the dynamic changes of oil and water in porous media during the following water flooding. It can be seen that after the SHC flooding system is injected, microprofile control is carried out on the

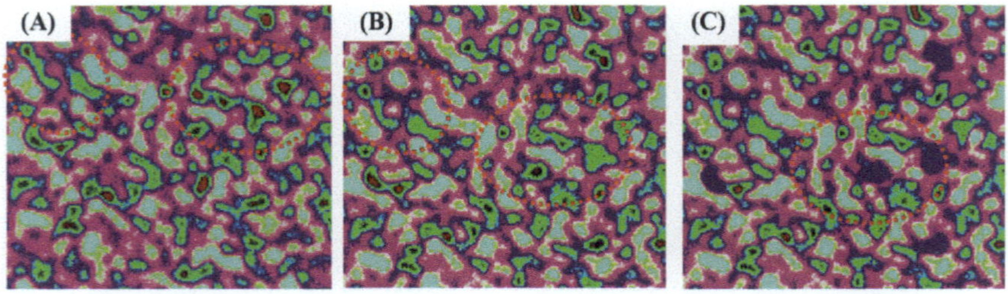

Figure 5.61 Dynamic changes of oil and water during following water flooding: (A) 5 min; (B) 10 min; (C) 15 min.

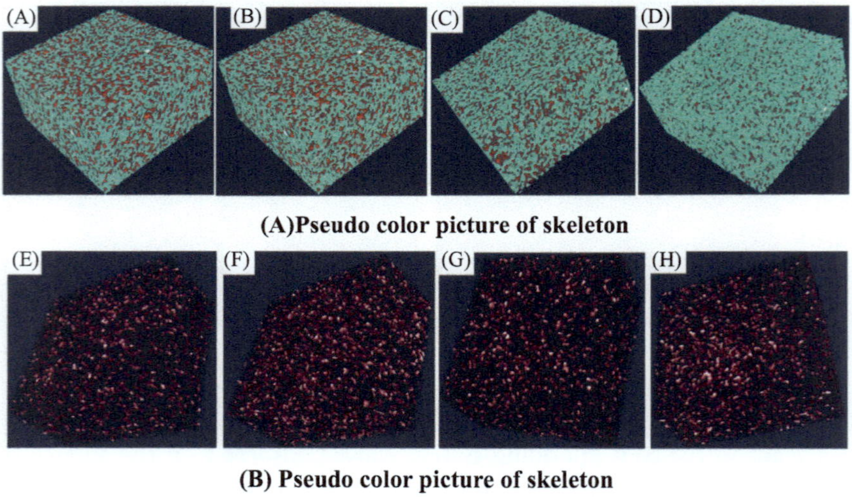

Figure 5.62 Dynamic variation of oil and water during SHC flooding: (A), (E) initial state, saturated oil; (B), (F) water flooding, 20 min; (C), (G) injection SHC flooding system, 15 min; (D), (H) following water flooding, 20 min.

seepage profile, forcing the following water flooding to move to the low-permeability zones in the core and driving out the remaining oil. In this way, the oil recovery enjoys an improvement.

5.5.2.4 Three-dimensional dynamic analysis of the displacement process

Three-dimensional skeleton and pore structure imaging during the SHC flooding process was carried out by *CT* scanning technology, for analysis of the dynamic variation of oil and water. The experimental results are shown in Fig. 5.62. According to the skeleton pseudocolor diagram in Fig. 5.62A and the pore pseudocolor diagram in Fig. 5.62E, the core shows high oil saturation in its initial state. After 20 minutes of water flooding, oil production generally stops increasing. This indicates the transformation of the core from homogeneous to heterogeneous, causing the injected water to move forward along the high-permeable part of the core and forming an inefficient cycle.

At this time, however, the remaining oil or residual oil (Fig. 5.62B) distributed in the small pores of the core is not driven out, resulting in low water flooding recovery. When the SHC flooding system is injected, the DPG particles preferentially enter high-permeability zones directly through or in a deformed way through deep into the core. Then, the reservoir on the microscale is regulated. The swept volume is expanded and the remaining oil is driven out through direct plugging, multiple particle bridging, or stranded in the pore plugging high-permeable zones. Meanwhile, the viscosity increasing effect of the DPG particles in the SHC flooding system improves the mobility of the surfactant and reduces its diffusion rate. In addition, the surfactant in the SHC flooding system can emulsify the remaining oil. Thus the viscosity of the displacement system is strengthened, so is the mobility controllability of the displacement system. The reservoir microscopic profile controllability of the SHC flooding system allows the surfactant to enter a pore that is inaccessible to the DPG particles. Then, the oil recovery is improved by stripping the residual oil from the rock surface through the wetting, emulsification, and capillary pressure reduction of the surfactant (Fig. 5.62C and D). The oil recovery also embraces a significant improvement through the microscopic profile control, the efficient oil displacement effect of superposition surfactant, and the synergistic effect of the composition of wettability alteration and emulsification.

5.5.3 Sweep efficiency of SHC flooding system in porous media

5.5.3.1 *Experimental principle*

The nuclear magnetic resonance (NMR) displacement experiment was used to explore the mobility controllability in a heterogeneous reservoir by the SHC flooding system. The principle of the nuclear magnetic resonance method is as follows. The nucleus of each element has a specific spin quantum number. If the spin quantum number is greater than 0, the nucleus generates a magnetic field when it spins. Through the interaction between the hydrogen nuclei and the magnetic field, the analysis of the amount and type of fluid in the pores of the rock can be determined by the characteristic that the nuclear magnetic resonance signal of the hydrogen atom is proportional to its porosity. The variation of the peak value with the transverse relaxation time T_2 was adopted to study the microcontrollability of the SHC flooding system. The relationship between the relaxation time T_2 and the magnetic field strength G, echo interval T_E, fluid diffusion coefficient D, and other factors is as follows (5−9):

$$\frac{1}{T_2} = \frac{D(\gamma G T_E)}{12} \tag{5.9}$$

The amplitude of T_2 is related to the fluid distribution in the pores. Due to the different hydrogen contents in the molecular structure of the water, oil, and plugging agent systems in the reservoir, the recovery rate of each fluid varies as well. Therefore the signal distribution with different attenuation amplitudes is available to infer the distribution information of oil, water, surfactant, and SHC flooding system.

5.5.3.2 Experimental method

In the experiment, the MacroMR12−150H-I nuclear magnetic displacement equipment produced by Suzhou Niumai Electronic Technology Co., Ltd. was used to scan the core. The T_2 spectrum was determined by the CPMG sequence. In order to reduce the interference of the hydrogen atom signal in the water, the manganese chloride solution was used for displacement. The specific experimental steps are as follows:

1. The core was cleaned with Soxhlet extraction and then dried. Next, the core was saturated with water to measure the permeability and pore volume of the core.
2. Using the gradient pressurization method, the water-saturated core was displaced with simulated oil to the irreducible water state. Then, NMR scanning was performed to determine the initial distribution state of oil and water in the core.
3. The core displacement experiment was next carried out. NMR scanning was performed on the core sample at a fixed time interval to obtain the oil−water saturation signal until the water-cut rate reached over 98%.
4. The SHC flooding system (0.1% DPG + 0.1% THSB surfactant) was injected. The core was scanned at a fixed time interval to collect the oil−water saturation signal.
5. In the follow-up water flooding experiment, the core was scanned by NMR at a fixed time interval to obtain oil−water saturation until the water-cut rate rises over 98%.

The SHC flooding system injected in step (4) was replaced with 0.1% surfactant. Then, the experiment steps (1)−(5) were repeated. The core permeability used in the experiment is 0.78 μm^2 and the pore volume 5.4 mL.

5.5.3.3 Displacement analysis of the SHC flooding system

In light of the T_2 spectrum (Fig. 5.63) and the oil saturation map (Fig. 5.64) during the displacement process by the SHC flooding system, the characteristic peaks in the initial water flooding stage are concentrated between 0.4−3, 10−14, and 100−1000 ms. When the water flooding starts, the peak in the smaller interval shifts rightward progressively. The longer the displacement time, the greater the rightward of the T_2 peak in the smaller interval. This indicates the gradual enlargement of core pores and the increase in the heterogeneity during the water flooding process. At this time, there is still some remaining oil in the core that has not been displaced (Fig. 5.64B). After the injection of the SHC flooding system, the characteristic peak of T_2 shifts leftward gradually. This movement suggests the profile control of the DPG particles in the SHC flooding system on the heterogeneity of the reservoir effectively. After the injection of the SHC flooding system for 20 minutes, the DPG particles in the SHC system can control the reservoir microscopically through direct plugging of a single particle or bridging and plugging by multiple particles. In this way, the reservoir is made heterogeneous. The qualitative decrease is

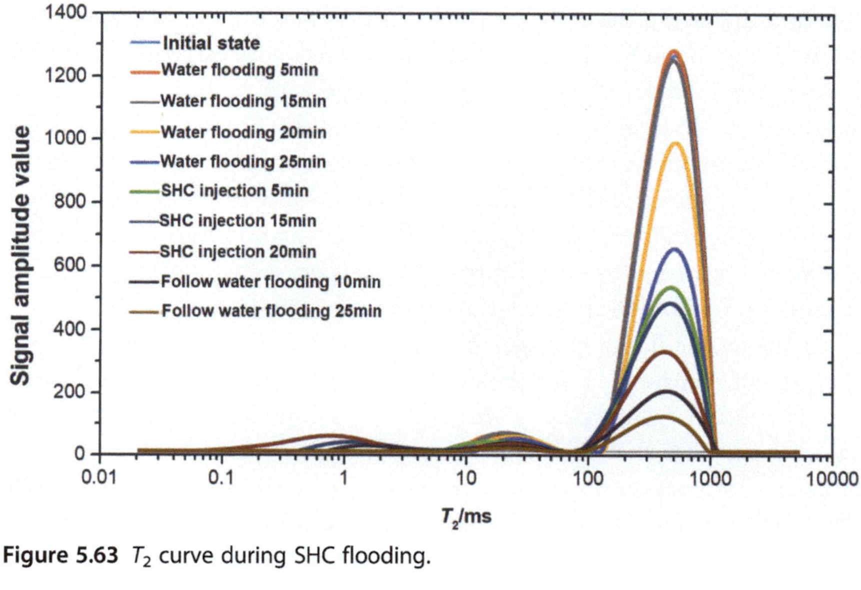

Figure 5.63 T_2 curve during SHC flooding.

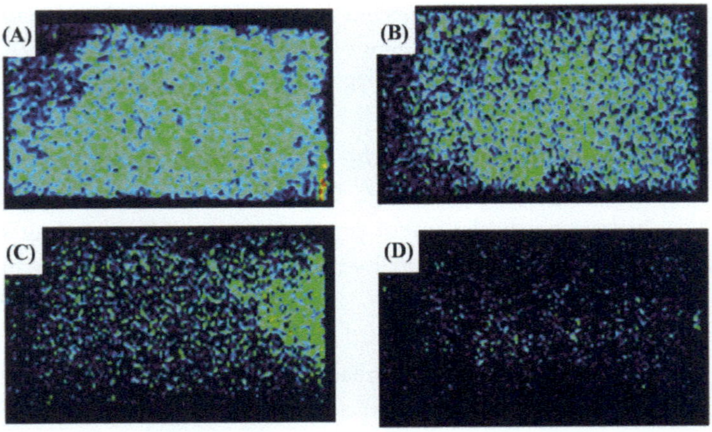

Figure 5.64 The oil saturation during SHC flooding stage: (A) initial state; (B) water flooding for 20 min; (C) injection of SHC flooding system for 20 min; (D) following water flooding for 25 min.

manifested in the microscopic view as the moving of the T_2 characteristic peak from 100–1000 ms, to 0.1–3 ms, and to 10–14 ms. This movement increases the swept volume of the SHC flooding system. The microcontrol of the SHC flooding system over the reservoir and the superimposed high-displacement efficiency improves the recovery factor (Fig. 5.64B and C). In the subsequent water flooding, the injected water turns to the low-permeability part of the core, enlarging the swept volume and displacing the remaining oil in it. As a result, the oil recovery is elevated (Fig. 5.64D). However, after the subsequent conversion to water flooding for multiple PVs, the characteristic peak of T_2 at 0.1–3 ms shifts rightward, indicating the successful control of the SHC flooding system over the reservoir, microscopically. However, its controllability is limited.

In order to further clarify the microcontrollability of the SHC flooding system on the reservoir, the T_2 curve and oil saturation during the surfactant flooding process were compared. The experimental results are shown in Figs. 5.65 and 5.66. It can be seen that the characteristic peak value of T_2 in the water flooding stage shifts from 7–13 to 100–1000 ms. This indicates an increase in both the pores and the core heterogeneity. When the conventional surfactant system is injected, the T_2 characteristic peak does not move to the left in the way it performs in the process of heterogeneous combination flooding. This performance shows the failure of the surfactant injection in controlling the reservoir. The increase in oil recovery in the process of surfactant flooding is ascribed only to the stripping of

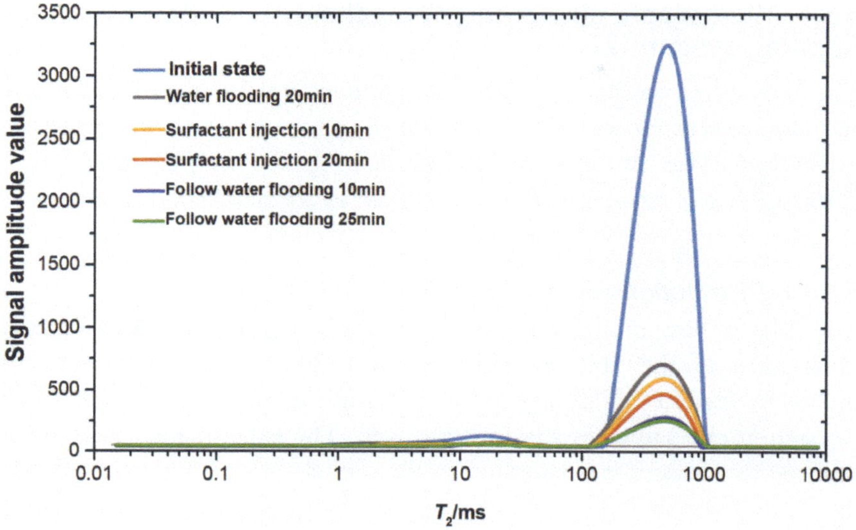

Figure 5.65 T_2 curve change during surfactant flooding.

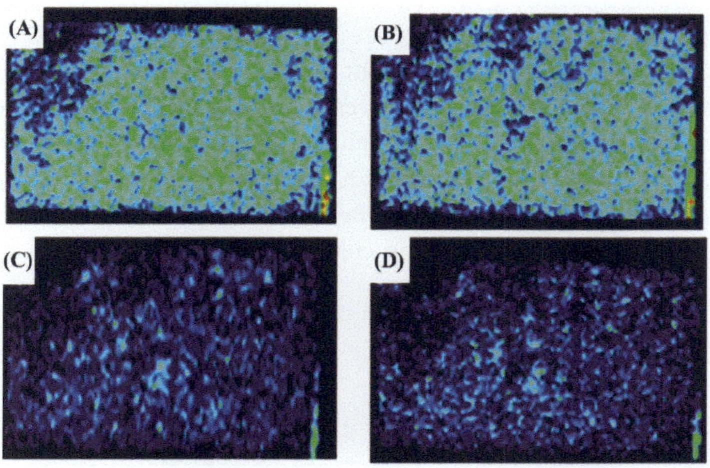

Figure 5.66 The oil saturation during surfactant flooding: (A) initial state; (B) water flooding for 20 min; (C) surfactant system injection for 20 min; (D) following water flooding for 25 min.

residual oil in the high-permeability part by the surfactant, which cannot improve the heterogeneity of the reservoir. Therefore in the subsequent water flooding process, there is still some remaining oil in the core that is not produced yet (Fig. 5.66D), resulting in much lower oil displacement efficiency than with the SHC flooding system.

5.5.4 Microscopic displacement mechanism of the SHC flooding system [2]

The microscopic visualization experiment can directly reflect the migration state and displacement process of the SHC flooding system in porous media. The experiment reveals its displacement mechanism by observing the activation of different forms of remaining oil and residual oil by the SHC flooding system.

5.5.4.1 Distribution of remaining oil

The glass etching model was used to study the microscopic displacement mechanism of the SHC flooding system. The experimental process is shown in Fig. 5.67. The injection end simulates an injection well while the output end simulates a production well. The experimental steps are as follows. First, water was saturated in the model, and oil and water flooding were saturated to 98%. Then, the SHC flooding system (0.1% DPG + 0.1% THSB surfactant) was injected, with the water flooding to 98%. In order to facilitate the observation of the controllability of the SHC flooding system on the reservoir, the DPG particles was dyed with tea polyphenols as dark blue.

Fig. 5.68 reflects the displacement effect of the SHC flooding system intuitively. The water flooding recovery rate calculated by the image analysis software is only 25%. After 25 minutes injection of the SHC flooding system, the recovery rate reaches 76% and the recovery rate increases by 51%. In the subsequent water flooding, the increase of the recovery rate is only 2.4%. It can be seen that the enhanced oil recovery mainly exists in

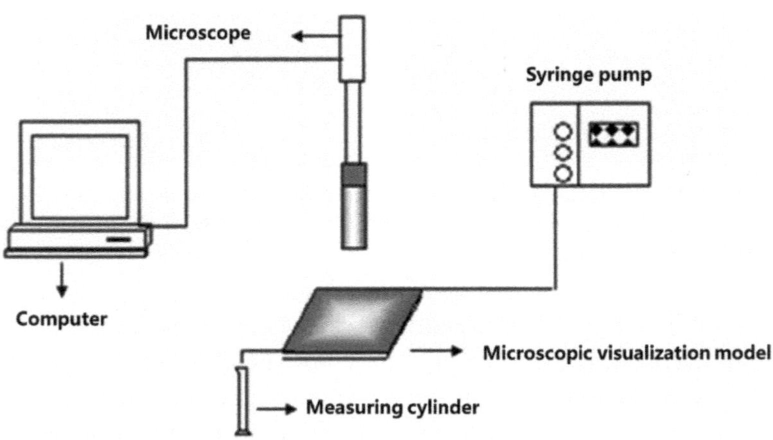

Figure 5.67 The flow chart of visual simulation experiment.

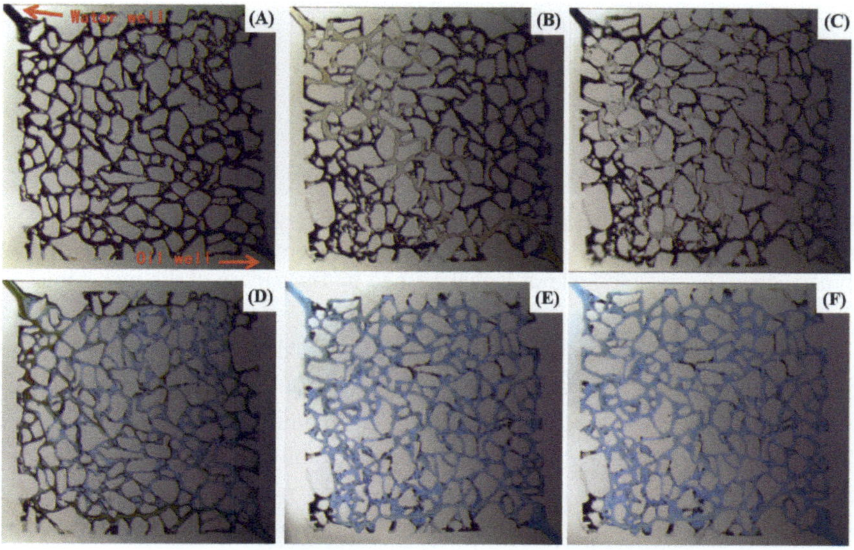

Figure 5.68 Visualization results of SHC flooding system: (A) initial state; (B) water flooding for 15 min; (C) injecting SHC flooding system for 5 min; (D) injecting SHC flooding system for 15 min; (E) injecting SHC flooding system for 25 min; (F) following water flooding for 20 min.

the displacement process of the SHC flooding system. In the initial stage of water flooding, the injected water drives the crude oil out of the model's high-permeability channel, thus increasing the heterogeneity of the model. A dominant channel is gradually formed between the water injection well and oil production well. Thus the injected water breaks through from the dominant channel to the oil well and results in limited recovery (Fig. 5.68B). When the SHC flooding system is injected, the DPG particles preferentially enters the high-permeability channel. Since the DPG particles are flexible particles, they can be deformed and transported into the depth of the porous media. The high-permeability channel can thus be controlled effectively by the direct plugging of a single particle or the bridging effect of multiple particles. At this point, the surfactant in the SHC flooding system enters the low-permeability part of the model, and the remaining oil is driven out through emulsification, wetting reversal, etc. (Fig. 5.68C and D). At the same time, the interaction of the components in the SHC flooding system enhances the emulsification of the remaining oil and wetting alteration. Both the microcontrol effect of superimposed DPG particles on the reservoir and the high displacement efficiency of the surfactant significantly improve the oil recovery.

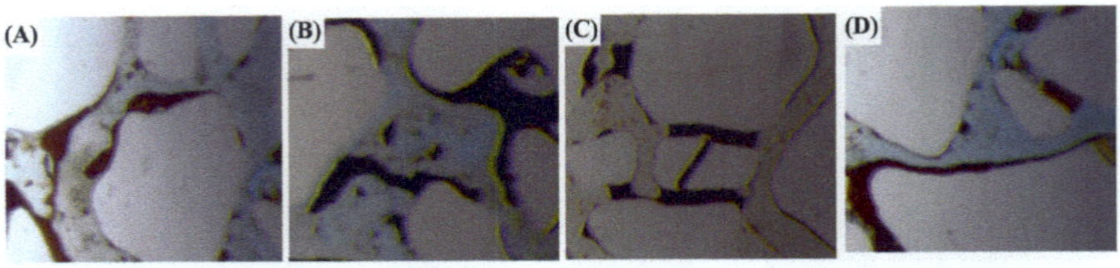

Figure 5.69 Distribution of remaining oil: (A) remaining oil formed by water channeling; (B) remaining oil on or inside thick oil layers; (C) remaining oil in the pores, fine throats, etc.; (D) remaining oil not highly washed on the surface of the rock particles.

According to the results of microscopic visualization experiments, the remaining oil that has not been displaced in the model are mainly in four forms after water flooding (Fig. 5.69): (A) remaining oil formed by water channeling; (B) remaining oil on or inside thick oil layers; (C) remaining oil in the pores, fine throats, etc.; (D) remaining oil not highly washed on the surface of the rock particles (mainly in the form of oil film). Also, after water flooding, the model heterogeneity intensifies and subsequent water flooding cannot drive the remaining oil of the above four forms. If the SHC flooding system is injected, the remaining oil is displaced through the synergistic effect of the microreservoir regulation effect of the DPG particles in the SHC flooding system and the efficient oil washing effect of the surfactant.

5.5.4.2 Reservoir microconformance controllability of SHC flooding system

In light of the results of previous studies, the SHC flooding system has the ability to microcontrol the reservoir. In order to further illustrate the microconformance ability of the SHC flooding system on the reservoir, the distribution state of the SHC system in the oil–water two phases was observed by a visualization device. The experimental results are shown in Fig. 5.70, in which Fig. 5.70A–D is enlarged by five times, Fig. 5.70E–I by 40 times, and Fig. 5.70J by 100 times. When the SHC flooding system migrates in the porous media, it generally passes through the pore throat in two forms. When the particle size of the DPG particle is smaller than the pore throat radius, it passes directly. When the particle radius is larger than the pore throat radius, the DPG particle deforms due to the increase of the pressure difference between the upper and lower flow of the pore throat. When approaching the size of the pore throat, DPG particles slowly pass through the pore throat and then returns to its original state. They are allowed to enter the deep core smoothly by the repulsive force between the SHC flooding system and the

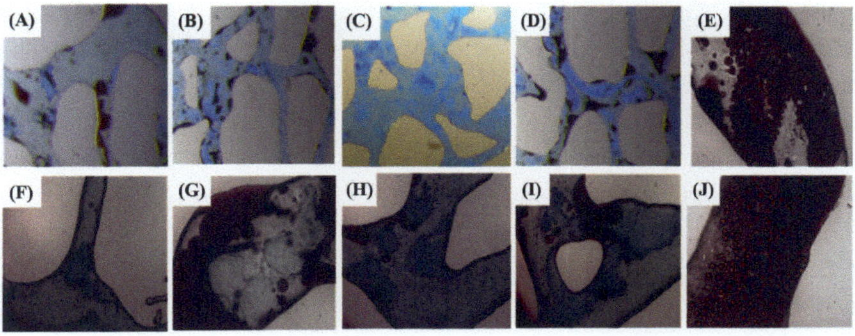

Figure 5.70 Microcontrollability of SHC flooding system on reservoir: (A), (F) single particle plugging; (B), (G) retention; (C), (H) multiple particles bridging and plugging; (D), (I) adsorption; (E), (J) emulsion plugging effect.

surface of the reservoir. After retention and adsorption on the surface of the strongly hydrophilic reservoir, the reservoir microcontrollability of the SHC flooding system can be enhanced. It can be seen from Fig. 5.70 that DPG particles of the SHC flooding system play a major role in the microcontrol of the reservoir. If the pore throat radius is smaller than the DPG particle size, the DPG particles are directly plugged when entering the model (Fig. 5.70A and E). If the pore throat radius is larger than the particle size of DPG, however, multiple particles are plugged by bridging or forming larger aggregates (Fig. 5.70B and F), forcing the subsequent injected water to turn to the untouched area. For large pores, multiple DPG particles stay in the pores, reducing the pore volume and increasing the flow resistance of subsequent injected water (Fig. 5.70C and G). The DPG particles can also be retained in the pore throat by adsorption, so that the subsequent injected water will flow around and drive out the remaining oil attached on the surface of the pore wall (Fig. 5.70D and H). The DPG particles are flexible and can be deformed through the pore throat. When the DPG particles pass through the pore throat, the negative pressure generated will also drive out the remaining oil. Fig. 5.70 further shows that when a single DPG particle or multiple DPG particles plug the pores, there is still a certain gap between the particles and the pore wall and between the particles and the particles. The difference in surface polarity between the DPG particles and the remaining oil causes the remaining oil to deform through the gap under the subsequent injection pressure, thereby improving the oil recovery. In addition, the viscosity-increasing effect of the DPG particles in the SHC flooding system can improve the fluidity ratio of the surfactant and reduce its diffusion rate. The microcontrol effect promotes the surfactant to shift to the middle and low-permeability zones, which improves the surface activity, increasing the microscopic oil washing efficiency of the agent. When the SHC flooding system is in contact with the remaining oil in the formation, the surfactant or the DPG particles adsorbing the surfactant emulsify the crude oil to form W/O or O/W emulsion particles. The concerned substance then has the effect of strengthening the microcontrol on the high-level permeability (Fig. 5.70E and J).

5.5.4.3 Microscopic oil displacement mechanism of SHC flooding system

The SHC flooding system is a binary system composed of DPG and surfactant. When the DPG microcontrols the reservoir, the surfactant improves the displacement efficiency of the SHC flooding system through the combination mechanisms (including oil−water interfacial tension reduction mechanism, emulsification mechanism, wetting reversal mechanism, and coalescence mechanism) to form an oil belt.

5.5.4.3.1 Interfacial tension reduction

Affected by capillary force, adhesion force, and cohesion force, part of the remaining oil adheres to the surface of the formation, reducing the oil displacement effect. The addition of surfactants in the SHC flooding system can be adsorbed on the oil−water interface, a way of effectively reducing the oil−water interfacial tension. Then, the above three forces are reduced, so as to displace the remaining oil adhering to the formation surface and improve the oil washing capacity. Capillary force, adhesion force, and cohesion force can be calculated by the following formula:

Capillary force:

$$P_c = \frac{2\varphi \cos\theta}{r} \tag{5.10}$$

where P_c is capillary force; φ is the oil−water interfacial tension; θ is water phase wetting contact angle; and r is capillary radius.

Adhesion energy:

$$W = \varphi(1 - \cos\theta) \tag{5.11}$$

where W is adhesion energy, the force of oil peeling off the rock surface.

Cohesion force:

$$A = 4\pi n r_2^2 \varphi \left(1 - \frac{r_2}{r_1}\right) \tag{5.12}$$

where A is cohesion force, the energy necessary to overcome cohesion; n is oil droplet numbers; r_1 is oil droplet radius before dispersion; and r_2 is oil droplet radius after dispersion.

5.5.4.3.2 Synergy-emulsification mechanism

The emulsification phenomenon during the displacement process of the SHC flooding system was observed by visualization devices. The experimental results are shown in Fig. 5.71, where Fig. 5.71A–D are magnified by five times, Fig. 5.71E by 40 times, and Fig. 5.71F and G by 100 times. The emulsified oil droplets have difficulty in adhering to the formation surface during the forward movement, thus improving the efficiency of oil washing. At the same time, the DPG particles adsorbing surfactants also have a certain surface activity. When coming into contact with crude oil, they emulsify the crude oil. The synergistic effect of the two behaviors enhances the remaining oil's emulsifying ability. The oil-in-water emulsion or multiple emulsions formed by emulsification produces a plugging effect at the high-permeability part (Fig. 5.71F and G). This performance plays a microcontrol function on the high-permeability part, enabling the subsequent injected fluid to be uniformly advanced and the sweep coefficient to be improved. In addition, the surfactant can emulsify the remaining oil, increase the viscosity of the displacement medium, and further enhance the fluidity controllability of the displacement medium.

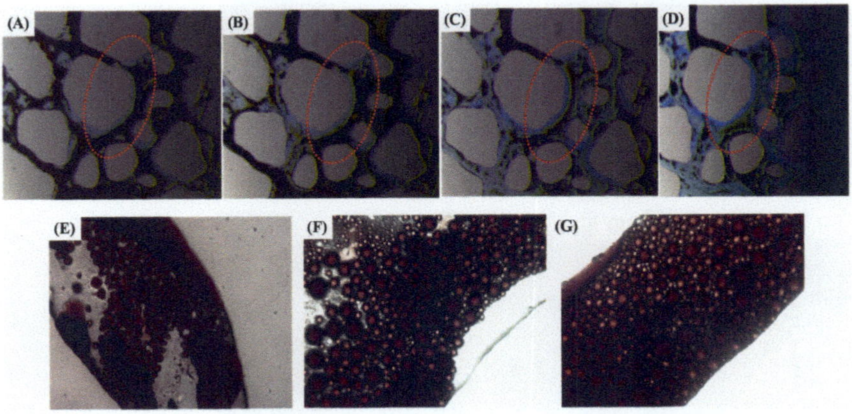

Figure 5.71 The synergistic-emulsification effect on the remaining oil: (A)–(C) 15 min, O/W emulsion and multiple; (D) initial state; (E) injection 5 min; (F) injection 15 min; (G) injection 25 min.

5.5.4.3.3 Strengthening-wetting reversal mechanism

The wettability of the porous media surface porous media plays an important role in improving the oil recovery. When the SHC flooding system flows in the porous media, the addition of the DPG weakens the negative charge of the SHC flooding system and reduces the electrostatic repulsion effect between the combination flooding system and the formation surface. This is favorable for surfactant molecules to adsorb on the surface of the formation and to strengthen the wetting ability of the combination flooding system. When the rock surface changes from lipophilic to moderate wettability, the capillary force gradually decreases, as shown in the Eq. (5.10). The adhesion energy between the oil and the rock surface is reduced, making detaching easy (as shown in Fig. 5.72) and thereby improving the oil washing efficiency of the SHC flooding system.

5.5.4.3.4 Mechanism of coalescence to form oil belts

During the flooding process of the SHC flooding system, the displaced remaining oil mainly migrates in the porous media in the form of oil droplets. When an increasing amount of crude oil is stripped from the rock surface, the oil droplets collide with each other during the migration process. If the collision energy overcomes the electrostatic repulsion between the oil droplets, the oil droplets can gather together and form large oil droplets. The large oil droplets gather with each other to form an oil belt (as shown in Fig. 5.73). Then, the oil belt moves in the porous media, continuously

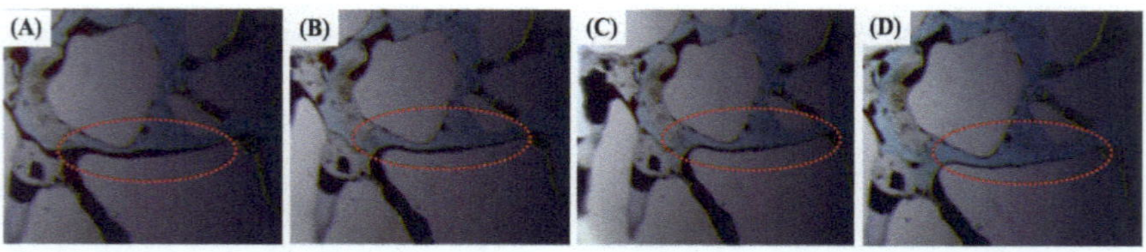

Figure 5.72 Enhanced wetting reversal of rock surface: (A) initial state; (B) 5 min; (C) 15 min; (D) 25 min.

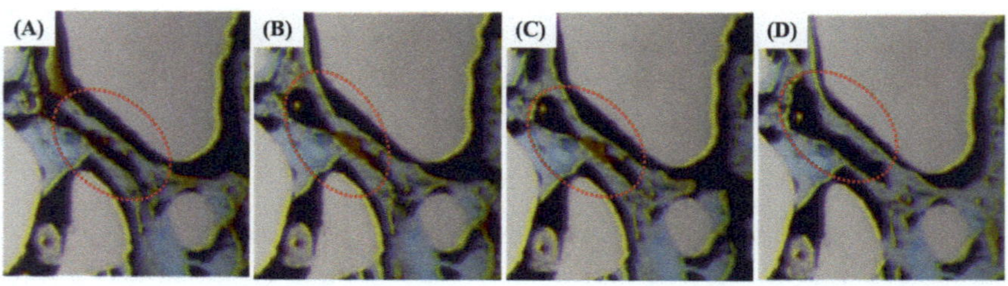

Figure 5.73 Coalescing and forming an oil belt: (A) initial state; (B) 5 min; (C) 10 min; (D) 20 min.

gathering and bringing in the oil droplets that it encounters, so as to continue enlarging. This is beneficial to the production of crude oil.

Section 5.6 Field application case in zhejiang oilfield

5.6.1 General situation of the targeted well

The reservoir of the block where well Ji-A stands is a typical high-temperature tight reservoir, with a porosity ranging from 3.89% to 13.03% and a permeability ranging from 0.03×10^{-3} to 2.13×10^{-3} μm^2. Its average porosity is 9.21% and its average permeability 0.74×10^{-3} m^2. The distribution of the pore throat radius measured by core mercury injection is shown in Fig. 5.74. The results indicate that the pore radius of the reservoir is mainly distributed between 0 and 0.4 μm, accounting for more than 90%. The formation temperature is roughly 130.0°C and the formation pressure about 38.0 MPa. The formation water is of $NaHCO_3$ type, with a salinity from 25,265 to 34,843 mg/L.

The crude oil in the block where the stimulated well is located contains a high pour point and a low sulfur content, so is the medium crude oil. The crude oil produced by the stimulated well is characterized by a surface density of 0.868 g/cm^3, a surface viscosity (60°C) of 18.45 mPa · s, a pour point of 42°C, and an initial boiling point of 225.1°C. The viscosity of

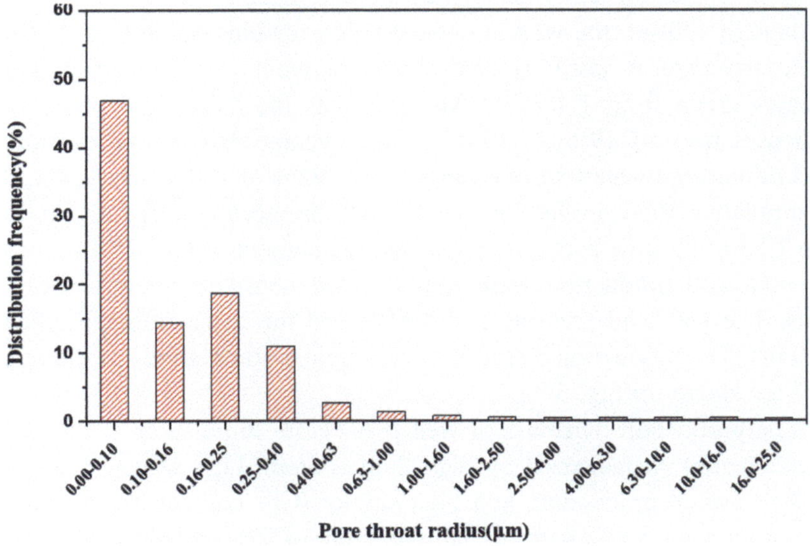

Figure 5.74 Pore throat radius distribution.

underground crude oil is generally lower than 10 mPa · s, whereas the fluidity of crude oil in the formation is relatively good.

Well Ji-A has been drilled since the beginning of 2016. The actual drill length of its horizontal section was 963.0 m and the penetration rate of the reservoir 100%. Oil and gas could be seen throughout the horizontal section. Subsequently, 13 stages of fracturing operations were completed. The cumulative number of proppants added was 950.4 m^3 and the cumulative fracturing fluid injected into the well was 12,229.7 m^3. Using the ground fractures monitoring method, the azimuth of the fractures was measured to be up to some 83 degrees north by east while the height was between 25 and 30 m and the lengths were between 230 and 340 m. The fracture treatments of this well meet the design requirements, allowing the creation of long fractures and controlling of the fracture height to realize an adequate reservoir stimulation. Blowout after fracturing operation of the well was constructed in September 2016.

5.6.2 Dynamic production parameters analysis of oil before operation

At the beginning of blowout after the fracturing of Ji-A, the nozzle with a size of "Φ2 mm → Φ3 mm → Φ5 mm → releasing freely" was adopted. The casing pressure releases down from 26 to 0 Mpa. The daily fluid production initially increases from 66.9 to 191.6 and then reduces to 18.6 m^3/d. The oil spots are visible during the blowout process. After 48 days of shut-in, the well is opened again for blowout and production while the gate is used to control the blowout. The casing pressure releases down from 1.4 to 0.8 MPa whereas the daily fluid production decreases from 45.19 to 11.19 m^3/d. The average daily oil production is 0.43 t/d and the water-cut rate ranges from 92% to 100%. Moreover, the accumulative fluid production is 6477 m^3, the accumulative oil production 12.3 t, the flow back rate 53%, and the content of Cl^- 18,147 mg/L. Blowout and production were implemented constantly before stimulations. The wellhead pressure is 3.4 MPa and the daily drainage volume 30.0 m^3. The production dynamic curves before stimulations of the target well are shown in Fig. 5.75.

The production dynamics of well Ji-A before stimulations demonstrate its production characterized by high fluid content, high water-cut (99%–100%), low oil production, and high salinity (Cl^- content 18,147 mg/L). The production is in a state of high water-cut of 99% and above, and basically no oil is produced during the process of flowback and production test.

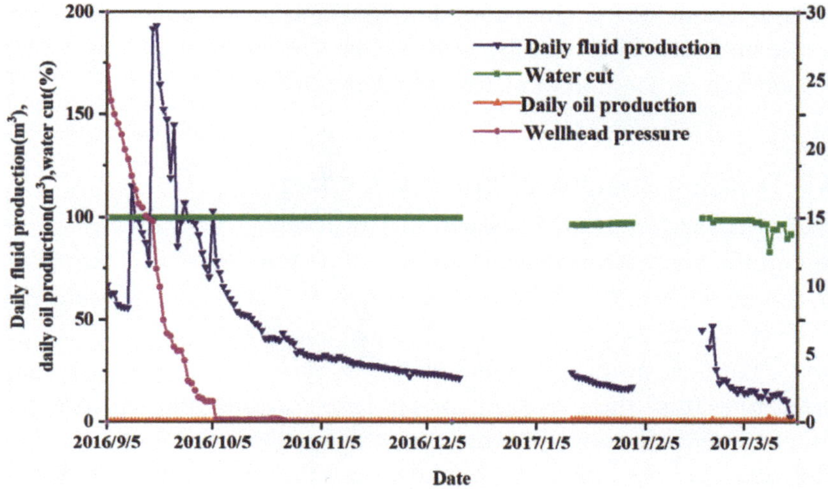

Figure 5.75 The production dynamic curves before operation.

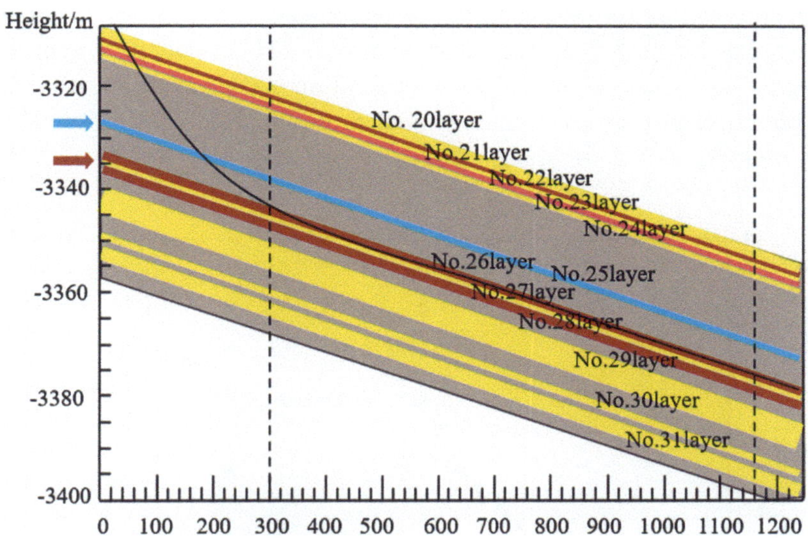

Figure 5.76 Profiles of the target reservoir.

It can be judged from the salinity of the produced fluid that formation water serves as the main source of flowback fluid. Combining with the oil reservoir profiles of the well Ji-A (Fig. 5.76), the No. 25 layer 5.5 m above the target layer (No. 26, 27, 28) is found to be the aquifer. Therefore it is possible that the cross-flow of the formation water in the No. 25 layer results from fracturing large fractures. The reservoir energy losses are because of the cross-flow through the large fractures. The reservoir pressure attenuates rapidly, resulting in the presence of low swept zones and the suppression of matrix oil expulsion.

5.6.3 Tracking analysis of operation effect

5.6.3.1 Measure ideas and design

Based on the produced water analysis of well Ji-A, it comes to the conclusion that it is necessary to control the large fractures for inhibiting the cross-flow of the large fractures and expanding the swept zone of the matrix. The self-developed DPG particles SHC flooding system was selected to control the cross-flow of the large fractures and enhance the oil displacement capacity of the matrix.

Because of the high temperature of the reservoir, the high-temperature DPG particles were prepared for the field test. The SHC flooding system (0.06% DPG + 0.4% surfactant) was designed, with an injection of 320 m^3 of fluid. In the early stage, 40.0 m^3 of formation water was injected as a pilot test; then, the SHC flooding system was injected to displace 80 m^3 of polymer solution and formation water. The construction process went smoothly (as shown in Fig. 5.77), after which the well was shut in for about 20 days.

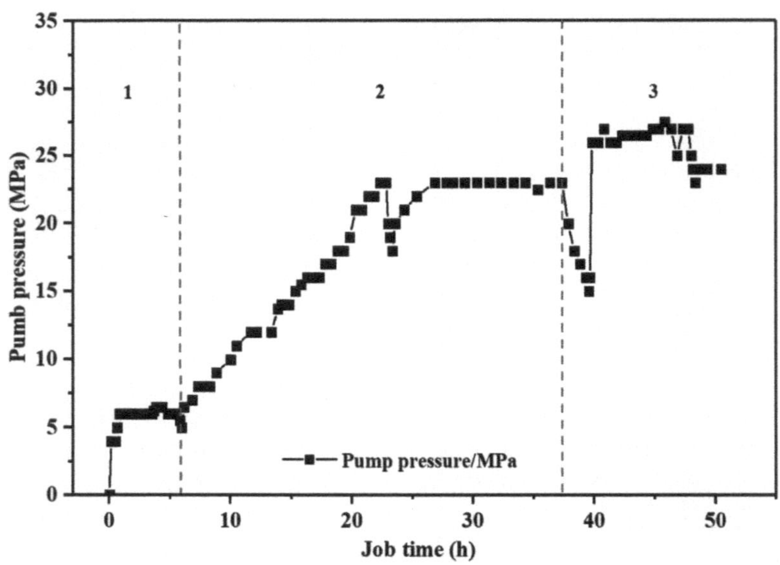

Figure 5.77 The pump pressure curves during pumping in the field.

5.6.3.2 Operation effect analysis

Production started in September 2017 after the injection of SHC flooding system in August 2017, to take effect after the initial water drainage. Then the water-cut decreased significantly from 100% to 78.5%, whereas the daily oil production increased sharply from 0 to 3.4 t/d. The cumulative increase in oil production exceeded 210 t in 6 months after the effect and the input–output ratio was higher than 5:1. The production dynamics before and after the stimulations are shown in Fig. 5.78.

In light of the comparison of the production before and after the stimulations, the cross-flow in large fractures of the stimulated well is controlled and the water-cut reduces in a significant manner after the injection of the SHC flooding system. Meanwhile, the significant weakening of the drainage capacity of the stimulated well indicates that the SHC flooding system achieves control of the large fractures without

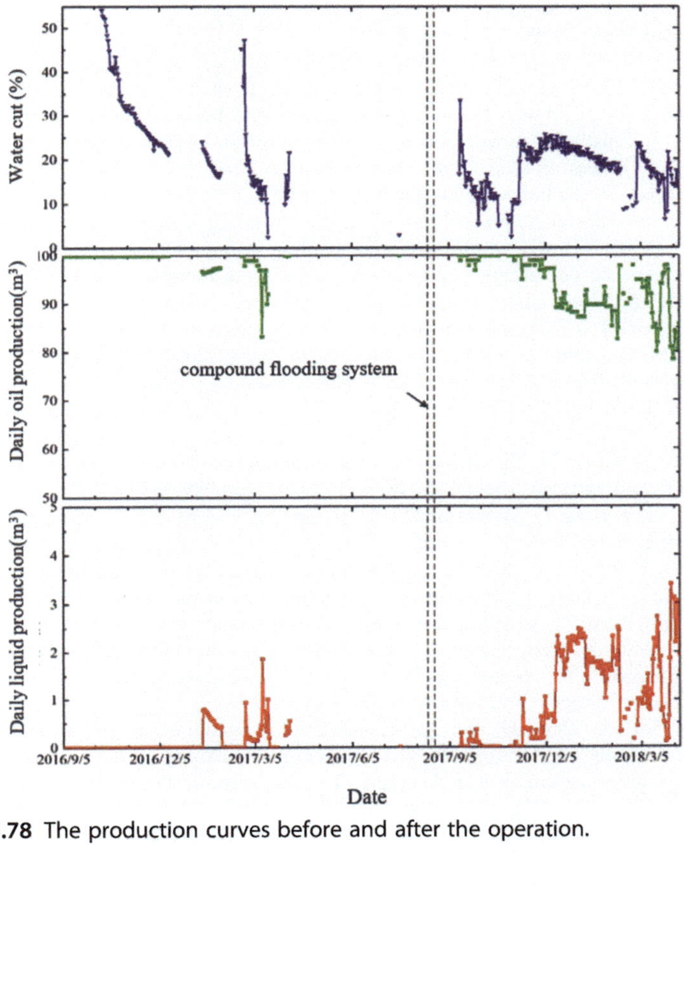

Figure 5.78 The production curves before and after the operation.

causing obvious pollution to the dense matrix. In addition, the periodic oil production increases significantly. This suggests that the SHC flooding system effectively broadens the operation range and improves the oil displacement efficiency of the matrix during the production process.

References

[1] Luo S, Yu Z, Niu G, et al. Study on the temperature-resistant and salt-tolerant surfactant BHJ-2 for EOR. Fine Chemicals 2016;33(1):98–104.

[2] Zhao G, Dai C, You Q. Characteristics and displacement mechanisms of the dispersed particle gel soft heterogeneous combination flooding system. Petroleum Exploration and Development 2018;45(3):464–73.

[3] Jang S, Lin S, Mai P, et al. Molecular dynamics study of a surfactant-mediated decane-water interface: effect of molecular architecture of alkyl benzene sulfonate. Journal of Physical Chemistry B 2004;108:12130–40.

[4] Dana A, Gocheva G, Millera R. Tensiometry and dilational rheology of mixed β-lactoglobulin/ionic surfactant adsorption layers at water/air and water/hexane interfaces. Journal of Colloid and Interface Science 2015;449:383–91.

[5] NY/T-2009. Colorimetric method for determination of betaine in betaine. China Standard Press; 2009.

[6] Zhao J, Dai C, Fang J, et al. Surface properties and adsorption behaviors of cocamidopropyl dimethyl amine oxide under high temperature and high salinity. Colloids and Surfaces A: Physicochemical and Engineering Aspects 2014;450:93–8.

[7] Boris A. Dilational surface rheology of polymer and polymer/surfactant solutions. Current Opinion in Colloid & Interface Science 2010;15:229–36.

[8] Yarranton H, Sztukowski D, Urrutia P. Effect of interfacial rheology on model emulsion coalescence. Journal of Colloid and Interface Science 2007;310:246–52.

[9] Lyu Y, Gu C, Fan X, et al. Interfacial rheology of a novel dispersed particle gel soft heterogeneous combination flooding system at the oil–water interface. Colloids and Surfaces A: Physicochemical and Engineering Aspects 2018;559:23–34.

[10] Israelachvili J, Min Y, Akbulut M, et al. Recent advances in the surface forces apparatus (SFA) technique. Reports on Progress in Physics 2001;73:036601–17.

[11] Omarjee P, Espert A, Mondain O. Polymer-induced repulsive forces at solid-liquid and at liquid-liquid interfaces. Langmuir: The ACS Journal of Surfaces and Colloids 2001;17:5693–5.

[12] Maeda N, Chen N, Tirrell M, et al. Adhesion and friction mechanisms of polymer-on-polymer surfaces. Science (New York, N.Y.) 2002;297:379–82.

[13] Zeng H, Hwang D, Israelachvili J, et al. Strong reversible Fe^{3+}-mediated bridging between dopa-containing protein films in water. Proceedings of the National Academy of Sciences of the United States of America 2010;107:12850–3.

[14] Leckband D. Surface force apparatus measurements of molecular forces in biological adhesion. New York: Springer Science; 2008. p. 1–9.

[15] Liu K, Yue X, Yang C, et al. Experimental study on chromatographic separation phenomenon of ternary combination oil displacement agent system. China Surfactant Detergent and Cosmetics 2014;44(12):671–5.

[16] Li D, Shi M, Wang D, et al. Chromatographic separation of chemicals in alkaline surfactant polymer flooding in reservoir rocks in the daqing oil field. In: SPE international symposium on oilfield chemistry, The Woodlands, Texas, April 20–22, 2009.

[17] Liu C, Wang K, Wang Y, et al. Polymer/surfactant binary flooding in A oilfield with high temperature and high salinity. Lithologic Reservoirs 2017;29(3):152–8.

[18] Yu Q, Guo S, Li Q, et al. Studies of adsorption and chromatographic separation of petroleum sulfonate/alkanolamide combined system. Speciality Petro-Chemicals 2015;32(5):68–71.
[19] Deng S, Lu W, Liu Q, et al. Research on oil displacement mechanism in conglomerate using *CT* scanning method. Petroleum Exploration and Development 2014;41:365–70.

CHAPTER 6

DPG-strengthened polymer/surfactant combination flooding technology

As one of the most popular tertiary oil recovery technologies, the polymer and surfactant combination flooding is widely popular in China's oilfields. It plays an important role to sustain oil production in the mature oilfields in China. Both its scale of application and annual increase of oil production in China rank at the top across the world [1–3]. Some drawbacks of this technology, however, have been exposed in the implementation of polymer flooding in oilfields. For example, the shear degradation of polymer caused by the injection device and the seepage shearing in porous media may lead to a sharp decrease in the polymer viscosity. Also, the dilution effect by the incompatibility of polymer with the reservoir, minerals, and fluid may lead to a significant reduction of the mobility control performance [4–7]. In particular, in the later water flooding stage, the rapid decline of well injection pressure makes the breakthrough of injection fluids much more rapid at the end of producing well. This will, however, greatly devalue the usage of injection fluid for enhanced oil recovery. In this chapter, a novel DPG-strengthened polymer/surfactant (DPS) combination flooding technology is first introduced as an improvement of the polymer/surfactant combination flooding technology. Through the synergistic effects DPG particles and polymer, the mobility control performance and the injection pressure in the later water flooding stage are improved in an ingenious way. The overall performance of enhanced oil recovery by the novel flooding system is then significantly enhanced.

Section 6.1 Construction of DPS combination flooding system

In this section, the compatibility, viscosity, and interfacial tension were used as evaluation indexes to design and evaluate the DPS combination flooding system. The polymer marked BD was applied at a hydrolysis degree of 17% and an average molecular weight of 18,000,000 mg/mol. The surfactant in

Theory and technology of multiscale dispersed particle gel for in-depth profile control
DOI: https://doi.org/10.1016/B978-0-323-99849-9.00006-0

use was petroleum sulfonate, codenamed CP. The average particle size of DPG particles was about 2.2 μm. The salinity of the simulated formation water was 6000 mg/L (i.e., Na^+, 5500 mg/L; Ca^{2+}, 400 mg/L; Mg^{2+}, 100 mg/L). The viscosity of simulated crude oil was 10 mPa·s at 45°C, a result that was collected in the north of Daqing Oilfield in China.

6.1.1 Compatibility

First of all, the DPS combination flooding system was prepared with the simulated formation water at room temperature. Then, the system was allowed to stand for an observation of whether the precipitation or flocculation occurs. In this case, the polymer concentration was 0.15% and 0.20%, the surfactant concentration ranged from 0.1% to 0.3%, and the DPG concentration ranged from 0.04% to 0.08%. The experimental results are shown in Table 6.1 and Fig. 6.1. It can be observed that no precipitation and flocculation are generated in the combination flooding system at either room temperature and 45°C. The results justify the excellent compatibility of DPG particles with the polymer/surfactant combination flooding system.

6.1.2 Interfacial tension

6.1.2.1 Effect of surfactant concentration

In this experiment, the concentration of BD polymer was fixed at 0.15%, the concentration of DPG particles at 0.06%, and the concentration of CP surfactant ranged from 0.05% to 0.4%. A TX-500C interfacial tensiometer was used to determine the interfacial tension (IFT) at the test temperature of 45°C. Through comparison testing, the capacity of both the

Table 6.1 Compatibility of DPG particles with polymer/surfactant combination flooding system.

No.	Formula			Precipitation or flocculation			
				Room temperature		45°C	
	BD polymer (%)	CP surfactant (%)	DPG (%)	1 day	5 day	1 day	5 day
A	0.15	0.1	0.04	No	No	No	No
B	0.15	0.2	0.06	No	No	No	No
C	0.15	0.3	0.08	No	No	No	No
D	0.20	0.1	0.04	No	No	No	No
E	0.20	0.2	0.06	No	No	No	No
F	0.20	0.3	0.08	No	No	No	No

Figure 6.1 Compatibility of DPG particles with polymer/surfactant combination flooding system: (A) 0.15% BD polymer + 0.1% CP surfactant + 0.04% DPG; (B) 0.15% BD polymer + 0.2% CP surfactant + 0.06% DPG; (C) 0.15% BD polymer + 0.3% CP surfactant + 0.08% DPG; (D) 0.20% BD polymer + 0.1% CP surfactant + 0.04% DPG; (E) 0.20% BD polymer + 0.2% CP surfactant + 0.06% DPG; (F) 0.20% BD polymer + 0.3% CP surfactant + 0.08% DPG.

conventional polymer/surfactant (PS) combination flooding system and the DPS combination flooding system to reduce oil–water interfacial tension were investigated. The results are shown in Fig. 6.2.

It can be observed from Fig. 6.2 that the interfacial tension using both the PS and DPS systems gradually decreases with the increase of surfactant concentration. When the concentration of CP surfactant exceeds 0.3%, no further changes in interfacial tension are observed in either combination flooding systems. This can be explained by when the surfactant concentration increases, more surfactant molecules are adsorbed on the oil–water interface, which then greatly reduces the oil–water interfacial tension. Notably, the adsorption and desorption behaviors of surfactant molecules simultaneously take place on the oil–water interface. When the two behaviors reach dynamic equilibrium, the interfacial tension does not change obviously even with the further increase of surfactant concentration. In addition, it can be concluded from Fig. 6.2 that the oil–water interfacial tension increases slightly after DPG particles are added to the polymer/surfactant combination flooding system. This phenomenon may result from the competitive adsorption behaviors between the DPG

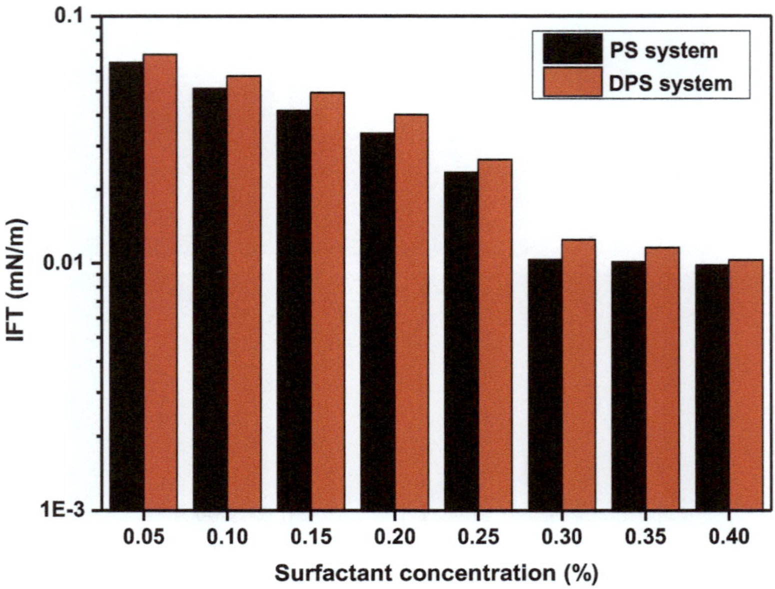

Figure 6.2 Effect of surfactant concentration on interfacial tension of combination flooding system.

particles and surfactant molecules onto the oil–water interface. This competitive adsorption behavior reduces the ability of the DPS combination flooding system, thereby weakening the oil–water interfacial tension. In addition, the surfactant molecules can be also adsorbed onto the surface of DPG particles through the hydrogen bond and dispersion force, etc. This may further reduce the concentration of surfactant molecules in the DPS system and lead to a slight increase in oil–water interfacial tension.

6.1.2.2 Effect of DPG particle concentration

In this experiment the concentration of BD polymer was fixed at 0.15%, the concentration of CP surfactant at 0.3%, and the concentration of DPG particles ranged from 0% to 0.1%. The oil–water interfacial tension of the combination flooding system was measured at 45°C. The results are shown in Fig. 6.3.

It can be observed from Fig. 6.3 that the ability of the DPS combination flooding system to reduce the oil–water interfacial tension shows a decreasing trend with the increase of the DPG particle concentration. The final stable value is about 1.412×10^{-2} mN/m, justifying the excellent performance of the DPS combination flooding system in reducing

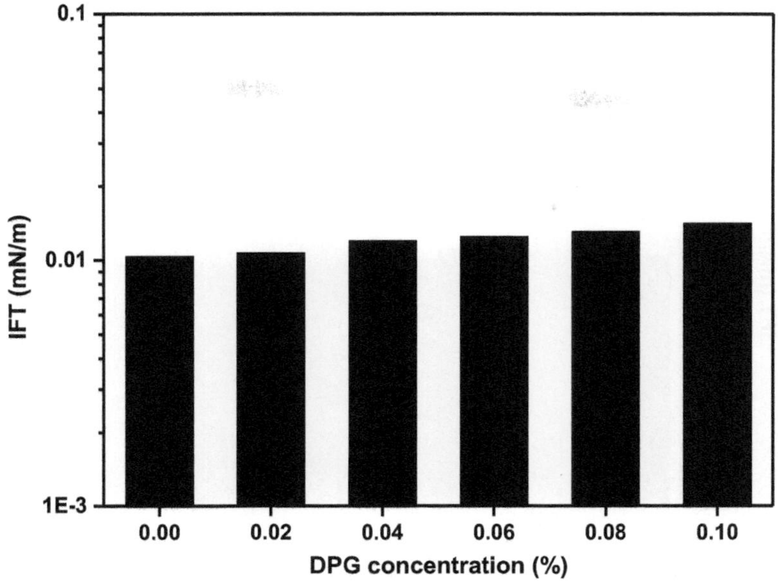

Figure 6.3 Effect of DPG particle concentration on interfacial tension of DPS combination flooding system.

the oil–water interfacial tension. When the DPG particles are added to the combination flooding system, the surfactant molecules are adsorbed onto the surface of the DPG particles through the hydrogen bond and dispersion force. On the one hand, this adsorption behavior consumes some of the surfactant molecules in both the solution and the oil–water interface. This leads to a slight increase in interfacial tension. On the other hand, the DPG particles have a good performance of oil displacement due to the adsorption of surfactant molecules. Thus the oil recovery of the DPS combination flooding system can be enhanced overall.

6.1.2.3 Effect of polymer concentration

In this experiment, the concentration of DPG particles was fixed at 0.06%, the concentration of CP surfactant was 0.3%, and the concentration of BD polymer ranged from 0% to 0.3%. The oil–water interfacial tension of the combination flooding system was measured at 45°C, for an evaluation on the ability of the combination flooding system to reduce the interfacial tension. The results are shown in Fig. 6.4.

It can be observed from Fig. 6.4 that the ability of the combination flooding system to reduce the oil–water interfacial tension slightly decreases around

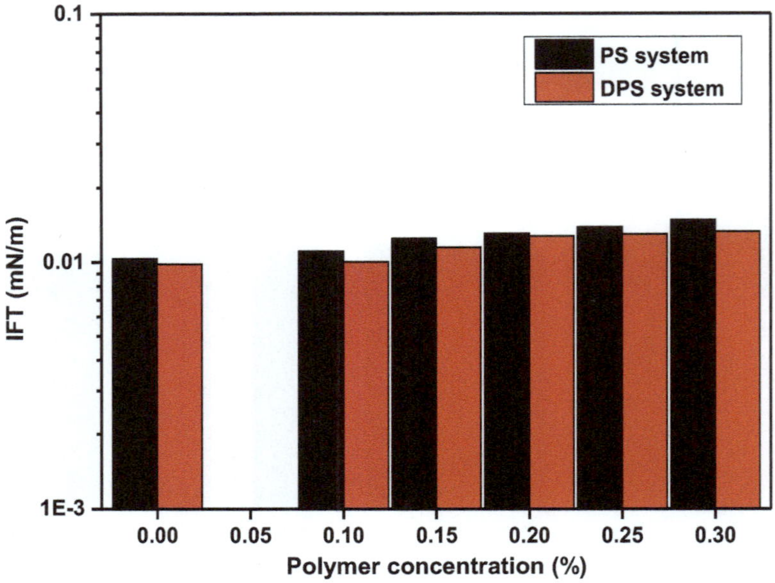

Figure 6.4 Effect of polymer concentration on interfacial tension (IFT) of combination flooding system.

10^{-2} mN/m even with the increase of the polymer concentration. This is brought about by the gradual increase in the viscosity of the combination flooding system along with the increase of polymer concentration. A decrease thus occurs in the surfactant molecules diffusion onto the oil–water interface and the weakened performance of interfacial tension reduction.

6.1.3 Viscosity

6.1.3.1 Effect of polymer concentration

The polymer, the main source to enhance the viscosity of the combination flooding system, determines the performance of oil displacement in the combination flooding process. In this case, the effect of polymer concentration on the viscosity of the combination flooding system was investigated. The concentration of DPG particles was fixed at 0.06%, the concentration of CP surfactant was 0.3%, and the concentration of BD polymer ranged from 0% to 0.3%. The Brookfield viscometer ($0^{\#}$ rotor; 6 rpm; 45°C) was used to test the viscosity of combination flooding system. The experimental results are shown in Fig. 6.5.

It can be observed from Fig. 6.5 that the viscosity of both PS and DPS systems keeps rising with the increase of polymer concentration. First, the

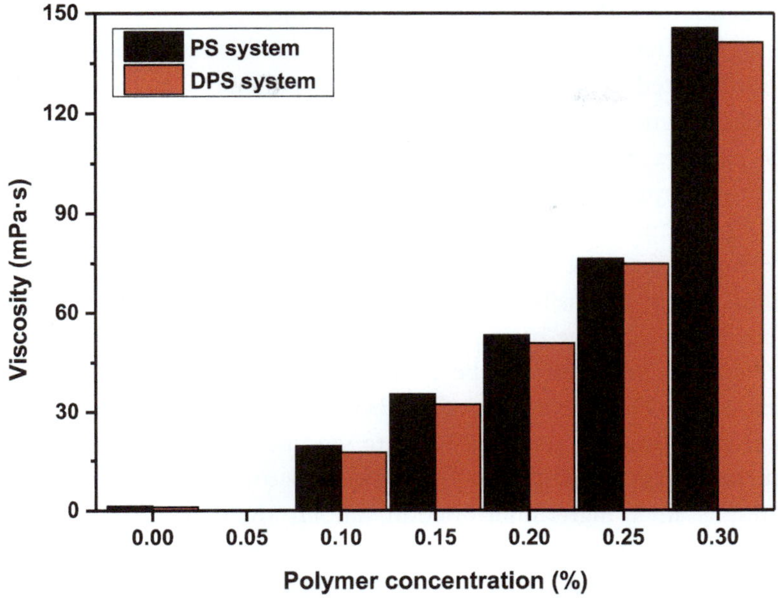

Figure 6.5 Effect of polymer concentration on viscosity of combination flooding system.

polymer molecules entangle with each other and then form a structure when the polymer concentration exceeds a certain critical value. As a result, the viscosity of an association structure is generated. Meanwhile, the solvent effect of the hydrophilic groups in the polymer chain can lead to an increase in the viscosity of the combination flooding system. Furthermore, according to Fig. 6.5, the viscosity of the DPS system is larger than that of the PS combination flooding system. On the one hand, the surfactant molecules are adsorbed onto the surface of DPG particles, helping to enlarge the hydraulic radius of the particles. Then, the viscosity of the combination flooding system increases as well. On the other hand, the DPG particles intersperse in the polymer network structure to strengthen the stability of the polymer through hydrogen bonding and dispersion forces. This could further strengthen the structure viscosity. Through comprehensively optimization of the viscosity increasing ability and cost, the BD polymer with a concentration of 0.15% is determined in the following experiments.

6.1.3.2 Effect of DPG particle concentration

DPG particles are viscoelastic particles. It can be speculated that the viscosity of the combination flooding system can increase by the introduction of DPG particles. In this experiment, the effect of DPG particles concentration on the viscosity of the combination flooding system was investigated. The concentration of BD polymer was fixed at 0.15%, the concentration of CP surfactant at 0.3%, and the concentration of DPG particles ranged from 0% to 0.1%. The Brookfield viscometer ($0^{\#}$ rotor; 6 rpm; 45°C) was used to test the viscosity of combination flooding system. The experimental results are shown in Fig. 6.6.

It can be observed from Fig. 6.6 that the viscosity of the combination flooding system gradually rises with the increase of DPG particle concentration. This means the presence of synergistic viscosity increases the effect of DPG particles, which can be explained as follows. First, a higher concentration of DPG particles usually produces a higher solid content of the combination flooding system. That is to say, the reduction of distance and the increase of collision probability between particles take place. This usually results in a high inner friction among these particles and finally leads to the increase in viscosity of the combination flooding system. Second, the surfactant molecules are adsorbed onto the surfaces of DPG particles to enlarge the hydraulic radius of the particles. This helps increase

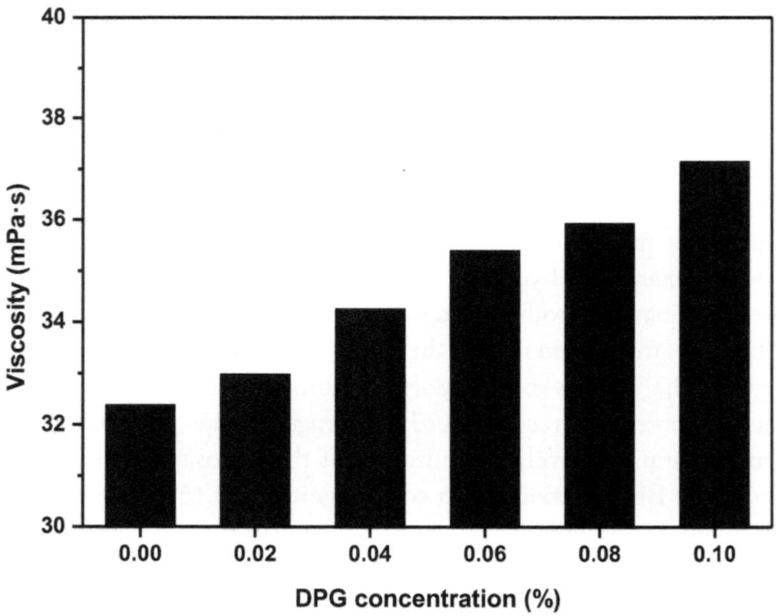

Figure 6.6 Effect of DPG particle concentration on viscosity of combination flooding system.

the collision probability among particles, thus making the viscosity of the combination flooding system further increase. Third, the DPG particles are interspersed in the network structure of the polymer, further synergistically increasing the viscosity of the combination flooding system.

6.1.3.3 Effect of surfactant concentration

In this experiment the effect of surfactant concentration on the viscosity of the combination flooding system is also investigated. The concentration of BD polymer was fixed at 0.15%, the concentration of DPG particles at 0.06%, and the concentration of CP surfactant ranged from 0% to 0.35%. The Brookfield viscometer ($0^{\#}$ rotor; 6 rpm; 45°C) was used to test the viscosity of the combination flooding system. The experimental results are shown in Fig. 6.7.

It can be observed from Fig. 6.7 that the viscosity of both PS and DPS systems increase slightly with the introduction of surfactant. These changes can be elaborated by the adsorption behavior of surfactants. The surfactant molecules are adsorbed onto the surfaces of DPG particles through the hydrogen bond and dispersion force. Then, an expansion emerges in the

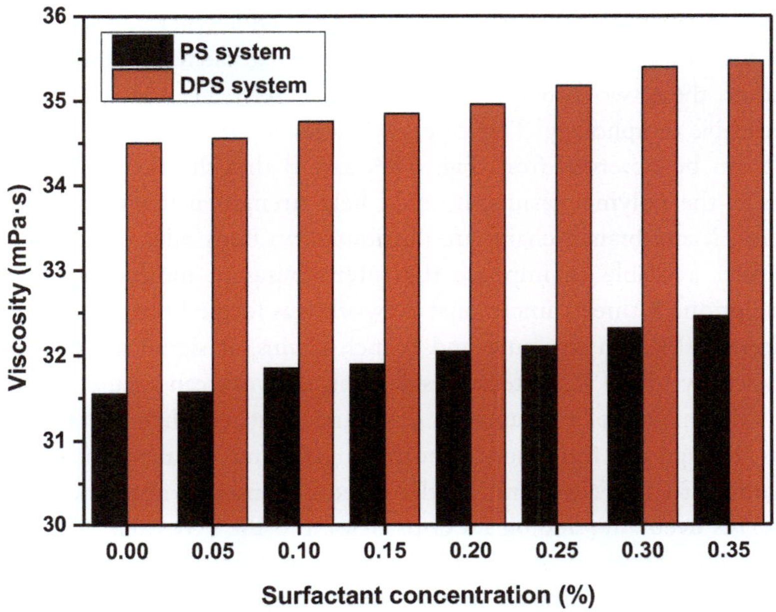

Figure 6.7 Effect of surfactant concentration on viscosity of combination flooding system.

hydraulic radius of the particles, so does the increase in the collision probability among particles [8,9]. In addition, the surfactant molecules enter the hydrophobic microdomains in the polymer molecules through the hydrophobic interaction, making the polymer chains more stretched. The expansion in the hydrodynamic sizes of the polymer molecules can also increase the viscosity of the combination flooding system.

Section 6.2 Characterization of DPS combination flooding system

In this section, modern including a viscometer, laser particle size analyzer, scanning electron microscope (SEM), and transmission electron microscope (TEM) were adopted to characterize the microscopic morphology, thermal stability, and coalescence expansion performance of the DPS combination flooding system.

6.2.1 Microscopic morphology of DPS combination flooding system

First of all, the PS combination flooding system (0.15% BD polymer + 0.3% CP surfactant, PS system) and DPS combination flooding system (0.15% BD polymer + 0.3% CP surfactant + 0.06% DPG, DPS system) were prepared with the simulated formation water. After letting them stand at 45°C for 24 hours, these two systems were engaged in SEM, in order to obtain the microscopic morphology. The experimental results are shown in Fig. 6.8.

It can be observed from Fig. 6.8A and B that the skeleton structure exists in the polymer chains. It could help strengthen the stability of the systems. Many branch chains are distributed on both sides of the skeleton structure, available to improve the intertwining of molecular chains in the solution. A three-dimensional network was formed through the intersection of skeleton structure and branch chains. It significantly increases the viscosity of the PS system. Besides, this network can ensure the excellent performance of resistance to shear and dilution while the PS system flows through the formation of porous media and enhances oil recovery. According to Fig. 6.8C and D, the three-dimensional network structure still exists even after adding DPG particles into the PS system. The skeleton structure with branched chains and the DPG particles are tightly locked together to form a compact and continuous network through the impact of hydrogen bond and dispersion force. Therefore both the viscosity and stability of DPS system are largely enhanced by the improved

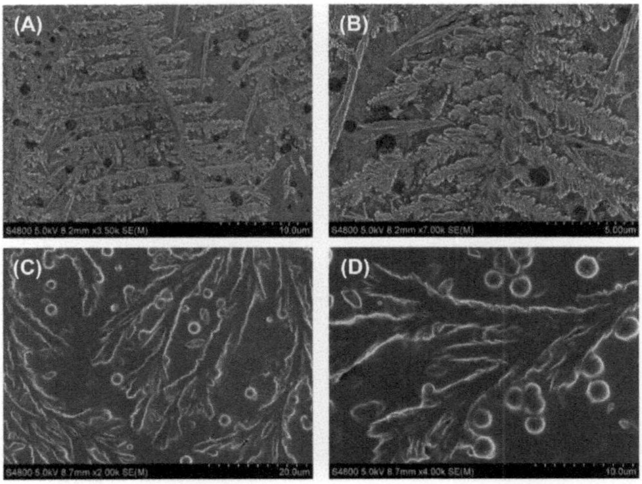

Figure 6.8 Microscopic morphology of the combination flooding system: (A, B) PS system, and (C, D) DPS system.

synergistic effects. The pressure during the injection process also increases while the DPS system flows through the porous media. In the following water flooding stage, the strengthening effect of dispersed particle gel further improves the mobility control performance of the DPS system and enhances oil recovery.

6.2.2 Thermal stability

6.2.2.1 Viscosity

The viscosity stability of DPS combination flooding system plays an important role in enhancing the oil recovery. In this case, the PS combination flooding system (0.15% BD polymer + 0.3% CP surfactant) and the DPS combination flooding system (0.15% BD polymer + 0.3% CP surfactant + 0.06% DPG) were continuously aged at 45°C. The viscosity changes of the two systems were investigated in the process of aging. The Brookfield viscometer ($0^{\#}$ rotor; 6 rpm; 45°C) was used to test the viscosity of the combination flooding system. The experimental results are shown in Fig. 6.9.

It can be observed from Fig. 6.9 that the viscosity of the two combination flooding systems shows a decreasing trend in the process of aging, and that the viscosity reduction of the DPS system is obviously weaker than that of the PS system. For the PS system, the degradation of polymer results in unequal positive and negative charge groups on the molecular chains in the

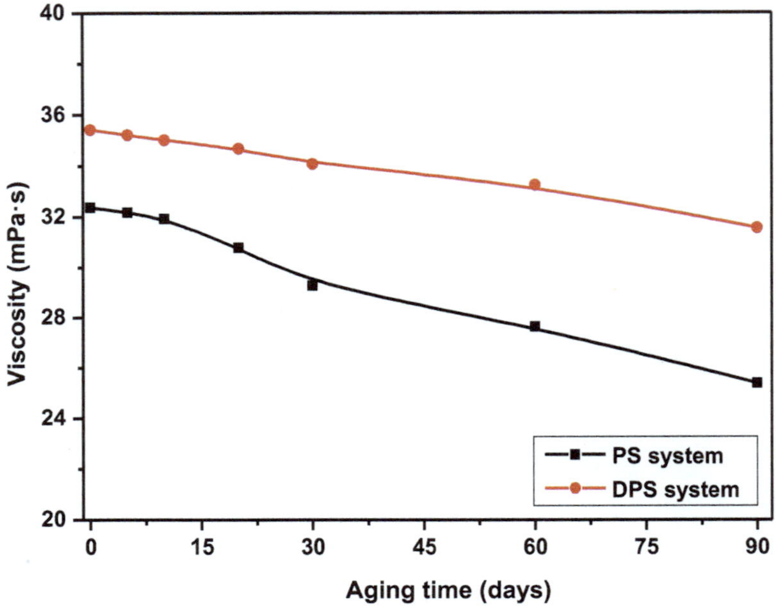

Figure 6.9 Viscosity changes of PS system and DPS system at different aging times.

process of aging. As a result, the degree of curling of the molecular chains increases and the viscosity of solution decreases [10,11]. For the DPS system, on the contrary, the adsorption of DPG particles onto the polymer chains improves the strength of the three-dimensional structure. Therefore the stability of the DPS system is significantly enhanced.

6.2.2.2 Interfacial tension

In this experiment, the PS combination flooding system (0.15% BD polymer + 0.3% CP surfactant, PS system) and the DPS combination flooding system (0.15% BD polymer + 0.3% CP surfactant + 0.06% DPG) were continuously aged at 45°C. The interfacial tension (IFT) changes of the two combination flooding systems were investigated in the process of aging. The TX-500C interfacial tensiometer was used to determine the interfacial tension at 45°C. The experimental results are shown in Fig. 6.10.

It can be observed from Fig. 6.10 that the interfacial tension of the two combination flooding systems slightly increases in the process of aging whereas the interfacial tension fluctuates in the range from 10^{-1} to 10^{-2} mN/m after 90 days of aging. These results indicate an excellent thermal stability of both systems. The interfacial tension of the DPS

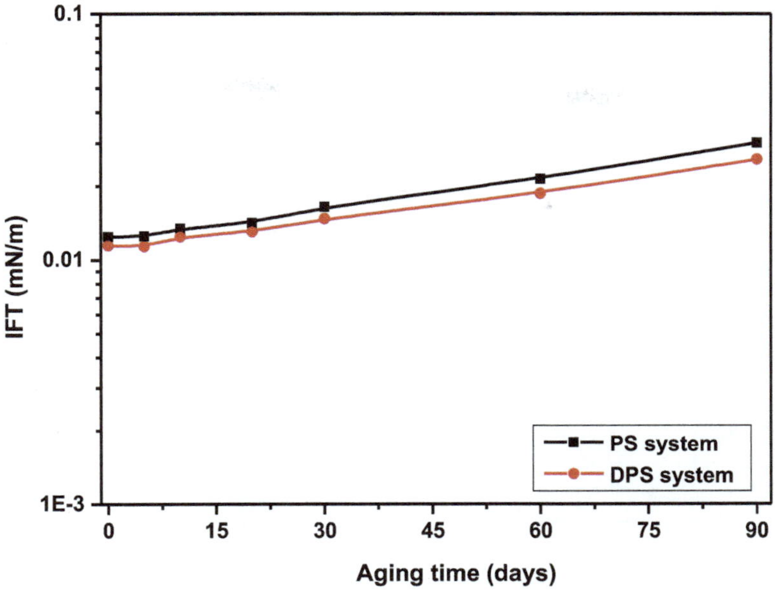

Figure 6.10 Interfacial tension changes of the PS system and the DPS system at with aging time.

system is slightly higher than that of the PS system. This may be due to the adsorption of surfactant molecules onto the surface of the dispersed particle gel, which leads to a decrease in the amount of surfactant molecules in the solution and on the oil—water interface.

6.2.3 Aggregation and expansion property

The DPG particles improve the profile control through their aggregation and expansion properties. As a consequence, the DPS combination flooding system harbors the characteristics of particle control. In this experiment, the DPS combination flooding system (0.06% DPG particles) and the DPS combination flooding system (0.15% BD polymer + 0.3% CP surfactant + 0.06% DPG) were continuously aged at 45°C in shaking water bath. The changes in the particle size of both flooding systems were investigated in the process of aging. The experimental results are shown in Fig. 6.11.

It can be observed from Fig. 6.11 that the particle size of the two combination flooding systems increases as the aging time increases. The aggregation and expansion ratio of dispersed particle gel is more than 20-fold after 90 days of aging. It indicates that both systems possess excellent aggregation and expansion property. For the DPS system with higher

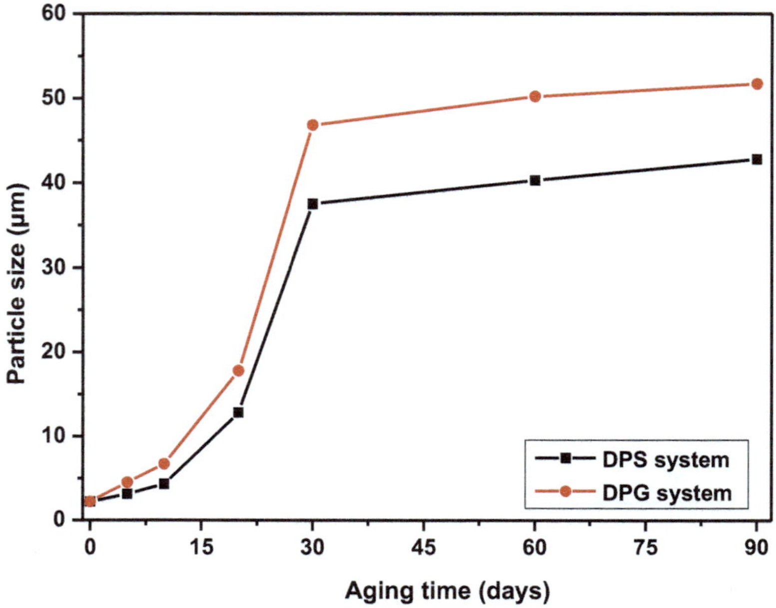

Figure 6.11 The aggregation and expansion property of the DPG system and the DPS system.

viscosity, the aggregation speed of DPG particles is hindered by the viscous force. Therefore the aggregation and expansion property of the DPS system is relatively weaker than that of the DPG system. It is worth noting that the aggregation and expansion property is still beneficial for the DPS system to improve the profile control. Particularly, in the following water flooding stage, the mobility control performance of the PS system is weakened due to the decline of viscosity. On the contrary, the DPS system still maintains an excellent performance in mobility control with the introduction of DPG particles.

Section 6.3 Enhanced oil recovery performance of DPS combination flooding system

In this section, the application performances of PS combination flooding system and DPS combination flooding system are comprehensively evaluated by different methods, including the mobility control performance, the profile improvement performance, and the enhanced oil recovery performance. The experimental results provide the theoretical foundation for the oilfield application.

6.3.1 Mobility control performance

In this experiment, the mobility control performance of the PS combination flooding system (0.15% BD polymer + 0.3% CP surfactant) and the DPS combination flooding system (0.15% BD polymer + 0.3% CP surfactant + 0.06% DPG) were investigated via single sand-pack flow experiments. All of the experiments were carried out at 45°C. The injection parameters are presented in Table 6.2, and the experimental results are presented in Figs. 6.12 and 6.13.

It can be observed from Figs. 6.12 and 6.13 that the injection of the two systems strengthens the resistance factor more than 20. It indicates an effective mitigation of the injection of the combination flooding system on the channel effects caused by the formation heterogeneity. In the water flooding stage, the long-term water flooding exacerbates the formation heterogeneity, thus leading to the reduction of resistance factor caused by the channel effect. In the combination flooding stage, the viscoelastic solution preferentially enters the high permeability channels. Here, the viscosity of the combination flooding system plays an important role in improving the mobility control. Through the viscosity increase in the water phase, the relative permeability of water phase is reduced effectively. Additionally, the polymer molecules of the combination flooding system are adsorbed onto the sand surfaces through the hydrogen bonds. This helps to reduce the effective radius of the pore-throat and further increase the flow resistance. Therefore a higher resistance factor is obtained during the combination flooding stage. It can be further observed from the Fig. 6.13 that the resistance factor of the DPS system is significantly higher than that of the PS system. This difference can be attributed to the plugging behavior of DPG particles in the DPS system. The DPG particles can directly plug or bridge across the pore-throats and then improve the profile control performance in reservoirs. Moreover, the synergistic thickening effect of DPG particles enhances the viscosity of the DPS system, helping contribute to a higher resistance factor.

However, the resistance factors of both systems have significant differences when the sand-packs are switched back to the following water flooding stage. The resistance factor of the PS system decreases more rapidly than that of the DPS system. For the PS system, the viscosity is the driving force for the mobility control. However, in the following water flooding stage, in the PS system the polymer viscosity gradually decreases as the polymer is easily diluted by contacting the injected water. Thus a

Table 6.2 Injection parameters.

Scheme	Core parameters					Injection volume (PV)
	Permeability (μm^2)	Length (cm)	Diameter (cm)	Porosity (%)	Pore volume (mL)	
PS system	1.031	10	2.5	26.3	13.06	1
DPS system	1.029	10	2.5	25.4	12.87	1

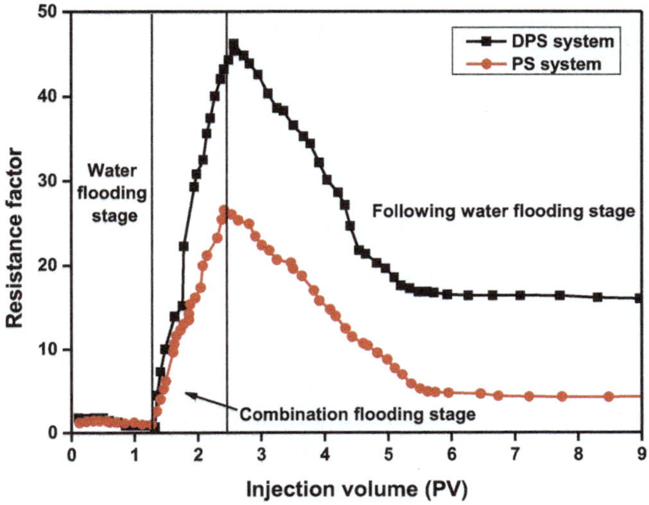

Figure 6.12 The mobility control performance of the PS system and the DPS system.

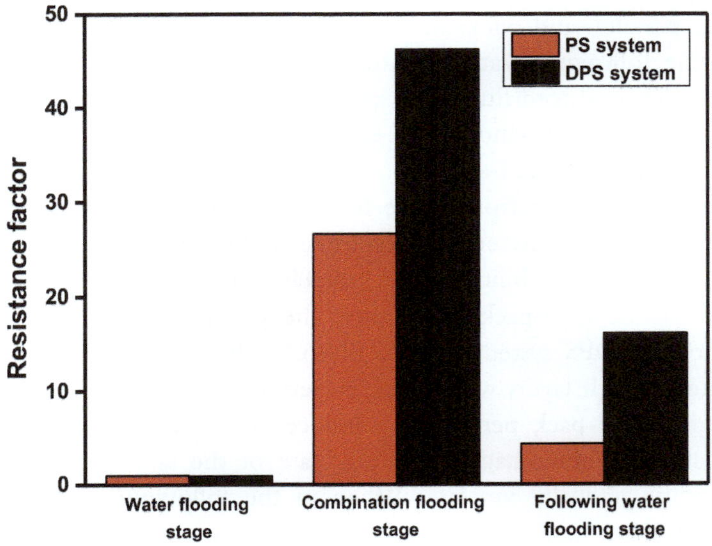

Figure 6.13 Resistance factors comparison of the PS system and DPS system.

lower resistance factor is generated. When the water flooding after DPS combination flooding is in operation, the viscosity of the DPS system and the plugging behaviors of the DPG particles produce high water flow resistance in the porous media. Although the viscosity of the DPS system gradually decreases as the water flooding continues, the DPG particles still effectively improve the profile control performance through the plugging behaviors of particles themselves. Therefore the DPS system can maintain a higher mobility control performance in the following water flooding stage.

6.3.2 Profile improvement performance

In this experiment, the shunt rate and profile improvement efficiency were used to evaluate the profile improvement performance of the DPS combination flooding system. All experiments were conducted using the parallel sand-packs model 20 cm in length and 2.5 cm in diameter. The experimental procedures are described as follows. First, the parallel sand-packs were prepared with different permeability ratio. Second, the water flooding was carried out at an injection rate of 0.5 mL/min until a stable pressure came into being. Third, a 1 PV DPS combination flooding system (0.15% BD polymer + 0.3% CP surfactant + 0.06% DPG) was injected into the parallel sand-packs. Last, the parallel sand-packs models were put to aging for 5 days at 45°C and the following water flooding was carried out. The changes of pressure and produced fluid rates were recorded to calculate the profile improvement performance. The experimental results are presented in Table 6.3 and Figs. 6.14–6.16.

It can be observed from Table 6.3 that the DPS system possesses an excellent profile improvement performance for the parallel sand-packs with different permeability ratios. Notably, the larger the permeability ratio of parallel sand-packs, the better the profile improvement performance of the DPS system. Figs. 6.14–6.16 show the changes of fluids flow rate through layers with different permeability. It can be confirmed that higher sand-pack permeability induces a greater shunt rate. In the combination flooding stage, the shunt rate of the sand-packs with low permeability gradually increase. When in the following water flooding stage, the shunt rate of the sand-packs with low permeability further increases. This indicates a significant mitigation of the side effects of formation heterogeneity by the DPS system. This may be related to the increasing effect of the viscosity by the polymer in the DPS system. The viscoelastic system can easily enter the sand-packs with high permeability and thus mitigate the damage to the nontarget formation. Therefore the

Table 6.3 Profile improvement performance of the DPS system with different permeability ratios.

Permeability ratio	Permeability type	Permeability (μm^2)	Shunt rate (%)		Profile improvement performance (%)
			Before injection	After injection	
1.14	Low permeability	0.524	57.45	16.32	85.56
	High permeability	0.597	42.55	83.68	
1.65	Low permeability	0.537	69.83	19.63	89.45
	High permeability	0.886	30.17	80.37	
4.38	Low permeability	0.574	76.52	20.84	91.92
	High permeability	2.514	23.48	79.16	

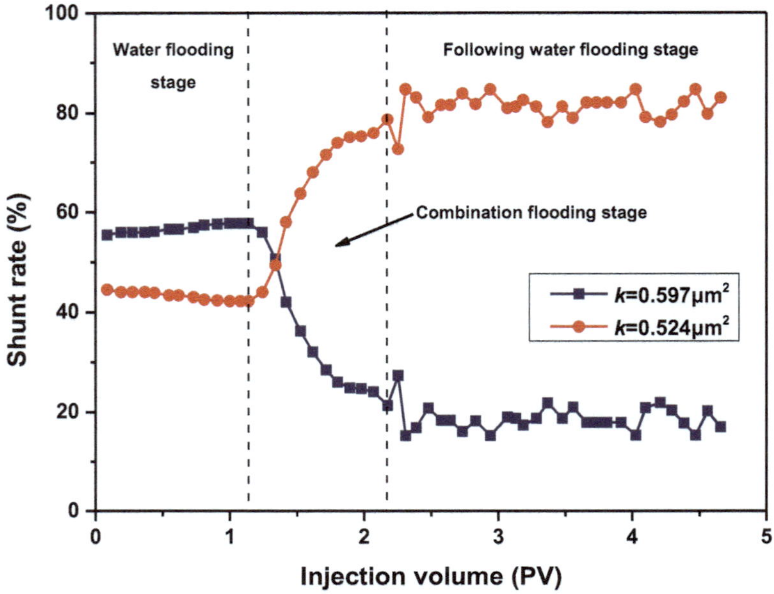

Figure 6.14 The shunt rate change of sand-pack model at permeability ratio of 1.14.

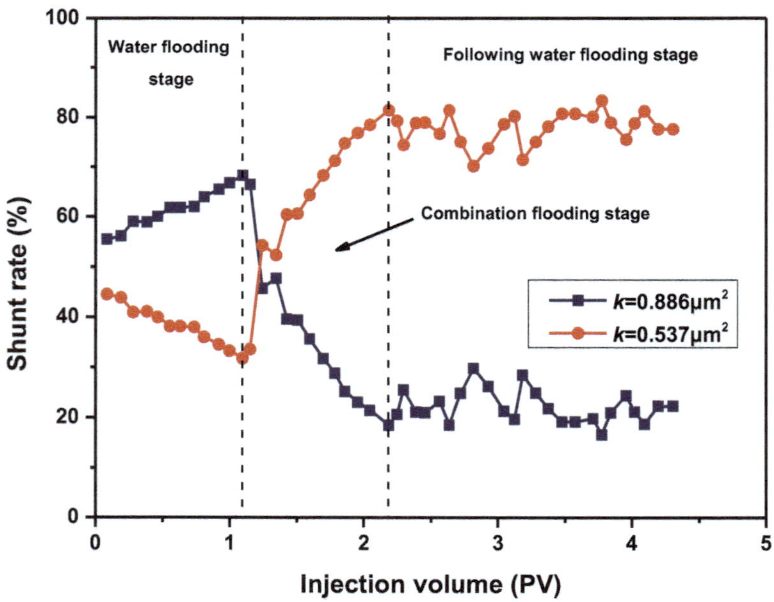

Figure 6.15 The shunt rate change of sand-pack model at permeability ratio of 1.65.

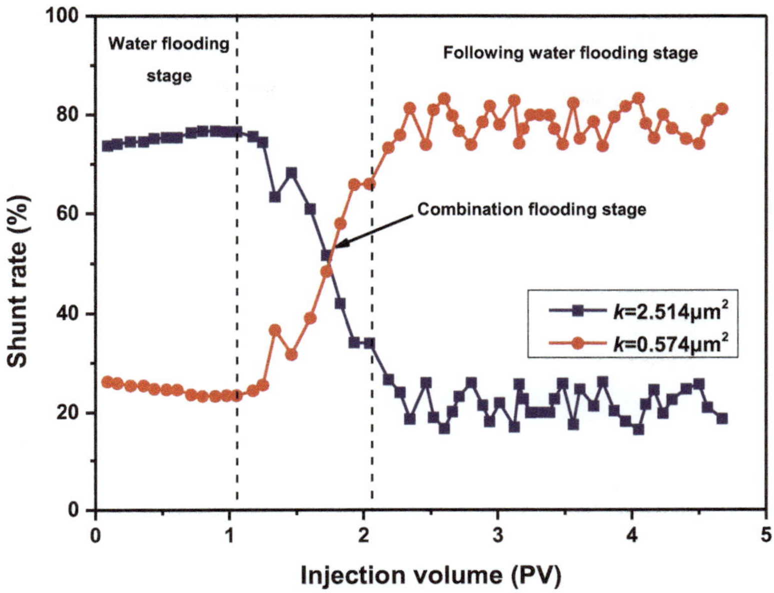

Figure 6.16 The shunt rate change of sand-pack model at permeability ratio of 4.38.

sand-packs with high permeability have a better profile improvement performance than those with low permeability.

6.3.3 Enhanced oil recovery performance

In this example, the single core flow model was used to compare the enhanced oil recovery performance of the PS system (0.15% BD polymer + 0.3% CP surfactant) and DPS system (0.15% BD polymer + 0.3% CP surfactant + 0.06% DPG). The experimental procedures are described as follows. First of all, the cores were dried to measure the permeability and the porosity. Then, they were successively saturated with the simulated formation water and simulated crude oil. After the saturation, they went through all the water flooding stage (until the water-cut reached 98%), the combination flooding stage, and the following water flooding stage (until the water-cut reached 98%). All of the experiments were conducted at 45°C. The changes in pressure, water production, and oil production rates were recorded to evaluate the enhanced oil recovery performance. The experimental results are presented in Table 6.4 and Figs. 6.17–6.19.

It can be observed from Table 6.4 that the oil recovery increases of the two systems are 18.34% and 34.88%, respectively. The oil recovery increase of the DPS system is significantly higher than that of the PS

Table 6.4 Enhanced oil recovery performance of the PS system and the DPS system.

Scheme	Permeability (μm^2)	Length × diameter (cm)	Pore volume (ml)	Oil saturation (%)	Recovery of water flooding (%)	Total recovery (%)	Recovery increment(%)
PS system	0.982	10 × 2.5	12.58	80.7	34.78	53.12	18.34
DPS system	0.976	10 × 2.5	12.39	81.5	35.95	70.83	34.88

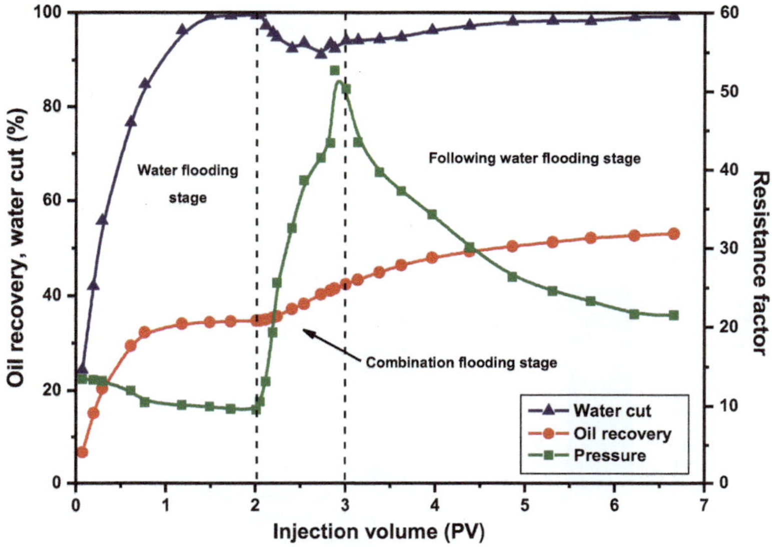

Figure 6.17 Enhanced oil recovery performance of the PS system.

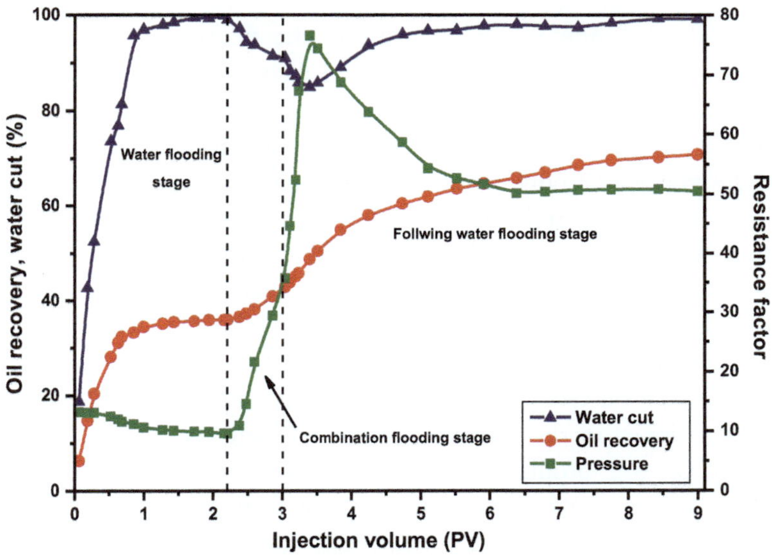

Figure 6.18 Enhanced oil recovery performance of the DPS system.

system. It means the DPS system has prominent enhanced oil recovery performance. According to Figs. 6.17–6.19, however, the enhanced oil recovery performances of these two systems are significantly different in

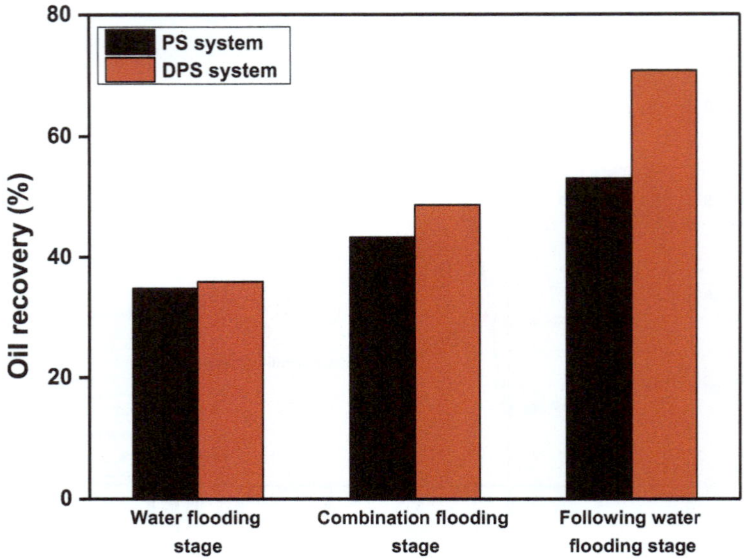

Figure 6.19 Oil recovery comparison of the PS system and DPS system at different displacement stages.

both the combination flooding stage and the following water flooding stage. For the PS system, the enhanced oil recovery mainly occurs in the combination flooding stage. Its viscoelastic and low interfacial tension is the primary driving force for enhancing the oil recovery. On the one hand, the viscoelastic polymer in the PS system can decrease the mobility ratio of water to oil and divert the flowing water into medium or low permeability zones, thus resulting in a rise in mobility control. On the other hand, the surfactant in the system can reduce the IFT and change the wettability of the sand surfaces, thus improving the swept efficiency. As a consequence, most of the residual oil in the unswept zones is driven out to enhance oil recovery. However, the viscosity of the PS system is gradually reduced in the following water flooding stage. This leads to a reduction of the profile control performance and mobility control. Therefore the oil recovery almost does not further increase during this stage. As an improvement of the PS system, the DPS system maintains all the above characteristics of the PS system. The DPS system has, additionally, its own unique features for enhancing oil recovery due to the introduction of the DPG particles. The plugging behavior of DPG particles and the synergistic thickening effect can strengthen the mobility control performance of the DPS system in both the combination flooding stage and the following water flooding stage. Consequently, the mobility control in terms of injection pressure of the DPS system maintains a higher level than that of the PS system in these two stages.

Section 6.4 Enhanced oil recovery mechanism of DPS combination flooding system

In this section, the scanning electron microscope (SEM) was used to investigate the morphologies and reveal the enhanced oil recovery mechanisms of the PS system and the DPS system in the porous media. The experimental procedures are as follows. At the beginning, the cores were dried to measure the permeability and the porosity. Then, the core flood was applied using the two combination flooding systems and aged for 5 days at 45°C. Next, the cores were freeze-dried for 24 hours and 1 × 1 cm fragments at the center of the cores were observed by the scanning electron microscope. The experimental results are presented in Fig. 6.20.

It can be observed from Fig. 6.20 that the distribution behaviors of the PS system and the DPS system in porous media are significantly different. A three-dimensional network of the PS system can be formed in small pores (Fig. 6.20A), while the network is not sufficiently compact and the bridging becomes more difficult as the pore size increases (Fig. 6.20B). Moreover, the network of the PS system gradually evolves into an

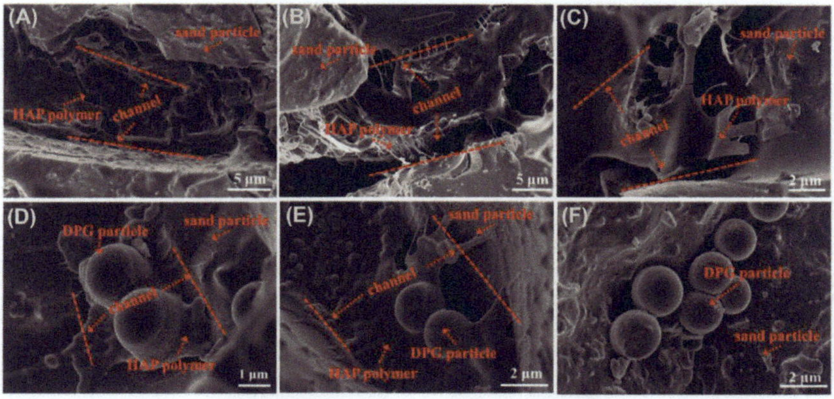

Figure 6.20 Morphologies of the PS system and the DPS system in porous media [12]: (A, B) combination flooding stage of the PS system; (C) following water flooding stage of the PS system; (D, E) combination flooding stage of the DPS system; (F) following water flooding stage of the DPS system.

instability system due to the dilution effect and the seepage shearing effect. Although the network is more or less observed in porous media in the following water flooding stage, the viscosity of the PS system is greatly reduced. As a result, the mobility control performance of the PS system is damaged, leading to a poorly enhanced oil recovery performance.

When the DPS combination flooding is in action, it is possible to observe the network (Fig. 6.20D and E). Given the existence of polymers and surfactants, the DPS system enjoys an improvement in its mobility control and profile control mechanisms, similar to those in the PS system. However, the existence of DPG particles enables an increase in the viscosity of the DPS combination flooding system by the synergistic effect on the DPG particles, polymers, and surfactants, when in the combination flooding stage. Additionally, the DPG particles are interspersed in the network structure, heightening the mobility ratio of water to oil in this stage (Fig. 6.20D and E). In the following water flooding stage, two or more DPG particles can bridge together and effectively plug into the pore-throats (Fig. 6.20F). Therefore the DPS system can easily reduce the water-phase permeability in the high permeability zones and drive out more injection fluids into the low permeability zones.

The DPS system presents through a synergistic thickening effect an excellent enhanced oil recovery performance, strengthened oil displacement efficiency, and an improved mobility control performance. With comprehensive consideration of its characteristics, its application properties and distribution behaviors in porous media, and its schematic diagram of strengthening mechanism are given in Fig. 6.21.

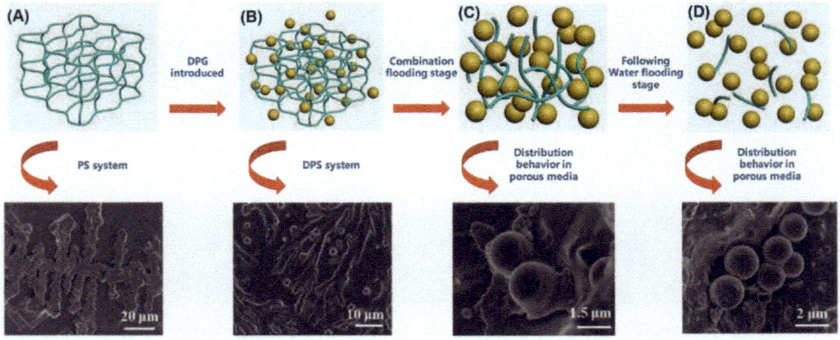

Figure 6.21 Schematic diagram of network structure changes of the DPS system [12]: (A) network structure of PS; (B) network structure of DPS; (C) loose network structure of DPS; (D) disappeared network structure of DPS.

6.4.1 Synergistic thickening effect

The synergistic thickening effect of the DPS system are mainly manifested in two aspects. On the one hand, the DPG particles are interspersed in the network formed by polymer, improving the strength of polymer chains and resulting in more difficulty of conformational transition of polymer. Therefore the viscosity of the DPS system is further increased. On the other hand, the surfactants are absorbed onto the surfaces of DPG particles through hydrogen bonds and electrostatic interaction. This leads to an expansion of the hydraulic radius of DPG particles and an increase in the system viscosity. In general, the synergistic thickening effect improves the mobility control performance and strengthens the resistance to seepage shear and dilution. As a result, the oil recovery of the DPS system is improved.

6.4.2 Strengthened oil displacement efficiency

The adsorption of surfactants onto the surfaces of DPG particles through hydrogen bonds and electrostatic interaction enables the DPG particles to perform interfacial surfactant-activeness. When contacting with the residual crude oil in formation, the DPG particles can reduce the oil—water interfacial tension, thus further strengthening oil displacement efficiency of the DPS system.

6.4.3 Improved mobility control performance

In the following water flooding stage, the viscosity and mobility control performance of the PS system decrease. Although the viscosity of the polymer decreases, the DPG particles are still present in the DPS system. In general, an effective profile control can be achieved by the plugging of single particles, bridges, and aggregates, as well as by the expanding behaviors of multiple particles across the pore throats. The above plugging behaviors of the DPG particles play a leading role in the following water flooding stage, thereby improving the mobility control performance of the DPS system.

References

[1] Ji B. Progress and prospects of enhanced oil recovery technologies at home and abroad. Oil & Gas Geology 2012;33(1):111—17.

[2] Wang D, Cheng J, Wu J, et al. Application of polymer flooding technology in daqing oilfield. Acta Petrolei Sinica 2005;26(1):74—8.

[3] Cheng J, Wang D, Li Q. Field test performance of alkaline surfactant polymer flooding in daqing oilfield. Acta Petrolei Sinica 2002;23(6):37–40.

[4] Argillier J, Dupas A, Tabary R, et al. Impact of polymer mechanical degradation on shear and extensional viscosities: toward better injectivity forecasts in polymer flooding operations. 2013.

[5] Mantia F, Morreale M, Botta L, et al. Degradation of polymer blends: a brief review. Polymer Degradation and Stability 2017;145:79–92.

[6] Zhang J. Study of crosslinking and transport plugging properties of polymer gel system in porous media. Petroleum Geology and Recovery Efficiency 2012;19 (2):54–6.

[7] Guo G, Ye Z, Shu Z. Effects of near wellbore shearing on the seepage characteristics of flooding polymer solutions. Oilfield Chemistry 2014;31(1):90–4.

[8] Chari K, Antalek B, Lin M, et al. The viscosity of polymer-surfactant mixtures in water. Journal of Chemical Physics 1994;100(7):5294–300.

[9] Guzman E, Llamas S, Maestro A, et al. Polymer-surfactant systems in bulk and at fluid interfaces. Advances in Colloid and Interface Science 2016;38–64.

[10] Sarsenbekuly B, Kang W, Fan H, et al. Study of salt tolerance and temperature resistance of a hydrophobically modified polyacrylamide based novel functional polymer for EOR. Colloids and Surfaces A: Physicochemical and Engineering Aspects 2017;514:91–7.

[11] Jamaloei B, Kharrat R, Asghari K. The influence of salinity on the viscous instability in viscous-modified low-interfacial tension flow during surfactant-polymer flooding in heavy oil reservoirs. Fuel 2012;97:174–85.

[12] Zhao G, Li J, Gu C, et al. Dispersed particle gel-strengthened polymer/surfactant as a novel combination flooding system for enhanced oil recovery. Energy & Fuels 2018;32(11):11317–27.

CHAPTER 7

Profile control technology of the DPG particles three-phase foam system

Foams can be formed by trapping pockets of air into a liquid or solid, where the air is the dispersed phase and the liquid or solid is the continuous phase. In the process of penetration during oil recovery, foams are successfully applied to oil recovery due to their excellent stability, good selectivity for pores, and their minimal damage to strata. They can be classified into two groups: conventional foam and enhanced foam. The stability of the former foam generated by a regular foaming agent is poor with limited plugging capability and short effective time period. To improve its stability, a reinforced foam system with polymers, gels, and particles as the continuous phase has been developed [1–3]. Thus the foam stability has been greatly improved either by increasing the viscosity of the liquid phase and the interface or by improving the particle adsorption effect on the interface. This chapter introduces a three-phase foam for profile control using DPG particles as the dispersed phase, with a discussion on the factors that affect the stability, the profile control capability, and the stabilization mechanism of foams, as well as the optimization of injection process parameters. We also explore the results of pilot tests in oil fields using this three-phase foam, in an attempt to pave the way for success in profile control application.

Section 7.1 Design of the DPG particles three-phase foam system

Using the modified Ross–Miles approach, we focus in this section on the design of the DPG particle three-phase foam system, optimizing the concentration of DPG particles, foaming agent, and gas/liquid volume ratio on the base of half-life, foaming volume, and foam comprehensive value.

Theory and technology of multiscale dispersed particle gel for in-depth profile control
DOI: https://doi.org/10.1016/B978-0-323-99849-9.00007-2

7.1.1 Modified Ross–Miles approach

A sand core was put at the bottom of the foam generator to simulate the situation when a three-phase foam was transferred through porous media (shown in Fig. 7.1). To investigate the foaming ability and the stability of the foam under different gas/liquid ratios, the amount of gas (nitrogen in this case) was carefully controlled by a gas flowmeter, as shown in Fig. 7.2. The foam comprehensive value is defined as the foam volume times half-life, as shown in Fig. 7.1:

$$F = V_0 \times t_{50} \tag{7.1}$$

Where F is foam comprehensive value (with a unit of mL min or mL s); V_0 is volume (mL); and t_{50} is half-life time (s).

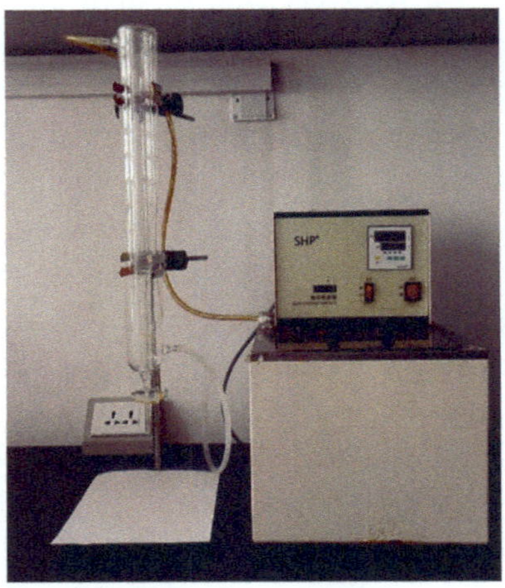

Figure 7.1 Modified Ross–Miles experiment setup.

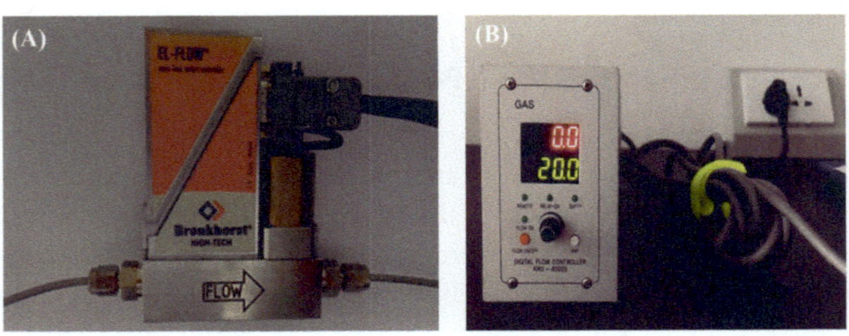

Figure 7.2 Gas flowmeter set up: (A) gas flowmeter; (B) gas flow control system.

Steps:

1. Prepare the DPG particle foaming agent solution: take 20 mL by pipette and add to the inlet, the turn on the circulating water bath at 80°C for 10 minutes;
2. Set gas injection speed at 20 mL/min by gas flowmeter;
3. Inject N_2, then shut down gas flowmeter and Ross–Miles, record the volume and bubble half-life time.

The size range of DPG particles is 600–2200 nm, and the foaming agent is glycine betaine foaming complex FD-1, where salinity: 4000 mg/L; Ca^2: 30 mg/L; Mg^{2+}: 30 mg/L; simulated oil: crude oil from Henan oilfield. The viscosity is 18 mPa · s at 50°C.

7.1.2 Optimization of foaming agent concentration

The concentration of the dispersed particle gel is 0.06%, the gas/liquid ratio is 3:1, and the temperature is 80°C. Fig. 7.3 shows the foaming performances of the conventional foam system and the DPG particle three-phase foam system under different foaming agent concentrations. The results indicate a direct impact of the concentration of the foaming agent on the foaming performance. As the concentration of the foaming agent increases, more and more surfactant molecules are adsorbed on the surface of the liquid film. This thickens the liquid film and slows down the drainage rate of the liquid film, thereby enhancing the foaming performance of the foam. However, the foaming performance of the DPG particle is significantly better if using the conventional approach. This can be attributed to the accumulation of DPG particles on the interfacial film that enhance its viscoelasticity and reduce the drainage speed, thus improving its ability as foam. Considering the foaming performance and cost factors, the optimized concentration of the foaming agent is 0.2%–0.4%.

7.1.3 Optimization of DPG particles concentration

We set the concentration of the foaming agent at 0.3%, the gas/liquid ratio at 3:1, and the temperature at 80°C, to optimize the DPG particles concentration. After the addition of DPG particles, the foaming performance is obviously improved. The foaming volume, half-life, and foam comprehensive value increase with increasing DPG particles concentration, as shown in Fig. 7.4. When the concentration of the DPG particles exceeds 0.06%, the foaming performance basically does not change. Thus the optimized concentration of the DPG particles is in the range of 0.06%–0.08%.

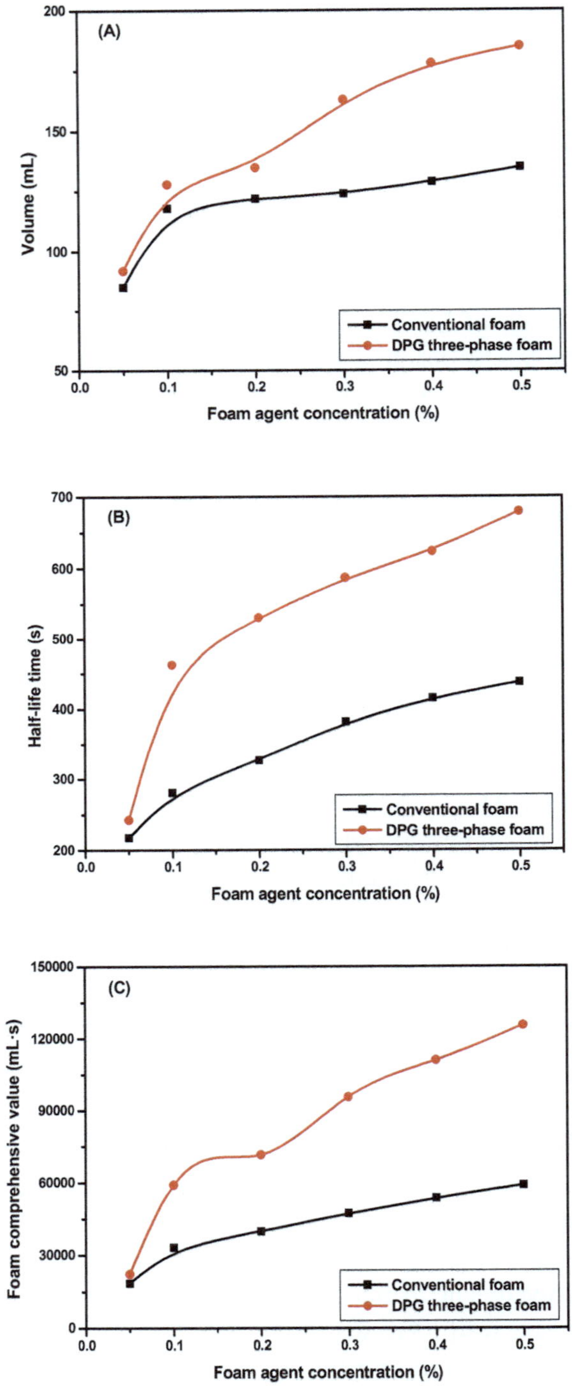

Figure 7.3 Influence of foaming agent on foaming performance: (A) foam volume; (B): half-life time; (C) foam comprehensive value.

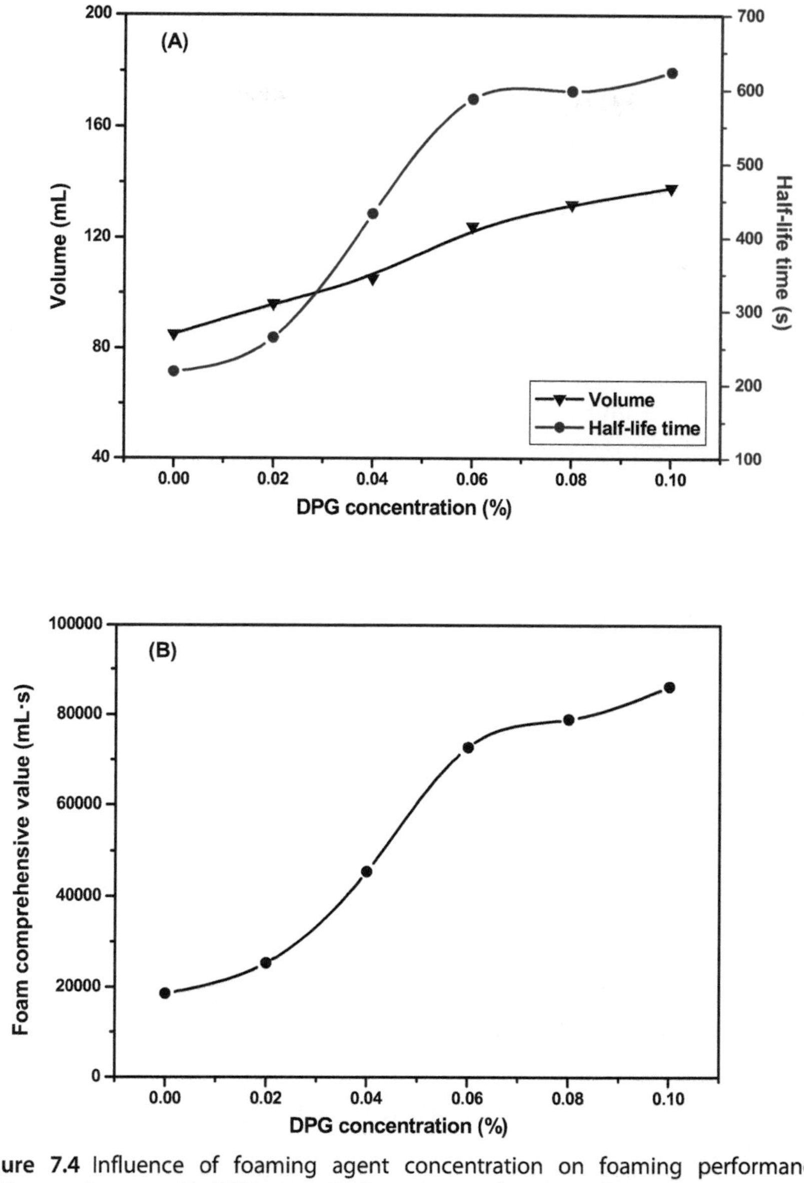

Figure 7.4 Influence of foaming agent concentration on foaming performance: (A) foam volume and half-life time; (B) foam comprehensive value.

Moreover, the addition of viscoelastic DPG particles increases the viscosity and enhances the viscoelasticity of the foam liquid film. The higher the concentration of the DPG particles, the greater the viscosity of the foaming agent, and the stronger the viscoelasticity of the interfacial film. In addition, the DPG particles can be adsorbed on the surface of the liquid film. This not only increases the thickness of the liquid film but also increases the repulsion between the foam liquid film, so that the stability of the foam is strengthened. Therefore the stability of the three-phase foam is largely improved. The foaming performance of the DPG particles does not, however, vary in a dramatic manner when the concentration of the DPG particles exceeds 0.06%. As the surfactant in the foaming agent solution can be absorbed on the DPG particles surface through dispersed force and hydrogen bonding, fewer surfactants may come up on the film surface of the foam, thus affecting the stability of the foam system. However, increasing the DPG particles concentration can reinforce the viscoelasticity of the film. When these two impacts reach a dynamic equilibrium state, the foam performance basically is maintained at the same condition.

7.1.4 Optimization of gas/liquid ratio

This part focuses on investigating the foaming performance of the DPG particles three-phase foam system (0.06% DPG particles + 0.3% foaming agent) under different gas/liquid volume ratios at 80°C. With the increase of the gas/liquid ratio, the foam volume of the three-phase foam of the dispersed particle gel gradually increases, whereas both the half-life and the foam comprehensive value gradually decrease, as shown in Fig. 7.5. Thus the optimized gas/liquid ratio is 2:1−3:1. At a low gas/liquid ratio, the injected nitrogen and the foaming liquid are fully mixed. The surfactant molecules are tightly arranged on the surface of the liquid film, which superimposes the viscoelasticity and particle adsorption effects to form a dense foam system [4,5]. At a high gas/liquid ratio, on the contrary, the numbers of surfactant molecules and DPG particles per unit area of the liquid film drop down as the foam is larger. Therefore the drainage speed is accelerated, increasing the instability of the foam and a decreased half-life time.

Section 7.2 Profile control ability of the DPG particles three-phase foam system

We investigated (1) the stability of the DPG particles three-phase foam system under different conditions, including age time, temperature, ionic strength, and crude oil based on the modified Ross−Miles approach; (2) the mechanism of profile control, the ability of profile control, and the ability for oil recovery through a core flooding experiment. All the works aim to give technical support for field application.

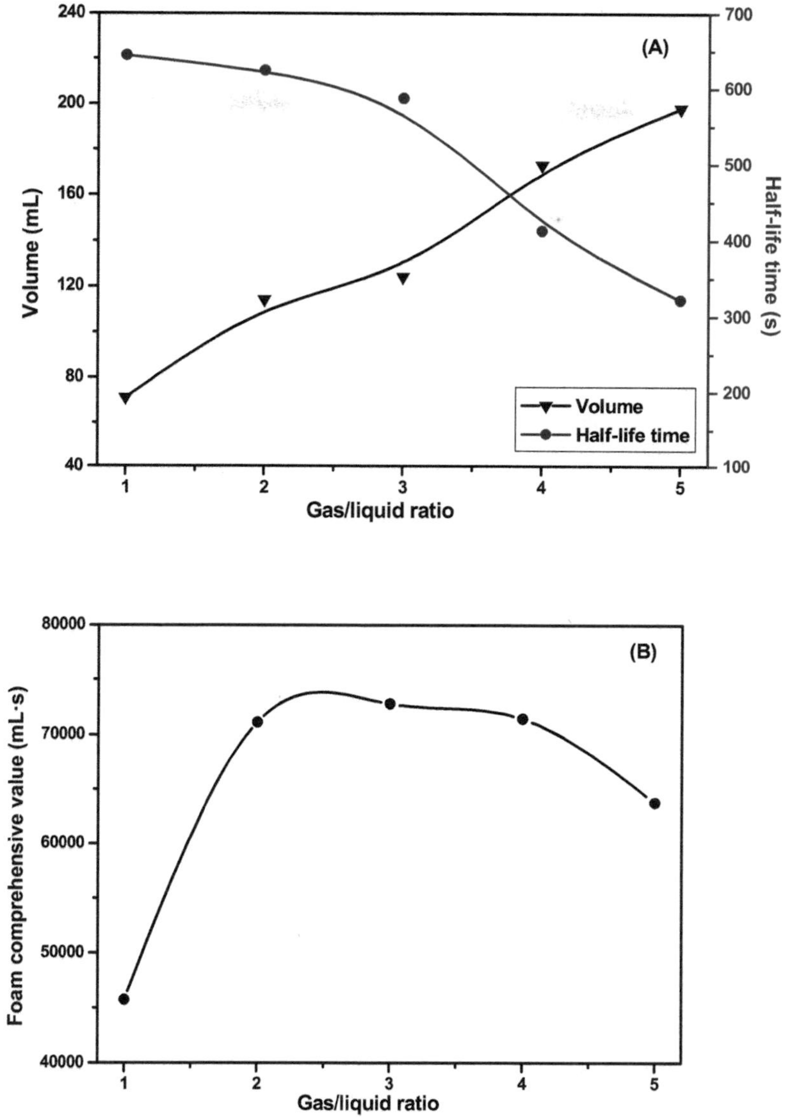

Figure 7.5 Influence of gas/liquid ratio on foam performance: (A) volume and half-life time; (B) foam comprehensive value.

7.2.1 Stability factors of DPG particles three-phase foam system

7.2.1.1 Aging time

The foam (0.06% DPG + 0.3% foaming agent) was put in an oven at 80°C for different times to investigate the thermal stability, with a gas/liquid ratio of 3:1. The results are shown in Fig. 7.6. As the aging time increases,

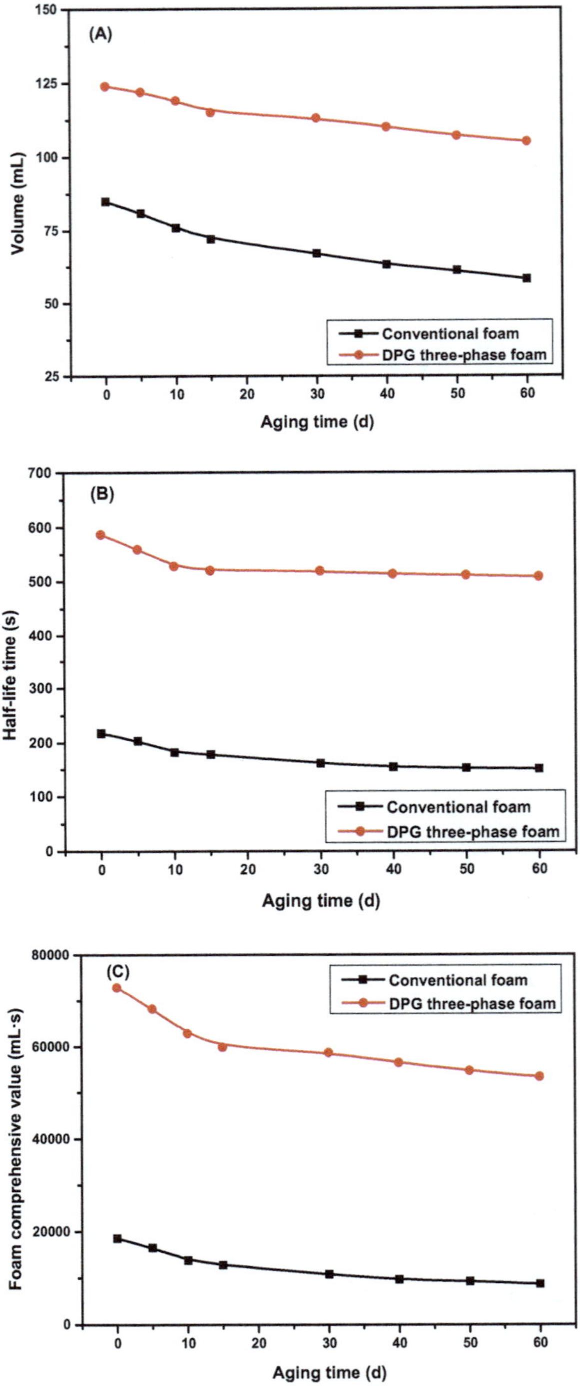

Figure 7.6 The effect of age time on foam performance: (A) foam volume; (B) half-life time; (C) foam comprehensive value.

the number of effective surfactant molecules decreases on the foaming liquid. This results in a decrease in the volume, the half-life, and the foam comprehensive value of both the conventional foam and the DPG particles three-phase foam. However, the three-phase foam shows an overall better performance than the conventional foam. High-temperature aging reduces the viscosity, the viscoelasticity of the foam film, and the thermal stability of the foam. Nevertheless, the DPG particles can be adsorbed on the surface of the liquid film, resulting in enhanced thickness and repulsive force and thus improving the thermal stability of the foam.

7.2.1.2 Temperature

Under formation conditions, the influence of temperature is of particular importance. We compared the foaming performances of the conventional foam system (0.3% foaming agent) and the DPG particles three-phase foam system (0.06% DPG + 0.3% foaming agent) under different temperature conditions, as shown in Fig. 7.7. As temperature increases, the foam volume of the two systems increases, but the half-life decreases. The high temperature induces an intensified thermal motion of the molecules, resulting in increased kinetic energy. Moreover, the intermolecular force becomes smaller and the viscosity of the liquid phase decreases, leading to larger foam volume. The influence of temperature on the half-life is more obvious. On the one hand, the increase in temperature strengthens the thermal motion of gas molecules. Thus the internal pressure of the bubble increases, inducing coalescence of bubbles. On the other hand, the water of the film evaporates increasingly as the temperature rises. This easily brings about raptures of bubbles. However, the DPG particles three-phase foam system still has a better foam stabilization ability at high temperatures due to the viscoelasticity of the DPG particles and the adsorption effect of the liquid film.

7.2.1.3 Salt ions

The formulation water and the formation water both contain salt but with different concentrations. If sensitive to salt ions, the three-phase foam is easy to burst in the formation but cannot effectively control the profile, thus resulting in a failure in profile control. Therefore the influence of the three typical salt ions (Na^+, Ca^{2+}, and Mg^{2+}) on the stability of the DPG particles three-phase foam system (0.06% DPG + 0.3% foaming agent) was investigated, with a gas/liquid ratio of 3:1, as shown in Fig. 7.8.

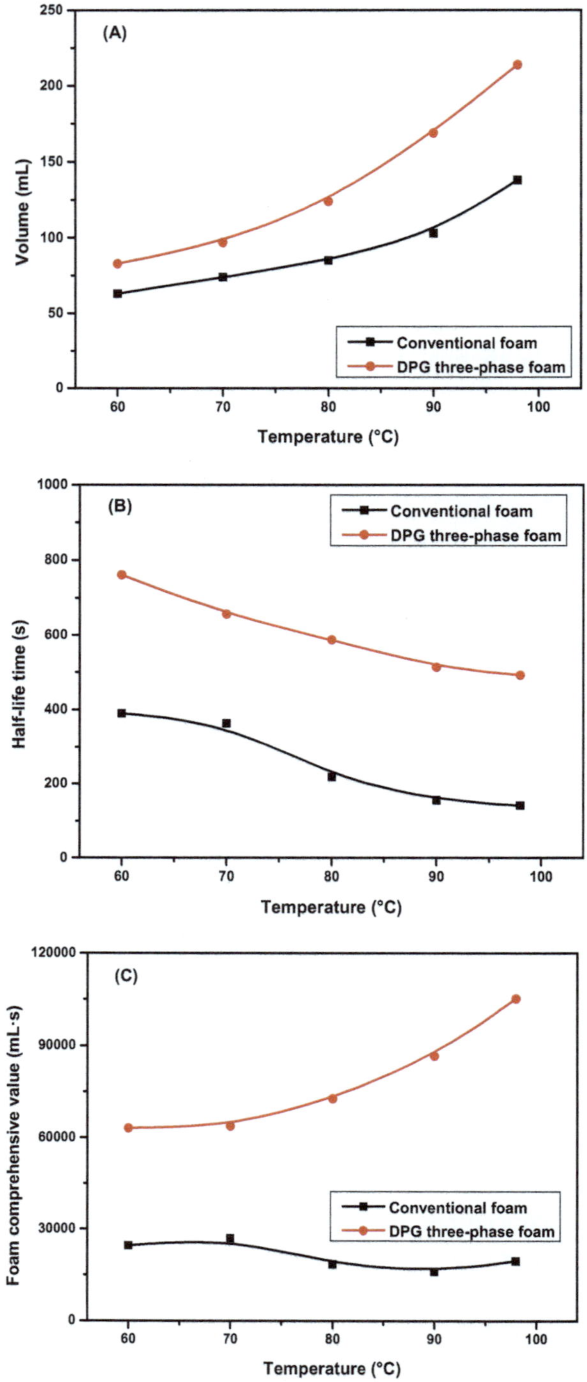

Figure 7.7 The effect of temperature on foam performance: (A) foam volume; (B) half-life time; (C) foam comprehensive value.

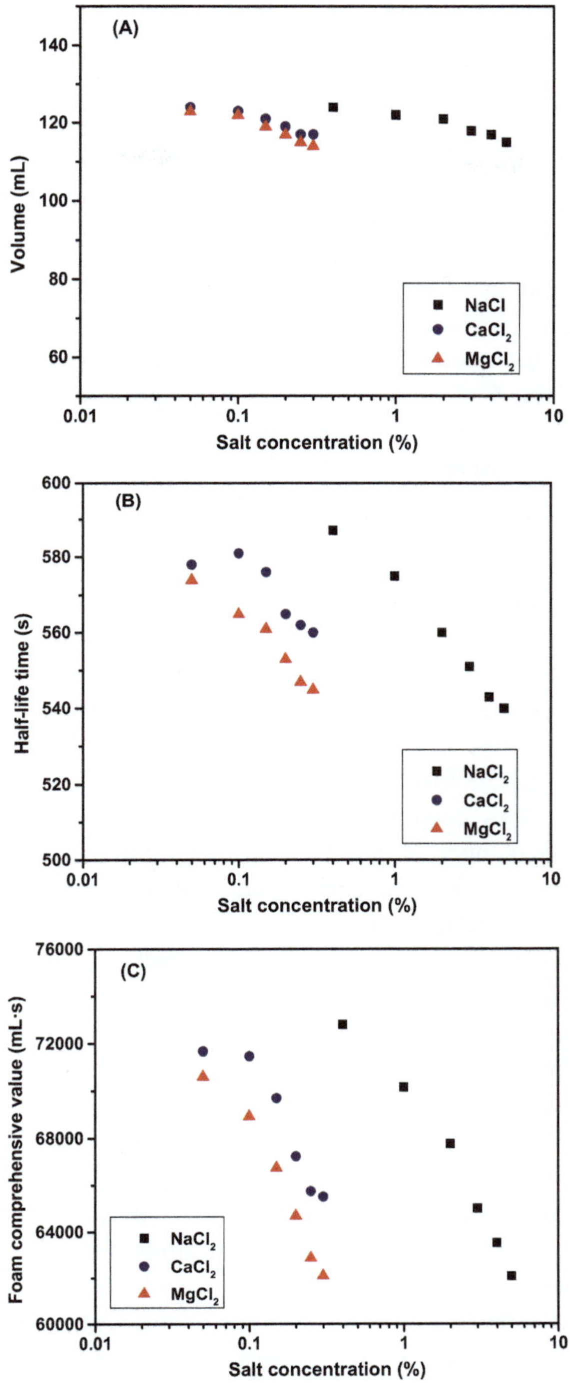

Figure 7.8 The effect of salt on foaming performance: (A) foam volume; (B) half-life time; (C) foam comprehensive value.

All the volume, the half-life time, and the foam comprehensive value of the three-phase foam system decrease with the increase of the salt ion concentration. This indicates the resistance of the foam system to salt ions, as shown in Fig. 7.8. After the addition of salt, the diffusive electric double layer in the surfactant is compressed. Also, a loose adsorption layer is formed due to the less attractive force between surfactant molecules, reducing the stability of the foam. Besides, higher cationic valence induces a stronger ability to compress the diffusive electric double layer. The foaming agent has good salt resistance with good absorption property on the surface of the film, however, the drainage speed of the bubble film is reduced. Therefore, the three-phase foam has good foaming ability and stability in a high-concentration salt solution.

7.2.1.4 Crude oil

After water flooding, lots of residual oil is left in the reservoir. While the foam is applied to profile control treatment, the foam system can preferentially penetrate into the high-permeability region due to its characteristics of being "stable with water, defoaming with oil" and the pressure selectivity of the reservoir. Though stable enough to exist in the water saturation region, the foam inevitably contacts with the remaining oil in the formation. This could affect its stability. Therefore the experiment investigates the stability of the DPG particles three-phase foam system (0.06% DPG + 0.3% foaming agent) under different crude oil concentration, with a gas/liquid ratio of 3:1. The results are shown in Fig. 7.9.

Fig. 7.9 shows the results concerning the effect of crude oil on the stability of conventional foam and three-phase foam systems. The addition of crude oil reduces the foaming performance of these two foam systems. The higher the crude oil concentration, the worse the foaming performance. The results prove that the crude oil weakens the stability of the foam, and these foam systems own the characteristics of "being stable in contact with water and defoaming in contact with oil." When the foam system contacts the oil phase, the oil droplets diffuse into the interfacial layer between gas and liquid. This can significantly reduce the surface tension while the surface tension around the foam barely changes. This surface tension difference causes the rupture of the foam [6,7]. When the DPG particles are added into the foaming agent, the foam film structure is changed into a viscoelastic film instead of an aqueous film. The mechanical property of the film improves, slowing down the diffusion rate of

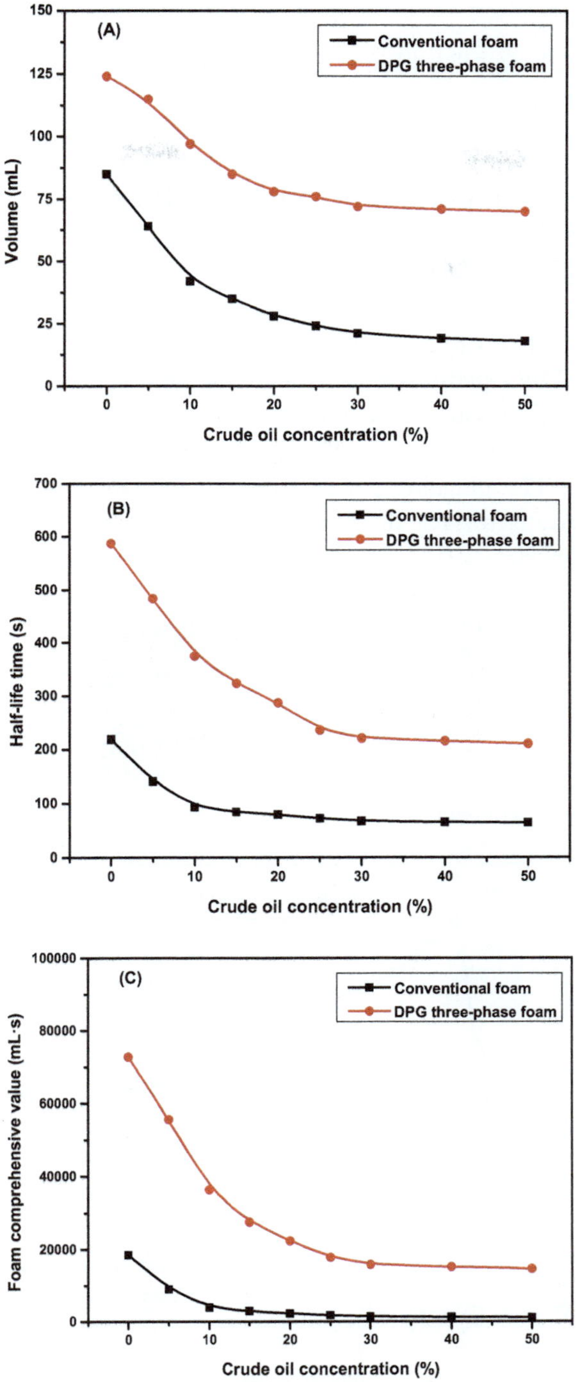

Figure 7.9 The effect of crude oil to foam performance: (A) foam volume; (B) half-life time; (C) foam comprehensive value.

crude oil into the gas/liquid interface. This leaves the three-phase foam as more stable than conventional foam.

7.2.2 Application performance of the DPG particles three-phase foam

7.2.2.1 Plugging performance

The single-core flow model was applied to compare foaming performance of conventional foam (0.3% foaming agent) and DPG particles three-phase foam (0.06% DPG particles + 0.3% foaming agent). Gas and liquid were injected sequentially into a core with a permeability of 1.05 μm^2 for 1PV at an injection speed of 1 mL/min. The gas/liquid ratio is 1:1 and the temperature is 80°C. The plugging performance of the foam system is evaluated by the resistance factor and the residual resistance factor. The foam generator is present in Fig. 7.10, and the results are shown in Fig. 7.11.

Figure 7.10 Foam generator.

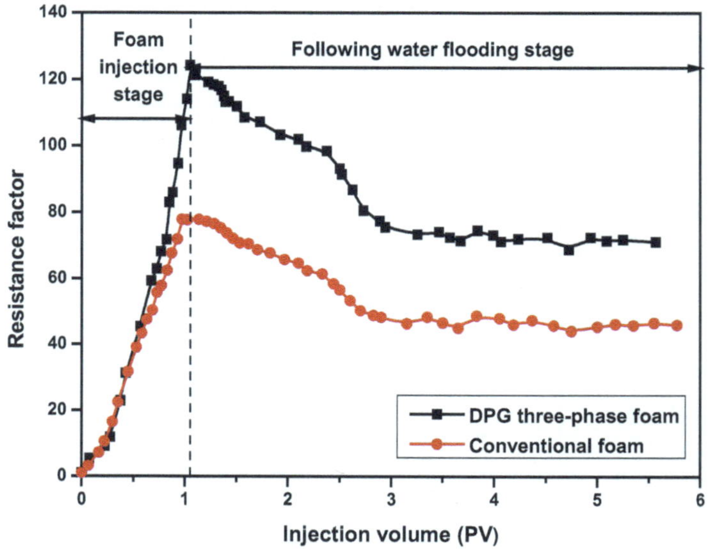

Figure 7.11 The plugging capability of DPG particles three-phase foam system.

The resistance factors of the two foam systems gradually increase with the increase of the injection volume, as shown in Fig. 7.11. The resistance factor reaches the maximum when the foam injection volume is 1PV. Switching to the water flooding stage, the residual resistance factor gradually decreases and still maintains a high residual resistance factor after 3PV continuous water injection. This indicates good plugging and erosion resistance performance of the foam system. It is found that the resistance factor and residual resistance factor of the three-phase foam are significantly higher than those of the conventional foam during the injection and the following water flooding stage. For the conventional foam system, the plugging ability is poor, relying on the Jamin effect or the superimposed Jamin effect of the foam due to its poor stability. For the DPG particles three-phase foam system, the stability is improved by the viscoelasticity of the DPG particles and adsorbed particles on the surface. Moreover, both the foam and DPG particles can plug the pore which can effectively control the profile of the reservoir. In the following water flooding stage, the DPG particles can still gather and bridge at the pore throats of the formation after the foam raptures, forming an effective secondary plugging and further improving control ability to the reservoir.

7.2.2.2 Recovery potential

The single-core flow model was applied to investigate the recovery potential of the three-phase foam system (0.06% DPG + 0.3% foaming agent). Here, the gas and liquid were injected simultaneously, at an injection speed of 1 mL/min and a temperature of 80°C, with an injection volume of 1PV and a gas/liquid ratio of 1:1. The results are shown in Table 7.1 and Fig. 7.12.

The increase in oil recovery for water flooding after gel injection is presented in Table 7.1 and Fig. 7.12. The reservoir becomes more heterogeneous after water flooding, leaving residual oil in the low-permeability region. After injecting three-phase foam, the pores in the reservoir can be selectively plugged in the high-permeability region. As continuously injecting foam, the foam is forced to flood into the low-permeability region due to Jamin effect and the recovery ratio can reach as high as 13.5%. After injecting gel foam, we perform another water flooding. The pressure of the reservoir maintains a high value during the pore has been selectively plugged in the high-permeability region, which further improves the oil recovery ratio. These results demonstrate the potential application of the three-phase foam in the oil recovery application.

Table 7.1 The recovery capacity of DPG particles three-phase foam system.

Permeability (μm^2)	Length (cm)	Diameter (cm)	Pore volume (mL)	Oil saturation (%)	Recovery ratio (%)			Recovery ratio (%)
					Water flooding	During Injection	Following water flooding	
1.82	10	2.5	14.7	82.7	29.69	43.26	61.33	31.64

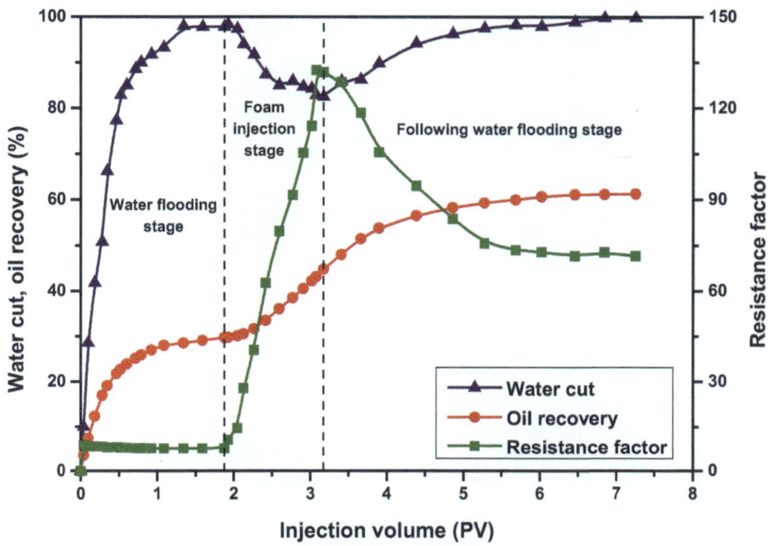

Figure 7.12 The recovery potential of DPG particles three-phase foam system.

Section 7.3 The stabilizing mechanism of the DPG particles three-phase foam system

The stability of the foam depends on the drainage rate of the liquid between the bubbles. The slower the drainage rate, the more stable the foam. This section discusses the stabilizing mechanism of a DPG particles three-phase foam system in terms of particle absorption characteristic of the interfacial film, interfacial film viscoelasticity, surface tension, and surface charge. A microchannel experimental model was established to investigate the bubble evolution process in porous media and the aggregation and rupture behavior of the sudden expansion and contraction structure. A foam general model, the rupture mechanism, and the mechanism of particle suppression of bubble aggregation were proposed to reveal the stabilizing mechanism of the DPG particles three-phase foam system.

7.3.1 Stabilizing mechanism of DPG particles three-phase foam

7.3.1.1 Particle absorption characteristic on interfacial film

We observed the particle distribution on the interfacial film between the conventional foam and the DPG particles three-phase foam system using an optical microscope. The conventional foam is heterogeneous in size, presenting a loose structure as a whole, as shown in Fig. 7.13. During

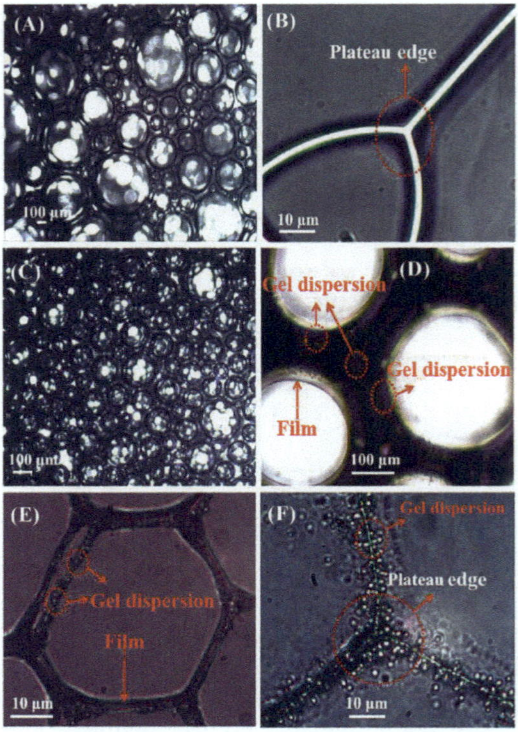

Figure 7.13 Microstructure of conventional foam system and DPG particles three-phase foam system [4]: (A) conventional foam; (B) Plateau edge of conventional foam; (C) DPG particles three-phase foam system; (D–E) distribution of DPG particles on film and in solution; (F) Plateau edge of DPG particles three-phase foam.

foam generation, bubbles tend to merge and form large bubbles. The newly formed generated bubbles possessing a thin film are easy to rupture, which results in poor stability, as shown in Fig. 7.13B. For the DPG particles three-phase foam system, the foam becomes compact and uniform in size (Fig. 7.13C). The DPG particles are absorbed on the surface of the foam, which can increase the film thickness and its viscoelasticity. This behavior prevents the foam from contracting and aggregating. Moreover, these DPG particles construct a compact layer conducive to enhancing the stability of the foam. Further, compared with the conventional foam, the interfacial film of the DPG three-phase foam has a thicker Plateau edge due to the compact backbone by particles, which further improves its stability, as shown in Fig. 7.13F.

7.3.1.2 Viscosity

A rheometer (HAAKE MARS 60, Germany) was applied to measure the viscosity and modulus of conventional foam system with 0.3% foaming agent and the DPG three-phase foam system with 0.06% DPG particles plus 0.3% foaming agent, as shown in Figs. 7.14 and 7.15.

As the shear rate increases, the foam viscosity gradually decreases, as shown in Fig. 7.14. The viscosity of conventional foam decreases from 200 to 20 mPa · s, a relatively sharp loss in viscosity, with the increase of frequency. The viscosity of the DPG three-phase foam is generally high. Even as the shear rate reaches 100 s^{-1}, the viscosity can be maintained at 200 mPa · s. This indicates that the viscosity of the foam system has increased after the addition of DPG particles. The increase of bulk viscosity would increase the resistance of the surfactant on the Plateau edge, slow down the drainage rate of the liquid film, and improve the stability of the foam.

As shown in Fig. 7.15, the bulk viscoelastic modulus of the foam rises exponentially when the frequency increases. This indicates that the foam would show a liquid-like behavior at a low frequency while showing a solid-like behavior at high frequency. The viscous and elastic modulus of the three-phase foam both increase after adding the DPG particles, in which the increase of elastic modulus is larger. When the frequency

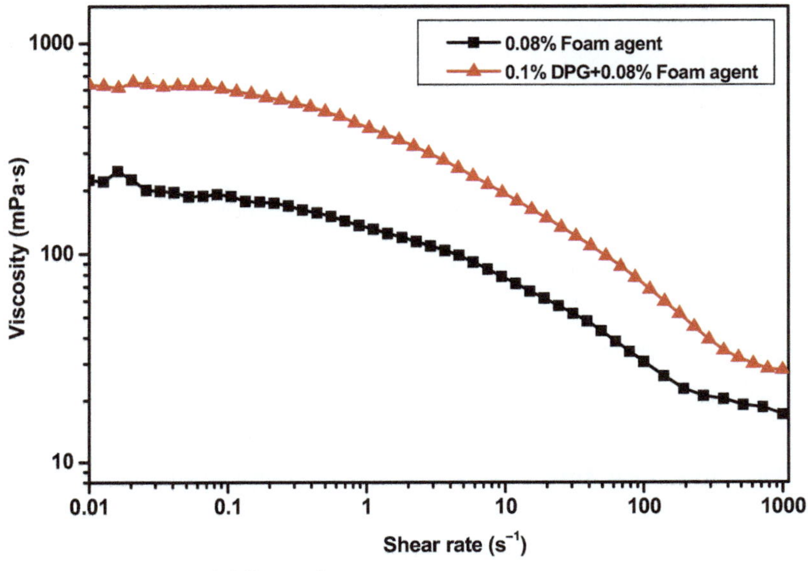

Figure 7.14 Viscosity of different foam systems.

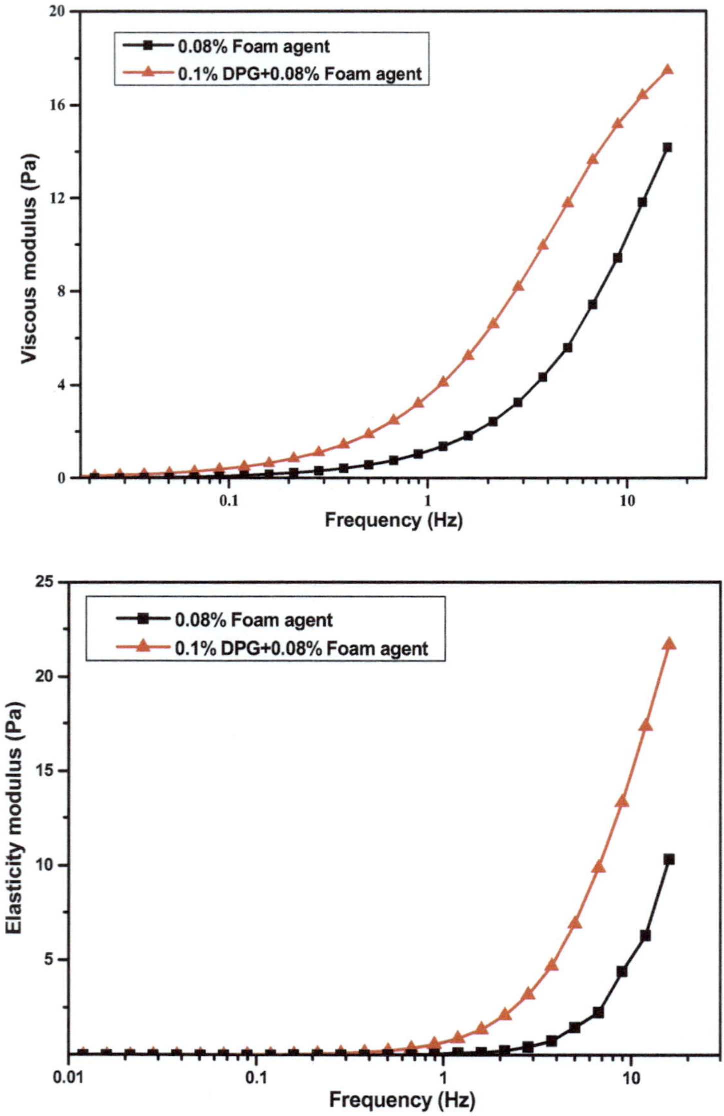

Figure 7.15 Viscosity modulus of different foam systems.

increases to 10 Hz, the elastic modulus of the three-phase foam reaches 16.7 Pa, more than three times that of the conventional foam. These results indicate that the dispersed particle gel three-phase foam has better resistance to external interference and mechanical strength, as well as better stability.

7.3.1.3 Surface tension

The surface energy always tends to stay at a low status based on Gibbs free energy theory. The low surface tension leads to a low capillary force and then reduces the foam drainage rate. The energy of a foam system reduces as well, but is beneficial to the stability of the foam. We measured the surface tension by interfacial rheometer (TECLIS, France) before and after adding DPG particles with a concentration of 0.06% at 30°C. The surface tension for both systems decreases with the increase in the concentration of the foaming agent. The surface tension of the conventional foam system is much lower than that of the DPG particles, as shown in Fig. 7.16. The stability of the DPG particles three-phase foam system can be greatly influenced by the addition of DPG particles. However, the DPG particles that accumulate in the interfacial film would increase the thickness and its viscoelasticity, compensating for the influence of the surface tension on the stability of the foam. Thus the DPG particles three-phase foam still possesses great stability even if the surface tension increases.

7.3.1.4 Surface charge

The DPG particles repel each other due to the surface charge, thus they can enhance the stability of the foam. We used zetasizer (NanoBrook Omni, Germany) to measure the zeta potential of the foam system at a value of

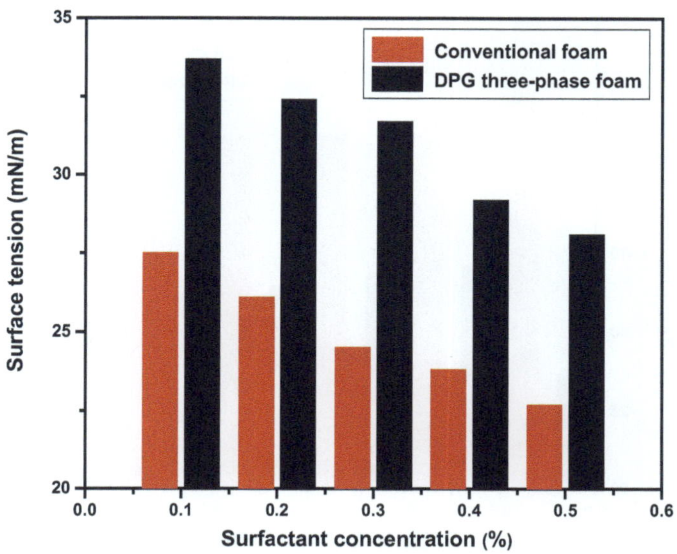

Figure 7.16 Surface tension comparison between conventional foam system and DPG particles three-phase foam.

0.06% at 30°C, as shown in Fig. 7.17. Both the DPG particles and the foaming agent are negatively charged, but the absolute value of the zeta potential of the DPG particles is significantly lower than that of the foaming agent. The absolute zeta potential value of a single DPG particle is lower than 30 mV, presenting a weak stable character. When DPG particles absorb the surfactants, the absolute value of the DPG zeta potential increases rapidly. The repulsive force increases accordingly between the DPG particles, preventing the coalescence of the particles. After generating bubbles, the DPG particles can be uniformly adsorbed at the gas/liquid interface. As the absolute value of the zeta potential increases, the electrostatic repulsion between the foam liquid films is increased, thereby preventing the aggregation between the foams and enhancing the stability of the foam.

7.3.2 The generation and migration of DPG particles three-phase foam in porous media [8]

7.3.2.1 The experiment design microchannel model

7.3.2.1.1 The design of microchannel model

The microchip used in the experiment was made by polypropylene methyl methacrylate (PMMA), a thermoplastic polymer with good light transmission property. The cross-section of the channel is 150 × 150 μm. The microchip

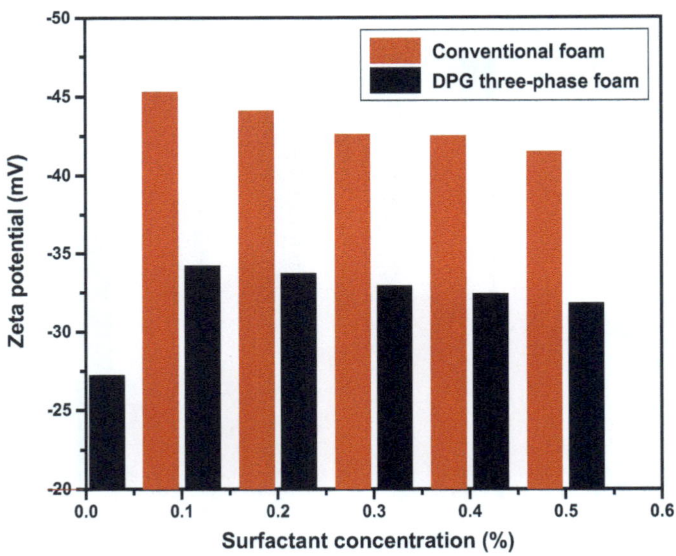

Figure 7.17 Zeta potential of conventional foam and DPG particles three-phase foam system.

is shown in Figs. 7.18 and 7.19. There is a cross-shaped bifurcation on the upstream of the microchannel for generating bubbles. A serpentine channel downstream serves for bubble collection. The length is roughly 3 cm. The final part of the microchannel system has five compression/expansion setups. The aim of this experiment is to investigate the bubble rapture process at the cross bifurcation, and the bursting and coalescence process of the bubble at the first compression/expansion setup.

7.3.2.1.2 Test procedures

The experimental equipment for studying bubble generation and flow characteristics is mainly divided into the fluid control system and image capturing and acquisition system. The experimental setup is shown in Fig. 7.20. The gas and liquid drives and flow control were controlled by a

Figure 7.18 Microfluidic devices used in the experiments.

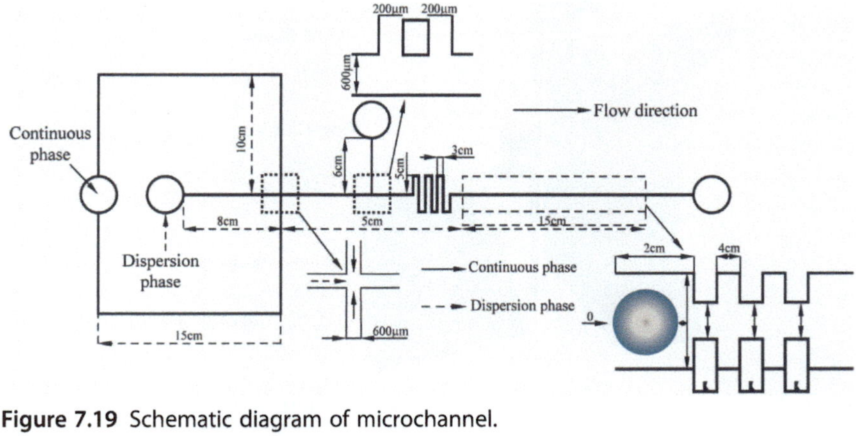

Figure 7.19 Schematic diagram of microchannel.

PHD2000 pump with a syringe connected to the channel with a polyethylene hose.

The flow and rupture of bubbles in the microchannel were captured by a high-speed camera (Fig. 7.21) and recorded in a computer.

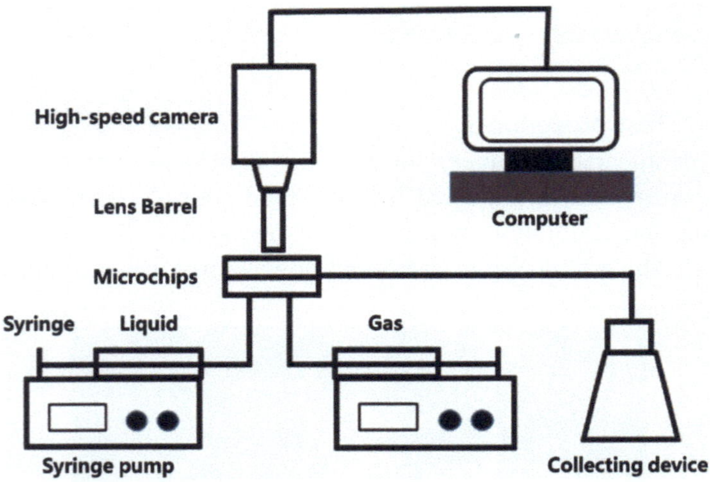

Figure 7.20 Schematic diagram of the experiment setup.

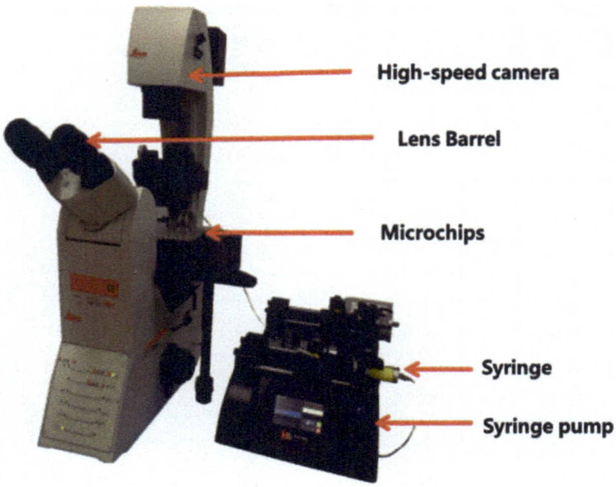

Figure 7.21 Experiment device of bubble generation and flow.

The steps are as follows:

1. Check the air tightness of the microchannel and syringes.
2. Fill the syringe with gas and liquid respectively, and connect the syringe and the microchannel as shown in Fig. 7.21.
3. Adjust the distance between the microchip and the objective lens to obtain a clear image.
4. Start the PHD2000 pump, set the gas and liquid flow rate to preset value to drive the gas and liquid.
5. After the generated bubble size reaches steady state, collect images and save them.
6. Stabilize for at least 5 minutes when the operating conditions are changed to ensure the stability of the bubble generation process.
7. After the experiments finished, use deionized water to clean the microchannel, use gas to drain out the remaining liquid in the channel for next use, turn off the syringe pump and the recording system.

The surfactant agent used in the experiments is sodium dodecyl sulfate (SDS), and the DPG is made by mechanical shearing using inorganic nanoparticle-strengthened gel.

7.3.2.2 The generation mechanism of three-phase foam

The bubble size affects the foam stability, so it is necessary to study the size evolution during the bubble generation process. The bubble size is usually characterized by its radius, length, and volume. Here we use the bubble volume V_b to characterize the size of the bubble. The bubble generation process is a stable periodic process, expressed as $V_b = Q_g T$, where Q_b is the volume flow rate of the gas phase and T the bubble generation period (when data is processed, five bubble generation periods are selected to obtain the average time).

7.3.2.2.1 The bubble generation process

The bubble generation processes can be observed and recorded with a high-speed camera. Fig. 7.22 shows the bubble generation process at different liquid phase flow rates under the same operating conditions. The bubble was squeezed by the walls, and the bubble width is almost equal to the channel width. After the bubble entered the cross channel, the bubblehead expands rapidly in a semicircular shape. With the advancement of the gas/liquid flow rate, after the bubblehead enters the main channel, the convex neck gradually shrinks, and finally it pinches off and forms a new bubble. Comparing Fig. 7.22A and B, it can be found that at a higher liquid phase flow rate, the bubble generation period is shorter

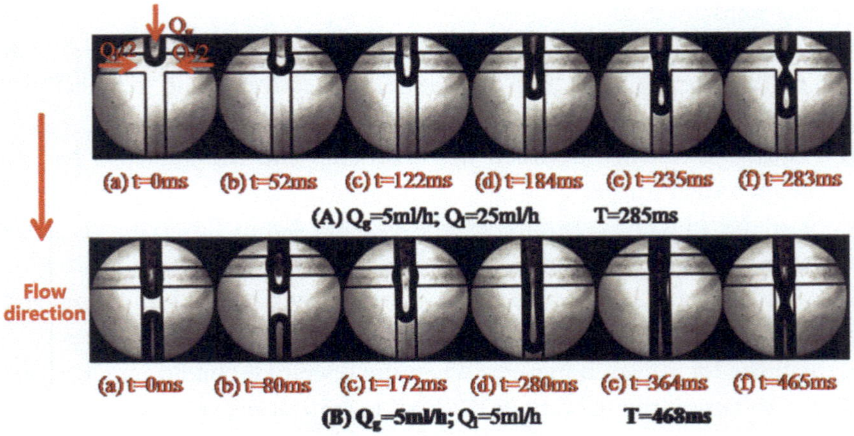

Figure 7.22 The generation process of bubbles at different liquid flow rates.

and the bubbles formed are smaller and the bubble pinch-off occurs in the middle of the cross channel. But at the lower liquid phase at a flow rate, bubble pinch-off occurs in the main channel, the bubble generation period is longer and the bubble formed is larger.

From the bubble generation processes, it can be known that at a certain gas flow rate, the liquid flow rate plays an important role in the bubble generation process. In the bubble generation process, with the increase of the liquid phase flow rate, the bubble generation period is shortened, the bubble volume decreases, and the generated bubble shape is longer.

Dietrich et al. believed that in the process of bubble formation and flow, the gas/liquid interface is dominated by inertial force, viscous force, and surface tension. Therefore, the increase of liquid phase flow rate endows the bubbles with a larger squeezing force, and the bubbles pinch off much more easily, which will result in changes in the bubble size, generation period, and shape during the bubble generation process.

7.3.2.2.2 Bubble size distribution

The variation of bubble size with liquid flow rate is shown in Fig. 7.23. When the fixed gas flow rate is 5 mL/h, the bubble volume increases with the increase of the gas/liquid ratio, and the bubble volume and the gas/liquid ratio conform to the power function relationship: $V_b \propto \varphi^{\gamma}$. The fitted results are plotted in logarithmic coordinates in Fig. 7.23B, where $\gamma = 0.61$. It can be seen that the change of bubble

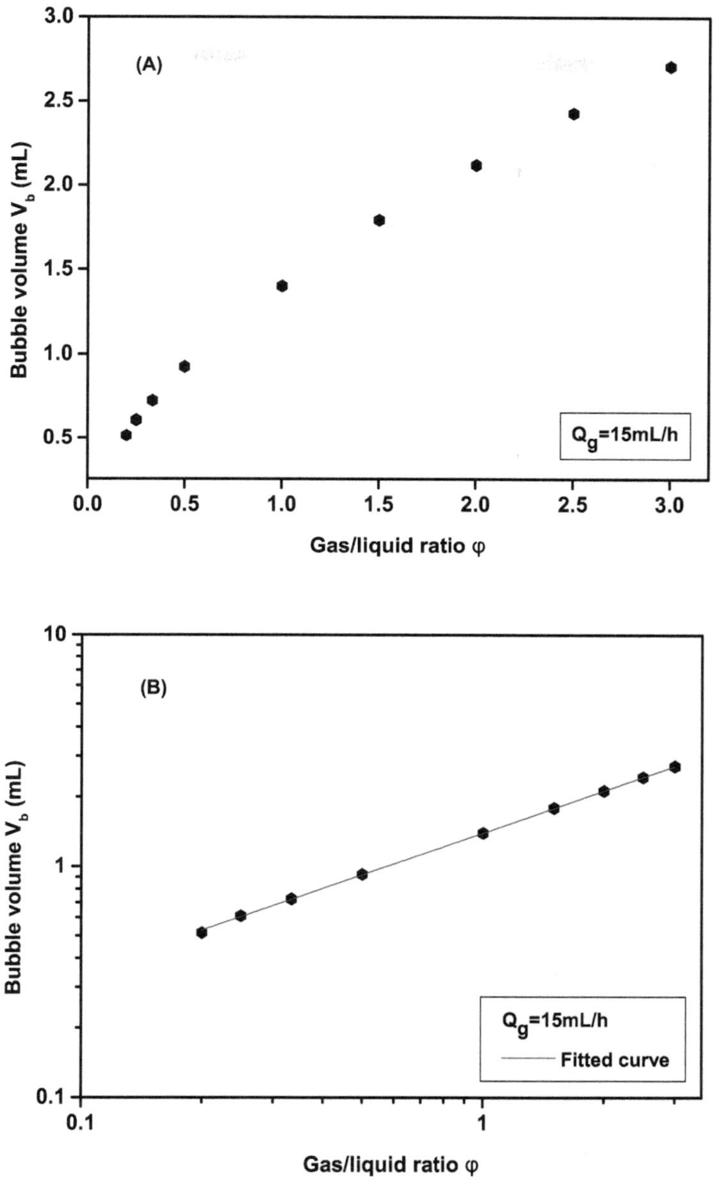

Figure 7.23 The change law of bubble volume V_b with gas/liquid ratio φ: (A) original data; (B) fitting results.

volume can span over an order of magnitude with the increase of liquid flow rate. The bubble volume can be controlled by changing the liquid flow rate.

The influence of gas flow velocity on bubble size is shown in Fig. 7.24. It can be seen from the figure that the bubble volume increases with the increase of the gas flow rate and the gas/liquid ratio, and that the bubble volume and the ratio conform to the power function relationship: $V_b \propto \varphi^{\gamma}$. The fitted results are plotted in logarithmic coordinates in Fig. 7.24B, where $\gamma = 0.54$. In addition, it can be seen from the figure that within the test range, the change in bubble volume is close to an order of magnitude. This indicates that the gas flow velocity has a greater effect on the bubble size.

It can be seen from Figs. 7.23 and 7.24 that the bubble size is affected by the gas/liquid two-phase flow rate, which is consistent with the extrusion mechanism: the bubble volume decreases with the increase of liquid flow rate and increases with the rising gas-phase flow rate. Therefore the bubble size can be controlled by adjusting the flow rate of gas and liquid, and then the density of foam can be regulated in practical applications.

7.3.2.2.3 Pinch-off rule of bubble interface

1. Interfacial pinch-off process of bubbles

   Fig. 7.25 shows the lifecycle of bubble generation. The foam generation occurs at different corresponding times, as shown in Fig. 7.25A–F. The moment when the bubble is first generated is defined as 0 second. The bubblehead expands in the focusing region (Fig. 7.25A–B). Fig. 7.25C shows that the bubblehead begins to enter the main channel and the protruding neck begins to pinch off. Fig. 7.25D–F shows that the bubblehead completely fills with the main channel, and the concavity of the neck which is pinched off is limited to the intersection.

2. Effect of addition of DPG particles on the interfacial pinch-off of bubbles

   The viscosity of 0.3% DPG solution was 9.1 mPa · s. The glycerin was gradually added into a certain amount of water to prepare a control group with the same viscosity, and then 0.3% foaming agent solution was prepared. The generation of bubbles was observed and recorded by microfluidic devices. The interfacial pinch-off law of three-phase bubbles with glycerol and DPG was obtained. The evolution pattern of width of the bubble neck with remaining time is shown in Fig. 7.26, where τ represents the remaining time, t the pinch-off time, and t_c the neck breaking up time, $\tau = t_c - t$. The slope of the curve in Fig. 7.26A represents the pinch off velocity of the bubble neck interface. In order to compare the interface pinch-off velocity of the bubble intuitively, the changes in velocity are plotted in

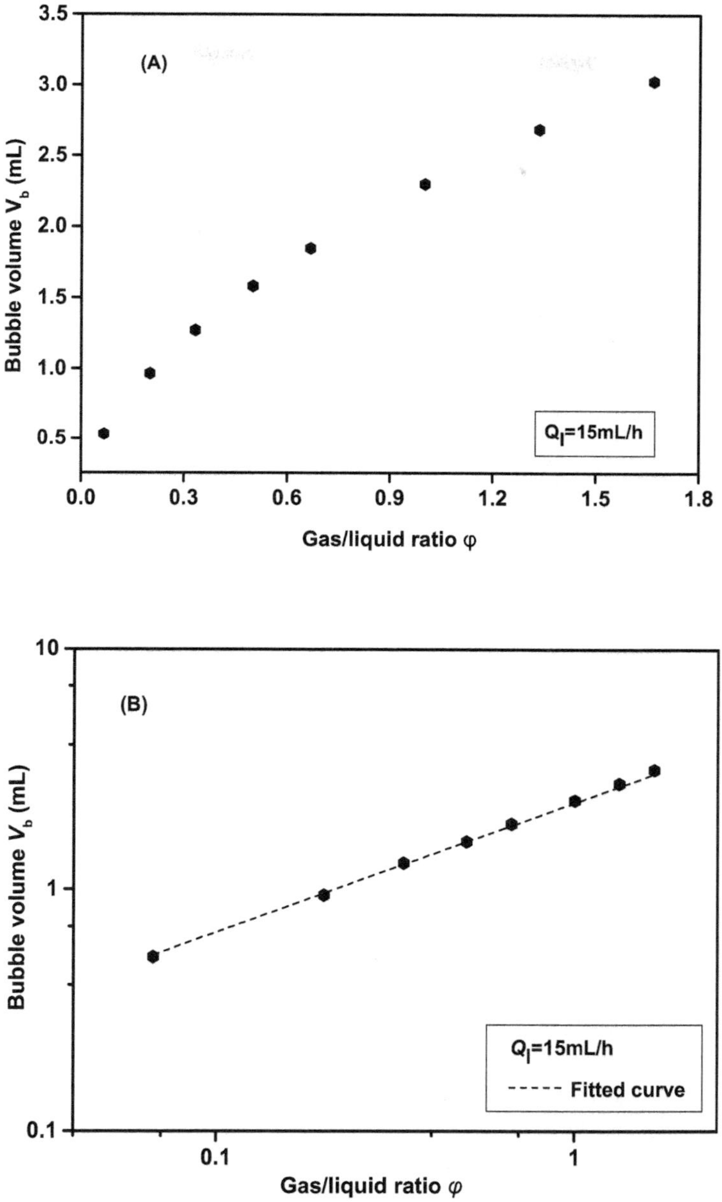

Figure 7.24 The change law of bubble volume V_b with gas/liquid flow ratio φ: (A) original data; (B) fitting results.

logarithmic coordinates (Fig. 7.26B and C). The interface pinch-off index α is obtained by fitting the data, and α represents the interface pinch-off velocity of the bubble. It can be seen from the comparison

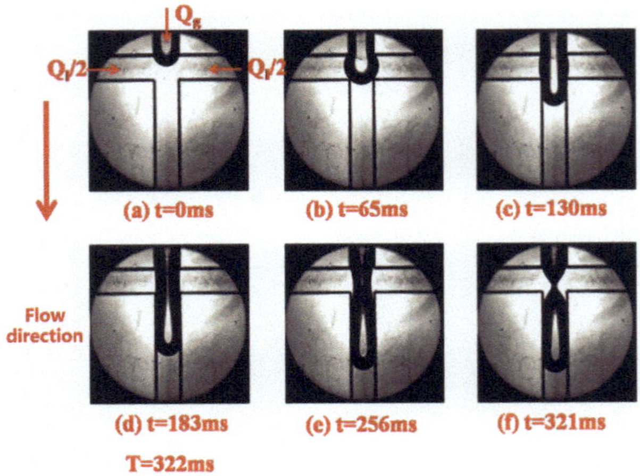

Figure 7.25 Schematic diagram of periodic foam generation process ($Q_l = Q_g = 15$ mL/h).

that the bubble pinch-off rule curve of the DPG particles increased faster than that of glycerol added, that is, the bubbles break up faster with the addition of DPG particles. This indicated that the DPG particles can promote the pinch-off and accelerate the generation of bubbles, which is beneficial to foam profile control. The acceleration of bubble generation is beneficial to prolong the half-life of the foam, making it stable in the pore-throat structure for a longer time and giving full play to the plugging and profile control.

3. Effect of flow velocity on the pinch-off law of bubble interface

A series of data in the process of bubble pinching off under different liquid phase flow rates were obtained through experiments. The data is plotted in Fig. 7.27. The results show that the liquid flow rate has a great influence on the initial pinch-off velocity of bubbles. With the increase of liquid flow rate, the initial pinch-off speed increases gradually. Under a certain liquid flow rate, the shrinking neck velocity speeds up with the evolution of the breakup process. At the same time, the curves of each flow rate gradually converge, indicating that the squeezing effect of the liquid gradually weakens. With the rapid increase of pinch off speed, the neck of the bubble is pinched off with the same rule under different flow rates. The effect of flow rate change is very weak.

In order to investigate the bubble rupture law, the data of neck width and remaining time are fitted with the method of Bergmann et al., assuming that $W_b = A\tau^{\alpha}$. The fitting results are plotted in logarithmic coordinates in Fig. 7.27B, and the discontinuities with

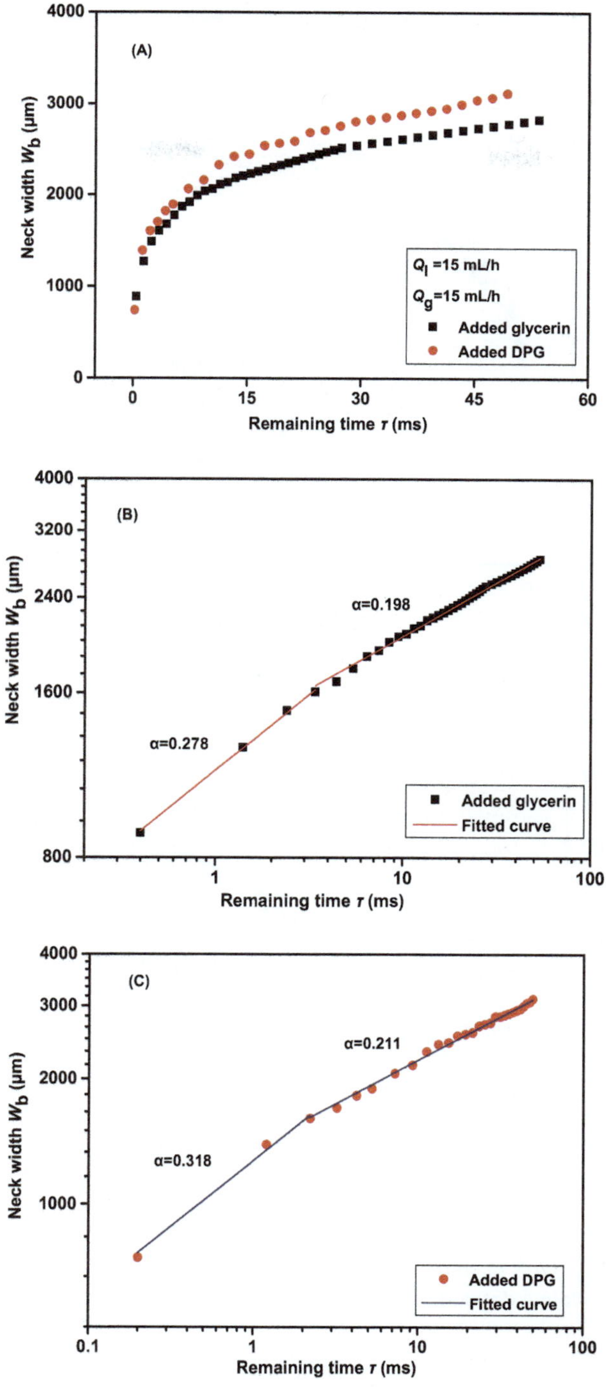

Figure 7.26 Evolution comparison of minimum neck width W_b with remaining time τ of bubble pinch-off: (A) original dates; (B) fitted results of adding glycerol; (C) the fitted results of adding three-phase foam of DPG particles.

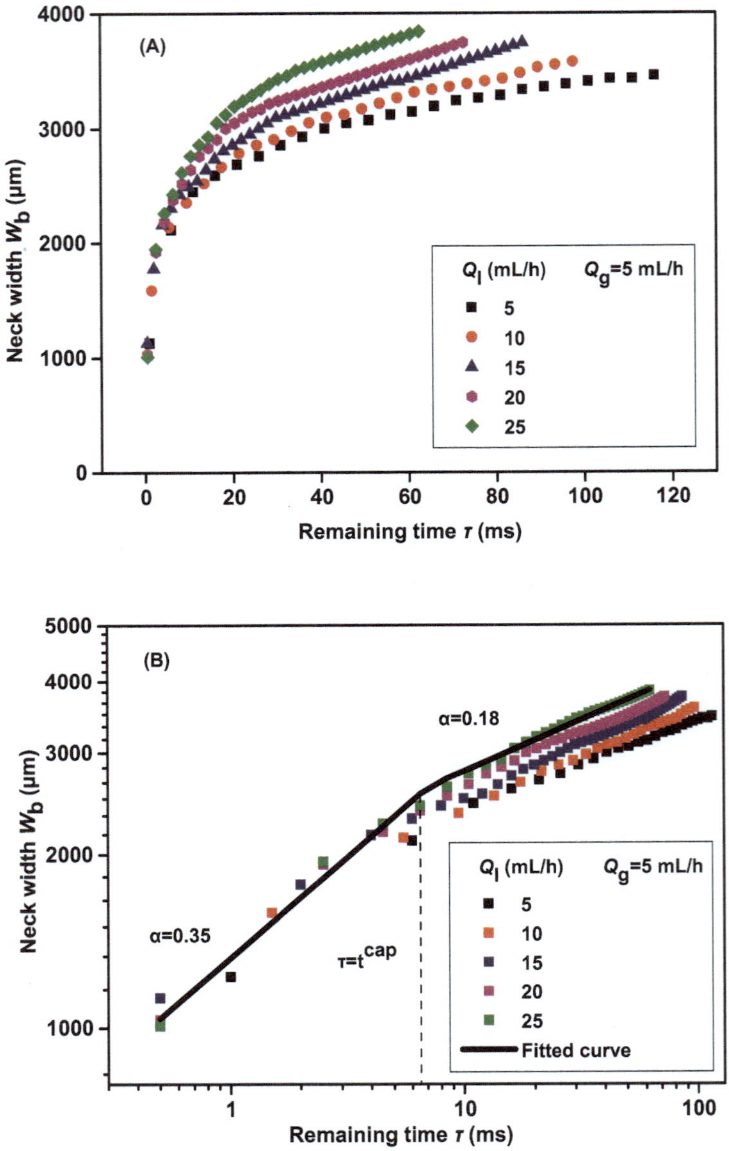

Figure 7.27 Evolution of minimum neck width W_b with remaining time τ of bubble clamping: (A) original data; (B) fitting results.

different slopes represent the pinch-off index of corresponding curve segments. Gartecki et al. [9] considered that when the deformation time of the neck interface is greater than the neck rupture time (i.e., when $t_{cap} > \tau$), the interface breaks rapidly and the pinch-off process is no longer affected by liquid extrusion. Lu Yutao et al. [10] believe that there is a simple size relationship between the remaining time and the deformation time of the neck interface: $t_{cap} < \tau$ before the sudden change of pinch-off speed; $t_{cap} \approx \tau$ when the sudden change of clamping speed occurs; and $t_{cap} > \tau$ after the sudden change of pinch-off speed. When $\tau > t_{cap}$, it is called the liquid fracture stage due to the strong effect of liquid velocity; when $\tau < t_{cap}$, the influence of liquid flow rate is small and is called free fracture stage [10].

In addition, the effect of gas velocity on the bubble pinch-off process was investigated. The results are shown in Fig. 7.28. It can be seen from the figure that the influence of gas velocity on the bubble formation process is very weak, and there is no obvious difference in the bubble rupture process under different gas velocity.

7.3.2.2.4 Bubble formation model

Fig. 7.29 shows the evolution law of the neck width W_b of the bubble over time. By analyzing the pictures, the evolution process of the bubble neck W_b is recorded (for the expansion stage, W_b is the maximum neck width, while for the pinch-off stage, W_b is the minimum neck width). As shown in Fig. 7.29, a complete bubble generation cycle consists of four

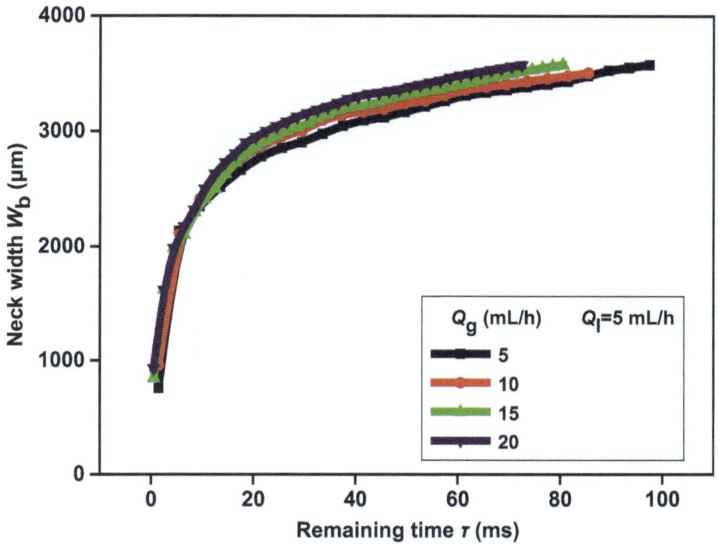

Figure 7.28 Evolution of minimum neck width W_b with residual time τ of bubble pinch-off.

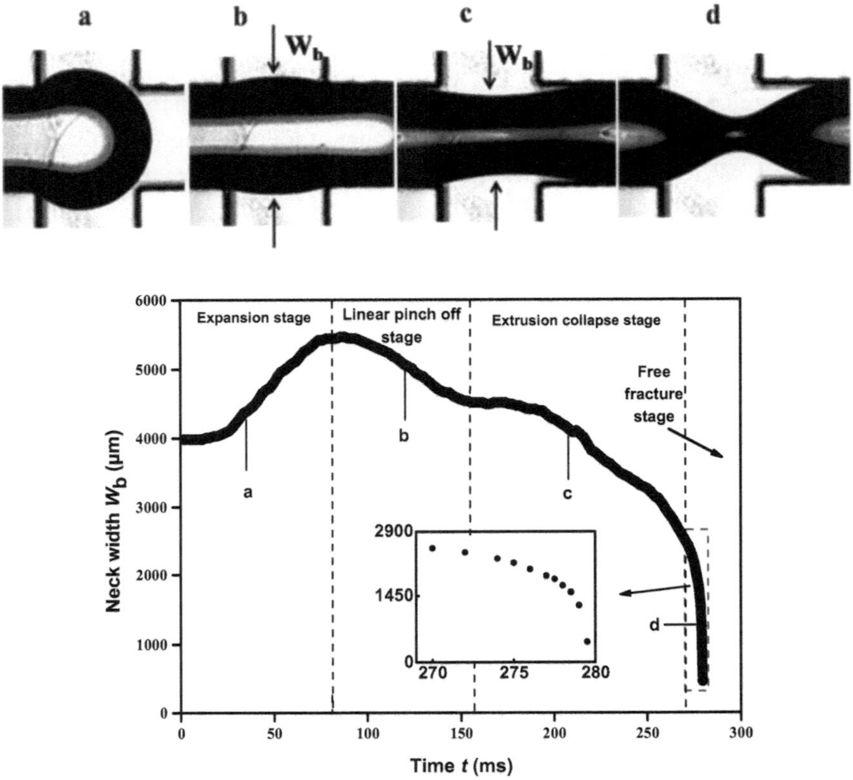

Figure 7.29 Evolution of neck width W_b with time and corresponding neck interface profile ($Q_l = Q_g = 15$ mL/h).

stages: expansion stage, linear pinch off stage, extrusion collapse stage, and free fracture stage. The shape of the interface is mainly controlled by the surface tension in the expansion stage and linear pinch-off stage. In the linear collapse stage, the neck width is mainly affected by the extrusion of the incoming liquid, and its velocity is directly proportional to the liquid flow rate. When the neck is separated from the channel wall, the neck gradually shrinks and the nonlinear interface pinches off. Also, the relationship between the bubble neck width and the remaining time satisfies the power function $r_o \propto \tau^{\alpha}$. The third stage is an extrusion collapse. Here, the neck is clamped under the joint action of liquid extrusion force and surface tension. The fourth stage is the free fracture stage. The phase interface begins to pinch off under the action of interfacial tension and breaks under the control of liquid inertial force, which is hardly affected by the initial liquid flow rate.

7.3.2.3 Migration mechanism of three-phase foam in microchannel with sudden expansion and contraction

7.3.2.3.1 The rupture and coalescence behavior of three-phase bubbles in sudden contraction and expansion structure

The aggregation of bubbles is random, and these experiments are carried out under specific experimental conditions. The concentration of the foaming agent was 0.3%. The foam breakdown and coalescence behavior under different DPG concentrations and glycerol–water solutions with the same viscosity were investigated. The volume of gas was $Q_g = 5$ mL/h and the volume of liquid $Q_l = 10$ mL/h.

The coalescence behavior of bubbles in the sudden contraction and expansion structure of the microchannel was compared between the two solutions. The number of bubbles coalesced in every 1000 bubbles through the structure of sudden contraction and expansion was counted and the final average value was taken after five interactions of the test to compare the aggregation of bubbles in SDS solution with DPG and glycerol at the same viscosity. The test results are shown in Fig. 7.30.

It is found that with the increase of glycerol and DPG particles concentration, the concentration of solution increases and the number of coalescence bubbles increases gradually. When adding glycerol and DPG particles at the same solution viscosity, the number of coalesced bubbles in the foaming agent solution with added glycerol is obviously more than that in the three-phase

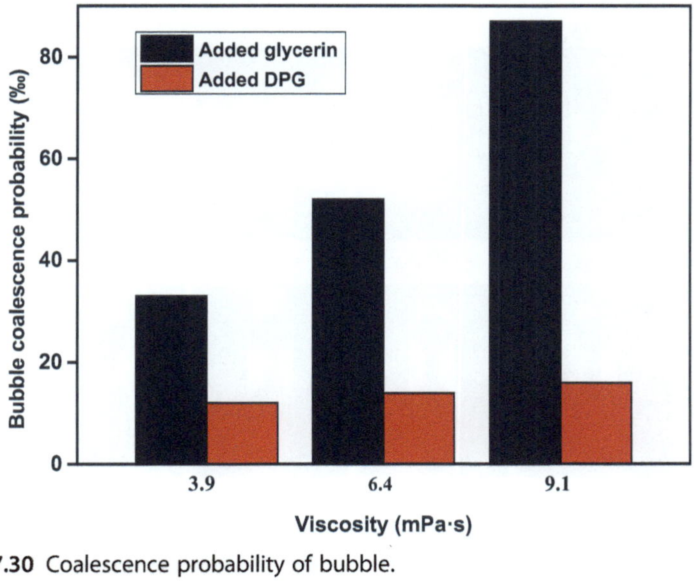

Figure 7.30 Coalescence probability of bubble.

foam system with DPG particles, indicating that the DPG could effectively inhibit the coalescence of bubbles. This is beneficial to the application of foam flooding in oilfield.

Fig. 7.31 shows the process of small bubbles breaking up when flowing through the pore-throat structure. The whole process lasts for 39 ms. In the first 30 ms, small bubbles are separated from the front big bubbles through the sudden contraction mouth. Then the small bubbles are connected with the large bubbles in the sudden expansion mouth and they converge quickly. The large bubble in the sudden expansion cannot move forward

Figure 7.31 The migration behavior diagram of bubbles generated by the solution with added glycerol through the sudden contraction and expansion structure.

due to the effect of the front and rear pressure. Therefore the small bubbles separated from the rear quickly catch up with the big bubbles, accumulating and gradually growing into a new round big bubble.

Fig. 7.32 shows the process where large bubbles generated in DPG fluids are continuously pinched to produce small bubbles when passing through the sudden contraction and expansion structure. It can be seen from the figure that the generated small bubbles move forward continuously with the advancement of the liquid, but without coalescence behavior.

7.3.2.3.2 The inhibiting bubble coalescence mechanism of DPG particles

The occurrence of bubble coalescence is the combination of the small bubbles separated from the main channel and the front bubble in the sudden expansion structure. Experiments show that the viscosity has a greater impact on the bubble coalescence. The greater the viscosity, the harder the bubbles are to coalesce. The development of the liquid film bridge between bubbles and the force it receives are the main factors for the bubble coalescence process. As viscoelastic particles, DPG increases the viscosity of the liquid phase to a certain extent. The particles can also enter between the two bubbles during the process of bubble migration, which effectively prevents the formation of liquid film bridges and reduces the coalescence behavior between bubbles.

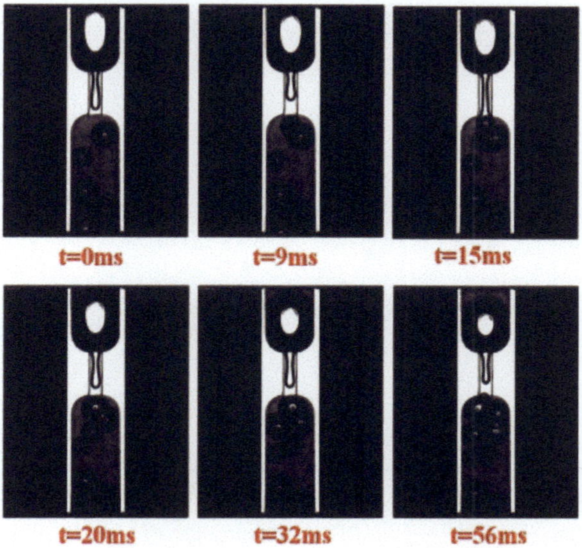

Figure 7.32 The process of the migration behavior of large bubbles generated by the DPG particles added to the solution through the sudden contraction and expansion structure.

7.3.2.3.3 The breakup mechanism of bubble

Burton et al. found that the viscosity of the liquid phase has a significant effect on the breakup of bubbles in the fluid [11]. In fact, in addition to the viscosity of the liquid phase, the bubble size is also affected by changing the operating conditions. The experiments investigated the influence of liquid viscosity and operating conditions on the bubble breakup process in the microchannel. The concentration of the foaming agent is fixed at 0.3%, and the concentration of the DPG is adjusted. The first 20 bubbles were statistically sized and averaged, and five groups were averaged.

The solution viscosity of the three-phase foam system of different concentrations DPG is shown in Table 7.2.

The experiments studied the effect of DPG concentrations on the breakup size of the bubble in the sudden expansion structure. The bubble diameter is shown in Fig. 7.33. It can be seen that the size of the bubble is almost not affected by the change in the viscosity of the solution. This

Table 7.2 Solution viscosity of three-phase foam system adding different concentrations DPG.

DPG concentration (%)	0.1	0.2	0.3	0.4	0.5
Viscosity (mPa · s)	3.9	5.3	6.4	8.1	9.1

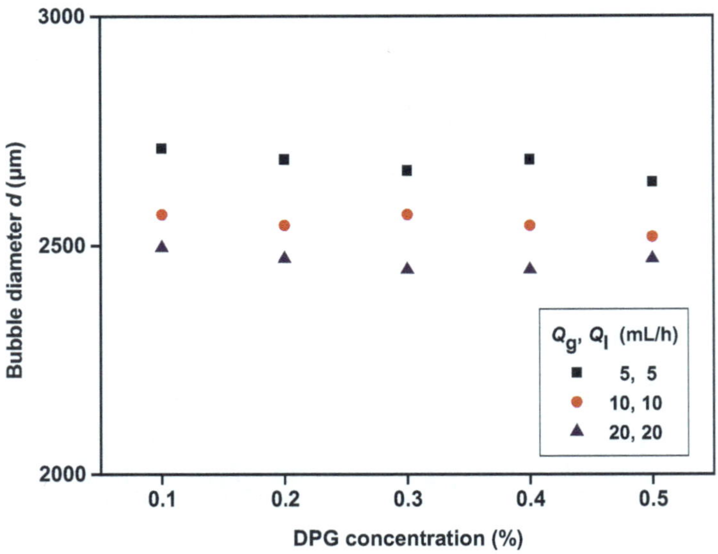

Figure 7.33 The variation of bubble diameter *d* with the viscosity of the three-phase foam solution of the DPG particles.

outcome is consistent with the conclusion drawn by Burton et al. that in the low viscosity range (<10 mPa · s), the breakup size of the bubble is not affected by the changing of viscosity.

Fig. 7.34 shows the change law of bubble size with the liquid flow rate through the sudden contraction and expansion structure. At a specific gas flow rate, the bubble size increases with the increase of the gas/liquid flow ratio, indicating that the size of the bubble formed in the sudden expansion structure can be controlled by changing the liquid flow rate.

Fig. 7.35 shows the effect of the change of gas flow on bubble size under different gas flow. It can be seen from the figure that the bubble size increases with the increase of the gas/liquid ratio. That is, the bubble size enlarges with the increase of the gas flow rate.

Based on the above analysis, it can be seen that the size of the bubble is affected by the gas/liquid flow, which is consistent with the squeezing mechanism: the bubble size increases with the increase of the gas flow, and decreases with the increase of the liquid flow.

In short, operating conditions and liquid viscosity play an important role in the breakup of bubbles. The squeezing mechanism considers that the squeezing action plays an important role in the bubble formation process and the size of the bubble depends on the gas/liquid flow ratio. The shear mechanism emphasizes that the shear force is dominant in the

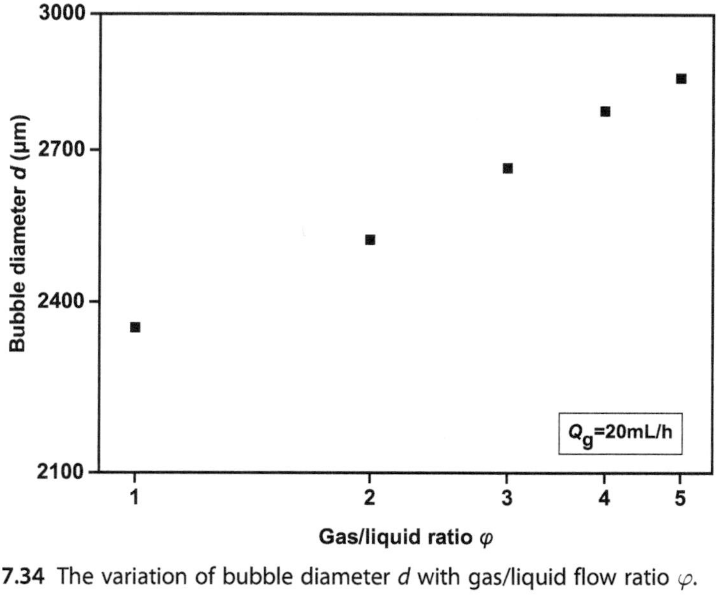

Figure 7.34 The variation of bubble diameter d with gas/liquid flow ratio φ.

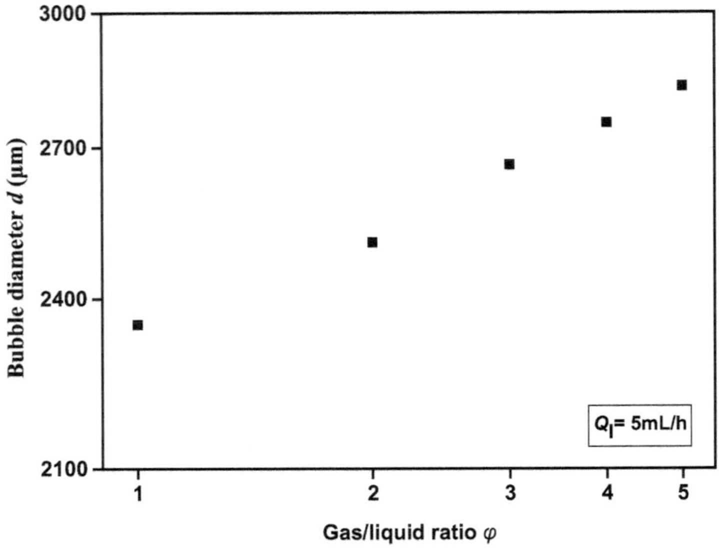

Figure 7.35 The variation of bubble diameter *d* with gas/liquid flow ratio φ.

bubble formation process, and the size of the bubble depends on the capillary number. The liquid used in this experiment is a low-viscosity liquid, so the squeezing force is the driving factor for bubble formation.

The breakup of the bubble in the sudden expansion structure can be explained by the classic Rayleigh–Plateau instability theory and the mechanism of the squeezing of the bubble neck by the accumulation pressure in the plugged channels. In the sudden expansion structure of the microchannel, due to the sudden increase in space, the bubblehead expands in the radial direction and the bubblehead takes an elliptical shape. The speed also decreases due to the increase in the volume of the head in the radial direction. The expanding bubblehead causes the fluid to be partially plugged, creating an accumulation of pressure. When the bubble neck is generated, the accumulation of pressure of the bubblehead and neck in the axial direction then develops in the radial direction. When the Rayleigh–Plateau instability condition is met (the length is greater than the interface circumference), the bubble is pinched off, resulting in two bubbles.

Section 7.4 Injection parameters optimization of DPG particles three-phase foam

A single sand-pack flow model was established in this section. With the resistance factor and residual resistance factor as evaluation indicators, the injection pattern, the gas injection rate, the injection volume, and the gas/liquid ratio of the DPG particles three-phase foam system was optimized. This section can provide guidance for site construction.

7.4.1 Optimization of injection pattern

Foam is mainly carried out in on-site construction through two patterns of mixed gas/liquid injection and alternate injection of segmented plugs. The resistance factor and residual resistance factor of the two injection patterns were compared in the laboratory. The formula of DPG particles three-phase foam is as follows: 0.06% DPG + 0.3% foaming agent, gas source as nitrogen, an injection gas/liquid ratio of 1:1, a gas injection rate of 0.5 mL/min, and at a test temperature of 85°C. The injection parameters are shown in Table 7.3, and the results are shown in Fig. 7.36.

It can be seen from Fig. 7.36 that the resistance factor and residual resistance factor of the gas/liquid mixed injection pattern are significantly higher than the gas/liquid alternate injection pattern. The mixed injection pattern adopts ground foaming; the gas and the foaming solution are fully contacted and mixed to generate more foam, which maximizes the foaming ability of the foaming solution. There is a strong Jamin effect which induces a higher resistance factor. The gas/liquid alternate injection pattern is likely to cause insufficient contact between the gas and the solution, and the underground foaming capability is not ideal, but the construction is convenient. It is mostly suitable for the profile control operations of reservoirs with high injection pressure and low permeability. Therefore, according to the heterogeneous development of the reservoir in the target block and the condition of the injection equipment, a suitable injection pattern can be selected to carry out site construction.

Table 7.3 Core parameters and steady resistance factor under different injection patterns.

Injection pattern	Core permeability (μm^2)	Pore volume (mL)	Injection volume (PV)	Steady resistance factor
Mixed gas/liquid injection	1.05	30	1	71.08
Alternate injection of segmented plugs	1.04	29	1	32.2

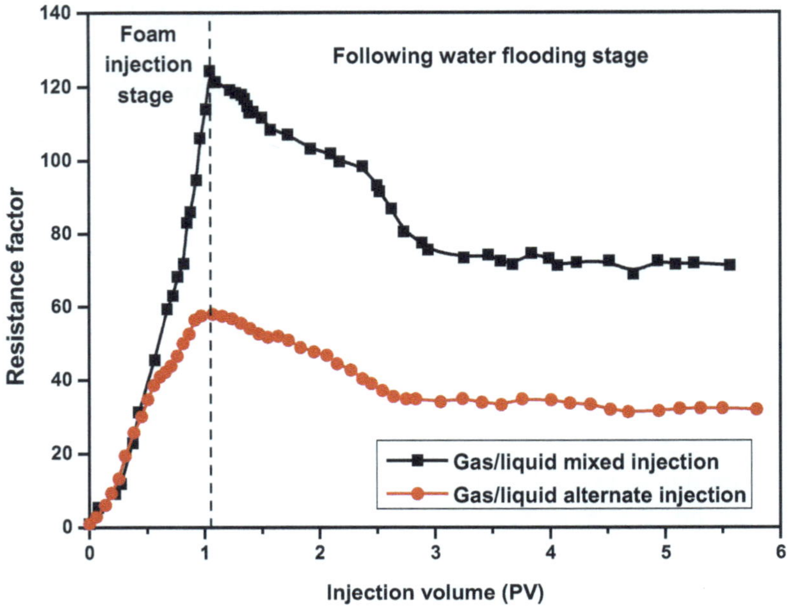

Figure 7.36 The influence of injection pattern on foam plugging performance.

7.4.2 Optimization of gas/liquid ratio

With the resistance factor and the residual resistance factor as the evaluation indicators, a single sand-pack flow model was established to optimize the gas/liquid ratio of the DPG particles three-phase foam system. The formula of the three-phase foam system for injection of DPG is as follows: 0.06% DPG + 0.3% foaming agent, gas source of nitrogen, injection volume of 1 PV, an injection speed of 0.5 mL/min, by a mixed gas/liquid injection mode, and at a test temperature of 85°C. The injection parameters are shown in Table 7.4, and the results are shown in Fig. 7.37.

According to Table 7.4 and Fig. 7.37, the plugging ability of the three-phase foam reaches the optimum at a gas/liquid ratio of 1:1. The larger the gas/liquid ratio, the larger the foam volume, but the more unstable the foam volume. When the foam enters the formation, it results in a poor sealing effect under the condition of high gas/liquid ratio. Therefore the gas/liquid ratio of 1:1 is the best for optimizing the three-phase foam.

7.4.3 Optimization of gas injection velocity

The gas injection velocity directly affects the stability of the foam in the porous media and construction schedule. If the gas injection rate is too

Table 7.4 Core parameters and steady resistance factor under different injection volume conditions.

Gas/liquid flow ratio	Core permeability (μm^2)	Pore volume (mL)	Steady resistance factor
1:1	1.04	29	71.08
2:1	1.05	29	50.86
4:1	0.98	33	45.95

fast, the volume of foam generated is large, and the requirements for the injection equipment are higher. On the contrary, the quality of foam generated is poor if the gas injection rate is slow. This affects the profile control effect. Evaluated by resistance factor and residual resistance factor, the gas injection rate of the DPG three-phase foam was optimized. The formula of the three-phase foam system for injection DPG is as follows: 0.06% DPG + 0.3% foaming agent, gas source of nitrogen, injection volume of 1 PV, a gas/liquid ratio of 1:1, by a mixed gas/liquid injection mode, and at a test temperature of 85°C. Injection parameters are shown in Table 7.5 and the optimization results are shown in Fig. 7.38.

As shown in Table 7.5 and Fig. 7.38, different gas injection velocities make a significant influence on the plugging performance of the DPG particles three-phase foam system. When the injection velocity is 0.5 mL/min, the plugging performance of the foam in the porous media and the following water flooding reaches its optimum. When the gas injection velocity is high, the foam produced by gas/liquid mixing with a large bubble size is not stable. In the porous media, the large bubble or air masses mostly pass through the pore throat in the form of bursting, which makes the resistance factor small. When the gas injection rate is low, nitrogen and foaming agent can be fully mixed to form a small and compact foam system with higher stability in porous media and better plugging performance in the injection process and the following water flooding stage.

7.4.4 Optimization of injection volume

With resistance factor and residual resistance factor as evaluation indicators, a single sand-pack flow model was established to optimize the injection volume. The formula of the three-phase foam system consists of 0.06% DPG and 0.3% foaming agent, nitrogen as the gas source, a gas/liquid ratio of 1:1, an injection speed of 0.5 mL/min, by a mixed gas injection at a test temperature of 85°C. The results were shown in Fig. 7.39.

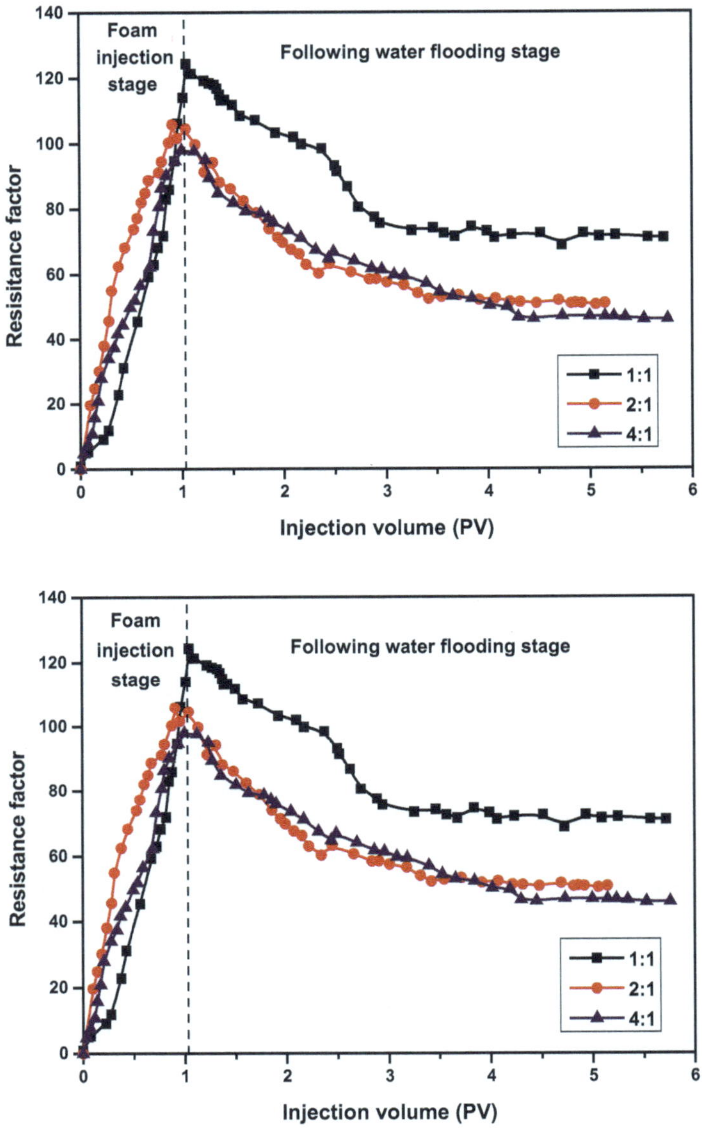

Figure 7.37 Influence of gas/liquid ratios on foam plugging performance.

As shown in Table 7.6 and Fig. 7.39, the resistance factor and residual resistance factor of the three-phase foam increases with the rising of the injection volume. When the injection volume is 1.5 PV, the steady resistance factor reaches the maximum. The larger the injection volume, the larger the Jamin effect of the three-phase foam in porous media. With the increase of

Table 7.5 Core parameters and resistance factor with different gas injection velocities.

Injection velocity (mL/min)	Permeability (μm^2)	Pore volume (mL)	Injected pore volume (PV)	Steady resistance factor
0.25	1.03	30	1	61.33
0.50	1.04	29	1	71.08
0.75	1.10	30	1	54.97
1.00	1.08	32	1	45.88

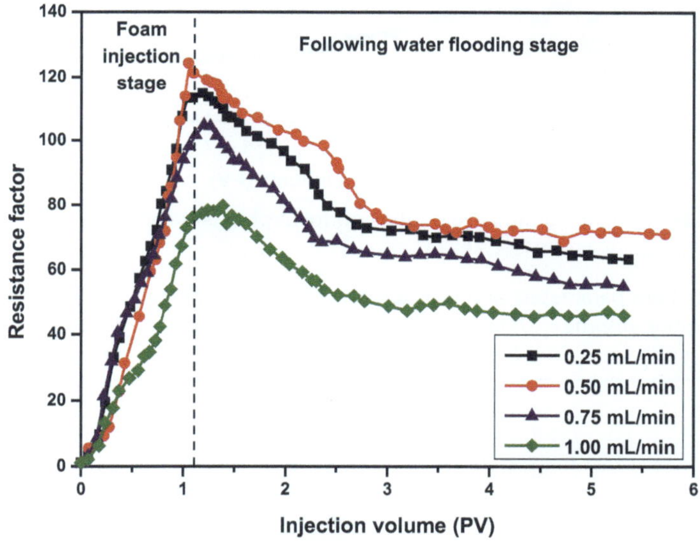

Figure 7.38 Influence of gas/liquid ratios on foam plugging performance.

injection volume, the number of DPG particles remains in the formation after defoaming, and the plugging effect of particles is enhanced. Considering the cost and plugging strength, the optimal injection volume is 1.0 PV.

Section 7.5 Application examples in oil fields

7.5.1 Geological survey of the test area

Spring 2 unit has 0.79 km^2 of oil-bearing area, with a porosity of 30.72%–38.46%, permeability of $1469-6438 \times 10^{-3}\ \mu m^2$, geological reserves of 49.0×10^4 t, calibrating recoverable reserves of 11.6×10^4 t, and recovery efficiency of 23.7%. The production horizon is $N_1S_1II_3$. The lithology of the reservoir is mainly fine sandstone with pebbly and loose cementation. It is a

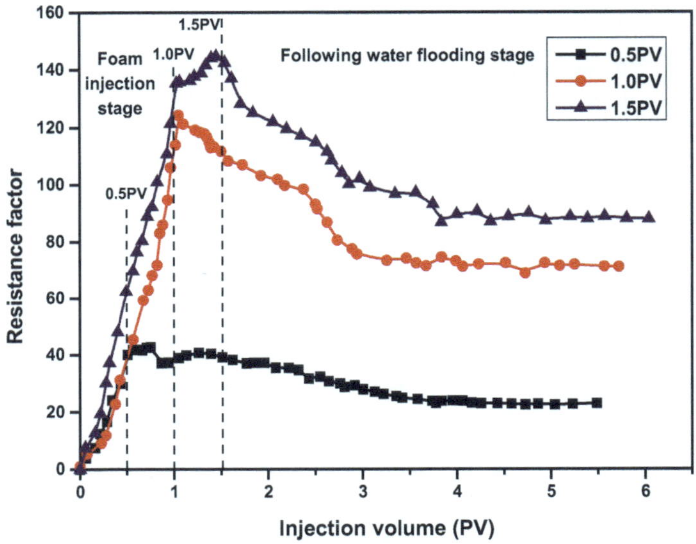

Figure 7.39 Influence of gas/liquid ratios on foam plugging capability.

Table 7.6 Core parameters and resistance factor under different injection volumes.

Injected pore volume (PV)	Permeability (μm^2)	Pore volume (mL)	Steady resistance factor
0.5	1.03	28	22.89
1.0	1.04	29	71.08
1.5	1.10	34	88.03

high-porosity and high-permeability reservoir with good reservoir conditions. It has grown seven development wells successively since July 2010. For now, three wells reach the yield of 18.66×10^4 t of liquid and 8.03×10^4 t of oil. The oil recovery rate is 0.8%, and the recovery degree is 16.57%.

The original formation pressure of Spring 2 unit is 9.31 MPa and the formation temperature is 41.55°C. Ground crude oil density is 0.9477 g/cm^3. The viscosity of degassed crude oil at reservoir temperature is 476.5 mPa · s. The oil/water viscosity ratio is 952. The edge water energy was sufficient and the early liquid recovery was quick, causing a rapid edge water fingering. The spread of edge water volume was small, and the oil displacement efficiency was low. The saturation figure shows that the fingering of edge water was significant. As shown in Fig. 7.40, Spring 2 unit has been in production for more than 6 years since November 2010, and most of the low area of the reservoir has been flooded, and oil saturation of the water flooded part is 25%–50%.

7.5.2 Development status of the pilot area

7.5.2.1 Influence of geological structure on the distribution of remaining oil

Spring 2 unit is a monoclinal structural sand, high in the north and low in the south, and edge water is located in the south of Chunguang Oilfield. The lower part of the structure is close to the edge water, and the degree of water flooding is relatively high. The higher part of the structure is far from the edge water, and the remaining oil is relatively enriched, as shown in Figs. 7.41 and 7.42.

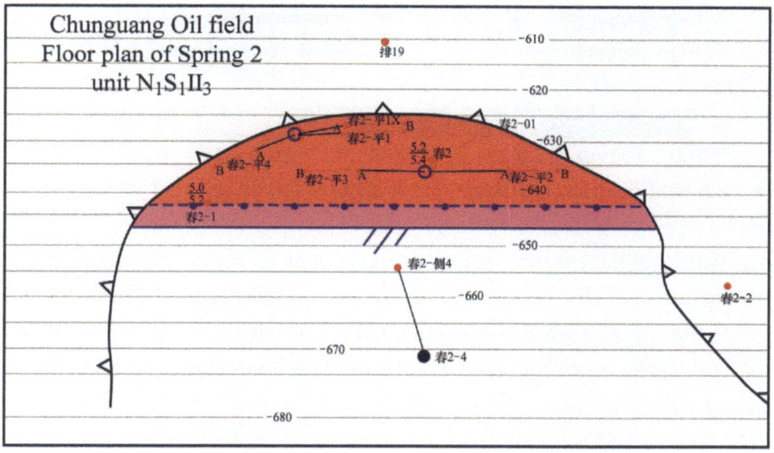

Figure 7.40 The plan of Spring 2 unit.

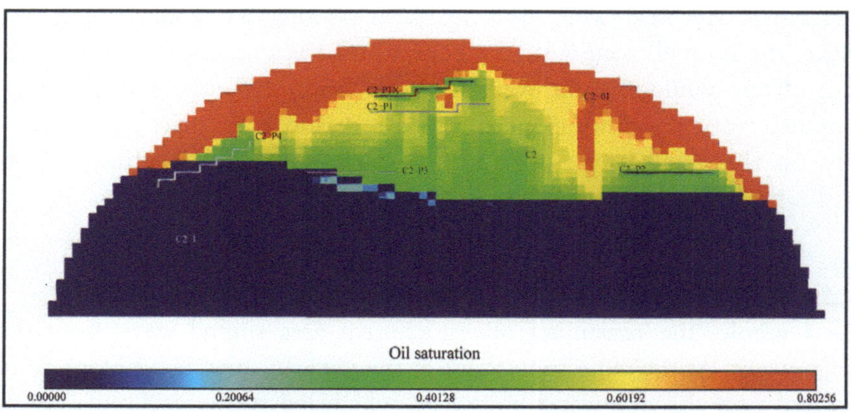

Figure 7.41 Remaining oil saturation of Spring 2 unit.

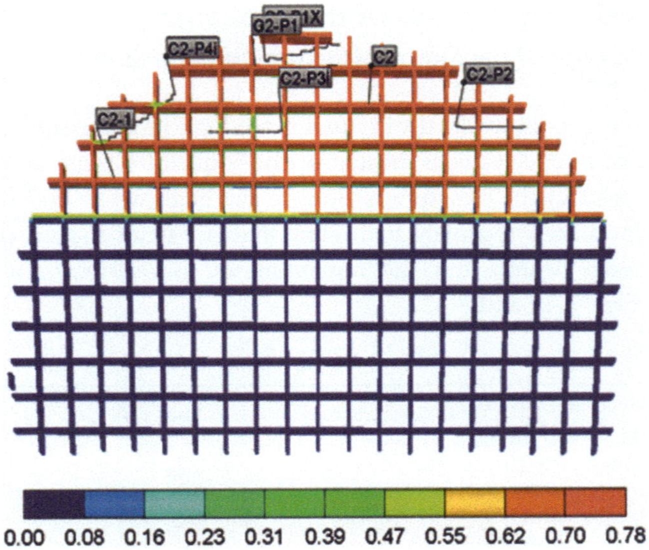

Figure 7.42 Grid diagram of remaining oil distribution of Spring 2 unit.

Longitudinally, the bottom of the oil layer is affected by the secondary bottom water and the combined action of edge water and secondary bottom water leads to a high water-cut, which is not conducive to the use of the top oil layer, and the top of the oil layer is relatively enriched with remaining oil.

7.5.2.2 Influence of pressure system on residual oil distribution

Due to the invasion of oil well production and edge water, the oil saturation in the near well zone decreases rapidly, while in the far well zone, the far well zone near the edge water has a relatively good flooding effect, while the far well zone far from the edge water has an unsatisfactory flooding effect. The main reason is that the energy supply on the side away from the waterside is limited, leading to low utilization.

7.5.2.3 Influence of well pattern on remaining oil distribution

As a whole, the remaining oil enrichment area shifted away from the edge water. At the same time, most of the wells in the Chun2 block were located near the oil–water boundary, which led to the accumulation of the remaining oil in the large no-well control zone.

7.5.2.4 Residual oil formed in interwell detention area

Pressure in some areas of the interwell zone is balanced, resulting in reservoir retention zones such as in the interwell zones of Chun2-Ping 1X and Chun2 wells. Meanwhile, long-term sand production in Chun2 well brought about low liquid production, poor utilization of well control range, and accumulation of residual oil around the well.

7.5.2.5 Influence of oil–water seepage channel on remaining oil

The overall block is high in porosity and permeability, and the oil flow degree of the reservoir is relatively large. After the partial breakthrough of boundary water in the horizontal well section, the dominant seepage channel is formed, which makes the boundary water flow into the wellbore along the dominant seepage channel, reducing the sweep coefficient, and forming the remaining oil.

7.5.3 On-site construction

The water absorption index was obtained by pressing, and test extrusion of deoiling effluent of the Spring 2-Ping4 well used 10 m^3 deoiled effluent by pressing and test extrusion on September 4, 2017. The following were obtained: the water absorption of 395 L/min at the pressure of 3 MPa, and the water adsorption of 450 L/min at the pressure of 4.2 MPa. The ground pipeline was tested for 18 MPa and passed the test on September 13. The site construction was carried out for DPG particles three-phase foam treatment, with nitrogen injection of 34,453 Nm^3, pressure of 2.25–10 MPa, and injection rate of 900 Nm^3/min on September 13–15. The mixed injection of DPG particles three-phase foam liquid slug was 111 m^3, the pressure was 6.8–9.3 MPa, and injection rate was 105 L/min on the ground. The injection volume of push fluid was 25 m^3 with an injection rate of 208 L/min and a pressure of 8.94–4.6 MPa on the ground. The shut-in was completed on September 15. Fig. 7.43 shows the pressure change curve and injection situation in the construction process.

The water absorption index was obtained by pressing and test extrusion of 10 m^3 deoiled effluent on September 5, 2017, in Spring 2-Ping 3 well. The water absorption index was as follows: the pressure of 0 MPa and the water absorption of 450 L/min. The ground pipeline was tested for 18 MPa and passed the test on September 5. The site construction of foam flooding was carried out with a nitrogen injection of 139,250 Nm^3, the pressure of 1.5–10.46 MPa and the injection rate of 900 Nm^3/min on September 5 to 12.

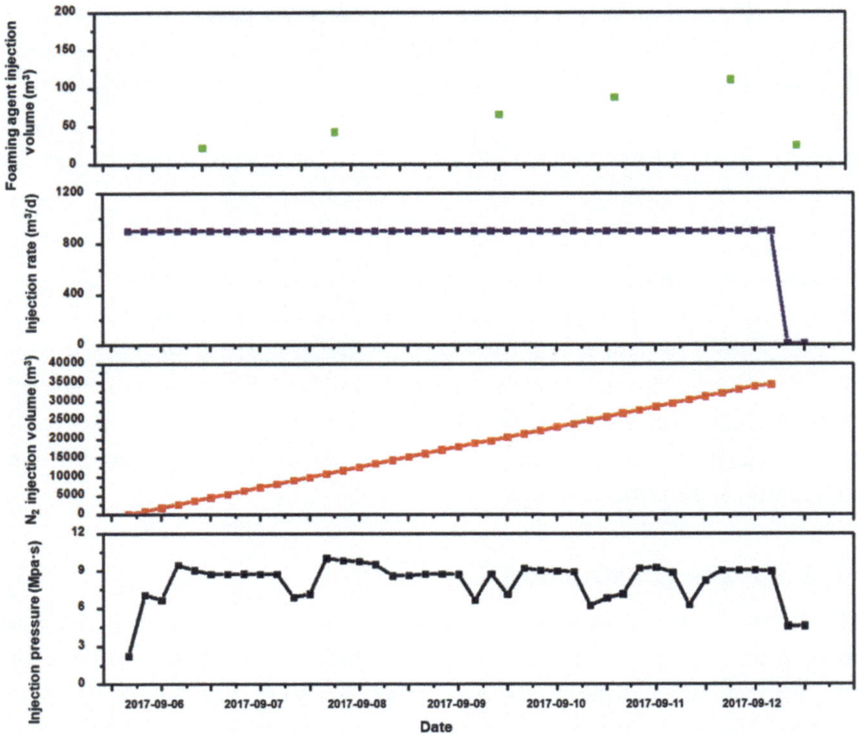

Figure 7.43 Construction curve of three-phase foam suppression in Spring 2-Ping 4 well.

The volume of mixed injection foam liquid was 410 m^3 and the volume of three-phase foam 42 m^3; the volume of pressure was 4–9.3 MPa; the injection rate was 105–120 L/min. The injection push fluid was extruding 30 m^3 with an injection rate of 215 L/min and pressure of 8.6–4.6 MPa on September 12. The shut-in was completed on September 12.

7.5.4 Construction effect analysis

The DPG particles three-phase foam system was implemented to plug the high-permeability channel and inhibit the planar interflow in Spring 2-Ping 3 and Spring 2-Ping 4 wells. No production was conducted in the wells after the construction. The high-position were Spring 2-Ping 1X and Spring 2–01 wells and the production curve is shown in Fig. 7.44. It can be seen from the Spring 2-Ping 1X production curve that the liquid production decreases while the daily oil production increases significantly after the completion of construction. The daily oil production rises from 3.5 t before the

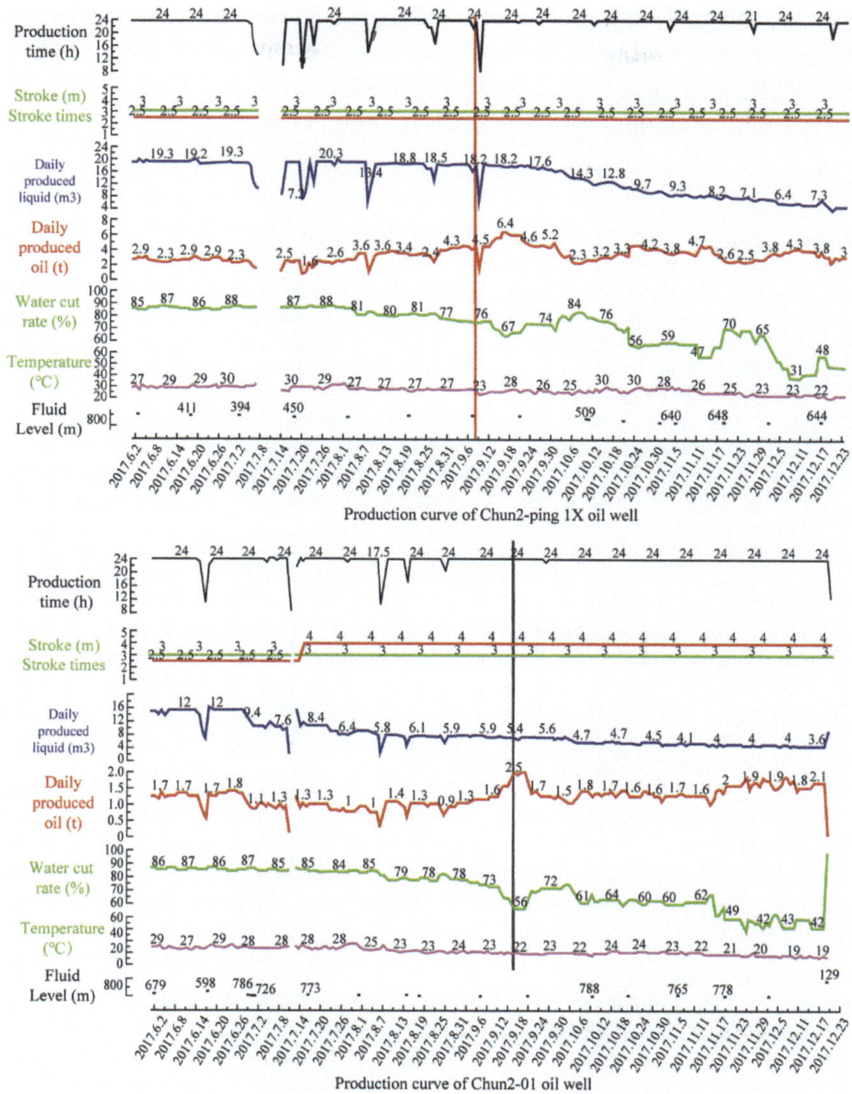

Figure 7.44 Production curves of Chun 2-Ping 1X and Chun 2–01 wells.

measures to 4.6 t, and the water-cut rate decreases from 81% to 73%, achieving an obvious effect of precipitation and oil increase with a cumulative increase of 106 t. According to Spring 2–01 well, the fluid production decreases and the daily oil production increases significantly after the completion of construction. The daily oil production rises from 1.2 t before the measures to 1.8 t and the water-cut decreases from 79% to 67%, achieving an obvious effect of precipitation and oil increase with an accumulative oil increase of 90.1 t. The oil production rises by 196.1 t and the water production of Chun2-Ping 1X and Chun2−01 wells decreases by 1703.2 t after the construction of Chun2-Ping 4 and Chun2-Ping 3 wells.

References

[1] Zhao G, Dai C, Zhang Y, et al. Enhanced foam stability by adding comb polymer gel for in-depth profile control in high temperature reservoirs. Colloids and Surfaces A: Physicochemical and Engineering Aspects 2015;482(3):115−24.

[2] Yao X, Sun N, Lyu Y, et al. Research progress of foam profile-controlling and flooding system. Oilfield Chemistry 2020;37(1):169−77.

[3] Sun Q, Li Z, Li S, et al. Utilization of surfactant-stabilized foam for enhanced oil recovery by adding nanoparticles. Energy & Fuels 2014;28:2384−94.

[4] Zhao G, Dai C, Wen D, et al. Stability mechanism of a novel three-phase foam by adding dispersed particle gel. Colloids and Surfaces A: Physico-chemical and Engineering Aspects 2016;497:214−24.

[5] Zhang Y, Xu G, Zou J, et al. Experimental study on water plugging effect and influence factors of three phase foam system. Reservoir Evaluation and Development 2019;9(2):44−9, 82.

[6] Harkins W. A general thermodynamic theory of the spreading of liquids to form duplex films and of liquids or solids to form monolayers. Journal of Chemical Physics 1941;9(7):552.

[7] Simjoo M, Rezaei T, Andrianov A, et al. Foam stability in the presence of oil: effect of surfactant concentration and oil type. Colloids and Surfaces A: Physicochemical and Engineering Aspects 2013;438:148−58.

[8] Yao X, Zhao G, Dai C, et al. Interfacial characteristics and the stability mechanism of a dispersed particle gel (DPG) three-phase foam. Journal of Molecular Liquids 2020;301.

[9] Garstecki P, Fuerstman MJ, Whitesides GM. Formation of droplets and bubbles in a microfluidic T-junction-scaling and mechanism of break-up. Lab on a Chip 2006;6:437−46.

[10] Lu Y, Fu T, Zhu C, et al. Dynamics of bubble breakup at a T junction. Physical Review E 2016;93:022802.

[11] Burton JC, Taborek P. Bifurcation from bubble to droplet behavior in inviscid pinch-off. Physical Review Letters 2008;101:214502.

CHAPTER 8

Prospect of multiscale DPG in-depth profile control technology

Since 2004, through almost 20 years' of R&D efforts, the research group lead by Prof. Caili Dai at China University of Petroleum (East China) has been devoted to strategy innovation, fundamental theory, industrial production, equipment development, and field applications of multiscale DPG particles, and have successfully developed the series of novel techniques, including the multiscale DPG in-depth profile control technique, DPG heterogeneous combination flooding technique, DPG-strengthened polymer/surfactant combination flooding technique, and DPG-enhanced three-phase foam flooding technique. The indoor experimental findings and field application experience have built the basic foundations for further industrialization promotion. As the worldwide oilfield exploration is now evolving from conventional to unconventional oilfields, from onshore to offshore oilfields, from shallow formation/water to deep formation/water, the challenges of production become more evident. This thus results in more demand for the advanced techniques and equipment. In the area of multiscale DPG technique, besides its further applications for water control for conventional water flooding, more advanced research and development are also demanded to meet with the national oil and gas development strategy.

8.1 Advancement of skid-mounted DPG online production/injection integration system

At the current moment of the oil market downturn, it has become more pressing to develop a more economically efficient profile control technique to further enhance the recovery of water flooding. Hence, to further save the operational cost of DPG profile control technique, we developed the skid-mounted DPG online production/injection integration system. This system effectively overcame the current problems of DPG application, including difficult transportation, costly packing, etc. This system has been certified by both the China Classification Society (CCS) and the French Classification Society

Theory and technology of multiscale dispersed particle gel for in-depth profile control
DOI: https://doi.org/10.1016/B978-0-323-99849-9.00008-4

Bureau Veritas, and has been piloted in Shengli oilfield, Xinjiang oilfield, and CNOOC Pengbo offshore oilfield. This system helped them increase the oil production by about 30t per day. It is our future endeavor to develop further both economic- and technique-efficient skid-mounted DPG online production/injection integration systems to offset the low oil price. To do this, the following tasks should be expected: to make the skid-mounted DPG online production/injection integration system more simple yet efficient, intelligent, and applicable to more complex application scenarios such as desert, tidal flats, hills, narrow sea platform, etc.; to make the core components of the equipment more standardized and sophisticated; thus, to make it possible to bring the skid-mounted DPG online production/injection integration system to more oilfields all over the world.

8.2 Further understanding on water control mechanisms of DPG chemical system

In this book, the water control mechanism of the DPG chemical system has been well clarified, and more fundamentals and the microscopic mechanism are explored. In our future endeavor, it is key to investigate the synergistic mechanisms of the DPG in-depth profile control technique combined with other enhanced oil recovery techniques, and perfect the mechanisms of water control mechanisms of the DPG chemical system.

8.3 Development of multiscale DPG profile control technique in ultradeep reservoirs

Deep oil and gas reservoirs have the characteristics of deep burial, high temperature, high salt, and high pressure, and they also have strong reservoir heterogeneity. Hence, it is growing to be a huge challenge to achieve effective profile control in such reservoirs. In our lab, the DPG in-depth profile control technique has been applied to a deep reservoir at a temperature of 150°C and salinity of 300,000 mg/L with initial success. However, there are still lots of problems yet to be solved to extend the DPG in-depth profile control technique more efficiently into the deep oil and gas reservoirs. It will be key to develop a new DPG that is resistant to ultrahigh temperature and ultrapressure, and thus improve the competitive advantage of the DPG technology and applicable scope for reservoirs. As for the polymer structure design of the bulk gel system, nanoinorganic particle materials (nanosilica, graphite, etc.) can be grafted onto the polymer branch chains to form a polymer system with larger bond energy and more stable structure. This may help to develop a more stable nanoparticle–inorganic strengthening polymer with good resistance to ultrahigh temperature and ultrahigh salt. In terms of cross-linking agent design, covalent bonds and noncovalent bonds can be designed to construct a more stable gel system. More efforts will also be made into investigating the stability mechanisms under harsh reservoir conditions.

Index

Note: Page numbers followed by "*f*" and "*t*" refer to figures and tables, respectively.

A

Acrylamide (AM), 4
Acrylic acid, 4
Adhesion energy, 248
Adsorption, 232
  behavior of surfactants, 188–195
  decreasing zone, 193
  equilibrium zone, 192–193
  rising zone, 192
  subequilibrium zone, 193
  time effect on SHC flooding system, 191–193
Aging time
  of DPG particles three-phase foam system, 293–295
  effect on dynamic interfacial tension, 199–200
  effect on dynamic surface expansion modulus, 202–203
Amide group ($-CONH_2$), 23, 35
Ammonia (NH_3), 34
Amylum–cadmium iodide method, 53
Atomic force microscopy, 205

B

Bettersize2000E laser particle size analyzer, 59
Boundary condition, 49
Breakup mechanism of bubble, 324–326
Bright Water overseas, 4
Bubble
  breakup mechanism of, 324–326
  coalescence mechanism of DPG particles, 323
  formation model, 319–320
  generation process, 311–312
  pinch-off rule of bubble interface, 314–319
  size distribution, 312–314
Bulk gels, 1–2, 165–166. *See also* Dispersed particle gel (DPG)
  chemical cross-linking reaction system, 92–93
  cross-linking reaction phase, 72–75
  formation mechanism of, 78
  system for low-temperature and low salinity reservoirs, 13–22
    cross-linking mechanism of zirconium-based gel, 13–14
    factors influencing formation of zirconium bulk gel, 14–22
  system for medium and high temperature reservoirs, 22–34
    cross-linking mechanism of resin bulk gel, 22–23
    influencing factors of resin bulk gel, 23–29
    temperature resistance of resin bulk gel, 29–32
  technical requirements for, 11
    reservoir type requirement, 11
    technical preparation requirements of, 11
  for ultra high temperature reservoirs, 34–45

C

Capability of self-growing and coalescence, 102–103
Capillary force, 248
Carman–Kozeny equation, 107
Changqing Oilfield, 128–134
  current development status of experimental area, 128–131
  geological profile of test area, 128
  on-site construction of DPG profile control treatment, 131–134
Chemical flooding, 155

China Classification Society (CCS), 339–340
Chromatographic separation effect of SHC flooding system, 225–233
  influence of DPG particle concentration, 227–229
  permeability effect on chromatographic separation effect, 229–233
    competitive adsorption, 232
    multipath migration, 233
    retention loss, 232
  test method and procedure, 225–226
Chromium ion, 1–2
Classic Rayleigh–Plateau instability theory, 326
Coalescence
  behavior of three-phase bubbles, 321–323
  mechanism to form oil belts, 250–251
Coaxial cylinder shear cross-linking method, 47–48
Cohesion force, 248–249
Colloid mill shearing method, 47–48
Colloid milling, 48
Colloidal dispersion gel (CDG), 2
Competitive adsorption, 232
Complexation, 13
Construction effect analysis, 336–338
Continuity equation, 49
Convenient dispersion injection system, 92–93
Conventional EOR methods, 155
Conventional foam, 287
Core oil saturation analysis, 235
Core scanning electron microscopy, 126
Covalent bonds, 340–341
Cross-linking agents, 1–2, 340–341
Cross-linking mechanism
  of zirconium-based gel, 13–14
  of resin bulk gel, 22–23
  of strengthened bulk gel, 34–36
Crude oil, 298–300

D

Deep profile control of multiscale DPG particles, 97
Degree of polymerization, 14
Design skid-mounted production equipment of multiscale DPG particles, 90–94
Differential scanning calorimeter (DSC), 23
Dispersed particle gel (DPG), 5, 47, 63–64. *See also* Bulk gels
  application examples in oil fields
    construction effect analysis, 336–338
    development status of pilot area, 333–335
    geological survey of test area, 331–332
    on-site construction, 335–336
  design of DPG particles three-phase foam system, 287–292
    modified Ross–Miles approach, 288–289
    optimization of DPG particles concentration, 289–292
    optimization of foaming agent concentration, 289
    optimization of gas/liquid ratio, 292
  development of multiscale DPG particles industrialized preparation software, 81–89
  efficiency of DPG preparation, 52–55
  establishment of DPG preparation technique, 47–52
    linear flow, 50*f*
    stator and rotor of colloid mill, 48*f*
    steady laminar flow, 50*f*
  flow chart of, 6*f*
  generation and migration of, 308–326
  industrialized production equipment design for DPG particles, 89–94
  injection parameters optimization of, 326–331
  method for testing DPG preparation efficiency, 53–55
    efficiency of DPG preparation, 54–55
    experimental principle, 53
    procedure of measurement, 54
    reagent preparation method, 53–54
  microscopic flooding mechanism of, 8*f*
  optimization of process parameters for DPG preparation, 52–70
  particles concentration influence on dynamic interfacial tension, 198–199

preparation mechanism of multiscale DPG particle, 70–80
preparation parameters effect on DPG particle size, 59–63
effect of bulk gel strength, 60
effect of bulk gel–water mixing ratio, 62–63
effect of shearing rate, 61
effect of shearing time, 61–62
preparation parameters effect on viscosity of DPG particle, 63–67
effect of bulk gel strength, 64
effect of bulk gel–water mixing ratio, 66–67
effect of shearing rate, 65
effect of shearing time, 65–66
preparation parameters effect on zeta potential of DPG particles, 67–70
effect of bulk gel strength, 67–68
effect of bulk gel–water mixing ratio, 70
effect of shearing rate, 68–69
effect of shearing time, 69–70
preparation parameters influence on microscopic morphology of, 56–59
effect of bulk gel strength, 56–57
effect of bulk gel–water mixing ratio, 59
effect of shearing rate, 57–58
effect of shearing time, 58
preparation procedure, 52
profile control ability of, 292–302
stabilizing mechanism of, 303–326
particle absorption characteristic on interfacial film, 303–304
surface charge, 307–308
surface tension, 307
viscosity, 305–306
Dispersed particle gel soft heterogeneous combination flooding technology (DPG SHC flooding technology), 155
characterization of DPG SHC system properties, 163–175
ability to reduce interfacial tension, 166–168
expansion capacity, 168–169
microscopic morphology of DPG SHC flooding system, 163–165
shear resistance, 172–174
viscosity stability, 165–166
wettability alteration, 169–171
zeta potential, 174–175
construction, 155–163
ability to reduce interfacial tension, 157–160
compatibility research, 155–157
viscosity characteristics, 160–163
displacement mechanism of SHC flooding system, 225–251
displacement performance of, 175–188
field application case in Zhejiang oilfield, 251–256
interaction mechanism of SHC flooding system, 188–225
Dispersed particle gel-strengthened polymer/surfactant (DPS), 259
characterization of, 268–272
aggregation and expansion property, 271–272
microscopic morphology, 268–269
thermal stability, 269–271
construction of DPS combination flooding system, 259–268
compatibility, 260
interfacial tension, 260–264
viscosity, 264–268
enhanced oil recovery mechanism of, 283–285
enhanced oil recovery performance of, 272–283
Displacement mechanism of SHC flooding system, 225–251
chromatographic separation effect of SHC flooding system, 225–233
distribution in porous media, 233–239
microscopic displacement mechanism of SHC flooding system, 244–251
Displacement performance of DPG SHC flooding system, 175–188
injectivity performance evaluation, 176–178
oil displacement potential evaluation, 182–188

Displacement performance of DPG SHC flooding system (*Continued*)
permeability reduction effect at different positions of sand-pack, 178–179
profile improvement of SHC flooding system, 179–182
Distribution behavior of DPG particles in porous media, 126–127
Dynamic interfacial tension, influencing factors of
aging time effect on dynamic interfacial tension, 199–200
DPG particles concentration influence on dynamic interfacial tension, 198–199
surfactant concentration effect on dynamic surface tension, 197–198

E

Emulsification phenomenon, 249
Emulsifier, 4
Enhanced foam, 287
Enhanced oil recovery (EOR), 1
mechanism of DPS combination flooding system, 283–285
improved mobility control performance, 285
strengthened oil displacement efficiency, 285
synergistic thickening effect, 285
performance of DPS combination flooding system, 272–283
mobility control performance, 273–276
profile improvement performance, 276–279
potential, 116–119
evaluation by sandpacks, 116–118
results by visualized model, 118–119
Environmental scanning electron microscope (ESEM), 33
Experiment design microchannel model
microchannel model design, 308–309
test procedures, 309–311

F

Fish-eye phenomenon, 92–93
Fluid diversion, 2–4
Foaming agent concentration, optimization of, 289
Foams, 287, 327
Force
characteristics of SHC flooding system, 205–225
force measurement scheme design, 206–207
interaction force characteristics analysis of DPG particles, 212–217
molecular structural analysis of SHC flooding system, 206–207
principle of force measurement, 205–206
roughness analysis of adsorption film in SHC flooding system, 210–212
interaction force characteristic analysis of SHC flooding system, 217–225
measurement
principle, 205–206
scheme design, 206–207
Formaldehyde (CH_2O), 34–35
French Classification Society Bureau Veritas, 339–340

G

Gas flooding technology, 155
Gas injection velocity optimization, 328–329
Gas/liquid ratio optimization, 292, 328
GE LIGHTSPEED 8-Slice Spiral CT Scanner, 234–235
Gelation
formation process, 33–34
performance
concentration influence on, 37–39
inorganic nanoparticles influence on, 39–40
temperature influence on, 40
time, 36–37
Geological structure influence on remaining oil distribution, 333–334

Glass etching model, 120
visualization results of glass etching model test, 122–126
visualization results of test in oil model, 122–126
visualization results of test in saturated water model, 122
Graphite, 340–341

H

HAAKE MARS 60 rheometer, 71
Hall curve of water injection well, 150
Hanging method, 14–15
Hexamethylenetetramine (HMTA), 34–35
High-speed mechanical shearing mechanical method, 47–48
Hydrolysis, 13–14
Hydrophilic groups, 2–4
Hydroxyl bridge action, 14
Hydroxyl bridging effect, 14
Hydroxyl group, 23

I

Indication curve of water injection well, 150–152
Industrialized production equipment design for DPG particles, 89–94
design industrialized workshop production equipment, 89
design skid-mounted production equipment of multiscale DPG particles, 90–94
Industrialized software of DPG preparation, 47
Injection parameters optimization of DPG particles three-phase foam, 326–331
gas injection velocity optimization, 328–329
gas/liquid ratio optimization, 328
injection pattern optimization, 327
injection volume optimization, 329–331
Injectivity performance evaluation, 176–178
Inorganic nanoparticles, 34–36
influence on gelation performance, 39–40
Interaction force characteristics analysis
of DPG particles, 212–217
of SHC flooding system, 217–225
Interaction mechanism of SHC flooding system, 188–225
adsorption behavior of surfactants on surface of DPG particles, 188–195
calculation of static adsorption capacity, 191
measurement principle, 189
measurement procedure, 189–190
preparation of chemicals, 189
effect of adsorption concentration, 194–195
effect of adsorption time, 191–193
adsorption decreasing zone, 193
adsorption equilibrium zone, 192–193
adsorption rising zone, 192
interface rheological behavior of SHC flooding system, 195–225
Interface rheological behavior of SHC flooding system, 195–225
theoretical foundation of interface rheological behavior, 196–197
influencing factors of dynamic interfacial tension, 197–200
influencing factors of expansion modulus of SHC flooding system, 200–203
Interfacial pinch-off process of bubbles, 314
Interfacial rheological behavior of SHC flooding system, 188
Interfacial tension (IFT), 157, 260–261, 270–271
ability to reduce, 166–168
influence of DPG particle concentration, 159–160
influence of surfactant concentration, 157–158
optimal concentration of components of SHC flooding system, 162–163
effect of
DPG particle concentration, 262–263
polymer concentration, 263–264
surfactant concentration, 260–262
Interwell detention area, residual oil formed in, 335

J

Jamin effect of foam, 301

K

Krieger–Dougherty equation, 97–98

L

Laser particle size analyzer, 97, 268
Linked polymer solution (LPS), 2

M

MacroMR12–150H-I nuclear magnetic displacement, 240–241
Macroscopic morphology, 104–105
Matching factor optimization, 107–111
Matching relationship between DPG particles and pore throat, 106–111
  establishing mathematical model of matching relationship, 106–107
  matching factor optimization, 107–111
Mathematics method, 47
Mechanical shear(ing) method, 47–48, 62
Medium–high salt oil reservoirs, 155
Medium–high temperature, 155
Microchannel model design, 308–309, 309*f*
Microscopic displacement mechanism of SHC flooding system, 244–251
  microscopic oil displacement mechanism of SHC flooding system, 248–251
  remaining oil distribution, 244–246
  reservoir microconformance controllability of SHC flooding system, 246–248
Microscopic morphology, 105–106
  of DPG SHC flooding system, 163–165
  of DPS combination flooding system, 268–269
Microscopic oil displacement mechanism of SHC flooding system, 248–251
  coalescence mechanism to form oil belts, 250–251
  interfacial tension reduction, 248–249
  strengthening-wetting reversal mechanism, 250
  synergy-emulsification mechanism, 249
Microstructure of strengthened bulk gel, 44–45
Migration mechanism of three-phase foam
  breakup mechanism of bubble, 324–326
  inhibiting bubble coalescence mechanism of DPG particles, 323
  rupture and coalescence behavior of three-phase bubbles, 321–323
Modern analytical methods, 97
Modified Ross–Miles approach, 287–289
Molecular structural analysis of SHC flooding system, 206–207
Momentum equation, 49
Multipath migration, 233
Multiscale DPG, 5–8
  characterization of DPG particles, 97–106
    capability of self-growing and coalescence, 102–103
    macroscopic morphology, 104–105
    microscopic morphology, 105–106
    shear resistance, 100–101
    thermal stability, 101–102
    viscosity characteristics of DPG particles, 97–99
    zeta potential analysis, 103–104
  in-depth profile control technology, 339
    advancement of skid-mounted DPG online production/injection integration system, 339–340
    development in ultradeep reservoirs, 340–341
    understanding on water control mechanisms of DPG chemical system, 340
  industrialized preparation software, 81–89
    development, 85–89
    establishment for multiscale DPG particles preparation model, 81–85
    functions introduction of software, 85
    login of software, 86
    operation of software, 86–88
    quit of software, 89
  matching relationship between DPG particles and pore throat, 106–111
    establishing mathematical model of matching relationship, 106–107

matching factor optimization, 107–111
oilfield application examples, 127–153
Changqing Oilfield, 128–134
Chunguang Oilfield, 333
Northwest Oilfield, 139–153
Shengli Oilfield, 134–139
Zhejiang Oilfield, 251–256
particles, 47–48
preparation mechanism of, 70–80
bulk gel cross-linking reaction phase, 72–75
multiscale DPG particles shaping shearing phase, 75–78
test principle, 71–72
preparation mechanism of DPG particles, 78–80
formation mechanism of bulk gels, 78
shaping shear mechanism of DPG particles, 78–80
preparation technology, 11
profile control mechanism of multiscale DPG particles, 120–127
distribution behavior of DPG particles in porous media, 126–127
profile control mode of multiscale DPG particles, 120–126
visualization results of glass etching model test, 122–126
profile control performance of DPG particles, 111–119
capability of mobility control, 111–112
enhanced oil recovery potential evaluation, 116–119
profile improvement performance, 113–115
profile control technique development in ultradeep reservoirs, 340–341

N

Nanoinorganic particle materials, 340–341
Nanosilica, 340–341
Newton inner friction law, 50–51
Noncovalent bonds, 340–341
Northwest Oilfield, 139–153. *See also* Changqing Oilfield
analysis of
fluid properties, 143–144
relationship between injection and production, 143
reservoir characteristics in YT2 well area, 141–143
effect evaluation of water injection wells, 150–153
identification of dominant channels, 147–149
pilot test of DPG treatment, 149
water absorption profile test, 144–147
Nuclear magnetic resonance (NMR), 239–240

O

Octadecyltrichlorosilane (OTS), 207
Oil displacement potential evaluation, 182–188
Oil model, visualization results of test in, 121–126
Oil phases
dynamic changes during
following water flooding, 237–238
SHC flooding, 236–237
water flooding, 235–236
Oil recovery, 287. *See also* Enhanced oil recovery (EOR)
increase technologies, 155
Oilfield application examples, 127–153
Changqing Oilfield, 128–134
Chunguang Oilfield, 333
Northwest Oilfield, 139–153
Shengli Oilfield, 134–139
Zhejiang Oilfield, 251–256
Oil–water seepage channel influence on remaining oil, 335
Optimal technical parameters, 47
Organic zirconium cross-linking agent, 13
Ostwald de Waele power law, 72

P

Parallel sand-packs models, 179
Particle absorption characteristic on interfacial film, 303–304
Particle size, 59, 81

Penetration process, 287
Phenolic resin, 1–2
  precondensation molecules, 35
Pinch-off rule of bubble interface, 314–319
Pipe flow shear cross-linking method, 47–48
Plate sand-pack model, 120
  visualization results of plate sand-pack model test, 120–121
    visualization results of test in oil model, 121
    visualization results of test in saturated water model, 120–121
Plugging
  agents, 5
  comparison of several typical, 7*t*
  performance of DPG particles three-phase foam, 300–301
Poiseuille equation, 51
Polyethylenimine, 1–2
Polymer/surfactant (PS), 260–261
  polymer–surfactant binary combination flooding, 155
  polymer–surfactant–alkaline ternary combination flooding, 155
Polymer(s), 1–2
  gels, 5, 11
  microspheres, 4
    plugging agents, 5
  polymer–based oil flooding system, 172
Polypropylene methyl methacrylate (PMMA), 308–309
Porous media
  displacement process analysis of SHC flooding system, 235–238
    core oil saturation analysis, 235
    dynamic changes of oil and water during following water flooding, 237–238
    dynamic changes of oil and water during SHC flooding, 236–237
    dynamic changes of oil and water phases during water flooding, 235–236
    three-dimensional dynamic analysis of displacement process, 238–239
  experiment method and procedures, 234
  experimental principle, 233
  generation and migration of DPG particles three-phase foam in, 308–326
    experiment design microchannel model, 308–311
    generation mechanism of three-phase foam, 311–320
    migration mechanism of three-phase foam in microchannel, 321–326
  sweep efficiency of, 239–243
Prepreparation gel particles (PPG), 2–5
Pressure system influence on residual oil distribution, 334
Production workshop, 47
Profile control
  ability of DPG particles three-phase foam system, 292–302
  application performance of DPG particles three-phase foam
    plugging performance, 300–301
    recovery potential, 301–302
  mechanism of multiscale DPG particles, 120–127
    distribution behavior of DPG particles in porous media, 126–127
    profile control mode of multiscale DPG particles, 120–126
    visualization results of glass etching model test, 122–126
  performance of DPG particles, 111–119
    capability of mobility control, 111–112
    enhanced oil recovery potential evaluation, 116–119
    profile improvement performance, 113–115
  stability factors of DPG particles three-phase foam system, 293–300
  and water plugging technology, 1–2
Profile improvement of SHC flooding system, 179–182
Python open-source language, 85

Q

Quanta 200
  ESEM, 33
  FEG scanning electron microscope, 20

R

Reagent preparation method, 53–54
Recovery potential, 301–302
Reservoir
geology, 134–135
microconformance controllability of SHC flooding system, 246–248
type requirement, 11
Residual oil
formed in interwell detention area, 335
pressure system influence on distribution, 334
Resin bulk gel
cross-linking mechanism of, 22–23
influencing factors of, 23–29
effect of cross-linking agent concentration, 25
effect of cross-linking coagulant concentration, 26
effect of salt concentration, 26–28
effect of shearing, 28–29
effect of temperature, 28
effect of type II functional polymer concentration, 24–25
temperature resistance of, 29–32
Resorcinol, 34–35
Retention loss, 232
Rheometer, 196–197, 305
Roughness analysis of adsorption film, 210–212
Rupture behavior of three-phase bubbles, 321–323

S

Salt ions of DPG particles three-phase foam system, 295–298
Salt tolerance of heat-resistant functional monomer AA, 4
Sandpacks, enhanced oil recovery potential evaluation by, 116–118
Saturated water model, visualization results of test in, 120–122
Scanning electron microscopy (SEM), 56, 97, 163–164, 268, 283
Shaping shear mechanism of DPG particles, 78–80
Shear
degradation, 1–2
of polymer, 259
resistance, 100–101, 172–174
Shearing degradation, 28–29
Shengli Oilfield, 134–139
on-site construction of DPG particles treatment, 137–139
operated well group production situation, 135–137
overview of test area, 134–135
Simulated water, 101–102
Single sand-pack flow(ing) model, 182–183, 326–327, 329
Single-core flow model, 300
Skid-mounted DPG online production/injection integration system, advancement of, 339–340
Skid-mounted equipment, 47
Sodium dodecyl sulfate (SDS), 311
Soft heterogeneous combination flooding system (SHC flooding system), 155
displacement mechanism of, 225–251
distribution in porous media, 233–239
influencing factors of expansion modulus of, 200–203
aging time effect on dynamic surface expansion modulus, 202–203
discussion on interface behavior, 203–205
DPG particles concentration influence on expansion modulus, 201–202
research on force characteristics of SHC flooding system, 205–225
surfactant concentration effect on expansion modulus, 200–201
interaction mechanism of, 188–225
interface rheological behavior of, 195–225
optimal concentration of components, 162–163
profile improvement of, 179–182
Stability factors of DPG particles three-phase foam system
aging time, 293–295
crude oil, 298–300

Stability factors of DPG particles three-phase foam system (*Continued*)
  salt ions, 295–298
  temperature, 295
Stabilizing mechanism of DPG particles three-phase foam system, 303–326
Static adsorption capacity calculation, 191
Strengthened bulk gel
  cross-linking mechanism of, 34–36
  factors influencing gelation of, 36–40
  microstructure of, 44–45
  thermal stability of, 40–44
Strengthened oil displacement efficiency, 285
Strengthening-wetting reversal mechanism, 250
Subequilibrium zone, 193
Surface charge of DPG particles three-phase foam system, 307–308
Surface force apparatus (SFA), 205
Surface tension of DPG particles three-phase foam system, 307
Surfactant concentration effect on dynamic surface tension, 197–198
Sweep efficiency of SHC flooding system, 239–243
  displacement analysis of SHC flooding system, 241–243
  experimental method, 240–241
  experimental principle, 239–240
Synergistic thickening effect of DPS system, 285
Synergy-emulsification mechanism, 249

T

Temperature
  of DPG particles three-phase foam system, 295
  influence on gelation performance, 40
  resistance of resin bulk gel
    analysis of bulk gel temperature resistance performance, 30–32
    stability investigation, 29–30
Tetradecyl hydroxysulfaine surfactant (THSB), 155–157
Thermal stability, 101–102
  of DPS combination flooding system
    interfacial tension, 270–271
    viscosity, 269–270
  of strengthened bulk gel, 40–44
  of zirconium bulk gel, 21–22
Thermally activated particles (TAP), 4
Three-dimensional
  dynamic analysis of displacement process, 238–239
  network, 268–269
Transmission electron microscope (TEM), 268
Type III functional polymer, 34–35

U

Ultra high temperature reservoirs
  bulk gel for, 34–45
  cross-linking mechanism of strengthened bulk gel, 34–36
  factors influencing gelation of strengthened bulk gel, 36–40
    concentration influence on gelation performance, 37–39
    inorganic nanoparticles influence on gelation performance, 39–40
    temperature influence on gelation performance, 40
  microstructure of strengthened bulk gel, 44–45
  thermal stability of strengthened bulk gel, 40–44
Ultradeep reservoirs, multiscale DPG profile control technique development in, 340–341
Ultralow-permeability microfractured well groups, 128–134

V

Viscometer, 97, 268
Viscosity
  characteristics of DPG particles, 97–99
    viscosity–concentration relationship, 97–98
    viscosity–temperature relationship, 98–99
  characteristics of DPG SHC flooding system

influence of DPG concentration, 161–162
influence of surfactant concentration, 160–161
of DPG particles, 63–67
three-phase foam system, 305–306
of DPS
combination flooding system, 269–270
effect of DPG particle concentration, 265–267
effect of polymer concentration, 264–265
effect of surfactant concentration, 267–268
viscosity–concentration relationship, 97–98
viscosity–temperature relationship, 98–99
Visualization results of plate sand-pack model test, 120–121
Visualized model, enhanced oil recovery potential results by, 118–119

W

Water
absorption
index, 335
profile test, 144–147
control
for conventional water flooding, 339
mechanisms of DPG chemical system, 340
drive characteristic curve, 152–153
phases
dynamic changes during following water flooding, 237–238
dynamic changes during SHC flooding, 236–237
dynamic changes during water flooding, 235–236
water–oil mobility ratio formula, 161–162
Well pattern influence on remaining oil distribution, 334
Wettability alteration of DPG SHC flooding technology, 169–171
experimental method, 170
experimental results and analysis, 170–171

Z

Zeta potential, 67, 174–175
analysis, 103–104
of DPG particles, 67–70
tester, 97
Zhejiang oilfield, field application case in
dynamic production parameters analysis of oil before operation, 252–254
general situation of targeted well, 251–252
tracking analysis of operation effect
measure ideas and design, 254
operation effect analysis, 255–256
Zirconium ion, 1–2
Zirconium-based gel
cross-linking mechanism of, 13–14
factors influencing formation of, 14–22
effect of concentration, 15–17
microstructure of zirconium gel, 20
effect of pH, 19–20
effect of salinity, 18–19
effect of temperature, 18
thermal stability of zirconium bulk gel, 21–22

FSC
www.fsc.org
MIX
Papier | Fördert gute Waldnutzung
FSC® C083411

Kolibri 360 GmbH
Ferdinand-Jühlke-Straße 7
99095 Erfurt, Deutschland
produktsicherheit@kolibri360.de